Chromatin Deregulation in Cancer

A subject collection from *Cold Spring Harbor Perspectives in Medicine*

OTHER SUBJECT COLLECTIONS FROM *COLD SPRING HARBOR PERSPECTIVES IN MEDICINE*

Malaria: Biology in the Era of Eradication

Antibiotics and Antibiotic Resistance

The p53 Protein: From Cell Regulation to Cancer

Aging: The Longevity Dividend

Epilepsy: The Biology of a Spectrum Disorder

Molecular Approaches to Reproductive and Newborn Medicine

The Hepatitis B and Delta Viruses

Intellectual Property in Molecular Medicine

Retinal Disorders: Genetic Approaches to Diagnosis and Treatment

The Biology of Heart Disease

Human Fungal Pathogens

Tuberculosis

The Skin and Its Diseases

MYC and the Pathway to Cancer

Bacterial Pathogenesis

Transplantation

Cystic Fibrosis: A Trilogy of Biochemistry, Physiology, and Therapy

Hemoglobin and Its Diseases

Addiction

Parkinson's Disease

SUBJECT COLLECTIONS FROM *COLD SPRING HARBOR PERSPECTIVES IN BIOLOGY*

Cilia

Microbial Evolution

Learning and Memory

DNA Recombination

Neurogenesis

Size Control in Biology: From Organelles to Organisms

Mitosis

Glia

Innate Immunity and Inflammation

The Genetics and Biology of Sexual Conflict

The Origin and Evolution of Eukaryotes

Endocytosis

Chromatin Deregulation in Cancer

A subject collection from *Cold Spring Harbor Perspectives in Medicine*

Scott A. Armstrong

Dana-Farber Cancer Institute

Steven Henikoff

Fred Hutchinson Cancer Research Center

Christopher R. Vakoc

Cold Spring Harbor Laboratory

COLD SPRING HARBOR LABORATORY PRESS
Cold Spring Harbor, New York • www.cshlpress.org

Chromatin Deregulation in Cancer

A subject collection from *Cold Spring Harbor Perspectives in Medicine*
Articles online at www.perspectivesinmedicine.org

Executive Editor	Richard Sever
Managing Editor	Maria Smit
Senior Project Manager	Barbara Acosta
Permissions Administrator	Carol Brown
Production Editor	Diane Schubach
Production Manager/Cover Designer	Denise Weiss
Publisher	John Inglis

Front cover artwork: The image depicts six broad categories of chromatin regulatory mechanisms, which are all known to be altered via mutations in the pathogenesis of human cancer. (Figure adapted by Meredith Cassuto, Cold Spring Harbor Laboratory, from the cover of *Epigenetics*, 2nd ed., 2015, with permission from Cold Spring Harbor Laboratory Press.)

Library of Congress Cataloging-in-Publication Data

Names: Armstrong, Scott A., editor. | Henikoff, Steve, editor. | Vakoc, Christopher R., editor.
Title: Chromatin deregulation in cancer / edited by Scott A. Armstrong, Dana-Farber Cancer Institute, Steven Henikoff, Fred Hutchinson Cancer Research Center and Christopher R. Vakoc, Cold Spring Harbor Laboratory.
Description: Cold Spring Harbor, New York : Cold Spring Harbor Laboratory Press, 2016. | Includes bibliographical references and index.
Identifiers: LCCN 2016027022 | ISBN 9781621821403 (hardback)
Subjects: LCSH: Cancer--Genetic aspects. | Chromosome abnormalities. | Mutation (Biology) | BISAC: SCIENCE / Life Sciences / Biology / Molecular Biology. | MEDICAL / Oncology. | SCIENCE / Life Sciences / Cytology.
Classification: LCC RC268.4 .C45 2016 | DDC 616.99/4042--dc23
LC record available at https://lccn.loc.gov/2016027022

10 9 8 7 6 5 4 3 2 1

All World Wide Web addresses are accurate to the best of our knowledge at the time of printing.

For a complete catalog of all Cold Spring Harbor Laboratory Press publications, visit our website at www.cshlpress.org.

Contents

Contents

Preface

THE STRUCTURAL STATE OF CHROMATIN, consisting of genomic DNA and its associated proteins, influences all DNA-templated processes, including transcription, replication, and repair. A vibrant area of discovery over the last several decades has been in revealing the repertoire of chromatin components, which includes the enzymes that covalently modify histones and DNA, the ATP-dependent remodelers of nucleosome structure, and the core histone variants that establish specialized chromatin domains. Because these basic chromatin components are ancestral to eukaryotic life, their study in model organisms has led to a deep understanding of roles that they play in mediating gene expression and chromosome integrity. Other chromatin components, especially sequence-specific DNA-binding proteins, drive development by activating or repressing transcription and altering the chromatin landscape to maintain cell-type identity during cell division.

Although mutation or aberrant expression of particular sequence-specific transcription factors has long been known to drive oncogenesis, only in the past several years have we come to appreciate the surprising involvement of chromatin regulators in cancer progression. This sea change can be largely attributed to cancer genome sequencing initiatives, which have revealed that loss- and gain-of-function mutations of genes encoding chromatin regulators are pervasive events in human oncogenesis. In contrast to the well-known oncogenes and tumor suppressor genes that were discovered and studied because of their cancer phenotypes, most chromatin oncoproteins were already very familiar from earlier genetic and biochemical studies. Thus, the study of chromatin deregulation in cancer has greatly profited from decades of research into basic mechanisms of gene regulation and development.

The impetus for this volume is the rapid expansion in our understanding in recent years of how perturbations of chromatin can result in the pathogenesis of human cancer. The major implication of this research is that many of the tumor-specific changes in DNA or histone methylation, which might have been previously interpreted as an epigenetic phenomena, are in fact the direct consequence of mutational changes in genes that encode chromatin regulators. Even histone genes themselves are found mutated in human tumors, a truly compelling example in which chromatin deregulation directly causes cancer. Based on these recently discovered paradigms, most of the chapters of this book are devoted to the chromatin regulatory machineries that are commonly mutated in human cancer and to our current understanding of underlying mechanisms. Other chapters are focused on the area of therapeutic modulation of chromatin, a field of study that has been invigorated by a new generation of small molecules that target chromatin reader, writer, and eraser functionalities. It is our hope that assembling this new body of knowledge in one volume will be of particular use to scientists just entering the chromatin field, as they seek to address the many unanswered questions that remain.

We wish to thank Richard Sever, Barbara Acosta, and their colleagues at the Cold Spring Harbor Laboratory Press office for their assistance in assembling this book. We would also like to express our gratitude to all of the authors, who generously sacrificed their time to the writing of the chapters of this volume.

SCOTT A. ARMSTRONG
STEVEN HENIKOFF
CHRISTOPHER R. VAKOC

Role of the Polycomb Repressive Complex 2 (PRC2) in Transcriptional Regulation and Cancer

Anne Laugesen,[1,2,3] **Jonas Westergaard Højfeldt,**[1,2,3] **and Kristian Helin**[1,2,3]

[1]Biotech Research and Innovation Centre (BRIC), University of Copenhagen, DK-2200 Copenhagen N, Denmark

[2]Centre for Epigenetics, University of Copenhagen, DK-2200 Copenhagen N, Denmark

[3]The Danish Stem Cell Center (DanStem), University of Copenhagen, DK-2200 Copenhagen N, Denmark

Correspondence: kristian.helin@bric.ku.dk

The chromatin environment is modulated by a machinery of chromatin modifiers, required for the specification and maintenance of cell fate. Many mutations in the machinery have been linked to the development and progression of cancer. In this review, we give a brief introduction to Polycomb group (PcG) proteins, their assembly into Polycomb repressive complexes (PRCs) and the normal physiological roles of these complexes with a focus on the PRC2. We review the many findings of mutations in the PRC2 coding genes, both loss-of-function and gain-of-function, associated with human cancers and discuss potential molecular mechanisms involved in the contribution of PRC2 mutations to cancer development and progression. Finally, we discuss some of the recent advances in developing and testing drugs targeting the PRC2 as well as emerging results from clinical trials using these drugs in the treatment of human cancers.

The genome of eukaryotic cells is organized into chromatin, consisting of DNA wrapped around an octamer of core histones to form nucleosomes. This organization serves both structural and functional purposes; organizing the genome into chromatin enables the linear genome to be packaged into the cell nucleus and protects the DNA strand from physical stresses. In addition, this organization allows specific areas of the genome to be condensed, thereby precluding the transcriptional machinery from gaining access to the underlying genes, while other areas in more open conformations can be actively transcribed, thereby facilitating cell-type specific gene expression patterns. Chroma-tin-associated proteins modulate the chromatin environment to help establish and maintain gene expression patterns over cell generations (Orkin and Hochedlinger 2011).

Deregulation of the chromatin environment can influence cell fate, and cancer cells often display disrupted chromatin environments with altered levels of various factors of the chromatin machinery. Recently, sequencing studies of human cancers have identified many somatic mutations in genes encoding chromatin-related factors, and intense research is ongoing to further decipher their involvement in cancer development and progression (You and Jones 2012).

The focus of this review is somatic mutations in genes coding for PRC2 subunits as well as recently discovered somatic mutations in the substrate for PRC2. We will discuss the potential roles of these mutations in the development of cancer. Before going into detail, however, we will give a general introduction to Polycomb group proteins, briefly outlining their biochemical and biological functions.

POLYCOMB GROUP PROTEINS

Polycomb group (PcG) proteins were originally identified in *Drosophila* as important regulators of fly development. The PcG proteins were shown to regulate the spatiotemporal expression pattern of important transcription factors (most notably, the *Hox* genes) during development, with PcG mutants showing characteristic phenotypes with defects in body segmentation. Orthologs have since been found in species ranging from plants to mammals, where they play key roles in establishing and maintaining correct gene expression patterns (Laugesen and Helin 2014).

POLYCOMB REPRESSIVE COMPLEXES

PcG proteins assemble into large multimeric protein complexes, of which the best characterized ones are Polycomb repressive complexes 1 and 2 (PRC1 and PRC2).

PRC1 complexes all contain either RING1A or RING1B with catalytic activity toward lysine 119 on histone H2A (H2AK119ub1), one of six Polycomb group RING finger (PCGF) subunits as well as additional subunits that define canonical PRC1 (CBX and PHC subunits) and noncanonical PRC1 (RYBP/YAP2) (Gao et al. 2012). The subunit composition affects recruitment to target genes and the catalytic activity of the complex (Blackledge et al. 2015). The mechanism by which PRC1 represses transcription is not entirely clear. Genetic studies have shown that it is only in part dependent on its ubiquitination activity (Illingworth et al. 2015; Pengelly et al. 2015), and, although PRC1 has been shown to promote chromatin compaction (Francis et al. 2004), the specific role of this

activity in transcriptional repression remains unclear.

PRC2 is a methyltransferase with activity toward lysine 27 on histone H3 (H3K27). The SET-domain-containing component (EZH1 or EZH2) is closely associated with several other subunits. The core complex necessary for catalytic function consists of EZH1/2, the Zinc-finger protein SUZ12, and the WD40 protein EED, which can be purified from cells with equimolar stoichiometry (Cao et al. 2002; Kuzmichev et al. 2002; Pasini et al. 2004; Smits et al. 2013). The core complex is associated with several additional proteins (RBBP4/7, JARID2, AEBP2, PCL1-3, C17orf96, and C10or12) (Smits et al. 2013) as well as ncRNAs (Blackledge et al. 2015), and the different interaction partners are thought to play roles in regulation of PRC2 activity or recruitment to target genes (Fig. 1) (Di Croce and Helin 2013).

RECRUITMENT OF POLYCOMB REPRESSIVE COMPLEXES

While the recruitment of PcG proteins in *Drosophila* to DNA stretches termed Polycomb response elements (PREs) is believed to involve transcription factors (Orsi et al. 2014), the mechanisms governing the recruitment of mammalian PRCs are still unclear. Mammalian cells lack distinct PREs, but mammalian PRCs preferentially bind in CpG-rich contexts and CpG-rich sequences can mediate recruitment of PRCs (Tanay et al. 2007; Ku et al. 2008; Mendenhall et al. 2010; Lynch et al. 2012).

Several studies show that PRC1 and PRC2 binding patterns overlap, that these complexes maintain the repression of common target genes (Boyer et al. 2006; Bracken et al. 2006), and that PRC1 recruitment is dependent on PRC2 (Rastelli et al. 1993; Cao et al. 2002; Wang et al. 2004). This, along with the fact that PRC1 contains a subunit with affinity for H3K27me3 (the CBX component) (Cao et al. 2002; Fischle et al. 2003; Min et al. 2003), has led to the proposal of a hierarchical recruitment model. According to this model, PRC2 is recruited to chromatin, where it trimethylates H3K27, facilitating PRC1 recruitment

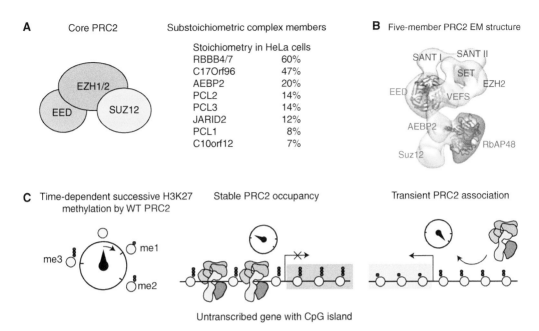

Figure 1. PRC2 catalyzes H3K27 methylation. (*A*) The core PRC2 complex and substoichiometric interactors measured in HeLa cells (Smits et al. 2013). (*B*) Electron micrograph (EM) structure of a PRC2 complex with RBBP4 (RBAP48) and AEBP2 (EMDataBank 2236, deposited image) (Ciferri et al. 2012). (*C*) Graphic representation of time-dependent successive H3K27 methylation (Sneeringer et al. 2010). A longer residence time at CpG islands at promoters of untranscribed genes allows for establishment of a trimethylated domain. WT, Wild-type.

and H2AK119 ubiquitination, leading to chromatin compaction and transcriptional repression (Blackledge et al. 2015). In recent years, however, this model has been challenged by several observations. First, in contrast to early studies from *Drosophila* and mammalian cells, PRC2 and H3K27me3 do not seem to be absolutely required for PRC1 recruitment and its activity (Leeb et al. 2010). This is partly explained by the discovery of "noncanonical" PRC1 lacking a CBX binding partner (Lagarou et al. 2008). Rather, recruitment of these complexes appears to rely on recognition of unmethylated CpG islands through the CXXC-domain of the KDM2B-component (Farcas et al. 2012; Gao et al. 2012; Qin et al. 2012; Tavares et al. 2012; Wu et al. 2013b). Second, recent results have shown that H2AK119ub1 is able to recruit and stimulate the activity of PRC2 (Blackledge et al. 2014; Cooper et al. 2014; Kalb et al. 2014), suggesting that the hierarchy under certain circumstances might be inverted. This suggestion still

lacks experimental support, and while H3K27 methylation is essential for normal development and gene repression (Muller et al. 2002; Pengelly et al. 2013), H2AK119ub1 is not required for "global" H3K27 methylation, gene repression and early development in *Drosophila* and in part in mouse (Illingworth et al. 2015; Pengelly et al. 2015). Thus, while it appears that noncanonical PRC1 can be recruited to chromatin through the binding of KDM2B to CpG-rich stretches and canonical PRC1 can be recruited through binding to H3K27me3, it is currently unclear how PRC2 is recruited to chromatin. Although several of the PRC2 subunits or associated proteins have been shown to possess weak DNA- or RNA-binding activities and other subunits are known to interact with histones, no study has been able to pinpoint one factor being solely responsible for PRC2 recruitment (Blackledge et al. 2015). The recruitment of PRC2 will be discussed further in the next section.

PRC2 AND H3K27 METHYLATION

A low resolution EM structure of a five-member human PRC2 has been solved, which is consistent with a structural composition proposed by biochemical studies (Fig. 1B) (Ketel et al. 2005; Ciferri et al. 2012). Recently, the first crystal structure of an active PRC2 (from the yeast species *Chaetomium thermophilum*) was solved (Jiao and Liu 2015). A striking feature of the crystal structure is that an amino-terminal region of EZH2 makes extensive contacts with and is wrapped around EED (Fig. 3D). This loop is followed by a region, immediately before the first SANT domain, that is sandwiched by EED and the catalytic SET domain of EZH2. This "pre-SANT" region appears to have important regulatory roles for the enzymatic activity. Previous data have shown that PRC2, more specifically the WD40 domains of EED, have an affinity for H3K27me3, suggesting a potential mechanism for the spreading of H3K27me3 over larger genomic regions (Hansen et al. 2008; Margueron et al. 2009; Xu et al. 2010). Based on the structure, it seems likely that binding of H3K27me3 to EED leads to a conformational change, which, through the pre-SANT region, affects the SET domain of EZH2 and could explain the increased activity of the PRC2 complex observed in the presence of H3K27me3 (Margueron et al. 2009; Xu et al. 2010; Jiao and Liu 2015). The large interaction surfaces observed from the structure also provide an explanation for the stable nature of the PRC2 complex—and why some of the individual components are unstable when not in the complex (Montgomery et al. 2005; Pasini et al. 2007).

The different methylation states of H3K27 (H3K27me1, H3K27me2, and H3K27me3) are found in different genomic contexts. ChIP-sequencing studies and mass spectrometry of histones isolated from embryonic stem cells show that around 5%–10% of histones carry H3K27me1, and this modification is primarily found within the gene bodies of actively transcribed genes (Ferrari et al. 2014). H3K27me2 is very abundant (about 50%–70% of histones have this modification) and is found in large domains, where it is proposed to exert a protective function against inappropriate transcription or enhancer activity (Jung et al. 2010, 2013; Ferrari et al. 2014). In contrast to H3K27me1 and H3K27me2, the distribution of H3K27me3 overlaps well with PRC2-binding patterns, and 5%–10% of histones carry this modification (Peters et al. 2003; Jung et al. 2010, 2013; Ferrari et al. 2014). These observations bring up several interesting points: The fact that PRC2 catalyzes all the different degrees of H3K27 methylation and H3K27me1 being associated with actively transcribed genes is hard to reconcile with the classic understanding of PRC2 as a transcriptional repressor. Furthermore, the fact that PRC2 binding is only found to overlap with H3K27me3 raises questions about the dynamics and rates of the deposition of each stage of H3K27 methylation, and might help us better understand PRC2 recruitment. With 70% of the genome being methylated on H3K27 by PRC2, it is unlikely that the bulk of PRC2 activity is directed by specific features in DNA sequence or composition. Mono- and dimethylation of H3K27 is reestablished very fast in newly incorporated histones in each cell cycle and show no sequence specificity. In contrast, the reestablishment of H3K27me3 is relatively slow (Zee et al. 2012; Alabert et al. 2015). Genome-wide location analyses have shown that PRC2 is highly enriched at CpG islands, whereas no enrichment of PRC2 is observed at H3K27me1/me2-positive regions. These results are in agreement with in vitro data showing that PRC2 is slower at converting H3K27me2 to H3K27me3 than catalyzing mono- and dimethylation of H3K27, and that it is therefore thought to need longer residence time to catalyze trimethylation (Sneeringer et al. 2010; Yap et al. 2011). Taken together, the current available data suggest that PRC2 interacts with chromatin in a manner independent of the underlying DNA sequence, facilitating mono- and dimethylation of H3K27, whereas conversion to H3K27me3 requires more stable binding of PRC2 (Fig. 1C). This might be achieved by interaction with sequence-specific or CpG island-associated proteins, which in the latter case might even be an intrinsic property of the

PRC2 complex. No PRC2 binding is observed at CpG islands by active genes, but appears at many such sites following drug-induced transcriptional blockage. This implies that the transcriptional machinery and associated proteins prevent stable binding of PRC2, which otherwise occurs by default (Riising et al. 2014).

PRC2 AND TRANSCRIPTIONAL REGULATION

While some studies indicate that PRC2 is bound to the stalled promoters of both coding and noncoding transcripts (Enderle et al. 2011; Brookes et al. 2012) with abortive transcripts potentially playing a role in recruitment (Kanhere et al. 2010), one study found the presence of PcG proteins to block transcriptional initiation altogether (Dellino et al. 2004; Kaneko et al. 2014), and it was recently demonstrated that PRC2 is recruited upon transcriptional repression rather than being required for setting up transcriptional repression (Riising et al. 2014). In agreement with the latter results, studies in *Drosophila* have shown that PRC2 is dispensable for initiating transcriptional repression of homeotic genes, but rather it is required for maintaining the repression of these genes during later stages of development (Struhl and Akam 1985; Jones and Gelbart 1990; Simon et al. 1992).

Whether loss of PRC2 function by itself leads to derepression of its target genes has been debated. Studies of embryonic stem cells lacking functional PRC2 initially showed a certain degree of derepression of PRC2 target genes (Montgomery et al. 2005; Pasini et al. 2007; Chamberlain et al. 2008; Shen et al. 2008). However, subsequent studies from our laboratory in more stable environments (ES cells grown in defined medium) showed that PRC2 loss does not lead to changes in transcription, indicating that the deregulation observed in earlier studies stems from outside stimuli (Riising et al. 2014). Based on these studies, we proposed that, rather than being recruited to target genes to promote their repression, lack of transcription allows a longer residence time of PRC2 at the promoters of repressed genes. Here, PRC2 contributes

to maintaining repression of these genes by setting thresholds for their activation, and in this way contributes to preserving cell identity. In accordance with this, loss of PRC2 function should not in itself lead to increased transcription. It would, however, make the chromatin environment more sensitive to stimuli, and abolishing PRC2 function, therefore, greatly influences cells undergoing changes to their transcriptional program, as occurring during cell-fate transitions.

PRC2 IN STEM CELLS AND DEVELOPMENT

As initially observed in *Drosophila*, PcG mutants display abnormal development in a range of species. Knockout of *Ezh2*, *Eed*, or *Suz12* in the mouse leads to early embryonic lethality with the embryos failing to undergo gastrulation (Schumacher et al. 1996; O'Carroll et al. 2001; Pasini et al. 2004). Studies of conditional and tissue-specific knockout mice have shown important roles of PRC2 in orchestrating cell-fate transitions in the development of a range of tissues (Laugesen and Helin 2014). Loss of PRC2 integrity impairs the stability of cell identity and makes the differentiation process more labile. With PRC2 being important for balancing differentiation versus proliferation, it is not surprising that PRC2 mutations or misregulation can skew this balance, leading to adverse outcomes including cancer development.

MISREGULATION OF PRC2 IN CANCER

The first observations implicating PRC2 in cancer came from the discoveries that *EZH2* is overexpressed and/or its locus amplified in several types of cancer, correlating with poor prognosis, and from knockdown-studies showing that PRC2 is required for proliferation of these cancer cells (Varambally et al. 2002; Bracken et al. 2003; Kleer et al. 2003). Subsequently, a plethora of studies have shown misregulation of PRC2 in human cancers, with PRC2 overexpression often correlating to poor prognosis (Deb et al. 2014; Jiang et al. 2015). Importantly, because of the cell-cycle-coupled expression of PRC2 components (Bracken et al. 2003), and PRC2 expression being high in the stem-cell compartment, increased PRC2 expression in cancer cells may stem from the increased proliferative capacity and/or dedifferentiated phenotype of cancer cells. It is, therefore, difficult to judge whether increased expression levels are a cause or a consequence of oncogenesis. It is clear, however, that many cancers depend on PRC2 expression for proliferation (discussed below). Along with the stronger evidence from *EZH2* amplifications (Bracken et al. 2003), the discovery of hyperactivating EZH2 mutants, and the fact that these mutations were found to be an early event through studies of clonality (Bodor et al. 2013), these observations collectively show that increased PRC2 activity is certainly acting as a driver in some cancers. This has recently been supported by numerous sequencing studies that have found the genes encoding the PRC2 members to be mutated in many different cancers. Surprisingly, however, both loss- and gain-of-function mutations have been identified, indicating that PRC2 acts as a tumor suppressor in some cancers and an oncogene in others (Table 1).

HYPERACTIVATING CATALYTIC MUTANTS OF EZH2

Several studies have identified recurring somatic point mutations in the sequence coding for the SET domain of EZH2 in follicular lymphoma (FL) and the germinal center B-cell (GCB) subset of diffuse large B-cell lymphoma (DLBCL) (Fig. 2). The most frequent of the hyperactivating mutations leads to a substitution of a tyrosine in the SET domain (Y646F/N/S/H/C, initially referred to as Y641), occurring in up to 25% of DLBCL and FLs (Morin et al. 2010, 2011; Bodor et al. 2013; Okosun et al. 2014). Although this mutant was originally reported to inactivate EZH2 activity as assessed by in vitro methylation assays (Morin et al. 2010), subsequent studies have demonstrated that it is a gain-of-function mutation with hyperactive conversion rate of H3K27 di- to trimethylation, and the mutant does indeed confer greatly increased levels of H3K27me3 (Sneeringer et al. 2010; Bodor et al. 2011, 2013; Yap et al. 2011).

The initial misperception concerning the biochemical properties of the EZH2-Y646 mutant stems from the fact that the mutant enzyme is very inefficient at the conversions of unmethylated H3K27 to H3K27me1 and of H3K27me1 to H3K27me2. If a wild-type EZH2 is present, however, allowing the mutant enzyme to use H3K27me2 as its substrate, they together very efficiently catalyze the conversion to H3K27me3 (Fig. 2C) (Swalm et al. 2014). Accordingly, the somatic mutation leading to the change of Y646 is always monoallelic and a wild-type allele of *EZH2* is always present in these cancers (Sneeringer et al. 2010). Interestingly, this does not seem to be the case for another characterized hyperactivating mutant (A682G), found in 1%–2% of FL, which catalyzes all the steps at a high rate and, consequently, does not seem to require wild-type EZH2 in vitro (Morin et al. 2011; McCabe et al. 2012a). A third recurrent point mutant in the SET domain of EZH2 (A692V) associated with lymphoma also leads to increased H3K27me3, but as it appears to prefer H3K27me1 as its substrate with little-to-no enhanced catalytic activity toward H3K27me2, it does not lead to a global depletion of H3K27me2 as seen in cells expressing Y646 or A682 mutants (Morin et al. 2011; Majer et al. 2012; Bodor et al. 2013; Ott et al. 2014). Interestingly, this mutant is also detected in a patient-derived B-cell acute lymphoblastic leukemia (B-ALL) cell line, and treatment with an EZH2 inhibitor (GSK126) leads to growth arrest and apoptosis (Ott et al. 2014). The same study found another B-cell ALL cell line to contain the Y646 substitution (Ott et al. 2014), and this mutant has also been found in sporadic parathyroid adenomas (Cromer et al. 2012) and in melanomas (Hodis et al. 2012), demonstrating that hyperactive *EZH2* mutants are not exclusive to lymphoma. Crystal structures of the human EZH2 SET domain (Antonysamy et al. 2013; Wu et al. 2013a) show that Y646 is placed so that the tyrosine hydroxyl group can make hydrogen bond coordination to the substrate's terminal amine nitrogen. This could impede free rotation needed to orient the nitrogen properly for transfer of the third methyl group from *S*-adenosylmethionine

Table 1. Mutations in genes coding for PRC2 subunits in cancer

Gene	Cancer	Notes	References
Hyperactivating mutations			
EZH2	DLBCL	Y646N/F/S/H/C A682G A692V	Morin et al. 2010, 2011; Lohr et al. 2012; McCabe et al. 2012a; Okosun et al. 2014
	FL	Y646N/F/S/H/C A682G A692V	Morin et al. 2010, 2011; Bodor et al. 2011, 2013; McCabe et al. 2012a; Okosun et al. 2014
	NHL ("other")	Y646F/S A692V	Morin et al. 2011
	B-ALL (patient-derived cell lines)	Y646N A692V	Ott et al. 2014
	Sporadic parathyroid adenomas	Y646N	Cromer et al. 2012
	Melanoma	Y646N/F/S	Hodis et al. 2012; Krauthammer et al. 2012; Alexandrov et al. 2013; Zingg et al. 2015
Loss-of-function mutations			
EZH2	Myeloid malignancies, including MDS, MPN, MF, CMML, and AML	Correlated with poor survival	Ernst et al. 2010; Nikoloski et al. 2010; Abdel- Wahab et al. 2011; Bejar et al. 2011; Guglielmelli et al. 2011; Jankowska et al. 2011; Score et al. 2012; Muto et al. 2013; Lindsley et al. 2015;
	T-ALL		Ntziachristos et al. 2012; Neumann et al. 2015
EED	MDS/MPN		Score et al. 2012
	MPNST		De Raedt et al. 2014; Lee et al. 2014
	GBM		De Raedt et al. 2014
	Melanoma		De Raedt et al. 2014
SUZ12	MDS/MPN		Score et al. 2012
	T-ALL		Ntziachristos et al. 2012; Simon et al. 2012; Neumann et al. 2015
	MPNST		De Raedt et al. 2014; Lee et al. 2014; Zhang et al. 2014
	GBM		De Raedt et al. 2014
	Melanoma		De Raedt et al. 2014
JARID2	MDS/MPN		Puda et al. 2012; Score et al. 2012
	T-ALL		Simon et al. 2012
AEBP2	MDS/MPN		Puda et al. 2012

AML, Acute myeloid leukemia; B-ALL, B-cell acute lymphoblastic leukemia; CMML, chronic myelomonocytic leukemia; DLBCL, diffuse large B-cell lymphoma; FL, follicular lymphoma; GBM, glioblastoma multiforme; MDS, myelodysplastic syndrome; MF, myelofibrosis; MPN, myeloproliferative neoplasm; MPNST, malignant peripheral nerve sheath tumor; NHL, non-Hodgkin lymphoma; T-ALL, T-cell acute lymphoblastic leukemia.

(SAM) (Fig. 2B). A682 likely holds Y646 in place, and mutation to glycine may loosen the pocket to decrease suppression of the third methylation step. A692 is also located in the active site, although it is not as clear what exact structural changes occur when it is mutated to valine.

In accordance with Ezh2 being specifically expressed at this stage of B-cell development (Velichutina et al. 2010), Ezh2 is required for the formation of germinal centers. Expression of hyperactive Ezh2 (Ezh2-Y646N) in GC-B cells induces GC hyperplasia, and this hyperplasia is reverted by treatment with an EZH2 inhibitor. Knockin of an *Ezh2* allele expressing Ezh2-Y646N alone does not give rise to lymphomas, but collaboration with Bcl2 and sustained activation of GC formation leads to

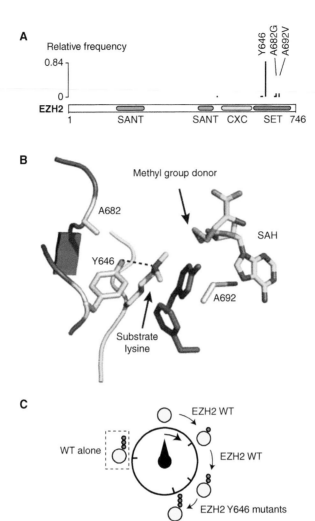

Figure 2. Hyperactive EZH2 mutants in DLBCL and FL. (*A*) Histogram of EZH2 mutation frequencies in FL and DLBCL. These map to three residues in the catalytic SET domain. (*B*) The mutated residues (light blue) are found in the active catalytic site of EZH2 near substrate and cofactor (yellow). Y646 coordinates the substrate lysine's side-chain amine and impedes rotation to allow transfer of the third methyl group, while A682 stabilizes the position of Y646. The crystal structure of the human EZH2 SET domain (PDB 4MI0) (Wu et al. 2013a) is used with substrate and cofactor from EHMT1 superimposed (PDB 2RFI) (Wu et al. 2010). (*C*) Combined activities of wild-type (WT) and Y646 mutant EZH2 achieves faster H3K27 trimethylation (Swalm et al. 2014).

GC-derived lymphomas (Beguelin et al. 2013). Similarly, expressing Ezh2-Y646F from a lymphocytic promoter does not in itself lead to tumor formation. However, crossing these mice with the Eμ-Myc transgenic mice markedly increases the rate of lymphoma formation in the offspring, and the resulting cancers can give rise to tumors in secondary recipients in transplantation experiments (Berg et al. 2014). Collec-

tively, these studies show that Ezh2-Y646N/F can collaborate with known oncogenes to accelerate tumorigenesis.

Both DLBCL and FL arise from GC B cells. As mentioned above, Ezh2 is required for GC formation with Ezh2 levels increasing as B cells enter the GC reaction and then decreasing upon exiting, enabling the expression of genes mediating terminal differentiation (Velichutina et al.

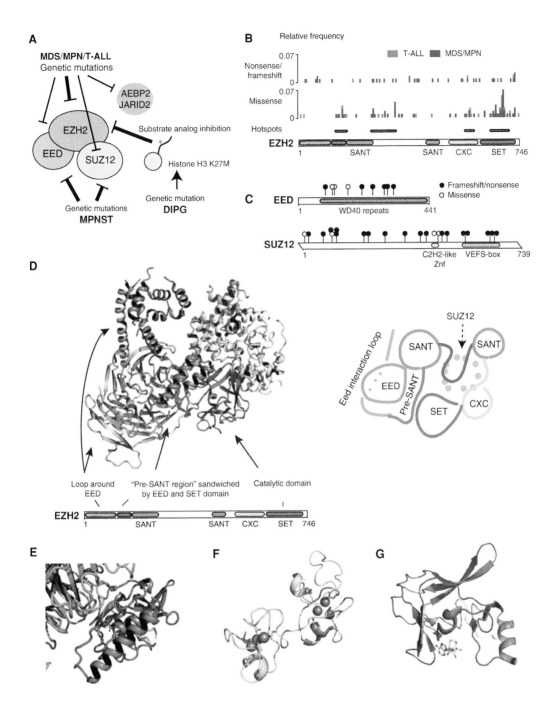

Figure 3. Loss-of-function PRC2 mutations in cancer. (*A*) Mutations conferring loss of PRC2 function are recurrent in several cancer types. The mutations are found in the genes coding for PRC2 core subunits or substoichiometric interaction partners, as well as in the genes coding for H3.1 and H3.3. (*B*) Histogram of *EZH2* mutants found in T-cell acute lymphoblastic leukemia (T-ALL) and myelodysplastic syndrome/myeloproliferative neoplasm (MDS/MPN) (Ernst et al. 2010; Nikoloski et al. 2010; Abdel-Wahab et al. 2011; Bejar et al. 2011; Guglielmelli et al. 2011; Jankowska et al. 2011; Score et al. 2012; Muto et al. 2013; Lindsley et al. 2015). (*Legend continues on following page.*)

2010; Beguelin et al. 2013). Hyperactive PRC2 in the GC appears to inhibit terminal differentiation, keeping cells in a more stem cell–like state, and allowing continuous rapid proliferation. Thus, the normal physiological role of PRC2 in GC B cells is related to keeping these cells from terminally differentiating—an observation that is mimicked in several other tissues, where PRC2 is often active in the stem-cell compartment and downregulated upon differentiation, seemingly guarding the balance between self-renewal and differentiation (Laugesen and Helin 2014).

LOSS-OF-FUNCTION MUTATIONS IN GENES ENCODING PRC2 MEMBERS

Loss-of-function mutations in *EZH2* in human cancers were first identified in myeloid malignancies (Ernst et al. 2010; Nikoloski et al. 2010). The observed mutations range from chromosomal loss over insertions/deletions to single nucleotide substitutions leading to mis- or nonsense amino acid substitutions, most of them believed to abolish PRC2 activity (Ernst et al. 2010). Subsequently, loss-of-function mutations in genes encoding core PRC2 members (*EZH2, SUZ12, EED*) and accessory factors (*JARID2, AEBP2*) have been identified in various myeloid malignancies (Ernst et al. 2010; Nikoloski et al. 2010; Jankowska et al. 2011; Puda et al. 2012; Score et al. 2012; Lindsley et al. 2015) with loss of *EZH2* correlating with poor prognosis (Bejar et al. 2011; Guglielmelli et al. 2011), as well as in T-cell lymphoblastic leukemia (T-ALL) (Ntziachristos et al. 2012; Si-

mon et al. 2012; Neumann et al. 2015), malignant peripheral nerve sheath tumor (MPNST) (De Raedt et al. 2014; Lee et al. 2014; Zhang et al. 2014), and melanoma and glioblastoma multiforme (GBM) (De Raedt et al. 2014).

In myelodysplastic syndrome/myeloproliferative neoplasm (MDS/MPN) and T-ALL, PRC2 is most frequently targeted through mutations of *EZH2*. This includes mutations leading to homozygous amino-terminal frame shifts, and thus complete loss of EZH2 is tolerated and maybe beneficial for these cancers. Missense mutations are also frequent, and are mostly found in the part of the gene coding for the catalytic domain, but three other regions of the gene appear to have a significant number of mutations clustering as well (Fig. 3B). The first hotspot region lies prior to the first SANT domain. Although sequence homology is poor between the crystallized PRC2 from *C. thermophilum* and human PRC2, it is plausible that an analogous pre-SANT region sandwiched between EED and the SET domain exists, and mutations may directly affect the SET domain or the allosteric communication from EED (Fig. 3E). The second hotspot region is immediately after the first SANT domain, but the crystal structure currently gives little clues as to how this can affect PRC2 function. Mutations in the CXC domain map to the two zinc-binding clusters and likely disrupt the (unknown) function of the CXC domain (Fig. 3F). In the SET domain mutants, the alterations map to the immediate proximity of the active site where catalysis takes place (Fig. 3G). In MDS/MPN

Figure 3. (*Continued*) The missense mutations cluster in four hotspot regions. (*C*) Mutations in *EED* and *SUZ12* found in malignant peripheral nerve sheath tumors (MPNSTs) (De Raedt et al. 2014; Lee et al. 2014; Zhang et al. 2014). (*D*) Crystal structure of Ezh2 and Suz12 VEFS domain from *C. thermophilum* (PDB 5CH1) (Jiao and Liu 2015) with Eed replaced by superimposed human EED (PDB3IIW) (Margueron et al. 2009). *Right* panel shows a schematic representation of the structure. (*E*) Loss-of-function mutations are frequently found in the part of *EZH2* coding for the 50 amino acid pre-SANT region. In the structure *D*, the pre-SANT region (red) is sandwiched between EED (green) and the catalytic SET domain of Ezh2 (blue), and mutations likely affect SET domain function directly or perturb a putative allosteric communication between EED and Ezh2. (*F*) Loss-of-function mutations are frequently found in the part of *EZH2* coding for the CXC domain. It has nine cysteines (light blue) that coordinate zinc ions in two separate clusters. All missense mutations (red) in this domain map to these clusters. Structure of the human CXC domain is used (PDB 4MI0) (Wu et al. 2013a). (*G*) The hotspot region for *EZH2* mutations in the SET domain maps to the active site and surrounding binding pockets for substrate and cofactor. Structure used is of human SET domain (PDB 4MI0) with substrate and cofactor from EHMT1 superimposed (PDB 2RFI) (Wu et al. 2010). DIPG, Diffuse intrinsic pontine glioma.

and T-ALL, mutations in *EED* and *SUZ12* are less frequent (1%–2%), but they do occur (Ntziachristos et al. 2012; Score et al. 2012; Simon et al. 2012; Neumann et al. 2015) and are further evidence that PRC2 disruption is the driving event. Missense mutations in *SUZ12* map to the VEFS domain, which mediates the binding to EZH2.

In some cancers, both alleles of a gene encoding a PRC2 member carry loss-of-function mutations, leading to complete abrogation of PRC2 activity, while other patients carry one wild-type allele. A recent study of MPNST patients found that 79% with and 34% without *NF1* microdeletions are homozygous for loss of either *SUZ12* or *EED* (De Raedt et al. 2014). In an $Nf1^{+/-}$; $Trp53^{+/-}$ mouse model, heterozygous loss of *Suz12* significantly accelerated development of MPNST and other cancer types, including high-grade glioma (De Raedt et al. 2014). MPNST appears to frequently be derived from a neurofibroma with monoallelic *NF1* deletions that include *SUZ12* (neighboring gene), and the second *SUZ12* mutation drives the event. This observation provides a possible explanation to why *SUZ12* is frequently the targeted PRC2 member in MPNST.

It was recently shown that knockout of *Suz12* or *Eed* leads to defects in hematopoietic stem cell (HSC) self-renewal, in part through derepression of the *Cdkn2a* locus, and that PRC2 function is required for lymphoid development (Xie et al. 2014; Lee et al. 2015). Conversely, partial loss of PRC2 function leads to increased proliferative capacity of HSCs (Lessard et al. 1999; Richie et al. 2002; Majewski et al. 2010; Lee et al. 2015), highlighting the sensitivity of these cells to PRC2 dosage.

Several mouse models have shown tumor suppressive roles of PRC2 in hematological malignancies, with *Ezh2* knockout leading to spontaneous development of T-ALL and MDS/MPN (Simon et al. 2012; Mochizuki-Kashio et al. 2015). Importantly, deletion of *Ezh1* alone does not cause any hematological malignancies, and concomitant loss of *Ezh1/Ezh2* abolishes the myoproliferative effects of *Ezh2* loss, indicating that the phenotype of *Ezh2* knockouts might rely on Ezh1 compensating to uphold

some residual PRC2 activity (Mochizuki-Kashio et al. 2015). In addition, loss of *Ezh2* was found to accelerate and exacerbate the development of MDS when deleted in conjunction with *Tet2* (Muto et al. 2013), and in a *Runx1* mouse model, *Ezh2* knockout promotes MDS development but prohibits malignant transformation to AML (Sashida et al. 2014). Interestingly, heterozygous loss of *Suz12* was found to accelerate lymphomagenesis in Eμ-myc mice (Lee et al. 2013), a striking observation, since hyperactive Ezh2 was found to induce a similar phenotype (Berg et al. 2014). Furthermore, several studies have shown that deletion or depletion of PRC2 members impedes MLL-AF9-driven AML (Neff et al. 2012; Tanaka et al. 2012; Shi et al. 2013; Danis et al. 2015), and overexpression of Ezh2 leads to myeloproliferative disorders (Herrera-Merchan et al. 2012). Taken together, this underlines the complex nature of PRC2 function in disease progression, and indicates that the consequence of PRC2 loss is highly dependent on the cellular context and exact developmental stage at which the loss occurs, compromising the intricate balance between self-renewal and differentiation. The various observations indicate that HCSs are dependent on a functional PRC2 to prevent activation of the *Cdkn2a* locus, whereas loss-of-function of one allele coding for a core component of PRC2 predisposes to the development of a hematopoietic malignancy. If these malignancies retain an intact *Cdkn2a* locus, they could therefore still be dependent on PRC2 activity and might thus respond to treatment with EZH2 inhibitors (see below).

THE SIGNIFICANCE OF H3K27 METHYLATION

Drosophila studies have shown that the catalytic activity of PRC2 is required to prevent derepression of target genes (Muller et al. 2002), and a histone mutant abolishing H3K27 methylation phenocopies of Polycomb mutants, strongly suggesting that H3K27 methylation is the essential physiological function of PRC2 (Pengelly et al. 2013). Additional data from *Drosophila* show that a mutant with an amino acid substitution (R741K) in the SET domain of *E(z)* (the

Drosophila homolog of EZH2) confers hyper-trimethylation activity of PRC2 with increased levels of H3K27me3 and decreased levels of H3K27me1/me2. Importantly, this increased trimethylation leads to inappropriate repression of PRC2 target genes with phenotypes mimicking those of loss-of-function mutations in *trithorax* (*trx*), encoding the *Drosophila* enzyme orthologous to the MLL H3K4 methyltransferases and key antagonist of Polycomb silencing (Stepanik and Harte 2012). Although the mutation conferring hypertrimethylation of H3K27 is distinct from the mutations observed in mammalian cancers, the data from *Drosophila* show that increased H3K27me3 can indeed cause inappropriate silencing of PRC2 target genes with consequences for cell-fate decisions.

The discovery of cancers with increased levels of H3K27me3 from hyperactivating mutations in *EZH2* point to a definite role of the catalytic activity of PRC2 and particularly H3K27me3 in cancers where PRC2 is playing an oncogenic role. In a case of PRC2 acting as a tumor suppressor, patients with benign neurofibromas retaining some PRC2 function, staining for H3K27me3 is lost upon progression to MPNST (Lee et al. 2014), indicating that complete loss of PRC2 function and H3K27me3 is contributing to malignant transformation in these cancers. In line with this, loss of H3K27me3 was recently shown to be a robust diagnostic marker for distinguishing sporadic and radiation-induced MPNSTs from other cancers in the differential diagnosis for MPNST (Prieto-Granada et al. 2015). In addition, the demethylases responsible for removing H3K27me3 are also deregulated in human cancers with KDM6A/*UTX* mutations in several cancer types (van Haaften et al. 2009; Gui et al. 2011; Jankowska et al. 2011; Ross et al. 2014; Van der Meulen et al. 2015) and KDM6B/JMJD3 overexpression in Hodgkin's lymphoma and T-ALL associated with loss of H3K27me3 (Anderton et al. 2011; Simon et al. 2012).

Seemingly direct implications of the histone substrate of PRC2 in cancer development have come from recent findings of somatic mutations, causing a lysine-to-methionine substitution at residue 27 in the H3 tail in genes encoding histone variants H3.1 and H3.3 in pediatric brain tumors (Schwartzentruber et al. 2012; Wu et al. 2012). Global levels of H3K27me3 are diminished in these tumors, putatively because of the inhibitory effect of H3K27M on PRC2 activity (Bender et al. 2013; Chan et al. 2013; Lewis et al. 2013; Venneti et al. 2013), and these mutants might primarily exert their tumorigenic function through their conferred loss-of-function of PRC2. Despite the global reduction in H3K27me3, PRC2 occupancy and H3K27me3 are specifically enriched at certain PRC2 target genes, correlating with lower expression of these genes. Conversely, PRC2 target genes with lower levels of H3K27me3 in cells carrying the H3K27M mutant were found to have reduced levels of DNA methylation and increased expression levels, indicating that the functional consequence of H3K27M mutation includes aberrant transcriptional activation as well as defective repression of PRC2 targets (Bender et al. 2013; Chan et al. 2013).

HOW DOES DEREGULATED PRC2 ACTIVITY CONTRIBUTE TO TUMORIGENESIS?

PRC2 is important for balancing proliferation versus differentiation, and PRC2 has been shown to promote a de-differentiated phenotype of several cancers (Richter et al. 2009; Tanaka et al. 2012; Beguelin et al. 2013). Increased PRC2 activity (be it through overexpression or hyperactive mutants) in stem and progenitor cells might promote self-renewal over differentiation by increasing the thresholds for transcriptional activation of differentiation-associated PRC2 target genes or genes controlling cell proliferation such as *CDKN2A*. A central role of *CDKN2A* in relation to PRC2 mutants is supported by findings that knockout of the *Cdkn2a* locus in combination with *Eed* knockout partially rescues the growth phenotype of both HSCs (Xie et al. 2014) and leukemic cells in an MLL-AF9 mouse model (Shi et al. 2013; Danis et al. 2015), as well as the co-occurrence of *CDKN2A* mutations and loss-of-function mutations in *EED* and *SUZ12* in MPNSTs (Lee et al. 2014) and the requirement of PRC2-

mediated repression of *CDKN2A* for proliferation of rhabdoid tumors (Kia et al. 2008).

Another plausible explanation is that deregulation of PRC2 activity in either direction alters thresholds for gene activation, promoting epigenetic instability and giving rise to transcriptional deregulation, thereby increasing the risk of cancer development (Brock et al. 2015). In addition to the direct misregulation of *CDKN2A* in some cancers with PRC2 alterations, this explanation might provide a more general model of how deregulated PRC2 activity promotes oncogenesis. This would also explain the difficulties with identifying key deregulated genes or conserved gene expression signatures in cancers dependent on PRC2 and with altered PRC2 activity.

TARGETING PRC2 IN CANCER

Knockdown studies in cancer cells with increased PRC2 levels show that many of these cancers depend on PRC2 for proliferation (Albert and Helin 2010). Along with the discovery of mutations in *EZH2* leading to hyperactive PRC2, this has prompted the hope of therapeutically targeting PRC2 in cancer. Several companies have developed small molecule inhibitors specific for EZH2, and their efficiency in targeting lymphoma cells harboring hyperactivated EZH2 has been demonstrated both in vitro and in preclinical mouse models (Knutson et al. 2012, 2014a; McCabe et al. 2012b; Qi et al. 2012; Beguelin et al. 2013; Garapaty-Rao et al. 2013; Bradley et al. 2014; Campbell et al. 2015). The inhibitors are all *S*-adenosylmethionine (SAM, cofactor) competitive inhibitors, and, perhaps because the hyperactivity-conferring mutations map in the substrate-binding pocket (Fig. 2B), these inhibitors are not selective for mutant versus wild-type EZH2.

Three companies (Epizyme, Cambridge, MA; GlaxoSmithKline, Parsippany, NJ; and Constellation Pharmaceuticals, Cambridge, MA) have moved their EZH2 inhibitors forward into phase 1 clinical trials in the last couple of years. Whereas the phase 1 clinical trials are primarily aimed at establishing acceptable safety profiles for the compounds, the three com-panies have all pointed at B-cell lymphomas with hyperactive mutant EZH2 as the primary indication to target. Recently, the FDA approved that Epizyme can initiate phase 2 clinical trials for the treatment of patients with relapsed DLBCL (Epizyme, press release). Interestingly, the Epizyme EZH2 inhibitor Tazemostat (also known as EPZ-6438) has shown promising antitumor activity in nine out of 15 DLBCL patients of which only one patient expressed hyperactive mutant EZH2 (Epizyme, press release). This intriguing finding is in line with several studies suggesting that targeting EZH2 might be a relevant therapeutic strategy in cancers without hyperactivated EZH2. In addition to EZH2 inhibition effectively targeting lymphomas without hyperactivated EZH2 (Beguelin et al. 2013; Bradley et al. 2014), these studies have shown that EZH2 inhibition shows synthetic lethality with mutations in components of the SWI/SNF remodeling complex (Wilson et al. 2010; Knutson et al. 2013; Bitler et al. 2015; Fillmore et al. 2015; Kim et al. 2015) BAP1 (La-Fave et al. 2015), and potentially also with mutations in *UTX* (Van der Meulen et al. 2015).

With PRC2 mutational status not being sufficient to decide whether a patient will respond to EZH2 inhibition, identification of robust biomarkers that can be used to stratify patients correctly becomes a major challenge. As mentioned above, PRC2 mutations may primarily contribute to oncogenesis through a destabilization of the chromatin environment by altering the thresholds for gene activation and increasing transcriptional noise. Thus, the gene expression changes ultimately promoting cancer development might well be diverse and unique to each patient. This is supported by the identification of very few common gene expression changes following EZH2 inhibition in patient-derived DLBCL cell lines (McCabe et al. 2012b; Beguelin et al. 2013). Analysis of gene expression signatures may thus not be a viable strategy for defining responders to treatment with EZH2 inhibitors.

Collectively, therapeutically targeting PRC2 in cancer through EZH2 inhibition has very promising prospects, and, with the particular requirement of PRC2 in stem and progenitor cells, might present a means of targeting the cancer stem cell compartment. However, the fact that PRC2 acts as a tumor suppressor in some cancer types along with the sensitivity of many cell types to alterations in PRC2 dosage raises some safety concerns and underlines the need to identify potential responders and minimize adverse effects of EZH2 inhibition in the clinic. It has already been demonstrated that secondary mutations can be acquired in tumors carrying either wild-type or mutated *EZH2* alleles under prolonged inhibitor treatment, thus making cancer cells refractory to current inhibitors (Baker et al. 2015; Gibaja et al. 2016). Such findings highlight the need to expand the range of EZH2 inhibitors to include a repertoire of inhibitors targeting potential resistance mutants in the clinic.

CONCLUDING REMARKS

In accordance with their crucial functions in specification and preservation of cell fate, the genes encoding the PRC2 components are often found mutated or deregulated in human cancers. The fact that both activating and inactivating mutations can contribute to the development of cancer is consistent with the general role of PRC2 in maintaining transcriptional programs and thereby guarding cell identity. Deregulation of PRC2 activity does not in itself lead to changes in transcriptional programs, but it alters the thresholds for gene activation. Therefore, the apparent role of PRC2 becomes context-dependent, that is, the phenotypic consequences of deregulated PRC2 activity are determined by the cellular environment and other mutations contributing to tumor development. With this in mind, it might not be surprising that efforts to identify common gene expression patterns in PRC2-dependent tumors have been futile. Apart from a crucial role of PRC2 in maintaining the transcriptional repression of the *CDKN2A* locus, genes have not been identified that indicate PRC2-dependency in tumors. However, despite this complexity, several companies have developed specific EZH2 inhibitors that have shown promising results in both preclinical and clinical trials. Thus far,

the clinical trials have focused on cancers overexpressing EZH2 (e.g., DLBCL and FL), cancers expressing hyperactivating mutants of EZH2 (e.g., DLBCL and FL), and cancers with mutations in genes coding for the SWI/SNF complex (e.g., rhabdoid tumors and synovial sarcomas). If further success is achieved in these clinical trials, they could be extended to other cancers with somatic mutations that show synthetic lethality with PRC2 inhibition (e.g., BAP1 mutated mesotheliomas [LaFave et al. 2015]). While EZH2 inhibitors have until now been used as single agents, several studies have suggested that combination therapies might be a viable strategy (Beguelin et al. 2013; De Raedt et al. 2014; Knutson et al. 2014b; Fillmore et al. 2015), which could be exploited.

ACKNOWLEDGMENTS

We thank members of the Helin laboratory for discussions. The work in the Helin laboratory is supported by The European Research Council (294666_DNAMET), the 7th framework program of the European Union (4DCellFate and INGENIUM), the Danish Cancer Society, the Danish National Research Foundation (DNRF 82), the Danish Council for Strategic Research, the Danish Medical Research Council, the Novo Nordisk Foundation, The Lundbeck Foundation, and through a center grant from the Novo Nordisk Foundation (The Novo Nordisk Foundation Section for Stem Cell Biology in Human Disease).

REFERENCES

Abdel-Wahab O, Pardanani A, Patel J, Wadleigh M, Lasho T, Heguy A, Beran M, Gilliland DG, Levine RL, Tefferi A. 2011. Concomitant analysis of EZH2 and ASXL1 mutations in myelofibrosis, chronic myelomonocytic leukemia and blast-phase myeloproliferative neoplasms. *Leukemia* **25:** 1200–1202.

Alabert C, Barth TK, Reveron-Gomez N, Sidoli S, Schmidt A, Jensen ON, Imhof A, Groth A. 2015. Two distinct modes for propagation of histone PTMs across the cell cycle. *Genes Dev* **29:** 585–590.

Albert M, Helin K. 2010. Histone methyltransferases in cancer. *Semin Cell Dev Biol* **21:** 209–220.

Alexandrov LB, Nik-Zainal S, Wedge DC, Aparicio SA, Behjati S, Biankin AV, Bignell GR, Bolli N, Borg A, Borresen-

Dale AL, et al. 2013. Signatures of mutational processes in human cancer. *Nature* **500:** 415–421.

Anderton JA, Bose S, Vockerodt M, Vrzalikova K, Wei W, Kuo M, Helin K, Christensen J, Rowe M, Murray PG, et al. 2011. The H3K27me3 demethylase, KDM6B, is induced by Epstein–Barr virus and over-expressed in Hodgkin's lymphoma. *Oncogene* **30:** 2037–2043.

Antonysamy S, Condon B, Druzina Z, Bonanno JB, Gheyi T, Zhang F, MacEwan I, Zhang A, Ashok S, Rodgers L, et al. 2013. Structural context of disease-associated mutations and putative mechanism of autoinhibition revealed by X-ray crystallographic analysis of the EZH2-SET domain. *PLoS ONE* **8:** e84147.

Baker T, Nerle S, Pritchard J, Zhao B, Rivera VM, Garner A, Gonzalvez F. 2015. Acquisition of a single EZH2 D1 domain mutation confers acquired resistance to EZH2-targeted inhibitors. *Oncotarget* **6:** 32646–32655.

Beguelin W, Popovic R, Teater M, Jiang Y, Bunting KL, Rosen M, Shen H, Yang SN, Wang L, Ezponda T, et al. 2013. EZH2 is required for germinal center formation and somatic EZH2 mutations promote lymphoid transformation. *Cancer Cell* **23:** 677–692.

Bejar R, Stevenson K, Abdel-Wahab O, Galili N, Nilsson B, Garcia-Manero G, Kantarjian H, Raza A, Levine RL, Neuberg D, et al. 2011. Clinical effect of point mutations in myelodysplastic syndromes. *N Engl J Med* **364:** 2496–2506.

Bender S, Tang Y, Lindroth AM, Hovestadt V, Jones DT, Kool M, Zapatka M, Northcott PA, Sturm D, Wang W, et al. 2013. Reduced H3K27me3 and DNA hypomethylation are major drivers of gene expression in K27M mutant pediatric high-grade gliomas. *Cancer Cell* **24:** 660–672.

Berg T, Thoene S, Yap D, Wee T, Schoeler N, Rosten P, Lim E, Bilenky M, Mungall AJ, Oellerich T, et al. 2014. A transgenic mouse model demonstrating the oncogenic role of mutations in the polycomb-group gene EZH2 in lymphomagenesis. *Blood* **123:** 3914–3924.

Bitler BG, Aird KM, Garipov A, Li H, Amatangelo M, Kossenkov AV, Schultz DC, Liu Q, Shih Ie M, Conejo-Garcia JR, et al. 2015. Synthetic lethality by targeting EZH2 methyltransferase activity in ARID1A-mutated cancers. *Nat Med* **21:** 231–238.

Blackledge NP, Farcas AM, Kondo T, King HW, McGouran JF, Hanssen LL, Ito S, Cooper S, Kondo K, Koseki Y, et al. 2014. Variant PRC1 complex-dependent H2A ubiquitylation drives PRC2 recruitment and Polycomb domain formation. *Cell* **157:** 1445–1459.

Blackledge NP, Rose NR, Klose RJ. 2015. Targeting Polycomb systems to regulate gene expression: Modifications to a complex story. *Nat Rev Mol Cell Biol* **16:** 643–649.

Bodor C, O'Riain C, Wrench D, Matthews J, Iyengar S, Tayyib H, Calaminici M, Clear A, Iqbal S, Quentmeier H, et al. 2011. EZH2 Y641 mutations in follicular lymphoma. *Leukemia* **25:** 726–729.

Bodor C, Grossmann V, Popov N, Okosun J, O'Riain C, Tan K, Marzec J, Araf S, Wang J, Lee AM, et al. 2013. EZH2 mutations are frequent and represent an early event in follicular lymphoma. *Blood* **122:** 3165–3168.

Boyer LA, Plath K, Zeitlinger J, Brambrink T, Medeiros LA, Lee TI, Levine SS, Wernig M, Tajonar A, Ray MK, et al. 2006. Polycomb complexes repress developmental regulators in murine embryonic stem cells. *Nature* **441:** 349–353.

Bracken AP, Pasini D, Capra M, Prosperini E, Colli E, Helin K. 2003. EZH2 is downstream of the pRB-E2F pathway, essential for proliferation and amplified in cancer. *EMBO J* **22:** 5323–5335.

Bracken AP, Dietrich N, Pasini D, Hansen KH, Helin K. 2006. Genome-wide mapping of Polycomb target genes unravels their roles in cell fate transitions. *Genes Dev* **20:** 1123–1136.

Bradley WD, Arora S, Busby J, Balasubramanian S, Gehling VS, Nasveschuk CG, Vaswani RG, Yuan CC, Hatton C, Zhao F, et al. 2014. EZH2 inhibitor efficacy in non-Hodgkin's lymphoma does not require suppression of H3K27 monomethylation. *Chem Biol* **21:** 1463–1475.

Brock A, Krause S, Ingber DE. 2015. Control of cancer formation by intrinsic genetic noise and microenvironmental cues. *Nat Rev Cancer* **15:** 499–509.

Brookes E, de Santiago I, Hebenstreit D, Morris KJ, Carroll T, Xie SQ, Stock JK, Heidemann M, Eick D, Nozaki N, et al. 2012. Polycomb associates genome-wide with a specific RNA polymerase II variant, and regulates metabolic genes in ESCs. *Cell Stem Cell* **10:** 157–170.

Campbell JE, Kuntz KW, Knutson SK, Warholic NM, Keilhack H, Wigle TJ, Raimondi A, Klaus CR, Rioux N, Yokoi A, et al. 2015. EPZ011989, a potent, orally-available EZH2 inhibitor with robust in vivo activity. *ACS Med Chem Lett* **6:** 491–495.

Cao R, Wang L, Wang H, Xia L, Erdjument-Bromage H, Tempst P, Jones RS, Zhang Y. 2002. Role of histone H3 lysine 27 methylation in Polycomb-group silencing. *Science* **298:** 1039–1043.

Chamberlain SJ, Yee D, Magnuson T. 2008. Polycomb repressive complex 2 is dispensable for maintenance of embryonic stem cell pluripotency. *Stem Cells* **26:** 1496–1505.

Chan KM, Fang D, Gan H, Hashizume R, Yu C, Schroeder M, Gupta N, Mueller S, James CD, Jenkins R, et al. 2013. The histone H3.3K27M mutation in pediatric glioma reprograms H3K27 methylation and gene expression. *Genes Dev* **27:** 985–990.

Ciferri C, Lander GC, Maiolica A, Herzog F, Aebersold R, Nogales E. 2012. Molecular architecture of human Polycomb repressive complex 2. *eLife* **1:** e00005.

Cooper S, Dienstbier M, Hassan R, Schermelleh L, Sharif J, Blackledge NP, De Marco V, Elderkin S, Koseki H, Klose R, et al. 2014. Targeting Polycomb to pericentric heterochromatin in embryonic stem cells reveals a role for H2AK119u1 in PRC2 recruitment. *Cell Rep* **7:** 1456–1470.

Cromer MK, Starker LF, Choi M, Udelsman R, Nelson-Williams C, Lifton RP, Carling T. 2012. Identification of somatic mutations in parathyroid tumors using whole-exome sequencing. *J Clin Endocrinol Metab* **97:** E1774–E1781.

Danis E, Yamauchi T, Echanique K, Haladyna J, Kalkur R, Riedel S, Zhu N, Xie H, Bernt KM, Orkin SH, et al. 2015. Inactivation of *Eed* impedes *MLL-AF9*-mediated leukemogenesis through *Cdkn2a*-dependent and *Cdkn2a*-independent mechanisms in a murine model. *Exp Hematol* **43:** 930–935.e6.

Deb G, Singh AK, Gupta S. 2014. EZH2: Not EZHY (easy) to deal. *Mol Cancer Res* **12:** 639–653.

Dellino GI, Schwartz YB, Farkas G, McCabe D, Elgin SC, Pirrotta V. 2004. Polycomb silencing blocks transcription initiation. *Mol Cell* **13:** 887–893.

De Raedt T, Beert E, Pasmant E, Luscan A, Brems H, Ortonne N, Helin K, Hornick JL, Mautner V, Kehrer-Sawatzki H, et al. 2014. PRC2 loss amplifies Ras-driven transcription and confers sensitivity to BRD4-based therapies. *Nature* **514:** 247–251.

Di Croce L, Helin K. 2013. Transcriptional regulation by Polycomb group proteins. *Nat Struct Mol Biol* **20:** 1147–1155.

Enderle D, Beisel C, Stadler MB, Gerstung M, Athri P, Paro R. 2011. Polycomb preferentially targets stalled promoters of coding and noncoding transcripts. *Genome Res* **21:** 216–226.

Ernst T, Chase AJ, Score J, Hidalgo-Curtis CE, Bryant C, Jones AV, Waghorn K, Zoi K, Ross FM, Reiter A, et al. 2010. Inactivating mutations of the histone methyltransferase gene *EZH2* in myeloid disorders. *Nat Genet* **42:** 722–726.

Farcas AM, Blackledge NP, Sudbery I, Long HK, McGouran JF, Rose NR, Lee S, Sims D, Cerase A, Sheahan TW, et al. 2012. KDM2B links the Polycomb repressive complex 1 (PRC1) to recognition of CpG islands. *eLife* **1:** e00205.

Ferrari KJ, Scelfo A, Jammula S, Cuomo A, Barozzi I, Stutzer A, Fischle W, Bonaldi T, Pasini D. 2014. Polycomb-dependent H3K27me1 and H3K27me2 regulate active transcription and enhancer fidelity. *Mol Cell* **53:** 49–62.

Fillmore CM, Xu C, Desai PT, Berry JM, Rowbotham SP, Lin YJ, Zhang H, Marquez VE, Hammerman PS, Wong KK, et al. 2015. EZH2 inhibition sensitizes BRG1 and EGFR mutant lung tumours to TopoII inhibitors. *Nature* **520:** 239–242.

Fischle W, Wang Y, Jacobs SA, Kim Y, Allis CD, Khorasanizadeh S. 2003. Molecular basis for the discrimination of repressive methyl-lysine marks in histone H3 by Polycomb and HP1 chromodomains. *Genes Dev* **17:** 1870–1881.

Francis NJ, Kingston RE, Woodcock CL. 2004. Chromatin compaction by a Polycomb group protein complex. *Science* **306:** 1574–1577.

Gao Z, Zhang J, Bonasio R, Strino F, Sawai A, Parisi F, Kluger Y, Reinberg D. 2012. PCGF homologs, CBX proteins, and RYBP define functionally distinct PRC1 family complexes. *Mol Cell* **45:** 344–356.

Garapaty-Rao S, Nasveschuk C, Gagnon A, Chan EY, Sandy P, Busby J, Balasubramanian S, Campbell R, Zhao F, Bergeron L, et al. 2013. Identification of EZH2 and EZH1 small molecule inhibitors with selective impact on diffuse large B cell lymphoma cell growth. *Chem Biol* **20:** 1329–1339.

Gibaja V, Shen F, Harari J, Korn J, Ruddy D, Saenz-Vash V, Zhai H, Rejtar T, Paris CG, Yu Z, et al. 2016. Development of secondary mutations in wild-type and mutant EZH2 alleles cooperates to confer resistance to EZH2 inhibitors. *Oncogene* **35:** 558–566.

Guglielmelli P, Biamonte F, Score J, Hidalgo-Curtis C, Cervantes F, Maffioli M, Fanelli T, Ernst T, Winkelman N, Jones AV, et al. 2011. *EZH2* mutational status predicts poor survival in myelofibrosis. *Blood* **118:** 5227–5234.

Gui Y, Guo G, Huang Y, Hu X, Tang A, Gao S, Wu R, Chen C, Li X, Zhou L, et al. 2011. Frequent mutations of chromatin remodeling genes in transitional cell carcinoma of the bladder. *Nat Genet* **43:** 875–878.

Hansen KH, Bracken AP, Pasini D, Dietrich N, Gehani SS, Monrad A, Rappsilber J, Lerdrup M, Helin K. 2008. A model for transmission of the H3K27me3 epigenetic mark. *Nat Cell Biol* **10:** 1291–1300.

Herrera-Merchan A, Arranz L, Ligos JM, de Molina A, Dominguez O, Gonzalez S. 2012. Ectopic expression of the histone methyltransferase Ezh2 in haematopoietic stem cells causes myeloproliferative disease. *Nat Commun* **3:** 623.

Hodis E, Watson IR, Kryukov GV, Arold ST, Imielinski M, Theurillat JP, Nickerson E, Auclair D, Li L, Place C, et al. 2012. A landscape of driver mutations in melanoma. *Cell* **150:** 251–263.

Illingworth RS, Moffat M, Mann AR, Read D, Hunter CJ, Pradeepa MM, Adams IR, Bickmore WA. 2015. The E3 ubiquitin ligase activity of RING1B is not essential for early mouse development. *Genes Dev* **29:** 1897–1902.

Jankowska AM, Makishima H, Tiu RV, Szpurka H, Huang Y, Traina F, Visconte V, Sugimoto Y, Prince C, O'Keefe C, et al. 2011. Mutational spectrum analysis of chronic myelomonocytic leukemia includes genes associated with epigenetic regulation: *UTX, EZH2,* and *DNMT3A. Blood* **118:** 3932–3941.

Jiang T, Wang Y, Zhou F, Gao G, Ren S, Zhou C. 2015. Prognostic value of high EZH2 expression in patients with different types of cancer: A systematic review with meta-analysis. *Oncotarget* **7:** 4584–4597.

Jiao L, Liu X. 2015. Structural basis of histone H3K27 trimethylation by an active Polycomb repressive complex 2. *Science* **350:** aac4383.

Jones RS, Gelbart WM. 1990. Genetic analysis of the enhancer of zeste locus and its role in gene regulation in *Drosophila melanogaster. Genetics* **126:** 185–199.

Jung HR, Pasini D, Helin K, Jensen ON. 2010. Quantitative mass spectrometry of histones H3.2 and H3.3 in Suz12-deficient mouse embryonic stem cells reveals distinct, dynamic post-translational modifications at Lys-27 and Lys-36. *Mol Cell Proteomics* **9:** 838–850.

Jung HR, Sidoli S, Haldbo S, Sprenger RR, Schwammle V, Pasini D, Helin K, Jensen ON. 2013. Precision mapping of coexisting modifications in histone H3 tails from embryonic stem cells by ETD-MS/MS. *Anal Chem* **85:** 8232–8239.

Kalb R, Latwiel S, Baymaz HI, Jansen PW, Muller CW, Vermeulen M, Muller J. 2014. Histone H2A monoubiquitination promotes histone H3 methylation in Polycomb repression. *Nat Struct Mol Biol* **21:** 569–571.

Kaneko S, Son J, Bonasio R, Shen SS, Reinberg D. 2014. Nascent RNA interaction keeps PRC2 activity poised and in check. *Genes Dev* **28:** 1983–1988.

Kanhere A, Viiri K, Araujo CC, Rasaiyaah J, Bouwman RD, Whyte WA, Pereira CF, Brookes E, Walker K, Bell GW, et al. 2010. Short RNAs are transcribed from repressed Polycomb target genes and interact with Polycomb repressive complex-2. *Mol Cell* **38:** 675–688.

Ketel CS, Andersen EF, Vargas ML, Suh J, Strome S, Simon JA. 2005. Subunit contributions to histone methyltransferase activities of fly and worm Polycomb group complexes. *Mol Cell Biol* **25:** 6857–6868.

Kia SK, Gorski MM, Giannakopoulos S, Verrijzer CP. 2008. SWI/SNF mediates Polycomb eviction and epigenetic reprogramming of the INK4b-ARF-INK4a locus. *Mol Cell Biol* **28:** 3457–3464.

Kim KH, Kim W, Howard TP, Vazquez F, Tsherniak A, Wu JN, Wang W, Haswell JR, Walensky LD, Hahn WC, et al. 2015. SWI/SNF-mutant cancers depend on catalytic and non-catalytic activity of EZH2. *Nat Med* **21:** 1491–1496.

Kleer CG, Cao Q, Varambally S, Shen R, Ota I, Tomlins SA, Ghosh D, Sewalt RG, Otte AP, Hayes DF, et al. 2003. EZH2 is a marker of aggressive breast cancer and promotes neoplastic transformation of breast epithelial cells. *Proc Natl Acad Sci* **100:** 11606–11611.

Knutson SK, Wigle TJ, Warholic NM, Sneeringer CJ, Allain CJ, Klaus CR, Sacks JD, Raimondi A, Majer CR, Song J, et al. 2012. A selective inhibitor of EZH2 blocks H3K27 methylation and kills mutant lymphoma cells. *Nat Chem Biol* **8:** 890–896.

Knutson SK, Warholic NM, Wigle TJ, Klaus CR, Allain CJ, Raimondi A, Porter Scott M, Chesworth R, Moyer MP, Copeland RA, et al. 2013. Durable tumor regression in genetically altered malignant rhabdoid tumors by inhibition of methyltransferase EZH2. *Proc Natl Acad Sci* **110:** 7922–7927.

Knutson SK, Kawano S, Minoshima Y, Warholic NM, Huang KC, Xiao Y, Kadowaki T, Uesugi M, Kuznetsov G, Kumar N, et al. 2014a. Selective inhibition of EZH2 by EPZ-6438 leads to potent antitumor activity in EZH2 mutant non-Hodgkin lymphoma. *Mol Cancer Ther* **13:** 842–854.

Knutson SK, Warholic NM, Johnston LD, Klaus CR, Wigle TJ, Iwanowicz D, Littlefield BA, Porter-Scott M, Smith JJ, Moyer MP, et al. 2014b. Synergistic anti-tumor activity of EZH2 inhibitors and glucocorticoid receptor agonists in models of germinal center non-Hodgkin lymphomas. *PLoS ONE* **9:** e111840.

Krauthammer M, Kong Y, Ha BH, Evans P, Bacchiocchi A, McCusker JP, Cheng E, Davis MJ, Goh G, Choi M, et al. 2012. Exome sequencing identifies recurrent somatic *RAC1* mutations in melanoma. *Nat Genet* **44:** 1006–1014.

Ku M, Koche RP, Rheinbay E, Mendenhall EM, Endoh M, Mikkelsen TS, Presser A, Nusbaum C, Xie X, Chi AS, et al. 2008. Genomewide analysis of PRC1 and PRC2 occupancy identifies two classes of bivalent domains. *PLoS Genet* **4:** e1000242.

Kuzmichev A, Nishioka K, Erdjument-Bromage H, Tempst P, Reinberg D. 2002. Histone methyltransferase activity associated with a human multiprotein complex containing the Enhancer of Zeste protein. *Genes Dev* **16:** 2893–2905.

LaFave LM, Beguelin W, Koche R, Teater M, Spitzer B, Chramiec A, Papalexi E, Keller MD, Hricik T, Konstantinoff K, et al. 2015. Loss of BAP1 function leads to EZH2-dependent transformation. *Nat Med* **21:** 1344–1349.

Lagarou A, Mohd-Sarip A, Moshkin YM, Chalkley GE, Bezstarosti K, Demmers JA, Verrijzer CP. 2008. dKDM2 couples histone H2A ubiquitylation to histone H3 demethylation during Polycomb group silencing. *Genes Dev* **22:** 2799–2810.

Laugesen A, Helin K. 2014. Chromatin repressive complexes in stem cells, development, and cancer. *Cell Stem Cell* **14:** 735–751.

Lee SC, Phipson B, Hyland CD, Leong HS, Allan RS, Lun A, Hilton DJ, Nutt SL, Blewitt ME, Smyth GK, et al. 2013. Polycomb repressive complex 2 (PRC2) suppresses Eμ-*myc* lymphoma. *Blood* **122:** 2654–2663.

Lee W, Teckie S, Wiesner T, Ran L, Prieto Granada CN, Lin M, Zhu S, Cao Z, Liang Y, Sboner A, et al. 2014. PRC2 is recurrently inactivated through EED or SUZ12 loss in malignant peripheral nerve sheath tumors. *Nat Genet* **46:** 1227–1232.

Lee SC, Miller S, Hyland C, Kauppi M, Lebois M, Di Rago L, Metcalf D, Kinkel SA, Josefsson EC, Blewitt ME, et al. 2015. Polycomb repressive complex 2 component Suz12 is required for hematopoietic stem cell function and lymphopoiesis. *Blood* **126:** 167–175.

Leeb M, Pasini D, Novatchkova M, Jaritz M, Helin K, Wutz A. 2010. Polycomb complexes act redundantly to repress genomic repeats and genes. *Genes Dev* **24:** 265–276.

Lessard J, Schumacher A, Thorsteinsdottir U, van Lohuizen M, Magnuson T, Sauvageau G. 1999. Functional antagonism of the *Polycomb*-Group genes *eed* and *Bmi1* in hemopoietic cell proliferation. *Genes Dev* **13:** 2691–2703.

Lewis PW, Muller MM, Koletsky MS, Cordero F, Lin S, Banaszynski LA, Garcia BA, Muir TW, Becher OJ, Allis CD. 2013. Inhibition of PRC2 activity by a gain-of-function H3 mutation found in pediatric glioblastoma. *Science* **340:** 857–861.

Lindsley RC, Mar BG, Mazzola E, Grauman PV, Shareef S, Allen SL, Pigneux A, Wetzler M, Stuart RK, Erba HP, et al. 2015. Acute myeloid leukemia ontogeny is defined by distinct somatic mutations. *Blood* **125:** 1367–1376.

Lohr JG, Stojanov P, Lawrence MS, Auclair D, Chapuy B, Sougnez C, Cruz-Gordillo P, Knoechel B, Asmann YW, Slager SL, et al. 2012. Discovery and prioritization of somatic mutations in diffuse large B-cell lymphoma (DLBCL) by whole-exome sequencing. *Proc Natl Acad Sci* **109:** 3879–3884.

Lynch MD, Smith AJ, De Gobbi M, Flenley M, Hughes JR, Vernimmen D, Ayyub H, Sharpe JA, Sloane-Stanley JA, Sutherland L, et al. 2012. An interspecies analysis reveals a key role for unmethylated CpG dinucleotides in vertebrate Polycomb complex recruitment. *EMBO J* **31:** 317–329.

Majer CR, Jin L, Scott MP, Knutson SK, Kuntz KW, Keilhack H, Smith JJ, Moyer MP, Richon VM, Copeland RA, et al. 2012. A687V EZH2 is a gain-of-function mutation found in lymphoma patients. *FEBS Lett* **586:** 3448–3451.

Majewski IJ, Ritchie ME, Phipson B, Corbin J, Pakusch M, Ebert A, Busslinger M, Koseki H, Hu Y, Smyth GK, et al. 2010. Opposing roles of Polycomb repressive complexes in hematopoietic stem and progenitor cells. *Blood* **116:** 731–739.

Margueron R, Justin N, Ohno K, Sharpe ML, Son J, Drury WJ III, Voigt P, Martin SR, Taylor WR, De Marco V, et al. 2009. Role of the Polycomb protein EED in the propagation of repressive histone marks. *Nature* **461:** 762–767.

McCabe MT, Graves AP, Ganji G, Diaz E, Halsey WS, Jiang Y, Smitheman KN, Ott HM, Pappalardi MB, Allen KE, et al. 2012a. Mutation of A677 in histone methyltransferase EZH2 in human B-cell lymphoma promotes hypertrimethylation of histone H3 on lysine 27 (H3K27). *Proc Natl Acad Sci* **109**: 2989–2994.

McCabe MT, Ott HM, Ganji G, Korenchuk S, Thompson C, Van Aller GS, Liu Y, Graves AP, Della Pietra A III, Diaz E, et al. 2012b. EZH2 inhibition as a therapeutic strategy for lymphoma with EZH2-activating mutations. *Nature* **492**: 108–112.

Mendenhall EM, Koche RP, Truong T, Zhou VW, Issac B, Chi AS, Ku M, Bernstein BE. 2010. GC-rich sequence elements recruit PRC2 in mammalian ES cells. *PLoS Genet* **6**: e1001244.

Min J, Zhang Y, Xu RM. 2003. Structural basis for specific binding of Polycomb chromodomain to histone H3 methylated at Lys 27. *Genes Dev* **17**: 1823–1828.

Mochizuki-Kashio M, Aoyama K, Sashida G, Oshima M, Tomioka T, Muto T, Wang C, Iwama A. 2015. Ezh2 loss in hematopoietic stem cells predisposes mice to develop heterogeneous malignancies in an Ezh1-dependent manner. *Blood* **126**: 1172–1183.

Montgomery ND, Yee D, Chen A, Kalantry S, Chamberlain SJ, Otte AP, Magnuson T. 2005. The murine Polycomb group protein Eed is required for global histone H3 lysine-27 methylation. *Curr Biol* **15**: 942–947.

Morin RD, Johnson NA, Severson TM, Mungall AJ, An J, Goya R, Paul JE, Boyle M, Woolcock BW, Kuchenbauer F, et al. 2010. Somatic mutations altering EZH2 (Tyr641) in follicular and diffuse large B-cell lymphomas of germinal-center origin. *Nat Genet* **42**: 181–185.

Morin RD, Mendez-Lago M, Mungall AJ, Goya R, Mungall KL, Corbett RD, Johnson NA, Severson TM, Chiu R, Field M, et al. 2011. Frequent mutation of histone-modifying genes in non-Hodgkin lymphoma. *Nature* **476**: 298–303.

Muller J, Hart CM, Francis NJ, Vargas ML, Sengupta A, Wild B, Miller EL, O'Connor MB, Kingston RE, Simon JA. 2002. Histone methyltransferase activity of a *Drosophila* Polycomb group repressor complex. *Cell* **111**: 197–208.

Muto T, Sashida G, Oshima M, Wendt GR, Mochizuki-Kashio M, Nagata Y, Sanada M, Miyagi S, Saraya A, Kamio A, et al. 2013. Concurrent loss of Ezh2 and Tet2 cooperates in the pathogenesis of myelodysplastic disorders. *J Exp Med* **210**: 2627–2639.

Neff T, Sinha AU, Kluk MJ, Zhu N, Khattab MH, Stein L, Xie H, Orkin SH, Armstrong SA. 2012. Polycomb repressive complex 2 is required for MLL-AF9 leukemia. *Proc Natl Acad Sci* **109**: 5028–5033.

Neumann M, Vosberg S, Schlee C, Heesch S, Schwartz S, Gokbuget N, Hoelzer D, Graf A, Krebs S, Bartram I, et al. 2015. Mutational spectrum of adult T-ALL. *Oncotarget* **6**: 2754–2766.

Nikoloski G, Langemeijer SM, Kuiper RP, Knops R, Massop M, Tonnissen ER, van der Heijden A, Scheele TN, Vandenberghe P, de Witte T, et al. 2010. Somatic mutations of the histone methyltransferase gene *EZH2* in myelodysplastic syndromes. *Nat Genet* **42**: 665–667.

Ntziachristos P, Tsirigos A, Van Vlierberghe P, Nedjic J, Trimarchi T, Flaherty MS, Ferres-Marco D, da Ros V, Tang Z, Siegle J, et al. 2012. Genetic inactivation of the Polycomb repressive complex 2 in T cell acute lymphoblastic leukemia. *Nat Med* **18**: 298–301.

O'Carroll D, Erhardt S, Pagani M, Barton SC, Surani MA, Jenuwein T. 2001. The *Polycomb*-group gene *Ezh2* is re-

quired for early mouse development. *Mol Cell Biol* **21**: 4330–4336.

Okosun J, Bodor C, Wang J, Araf S, Yang CY, Pan C, Boller S, Cittaro D, Bozek M, Iqbal S, et al. 2014. Integrated genomic analysis identifies recurrent mutations and evolution patterns driving the initiation and progression of follicular lymphoma. *Nat Genet* **46**: 176–181.

Orkin SH, Hochedlinger K. 2011. Chromatin connections to pluripotency and cellular reprogramming. *Cell* **145**: 835–850.

Orsi GA, Kasinathan S, Hughes KT, Saminadin-Peter S, Henikoff S, Ahmad K. 2014. High-resolution mapping defines the cooperative architecture of Polycomb response elements. *Genome Res* **24**: 809–820.

Ott HM, Graves AP, Pappalardi MB, Huddleston M, Halsey WS, Hughes AM, Groy A, Dul E, Jiang Y, Bai Y, et al. 2014. A687V EZH2 is a driver of histone H3 lysine 27 (H3K27) hypertrimethylation. *Mol Cancer Ther* **13**: 3062–3073.

Pasini D, Bracken AP, Jensen MR, Lazzerini Denchi E, Helin K. 2004. Suz12 is essential for mouse development and for EZH2 histone methyltransferase activity. *EMBO J* **23**: 4061–4071.

Pasini D, Bracken AP, Hansen JB, Capillo M, Helin K. 2007. The Polycomb group protein Suz12 is required for embryonic stem cell differentiation. *Mol Cell Biol* **27**: 3769–3779.

Pengelly AR, Copur O, Jackle H, Herzig A, Muller J. 2013. A histone mutant reproduces the phenotype caused by loss of histone-modifying factor Polycomb. *Science* **339**: 698–699.

Pengelly AR, Kalb R, Finkl K, Muller J. 2015. Transcriptional repression by PRC1 in the absence of H2A monoubiquitylation. *Genes Dev* **29**: 1487–1492.

Peters AH, Kubicek S, Mechtler K, O'Sullivan RJ, Derijck AA, Perez-Burgos L, Kohlmaier A, Opravil S, Tachibana M, Shinkai Y, et al. 2003. Partitioning and plasticity of repressive histone methylation states in mammalian chromatin. *Mol Cell* **12**: 1577–1589.

Prieto-Granada CN, Wiesner T, Messina JL, Jungbluth AA, Chi P, Antonescu CR. 2015. Loss of H3K27me3 expression is a highly sensitive marker for sporadic and radiation-induced MPNST. *Am J Surg Pathol* **40**: 479–489.

Puda A, Milosevic JD, Berg T, Klampfl T, Harutyunyan AS, Gisslinger B, Rumi E, Pietra D, Malcovati L, Elena C, et al. 2012. Frequent deletions of JARID2 in leukemic transformation of chronic myeloid malignancies. *Am J Hematol* **87**: 245–250.

Qi W, Chan H, Teng L, Li L, Chuai S, Zhang R, Zeng J, Li M, Fan H, Lin Y, et al. 2012. Selective inhibition of Ezh2 by a small molecule inhibitor blocks tumor cells proliferation. *Proc Natl Acad Sci* **109**: 21360–21365.

Qin J, Whyte WA, Anderssen E, Apostolou E, Chen HH, Akbarian S, Bronson RT, Hochedlinger K, Ramaswamy S, Young RA, et al. 2012. The Polycomb group protein L3mbtl2 assembles an atypical PRC1-family complex that is essential in pluripotent stem cells and early development. *Cell Stem Cell* **11**: 319–332.

Rastelli L, Chan CS, Pirrotta V. 1993. Related chromosome binding sites for *zeste*, suppressors of *zeste* and Polycomb group proteins in *Drosophila* and their dependence on *Enhancer of zeste* function. *EMBO J* **12**: 1513–1522.

Richie ER, Schumacher A, Angel JM, Holloway M, Rinchik EM, Magnuson T. 2002. The *Polycomb*-group gene *eed* regulates thymocyte differentiation and suppresses the development of carcinogen-induced T-cell lymphomas. *Oncogene* **21**: 299–306.

Richter GH, Plehm S, Fasan A, Rossler S, Unland R, Bennani-Baiti IM, Hotfilder M, Lowel D, von Luettichau I, Mossbrugger I, et al. 2009. EZH2 is a mediator of EWS/FLI1 driven tumor growth and metastasis blocking endothelial and neuro-ectodermal differentiation. *Proc Natl Acad Sci* **106**: 5324–5329.

Riising EM, Comet I, Leblanc B, Wu X, Johansen JV, Helin K. 2014. Gene silencing triggers Polycomb repressive complex 2 recruitment to CpG islands genome wide. *Mol Cell* **55**: 347–360.

Ross JS, Wang K, Al-Rohil RN, Nazeer T, Sheehan CE, Otto GA, He J, Palmer G, Yelensky R, Lipson D, et al. 2014. Advanced urothelial carcinoma: Next-generation sequencing reveals diverse genomic alterations and targets of therapy. *Mod Pathol* **27**: 271–280.

Sashida G, Harada H, Matsui H, Oshima M, Yui M, Harada Y, Tanaka S, Mochizuki-Kashio M, Wang C, Saraya A, et al. 2014. Ezh2 loss promotes development of myelodysplastic syndrome but attenuates its predisposition to leukaemic transformation. *Nat Commun* **5**: 4177.

Schumacher A, Faust C, Magnuson T. 1996. Positional cloning of a global regulator of anterior–posterior patterning in mice. *Nature* **384**: 648.

Schwartzentruber J, Korshunov A, Liu XY, Jones DT, Pfaff E, Jacob K, Sturm D, Fontebasso AM, Quang DA, Tonjes M, et al. 2012. Driver mutations in histone H3.3 and chromatin remodelling genes in paediatric glioblastoma. *Nature* **482**: 226–231.

Score J, Hidalgo-Curtis C, Jones AV, Winkelmann N, Skinner A, Ward D, Zoi K, Ernst T, Stegelmann F, Dohner K, et al. 2012. Inactivation of Polycomb repressive complex 2 components in myeloproliferative and myelodysplastic/myeloproliferative neoplasms. *Blood* **119**: 1208–1213.

Shen X, Liu Y, Hsu YJ, Fujiwara Y, Kim J, Mao X, Yuan GC, Orkin SH. 2008. EZH1 mediates methylation on histone H3 lysine 27 and complements EZH2 in maintaining stem cell identity and executing pluripotency. *Mol Cell* **32**: 491–502.

Shi J, Wang E, Zuber J, Rappaport A, Taylor M, Johns C, Lowe SW, Vakoc CR. 2013. The Polycomb complex PRC2 supports aberrant self-renewal in a mouse model of MLL-AF9;Nras^{G12D} acute myeloid leukemia. *Oncogene* **32**: 930–938.

Simon J, Chiang A, Bender W. 1992. Ten different *Polycomb* group genes are required for spatial control of the *abdA* and *AbdB* homeotic products. *Development* **114**: 493–505.

Simon C, Chagraoui J, Krosl J, Gendron P, Wilhelm B, Lemieux S, Boucher G, Chagnon P, Drouin S, Lambert R, et al. 2012. A key role for *EZH2* and associated genes in mouse and human adult T-cell acute leukemia. *Genes Dev* **26**: 651–656.

Smits AH, Jansen PW, Poser I, Hyman AA, Vermeulen M. 2013. Stoichiometry of chromatin-associated protein complexes revealed by label-free quantitative mass spectrometry-based proteomics. *Nucleic Acids Res* **41**: e28.

Sneeringer CJ, Scott MP, Kuntz KW, Knutson SK, Pollock RM, Richon VM, Copeland RA. 2010. Coordinated activities of wild-type plus mutant EZH2 drive tumor-associated hypertrimethylation of lysine 27 on histone H3 (H3K27) in human B-cell lymphomas. *Proc Natl Acad Sci* **107**: 20980–20985.

Stepanik VA, Harte PJ. 2012. A mutation in the E(Z) methyltransferase that increases trimethylation of histone H3 lysine 27 and causes inappropriate silencing of active Polycomb target genes. *Dev Biol* **364**: 249–258.

Struhl G, Akam M. 1985. Altered distributions of Ultrabithorax transcripts in extra sex combs mutant embryos of *Drosophila*. *EMBO J* **4**: 3259–3264.

Swalm BM, Knutson SK, Warholic NM, Jin L, Kuntz KW, Keilhack H, Smith JJ, Pollock RM, Moyer MP, Scott MP, et al. 2014. Reaction coupling between wild-type and disease-associated mutant EZH2. *ACS Chem Biol* **9**: 2459–2464.

Tanaka S, Miyagi S, Sashida G, Chiba T, Yuan J, Mochizuki-Kashio M, Suzuki Y, Sugano S, Nakaseko C, Yokote K, et al. 2012. Ezh2 augments leukemogenicity by reinforcing differentiation blockage in acute myeloid leukemia. *Blood* **120**: 1107–1117.

Tanay A, O'Donnell AH, Damelin M, Bestor TH. 2007. Hyperconserved CpG domains underlie Polycomb-binding sites. *Proc Natl Acad Sci* **104**: 5521–5526.

Tavares L, Dimitrova E, Oxley D, Webster J, Poot R, Demmers J, Bezstarosti K, Taylor S, Ura H, Koide H, et al. 2012. RYBP-PRC1 complexes mediate H2A ubiquitylation at Polycomb target sites independently of PRC2 and H3K27me3. *Cell* **148**: 664–678.

Van der Meulen J, Sanghvi V, Mavrakis K, Durinck K, Fang F, Matthijssens F, Rondou P, Rosen M, Pieters T, Vandenberghe P, et al. 2015. The H3K27me3 demethylase UTX is a gender-specific tumor suppressor in T-cell acute lymphoblastic leukemia. *Blood* **125**: 13–21.

van Haaften G, Dalgliesh GL, Davies H, Chen L, Bignell G, Greenman C, Edkins S, Hardy C, O'Meara S, Teague J, et al. 2009. Somatic mutations of the histone H3K27 demethylase gene *UTX* in human cancer. *Nat Genet* **41**: 521–523.

Varambally S, Dhanasekaran SM, Zhou M, Barrette TR, Kumar-Sinha C, Sanda MG, Ghosh D, Pienta KJ, Sewalt RG, Otte AP, et al. 2002. The Polycomb group protein EZH2 is involved in progression of prostate cancer. *Nature* **419**: 624–629.

Velichutina I, Shaknovich R, Geng H, Johnson NA, Gascoyne RD, Melnick AM, Elemento O. 2010. EZH2-mediated epigenetic silencing in germinal center B cells contributes to proliferation and lymphomagenesis. *Blood* **116**: 5247–5255.

Venneti S, Garimella MT, Sullivan LM, Martinez D, Huse JT, Heguy A, Santi M, Thompson CB, Judkins AR. 2013. Evaluation of histone 3 lysine 27 trimethylation (H3K27me3) and enhancer of Zest 2 (EZH2) in pediatric glial and glioneuronal tumors shows decreased H3K27me3 in H3F3A K27M mutant glioblastomas. *Brain Pathol* **23**: 558–564.

Wang L, Brown JL, Cao R, Zhang Y, Kassis JA, Jones RS. 2004. Hierarchical recruitment of Polycomb group silencing complexes. *Mol Cell* **14**: 637–646.

Wilson BG, Wang X, Shen X, McKenna ES, Lemieux ME, Cho YJ, Koellhoffer EC, Pomeroy SL, Orkin SH, Roberts CW. 2010. Epigenetic antagonism between Polycomb and SWI/SNF complexes during oncogenic transformation. *Cancer Cell* **18:** 316–328.

Wu H, Min J, Lunin VV, Antoshenko T, Dombrovski L, Zeng H, Allali-Hassani A, Campagna-Slater V, Vedadi M, Arrowsmith CH, et al. 2010. Structural biology of human H3K9 methyltransferases. *PLoS ONE* **5:** e8570.

Wu G, Broniscer A, McEachron TA, Lu C, Paugh BS, Becksfort J, Qu C, Ding L, Huether R, Parker M, et al. 2012. Somatic histone H3 alterations in pediatric diffuse intrinsic pontine gliomas and non-brainstem glioblastomas. *Nat Genet* **44:** 251–253.

Wu H, Zeng H, Dong A, Li F, He H, Senisterra G, Seitova A, Duan S, Brown PJ, Vedadi M, et al. 2013a. Structure of the catalytic domain of EZH2 reveals conformational plasticity in cofactor and substrate binding sites and explains oncogenic mutations. *PLoS ONE* **8:** e83737.

Wu X, Johansen JV, Helin K. 2013b. Fbxl10/Kdm2b recruits Polycomb repressive complex 1 to CpG islands and regulates H2A ubiquitylation. *Mol Cell* **49:** 1134–1146.

Xie H, Xu J, Hsu JH, Nguyen M, Fujiwara Y, Peng C, Orkin SH. 2014. Polycomb repressive complex 2 regulates normal hematopoietic stem cell function in a developmental-stage-specific manner. *Cell Stem Cell* **14:** 68–80.

Xu C, Bian C, Yang W, Galka M, Ouyang H, Chen C, Qiu W, Liu H, Jones AE, MacKenzie F, et al. 2010. Binding of different histone marks differentially regulates the activity and specificity of Polycomb repressive complex 2 (PRC2). *Proc Natl Acad Sci* **107:** 19266–19271.

Yap DB, Chu J, Berg T, Schapira M, Cheng SW, Moradian A, Morin RD, Mungall AJ, Meissner B, Boyle M, et al. 2011. Somatic mutations at EZH2 Y641 act dominantly through a mechanism of selectively altered PRC2 catalytic activity, to increase H3K27 trimethylation. *Blood* **117:** 2451–2459.

You JS, Jones PA. 2012. Cancer genetics and epigenetics: Two sides of the same coin? *Cancer Cell* **22:** 9–20.

Zee BM, Britton LM, Wolle D, Haberman DM, Garcia BA. 2012. Origins and formation of histone methylation across the human cell cycle. *Mol Cell Biol* **32:** 2503–2514.

Zhang M, Wang Y, Jones S, Sausen M, McMahon K, Sharma R, Wang Q, Belzberg AJ, Chaichana K, Gallia GL, et al. 2014. Somatic mutations of *SUZ12* in malignant peripheral nerve sheath tumors. *Nat Genet* **46:** 1170–1172.

Zingg D, Debbache J, Schaefer SM, Tuncer E, Frommel SC, Cheng P, Arenas-Ramirez N, Haeusel J, Zhang Y, Bonalli M, et al. 2015. The epigenetic modifier EZH2 controls melanoma growth and metastasis through silencing of distinct tumour suppressors. *Nat Commun* **6:** 6051.

The Role of Nuclear Receptor–Binding SET Domain Family Histone Lysine Methyltransferases in Cancer

Richard L. Bennett, Alok Swaroop, Catalina Troche, and Jonathan D. Licht

Departments of Medicine, Biochemistry and Molecular Biology and University of Florida Health Cancer Center, The University of Florida, Gainesville, Florida 32610

Correspondence: jdlicht@ufl.edu

The nuclear receptor–binding SET Domain (NSD) family of histone H3 lysine 36 methyltransferases is comprised of NSD1, NSD2 (MMSET/WHSC1), and NSD3 (WHSC1L1). These enzymes recognize and catalyze methylation of histone lysine marks to regulate chromatin integrity and gene expression. The growing number of reports demonstrating that alterations or translocations of these genes fundamentally affect cell growth and differentiation leading to developmental defects illustrates the importance of this family. In addition, overexpression, gain of function somatic mutations, and translocations of NSDs are associated with human cancer and can trigger cellular transformation in model systems. Here we review the functions of NSD family members and the accumulating evidence that these proteins play key roles in tumorigenesis. Because epigenetic therapy is an important emerging anticancer strategy, understanding the function of NSD family members may lead to the development of novel therapies.

Histone lysine methyltransferases (HMTases) catalyze the transfer of up to three methyl groups to specific lysine (K) residues on the tails of histones H3 and H4 that are critical for chromatin maintenance and the fine regulation of gene expression. Histone marks created by lysine HMTases are associated with either active transcription (such as H3K4me or H3K36me2) or repressed transcription (such as H3K27me or H2K9me) (Lachner et al. 2003; Barski et al. 2007). The global activities of these enzymes keep genes in a state poised for rapid activation or repression. Methylated histones are sensed and linked to downstream biological functions by an array of methyllysine-binding proteins. Thus, lysine HMTases play important roles in many downstream cellular processes, such as DNA replication, DNA damage response, cell-cycle progression, cytokinesis, and transcriptional regulation of important developmental and tumor-suppressor genes. The human genome encodes more than 50 predicted lysine methyltransferases, and for many of these genes dysregulation has been reported to play a causative role in human disease (Kouzarides 2007; Morishita and di Luccio 2011). Here we review the most recently reported functions of nuclear receptor–binding SET domain (NSD) family members with particular attention paid to their roles in cancer.

NSD FAMILY MEMBERS ARE STRUCTURALLY CONSERVED HISTONE H3 LYSINE 36 MONO- AND DIMETHYLTRANSFERASES

The NSD family of HMTases are a phylogenetically distinct subfamily of lysine-HMTases comprised of: NSD1, NSD2 (MMSET/WHSC1), and NSD3 (WHSC1L1) (Morishita and di Luc-

cio 2011). The full-length members of the NSD family are large multidomain proteins that contain the evolutionarily conserved catalytic SET [Su(var)3-9, Enhancer-of-zeste, Trithorax] domain that is further subdivided into pre-SET, SET, and post-SET domains (Fig. 1) (Dillon et al. 2005; Herz et al. 2013). In addition, full-length NSD family members have two PWWP (proline–tryptophan–tryptophan–proline) domains that are critical for binding to methylated histone H3 as well as DNA and plant homeo-

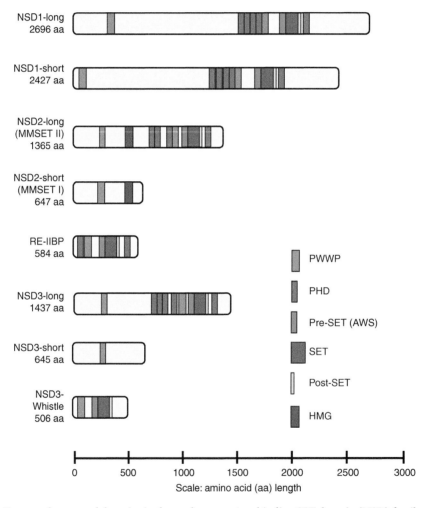

Figure 1. Conserved structural domains in the nuclear receptor–binding SET domain (NSD) family of histone lysine methyltransferases. Protein domain assignments were calculated using a simple modular architecture research tool (SMART) and the following UniProtKB entries: NSD1-long, Q96L73-1; NSD1-short, Q96L73-2; NSD2-long, O96028-1; NSD2-short, O96028-3; RE-IIBP, O96028-4; NSD3-long, Q9BZ95-1; and NSD3-short, Q9BZ95-3. NSD3-WHISTLE domain assignments are from data in Kim et al. (2006).

domain (PHD) zinc fingers important for interactions with other methylated histones (Fig. 1) (Baker et al. 2008; Pasillas et al. 2011; Sankaran et al. 2016).

NSD1, NSD2, and NSD3 function to mono- and dimethylate histone H3 on lysine 36 (H3K36) (Kim et al. 2006, 2008; Marango et al. 2008; Li et al. 2009b; Nimura et al. 2009; Lucio-Eterovic et al. 2010; Kuo et al. 2011; Qiao et al. 2011; Rahman et al. 2011). Histone H3K36 is found in nonmethylated and mono-, di-, and trimethylated forms (me1, me2, me3, respectively), and methylation of H3K36 is associated with transcription of active euchromatin (Rao et al. 2005; Wagner and Carpenter 2012). To date, eight different mammalian H3K36 HMTases have been identified with distinct preferences for which methylation state of H3K36 they recognize and modify in vitro and/or in vivo (Wagner and Carpenter 2012). In addition, the various methylated forms of H3K36 may have different biological functions depending on the organism or cellular context. In worms and humans, K36me3 has been shown to link transcription with splicing (Kolasinska-Zwierz et al. 2009; Sims and Reinberg 2009; Spies et al. 2009). In flies, H3K36me3 functions in dosage compensation (Larschan et al. 2007). In yeast, H3K36me2 and H3K36me3 have been implicated in transcription frequency and elongation that is coupled to histone acetylation (Kaplan et al. 2003; Carrozza et al. 2005; Joshi and Struhl 2005; Kizer et al. 2005; Morris et al. 2005; Shilatifard 2006; Lee and Shilatifard 2007; Xu et al. 2008b; Li et al. 2009a). In *Arabidopsis* and chicken, H3K36me2 and H3K36me3 mark actively transcribed chromatin (Bannister et al. 2005; Xu et al. 2008a). In humans, there is a preference for K36me1 at active promoters, and this mark is detected in active regions of the β-globin locus (Kim et al. 2007). Furthermore, at the human globin genes, H3K36me3 is broadly associated with transcription (Kim et al. 2007).

Evidence suggests that NSD1, NSD2, and/or NSD3 normally play nonredundant roles during development because genetic deletion of either *NSD1* or *NSD2* is lethal in mice (Rayasam et al. 2003; Nimura et al. 2009). Thus, alterations or amplifications of *NSD1*, *NSD2*, and/ or *NSD3* that dysregulate H3K36 methylation marks have profound effects on cell growth and differentiation and are linked to numerous developmental defects. In addition, overexpression, gain of function somatic mutations, and translocation of NSDs have been reported to frequently occur in a variety of cancers.

IDENTIFICATION AND FUNCTION OF NSD

The human nuclear receptor SET domain-containing 1 (*NSD1, KMT3B*) gene is comprised of 24 exons on chromosome 5q35. NSD1 has two protein isoforms: a predominant 2427 amino acid (aa) short form and a less abundant 2696 aa long form that occurs by retention of the intron between exons 2 and 3 (Fig. 1) (Lucio-Eterovic et al. 2010). NSD1 was originally identified to contain two nuclear steroid receptor interaction domains that regulate the function of retinoic acid, thyroid, retinoid X, and estrogen nuclear receptors (Huang et al. 1998). Subsequent studies revealed that the SET domain of NSD1 methylated H3K36 and H4K20 in vitro (Rayasam et al. 2003). However, more recent experiments suggest that the enzyme is specific for H3K36 (Bender et al. 2006; Bell et al. 2007; Stabell et al. 2007; Li et al. 2009b). Importantly, reports using defined nucleosome substrates with various forms of methylated histone H3 at lysine 36 show that NSD1 is a dimethylase specific for H3K36 (Li et al. 2009b; Qiao et al. 2011). Furthermore, depletion of NSD1 reduced the levels of H3K36me1, 2, and 3, suggesting that NSD1 is a mono/dimethylase and that this modification serves as a substrate for trimethylation by the HMTase SETD2 (Lucio-Eterovic et al. 2010). In addition, depletion of NSD1 reduced RNA polymerase II (RNAPII) promoter occupancy and inhibited the transition of RNAPII from an initiation to an elongation-competent state (Lucio-Eterovic et al. 2010).

The crystal structure for NSD1 has been solved and serves as a model for the other NSD family proteins (Qiao et al. 2011; Graham et al. 2016). This structure reveals that an auto-inhibitory loop blocks binding of NSD1 to the histone peptide as well as the entrance to the lysine-binding channel. As nucleosomal DNA contacts the NSD1 post-SET loop, the active conformation is stabilized. This may explain the preference of NSD family members for nucleosomal H3K36 substrates compared to octamers and why NSD family HMTases methylate histone lysine residues other than H3K36 when octamers, recombinant histones, or peptides are used as substrates (Li et al. 2009b; Kudithipudi et al. 2014; Morishita et al. 2014).

NSD1 is critical for normal murine embryonic development. Homozygous loss-of-function embryos are able to gastrulate and initiate mesoderm formation but fail at embryonic day 6.5 (Rayasam et al. 2003). Defects of *NSD1* are linked to Sotos syndrome (aberrant *NSD1* expression is present in 80% of patients) as well as unique variants of Weaver or Beckwith–Wiedemann syndrome (Kurotaki et al. 2002; Douglas et al. 2003; Baujat et al. 2004; Gibson et al. 2012). Characterized as overgrowth disorders, these patients experience pre-/postnatal overgrowth, enhanced bone age, neurodevelopmental delay, and an enhanced risk for cancer (Rahman 2005). A genomic analysis of 435 Sotos patient samples revealed nonsense, deletion/insertion/duplication, splice site, and missense *NSD1* mutations (Waggoner et al. 2005). Although half of the point mutations (nonsense and missense) occurred at CpG sites, missense mutations were primarily in exons that code for functional domains (i.e., PHD, PWWP, and SET domains) (Waggoner et al. 2005). Although not experimentally confirmed, there is evidence of *NSD1* gene-dosage effects linked to significant phenotypic outcomes beyond Sotos syndrome. Microdeletions lead to bone overgrowth/macrocephaly, whereas microduplications are mirrored by stunted growth and microcephaly (Dikow et al. 2013; Rosenfeld et al. 2013). *NSD1* knockout mouse are embryonic lethal (Rayasam et al. 2003). However, mice carrying a heterozygous 1.5-Mb deletion of 36 genes on mouse chromosome 13 that corresponds to the human chromosome 5q35.2-q35.3 region where *NSD1* is located have been characterized (Migdalska et al. 2012). Although these mice did not show the phenotypic overgrowth observed in Sotos syndrome, they did display impaired long-term memory and renal abnormalities (Migdalska et al. 2012).

In addition to histone H3, nonhistone targets of NSD1 have been reported. The carboxyl-terminus of NSD1 contains a unique PHD finger region termed the PHDvC5HCH domain that interacts with the transcription repressor NIzp1, and mutations in the NSD1 PHDvC5HCH domain may interfere with NIsp1 transcription repression (Nielsen et al. 2004; Berardi et al. 2016). Also, on activation of NF-κB, NSD1 has been reported to coimmunoprecipitate with the NF-κB p65 subunit, and be required for inducible p65 methylation at lysine residues K218 and K221 (Lu et al. 2010). However, whether p65 is a direct substrate of NSD1 has yet to be confirmed in vitro. These functions may contribute to the overgrowth phenotype observed when *NSD1* is mutated in Sotos syndrome.

NSD1 ROLE IN CANCER

Aberrant *NSD1* expression has been associated with many cancer pathologies, and tumors occur in 3% of patients diagnosed with Sotos syndrome (Tatton-Brown et al. 2005). In addition, *NSD1* was inactivated via CpG island–promoter hypermethylation in neuroblastomas and gliomas (Berdasco et al. 2009). This transcriptional silencing was associated with diminished methylation of H3K36 and H4K20. In prostate tumors, enhanced *NSD1* expression was associated with metastases, whereas reduced expression was associated with cancers having biochemical recurrence (Bianco-Miotto et al. 2010). Also, *NSD1* expression was up-regulated in patient-derived metastatic melanoma cell lines compared to primary epidermal melanocytes. However, during the progression from nonmetastatic to metastatic melanoma, expression of *NSD1* was down-regulated (de Souza et al. 2012). Furthermore, a transposon screen for frequently occurring mutations in skin tumors of mice revealed that *NSD1* is among those genes with significantly decreased expression during tumor development (Quintana et al. 2013). Taken together, these studies suggest that increased *NSD1* expression may promote oncogenic initiation through enhanced methylation of H3K36, which activates genes normally silenced by H3K27me3 (e.g., *MEIS1* and *Hox*) (Wang et al. 2007; Berdasco et al. 2009). Subsequently, metastatic progression may activate negative feedback of NSD1 in a tissue and/or stage-specific manner. Further studies are required to elucidate these potential roles of NSD1 in cancer.

An NUP98-NSD1 fusion protein has been observed in childhood acute myeloid leukemia (AML) (Jaju et al. 2001; Cerveira et al. 2003; La Starza et al. 2004). This recurrent translocation at t(5;11)(q35;p15.5) contains the FG-repeat domain of NUP98, a nucleoporin protein family member that can interact with the histone acetyltransferase CBP/p300 and the carboxyl terminal of NSD1 that retains the five PHD fingers, the Cys-His-rich domain (C5HCH), one PWWP domain, and the catalytic SET domain. NUP98-NSD1 is the most frequent (4%–5% of cases) fusion reported in pediatric AML and is associated with poor prognosis (Shiba et al. 2013). Retroviral infection of NUP98-NSD1 enhanced expression of *HoxA7*, *HoxA7*, *HoxA9*, and *Meis1* proto-oncogenes (Wang et al. 2007). In addition, marrow progenitors transduced with NUP98-NSD1-induced AML when transplanted into lethally irradiated mice (Wang et al. 2007). Mechanistically, the NUP98-NSD1 fusion drives methylation of H3K36 and histone acetylation. This persistent methylation prevents transcriptional repression of the Hox-A locus by EZH2 causing progenitor immortalization (Wang et al. 2007). Although the NUP98-NSD1 fusion protein occurs in a low percentage of patients, the incidence of concurrence between NUP98-NSD1 and FLT3 (FMS-like tyrosine kinase 3) internal tandem duplication (ITD) mutants is high (70%) in human AML cases (Thanasopoulou et al. 2014). Coinfection of progenitor cells expressing both NUP98-NSD1 and FLT3-ITD mutation resulted in a strikingly decreased latency of development of AML in mice compared to those reconstituted with NUP98-NSD1 alone (Thanasopoulou et al. 2014). Furthermore, experiments using immortalized bone marrow progenitor cells show that NUP98-NSD1 pro-

motes the expression and activation of FLT3-ITD, suggesting a potent cooperation between NUP98-NSD1 and FLT3-ITD during leukemic transformation that may be therapeutically targeted with SET domain or FLT3 inhibitors (Thanasopoulou et al. 2014).

In summary, NSD1 is an H3K36-specific mono- and dimethyltransferase that promotes transcription and is critical for normal growth and development. Aberrant expression of *NSD1* drives the pathobiology of Sotos syndrome and tumorigenesis. Although much work has been performed regarding the structure and function of NSD1, questions remain. For instance, whereas normal and cancer cell lines endogenously express both the short and long form of NSD1 the specific functions of these protein isoforms have not been well characterized. In addition, work remains to determine what cellular context or mechanisms regulate the oncogenic properties of NSD1 and how transcription of *NSD1* may be altered during metastasis.

IDENTIFICATION AND FUNCTION OF NSD2

NSD2 (also known as WHSC1 and MMSET) was identified by its presence in a chromosomal region found to be deleted in the Wolf–Hirschhorn syndrome (WHS) (Stec et al. 1998), and separately by its rearrangement with the immunoglobulin locus in a subset of cases of the plasma cell malignancy, multiple myeloma (Chesi et al. 1998). Since then, *NSD2* has been found to play a role in many facets of development and malignancy. This 90-kb, 25-exon gene on chromosome 4p16.3 encodes two main isoforms, *NSD2*-short (MMSET-I) and NSD2-long (MMSET-II), and an intronic transcript that encodes response element II–binding protein (RE-IIBP). The 1365 amino acid full-length species is composed of two PWWP domains, a high-mobility group (HMG) DNA-binding domain, four PHD zinc fingers, and a SET domain (Fig. 1), whereas the 647 amino acid short species lacks all but the initial PWWP and HMG domains (Stec et al. 1998).

NSD2 catalyzes the mono- and dimethylation of H3K36 (Martinez-Garcia et al. 2011; Morishita et al. 2014). The first amino-terminal

PWWP domain of NSD2 specifically binds to H3K36me2 to stabilize NSD2 at chromatin, and the catalytic SET domain of NSD2 propagates this gene-activating mark to adjacent nucleosomes (Waggoner et al. 2005; Kuo et al. 2011; Martinez-Garcia et al. 2011; Morishita et al. 2014; Sankaran et al. 2016). Similar to NSD1, the NSD2 post-SET domain is attached to the catalytic SET domain via an autoinhibitory loop region and inhibition is relieved on nucleosome binding (Qiao et al. 2011; Poulin et al. 2016a). Furthermore, NSD2 has been reported to preferentially catalyze H3K36 dimethylation compared to H3K36 monomethylation (Li et al. 2009b; Kuo et al. 2011; Poulin et al. 2016b).

NSD2 is broadly expressed and its importance in development is highlighted by its involvement in Wolf–Hirschhorn malformation syndrome (WHS). The full syndrome is characterized by brain defects associated with developmental delay and epilepsy as well as craniofacial anomalies, growth delay, heart defects, and midline fusion abnormalities (Stec et al. 1998; Bergemann et al. 2005). Variable deletions in the short arm of chromosome 4 (4p16.3) are typical of WHS and *NSD2* is the only gene in this region that is deleted in almost every case. Partial or full hemizygosity of *NSD2* appears to be necessary but not sufficient for the development of WHS, as the deletion of other genes nearby contributes to the constellation of abnormalities that make up the syndrome (Bergemann et al. 2005; Andersen et al. 2013). Mice with a homozygous NSD2 SET domain deletion do not survive past 10 days of age, and heterozygous mice develop significant developmental defects that imitate WHS (Nimura et al. 2009). Furthermore, chromatin immunoprecipitation (ChIP) experiments on embryonic stem (ES) cells of these mice revealed that NSD2 binds to several genes associated with development including *Sall1*, *Sall4*, and *Nanog*. Additionally, WHS patients often have antibody deficiencies, suggesting a role for NSD2 in B-cell development. This theory is supported by the ability of NSD2 to recruit the DNA-damage-responsive, p53-binding protein 1 (53BP1), which is critical for class switch recombination and also implicates NSD2 in DNA repair (Hajdu et al.

2011; Pei et al. 2011). Indeed, it was recently reported that NSD2 regulates the expression of DNA repair genes and may play a critical function in multiple myeloma chemoresistance (Shah et al. 2016). Cancer cells expressing high levels of NSD2 repaired DNA damage at a much faster rate than cells with low levels of NSD2, allowing them to survive and proliferate even while being treated with DNA-damaging chemotherapy.

NSD2 ROLE IN CANCER

NSD2 was initially described as a gene rearranged and linked to regulatory sequences of the immunoglobulin heavy chain gene in t(4;14) multiple myeloma (Chesi et al. 1998; Stec et al. 1998); *NSD2* and its translocations, amplifications, and mutations were subsequently identified in a wide spectrum of malignancies. In multiple myeloma, the t(4;14) translocation is present in 15%–20% of cases, resulting in overexpression of *NSD2* and *FGFR3* (Chesi et al. 1998; Stec et al. 1998; Finelli et al. 1999). However, NSD2 is purported to be the primary oncogenic driver, as approximately 30% of cases harboring this translocation have normal expression of *FGFR3* alongside *NSD2* overexpression (Santra et al. 2003). Furthermore, knockdown of *NSD2* expression in t(4;14)$^+$ multiple myeloma cell lines reduces proliferation, cell-cycle progression, and DNA repair, while increasing apoptosis and adhesion (Lauring et al. 2008; Brito et al. 2009; Martinez-Garcia et al. 2011; Huang et al. 2013; Shah et al. 2016). These phenotypic changes are driven by redistribution of activating and repressive chromatin marks that, in turn, affect gene expression. Overexpression of *NSD2* globally increases levels of H3K36 dimethylation (Martinez-Garcia et al. 2011; Zheng et al. 2012; Popovic et al. 2014; Sankaran et al. 2016). Interestingly, NSD2 also affects enhancer of zeste homolog 2 (EZH2), which is responsible for the reciprocal repressive chromatin mark, histone 3 lysine 27 trimethylation (H3K27me3). The spread of H3K36me2 from *NSD2* overexpression leads to global reduction in H3K27me3 and restricts EZH2 to small islands of chromatin where it then hypermethylates H3K27me3 (Zheng et al. 2012; Popovic et al. 2014). The transcriptional disturbance that results from *NSD2* overexpression primarily involves inappropriate activation of genes, but there are some genes that are inappropriately repressed because of these pockets of EZH2. This effect appears to be important for the survival of cells overexpressing *NSD2*, as they are sensitive to inhibition of EZH2. One contribution to this sensitivity is that inhibition of EZH2 decreases c-MYC protein levels, a fundamentally up-regulated gene in multiple myeloma (Popovic et al. 2014). In t(4;14)$^+$, multiple myeloma cells EZH2 represses miR-126, a microRNA that targets the MYC transcript (Min et al. 2012). Therefore, EZH2 inhibitors derepress miR-126, allowing it to reduce c-MYC protein levels and slow proliferation. However, c-MYC is not the sole cause of high NSD2-mediated oncogenesis. Transcriptional profiling of t(4;14)$^+$ multiple myeloma indicates that NSD2 regulates the expression of genes in apoptosis, DNA repair, cell-cycle control, and cell motility.

In addition to multiple myeloma, *NSD2* overexpression is detected in gastric, colon, lung, and skin cancer (Hudlebusch et al. 2011a). It may play a major role in neuroblastoma and breast, bladder, and prostate tumors, where overexpression is associated with worse prognosis (Hudlebusch et al. 2011b). In many of these cancers, *NSD2* expression is positively correlated with *EZH2* expression, and as opposed to the situation in t(4;14)$^+$ multiple myeloma, global levels of H3K27me3 and H3K36me2 are both increased (Asangani et al. 2013). In prostate cancer, EZH2 functions upstream of NSD2, repressing several microRNAs, including miR-203, miR-26, and miR-31, that target NSD2 (Asangani et al. 2013). However, it is the knockdown of NSD2 that abrogates the ability of prostate cancer cells to proliferate, form colonies, migrate, and invade (Ezponda et al. 2012). Furthermore, enforced expression of *NSD2* in nontransformed prostate epithelial cells promotes migration and invasion and leads to epithelial–mesenchymal transition (EMT). NSD2 directly binds the *TWIST1* locus and up-regulates expression of this EMT factor, which plays a key role in the aggressive biological behavior of advanced prostate cancer.

THE NSD2 E1099K POINT MUTATION IN LYMPHOID MALIGNANCIES

In cases of t(4;14)$^+$ multiple myeloma and some acute lymphoblastic leukemia (ALL) in which *TWIST1* is also overexpressed, *NSD2* harbors a recurrent point mutation in the SET domain (Oyer et al. 2014). The significance of the guanine to alanine substitution that results in a glutamic acid to lysine switch at amino acid 1099 (E1099K) of NSD2 was first noted following examination of the Broad Institute's Cancer Cell Line Encyclopedia (CCLE) that revealed numerous ALL cell lines with the mutation, which has since been identified in 10%–20% of relapsed pediatric ALL, 10% of mantle cell lymphoma (MCL), and 10% of chronic lymphocytic leukemia (CLL) (Fabbri et al. 2011; Barretina et al. 2012; Beà et al. 2013; Loh et al. 2013; Oyer et al. 2014). Rare mutations are found in glioblastoma, lung cancer, and multiple myeloma. Although the mechanism remains to be determined, the E1099K mutation was revealed to result in hyperactive NSD2, leading to increased H3K36me2 and decreased H3K27me3, similar to the epigenetic profile of t(4;14)$^+$ multiple myeloma (Oyer et al. 2014). Because of the common occurrence of this mutation in relapsed pediatric ALL, especially cases harboring other oncogenic lesions, such as the TEL-AML1 and E2A-PBX1 fusions, E1099K appears to be an important factor in progression of these malignancies rather than initiation (Jaffe et al. 2013; Loh et al. 2013).

In summary, NSD2 plays a significant role in normal development and malignancy. Haploinsufficiency of *NSD2* leads to severe developmental defects, such as cardiac lesions and midline abnormalities associated with WHS. In malignancy, NSD2 has prolific effects resulting from translocation, overexpression, and activating mutations of the gene. The t(4;14) translocation in multiple myeloma has been extensively characterized and indicates that high levels of NSD2 drive oncogenic phenotypes by spreading H3K36me2 throughout the genome and altering gene-expression profiles. The relationship between NSD2 and EZH2 is crucial in myeloma as the reciprocal H3K27me3 mark deposited by EZH2 is reduced and restricted to certain regions by *NSD2* overexpression. In other malignancies, such as prostate cancer, this relationship is different as EZH2 appears to function upstream of NSD2, although it is no less important. Inhibition or knockdown of NSD2 or EZH2 in both situations abrogates critical oncogenic pathways and phenotypes. However, there is still much to learn about NSD2 in these malignancies. Although NSD2 itself is already an attractive therapeutic candidate, thoroughly characterizing its binding partners will tell us how NSD2 functions both endogenously and in malignancy and provide more targets for designing therapies for NSD2-misregulated cancers. The recurrent E1099K mutation that results in a hyperactive NSD2 is also not fully understood. Its global effects on chromatin mimic *NSD2* overexpression, but it must be investigated further to identify its local effects on chromatin, gene expression, and oncogenicity. Additionally, as it appears that this mutation is more likely to contribute to relapse after therapy, it will be important to understand its role in drug resistance. Finally, we must understand the mechanism by which E1099K and other activating mutations alter the function of NSD2 to design directed therapies.

IDENTIFICATION AND FUNCTION OF NSD3

Using database searches for sequences similar to *NSD1* and *NSD2*, *NSD3* (also named Wolf–Hirschhorn syndrome candidate 1–like 1, WHSC1L1) was independently identified by two groups in 2001 (Angrand et al. 2001; Stec et al. 2001). The *NSD3* gene contains 24 exons that span approximately 112 kb of genomic DNA on chromosome 8p11.2 and is predicted to have 12 transcript variants. Three protein products of *NSD3* have been reported and characterized to date: Long, Short, and WHISTLE.

The NSD3-long transcript encodes a 1437–amino acid protein containing two PWWP domains, five PHD-type zinc-finger motifs, a SET-

associated Cys-rich (SAC) domain, and a SET domain (Fig. 1) (Angrand et al. 2001; Stec et al. 2001). NSD3-long is highly expressed in brain, heart, and skeletal muscle and to a lesser degree in liver and lung (Angrand et al. 2001). The sequence of NSD3-long from amino acids 703 to 1409 is highly conserved between NSD1 (68% identical) and NSD2 (55% identical) and contains the PHD, second PWWP, and SET domain (Angrand et al. 2001). Similar to NSD1 and NSD2, the carboxy-terminal region of NSD3-long that contains the catalytic pre-SET, SET, and post-SET domains is able to recognize and methylate histone H3 and H4 targets in vitro (Allali-Hassani et al. 2014; Morishita et al. 2014). However, unlike NSD2, a clear HMG box is missing from NSD3-long and the fourth PHD-type zinc finger contains an insertion of 49 amino acids (Stec et al. 2001). In addition, evidence suggests that the conserved PHD5 domain of NSD3-long is functionally distinct in its histone-binding properties from the PHD5 domain of NSD1 or 2 and targets NSD3 to specific genomic regions in vivo (He et al. 2013). The NSD3-short transcript encodes a protein of 645 amino acids (Stec et al. 2001). NSD3-long and NSD3-short have an identical amino-terminal 620–amino acid sequence, but NSD3-short lacks a catalytic SET domain and only contains the amino-terminal PWWP domain that binds to histone H3 when it is methylated on lysine 36 (Stec et al. 2001; Vermeulen et al. 2010; Wu et al. 2011). The NSD3-long and NSD3-short transcripts are coexpressed in many tissues (Stec et al. 2001).

WHISTLE (WHSC1-like 1 isoform 9 with methyltransferase activity to lysine) was identified by homology searching and functional assay to be the shortest isoform of NSD3 to retain a SET domain and methyltransferase activity (Kim et al. 2006). WHISTLE consists of 506 amino acids and was found expressed in testis and in bone marrow mononuclear cells of AML and ALL patients (Kim et al. 2006). It contains the PWWP, SET, and post-SET domains and was reported to facilitate transcription repression by promoting methylation of histone H3K4 and H3K27 (Kim et al. 2006; Allali-Hassani et al. 2014).

 Cite this article as *Cold Spring Harb Perspect Med* doi: 10.1101/cshperspect.a026708

NSD3 has been identified as an essential methyltransferase for neural crest gene expression during specification (Jacques-Fricke and Gammill 2014). NSD3 is expressed in premigratory and migratory neural crest cells and is necessary for expression of the neural plate border gene *Msx1*, as well as the key neural crest transcription factors Sox10, Snail2, Sox9, and FoxD3 (Jacques-Fricke and Gammill 2014). In addition, recent reports indicate that neuronal ten-eleven translocation (TET) hydroxylase 3 interacts with and activates NSD3 to stimulate H3K36 trimethylation and transcription of neuronal genes in retinal cells (Perera et al. 2015). Thus, NSD3 may promote context-specific chromatin remodeling and gene activation that is necessary for neural crest migration and retinal network formation during development.

NSD3 ROLE IN CANCER

Since its discovery, amplification and overexpression of *NSD3* has been a consistently identified feature in many cancer types (Angrand et al. 2001; Mahmood et al. 2013). For example, integrated DNA-RNA analyses of regional amplifications and deletions coupled with gene-expression profiling and have identified the 8p11-12 *NSD3* locus as an amplicon commonly expressed in both pancreatic ductal adenocarcinoma and non-small-cell lung cancer (Tonon et al. 2005). In addition, a bioinformatics screen to identify putative cancer driver genes amplified across TCGA datasets discovered frequent *NSD3* amplifications in bladder, breast, liver, lung, ovarian, head and neck, and colorectal cancer samples (Chen et al. 2014). Amplification of the 8p11-12 region has also been reported to be present in about 15% of primary human breast cancer samples, which correlates with histological grade and is associated with poor prognosis (Angrand et al. 2001; Yang et al. 2010; Chen et al. 2014). Furthermore, NSD3 expression and protein level are increased in breast, lung, pancreatic, and colorectal cell lines, and immunohistochemical analysis indicates that NSD3 is increased in primary breast carcinoma, bladder cancer, lung cancer, and liver cancer (Kang et al. 2013; Mahmood et al. 2013; Chen et al. 2014).

Mounting evidence suggests that NSD3 activates signaling pathways that promote cell-cycle progression and proliferation. Although an *NSD3* knockout mouse has not yet been reported, NSD3 has been identified to cooperate with oncogenic KRAS to drive tumorigenesis in a pancreatic cancer model mouse (Mann et al. 2012). Furthermore, transduction of either NSD3-short or NSD3-long into the mammary epithelial cell line MCF10A cells has been reported to increase proliferation, soft agar colony formation, and cause abnormal acini formation (Yang et al. 2010). In contrast, knockdown of *NSD3* reduced proliferation and promoted cell death in 8p11-12-amplified breast cancer cells and significantly decreased anchorage-dependent and anchorage-independent growth of pancreatic adenocarcinoma and small-cell lung cancer–derived cell lines (Yang et al. 2010; Mahmood et al. 2013). In addition, knockdown of *NSD3*-expression induced G_2/M cell-cycle arrest and suppressed proliferation of breast, bladder, and lung cancer cell lines (Zhou et al. 2010; Kang et al. 2013). Expression profile analysis showed that *NSD3* affected the expression of a number of genes known to play crucial roles in cell-cycle progression (Kang et al. 2013).

Although the mechanism by which NSD3 regulates the mechanism governing the cell cycle and proliferation is not well characterized, evidence suggests that *NSD3* amplification may promote transcription of a subset of genes involved in proliferation pathways. NSD3 can form complexes with H3K4-specific demethylase LSD2 and H3K9-specific methyltransferase G9a and may cooperate with these enzymes to coordinate the dynamics of H3K4 and H3K36 methylation to promote transcription elongation for a subset of genes (Fang et al. 2010). NSD3 promoted expression of transcription factors Iroquois homeobox 3 (IRX3) and TBIL1X known to regulate WNT signaling (Yang et al. 2010). Furthermore, knockdown of *NSD3* in MCF10A cells or breast cancer cell lines with amplified 8p11-12 resulted in increased expression of *SRFp1*, a negative regulator of WNT-signaling, and decreased expression of *TGFBI* leading to profound loss of growth and survival of these cells (Yang et al. 2010).

Taken together, these results suggest that *NSD3* is a putative cancer driver gene.

In addition to amplification, chromosomal translocations resulting in NSD3 fusion products have been described in myelodysplastic syndrome (MDS), AML, and nuclear protein in testis (NUT) carcinoma. NUP98-NSD3 fusion transcripts associated with t(8;11)(p11.2;p15) have been reported in one patient with AML, one patient with therapy-related MDS, and one patient with radiation-associated MDS, and both NSD3-long and NSD3-short isoforms have been detected to be fused to NUP98 in leukemic cell lines (Rosati et al. 2002; Romana et al. 2006; Taketani et al. 2009). In addition, NSD3 was detected as a fusion oncoprotein with the *NUT* gene in the rare and aggressive NUT midline carcinoma (NMC) (French et al. 2014; Harms et al. 2015; Kuroda et al. 2015; Suzuki et al. 2015). A t(8;15)(p12;q15) translocation was identified as responsible for the NSD3-NUT fusion in a patient-derived NMC cell line, and knockdown of the NSD3-NUT fusion revealed that this fusion protein functioned to block differentiation and promote proliferation (French et al. 2014).

NSD3 AS AN EFFECTOR OF BRD4 IN CANCER

The bromodomain and extraterminal domain (BET) family of transcriptional activators are promising therapeutic targets for cancers, particularly AML, because of their role in maintaining the expression of key oncogenes (Dawson et al. 2011; Mertz et al. 2011; Zuber et al. 2011). NSD3 can bind the ET domain of BET proteins and associates with BRD4 in nuclear lysates (Rahman et al. 2011; French et al. 2014; Crowe et al. 2016). The ET domain of BRD4 interacts with amino acids 100–263 of NSD3, a region immediately before the amino-terminal PWWP domain located at amino acids 270–333 (Fig. 2) (Shen et al. 2015). The corresponding regions in NSD1 and NSD2 have significant homology with NSD3, suggesting this may be a common element responsible for BRD4 interaction with NSD family members (Fig. 2). In addition, the NSD3-NUT fusion binds to BRD4 and BRD bromodomain inhib-

itors induce differentiation and arrest proliferation of t(8;15)(p12;q15) cells (French et al. 2014). Recent evidence suggests that interaction with NSD3-short may be required for the AML maintenance function of BRD4 (Shen et al. 2015). The NSD3-short isoform has been reported to be an adaptor protein that links BRD4 to the CHD8 chromatin remodeler. BRD4, NSD3, and CHD8 colocalize across the AML genome, and each is released from super-enhancer regions on treatment with BET inhibitors such as JQ1 (Shen et al. 2015). Furthermore, genetic targeting of *NSD3* or *CHD8* mimics the phenotypic and transcriptional effects of pharmacological BRD4 inhibition (Shen et al. 2015). Thus, BET inhibitors may function by evicting BRD4–NSD3–CHD8 complexes from chromatin to suppress transcription.

CONCLUDING REMARKS

The NSD HMTases are key regulators of development. Mutations, amplifications, or translocations of these genes lead to developmental defects and cancer. Differences between the function and tissue specificity of the many isoforms of each NSD family member are poorly understood, and additional research remains critical to improve our understanding of how dysregulation of NSD family members may lead to cancer. For instance, NSD family members have no known sequence-specific DNA-binding properties, and the precise mechanism that directs these lysine-HMTs to specific chromatin loci is unclear. Thus, identification of partner proteins that may guide NSD family members to specific promoter/enhancer regions is important to understanding how these proteins may be directed to specific functional contexts. The recent report that BRD4 may direct NSD3 to chromatin as well as other reports indicating NSD2 associates with BRD4 raises the intriguing possibility that bromodomains (acetyllysine side chains) on histone H3 may recruit BET family members such as BRD4 to enhancer and promoter regions that, in turn, recruit NSD family proteins to direct H3 K36 methylation, inhibit EZH2-mediated H3 K27

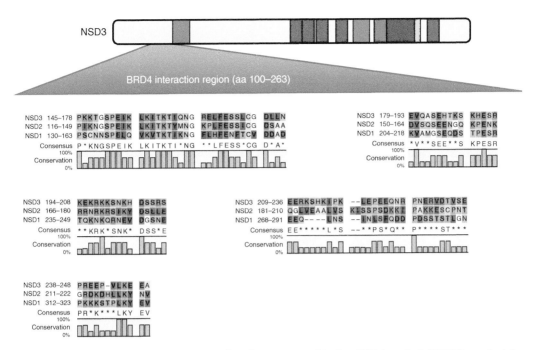

Figure 2. Sequence alignment of the region of nuclear receptor–binding SET domain 3 (NSD3) required for BRD4 interaction with NSD1 and NSD2 reveals conserved regions amino terminal to the first PWWP domain. The region in NSD3 identified as being required for interaction with BRD4 (amino acids 100-263) was found to have significant homology with the corresponding region in NSD1 and NSD2. Sequence analysis was performed using CLC Sequence Viewer (Qiagen).

methylation, and promote transcription initiation and elongation (Fig. 3) (Min et al. 2012; Shen et al. 2015). In addition, identifying how activating mutations such as E1099K in NSD2 may alter enzymatic activity to drive increased H3K36 dimethylation is critical to understanding the biological effect of this mutation in oncogenesis. Preliminary epigenetic and transcriptional profiling of E1099K in ALL cells indicates there are both similarities and differences between E1099K NSD2 and overexpression of *NSD2* (unpublished data). Elucidating the downstream effects of E1099K may provide new insight into the mechanism by which aberrant lysine HMTase activity promotes tumorigenesis.

The prominent role of translocations, upregulation, and activating mutations of NSD family members in driving tumor progression and aggressiveness, suggests that specific lysine-HMTase inhibitors are a promising therapeutic approach to suppress cancer growth. In support of this, abolition of the methyltransferase activity of NUP98-NSD1 by point mutation suppresses Hox-A gene activation and myeloid progenitor immortalization (Wang et al. 2007). In addition, knockdown of *NSD2* in several model systems has indicated the potential utility of NSD2 inhibitors for down-regulating critical tumor cell phenotypes. Targeting the SET domain is an attractive strategy for development of NSD family inhibitors, because the lysine HMTase activity is considered most likely to drive oncogenic reprogramming (Kuo et al. 2011; Martinez-Garcia et al. 2011). Recently, a high throughput luminescence-based assay using the NSD1 SET domain was used to screen for specific inhibitors of NSD1 (Drake et al. 2014). The HMTase inhibitor suramin was identified in this screen but had minimal selectivity for NSD1 over other histone methyltransferases and other compounds identified by this screen inactivated

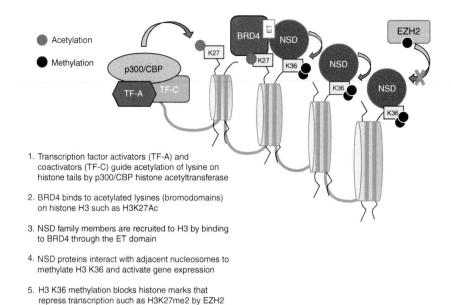

1. Transcription factor activators (TF-A) and coactivators (TF-C) guide acetylation of lysine on histone tails by p300/CBP histone acetyltransferase

2. BRD4 binds to acetylated lysines (bromodomains) on histone H3 such as H3K27Ac

3. NSD family members are recruited to H3 by binding to BRD4 through the ET domain

4. NSD proteins interact with adjacent nucleosomes to methylate H3 K36 and activate gene expression

5. H3 K36 methylation blocks histone marks that repress transcription such as H3K27me2 by EZH2

Figure 3. Proposed mechanism for targeting of nuclear receptor–binding SET domain (NSD) family proteins to specific chromatin loci.

NSD1 through nonspecific mechanisms. Alternatively, targeting domains outside of the catalytic SET domain may be a useful strategy for developing NSD inhibitors. The PHD zinc fingers 2, 3, and 4, as well as PWWP domain 2 of NSD2 are critical for the H3K36me2 up-regulation in t(4;14)$^+$ multiple myeloma (Popovic et al. 2014). The fourth PHD finger plays a role in the NSD2/EZH2 interplay, as its deletion abrogates the reduction of H3K27me3 typically observed with *NSD2* overexpression in myeloma. Furthermore, recent reports that BET inhibitors evict NSD3-BRD4 from chromatin raise the attractive hypothesis that these compounds may also inhibit interactions between other BRD and NSD family members to suppress aberrant gene activation and limit tumor growth. Although evidence suggests that genetic and epigenetic alterations cooperate in the stepwise initiation and progression of cancers, only epigenetic aberrations can be reversed. Thus, as we work to further our understanding of NSD family members, we are hopeful that inhibition of these epigenetic regulators will play a central role in the next generation of cancer therapy.

REFERENCES

Allali-Hassani A, Kuznetsova E, Hajian T, Wu H, Dombrovski L, Li Y, Graslund S, Arrowsmith CH, Schapira M, Vedadi M. 2014. A basic post-SET extension of NSDs is essential for nucleosome binding in vitro. *J Biomol Screen* **19:** 928–935.

Andersen EF, Carey JC, Earl DL, Corzo D, Suttie M, Hammond P, South ST. 2013. Deletions involving genes WHSC1 and LETM1 may be necessary, but are not sufficient to cause Wolf-Hirschhorn Syndrome. *Eur J Hum Genet* **22:** 464–470.

Angrand PO, Apiou F, Stewart AF, Dutrillaux B, Losson R, Chambon P. 2001. NSD3, a new SET domain-containing gene, maps to 8p12 and is amplified in human breast cancer cell lines. *Genomics* **74:** 79–88.

Asangani IA, Ateeq B, Cao Q, Dodson L, Pandhi M, Kunju LP, Mehra R, Lonigro RJ, Siddiqui J, Palanisamy N, et al. 2013. Characterization of the EZH2-MMSET histone methyltransferase regulatory axis in cancer. *Mol Cell* **49:** 80–93.

Baker LA, Allis CD, Wang GG. 2008. PHD fingers in human diseases: Disorders arising from misinterpreting epigenetic marks. *Mutat Res* **647:** 3–12.

Bannister AJ, Schneider R, Myers FA, Thorne AW, Crane-Robinson C, Kouzarides T. 2005. Spatial distribution of di- and tri-methyl lysine 36 of histone H3 at active genes. *J Biol Chem* **280:** 17732–17736.

Barretina J, Caponigro G, Stransky N, Venkatesan K, Margolin AA, Kim S, Wilson CJ, Lehár J, Kryukov GV, Sonkin D, et al. 2012. The Cancer Cell Line Encyclopedia enables predictive modelling of anticancer drug sensitivity. *Nature* **483:** 603–607.

Barski A, Cuddapah S, Cui K, Roh TY, Schones DE, Wang Z, Wei G, Chepelev I, Zhao K. 2007. High-resolution profiling of histone methylations in the human genome. *Cell* **129:** 823–837.

Baujat G, Rio M, Rossignol S, Sanlaville D, Lyonnet S, Le Merrer M, Munnich A, Gicquel C, Cormier-Daire V, Colleaux L. 2004. Paradoxical NSD1 mutations in Beckwith–Wiedemann syndrome and 11p15 anomalies in Sotos syndrome. *Am J Hum Genet* **74:** 715–720.

Beà S, Valdés-Mas R, Navarro A, Salaverria I, Martín-Garcia D, Jares P, Giné E, Pinyol M, Royo C, Nadeu F, et al. 2013. Landscape of somatic mutations and clonal evolution in mantle cell lymphoma. *Proc Natl Acad Sci* **110:** 18250–18255.

Bell O, Wirbelauer C, Hild M, Scharf AN, Schwaiger M, MacAlpine DM, Zilbermann F, van Leeuwen F, Bell SP, Imhof A, et al. 2007. Localized H3K36 methylation states define histone H4K16 acetylation during transcriptional elongation in Drosophila. *EMBO J* **26:** 4974–4984.

Bender LB, Suh J, Carroll CR, Fong Y, Fingerman IM, Briggs SD, Cao R, Zhang Y, Reinke V, Strome S. 2006. MES-4: An autosome-associated histone methyltransferase that participates in silencing the X chromosomes in the *C. elegans* germ line. *Development* **133:** 3907–3917.

Berardi A, Quilici G, Spiliotopoulos D, Corral-Rodriguez MA, Martin-Garcia F, Degano M, Tonon G, Ghitti M, Musco G. 2016. Structural basis for PHDVC5HCHNSD1-C2HRNizp1 interaction: Implications for Sotos syndrome. *Nucleic Acids Res* **44:** 3448–3463.

Berdasco M, Ropero S, Setien F, Fraga MF, Lapunzina P, Losson R, Alaminos M, Cheung NK, Rahman N, Esteller M. 2009. Epigenetic inactivation of the Sotos overgrowth syndrome gene histone methyltransferase NSD1 in human neuroblastoma and glioma. *Proc Natl Acad Sci* **106:** 21830–21835.

Bergemann AD, Cole F, Hirschhorn K. 2005. The etiology of Wolf–Hirschhorn syndrome. *Trends Genet* **21:** 188–195.

Bianco-Miotto T, Chiam K, Buchanan G, Jindal S, Day TK, Thomas M, Pickering MA, O'Loughlin MA, Ryan NK, Raymond WA, et al. 2010. Global levels of specific histone modifications and an epigenetic gene signature predict prostate cancer progression and development. *Cancer Epidemiol Biomarkers Prev* **19:** 2611–2622.

Brito JLR, Walker B, Jenner M, Dickens NJ, Brown NJM, Ross FM, Avramidou A, Irving JaE, Gonzalez D, Davies FE, et al. 2009. MMSET deregulation affects cell cycle progression and adhesion regulons in t(4;14) myeloma plasma cells. *Haematologica* **94:** 78–86.

Carrozza MJ, Li B, Florens L, Suganuma T, Swanson SK, Lee KK, Shia WJ, Anderson S, Yates J, Washburn MP, et al. 2005. Histone H3 methylation by Set2 directs deacetylation of coding regions by Rpd3S to suppress spurious intragenic transcription. *Cell* **123:** 581–592.

Cerveira N, Correia C, Doria S, Bizarro S, Rocha P, Gomes P, Torres L, Norton L, Borges BS, Castedo S, et al. 2003. Frequency of NUP98-NSD1 fusion transcript in childhood acute myeloid leukaemia. *Leukemia* **17:** 2244–2247.

Chen Y, McGee J, Chen X, Doman TN, Gong X, Zhang Y, Hamm N, Ma X, Higgs RE, Bhagwat SV, et al. 2014. Identification of druggable cancer driver genes amplified across TCGA datasets. *PloS ONE* **9:** e98293.

Chesi M, Nardini E, Lim RS, Smith KD, Kuehl WM, Bergsagel PL. 1998. The t(4;14) translocation in myeloma dysregulates both FGFR3 and a novel gene, MMSET, resulting in IgH/MMSET hybrid transcripts. *Blood* **92:** 3025–3034.

Crowe BL, Larue RC, Yuan C, Hess S, Kvaratskhelia M, Foster MP. 2016. Structure of the Brd4 ET domain bound to a C-terminal motif from γ-retroviral integrases reveals a conserved mechanism of interaction. *Proc Natl Acad Sci* **113:** 2086–2091.

Dawson MA, Prinjha RK, Dittmann A, Giotopoulos G, Bantscheff M, Chan WI, Robson SC, Chung CW, Hopf C, Savitski MM, et al. 2011. Inhibition of BET recruitment to chromatin as an effective treatment for MLL-fusion leukaemia. *Nature* **478:** 529–533.

de Souza CF, Xander P, Monteiro AC, Silva AG, da Silva DC, Mai S, Bernardo V, Lopes JD, Jasiulionis MG. 2012. Mining gene expression signature for the detection of premalignant melanocytes and early melanomas with risk for metastasis. *PloS ONE* **7:** e44800.

Dikow N, Maas B, Gaspar H, Kreiss-Nachtsheim M, Engels H, Kuechler A, Garbes L, Netzer C, Neuhann TM, Koehler U, et al. 2013. The phenotypic spectrum of duplication 5q35.2-q35.3 encompassing NSD1: Is it really a reversed Sotos syndrome? *Am J Med Genet A* **161A:** 2158–2166.

Dillon SC, Zhang X, Trievel RC, Cheng X. 2005. The SET-domain protein superfamily: Protein lysine methyltransferases. *Genome Biol* **6:** 227.

Douglas J, Hanks S, Temple IK, Davies S, Murray A, Upadhyaya M, Tomkins S, Hughes HE, Cole TR, Rahman N. 2003. NSD1 mutations are the major cause of Sotos syndrome and occur in some cases of Weaver syndrome but are rare in other overgrowth phenotypes. *Am J Hum Genet* **72:** 132–143.

Drake KM, Watson VG, Kisielewski A, Glynn R, Napper AD. 2014. A sensitive luminescent assay for the histone methyltransferase NSD1 and other SAM-dependent enzymes. *Assay Drug Dev Technol* **12:** 258–271.

Ezponda T, Popovic R, Shah MY, Martinez-Garcia E, Zheng Y, Min D-J, Will C, Neri A, Kelleher NL, Yu J, et al. 2012. The histone methyltransferase MMSET/WHSC1 activates TWIST1 to promote an epithelial–mesenchymal transition and invasive properties of prostate cancer. *Oncogene* **32:** 2882–2890.

Fabbri G, Rasi S, Rossi D, Trifonov V, Khiabanian H, Ma J, Grunn A, Fangazio M, Capello D, Monti S, et al. 2011. Analysis of the chronic lymphocytic leukemia coding genome: Role of NOTCH1 mutational activation. *J Exp Med* **208:** 1389–1401.

Fang R, Barbera AJ, Xu Y, Rutenberg M, Leonor T, Bi Q, Lan F, Mei P, Yuan GC, Lian C, et al. 2010. Human LSD2/KDM1b/AOF1 regulates gene transcription by modulating intragenic H3K4me2 methylation. *Mol Cell* **39:** 222–233.

Finelli P, Fabris S, Zagano S, Baldini L, Intini D, Nobili L, Lombardi L, Maiolo AT, Neri A. 1999. Detection of t(4;14)(p16.3;q32) chromosomal translocation in multiple myeloma by double-color fluorescent in situ hybridization. *Blood* **94:** 724–732.

French CA, Rahman S, Walsh EM, Kuhnle S, Grayson AR, Lemieux ME, Grunfeld N, Rubin BP, Antonescu CR, Zhang S, et al. 2014. NSD3-NUT fusion oncoprotein in NUT midline carcinoma: Implications for a novel oncogenic mechanism. *Cancer Discov* **4:** 928–941.

Gibson WT, Hood RL, Zhan SH, Bulman DE, Fejes AP, Moore R, Mungall AJ, Eydoux P, Babul-Hirji R, An J, et al. 2012. Mutations in EZH2 cause Weaver syndrome. *Am J Hum Genet* **90:** 110–118.

Graham SE, Tweedy SE, Carlson HA. 2016. Dynamic behavior of the post-SET loop region of NSD1: Implications for histone binding and drug development. *Protein Sci* **25:** 1021–1029.

Hajdu I, Ciccia A, Lewis SM, Elledge SJ. 2011. Wolf–Hirschhorn syndrome candidate 1 is involved in the cellular response to DNA damage. *Proc Natl Acad Sci* **108:** 13130–13134.

Harms A, Herpel E, Pfarr N, Penzel R, Heussel CP, Herth FJ, Dienemann H, Weichert W, Warth A. 2015. NUT carcinoma of the thorax: Case report and review of the literature. *Lung Cancer* **90:** 484–491.

He C, Li F, Zhang J, Wu J, Shi Y. 2013. The methyltransferase NSD3 has chromatin-binding motifs, PHD5-C5HCH, that are distinct from other NSD (nuclear receptor SET domain) family members in their histone H3 recognition. *J Biol Chem* **288:** 4692–4703.

Herz HM, Garruss A, Shilatifard A. 2013. SET for life: Biochemical activities and biological functions of SET domain-containing proteins. *Trends Biochem Sci* **38:** 621–639.

Huang N, vom Baur E, Garnier JM, Lerouge T, Vonesch JL, Lutz Y, Chambon P, Losson R. 1998. Two distinct nuclear receptor interaction domains in NSD1, a novel SET protein that exhibits characteristics of both corepressors and coactivators. *EMBO J* **17:** 3398–3412.

Huang Z, Wu H, Chuai S, Xu F, Yan F, Englund N, Wang Z, Zhang H, Fang M, Wang Y, et al. 2013. NSD2 is recruited through its PHD domain to oncogenic gene loci to drive multiple myeloma. *Cancer Res* **73:** 6277–6288.

Hudlebusch HR, Santoni-Rugiu E, Simon R, Ralfkiær E, Rossing HH, Johansen JV, Jørgensen M, Sauter G, Helin K. 2011a. The histone methyltransferase and putative oncoprotein MMSET is overexpressed in a large variety of human tumors. *Clin Cancer Res* **17:** 2919–2933.

Hudlebusch HR, Skotte J, Santoni-Rugiu E, Zimling ZG, Lees MJ, Simon R, Sauter G, Rota R, De Ioris MA, Quarto M, et al. 2011b. MMSET is highly expressed and associated with aggressiveness in neuroblastoma. *Cancer Res* **71:** 4226–4235.

Jacques-Fricke BT, Gammill LS. 2014. Neural crest specification and migration independently require NSD3-related lysine methyltransferase activity. *Mol Biol Cell* **25:** 4174–4186.

Jaffe JD, Wang Y, Chan HM, Zhang J, Huether R, Kryukov GV, Bhang H-eC, Taylor JE, Hu M, Englund NP, et al. 2013. Global chromatin profiling reveals NSD2 mutations in pediatric acute lymphoblastic leukemia. *Nat Genet* **45:** 1386–1391.

Jaju RJ, Fidler C, Haas OA, Strickson AJ, Watkins F, Clark K, Cross NCP, Cheng JF, Aplan PD, Kearney L, et al. 2001. A novel gene, NSD1, is fused to NUP98 in the t(5;11)(q35;p15.5) in de novo childhood acute myeloid leukemia. *Blood* **98:** 1264–1267.

Joshi AA, Struhl K. 2005. Eaf3 chromodomain interaction with methylated H3-K36 links histone deacetylation to Pol II elongation. *Mol Cell* **20:** 971–978.

Kang D, Cho HS, Toyokawa G, Kogure M, Yamane Y, Iwai Y, Hayami S, Tsunoda T, Field HI, Matsuda K, et al. 2013. The histone methyltransferase Wolf–Hirschhorn syndrome candidate 1-like 1 (WHSC1L1) is involved in human carcinogenesis. *Genes Chromosomes Cancer* **52:** 126–139.

Kaplan CD, Laprade L, Winston F. 2003. Transcription elongation factors repress transcription initiation from cryptic sites. *Science* **301:** 1096–1099.

Kim SM, Kee HJ, Eom GH, Choe NW, Kim JY, Kim YS, Kim SK, Kook H, Kook H, Seo SB. 2006. Characterization of a novel WHSC1-associated SET domain protein with H3K4 and H3K27 methyltransferase activity. *Biochem Biophys Res Commun* **345:** 318–323.

Kim A, Kiefer CM, Dean A. 2007. Distinctive signatures of histone methylation in transcribed coding and noncoding human β-globin sequences. *Mol Cell Biol* **27:** 1271–1279.

Kim JY, Kee HJ, Choe NW, Kim SM, Eom GH, Baek HJ, Kook H, Kook H, Seo SB. 2008. Multiple-myeloma-related WHSC1/MMSET isoform RE-IIBP is a histone methyltransferase with transcriptional repression activity. *Mol Cell Biol* **28:** 2023–2034.

Kizer KO, Phatnani HP, Shibata Y, Hall H, Greenleaf AL, Strahl BD. 2005. A novel domain in Set2 mediates RNA polymerase II interaction and couples histone H3 K36 methylation with transcript elongation. *Mol Cell Biol* **25:** 3305–3316.

Kolasinska-Zwierz P, Down T, Latorre I, Liu T, Liu XS, Ahringer J. 2009. Differential chromatin marking of introns and expressed exons by H3K36me3. *Nat Genet* **41:** 376–381.

Kouzarides T. 2007. SnapShot: Histone-modifying enzymes. *Cell* **128:** 802.

Kudithipudi S, Lungu C, Rathert P, Happel N, Jeltsch A. 2014. Substrate specificity analysis and novel substrates of the protein lysine methyltransferase NSD1. *Chem Biol* **21:** 226–237.

Kuo AJ, Cheung P, Chen K, Zee BM, Kioi M, Lauring J, Xi Y, Park BH, Shi X, Garcia BA, et al. 2011. NSD2 links dimethylation of histone H3 at lysine 36 to oncogenic programming. *Mol Cell* **44:** 609–620.

Kuroda S, Suzuki S, Kurita A, Muraki M, Aoshima Y, Tanioka F, Sugimura H. 2015. Cytological features of a variant NUT midline carcinoma of the lung harboring the NSD3-NUT fusion gene: A case report and literature review. *Case Rep Pathol* **2015:** 572951.

Kurotaki N, Imaizumi K, Harada N, Masuno M, Kondoh T, Nagai T, Ohashi H, Naritomi K, Tsukahara M, Makita Y, et al. 2002. Haploinsufficiency of NSD1 causes Sotos syndrome. *Nat Genet* **30:** 365–366.

Lachner M, O'Sullivan RJ, Jenuwein T. 2003. An epigenetic road map for histone lysine methylation. *J Cell Sci* **116:** 2117–2124.

Larschan E, Alekseyenko AA, Gortchakov AA, Peng S, Li B, Yang P, Workman JL, Park PJ, Kuroda MI. 2007. MSL complex is attracted to genes marked by H3K36 tri-methylation using a sequence-independent mechanism. *Mol Cell* **28**: 121–133.

La Starza R, Gorello P, Rosati R, Riezzo A, Veronese A, Ferrazzi E, Martelli MF, Negrini M, Mecucci C. 2004. Cryptic insertion producing two NUP98/NSD1 chimeric transcripts in adult refractory anemia with an excess of blasts. *Genes Chromosomes Cancer* **41**: 395–399.

Lauring J, Abukhdeir AM, Konishi H, Garay JP, Gustin JP, Wang Q, Arceci RJ, Matsui W, Park BH. 2008. The multiple myeloma associated MMSET gene contributes to cellular adhesion, clonogenic growth, and tumorigenicity. *Blood* **111**: 856–864.

Lee JS, Shilatifard A. 2007. A site to remember: H3K36 methylation a mark for histone deacetylation. *Mutat Res* **618**: 130–134.

Li B, Jackson J, Simon MD, Fleharty B, Gogol M, Seidel C, Workman JL, Shilatifard A. 2009a. Histone H3 lysine 36 dimethylation (H3K36me2) is sufficient to recruit the Rpd3s histone deacetylase complex and to repress spurious transcription. *J Biol Chem* **284**: 7970–7976.

Li Y, Trojer P, Xu CF, Cheung P, Kuo A, Drury WJ, Qiao Q, Neubert TA, Xu RM, Gozani O, et al. 2009b. The target of the NSD family of histone lysine methyltransferases depends on the nature of the substrate. *J Biol Chem* **284**: 34283–34295.

Loh ML, Ma X, Rusch M, Wu G, Harvey RC, Wheeler DA, Hampton OA, Carroll WL, Chen IM, Gerhard DS, et al. 2013. Comparison of mutational profiles of diagnosis and relapsed pediatric B-acute lymphoblastic leukemia: A report from the COG ALL target project. *Blood* **122**: 824.

Lu T, Jackson MW, Wang B, Yang M, Chance MR, Miyagi M, Gudkov AV, Stark GR. 2010. Regulation of NF-κB by NSD1/FBXL11-dependent reversible lysine methylation of p65. *Proc Natl Acad Sci* **107**: 46–51.

Lucio-Eterovic AK, Singh MM, Gardner JE, Veerappan CS, Rice JC, Carpenter PB. 2010. Role for the nuclear receptor-binding SET domain protein 1 (NSD1) methyltransferase in coordinating lysine 36 methylation at histone 3 with RNA polymerase II function. *Proc Natl Acad Sci* **107**: 16952–16957.

Mahmood SF, Gruel N, Nicolle R, Chapeaublanc E, Delattre O, Radvanyi F, Bernard-Pierrot I. 2013. PPAPDC1B and WHSC1L1 are common drivers of the 8p11-12 amplicon, not only in breast tumors but also in pancreatic adenocarcinomas and lung tumors. *Am J Pathol* **183**: 1634–1644.

Mann KM, Ward JM, Yew CC, Kovochich A, Dawson DW, Black MA, Brett BT, Sheetz TE, Dupuy AJ, Australian Pancreatic Cancer Genome I, et al. 2012. Sleeping Beauty mutagenesis reveals cooperating mutations and pathways in pancreatic adenocarcinoma. *Proc Natl Acad Sci* **109**: 5934–5941.

Marango J, Shimoyama M, Nishio H, Meyer JA, Min DJ, Sirulnik A, Martinez-Martinez Y, Chesi M, Bergsagel PL, Zhou MM, et al. 2008. The MMSET protein is a histone methyltransferase with characteristics of a transcriptional corepressor. *Blood* **111**: 3145–3154.

Martinez-Garcia E, Popovic R, Min DJ, Sweet SMM, Thomas PM, Zamdborg L, Heffner A, Will C, Lamy L, Staudt LM, et al. 2011. The MMSET histone methyl transferase switches global histone methylation and alters gene expression in t(4;14) multiple myeloma cells. *Blood* **117**: 211–220.

Mertz JA, Conery AR, Bryant BM, Sandy P, Balasubramanian S, Mele DA, Bergeron L, Sims RJ III. 2011. Targeting MYC dependence in cancer by inhibiting BET bromodomains. *Proc Natl Acad Sci* **108**: 16669–16674.

Migdalska AM, Van Der Weyden L, Ismail O, Rust AG, Rashid M, White JK, Sánchez-Andrade G, Lupski JR, Logan DW, Arends MJ, et al. 2012. Generation of the Sotos syndrome deletion in mice. *Mamm Genome* **23**: 749–757.

Min D-J, Ezponda T, Kim MK, Will CM, Martinez-Garcia E, Popovic R, Basrur V, Elenitoba-Johnson KS, Licht JD. 2012. MMSET stimulates myeloma cell growth through microRNA-mediated modulation of c-MYC. *Leukemia* **27**: 686–694.

Morishita M, di Luccio E. 2011. Cancers and the NSD family of histone lysine methyltransferases. *Biochim Biophys Acta* **1816**: 158–163.

Morishita M, Mevius D, di Luccio E. 2014. In vitro histone lysine methylation by NSD1, NSD2/MMSET/WHSC1 and NSD3/WHSC1L. *BMC Struct Biol* **14**: 25.

Morris SA, Shibata Y, Noma K, Tsukamoto Y, Warren E, Temple B, Grewal SI, Strahl BD. 2005. Histone H3 K36 methylation is associated with transcription elongation in *Schizosaccharomyces pombe*. *Eukaryot Cell* **4**: 1446–1454.

Nielsen AL, Jørgensen P, Lerouge T, Cerviño M, Chambon P, Losson R. 2004. Nizp1, a novel multitype zinc finger protein that interacts with the NSD1 histone lysine methyltransferase through a unique C2HR motif. *Mol Cell Biol* **24**: 5184–5196.

Nimura K, Ura K, Shiratori H, Ikawa M, Okabe M, Schwartz RJ, Kaneda Y. 2009. A histone H3 lysine 36 trimethyltransferase links Nkx2-5 to Wolf–Hirschhorn syndrome. *Nature* **460**: 287–291.

Oyer Ja, Huang X, Zheng Y, Shim J, Ezponda T, Carpenter Z, Allegretta M, Okot-Kotber CI, Patel JP, Melnick A, et al. 2014. Point mutation E1099K in MMSET/NSD2 enhances its methyltranferase activity and leads to altered global chromatin methylation in lymphoid malignancies. *Leukemia* **28**: 198–201.

Pasillas MP, Shah M, Kamps MP. 2011. NSD1 PHD domains bind methylated H3K4 and H3K9 using interactions disrupted by point mutations in human sotos syndrome. *Hum Mutat* **32**: 292–298.

Pei H, Zhang L, Luo K, Qin Y, Chesi M, Fei F, Bergsagel PL, Wang L, You Z, Lou Z. 2011. MMSET regulates histone H4K20 methylation and 53BP1 accumulation at DNA damage sites. *Nature* **470**: 124–128.

Perera A, Eisen D, Wagner M, Laube SK, Kunzel AF, Koch S, Steinbacher J, Schulze E, Splith V, Mittermeier N, et al. 2015. TET3 is recruited by REST for context-specific hydroxymethylation and induction of gene expression. *Cell Rep* **11**: 283–294.

Popovic R, Martinez-Garcia E, Giannopoulou EG, Zhang Q, Zhang Q, Ezponda T, Shah MY, Zheng Y, Will CM, Small EC, et al. 2014. Histone methyltransferase MMSET/NSD2 alters EZH2 binding and reprograms the myeloma epigenome through global and focal changes in H3K36 and H3K27 methylation. *PLoS Genet* **10**: e1004566.

Poulin MB, Schneck JL, Matico RE, Hou W, McDevitt PJ, Holbert M, Schramm VL. 2016a. Nucleosome binding alters the substrate bonding environment of histone H3 lysine 36 methyltransferase NSD2. *J Am Chem Soc* **138**: 6699–6702.

Poulin MB, Schneck JL, Matico RE, McDevitt PJ, Huddleston MJ, Hou W, Johnson NW, Thrall SH, Meek TD, Schramm VL. 2016b. Transition state for the NSD2-catalyzed methylation of histone H3 lysine 36. *Proc Natl Acad Sci* **113**: 1197–1201.

Qiao Q, Li Y, Chen Z, Wang M, Reinberg D, Xu RM. 2011. The structure of NSD1 reveals an autoregulatory mechanism underlying histone H3K36 methylation. *J Biol Chem* **286**: 8361–8368.

Quintana RM, Dupuy AJ, Bravo A, Casanova ML, Alameda JP, Page A, Sanchez-Viera M, Ramirez A, Navarro M. 2013. A transposon-based analysis of gene mutations related to skin cancer development. *J Invest Dermatol* **133**: 239–248.

Rahman N. 2005. Mechanisms predisposing to childhood overgrowth and cancer. *Curr Opin Genet Dev* **15**: 227–233.

Rahman S, Sowa ME, Ottinger M, Smith JA, Shi Y, Harper JW, Howley PM. 2011. The Brd4 extraterminal domain confers transcription activation independent of pTEFb by recruiting multiple proteins, including NSD3. *Mol Cell Biol* **31**: 2641–2652.

Rao B, Shibata Y, Strahl BD, Lieb JD. 2005. Dimethylation of histone H3 at lysine 36 demarcates regulatory and non-regulatory chromatin genome-wide. *Mol Cell Biol* **25**: 9447–9459.

Rayasam GV, Wendling O, Angrand PO, Mark M, Niederreither K, Song L, Lerouge T, Hager GL, Chambon P, Losson R. 2003. NSD1 is essential for early post-implantation development and has a catalytically active SET domain. *EMBO J* **22**: 3153–3163.

Romana SP, Radford-Weiss I, Ben Abdelali R, Schluth C, Petit A, Dastugue N, Talmant P, Bilhou-Nabera C, Mugneret F, Lafage-Pochitaloff M, et al. 2006. NUP98 rearrangements in hematopoietic malignancies: A study of the Groupe Francophone de Cytogenetique Hematologique. *Leukemia* **20**: 696–706.

Rosati R, La Starza R, Veronese A, Aventin A, Schwienbacher C, Vallespi T, Negrini M, Martelli MF, Mecucci C. 2002. NUP98 is fused to the NSD3 gene in acute myeloid leukemia associated with t(8;11)(p11.2;p15). *Blood* **99**: 3857–3860.

Rosenfeld JA, Kim KH, Angle B, Troxell R, Gorski JL, Westemeyer M, Frydman M, Senturias Y, Earl D, Torchia B, et al. 2013. Further evidence of contrasting phenotypes caused by reciprocal deletions and duplications: Duplication of NSD1 causes growth retardation and microcephaly. *Mol Syndromol* **3**: 247–254.

Sankaran SM, Wilkinson AW, Elias JE, Gozani O. 2016. A PWWP domain of histone-lysine N-methyltransferase NSD2 binds to dimethylated Lys-36 of histone H3 and regulates NSD2 function at chromatin. *J Biol Chem* **291**: 8465–8474.

Santra M, Zhan F, Tian E, Barlogie B, Shaughnessy JS Jr. 2003. Brief report A subset of multiple myeloma harboring the t(4; 14)(p16; q32) translocation lacks FGFR3

expression but maintains an IGH/MMSET fusion transcript. *Blood* **101**: 2374–2376.

Shah MY, Martinez-Garcia E, Phillip JM, Chambliss AB, Popovic R, Ezponda T, Small EC, Will C, Phillip MP, Neri P, et al. 2016. MMSET/WHSC1 enhances DNA damage repair leading to an increase in resistance to chemotherapeutic agents. *Oncogene* doi: 10.1038/onc .2016.116.

Shen C, Ipsaro JJ, Shi J, Milazzo JP, Wang E, Roe JS, Suzuki Y, Pappin DJ, Joshua-Tor L, Vakoc CR. 2015. NSD3-short is an adaptor protein that couples BRD4 to the CHD8 chromatin remodeler. *Mol Cell* **60**: 847–859.

Shiba N, Ichikawa H, Taki T, Park MJ, Jo A, Mitani S, Kobayashi T, Shimada A, Sotomatsu M, Arakawa H, et al. 2013. NUP98-NSD1 gene fusion and its related gene expression signature are strongly associated with a poor prognosis in pediatric acute myeloid leukemia. *Genes Chromosomes Cancer* **52**: 683–693.

Shilatifard A. 2006. Chromatin modifications by methylation and ubiquitination: Implications in the regulation of gene expression. *Annu Rev Biochem* **75**: 243–269.

Sims RJ III, Reinberg D. 2009. Processing the H3K36me3 signature. *Nat Genet* **41**: 270–271.

Spies N, Nielsen CB, Padgett RA, Burge CB. 2009. Biased chromatin signatures around polyadenylation sites and exons. *Mol Cell* **36**: 245–254.

Stabell M, Larsson J, Aalen RB, Lambertsson A. 2007. *Drosophila* dSet2 functions in H3-K36 methylation and is required for development. *Biochem Biophys Res Commun* **359**: 784–789.

Stec I, Wright TJ, van Ommen GJ, de Boer PA, van Haeringen A, Moorman AF, Altherr MR, den Dunnen JT. 1998. WHSC1, a 90 kb SET domain-containing gene, expressed in early development and homologous to a Drosophila dysmorphy gene maps in the Wolf–Hirschhorn syndrome critical region and is fused to IgH in t(4;14) multiple myeloma. *Hum Mol Genet* **7**: 1071–1082.

Stec I, van Ommen GJ, den Dunnen JT. 2001. WHSC1L1, on human chromosome 8p11.2, closely resembles WHSC1 and maps to a duplicated region shared with 4p16.3. *Genomics* **76**: 5–8.

Suzuki S, Kurabe N, Ohnishi I, Yasuda K, Aoshima Y, Naito M, Tanioka F, Sugimura H. 2015. NSD3-NUT-expressing midline carcinoma of the lung: First characterization of primary cancer tissue. *Pathol Res Pract* **211**: 404–408.

Taketani T, Taki T, Nakamura H, Taniwaki M, Masuda J, Hayashi Y. 2009. NUP98-NSD3 fusion gene in radiation-associated myelodysplastic syndrome with t(8;11)(p11;p15) and expression pattern of NSD family genes. *Cancer Genet Cytogenet* **190**: 108–112.

Tatton-Brown K, Douglas J, Coleman K, Baujat G, Chandler K, Clarke a, Collins a, Davies S, Faravelli F, Firth H, et al. 2005. Multiple mechanisms are implicated in the generation of 5q35 microdeletions in Sotos syndrome. *J Med Genet* **42**: 307–313.

Thanasopoulou A, Tzankov A, Schwaller J. 2014. Potent cooperation between the NUP98-NSD1 fusion and the FLT3-ITD mutation in acute myeloid leukemia induction. *Haematologica* **99**: 1465–1471.

Tonon G, Wong KK, Maulik G, Brennan C, Feng B, Zhang Y, Khatry DB, Protopopov A, You MJ, Aguirre AJ, et al.

2005. High-resolution genomic profiles of human lung cancer. *Proc Natl Acad Sci* **102:** 9625–9630.

Vermeulen M, Eberl HC, Matarese F, Marks H, Denissov S, Butter F, Lee KK, Olsen JV, Hyman AA, Stunnenberg HG, et al. 2010. Quantitative interaction proteomics and genome-wide profiling of epigenetic histone marks and their readers. *Cell* **142:** 967–980.

Waggoner DJ, Raca G, Welch K, Dempsey M, Anderes E, Ostrovnaya I, Alkhateeb A, Kamimura J, Matsumoto N, Schaeffer GB, et al. 2005. NSD1 analysis for Sotos syndrome: Insights and perspectives from the clinical laboratory. *Genet Med* **7:** 524–533.

Wagner EJ, Carpenter PB. 2012. Understanding the language of Lys36 methylation at histone H3. *Nat Rev Mol Cell Biol* **13:** 115–126.

Wang GG, Cai L, Pasillas MP, Kamps MP. 2007. NUP98-NSD1 links H3K36 methylation to Hox-A gene activation and leukaemogenesis. *Nat Cell Biol* **9:** 804–812.

Wu H, Zeng H, Lam R, Tempel W, Amaya MF, Xu C, Dombrovski L, Qiu W, Wang Y, Min J. 2011. Structural and histone binding ability characterizations of human PWWP domains. *PloS ONE* **6:** e18919.

Xu L, Zhao Z, Dong A, Soubigou-Taconnat L, Renou JP, Steinmetz A, Shen WH. 2008a. Di- and tri- but not monomethylation on histone H3 lysine 36 marks active transcription of genes involved in flowering time regulation and other processes in Arabidopsis thaliana. *Mol Cell Biol* **28:** 1348–1360.

Xu LN, Wang X, Zou SQ. 2008b. Effect of histone deacetylase inhibitor on proliferation of biliary tract cancer cell lines. *World J Gastroenterol* **14:** 2578–2581.

Yang ZQ, Liu G, Bollig-Fischer A, Giroux CN, Ethier SP. 2010. Transforming properties of 8p11-12 amplified genes in human breast cancer. *Cancer Res* **70:** 8487–8497.

Zheng Y, Sweet SMM, Popovic R, Martinez-Garcia E, Tipton JD, Thomas PM, Licht JD, Kelleher NL. 2012. Total kinetic analysis reveals how combinatorial methylation patterns are established on lysines 27 and 36 of histone H3. *Proc Natl Acad Sci* **109:** 13549–13554.

Zhou Z, Thomsen R, Kahns S, Nielsen AL. 2010. The NSD3L histone methyltransferase regulates cell cycle and cell invasion in breast cancer cells. *Biochem Biophys Res Commun* **398:** 565–570.

Zuber J, Shi J, Wang E, Rappaport AR, Herrmann H, Sison EA, Magoon D, Qi J, Blatt K, Wunderlich M, et al. 2011. RNAi screen identifies Brd4 as a therapeutic target in acute myeloid leukaemia. *Nature* **478:** 524–528.

Mixed-Lineage Leukemia Fusions and Chromatin in Leukemia

Andrei V. Krivtsov, Takayuki Hoshii, and Scott A. Armstrong

Department of Pediatric Oncology, Dana-Farber Cancer Institute, and Division of Hematology/Oncology, Boston Children's Hospital, Boston, Massachusetts 02215

Correspondence: scott_armstrong@dfci.harvard.edu

Recent studies have shown the importance of chromatin-modifying complexes in the maintenance of developmental gene expression and human disease. The mixed lineage leukemia gene (*MLL1*) encodes a chromatin-modifying protein and was discovered as a result of the cloning of translocations involved in human leukemias. MLL1 is a histone lysine 4 (H3K4) methyltransferase that supports transcription of genes that are important for normal development including *homeotic* (*Hox*) genes. *MLL1* rearrangements result in expression of fusion proteins without H3K4 methylation activity but may gain the ability to recruit other chromatin-associated complexes such as the H3K79 methyltransferase DOT1L and the super elongation complex. Therefore, chromosomal translocations involving *MLL1* appear to directly perturb the regulation of multiple chromatin-associated complexes to allow inappropriate expression of developmentally regulated genes and thus drive leukemia development.

IDENTIFICATION AND STRUCTURE OF MLL FUSIONS

Studies of acute leukemias carrying chromosomal translocation of the long arm (q) of chromosome 11 at band q23 (11q23) (Rowley 1993) resulted in identification of the gene lysine methyltransferase 2A (*KMT2A*; also known as *MLL, MLL1, HTRX, HRX, TRX1, ALL-1*) (Ziemin-van der Poel et al. 1991; Tkachuk et al. 1992) as the recurrent target for these translocations. Subsequent studies have shown that *MLL1/KMT2A* mutations can result either in partial tandem duplications (MLL-PTDs) or in fusions with proteins functionally associated with active transcription (MLL fusions) (Domer et al. 1993; Gu et al. 1993; Prasad et al. 1994; Schichman et al. 1994; Chaplin et al.

1995). In this review, we will use *MLL1/KMT2A* as the name for the gene and the more conventional and historical name "MLL fusions" or "MLL-PTDs" when referring to the protein generated by these chromosomal abnormalities.

More than 98% of breakpoints within the *MLL1* gene are located within an 8.3-kb breakpoint cluster region between exons 8 and 13 (Meyer et al. 2013) that has multiple topoisomerase II cleavage sites as well as nuclear matrix attachment regions. Interference with normal cellular processes such as therapeutic inhibition of topoisomerase II leads to the formation of chromosomal rearrangements found in leukemia (Strissel et al. 1998). The rearrangements always result in either expression of an in-frame chimeric protein encoded by the first eight to 13 exons of *KMT2A/MLL* and a variable number of exons from the fusion partner gene or an aberrant KMT2A/MLL protein encoded by duplication of the first five to 12 exons of *KMT2A/MLL* inserted into exons 11 and 12 (Fig. 1) (Krivtsov and Armstrong 2007). Currently, more than 70 translocation partner genes have been reported; however, four (*AF4, AF9, AF10, ENL*) account for >75% of cases (Meyer et al. 2013).

MLL FUSIONS AT THE FOREFRONT OF LEUKEMIA RESEARCH

Since their discovery, MLL fusions have been at the frontier of the leukemia research, likely because *MLL1/KMT2A* rearrangements are quite potent oncogenes and require very few other, if any, secondary mutations to generate leukemia. This is in line with the natural history of *MLL1*-rearranged leukemias that are one of the few mutations that generate leukemia in very young children including as early as a few days of life. Genetic "knockin" of *AF9*, a gene commonly found rearranged with *MLL1,* into the *Mll* locus generated transgenic mice expressing the Mll-AF9 fusion protein product of t(9;11) chromosomal translocation from the endogenous *Mll1* promoter and formally proved that this oncogene can initiate acute myeloid leukemia (AML) and B-lymphoblastic leukemia (B-ALL), mimicking the human disease (Corral et al. 1996). Later, the same laboratory pioneered an elegant system in which interchromosomal *Mll1* translocations could be induced in mice de novo and led to the development of acute leukemia (Collins et al. 2000; Forster et al. 2003).

A new avenue in the leukemia research was opened by ectopic expression of an MLL-ENL fusion protein in prospectively isolated hematopoietic stem cells (HSCs) and committed myeloid progenitor cells that resulted in similar AMLs (Cozzio et al. 2003). This experiment showed that MLL-ENL can reactivate hematopoietic self-renewal in committed myeloid progenitor cells called granulocyte-macrophage progenitors (GMPs). Previously, it was widely accepted that leukemogenic transformation originates only in HSCs (Dick 2005). Further-

more, conditional expression of an Mll-Cbp fusion protein in hematopoietic cells led to expansion of GMPs before progression to fatal AML (Wang et al. 2005). Ectopic expression of MLL-AF9 in GMPs followed by transplantation into syngeneic recipients resulted in AML with a relatively high leukemia stem cell frequency with approximately 1 in 150 cells being able to initiate disease in secondary recipient mice. Prospective isolation of leukemic cells (L-GMPs) with an immunophenotype similar to normal myeloid progenitors followed by transplantation into syngeneic recipients showed that leukemia initiating activity of this cell fraction was increased to ~1 in 6. Microarray-based gene expression analyses showed up-regulation of a subset of genes in L-GMPs that are normally expressed in self-renewing HSC and suppressed during normal differentiation to GMPs. These genes are also highly expressed in human *MLL*-rearranged AML as compared with other subtypes of AML, and we now know that many of these genes are direct binding targets of MLL-fusion proteins. These studies showed that leukemia initiating cells do not necessarily have an immunophenotype similar to HSC and suggest that leukemia may be the result of expression of stem cell–associated genes in developmental cell types in which they should not be highly expressed (Krivtsov et al. 2006; Somervaille and Cleary 2006). A subset of the self-renewal-associated genes that are highly expressed in the majority of mouse and human MLL-fusion leukemias are the posterior *HOXA* cluster genes *HOXA5-HOXA10* and the heterodimerizing partners of HOX proteins, MEIS1 and PBX3. The only known exceptions are the ~10% of *t(4;11)* rearranged B-ALL that do not express *HOXA* cluster genes (Krivtsov and Armstrong 2007; Krivtsov et al. 2008).

Overall, these studies interrogating the biology of *MLL1*-rearranged leukemia have made a significant impact on the both the leukemia and hematology fields. In particular, the mouse model systems developed to study this leukemia continue to be used to study many aspects of leukemia biology. Furthermore, the detailed characterization of the gene expression programs driven by MLL-fusion proteins has pro-

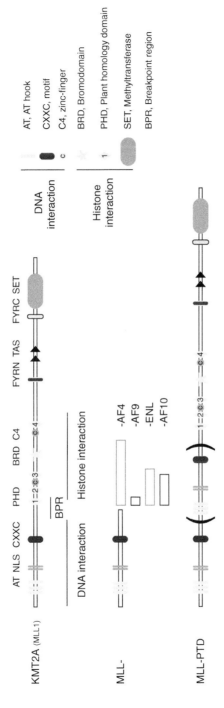

Figure 1. Schematic drawing of KMT2A, MLL fusions, and MLL-PTD with indicated domains. KMT2A, Lysine methyl-transferase 2A; MLL, mixed-lineage leukemia; PTD, partial tandem duplication; NLS, nuclear localization signal; FYRN, FY-rich amino terminal; TAS, transcription activation domain; FYRC, FY-rich carboxyl terminal; SET, Su(Var)39, en-hancer of zeste, and trithorax; AF4(9)(10), ALL1-Fused gene from chromosome 4(9)(10) protein; ENL, eleven nineteen leukemia.

vided a foundation for studies that have focused on the molecular mechanisms of MLL1/KMT2A and MLL-fusion protein function.

STRUCTURE OF THE MLL/KMT2 FAMILY

MLL1 (or KMT2A) is a member of the lysine (K) methyltransferase (KMT) 2 family of proteins that are highly conserved in eukaryotes most of which possess (Su(var)3−9, enhancer of zeste and trithorax) SET domains. Based on its protein domain composition, KMT2A's closest homolog is KMT2B (Fig. 2).

The MLL1 protein is a large nuclear protein that has multiple important well-defined domains. The amino terminus of MLL1 interacts with menin, a critical interaction for the MLL-fusion protein function. This is followed by three AT-hook domains that can bind to the minor grove of DNA and a CXXC domain that binds to unmethylated CpGs (Rao and Dou 2015). The CXXC domain is followed by a set of domains that can interact with modified histone tails and includes four plant homology domains (PHDs) and a bromodomain (BRD). All MLL fusions include the MLL1 sequence up to, but not including, the PHDs. In fact, inclusion of PHDs within the MLL fusions negates the transformation activity of these fusions (Muntean et al. 2008). Full-length MLL1 is cleaved by Taspase1 and reassembled through interaction of FYRN with FYRC domains (Rao and Dou 2015). For almost a decade, studies were unable to show enzymatic activity of MLL1 (Rea et al. 2000). However, in 2002 it was shown that a bacterially expressed SET domain of MLL1 has histone H3 lysine 4 (H3K4) methyltransferase activity on peptides resembling histone tails and recombinant nucleosomes and this catalytic activity was increased if substrates were preacetylated on nearby lysine residues (Milne et al. 2002; Nakamura et al. 2002; Zhang et al. 2015). Later, it was established that the SET domain of MLL1 can catalyze monomethyl and dimethyl and to lesser extent the trimethyl modification of H3K4 (Wu et al. 2013). These studies showed enzymatic activity of the SET domain, but we now know that the full MLL1 complex is required to support efficient modification of H3K4 (Rao and Dou 2015). Genome-scale occupancy studies using chromatin immunoprecipitation followed by sequencing (ChIP-seq) revealed association of MLL1 with a subset of active promoters (Milne et al. 2005; Wang et al. 2009) as well as certain enhancers (Wang et al. 2011; Yang et al. 2014). Although the exact mechanisms that control the appropriate association of MLL1 with chromatin are still not completely understood, it is plausible to suggest that MLL1 interacts with DNA and chromatin through multiple protein domains with amino-terminal regions such as the CXXC domain interacting with DNA, which is further supported through binding of four PHDs and possibly the bromodomain to modified histone tails (reviewed in Rao and Dou 2015).

Intriguingly, the catalytic activity of MLL1 does not seem to be essential for cell survival or function, as compared with the MLL1 protein itself. Insertion of a stop codon after either exon 3 or 12 leading to loss of MLL1 results in early embryonic death and severe hematopoietic abnormalities (Yu et al. 1995; Yagi et al. 1998). Surprisingly, mice with genetic deletion of only the MLL1 SET domain are born fertile and only show modest skeletal abnormalities and no gross hematopoietic defects (Terranova et al. 2006; Mishra et al. 2014). Moreover, loss of MLL1 does not result in global changes of H3K4 methylation, suggesting MLL1 is not the major H3K4 methyltransferase in most cells (Wang et al. 2009). These data prompt questions as to the specific function of H3K4 methylation during development. Given the pattern of H3K4 methylation throughout the genome and multiple studies showing that perturbation of H3K4 methylation leads to relevant changes in gene expression, there is little doubt that H3K4 methylation is playing a role in the control of gene expression, but its exact mechanistic contribution remains an open question.

MLL1/KMT2A SUPER COMPLEXES

The primary structure of MLL1 suggests multiple protein–DNA and protein–protein interactions, many of which have been validated. Indeed, the KMT2 family of proteins form large

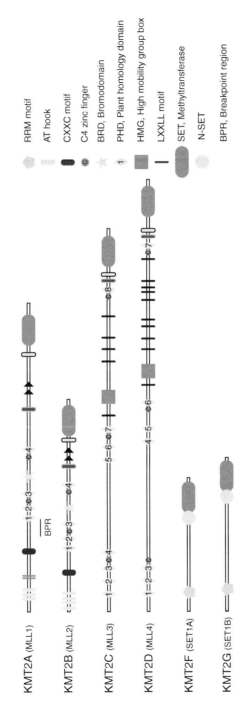

Figure 2. Schematic representation of domain composition of the KMT2 (lysine methyltransferase 2) family of proteins.

multiprotein complexes with common and subgroup specific proteins. For instance, the retinoblastoma binding protein 5 (RbBP5), ASH2 like histone lysine methyltransferase complex subunit (ASH2L), WD repeat 5 (WDR5), and DPY30 are found in complexes with all KMT2 family proteins and are in some cases required for full catalytic activation. Biochemical reconstitution assays show that WDR5, ASH2L, and RbBP5 form a stable core that interacts with MLL1 and potently activate the SET domain catalytic activity (Dou et al. 2006; Patel et al. 2009). At the same time, association of MLL1 with DPY30 only slightly activates SET activity (Patel et al. 2009). Blockage of association of MLL1 with WDR5 using WDR5 interacting (WIN) peptides leads to disintegration of the MLL1 complex in vitro likely leading to the reduction in SET domain catalytic activity (Karatas et al. 2010). In addition to the proteins that are found in all KMT2 protein complexes, there are also subgroup specific subunits. For instance, multiple endocrine neoplasia type 1 (menin or MEN1) (Hughes et al. 2004) and lens epithelium–derived growth factor (LEDGF) (Yokoyama and Cleary 2008) are unique subunits that are found only in the MLL1 and MLL2/KMT2B complexes. Both MEN1 and LEDGF are believed to be necessary for MLL1 and MLL-fusion protein recruitment to target genes. KMT2A can also physically associate with other enzymes that modify chromatin. For instance, MLL1 has been reported to be associated with the histone H3K9/H4K16 acetyltransferase MOF, possibly through WDR5 (Dou et al. 2005). It was observed that active promoters are decorated with H4K16 acetyl modifications in addition to H3K4me3 (Wang et al. 2008). However, another group argues that both MLL1 and MOF associate with WDR5 independently, and are present in different complexes (Dias et al. 2014). Additionally, in specific cellular contexts KMT2A/B were reported to interact with transcription factors such as OCT4 (Ang et al. 2011), E2F (Tyagi et al. 2007), PAX7 (Kawabe et al. 2012), and NF-Y (Fossati et al. 2011); however, the functional meaning of these interactions remains to be studied. Furthermore, it may be the case that MLL1 methylates these

transcription factors to modulate their activity as the full complement of MLL1 substrates has not been defined. This bouquet of MLL1 interactions with other proteins opens the possibility that the various complexes may play different roles in the control of gene expression.

MECHANISMS OF MLL-FUSION ACTION

At least two potentially complementary mechanisms for MLL-fusion function have been identified and studied. The first includes recruitment of the super elongation complex (SEC) by MLL-fusion proteins. ELL, one of the more frequent MLL fusion partners (Thirman et al. 1994), was shown to be required for elongation by RNA Pol II (Shinobu et al. 1999). This finding initiated the concept that MLL fusions may function to activate transcription of target genes via directly influencing transcriptional elongation by RNA Pol II. Subsequent characterization of the proteins encoded by the most frequent fusion partner genes AF4, AF9, and ENL were found to be members of a large multiprotein complex that has been isolated and given multiple names including the elongation-assisting multiprotein complex (EAP) and the SEC (Mueller et al. 2007; Lin et al. 2010). These complexes include at least ENL or AF9, AF4, AF5 (all known MLL-fusion partners) and members of the transcriptional elongation complex b (pTEFb). P-TEFb is a heterodimer of CDK9 and CyclinT1 or T2 that phosphorylates serine 2 within the carboxy-terminal repeat domain of RNA polymerase II which releases RNA Pol II from the paused state (Jonkers and Lis 2015). The polymerase-associated factor elongation complex (PAF1C) can also be coimmunoprecipitated (co-IP) with MLL1 as well as MLL-ENL (Mueller et al. 2007). PAF1C is associated with elongating RNA polymerase II and can influence proper methylation of H3K4 (Wood et al. 2003). It appears that PAF1 complex may serve as a dock for interaction of elongating RNA Pol II and other histone chromatin remodeling complexes (Wang et al. 2008). Moreover, PAF1 appears to be involved in regulation of methylation on H3K4 and H3K79 residues through recruitment of E2/E3 ubiquitin ligase complex (Wood et al. 2003). Therefore, MLL fusions may act to bring together two independently regulated complexes, MLL1 and SEC, into a single and potentially unregulated complex that allows unrestricted transcription of genes directly bound by the MLL-fusion proteins.

The second potential mechanism stems from the seminal identification of the interaction between one of the common MLL fusion partners, AF10, with the histone H3 lysine 79 (H3K79) methyltransferase DOT1L (Okada et al. 2005). Cells transformed with MLL-AF10 were shown to have elevated levels of H3K79 methylation associated with the MLL-fusion target gene *HoxA9* gene in vitro. Coincidentally, SEC complex members AF9 and ENL (both MLL fusion partners) can bind DOT1L directly in addition to the SEC (Erfurth et al. 2004; Zhang et al. 2006; Bitoun et al. 2007; Kuntimaddi et al. 2015). However, the regions of the ENL and AF9 proteins that interact with DOT1L and SEC are overlapping and therefore it appears that the SEC complex and DOT1L complex are two independent complexes (Shen et al. 2013). Given that H3K79me2 is also broadly correlated with actively transcribed genes, it is attractive to hypothesize that the MLL fusions recruit DOT1L to target genes to enhance H3K79 methylation and therefore positively regulate expression MLL-fusion target genes (Krivtsov and Armstrong 2007). This hypothesis was confirmed by the finding that mouse and human B-ALLs expressing an MLL-AF4 fusion protein have significantly increased H3K79 di-methylation on MLL-AF4-bound genes (Guenther et al. 2008; Krivtsov et al. 2008). Furthermore, elevated H3K79 methylation has been found at regions of chromatin bound by MLL-fusion proteins in all MLL-fusion leukemia models that have been assessed including fusions such as MLL-AF6 that do not appear to interact with either DOT1L or SEC (Yun et al. 2011; Deshpande et al. 2013). Therefore, it may be that elevated H3K79 methylation across MLL-fusion target genes is common feature among all *MLL*-rearranged acute leukemias. However, mice transgenic for an *Mll-LacZ* fusion also develop AML with similar latency as

mice transgenic for *Mll-AF9* (Dobson et al. 2000.). This has been hypothesized to be a result of oligomerization of the amino terminus of Mll. Therefore, it may be that both the DOT1L complexes and SECs are aberrantly targeted to MLL-fusion target genes in many cases, but this is unlikely to completely account for transformation. Furthermore, DOT1L is a physiologic regulator of *Hox* gene expression during normal hematopoietic development. Thus, the details defining how the SEC and DOT1L complexes influence leukemic gene expression remain a work in progress.

Perhaps, other mechanisms exist to recruit or influence DOT1L and/or SEC at MLL-fusion target genes. Leukemias with an MLL-PTD retain a SET domain. MLL-PTDs have not been reported to interact with proteins outside the normal MLL1 complexes. It is interesting to note the only difference between the MLL-PTD and MLL1 proteins is duplication of the AT-hook and CXXC domains. Therefore, it is not immediately clear how this would influence either the SEC or DOT1L complexes. ChIP-seq experiments have, however, detected extensive H3K79me2 hyper methylation of several genes in cells expressing an MLL-PTD, and these leukemia cells are quite sensitive to DOT1L inhibition, again suggesting a role for DOT1L in this form of MLL1/KMT2A leukemia (Kühn et al. 2015).

The exquisite sensitivity of leukemias with *MLL1/KMT2A* rearrangements to DOT1L inhibition has prompted the question as to the function of DOT1L and H3K79 methylation. H3K79 methylation is somewhat unique compared with many histone modifications and therefore may also function differently. Lysine 79 on histone 3 is found in the globular region of H3 as opposed to other amino acid residues that are targets of modifications, which most often reside in histone tails. There are currently no known proteins or domains that can interact with methylated H3K79. This is also different from the more highly studied lysine 4 and lysine 27 residues that, when methylated, are "read" by specific protein domains (Yun et al. 2011; Vlaming and van Leeuwen 2016). There is also no known demethylase that removes monomethylation, dimethylation, or trimethylation from H3K79. All of these characteristics suggest that H3K79 may play a unique role in cellular biology. Consistent with the lack of a demethylase, when DOT1L is inhibited, the kinetics of H3K79me2/3 loss is relatively slow and more consistent with dilution during cell division rather than active removal by a demethylase (Bernt et al. 2011). At the same time, H3K79me2 is present on most genes with high-level expression (Steger et al. 2008). Recently, a mechanism describing how H3K79 methylation may influence transcription was discovered in a genome-scale shRNA "rescue" screen. It appears that H3K79 methylation in leukemia cells prevents association of the NAD-dependent deacetylase sirtuin-1 (SIRT1) complex. Upon inhibition of DOT1L, SIRT1 localizes to regions of chromatin that were previously modified with H3K79 methylation. This leads to loss of H3K9 and H4K16 acetylation and subsequent histone H3 lysine 9 methylation (H3K9me3) via the SUV39 enzymes. This leads to suppression of gene transcription (Chen et al. 2015). Therefore, H3K79 methylation may prevent repressive mechanisms from converting chromatin from a state that allows gene expression to a state that does not allow access of the transcriptional machinery. This is consistent with pioneering work in budding yeast in which it has been shown that H3K79 methylation inhibits Sir protein spreading on chromatin (Sneppen and Dodd 2015).

POTENTIAL THERAPEUTIC STRATEGIES THAT TARGET CHROMATIN IN MLL-FUSION LEUKEMIA

Traditionally, patients with *MLL* rearrangements are assigned to intermediate- or poor-risk groups with long-term-survival rates at or below 50%. This group of patients would greatly benefit from development of novel therapeutic approaches. Because MLL fusions seem to act in concert with a number of other enzymes, those may potentially represent therapeutic targets. Recent efforts have made great progress toward identification of small molecule inhibitors that target MLL-fusion-associated mechanisms as well as other pathways the suppression of which might lead to synthetic lethality with the MLL-fusion protein.

Aberrant recruitment of the DOT1L enzyme by MLL fusions provided a rationale for development and testing of small molecule inhibitors. DOT1L inhibitors efficiently and specifically eradicated leukemia cells in vitro and in vivo model systems (Daigle et al. 2011, 2013). Phase 1 clinical studies assessing a small molecule DOT1L inhibitor in adults and children with relapsed or refractory leukemias have been completed. Approximately 20% of patients showed some type of clinical or biological response, two of which went into remission. Of importance, no side effects of the inhibitor, EPZ-5676, have been reported despite initial concerns of potential transcriptional suppression of the $2-5 \times 10^3$ genes that are normally marked by H3K79me2 (Stein and Tallman 2016). The initial successful development of DOT1L inhibitors was followed by small molecule discovery by other groups and there are currently multiple DOT1L inhibitors with varying pharmacokinetic properties available for research (Yu et al. 2012; Yi et al. 2015; Chen et al. 2016; Dafflon et al. 2016; Spurr et al. 2016). Although these initial results are encouraging, there is no doubt these approaches will need to be combined with other therapies to realize the maximal therapeutic benefit. Further development of DOT1L inhibitors will proceed with combination approaches as they are identified.

A second interesting potential therapeutic approach is disruption of the MLL-fusion complex via inhibition of the interaction of the amino terminus of MLL with the MLL-fusion complex member menin1 (MEN). MLL-fusion-driven gene expression and leukemia cell proliferation is dependent on interaction with MEN (Yokoyama and Cleary 2008). Several small molecules that disrupt interaction of menin with MLL fusions have been identified (Grembecka et al. 2012; Borkin et al. 2015; Xu et al. 2016). Treatment of leukemia cells expressing various fusion such as MLL-ENL, MLL-AF9, MLL-CBP, MLL-AF6, MLL-GAS7, and MLL-AF1P with MI-2-2 or MI-503 at nanomolar concentrations results in strong growth inhibition and MI-503 has antileukemia activity in vivo. Menin–MLL fusion interaction inhibition shows some degree of selectivity as cells expressing either Hoxa9-Meis1 or E2A-HLF oncogenes are not sensitive to the same concentrations of MI-2-2 in vitro. Moreover, disruption of MLL/MEN1 interaction results in rapid differentiation of the cells in vitro as assessed by morphology. Of potential therapeutic importance, combined inhibition of the MLL fusion–MEN1 interaction and DOT1L activity enhances leukemia cell differentiation (He et al. 2016). The advantage of this approach is relative specificity because menin1 interacts only with MLL fusion, MLL1, and MLL2/KMT2B and does not affect other members of the KMT2 family. Other approaches to disrupt the MLL-fusion complex have also been developed. One potentially interesting approach in inhibition of the interaction between WDR5 and MLL1, which leads to profound biochemical changes in the complexes similar to loss of wild-type *Mll* in mice, including global gene expression changes. Tool molecules, including the WDR5/MLL1 inhibitor MM-401, induced growth arrest followed by differentiation of AML cells expressing MLL fusions (Cao et al. 2014). Although these initial data are of interest, these molecules are not particularly potent so the full potential of this approach awaits development of better small-molecule inhibitors. Whether or not WDR5 inhibition will show selectivity toward MLL-fusion leukemias remains to be fully addressed and is a concern because WDR5 is found in all KMT2 complexes and thus could have profound cellular effects. Based on this progress, the development of inhibitors of these protein–protein interactions remains an active area of investigation.

Another potential approach that has garnered much interest is inhibition of the BET bromodomains. BRD4, a member of BET family of proteins, was also discovered as an interesting cancer dependency in *MLL*-rearranged AML using genome-wide shRNA screening (Dawson et al. 2011; Zuber et al. 2011). BRD4 is a "chromatin reader" protein that binds acetylated lysines in the H3 and H4 tails through its acetyllysine binding bromodomain. BRD4 is

found associated with many active enhancers throughout the genome, but particular excitement has focused on enhancers with unusually high levels of H3K27 acetylation and BRD4 binding. The enhancers, termed super-enhancers, may be particularly important as removal of BRD4 from super-enhancers like the *c-Myc* enhancer leads to rapid down-regulation of *Myc* gene expression. Genetic and shRNA-mediated silencing of BRD4 expression also resulted in differentiation of MLL-AF9 expressing leukemia cells and suppression of leukemogenesis in vitro and in vivo. However, the antiproliferative effects of the bromodomain inhibitors were broadly observed in other genetically defined types of AML as well as in other cancers (Delmore et al. 2011). Therefore, it is likely that bromodomain inhibitors are targeting a myeloid lineage program rather than the MLL-fusion driven program per se. Clinical trials with bromodomain inhibitors are ongoing and we should get an assessment of their potential clinical activity in the near future.

Another recent chromatin-targeted approach came from more focused studies on small molecules that inhibit the mediator component cyclin-dependent kinase 8 (CDK8). MLL-fusion-expressing cell lines are particularly sensitive to CDK8 inhibition. CDK8 is colocalized with MED1, BRD4, and H3K27Ac on super-enhancers providing a potential explanation for this sensitivity. However, inhibition of CDK8 enzymatic activity resulted in disproportionate activation of transcription of genes associated with super-enhancers, which is the inverse of the effects of BET inhibitors. The overexpression of such super-enhancer-associated tumor suppressors as IRF1, IRF8, CEBPA, and ETV6 seems to contribute to the antiproliferative effects of CDK8 inhibition. This study suggests that viability of leukemia cells might depend on a precise "dosage" of super-enhancer-associated gene expression to maintain proliferation and the undifferentiated state (Pelish et al. 2015). These studies suggest that either suppression of a gene expression program that inhibits differentiation or enhancement of a gene expression program that drives differentiation might be therapeutic approaches in

AML. Further studies will help clarify this possibility.

CONCLUSIONS

It is now clear that disruption of the normal cellular processes that control chromatin state is common in human cancers. The best studied of the mutations in chromatin regulators is the MLL fusion that drives expression of developmental gene expression in cells that should not express these programs. Studies have now shown that this "aberrant" gene expression can be reversed by inhibition of chromatin-based mechanism leading to the hope that this will allow new therapeutic development. We are still in the early days of assessment of such approaches, but the initial clinical trials suggest these approaches do have clinical activity. Studying acute leukemias bearing *MLL* rearrangements has provided valuable insight into the mechanisms of this disease as well as led the way for development of small molecule inhibitors targeting epigenetic mechanisms. However, it is also clear that much work needs to be performed to fully understand chromatin biology such that we can use these new approaches for the greatest benefit in patients with cancer.

ACKNOWLEDGMENTS

This work was supported by National Institutes of Health (NIH) Grants PO1 CA66996, R01 CA140575, the Leukemia & Lymphoma society, and Gabrielle's Angel Research Foundation (to S.A.A.). S.A.A. is a consultant for Epizyme Inc., Vitae Inc., and Imago Biosciences.

REFERENCES

Ang YS, Tsai SY, Lee DF, Monk J, Su J, Ratnakumar K, Ding J, Ge Y, Darr H, Chang B, et al. 2011. Wdr5 mediates self-renewal and reprogramming via the embryonic stem cell core transcriptional network. *Cell* **145:** 183–197.

Bernt KM, Zhu N, Sinha AU, Vempati S, Faber J, Krivtsov AV, Feng Z, Punt N, Daigle A, Bullinger L, et al. 2011. *MLL*-rearranged leukemia is dependent on aberrant H3K79 methylation by DOT1L. *Cancer Cell* **20:** 66–78.

Bitoun E, Oliver PL, Davies KE. 2007. The mixed-lineage leukemia fusion partner AF4 stimulates RNA polymerase II transcriptional elongation and mediates coordinated chromatin remodeling. *Hum Mol Genet* **16:** 92–106.

Borkin D, He S, Miao H, Kempinska K, Pollock J, Chase J, Purohit T, Malik B, Zhao T, Wang J, et al. 2015. Pharmacologic inhibition of the menin-MLL interaction blocks progression of MLL leukemia in vivo. *Cancer Cell* **27:** 589–602.

Cao F, Townsend EC, Karatas H, Xu J, Li L, Lee S, Liu L, Chen Y, Ouillette P, Zhu J, et al. 2014. Targeting MLL1 H3K4 methyltransferase activity in mixed-lineage leukemia. *Mol Cell* **53:** 247–261.

Chaplin T, Bernard O, Beverloo HB, Saha V, Hagemeijer A, Berger R, Young BD. 1995. The t(10;11) translocation in acute myeloid leukemia (M5) consistently fuses the leucine zipper motif of AF10 onto the HRX gene. *Blood* **86:** 2073–2076.

Chen CW, Koche RP, Sinha AU, Deshpande AJ, Zhu N, Eng R, Doench JG, Xu H, Chu SH, Qi J, et al. 2015. DOT1L inhibits SIRT1-mediated epigenetic silencing to maintain leukemic gene expression in *MLL*-rearranged leukemia. *Nat Med* **21:** 335–343.

Chen C, Zhu H, Stauffer F, Caravatti G, Vollmer S, Machauer R, Holzer P, Möbitz H, Scheufler C, Klumpp M, et al. 2016. Discovery of novel Dot1L inhibitors through a structure-based fragmentation approach. *ACS Med Chem Lett* **7:** 735–740.

Collins EC, Pannell R, Simpson EM, Forster A, Rabbitts TH. 2000. Inter-chromosomal recombination of *Mll* and *Af9* genes mediated by cre-*loxP* in mouse development. *EMBO Rep* **1:** 127–132.

Corral J, Lavenir I, Impey H, Warren AJ, Forster A, Larson TA, Bell S, McKenzie AN, King G, Rabbitts TH. 1996. An *Mll–AF9* fusion gene made by homologous recombination causes acute leukemia in chimeric mice: A method to create fusion oncogenes. *Cell* **85:** 853–861.

Cozzio A, Passegué E, Ayton PM, Karsunky H, Cleary ML, Weissman IL. 2003. Similar MLL-associated leukemias arising from self-renewing stem cells and short-lived myeloid progenitors. *Genes Dev* **17:** 3029–3035.

Dafflon C, Craig VJ, Méreau H, Gräsel J, Schacher Engstler B, Hoffman G, Nigsch F, Gaulis S, Barys L, Ito M, et al. 2016. Complementary activities of DOT1 and menin inhibitors in MLL-rearranged leukemia. *Leukemia.* doi: 10.1038/leu.2016.327.

Daigle SR, Olhava EJ, Therkelsen CA, Majer CR, Sneeringer CJ, Song J, Johnston LD, Scott MP, Smith JJ, Xiao Y, et al. 2011. Selective killing of mixed lineage leukemia cells by a potent small-molecule DOT1L inhibitor. *Cancer Cell* **20:** 53–65.

Daigle SR, Olhava EJ, Therkelsen CA, Basavapathruni A, Jin L, Boriack-Sjodin PA, Allain CJ, Klaus CR, Raimondi A, Scott MP, et al. 2013. Potent inhibition of DOT1L as treatment of MLL-fusion leukemia. *Blood* **122:** 1017–1025.

Dawson MA, Prinjha RK, Dittmann A, Giotopoulos G, Bantscheff M, Chan WI, Robson SC, Chung CW, Hopf C, Savitski MM, et al. 2011. Inhibition of BET recruitment to chromatin as an effective treatment for MLL-fusion leukaemia. *Nature* **478:** 529–533.

Delmore JE, Issa GC, Lemieux ME, Rahl PB, Shi J, Jacobs HM, Kastritis E, Gilpatrick T, Paranal RM, Qi J, et al.

2011. BET bromodomain inhibition as a therapeutic strategy to target c-Myc. *Cell* **146:** 904–917.

Deshpande AJ, Chen L, Fazio M, Sinha AU, Bernt KM, Banka D, Dias S, Chang J, Olhava EJ, Daigle SR, et al. 2013. Leukemic transformation by the MLL–AF6 fusion oncogene requires the H3K79 methyltransferase Dot1l. *Blood* **121:** 2533–2541.

Dias J, Van Nguyen N, Georgiev P, Gaub A, Brettschneider J, Cusack S, Kadlec J, Akhtar A. 2014. Structural analysis of the KANSL1/WDR5/KANSL2 complex reveals that WDR5 is required for efficient assembly and chromatin targeting of the NSL complex. *Genes Dev* **28:** 929–942.

Dick JE. 2005. Acute myeloid leukemia stem cells. *Ann NY Acad Sci* **1044:** 1–5.

Dobson CL, Warren AJ, Pannell R, Forster A, Rabbitts TH. 2000. Tumorigenesis in mice with a fusion of the leukaemia oncogene *Mll* and the bacterial *lacZ* gene. *EMBO J* **19:** 843–851.

Domer PH, Fakharzadeh SS, Chen CS, Jockel J, Johansen L, Silverman GA, Kersey JH, Korsmeyer SJ. 1993. Acute mixed-lineage leukemia t(4;11)(q21;q23) generates an *MLL–AF4* fusion product. *Proc Natl Acad Sci* **90:** 7884–7888.

Dou Y, Milne TA, Tackett AJ, Smith ER, Fukuda A, Wysocka J, Allis CD, Chait BT, Hess JL, Roeder RG. 2005. Physical association and coordinate function of the H3 K4 methyltransferase MLL1 and the H4 K16 acetyltransferase MOF. *Cell* **121:** 873–885.

Dou Y, Milne TA, Ruthenburg AJ, Lee S, Lee JW, Verdine GL, Allis CD, Roeder RG. 2006. Regulation of MLL1 H3K4 methyltransferase activity by its core components. *Nat Struct Mol Biol* **13:** 713–719.

Erfurth F, Hemenway CS, de Erkenez AC, Domer PH. 2004. MLL fusion partners AF4 and AF9 interact at subnuclear foci. *Leukemia* **18:** 92–102.

Forster A, Pannell R, Drynan LF, McCormack M, Collins EC, Daser A, Rabbitts TH. 2003. Engineering de novo reciprocal chromosomal translocations associated with *Mll* to replicate primary events of human cancer. *Cancer Cell* **3:** 449–458.

Fossati A, Dolfini D, Donati G, Mantovani R. 2011. NF-Y recruits Ash2L to impart H3K4 trimethylation on CCAAT promoters. *PLoS ONE* **6:** e17220.

Grembecka J, He S, Shi A, Purohit T, Muntean AG, Sorenson RJ, Showalter HD, Murai MJ, Belcher AM, Hartley T, et al. 2012. Menin–MLL inhibitors reverse oncogenic activity of MLL fusion proteins in leukemia. *Nat Chem Biol* **8:** 277–284.

Gu Y, Nakamura T, Alder H, Prasad R, Canaani O, Cimino G, Croce CM, Canaani E. 1993. The t(4;11) chromosome translocation of human acute leukemias fuses the *ALL-1* gene, related to *Drosophila* trithorax, to the *AF-4* gene. *Cell* **71:** 701–708.

Guenther MG, Lawton LN, Rozovskaia T, Frampton GM, Levine SS, Volkert TL, Croce CM, Nakamura T, Canaani E, Young RA. 2008. Aberrant chromatin at genes encoding stem cell regulators in human mixed-lineage leukemia. *Genes Dev* **22:** 3403–3408.

He S, Malik B, Borkin D, Miao H, Shukla S, Kempinska K, Purohit T, Wang J, Chen L, Parkin B, et al. 2016. Menin–MLL inhibitors block oncogenic transformation by MLL-

fusion proteins in a fusion partner-independent manner. *Leukemia* **30:** 508–513.

Hughes CM, Rozenblatt-Rosen O, Milne TA, Copeland TD, Levine SS, Lee JC, Hayes DN, Shanmugam KS, Bhattacharjee A, Biondi CA, et al. 2004. Menin associates with a trithorax family histone methyltransferase complex and with the *Hoxc8* locus. *Mol Cell* **13:** 587–597.

Jonkers I, Lis JT. 2015. Getting up to speed with transcription elongation by RNA polymerase II. *Nat Rev Mol Cell Biol* **16:** 167–177.

Karatas H, Townsend EC, Bernard D, Dou Y, Wang S. 2010. Analysis of the binding of mixed lineage leukemia 1 (MLL1) and histone 3 peptides to WD repeat domain 5 (WDR5) for the design of inhibitors of the MLL1–WDR5 interaction. *J Med Chem* **53:** 5179–5185.

Kawabe Y, Wang YX, McKinnell IW, Bedford MT, Rudnicki MA. 2012. Carm1 regulates Pax7 transcriptional activity through MLL1/2 recruitment during asymmetric satellite stem cell divisions. *Cell Stem Cell* **11:** 333–345.

Krivtsov AV, Armstrong SA. 2007. *MLL* translocations, histone modifications and leukaemia stem-cell development. *Nat Rev Cancer* **7:** 823–833.

Krivtsov AV, Twomey D, Feng Z, Stubbs MC, Wang Y, Faber J, Levine JE, Wang J, Hahn WC, Gilliland DG, et al. 2006. Transformation from committed progenitor to leukaemia stem cell initiated by MLL–AF9. *Nature* **442:** 818–822.

Krivtsov AV, Feng Z, Lemieux ME, Faber J, Vempati S, Sinha AU, Xia X, Jesneck J, Bracken AP, Silverman LB, et al. 2008. H3K79 methylation profiles define murine and human MLL–AF4 leukemias. *Cancer Cell* **14:** 355–368.

Kühn MW, Hadler MJ, Daigle SR, Koche RP, Krivtsov AV, Olhava EJ, Caligiuri MA, Huang G, Bradner JE, Pollock RM, et al. 2015. *MLL* partial tandem duplication leukemia cells are sensitive to small molecule DOT1L inhibition. *Haematologica* **100:** e190–193.

Kuntimaddi A, Achille NJ, Thorpe J, Lokken AA, Singh R, Hemenway CS, Adli M, Zeleznik-Le NJ, Bushweller JH. 2015. Degree of recruitment of DOT1L to MLL–AF9 defines level of H3K79 di- and tri-methylation on target genes and transformation potential. *Cell Rep* **11:** 808–820.

Lin C, Smith ER, Takahashi H, Lai KC, Martin-Brown S, Florens L, Washburn MP, Conaway JW, Conaway RC, Shilatifard A. 2010. AFF4, a component of the ELL/P-TEFb elongation complex and a shared subunit of MLL chimeras, can link transcription elongation to leukemia. *Mol Cell* **37:** 429–437.

Meyer C, Hofmann J, Burmeister T, Gröger D, Park TS, Emerenciano M, Pombo de Oliveira M, Renneville A, Villarese P, Macintyre E, et al. 2013. The *MLL* recombinome of acute leukemias in 2013. *Leukemia* **27:** 2165–2176.

Milne TA, Briggs SD, Brock HW, Martin ME, Gibbs D, Allis CD, Hess JL. 2002. MLL targets SET domain methyltransferase activity to *Hox* gene promoters. *Mol Cell* **10:** 1107–1117.

Milne TA, Dou Y, Martin ME, Brock HW, Roeder RG, Hess JL. 2005. MLL associates specifically with a subset of transcriptionally active target genes. *Proc Natl Acad Sci* **102:** 14765–14770.

Mishra BP, Zaffuto KM, Artinger EL, Org T, Mikkola HK, Cheng C, Djabali M, Ernst P. 2014. The histone methyltransferase activity of MLL1 is dispensable for hematopoiesis and leukemogenesis. *Cell Rep* **7:** 1239–1247.

Mueller D, Bach C, Zeisig D, Garcia-Cuellar MP, Monroe S, Sreekumar A, Zhou R, Nesvizhskii A, Chinnaiyan A, Hess JL, et al. 2007. A role for the MLL fusion partner ENL in transcriptional elongation and chromatin modification. *Blood* **110:** 4445–4454.

Muntean AG, Giannola D, Udager AM, Hess JL. 2008. The PHD fingers of MLL block MLL fusion protein-mediated transformation. *Blood* **112:** 4690–4693.

Nakamura T, Mori T, Tada S, Krajewski W, Rozovskaia T, Wassell R, Dubois G, Mazo A, Croce CM, Canaani E. 2002. ALL-1 is a histone methyltransferase that assembles a supercomplex of proteins involved in transcriptional regulation. *Mol Cell* **10:** 1119–1128.

Okada Y, Feng Q, Lin Y, Jiang Q, Li Y, Coffield VM, Su L, Xu G, Zhang Y. 2005. hDOT1L links histone methylation to leukemogenesis. *Cell* **121:** 167–178.

Patel A, Dharmarajan V, Vought VE, Cosgrove MS. 2009. On the mechanism of multiple lysine methylation by the human mixed lineage leukemia protein-1 (MLL1) core complex. *J Biol Chem* **284:** 24242–24256.

Pelish HE, Liau BB, Nitulescu II, Tangpeerachaikul A, Poss ZC, Da Silva DH, Caruso BT, Arefolov A, Fadeyi O, Christie AL, et al. 2015. Mediator kinase inhibition further activates super-enhancer-associated genes in AML. *Nature* **526:** 273–276.

Prasad R, Leshkowitz D, Gu Y, Alder H, Nakamura T, Saito H, Huebner K, Berger R, Croce CM, Canaani E. 1994. Leucine-zipper dimerization motif encoded by the *AF17* gene fused to *ALL-1 (MLL)* in acute leukemia. *Proc Natl Acad Sci* **91:** 8107–8111.

Rao RC, Dou Y. 2015. Hijacked in cancer: The KMT2 (MLL) family of methyltransferases. *Nat Rev Cancer* **15:** 334–346.

Rea S, Eisenhaber F, O'Carroll D, Strahl BD, Sun ZW, Schmid M, Opravil S, Mechtler K, Ponting CP, Allis CD, et al. 2000. Regulation of chromatin structure by site-specific histone H3 methyltransferases. *Nature* **406:** 593–599.

Rowley JD. 1993. Rearrangements involving chromosome band 11Q23 in acute leukaemia. *Semin Cancer Biol* **4:** 377–385.

Schichman SA, Caligiuri MA, Gu Y, Strout MP, Canaani E, Bloomfield CD, Croce CM. 1994. *ALL-1* partial duplication in acute leukemia. *Proc Natl Acad Sci* **91:** 6236–6239.

Shen C, Jo SY, Liao C, Hess JL, Nikolovska-Coleska Z. 2013. Targeting recruitment of disruptor of telomeric silencing 1-like (DOT1L): Characterizing the interactions between DOT1L and mixed lineage leukemia (MLL) fusion proteins. *J Biol Chem* **288:** 30585–30596.

Shinobu N, Maeda T, Aso T, Ito T, Kondo T, Koike K, Hatakeyama M. 1999. Physical interaction and functional antagonism between the RNA polymerase II elongation factor ELL and p53. *J Biol Chem* **274:** 17003–17010.

Sneppen K, Dodd IB. 2015. Cooperative stabilization of the SIR complex provides robust epigenetic memory in a model of SIR silencing in *Saccharomyces cerevisiae*. *Epigenetics* **10:** 293–302.

Somervaille TC, Cleary ML. 2006. Identification and characterization of leukemia stem cells in murine MLL-AF9 acute myeloid leukemia. *Cancer Cell* **10:** 257–268.

Spurr SS, Bayle ED, Yu W, Li F, Tempel W, Vedadi M, Schapira M, Fish PV. 2016. New small molecule inhibitors of histone methyl transferase DOT1L with a nitrile as a nontraditional replacement for heavy halogen atoms. *Bioorg Med Chem Lett* **26:** 4518–4522.

Steger DJ, Lefterova MI, Ying L, Stonestrom AJ, Schupp M, Zhuo D, Vakoc AL, Kim JE, Chen J, Lazar MA, et al. 2008. DOT1L/KMT4 recruitment and H3K79 methylation are ubiquitously coupled with gene transcription in mammalian cells. *Mol Cell Biol* **28:** 2825–2839.

Stein EM, Tallman MS. 2016. Emerging therapeutic drugs for AML. *Blood* **127:** 71–78.

Strissel PL, Strick R, Rowley JD, Zeleznik-Le NJ. 1998. An in vivo topoisomerase II cleavage site and a DNase I hypersensitive site colocalize near exon 9 in the *MLL* breakpoint cluster region. *Blood* **92:** 3793–3803.

Terranova R, Agherbi H, Boned A, Meresse S, Djabali M. 2006. Histone and DNA methylation defects at *Hox* genes in mice expressing a SET domain-truncated form of *Mll*. *Proc Natl Acad Sci* **103:** 6629–6634.

Thirman MJ, Levitan DA, Kobayashi H, Simon MC, Rowley JD. 1994. Cloning of *ELL*, a gene that fuses to *MLL* in a t(11;19)(q23;p13.1) in acute myeloid leukemia. *Proc Natl Acad Sci* **91:** 12110–12114.

Tkachuk DC, Kohler S, Cleary ML. 1992. Involvement of a homolog of *Drosophila* trithorax by 11q23 chromosomal translocations in acute leukemias. *Cell* **71:** 691–700.

Tyagi S, Chabes AL, Wysocka J, Herr W. 2007. E2F activation of S phase promoters via association with HCF-1 and the MLL family of histone H3K4 methyltransferases. *Mol Cell* **27:** 107–119.

Vlaming H, van Leeuwen F. 2016. The upstreams and downstreams of H3K79 methylation by DOT1L. *Chromosoma* **125:** 593–605.

Wang J, Iwasaki H, Krivtsov A, Febbo PG, Thorner AR, Ernst P, Anastasiadou E, Kutok JL, Kogan SC, Zinkel SS, et al. 2005. Conditional MLL-CBP targets GMP and models therapy-related myeloproliferative disease. *EMBO J* **24:** 368–381.

Wang P, Bowl MR, Bender S, Peng J, Farber L, Chen J, Ali A, Zhang Z, Alberts AS, Thakker RV, et al. 2008. Parafibromin, a component of the human PAF complex, regulates growth factors and is required for embryonic development and survival in adult mice. *Mol Cell Biol* **28:** 2930–2940.

Wang Z, Zang C, Rosenfeld JA, Schones DE, Barski A, Cuddapah S, Cui K, Roh TY, Peng W, Zhang MQ, et al. 2008. Combinatorial patterns of histone acetylations and methylations in the human genome. *Nat Genet* **40:** 897–903.

Wang P, Lin C, Smith ER, Guo H, Sanderson BW, Wu M, Gogol M, Alexander T, Seidel C, Wiedemann LM, et al. 2009. Global analysis of H3K4 methylation defines MLL family member targets and points to a role for MLL1-mediated H3K4 methylation in the regulation of transcriptional initiation by RNA polymerase II. *Mol Cell Biol* **29:** 6074–6085.

Wang KC, Yang YW, Liu B, Sanyal A, Corces-Zimmerman R, Chen Y, Lajoie BR, Protacio A, Flynn RA, Gupta RA, et al.

2011. A long noncoding RNA maintains active chromatin to coordinate homeotic gene expression. *Nature* **472:** 120–124.

Wood A, Schneider J, Dover J, Johnston M, Shilatifard A. 2003. The Paf1 complex is essential for histone mono-ubiquitination by the Rad6-Bre1 complex, which signals for histone methylation by COMPASS and Dot1p. *J Biol Chem* **278:** 34739–34742.

Wu L, Lee SY, Zhou B, Nguyen UT, Muir TW, Tan S, Dou Y. 2013. ASH2L regulates ubiquitylation signaling to MLL: *Trans*-regulation of H3 K4 methylation in higher eukaryotes. *Mol Cell* **49:** 1108–1120.

Xu Y, Yue L, Wang Y, Xing J, Chen Z, Shi Z, Liu R, Liu YC, Luo X, Jiang H, et al. 2016. Discovery of novel inhibitors targeting the menin-mixed lineage leukemia interface using pharmacophore- and docking-based virtual screening. *J Chem Inf Model* **56:** 1847–1855.

Yagi H, Deguchi K, Aono A, Tani Y, Kishimoto T, Komori T. 1998. Growth disturbance in fetal liver hematopoiesis of *Mll*-mutant mice. *Blood* **92:** 108–117.

Yang YW, Flynn RA, Chen Y, Qu K, Wan B, Wang KC, Lei M, Chang HY. 2014. Essential role of lncRNA binding for WDR5 maintenance of active chromatin and embryonic stem cell pluripotency. *Elife* **3:** e02046.

Yi JS, Federation AJ, Qi J, Dhe-Paganon S, Hadler M, Xu X, St Pierre R, Varca AC, Wu L, Marineau JJ, et al. 2015. Structure-guided DOT1L probe optimization by label-free ligand displacement. *ACS Chem Biol* **10:** 667–674.

Yokoyama A, Cleary ML. 2008. Menin critically links MLL proteins with LEDGF on cancer-associated target genes. *Cancer Cell* **14:** 36–46.

Yu BD, Hess JL, Horning SE, Brown GA, Korsmeyer SJ. 1995. Altered *Hox* expression and segmental identity in *Mll*-mutant mice. *Nature* **378:** 505–508.

Yu W, Chory EJ, Wernimont AK, Tempel W, Scopton A, Federation A, Marineau JJ, Qi J, Barsyte-Lovejoy D, Yi J, et al. 2012. Catalytic site remodelling of the DOT1L methyltransferase by selective inhibitors. *Nat Commun* **3:** 1288.

Yun M, Wu J, Workman JL, Li B. 2011. Readers of histone modifications. *Cell Res* **21:** 564–578.

Zhang W, Xia X, Reisenauer MR, Hemenway CS, Kone BC. 2006. Dot1a-AF9 complex mediates histone H3 Lys-79 hypermethylation and repression of *ENaCα* in an aldosterone-sensitive manner. *J Biol Chem* **281:** 18059–18068.

Zhang Y, Mittal A, Reid J, Reich S, Gamblin SJ, Wilson JR. 2015. Evolving catalytic properties of the MLL family SET domain. *Structure* **23:** 1921–1933.

Ziemin-van der Poel S, McCabe NR, Gill HJ, Espinosa RIII, Patel Y, Harden A, Rubinelli P, Smith SD, LeBeau MM, Rowley JD, et al. 1991. Identification of a gene, *MLL*, that spans the breakpoint in 11q23 translocations associated with human leukemias. *Proc Natl Acad Sci* **88:** 10735–10739.

Zuber J, Shi J, Wang E, Rappaport AR, Herrmann H, Sison EA, Magoon D, Qi J, Blatt K, Wunderlich M, et al. 2011. RNAi screen identifies Brd4 as a therapeutic target in acute myeloid leukaemia. *Nature* **478:** 524–528.

Oncogenic Mechanisms of Histone H3 Mutations

Daniel N. Weinberg, C. David Allis, and Chao Lu

Laboratory of Chromatin Biology and Epigenetics, The Rockefeller University, New York, New York 10065

Correspondence: alliscd@rockefeller.edu; clu01@mail.rockefeller.edu

Recurrent missense mutations in histone H3 were recently reported in pediatric gliomas and soft tissue tumors. Strikingly, these mutations only affected a minority of the total cellular H3 proteins and occurred at or near lysine residues at positions 27 and 36 on the amino-terminal tail of H3 that are subject to well-characterized posttranslational modifications. Here we review recent progress in elucidating the mechanisms by which these mutations perturb the chromatin landscape in cells through their effects on chromatin-modifying machinery, particularly through inhibition of specific histone lysine methyltransferases. One common feature of histone mutations is their ability to arrest cells in a primitive state refractory to differentiation induction, highlighting the importance of studying these mutations in their proper developmental context.

Chromatin, the combination of DNA and its interacting proteins, is the physiologically relevant form of eukaryotic genomes. The basic repeating unit of chromatin is the nucleosome, comprised of two copies of the core histone proteins H2A, H2B, H3, and H4 that together form an octamer wrapped by two superhelical turns of DNA (Luger et al. 1997). Historically, nucleosomes were thought to mainly provide structural support for genome packaging. However, research from the past two decades has revealed a remarkable role of nucleosome composition, modification, and positioning in virtually all DNA-based processes, including replication, transcription, and damage repair (Jenuwein and Allis 2001; Ernst and Kellis 2010).

Posttranslational modifications (PTMs) of histones are critically involved in chromatin-mediated gene regulation (Jenuwein and Allis 2001). It is believed that histone PTMs exert their effects through direct physical modulation of nucleosome–DNA contacts and/or recruitment of downstream "reader" protein complexes. To date more than 100 histone PTMs have been identified (Huang et al. 2014), many of which are dynamically controlled by enzymes catalyzing their addition ("writers") or removal ("erasers"). Among them, PTMs of several lysine residues located at the amino-terminal tail of histone H3 have been extensively characterized (Fig. 1A). For example, H3 lysine 27 (H3K27) can be acetylated by p300/CBP and H3K27ac is preferentially located at promoters and/or enhancers of genes that are actively transcribed (Ogryzko et al. 1996). In contrast, methylation of H3K27, catalyzed by the Polycomb repressive complex 2 (PRC2) and removed by KDM6 family demethylases, is a mark associated with gene

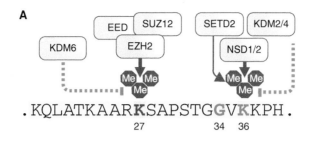

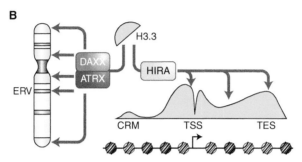

Figure 1. Posttranslational modification of the histone variant H3.3 and its chaperone/deposition machinery. (A) The amino-terminal tail of histone H3.3 (shown here) and other histone H3 proteins are subject to methylation of lysine residue 27 by the Polycomb repressive complex 2 (PRC2), containing core subunits EED, SUZ12, and EZH2. Removal of H3K27 methylation is performed by KDM6 family demethylases, including JMJD3 and UTX. Lysine residue 36 is subject to methylation by multiple enzymes, including the NSD family enzymes and SETD2. Removal of H3K36 methylation is performed by the KDM2 and KDM4 family demethylases. (B) Histone variant H3.3 is deposited at pericentric heterochromatin, telomeres, and certain endogenous retroviral elements (ERV) by the ATRX/DAXX heterodimeric complex. In contrast, H3.3 is deposited at euchromatin regions such as promoters and gene bodies by the histone chaperone HIRA (see Banaszynski et al. 2010 and Maze et al. 2014 for details and references).

silencing (Margueron and Reinberg 2011). Methylation of H3 lysine 36 (H3K36), depending on the context, can regulate transcriptional elongation, RNA processing, and DNA damage sensing (Kuo et al. 2011; Carvalho et al. 2014; Simon et al. 2014; Wen et al. 2014). In mammals, several enzymes targeting H3K36 have been reported (Wagner and Carpenter 2012). Although SETD2 is the only methyltransferase that can generate trimethylation of H3K36 (H3K36me3), multiple methyltransferases catalyze mono- and dimethylation of H3K36 (H3K36me1/2), including NSD1/2/3 and ASH1L. Conversely, members of the KDM2 and KDM4 families act as H3K36-specific demethylases.

More recently, it was appreciated that changes to the cellular epigenetic state could also result from expression and incorporation of histone variants (Banaszynski et al. 2010; Maze et al. 2014), further increasing the complexity of chromatin regulation. With the exception of H4, all histones are expressed in variant forms differing in primary amino acid sequence, leading to minor or major structural dissimilarities. Importantly, such variations were shown to have functional relevance. For example, in metazoans, variants of H3 include H3.1, H3.2, H3.3, and CENPA (Hake and Allis 2006). CENPA is centromere-specific and structurally dissimilar to the other H3 variants, whereas H3.3 differs from the "canonical" H3.1 and H3.2 by only a few amino acids. However, several lines of evidence suggest that H3.3 plays a distinct role in chromatin biology from canonical H3 (Fig. 1B). First, although canonical H3 is expressed and incorporated into chromatin in a DNA replication-dependent manner,

Cite this article as *Cold Spring Harb Perspect Med* doi: 10.1101/cshperspect.a026443

the expression, assembly, and deposition of H3.3-containing nucleosomes are cell-cycle-independent (Tagami et al. 2004). Second, in proliferating cells, H3.3 is enriched at selected genomic regions, including promoters and gene bodies of highly transcribed or transcriptionally "poised" genes in euchromatin (McKittrick et al. 2004), pericentric heterochromatin and telomeres (Goldberg et al. 2010), and certain classes of endogenous retroviral elements (ERVs) (Elsässer et al. 2015). Third, compared with canonical nucleosomes, distinct sets of factors facilitate the assembly and deposition of H3.3-containing nucleosomes. Deposition of H3.3 to euchromatic regions is mainly mediated by the HIRA complex (Tagami et al. 2004), whereas ATRX/DAXX were identified as the complex that specifically incorporates H3.3 to heterochromatic regions (Lewis et al. 2010; Elsässer et al. 2012, 2015). Depletion of H3.3 in embryonic stem cells leads to aberrant PRC2 binding at developmentally regulated genes, derepression of ERVs, and abnormal cell differentiation (Banaszynski et al. 2013; Elsässer et al. 2015). In mammals, two genes (*H3F3A* and *H3F3B*) encode H3.3 and mice deficient for either H3.3 gene show postnatal death, growth retardation, and infertility (Couldrey et al. 1999; Bush et al. 2013). Therefore, although the specific mechanisms remain to be fully elucidated, it appears that H3.3 is required to establish the proper chromatin states at specific genomic regions to maintain cell identity during development.

As the precise regulation of chromatin is essential for many cellular events, including proliferation and differentiation, not surprisingly chromatin misregulation has been linked to various human diseases, notably cancer. A major finding from recent tumor genome sequencing studies was the discovery that chromatin regulators, including writers, erasers, and readers of histone and DNA modifications and nucleosome remodelers, are frequently altered in malignancies (Shen and Laird 2013). Although aberrant chromatin states are increasingly recognized as an emerging hallmark of cancer, few researchers in the cancer epigenetics field anticipated the reports of recurrent mutations in histone H3 themselves. As detailed

in the following sections, these mutations are highly clustered missense mutations of residues at or near well-studied PTM sites. Furthermore, the mutations are always monoallelic and affect only one of the 16 genes encoding H3 in humans. These interesting features have attracted considerable attention from the oncology and chromatin biology communities and recently the mechanisms underlying these so-called "oncohistones" have begun to be unraveled. In this review, we will summarize and discuss efforts to identify and understand histone H3 mutations.

HISTONE MUTATIONS IN CANCER

Recurrent mutations in histone H3 were first reported in pediatric high-grade gliomas (pHGGs) (Fig. 2). Simultaneous reports described H3K27M mutations in the majority of diffuse intrinsic pontine gliomas (DIPGs), a type of pHGG associated with dismal prognosis owing to its location in the brainstem, as well as

Hemispheric pHGG
H3.3 G34R/V

Midline pHGG
H3.1 K27M
H3.3 K27M/I

GCT of the bone
H3.3 G34W/L

**Chondroblastoma/
pediatric sarcoma**
H3.1 K36M/I
H3.3 K36M

**Diffuse large B-cell/
follicular lymphoma**
H1 mutations

Figure 2. Recurrent histone mutations in human cancer (see text and Fontebasso et al. 2014a; Kallappagoudar et al. 2015). pHGG, Pediatric high-grade glioma; GCT, giant cell tumor.

in thalamic gliomas (Schwartzentruber et al. 2012; Wu et al. 2012). H3K27M mutations were not detected in other pediatric brain tumors including medulloblastomas and ependymomas (Wu et al. 2012) and seemed to be specific for midline pHGGs as they were subsequently found in pHGGs involving the cerebellum and the spinal cord (Sturm et al. 2012). Overall, pHGGs were characterized by additional mutations affecting growth factor signaling (e.g., RAS and PI3K) and the RB1 and TP53 pathways for cell-cycle regulation (Buczkowicz et al. 2014; Fontebasso et al. 2014b; Taylor et al. 2014; Wu et al. 2014). Evolutionary reconstruction of individual DIPG tumors revealed that H3K27M mutations arose early and were accompanied by an obligate partner mutation in a member of either of these pathways (e.g., *PIK3R1* or *TP53*), suggesting that both are needed for tumorigenesis (Nikbakht et al. 2016). H3K27M mutations were also reported in rare cases of pediatric low-grade gliomas (Jones et al. 2013; Zhang et al. 2013) as well as in thalamic gliomas in adults under the age of 50 (Aihara et al. 2014). Notably, DIPG patients carrying the H3K27M mutation had worse overall survival compared with patients lacking the mutation (Khuong-Quang et al. 2012; Feng et al. 2015). Because of its prevalence and association with patient outcomes, the detection of H3K27M mutations by immunohistochemistry is increasingly being considered for diagnostic purposes in pediatric gliomas. Antibodies directed against H3K27M have been developed and were shown to be 100% sensitive and specific for the mutation (Bechet et al. 2014; Venneti et al. 2014).

The majority of pHGG-associated H3K27M mutations affect the variant histone H3.3, whereas others are found in canonical histone H3.1/2. Intriguingly, H3.3K27M mutations differ from H3.1/2K27M mutations in several clinical features. H3.1K27M-mutant gliomas were restricted to the brainstem, unlike H3.3K27M-mutant gliomas, which were found in the brainstem in addition to other midline locations (Fontebasso et al. 2014b; Castel et al. 2015). DIPG patients with H3.1K27M were, on average, 2 years younger than those with H3.3K27M (Castel et al. 2015) and had a distinctive set of co-occurring mutations. Missense mutations in *ACVR1*, which encodes the bone morphogenetic protein (BMP) type I receptor ALK2, were significantly associated only with H3.1K27M but not H3.3K27M (Buczkowicz et al. 2014; Fontebasso et al. 2014b; Taylor et al. 2014; Wu et al. 2014). Many of the *ACVR1* mutations were identical to those found in the autosomal dominant syndrome fibrodysplasia ossificans progressiva in which gain-of-function mutations in *ACVR1* lead to heterotopic ossification (Shore et al. 2006; Chaikuad et al. 2012). Phosphorylation of SMAD1/5/8, downstream events of ALK2 activation, were increased in *ACVR1*-mutant DIPG tumors (Buczkowicz et al. 2014; Fontebasso et al. 2014b), and expression of mutant *ACVR1* transgenes was sufficient to increase phosphorylation of SMAD1/5/8 in astrocytes (Wu et al. 2014) and DIPG cell lines (Taylor et al. 2014). Therefore, ALK2 inhibitors currently in development to treat fibrodysplasia ossificans progressiva could be therapeutically effective against a subset of pHGGs. In contrast, amplification of *PDGFRA* was significantly associated only with H3.3K27M (Buczkowicz et al. 2014; Castel et al. 2015). Other H3.3K27M-mutant gliomas, particularly in the thalamus, were associated with mutations in both *FGFR1* and *NF1* (Jones et al. 2013). H3.1K27M and H3. 3K27M DIPGs also have distinct gene expression profiles, although it remains unclear to what extent this is attributable to differences in the H3 variants as opposed to the unique co-occurring mutations in each subgroup or the possibility that each subgroup arises from a distinct cell of origin (Castel et al. 2015).

Unlike midline pHGGs, many pHGGs located in the cerebral hemispheres carried H3.3G34R or H3.3G34V mutations (Fig. 2) (Schwartzentruber et al. 2012; Wu et al. 2012). H3.3G34R/V-mutant tumors were typically diagnosed during adolescence, rather than in early childhood like most H3K27M-mutant tumors. They had distinct gene expression and DNA methylation signatures from H3K27M-mutant tumors and H3 mutational status could be predicted based on the expression of FOXG1 exclusively in H3.3G34-mutant gliomas and OLIG2 exclusively in H3K27M-mutant gliomas (Sturm et al. 2012). Mutations in ATRX, which heterodimerizes with the histone chaperone DAXX to specifically incorporate H3.3 at pericentric heterochromatin and telomeres, were significantly associated with H3.3G34-mutant pHGGs although they were also observed at a lower frequency in wild-type (WT) and H3K27M-mutant pHGGs. Similar to other cancer types, ATRX-mutant pHGGs show alternative lengthening of telomeres (Schwartzentruber et al. 2012). Notably, mutations in the H3K36 methyltransferase SETD2 were also seen in high-grade gliomas involving the cerebral hemispheres in adolescents and were mutually exclusive with H3 mutations (Fontebasso et al. 2013).

Recurrent mutations in histone H3 were subsequently reported in specific types of bone and cartilage tumors (Fig. 2). H3.3K36M mutations were found in nearly all chondroblastomas, a benign tumor of the active growth plate of bones, and rarely in malignant conventional and clear-cell chondrosarcomas (Behjati et al. 2013). Antibodies directed against H3K36M were 100% specific and sensitive for the mutation and may prove useful for diagnosis (Amary et al. 2016; Lu et al. 2016). Interestingly, although H3.3K27M mutations occurred exclusively in *H3F3A*, H3.3K36M mutations predominately were found in *H3F3B* despite similar expression levels and identical amino acid sequence of both isoforms (Behjati et al. 2013). In addition, nearly all giant cell tumors of the bone had H3.3G34 mutations mostly to tryptophan (and in one case, to leucine), whereas osteosarcomas infrequently had the H3.3G34R mutation originally identified in pHGGs (Behjati et al. 2013; Joseph et al. 2014; Sarungbam et al. 2016). Two patients with postzygotic H3.3G34W mutations presented with paragangliomas and recurrent giant cell tumors of the bone, suggesting that somatic mosaic mutation of H3.3G34 may be the basis for a new nonhereditary cancer syndrome (Toledo et al. 2015). No H3 mutations were detected in other types of bone and cartilage tumors including chondromyxoid fibromas, chordomas, and chondromas (Behjati et al. 2013).

Recurrent mutations in the linker histone H1 have also been reported in diffuse large B-cell lymphomas (Lohr et al. 2012) and follicular lymphomas (Okosun et al. 2014). Many of these mutations are believed to be loss-of-function (Okosun et al. 2014), but their contribution to oncogenesis remains unexplored and therefore will not be discussed in this review.

MECHANISM OF H3K27M-MEDIATED ONCOGENESIS

Initial characterization of the histone PTMs in H3K27M-mutant DIPG patient tumors and cell lines revealed a global reduction of H3K27me3 compared with H3 WT DIPG tumors (Chan et al. 2013; Lewis et al. 2013; Venneti et al. 2013), despite similar expression of EZH2 (Venneti et al. 2013). K27M-mutant H3 contributed to only approximately 3%–17% of the total H3 proteins in DIPG samples (Lewis et al. 2013), which is consistent with the fact that only one of the 32 H3-encoding alleles harbors the mutation and suggests that the global reduction in H3K27me3 was caused by a dominant effect of the H3K27M mutation. Indeed, expression of an H3K27M transgene in 293T cells was sufficient to reduce H3K27me2 and H3K27me3 globally despite the mutant histone accounting for only 1% of total H3 protein (Lewis et al. 2013). Similar effects were seen regardless of whether the transgene was H3.1 or H3.3 (Chan et al. 2013) and were consistent across a range of different cell types including astrocyte, fibroblast, and glioma cell lines (Bender et al. 2013; Chan et al. 2013). This effect was restricted to the mutation of the lysine residue at position 27 to methionine (and to a lesser extent isoleucine), but not to other amino acids (Lewis et al. 2013). The dominant nature of the mutation was further shown by the loss of H3K27me3 on endogenous WT H3 proteins (Bender et al. 2013; Lewis et al. 2013).

In-depth biochemical and molecular work has revealed that the H3K27M mutation achieves global reductions in H3K27 methylation through inhibition of the PRC2 complex in *trans* (Fig. 3A,B). Addition of H3K27M peptides (Lewis et al. 2013) or mononucleosomes

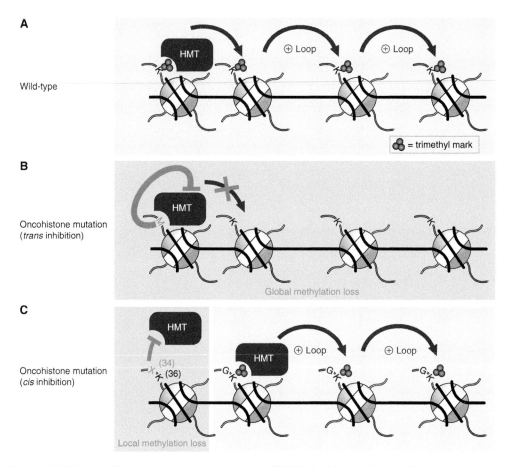

Figure 3. Inhibition of histone lysine methyltransferases (HMTs) by histone H3 oncohistone mutations. (*A*) Unimpeded activity of HMTs in the absence of a mutation (wild-type). (*B*) H3K9M, H3K27M, and H3K36M mutations dominantly inhibit their respective HMTs in *trans*. Consequently, nonmutant nucleosomes are hypomethylated (global methylation loss). (*C*) H3G34 mutations inhibit H3K36 HMTs in *cis* such that only nucleosomes containing the mutation are unable to be methylated (localized methylation loss).

(Brown et al. 2014), or heterotypic nucleosome arrays containing three WT nucleosomes and one spatially defined H3K27M mutant nucleosome (Brown et al. 2015), inhibited recombinant PRC2 activity in a dose-dependent and competitive manner. The IC$_{50}$ for H3K27M inhibition of PRC2 calculated in these in vitro assays is less than the estimated nuclear concentration of a 3%–5% fraction of the total H3 pool (Lewis and Allis 2013), suggesting that such a mechanism could be responsible for the global reduction of H3K27me3 in cells. Although nucleosomes isolated from H3K27M-expressing cells were able to inhibit the activity of recombinant PRC2, their effect on the activity of H3K27 demethylases JMJD3 and UTX was relatively minimal (Bender et al. 2013).

Insights into the interaction between H3K27M and PRC2 have begun to elucidate the structural basis of the inhibition. H3K27M peptides with photo-cross-linkers inserted at or near the mutated residue both pulled down EZH2 from the recombinant PRC2 (Lewis et al. 2013). EZH2 is also enriched on mononucleosomes containing H3K27M isolated from cells (Bender et al. 2013; Chan et al. 2013). Screening of H3K27-mutant peptides found that mutation of K27 to amino acids that have a long, hydrophobic side chain with minimal branching such as methionine, isoleucine, and the unnatural amino acid norleucine were among the most potent inhibitors of PRC2 activity (Lewis et al. 2013; Brown et al. 2014). The inhibitory properties of isoleucine led to the prediction of H3K27I mutation in pHGGs (Lewis and Allis 2013), which was indeed recently identified in a DIPG patient tumor (Castel et al. 2015). These three side chains are thought to be compatible with binding to the active site of the SET domain of EZH2, which normally uses an aromatic cage to bind the unbranched, hydrophobic side chain of lysine. Mutation of one of these aromatic residues in EZH2 rendered the protein less sensitive to inhibition by H3K27M (Lewis et al. 2013). Recently, the crystal structure of *Chaetomium thermophilum* PRC2 bound to a H3K27M peptide in the presence of *S*-adenosyl-L-homocysteine (SAH) was solved and suggests that a residue adjacent to the missense mutation, H3 arginine 26 (H3R26), occupies the active site and prevents substrate binding (Jiao and Liu 2015). Interestingly, in the human PRC2:H3K27M complex the methionine 27 side chain was positioned in the lysine 27 access channel (Justin et al. 2016). Further structural and biochemical studies of the PRC2-H3K27M interaction will be important to uncover potential ways to reverse the inhibition of PRC2 (Brown et al. 2014).

Based on the strong degree of conservation among SET domain–containing proteins, the widely adaptable inhibition of histone methyltransferases by H3 "K-to-M" mutations was appreciated even before reports surfaced of H3K36M mutations in patients (Behjati et al. 2013; Lewis et al. 2013). Analogous to H3K27M, the introduction of exogenous H3K9M and H3K36M into 293T cells was able to reduce global levels of H3K9me3 and H3K36me3, respectively (Lewis et al. 2013). Recombinant H3K9-specific methyltransferases, SUV39H1 and G9a, were also each shown to be inhibited by the addition of H3K9M peptides in *trans* and in an *S*-adenosyl methionine (SAM)-dependent manner (Lewis et al. 2013; Jayaram et al. 2016; Justin et al. 2016). In contrast, introduction of H3K4M had minimal effect on global H3K4me3 levels in cells, a finding Lewis and colleagues suggest may be a result of inhibition

of the H3K4 demethylases LSD1 and LSD2 by H3K4M counterbalancing any potential inhibition of H3K4 methyltransferases by H3K4M (Karytinos et al. 2009).

Although much interest has been directed toward understanding the inhibition of methyltransferases by H3 "K-to-M" mutations, these analyses may overlook additional cellular targets of these mutations. Indeed, histone demethylases have alternatively been implicated as being responsible for the alteration of H3K9 methylation in cells expressing H3K9M. Unbiased mass-spectrometry analysis of immunoprecipitated mononucleosomes containing H3K9M revealed an enrichment for the K9 demethylase KDM3B but a depletion for HP1, a reader of H3K9me3, in native conditions (Herz et al. 2014). In this model, H3K9M serves to recruit KDM3B to genomic loci where it demethylates adjacent WT H3K9, thereby releasing HP1 and decompacting the surrounding chromatin (Herz et al. 2014). It will be important to further validate this model by directly testing the ability of H3K9M to potentiate the activity of KDM3B in *cis* on nucleosomal templates as well as to assess whether deletion of KDM3B can blunt local and/or global changes in H3K9 methylation induced by the H3K9M mutation. Nonetheless, these findings highlight that multiple non–mutually exclusive mechanisms may shape how H3 "K-to-M" mutations alter the chromatin landscape in cells.

Despite global depletion of H3K27 methylation in H3K27M-expressing cells, genome-wide profiling of the chromatin landscape with high-throughput sequencing revealed a striking gain of H3K27me3 at selected genomic loci. The existence of these regions was first observed when comparing H3K27M-mutant DIPG cell lines with human neural stem cells (NSCs) (Chan et al. 2013) and also when comparing H3K27M and H3 WT DIPG tumor samples (Bender et al. 2013). Although H3K27me3 was reduced in mutant DIPG cell lines at every type of genetic element examined (including promoters, 5′ UTRs, coding exons, introns, 3′ UTRs, and intergenic regions), ~60% of the H3K27me3 peaks that remained were not present in NSCs (Chan et al. 2013). A similar reduction in overall H3K27me3 peaks was observed in H3K27M-mutant DIPG tumors compared with H3 WT tumors, with gain of H3K27me3 at limited loci that were significantly enriched for intergenic regions (Bender et al. 2013). Integrated RNA-seq and ChIP-seq analyses in both cell lines and tumors revealed that expression of genes that gained H3K27me3 in H3K27M-mutant samples was down-regulated (e.g., the long isoform of CDK6, p16Ink4a, MICA). Conversely, genes that lost H3K27me3 showed enhanced expression (Bender et al. 2013; Chan et al. 2013). In addition, H3K27M-mutant DIPG tumors showed global DNA hypomethylation, which was associated with the loss of H3K27me3 at promoter regions (Bender et al. 2013). A better understanding of the mechanism through which H3K27me3 is established and maintained at certain regions will likely require assessing how EZH2 localization and H3K27 methylation change genome-wide following the introduction and removal of H3K27M in the same cell type.

New models to study DIPG biology have been developed in light of the discovery of H3K27M mutations and the recognition of their ability to perturb the chromatin landscape in cells. Human embryonic stem cell–derived NSCs have been used to elucidate the oncogenic nature of H3K27M based on their role as the presumed cells-of-origin for DIPG. Introduction of H3K27M selectively increased cell proliferation in NSCs, but not in embryonic stem cells (ESCs), ESC-derived astrocytes, or human fibroblasts (Funato et al. 2014). In the setting of TP53 knockdown and the presence of a constitutively active form of PDGFRa, addition of the H3K27M mutation enabled NSCs to acquire several neoplastic properties (Funato et al. 2014). These included the ability to suppress apoptosis on growth factor withdrawal, sustained proliferation following irradiation, increased invasiveness, impaired differentiation into astrocytes and oligodendrocytes, and formation of tumors when transplanted in vivo (Funato et al. 2014). None of these attributes were observed with H3K27M expression alone, or when the combination of PDGFRa activation and p53 knockdown was not paired with H3K27M, consistent with the observations that these mutations tend to co-occur in patient tumors. Leveraging this model, a small molecule screen was performed, which identified the menin inhibitor MI-2 to selectively decrease the proliferation of transformed H3K27M-expressing NSCs and an H3K27M DIPG cell line, but not H3 WT NSCs (Funato et al. 2014). Treatment with MI-2 also led to the removal of the differentiation block and slowed in vivo growth of orthotopic tumor xenografts, indicating that menin inhibition might be a potential therapeutic option for DIPG patients (Funato et al. 2014).

Other approaches taken to therapeutically target H3K27M-mutant DIPG focused on reversing the reduction of H3K27me3 through manipulating various forms of chromatin-modifying machinery. For example, a small molecule inhibitor of the H3K27 demethylases JMJD3 and UTX known as GSKJ4 was able to increase global levels of H3K27me2/3 in pediatric glioma cell lines expressing H3K27M (Hashizume et al. 2014). GSKJ4 more potently inhibited the growth of these lines both in vitro and in vivo compared with other glioma cell lines that were H3 WT (Hashizume et al. 2014). The efficacy of GSKJ4 may be largely through its effect on JMJD3, as knockdown of JMJD3 but not UTX slowed the growth of H3K27M-expressing cell lines (Hashizume et al. 2014). An alternative method of restoring H3K27me3 levels in cells involves "detoxifying" the H3K27M nucleosomes by weakening their interaction with PRC2 (Brown et al. 2014). PRC2 interacts with the entire H3 N-terminal tail, and its inhibition by peptides containing K27Nle was substantially weakened when residues next to position 27 were deleted or altered (Brown et al. 2014). PTMs, including K4 methylation, polyacetylation, and S28 phosphorylation, all reduced inhibition of PRC2 activity by H3K27Nle peptides in vitro (Brown et al. 2014). Furthermore, the expression of an H3K27M, S28E (mimicking a phosphoserine) double mutant transgene partially rescued the global reduction of H3K27me2/3 in cells (Brown et al. 2014). This strategy has now been taken one step closer to the clinic as H3K27M DIPG cell lines were found to be selectively sensitive to histone deacetylase (HDAC) inhibitors (Grasso et al. 2015). Knockdown of either HDAC1 or HDAC2 showed a similar effect. The clinically approved multi-HDAC inhibitor panobinostat led to a dose-dependent increase in both global H3 acetylation and H3K27me3, likely reflecting diminished inhibition of PRC2 and thereby contributing to a normalization of the H3K27M-induced aberrant gene expression signature (Grasso et al. 2015). Several clinical trials are ongoing to test the safety and efficacy of panobinostat and other HDAC inhibitors in patients with DIPG.

MECHANISM OF H3K36M-MEDIATED ONCOGENESIS

The discovery of H3.3K36M mutations in the vast majority of chondroblastomas and in rare cases of chondrosarcomas (Behjati et al. 2013) prompted the initial description of global reduction in H3K36me3 in 293T cells (Lewis et al. 2013) to be revisited in a more developmentally relevant context. The introduction of an H3K36M transgene into murine mesenchymal progenitor cells blocked their ability to differentiate into chondrocytes and promoted tumor formation when the cells were subcutaneously transplanted into mice (Lu et al. 2016). The tumors did not appear by histology as chondroblastomas or chondrosarcomas but rather as undifferentiated sarcomas and indeed H3K36M was found to also block the differentiation of these cells into adipocytes and osteocytes. Screening a small panel of 10 pediatric undifferentiated soft tissue sarcomas identified one tumor with an H3.1K36M mutation and another with an H3.1K36I mutation. H3K36I similarly blocked the differentiation of murine mesenchymal progenitor cells and its presence in patient tumors mirrors the existence of H3K27I mutations in pediatric gliomas (Castel et al. 2015). Consistent with these findings, CRISPR-Cas9-mediated knock-in of the K36M mutation to the endogenous *H3F3B* allele in immortalized human chondrocytes bestowed similar oncogenic properties, including impaired differentiation and suppressed alkaloid-induced apoptosis (Fang et al. 2016).

Expression of either H3K36M or H3K36I resulted in global reductions in H3K36me2/3 in murine mesenchymal progenitor cells, an effect that appeared to be independent of whether the mutation occurred in H3.1 or H3.3 (Lu et al. 2016). Interestingly, reductions in H3K36me2/3 levels were positively correlated to enrichment of H3.3K36M, suggesting that in addition to its global effects, H3K36M may also exert a localized inhibitory influence (Fang et al. 2016). Nucleosomes containing H3K36M inhibited purified SETD2 and NSD2 activity in vitro, and knockdown of H3K36 methyltransferases recapitulated the impact of H3K36M on the epigenome, transcriptome, and cellular differentiation (Fang et al. 2016; Lu et al. 2016). These results lend support for a critical role of methyltransferase inhibition downstream of the H3K36M mutation.

Profiling the perturbations in the chromatin landscape induced by expression of H3K36M revealed one mechanism by which differentiation blockade is achieved. H3K36me2 was found to be significantly reduced at intergenic domains, which was associated with an increase in H3K27me3 at the same regions. This is consistent with the known inhibition of PRC2 methyltransferase activity by nucleosomes carrying H3K36me2/3 in vitro and the mutual exclusivity of H3K27me3 and H3K36me2/3 genome-wide (Schmitges et al. 2011; Yuan et al. 2011). The increase in intergenic H3K27me3 served to compete with gene-associated H3K27me3 for the recruitment of H3K27me3 "readers" and led to a dilution of the H3K27me3-binding canonical PRC1 complex away from its target genes where it normally represses gene expression. As a consequence, genes that were normally silenced by PRC1 became derepressed on expression of H3K36M. These genes were highly enriched for mesenchymal tissue development regulators and consistently, knockdown of PRC1 components significantly blocked mesenchymal progenitor cell differentiation (Lu et al. 2016).

As H3K36 methylation has also been implicated in the regulation of transcriptional elongation (Wen et al. 2014), RNA processing (Simon et al. 2014), and the DNA damage response (Carvalho et al. 2014), the H3K36M mutation may rely on additional oncogenic mechanisms besides dilution of PRC1. Indeed, defective homologous recombination was noted in chondrocytes expressing H3K36M (Fang et al. 2016). Further study is needed to address whether dysregulation of these processes contributes to H3K36M-mediated oncogenesis.

MECHANISM OF H3G34R/V-MEDIATED ONCOGENESIS

Less is known about H3G34 mutations in cancer, but they appear to work through a different mechanism than H3 "K-to-M" mutations. Expression of H3G34R or H3G34V in 293T cells both had no effect on global H3K27me2/3 or H3K36me3 levels (Lewis et al. 2013). Instead, H3K36me2/3 was exclusively reduced on exogenous H3G34-mutant nucleosomes but not on endogenous WT nucleosomes (Lewis et al. 2013). Consistent with this finding, recombinant SETD2 was less efficient at methylating H3K36 using H3G34-mutant nucleosomes in vitro (Lewis et al. 2013). Thus, H3G34 mutations appear to exert their effect in *cis*, in contrast to H3K27M and H3K36M mutations, which are able to inhibit SET-domain methyltransferases in *trans* (Fig. 3C). Of note, whereas H3K27M and H3K36M mutations have been reported in H3.1 and H3.3, H3G34 mutations have only been identified in H3.3, further suggesting that they may result in local chromatin changes in genomic regions enriched for H3.3. How H3G34 mutations contribute to oncogenesis remains largely unexplored, although an initial report points to MYCN as being the most highly differentially expressed gene in a H3G34-mutant glioma cell line (Bjerke et al. 2013). Knockdown of MYCN decreased viability of H3G34-mutant cells and expression of H3G34-mutant transgenes in astrocyte and glial cells was sufficient to upregulate MYCN (Bjerke et al. 2013). Further work is needed to assess how H3G34 mutations act to regulate MYCN expression as well as other genes associated with early brain development (Bjerke et al. 2013).

CONCLUDING REMARKS

The application of next-generation sequencing to an increasing number and array of human cancers is revealing mutations where no one expected—perhaps most surprisingly in histones, the fundamental building blocks of chromatin. Strikingly, these mutations occur at or near residues on the amino-terminal tail of histone H3 that have well-characterized writers, readers, and erasers. As more types of cancer are sequenced and as appreciation for the dominant nature of these "oncohistone" mutations grows, it will be interesting to determine if new histone mutations, "K-to-M" or otherwise, are identified. One common feature of histone mutations is their ability to arrest cells in a primitive state refractory to differentiation induction. This is in line with the exquisite tissue and lineage specificity of these mutations and highlights the importance of modeling and studying oncohistones in the appropriate developmental context.

Despite rapid progress in elucidating the molecular mechanism of H3K27M, H3K36M, and H3G34R/V mutations, many questions remain. Although the effects of H3K27M and H3K36M mutations seemed to be independent of the H3 isoforms in cell culture models, patient tumors carrying H3.1/2 "K-to-M" mutations show clear pathological and clinical distinctions from H3.3 "K-to-M" mutant tumors. Further research is warranted to resolve this discrepancy. In addition, more work is needed to understand the persistence of H3K27me3 enrichment at limited genomic loci in H3K27M-mutant gliomas and whether or how it may contribute to oncogenesis. This will likely require a renewed focus on changes to the chromatin landscape beyond genic regions to include intergenic loci. Why specific oncogenes and tumor suppressors co-occur with different oncohistones, such as H3K27M with specific oncogenic signaling pathways or H3G34R/V with ATRX mutations, is also unknown. At another level, the remarkable anatomical and age specificity of oncohistone-associated cancers begs for studies aimed at determining underlying cells-of-origin and developmental timing issues. If our current knowledge of histone mutations is any indication, answering these questions and making further progress in developing new therapies for patients will require integrating our understanding of chromatin structure and biochemistry with developmental pathways and the hallmarks of cancer.

ACKNOWLEDGMENTS

We apologize to colleagues whose work could not be cited because of space limitations. We thank Alexey Soshnev for help in the preparation of the figures. C.D.A. is supported by funding from The Rockefeller University and the National Institutes of Health (NIH) Grant P01CA196539. C.L. is the Kandarian Family Fellow supported by the Damon Runyon Cancer Research Foundation (DRG-2195-14). D.N.W. is supported by a Medical Scientist Training Program grant from the National Institute of General Medical Sciences of the NIH under award number T32GM007739 to the Weill Cornell/Rockefeller/Sloan Kettering Tri-Institutional MD-PhD Program.

REFERENCES

Aihara K, Mukasa A, Gotoh K, Saito K, Nagae G, Tsuji S, Tatsuno K, Yamamoto S, Takayanagi S, Narita Y, et al. 2014. H3F3A K27M mutations in thalamic gliomas from young adult patients. *Neuro Oncol* 16: 140–146.

Amary MF, Berisha F, Mozela R, Gibbons R, Guttridge A, O'Donnell P, Baumhoer D, Tirabosco R, Flanagan AM. 2016. The H3F3 K36M mutant antibody is a sensitive and specific marker for the diagnosis of chondroblastoma. *Histopathology* 69: 121–127.

Banaszynski LA, Allis CD, Lewis PW. 2010. Histone variants in metazoan development. *Dev Cell* 19: 662–674.

Banaszynski LA, Wen D, Dewell S, Whitcomb SJ, Lin M, Diaz N, Elsässer SJ, Chapgier A, Goldberg AD, Canaani E, et al. 2013. Hira-dependent histone H3.3 deposition facilitates PRC2 recruitment at developmental loci in ES Cells. *Cell* 155: 107–120.

Bechet D, Gielen G, Korshunov A, Pfister SM, Rousso C, Faury D, Fiset PO, Benlimane N, Lewis PW, Lu C, et al. 2014. Specific detection of methionine 27 mutation in histone 3 variants (H3K27M) in fixed tissue from high-grade astrocytomas. *Acta Neuropathol* 128: 733–741.

Behjati S, Tarpey PS, Presneau N, Scheipl S, Pillay N, Van Loo P, Wedge DC, Cooke SL, Gundem G, Davies H, et al. 2013. Distinct H3F3A and H3F3B driver mutations define chondroblastoma and giant cell tumor of bone. *Nat Genet* 45: 1479–1482.

Bender S, Tang Y, Lindroth AM, Hovestadt V, Jones DT, Kool M, Zapatka M, Northcott PA, Sturm D, Wang W, et al. 2013. Reduced H3K27me3 and DNA hypomethylation are major drivers of gene expression in K27M mutant pediatric high-grade gliomas. *Cancer Cell* 24: 660–672.

Bjerke L, Mackay A, Nandhabalan M, Burford A, Jury A, Popov S, Bax DA, Carvalho D, Taylor KR, Vinci M, et al. 2013. Histone H3.3 mutations drive pediatric glioblastoma through upregulation of MYCN. *Cancer Discov* 3: 512–519.

Brown ZZ, Müller MM, Jain SU, Allis CD, Lewis PW, Muir TW. 2014. Strategy for "fetoxification" of a cancer-derived histone mutant based on mapping its interaction with the methyltransferase PRC2. *J Am Chem Soc* 136: 13498–13501.

Brown ZZ, Müller MM, Kong HE, Lewis PW, Muir TW. 2015. Targeted histone peptides: Insights into the spatial regulation of the methyltransferase PRC2 by using a surrogate of heterotypic chromatin. *Angew Chem Int Ed Engl* 54: 6457–6461.

Buczkowicz P, Hoeman C, Rakopoulos P, Pajovic S, Letourneau L, Dzamba M, Morrison A, Lewis P, Bouffet E, Bartels U, et al. 2014. Genomic analysis of diffuse intrinsic pontine gliomas identifies three molecular subgroups and recurrent activating *ACVR1* mutations. *Nat Genet* 46: 451–456.

Bush KM, Yuen BT, Barrilleaux BL, Riggs JW, O'Geen H, Cotterman RF, Knoepfler PS. 2013. Endogenous mammalian histone H3.3 exhibits chromatin-related functions during development. *Epigenet Chrom* 6: 7.

Carvalho S, Vítor AC, Sridhara SC, Filipa BM, Ana CR, Desterro JMP, Ferreira J, de Almeida SF. 2014. SETD2 is required for DNA double-strand break repair and activation of the p53-mediated checkpoint. *eLife* 2014: 1–19.

Castel D, Philippe C, Calmon R, Le Dret L, Truffaux N, Boddaert N, Pagès M, Taylor KR, Saulnier P, Lacroix L, et al. 2015. Histone H3F3A and HIST1H3B K27M mutations define two subgroups of diffuse intrinsic pontine gliomas with different prognosis and phenotypes. *Acta Neuropathol* 130: 815–827.

Chaikuad A, Alfano I, Kerr G, Sanvitale CE, Boergermann JH, Triffitt JT, Von Delft F, Knapp S, Knaus P, Bullock AN. 2012. Structure of the bone morphogenetic protein receptor ALK2 and implications for fibrodysplasia ossificans progressiva. *J Biol Chem* 287: 36990–36998.

Chan K, Fang D, Gan H, Hashizume R, Yu C, Schroeder M, Gupta N, Mueller S, James CD, Jenkins R, et al. 2013. The histone H3.3K27M mutation in pediatric glioma reprograms H3K27 methylation and gene expression. *Genes Dev* 27: 985–990.

Couldrey C, Carlton MB, Nolan PM, Colledge WH, Evans MJ. 1999. A retroviral gene trap insertion into the histone 3.3A gene causes partial neonatal lethality, stunted growth, neuromuscular deficits and male sub-fertility in transgenic mice. *Hum Mol Genet* 8: 2489–2495.

Elsässer SJ, Huang H, Lewis PW, Chin JW, Allis CD, Patel DJ. 2012. DAXX envelops a histone H3.3–H4 dimer for H3.3-specific recognition. *Nature* 491: 560–565.

Elsässer SJ, Noh KM, Diaz N, Allis CD, Banaszynski LA. 2015. Histone H3.3 is required for endogenous retroviral element silencing in embryonic stem cells. *Nature* 522: 240–244.

Ernst J, Kellis M. 2010. Discovery and characterization of chromatin states for systematic annotation of the human genome. *Nat Biotechnol* **28:** 817–825.

Fang D, Gan H, Lee JH, Han J, Wang Z, Riester SM, Jin L, Chen J, Zhou H, Wang J, et al. 2016. The histone H3.3K36M mutation reprograms the epigenome of chondroblastomas. *Science* **352:** 1344–1348.

Feng J, Hao S, Pan C, Wang Y, Wu Z, Zhang J, Yan H, Zhang L, Wan H. 2015. The H3.3 K27M mutation results in a poorer prognosis in brainstem gliomas than thalamic gliomas in adults. *Hum Pathol* **46:** 1626–1632.

Fontebasso AM, Schwartzentruber J, Khuong-Quang DA, Liu XY, Sturm D, Korshunov A, Jones DT, Witt H, Kool M, Albrecht S, et al. 2013. Mutations in *SETD2* and genes affecting histone H3K36 methylation target hemispheric high-grade gliomas. *Acta Neuropathol* **125:** 659–669.

Fontebasso AM, Gayden T, Nikbakht H, Neirinck M, Papillon-Cavanagh S, Majewski J, Jabado N. 2014a. Epigenetic dysregulation: A novel pathway of oncogenesis in pediatric brain tumors. *Acta Neuropathol* **128:** 615–627.

Fontebasso AM, Papillon-Cavanagh S, Schwartzentruber J, Nikbakht H, Gerges N, Fiset PO, Bechet D, Faury D, De Jay N, Ramkissoon LA, et al. 2014b. Recurrent somatic mutations in *ACVR1* in pediatric midline high-grade astrocytoma. *Nat Genet* **46:** 462–466.

Funato K, Major T, Lewis PW, Allis CD, Tabar V. 2014. Use of human embryonic stem cells to model pediatric gliomas with H3.3K27M histone mutation. *Science* **346:** 1529–1533.

Goldberg AD, Banaszynski LA, Noh KM, Lewis PW, Elsässer SJ, Stadler S, Dewell S, Law M, Guo X, Li X, et al. 2010. Distinct factors control histone variant H3.3 localization at specific genomic regions. *Cell* **140:** 678–691.

Grasso CS, Tang Y, Truffaux N, Berlow NE, Liu L, Debily MA, Quist MJ, Davis LE, Huang EC, Woo PJ, et al. 2015. Functionally defined therapeutic targets in diffuse intrinsic pontine glioma. *Nat Med* **21:** 555–559.

Hake SB, Allis CD. 2006. Histone H3 variants and their potential role in indexing mammalian genomes: The "H3 barcode hypothesis." *Proc Natl Acad Sci* **103:** 6428–6435.

Hashizume R, Andor N, Ihara Y, Lerner R, Gan H, Chen X, Fang D, Huang X, Tom MW, Ngo V, et al. 2014. Pharmacologic inhibition of histone demethylation as a therapy for pediatric brainstem glioma. *Nat Med* **20:** 1394–1396.

Herz HM, Morgan M, Gao X, Jackson J, Rickels R, Swanson SK, Florens L, Washburn MP, Eissenberg JC, Shilatifard A. 2014. Histone H3 lysine-to-methionine mutants as a paradigm to study chromatin signaling. *Science* **345:** 1065–1070.

Huang H, Sabari BR, Garcia BA, Allis CD, Zhao Y. 2014. SnapShot: Histone modifications. *Cell* **159:** 458–458.e1.

Jayaram H, Hoelper D, Jain SU, Cantone N, Lundgren SM, Poy F, Allis CD, Cummings R, Bellon S, Lewis PW. 2016. S-adenosyl methionine is necessary for inhibition of the methyltransferase G9a by lysine 9 to methionine mutation on histone H3. *Proc Natl Acad Sci* **113:** 6182–6187.

Jenuwein T, Allis CD. 2001. Translating the histone code. *Science* **293:** 1074–1080.

Jiao L, Liu X. 2015. Structural basis of histone H3K27 trimethylation by an active Polycomb repressive complex 2. *Science* **350:** 291.

Jones DT, Hutter B, Jäger N, Korshunov A, Kool M, Warnatz HJ, Zichner T, Lambert SR, Ryzhova M, Quang DA, et al. 2013. Recurrent somatic alterations of FGFR1 and NTRK2 in pilocytic astrocytoma. *Nat Genet* **45:** 927–932.

Joseph CG, Hwang H, Jiao Y, Wood LD, Kinde I, Wu J, Mandahl N, Luo J, Hruban RH, Diaz LA Jr, et al. 2014. Exomic analysis of myxoid liposarcomas, synovial sarcomas, and osteosarcomas. *Genes Chromosomes Cancer* **53:** 15–24.

Justin N, Zhang Y, Tarricone C, Martin SR, Chen S, Underwood E, De Marco V, Haire LF, Walker PA, Reinberg D, et al. 2016. Structural basis of oncogenic histone H3K27M inhibition of human Polycomb repressive complex 2. *Nat Commun* **7:** 11316.

Kallappagoudar S, Yadav RK, Lowe BR, Partridge JF. 2015. Histone H3 mutations—A special role for H3.3 in tumorigenesis? *Chromosoma* **124:** 177–189.

Karytinos A, Forneris F, Profumo A, Ciossani G, Battaglioli E, Binda C, Mattevi A. 2009. A novel mammalian flavin-dependent histone demethylase. *J Biol Chem* **284:** 17775–17782.

Khuong-Quang DA, Buczkowicz P, Rakopoulos P, Liu XY, Fontebasso AM, Bouffet E, Bartels U, Albrecht S, Schwartzentruber J, Letourneau L, et al. 2012. K27M mutation in histone H3.3 defines clinically and biologically distinct subgroups of pediatric diffuse intrinsic pontine gliomas. *Acta Neuropathol* **124:** 439–447.

Kuo AJ, Cheung P, Chen K, Zee BM, Kioi M, Lauring J, Xi Y, Park BH, Shi X, Garcia BA, et al. 2011. NSD2 links dimethylation of histone H3 at lysine 36 to oncogenic programming. *Mol Cell* **44:** 609–620.

Lewis PW, Allis CD. 2013. Poisoning the "histone code" in pediatric gliomagenesis. *Cell Cycle* **12:** 3241–3242.

Lewis PW, Elsässer SJ, Noh KM, Stadler SC, Allis CD. 2010. Daxx is an H3.3-specific histone chaperone and cooperates with ATRX in replication-independent chromatin assembly at telomeres. *Proc Natl Acad Sci* **107:** 14075–14080.

Lewis PW, Muller MM, Koletsky MS, Cordero F, Lin S, Banaszynski LA, Garcia BA, Muir TW, Becher OJ, Allis CD. 2013. Inhibition of PRC2 activity by a gain-of-function H3 mutation found in pediatric glioblastoma. *Science* **340:** 857–861.

Lohr JG, Stojanov P, Lawrence MS, Auclair D, Chapuy B, Sougnez C, Cruz-Gordillo P, Knoechel B, Asmann YW, Slager SL, et al. 2012. Discovery and prioritization of somatic mutations in diffuse large B-cell lymphoma (DLBCL) by whole-exome sequencing. *Proc Natl Acad Sci* **109:** 3879–3884.

Lu C, Jain SU, Hoelper D, Bechet D, Molden R, Ran L, Murphy D, Venneti S, Hameed M, Pawel B, et al. 2016. Histone H3K36 mutations promote sarcomagenesis through altered histone methylation landscape. *Science* **352:** 844–849.

Luger K, Mäder AW, Richmond RK, Sargent DF, Richmond TJ. 1997. Crystal structure of the nucleosome core particle at 2.8 Å resolution. *Nature* **389:** 251–260.

Margueron R, Reinberg D. 2011. The Polycomb complex PRC2 and its mark in life. *Nature* **469:** 343–349.

Maze I, Noh KM, Soshnev AA, Allis CD. 2014. Every amino acid matters: Essential contributions of histone variants to mammalian development and disease. *Nat Rev Genet* **15:** 259–271.

McKittrick E, Gafken PR, Ahmad K, Henikoff S. 2004. Histone H3.3 is enriched in covalent modifications associated with active chromatin. *Proc Natl Acad Sci* **101:** 1525–1530.

Nikbakht H, Panditharatna E, Mikael LG, Li R, Gayden T, Osmond M, Ho CY, Kambhampati M, Hwang EI, Faury D, et al. 2016. Spatial and temporal homogeneity of driver mutations in diffuse intrinsic pontine glioma. *Nat Commun* **7:** 11185.

Ogryzko V, Schiltz RL, Russanova V, Howard BH, Nakatani Y. 1996. The transcriptional coactivators p300 and CBP are histone acetyltransferases. *Cell* **87:** 953–959.

Okosun J, Bödör C, Wang J, Araf S, Yang CY, Pan C, Boller S, Cittaro D, Bozek M, Iqbal S, et al. 2014. Integrated genomic analysis identifies recurrent mutations and evolution patterns driving the initiation and progression of follicular lymphoma. *Nat Genet* **46:** 176–181.

Sarungbam J, Agaram N, Hwang S, Lu C, Wang L, Healey J, Hameed M. 2016. Symplastic/pseudoanaplastic giant cell tumor of the bone. *Skeletal Radiol* doi: 10.1007/s00256-016-2373-z.

Schmitges FW, Prusty AB, Faty M, Stützer A, Lingaraju GM, Aiwazian J, Sack R, Hess D, Li L, Zhou S, et al. 2011. Histone methylation by PRC2 is inhibited by active chromatin marks. *Mol Cell* **42:** 330–341.

Schwartzentruber J, Korshunov A, Liu XY, Jones DT, Pfaff E, Jacob K, Sturm D, Fontebasso AM, Quang DA, Tönjes M, et al. 2012. Driver mutations in histone H3.3 and chromatin remodelling genes in paediatric glioblastoma. *Nature* **482:** 226–231.

Shen H, Laird PW. 2013. Interplay between the cancer genome and epigenome. *Cell* **153:** 38–55.

Shore EM, Xu M, Feldman GJ, Fenstermacher DA, Cho TJ, Choi IH, Connor JM, Delai P, Glaser DL, LeMerrer M, et al. 2006. A recurrent mutation in the BMP type I receptor ACVR1 causes inherited and sporadic fibrodysplasia ossificans progressiva. *Nat Genet* **38:** 525–527.

Simon JM, Hacker KE, Singh D, Brannon AR, Parker JS, Weiser M, Ho TH, Kuan PF, Jonasch E, Furey TS, et al. 2014. Variation in chromatin accessibility in human kidney cancer links H3K36 methyltransferase loss with widespread RNA processing defects. *Genome Res* **24:** 241–250.

Sturm D, Witt H, Hovestadt V, Khuong-Quang DA, Jones DT, Konermann C, Pfaff E, Tönjes M, Sill M, Bender S, et al. 2012. Hotspot mutations in H3F3A and IDH1 define distinct epigenetic and biological subgroups of glioblastoma. *Cancer Cell* **22:** 425–437.

Tagami H, Ray-Gallet D, Almouzni G, Nakatani Y. 2004. Histone H3.1 and H3.3 complexes mediate nucleosome assembly pathways dependent or independent of DNA synthesis. *Cell* **116:** 51–61.

Taylor KR, Mackay A, Truffaux N, Butterfield YS, Morozova O, Philippe C, Castel D, Grasso CS, Vinci M, Carvalho D, et al. 2014. Recurrent activating *ACVR1* mutations in diffuse intrinsic pontine glioma. *Nat Genet* **46:** 457–461.

Toledo RA, Qin Y, Cheng ZM, Gao Q, Iwata S, Silva GM, Prasad ML, Ocal IT, Rao S, Aronin N, et al. 2015. Recurrent mutations of chromatin-remodeling genes and kinase receptors in pheochromocytomas and paragangliomas. *Clin Cancer Res* **22:** 1–11.

Venneti S, Garimella MT, Sullivan LM, Martinez D, Huse JT, Heguy A, Santi M, Thompson CB, Judkins AR. 2013. Evaluation of histone 3 lysine 27 trimethylation (H3K27me3) and enhancer of zest 2 (EZH2) in pediatric glial and glioneuronal tumors shows decreased H3K27me3 in H3F3A K27M mutant glioblastomas. *Brain Pathol* **23:** 558–564.

Venneti S, Santi M, Felicella MM, Yarilin D, Phillips JJ, Sullivan LM, Martinez D, Perry A, Lewis PW, Thompson CB, et al. 2014. A sensitive and specific histopathologic prognostic marker for H3F3A K27M mutant pediatric glioblastomas. *Acta Neuropathol* **128:** 743–753.

Wagner EJ, Carpenter PB. 2012. Understanding the language of Lys36 methylation at histone H3. *Nat Rev Mol Cell Biol* **13:** 115–126.

Wen H, Li Y, Xi Y, Jiang S, Stratton S, Peng D, Tanaka K, Ren Y, Xia Z, Wu J, et al. 2014. ZMYND11 links histone H3.3K36me3 to transcription elongation and tumour suppression. *Nature* **508:** 263–268.

Wu G, Broniscer A, McEachron TA, Lu C, Paugh BS, Becksfort J, Qu C, Ding L, Huether R, Parker M, et al. 2012. Somatic histone H3 alterations in pediatric diffuse intrinsic pontine gliomas and non-brainstem glioblastomas. *Nat Genet* **44:** 251–253.

Wu G, Diaz AK, Paugh BS, Rankin SL, Ju B, Li Y, Zhu X, Qu C, Chen X, Zhang J, et al. 2014. The genomic landscape of diffuse intrinsic pontine glioma and pediatric non-brainstem high-grade glioma. *Nat Genet* **46:** 444–450.

Yuan W, Xu M, Huang C, Liu N, Chen S, Zhu B. 2011. H3K36 methylation antagonizes PRC2-mediated H3K27 methylation. *J Biol Chem* **286:** 7983–7989.

Zhang J, Wu G, Miller CP, Tatevossian RG, Dalton JD, Tang B, Orisme W, Punchihewa C, Parker M, Qaddoumi I, et al. 2013. Whole-genome sequencing identifies genetic alterations in pediatric low-grade gliomas. *Nat Genet* **45:** 602–612.

MLL3/MLL4/COMPASS Family on Epigenetic Regulation of Enhancer Function and Cancer

Christie C. Sze and Ali Shilatifard

Department of Biochemistry and Molecular Genetics and Robert H. Lurie NCI Comprehensive Cancer Center, Northwestern University Feinberg School of Medicine, Chicago, Illinois 60611

Correspondence: ash@northwestern.edu

During development, precise spatiotemporal patterns of gene expression are coordinately controlled by *cis*-regulatory modules known as enhancers. Their crucial role in development helped spur numerous studies aiming to elucidate the functional properties of enhancers within their physiological and disease contexts. In recent years, the role of enhancer malfunction in tissue-specific tumorigenesis is increasingly investigated. Here, we direct our focus to two primary players in enhancer regulation and their role in cancer pathogenesis: MLL3 and MLL4, members of the COMPASS family of histone H3 lysine 4 (H3K4) methyltransferases, and their complex-specific subunit UTX, a histone H3 lysine 27 (H3K27) demethylase. We review the most recent evidence on the underlying roles of MLL3/MLL4 and UTX in cancer and highlight key outstanding questions to help drive future research and contribute to our fundamental understanding of cancer and facilitate identification of therapeutic opportunities.

Enhancers are noncoding DNA regulatory sequences that govern the complex spatiotemporal patterns of gene expression throughout development by heightening the rate of transcription of target genes (Banerji et al. 1981; Smith and Shilatifard 2014). These DNA elements can span several hundred base pairs (bp) to a few kilobases (kb) and contain arrays of short DNA modules that serve as binding sites for sequence-specific transcription factors, which recruit a combination of factors that together dictate the function of the enhancer (Maniatis et al. 1987). A key attribute of enhancers is that they act independently of orientation and distance to their target promoter(s), and are littered throughout the genome within intragenic and intergenic regions (Smith and Shila-tifard 2014). Despite several decades of extensive research, the precise mechanism of action of enhancers is still poorly understood. Studies have shown that enhancers can exert their activity over long distances, bypassing neighboring genes, and communicate with a specific distal promoter (Blackwood and Kadonaga 1998; Bulger and Groudine 2011; Levine et al. 2014). Such enhancer–promoter communication is established via a looping mechanism mediated by the cohesin and mediator complexes and other associated proteins (Dorsett 1999; Kagey et al. 2010; Dorsett and Merkenschlager 2013).

The identification and functional annotation of enhancers in the metazoan genome have been challenging; however, the development of high-throughput sequencing in recent years has facilitated the discovery of tissue-specific enhancers. Genome-wide chromatin immunoprecipitation (ChIP) analyses of histone modifications led to the identification of chromatin signatures for enhancers. The epigenetic mark histone H3 lysine 4 (H3K4) monomethylation is particularly enriched on these regulatory elements, both active and inactive/poised (Heintzman et al. 2007, 2009; Smith and Shilatifard 2014). Additionally, the presence of acetylated histone H3 lysine 27 (H3K27ac) and trimethylated histone H3 lysine 27 (H3K27me) can further distinguish active from inactive enhancers, respectively (Creyghton et al. 2010; Rada-Iglesias et al. 2011; Zentner et al. 2011; Smith and Shilatifard 2014).

Given their essential role in transcriptional regulation and gene expression, it is not surprising that disruption of enhancers can lead to disease. In recent years, the role of enhancer malfunction in tumorigenesis is increasingly studied (Akhtar-Zaidi et al. 2012; Kurdistani 2012; Sur et al. 2012; Aran et al. 2013; Loven et al. 2013; Herz et al. 2014; Morgan and Shilatifard 2015). In this review, we focus our attention on the proteins and factors that mediate changes in enhancer chromatin states and their role in cancer pathogenesis.

THE COMPASS FAMILY OF HISTONE H3K4 METHYLASES

The mixed lineage leukemia (MLL) gene was first discovered as an oncogenic fusion resulting from seemingly random translocations in patients with hematological malignancies (Ziemin-van der Poel et al. 1991; Djabali et al. 1992; Gu et al. 1992; Tkachuk et al. 1992; Shilatifard 2012). To investigate the role of MLL in leukemia, initial efforts were focused on isolating MLL-containing complexes to understand the fundamental biochemical properties, functions, and regulation of MLL under normal conditions. An ancestral homolog of MLL, Set1, was identified in the budding yeast *Saccharomyces cerevisiae* and was found to exist within a macromolecular complex named COMPASS (complex of proteins associated with Set1) (Miller et al. 2001; Roguev et al. 2001; Krogan

et al. 2002). Set1, together with other subunits within COMPASS, is capable of catalyzing mono-, di-, and trimethylation on histone H3K4 in yeast (Schneider et al. 2005). Subsequent studies revealed a diverse family of COMPASS in metazoans. Although yeast only has a single Set1/COMPASS that can mediate all three H3K4 methylation patterns, *Drosophila melanogaster* has three H3K4 methyltransferases, named dSet1, trithorax (Trx), and trithorax-related (Trr) (Eissenberg and Shilatifard 2010; Mohan et al. 2011). For mammals, there are two paralogs corresponding to each of the three *Drosophila* members: Set1a (also known as KMT2F) and Set1b (or KMT2G), orthologous to dSet1; MLL1 (or KMT2A; GeneID 4297) and MLL2 (or KMT2B; GeneID 9757), orthologous to Trx; and MLL3 (or KMT2C; GeneID 58508) and MLL4 (or KMT2D; GeneID 8085), orthologous to Trr (Allis et al. 2007; Shilatifard 2012). The *Drosophila* and mammalian methylases also reside in COMPASS-like complexes, which were shown through ensuing studies to contain (1) core subunits critical for the enzymatic activity (Fig. 1, highlighted in dark blue), and (2) specific components that may confer functional uniqueness to each complex (Fig. 1, highlighted in green) (Mohan et al. 2011; Shilatifard 2012).

The methylase subunits of the COMPASS family all possess a catalytic 130-amino-acid-long carboxy-terminal motif called the SET domain, named after the *Drosophila* proteins Su(var)3-9, enhancer of zeste [E(z)], and trithorax (Trx) (Tschiersch et al. 1994; Stassen et al. 1995). In contrast, regions amino-terminal to the SET domain differ across the family members. In brief, mammalian Set1a and Set1b each have an amino-terminal RNA recognition motif (RRM) and an N-SET domain adjacent to the SET domain, whereas mammalian MLL1-4 methylases contain varying numbers of plant homeodomain (PHD) fingers, FY-rich (FYR) domains, and DNA-binding motifs such as AT-hooks, high mobility group (HMG) boxes, and CXXC domains (Fig. 2) (Herz et al. 2013). The domain architectural variability across the methylases denotes the binding and functional diversity of the COMPASS family.

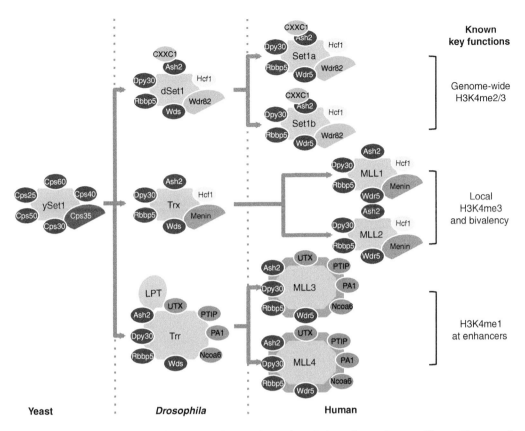

Figure 1. The COMPASS family of histone H3 lysine 4 (H3K4) methyltransferases in yeast, flies, and humans. In yeast, there is only one Set1 methyltransferase capable of methylating histone H3K4. Flies have three COMPASS family members: dSet1, trithorax (Trx), and trithorax-related (Trr). Mammals have two paralogs for each of the three fly members for a total of six COMPASS members. Core subunits found in all COMPASS complexes are highlighted in dark blue, whereas subunits specific to the complex are marked in green. Hcf1 (light blue) is reportedly specific to be in Set1 and Trx branches but not in the Trr complex (van Nuland et al. 2013). Mixed lineage leukemia (MLL)3 and MLL4, the focus of this review, are outlined in orange. Key functions known to date for each branch of COMPASS are noted.

THE MLL3 AND MLL4/COMPASS FAMILY AS ENHANCER MONOMETHYLASES

A growing body of evidence points to a model in which the responsibilities of H3K4 methylation are divided among the COMPASS family members to ensure proper transcriptional modulation. Several studies have shown that dSet1 and mammalian Set1a/b are responsible for bulk H3K4 di- and trimethylation across the genome (Wu et al. 2008; Ardehali et al. 2011; Mohan et al. 2011; Hallson et al. 2012), whereas Trx and MLL1/MLL2 are necessary for gene-specific H3K4 trimethylation, including *Hox* gene

promoters (Wang et al. 2009) and bivalent promoters (promoters marked by concurrent trimethylation of H3K4 and H3K27 and poised to express developmental genes) in mouse embryonic stem (mES) cells (Hu et al. 2013b). Trx was initially discovered as a regulator of the developmental expression of *Hox* genes in *Drosophila*, specifically being required for maintaining *Hox* gene activation (Breen and Harte 1991; Pirrotta 1998; Mahmoudi and Verrijzer 2001; Poux et al. 2002; Klymenko and Muller 2004; Shilatifard 2012). It is through shared protein homology with Trx that Trr was cloned (Sedkov et al. 1999).

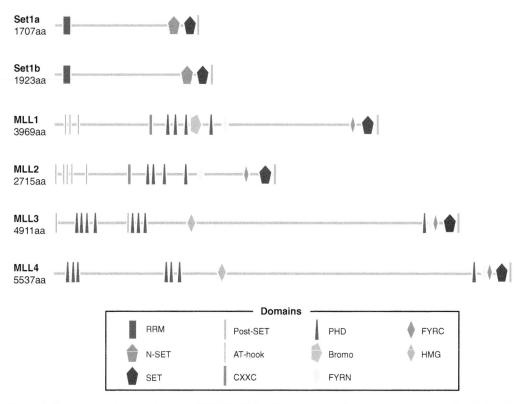

Figure 2. Known domain organization of COMPASS family members in humans. Annotation of each domain structure follows SMART (see smart.embl-heidelberg.de) (Schultz et al. 1998; Letunic et al. 2015) using protein sequences obtained from the National Center for Biotechnology Information (NCBI) as accessed on January 8, 2016. Names for illustrated domains are specified in the box labeled "Domains." All COMPASS family members possess the highly conserved SET and post-SET domains at the carboxyl terminus. Meanwhile, the amino terminus vastly varies across the subfamilies. Mammalian Set1a and Set1b have amino-terminal RNA recognition motifs (RRM) and an N-SET domain juxtaposing the SET domain. Mammalian MLL-related proteins have several plant homeodomain (PHD) fingers and other domains associated with chromatin binding (e.g., AT-hooks, high-mobility group [HMG] boxes, and CXXC domains). MLL1-4 methylases also have FY-rich (FYR) motifs, in which MLL1 and MLL2 have FYRN and FYRC regions distant from each other, whereas MLL3 and MLL4 have such regions adjacent to each other. The diversity in domains contributes to the binding and functional properties of the COMPASS complexes.

Trr and its mammalian homologs MLL3/MLL4 have been accredited as key H3K4 monomethyltransferases at enhancers, primarily implementing H3K4 monomethylation at intergenic and intragenic regions (Herz et al. 2012; Hu et al. 2013a; Lee et al. 2013). In *Drosophila*, Trr, which contains the SET domain, corresponds to the carboxy-terminal portion of MLL3/MLL4, whereas another protein LPT (lost plant homeodomains of Trr), which contains several PHD domains and an HMG box, is homologous to the amino-terminal region of MLL3/MLL4 (Mohan et al. 2011; Herz et al. 2012). Together, Trr and LPT serve an analogous role to mammalian MLL3/MLL4. ChIP-seq studies revealed that Trr and MLL3/MLL4 bind to enhancer regions as well as transcription start sites (Herz et al. 2012; Hu et al. 2013a). Depletion of Trr and MLL3/MLL4 resulted in a striking genome-wide reduction of H3K4 monomethylation, primarily occurring at enhancer regions (Herz et al. 2012; Hu et al. 2013a). As histone H3K27ac decreases, the H3K27me3 levels increase at putative enhancers upon loss of Trr in *Drosophila* S2 cells or loss of MLL3/MLL4 in mouse embryonic fibroblasts (MEFs) (Herz et al. 2012; Hu et al. 2013a). Because H3K27ac and H3K27me marks are part of different enhancer chromatin signatures, these findings further implicated Trr/MLL3/ MLL4 in enhancer regulation (Herz et al. 2014). The role of Trr/MLL3/MLL4 as enhancer monomethylases regulating enhancer/promoter communication has also been confirmed by other studies of mammalian myogenesis and adipogenesis, macrophage activation, cardiac development, and B-cell lymphomagenesis (Kaikkonen et al. 2013; Lee et al. 2013; Ortega-Molina et al. 2015; Ang et al. 2016).

MLL3 AND MLL4/COMPASS IN CANCER

The Mutational Landscape

Advancement of high-throughput genome sequencing in recent years resulted in the identification of a myriad of somatic mutations of MLL3 and MLL4 across different malignancies, which include but are not limited to non-Hodgkin's lymphoma (NHL), bladder cancer, breast cancer, medulloblastoma, prostate cancer, colorectal cancer, esophageal squamous cell carcinoma, acute myeloid leukemia (AML), and cutaneous T-cell lymphoma (Ruault et al. 2002; Ashktorab et al. 2010; Gui et al. 2011; Morin et al. 2011; Parsons et al. 2011; Pasqualucci et al. 2011; Akhtar-Zaidi et al. 2012; Ellis et al. 2012; Grasso et al. 2012; Jones et al. 2012; Pugh et al. 2012; Gao et al. 2014; Lin et al. 2014; da Silva Almeida et al. 2015; Tan et al. 2015). In fact, extensive genomic analyses of these sequencing data revealed that MLL3 and MLL4/ COMPASS family mutations to be among the most frequent in human cancer (Kandoth et al. 2013; Lawrence et al. 2014). Missense and nonsense mutations of MLL3 and MLL4 are distributed along the whole length of the protein, rendering the enzyme inactive or truncated (Fig. 3) (Forbes et al. 2015). However, closer observation of the "Catalogue of Somatic Mutations in Cancer" (COSMIC) data reveals a higher density of mutations in the amino-terminal region of MLL3 containing clusters of PHD fingers,

whereas mutations of MLL4, although abundant, are relatively more dispersed throughout the protein (Fig. 3) (see cancer.sanger.ac.uk/ cosmic; Forbes et al. 2015). The enrichment of mutations over the highly conserved amino-terminal region of MLL3 indicates the potential importance of these domains in cancer and justifying the need for further molecular understanding of these functional motifs (Fig. 3) (see cancer.sanger.ac.uk/cosmic; Forbes et al. 2015).

TUMOR SUPPRESSOR AND ONCOGENIC ACTIVITIES

Accumulating evidence suggests that MLL3 and MLL4/COMPASS family are tumor suppressors, such that their mutations result in abrogation of their tumor suppressing activity and promote cancer in a tissue-specific manner. Trr, the *Drosophila* homolog of MLL3/MLL4, was reported to negatively regulate cell proliferation, for *Trr* mutant clones resulted in tissue overgrowth compared with their wild-type (WT) counterparts during *Drosophila* eye development (Kanda et al. 2013). Another in vivo study linked MLL3/MLL4 to cancer by serving as a coactivator for p53 to induce expression of p53 target genes involved in DNA damage response (Lee et al. 2009). The same team reported that mice with deleted SET domain of MLL3 developed ureter epithelial tumors, which was exacerbated in a $p53^{+/-}$ background, hinting that the enzymatic activity may be required in inhibiting tumorigenesis (Lee et al. 2009). Newly published evidence has implicated MLL4 in genomic instability (Kantidakis et al. 2016), one of the essential enabling characteristics underlying oncogenesis as described by Hanahan and Weinberg in 2011 (Hanahan and Weinberg 2011). By targeting both alleles of MLL4 in immortalized MEFs and using human MLL4-KO HCT116 cells, Kantidakis et al. (2016) reported that such MLL4 deficiency results in significant perturbation of genomic integrity, including increase sister chromatid exchange and/or chromosomal abnormalities, and induced transcription stress, that is, slow elongation. A recent study has provided a dif-

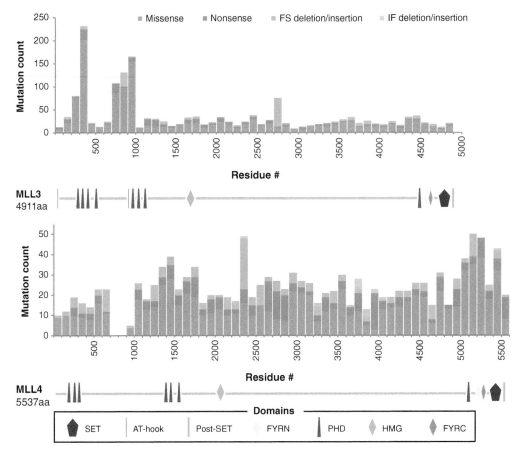

Figure 3. Mutations of MLL3/MLL4 identified in cancer patient samples. Missense and nonsense mutations, frameshift (FS) deletions and insertions, and in-frame (IF) deletions and insertions of MLL3 and MLL4 were obtained from the Catalogue of Somatic Mutations in Cancer (COSMIC) database (see cancer.sanger.ac.uk/cosmic; Forbes et al. 2015). Note that COSMIC reports sites of MLL4 mutations aligned to an alternatively spliced transcript encoding a shorter MLL4 protein (5268 amino acids). In the present schematic, the positions of these mutations were adjusted to match the 5537-amino-acid protein most commonly reported in the literature. To generate the bar plot, binning of 100 residues across each protein was performed, and mutations were categorized by mutation type. The plot shows the number of unique mutated patient samples for a specific mutation ("mutation count"), as documented by COSMIC. Domain schematic of MLL3 and MLL4 to serve as reference, and names for illustrated domains are specified in the box labeled "Domains."

ferent angle of MLL4 involvement in cancer: gain-of-function p53, an oncogenic p53 mutant resulting from a missense mutation in its DNA-binding domain, interacts with and up-regulates a set of chromatin regulators that include MLL1 and MLL4, resulting in altered genome-wide methylation patterns, whereas the WT p53 did not bind these genes or affect their expression (Zhu et al. 2015). The same study showed that knockdown of MLL1 or MLL4 in gain-of-

function p53 breast cancer cells severely reduced cell growth, phenocopying knockdown of gain-of-function p53 (Zhu et al. 2015). Together, these studies clearly indicate that role of MLL4 in cancer is context dependent.

Several studies point to a role for MLL3 and MLL4/COMPASS as tumor suppressors against a wide variety of neoplasms. MLL3 was identified as a novel haploinsufficient tumor suppressor in 7q-deficient myelodysplastic syndrome

(MDS) and AML (Chen et al. 2014). Chromosomal deletions of 7q occur frequently in MDS and AML and correlate with poor patient prognosis. It turns out that 7q deletion results in MLL3 loss, which frequently co-occurs with neurofibromin-1 (NF1) suppression and p53 inactivation (Chen et al. 2014). The investigators found that transplantation of *p53*-deficient hematopoietic stem and progenitor cells (HSPCs), with simultaneous knockdown of MLL3 and NF1, resulted in overt myeloid leukemia in mice. In contrast, suppression of MLL3 and NF1 in WT HSPCs did not induce leukemogenesis, and MLL3-only knockdown in *p53*-deficient HSPCs did not accelerate $p53^{-/-}$-induced thymic lymphoma (Chen et al. 2014). To support these findings, the team executed the CRISPR-Cas9 strategy to disrupt MLL3 in *p53*-deficient HSPCs with NF1-knockdown, which similarly led to AML pathogenesis. Sequencing of resulting AML clones confirmed that the clones contained heterozygous WT MLL3, a compelling finding signifying that MLL3 haploinsufficiency in coordination with NF1 and p53 suppression promotes AML (Chen et al. 2014).

MLL3 was also shown to exert tumor suppressive function in an aggressive form of AML with FLT3-ITD mutation (Garg et al. 2015). In this form of AML, internal tandem duplication (ITD) of Fms-like tyrosine kinase 3 (FLT3) renders the kinase constitutively active, and FLT3-ITD AML patients eventually face tumor relapse or drug resistance. In this study, whole exome sequencing followed by targeted deep sequencing in FLT3-ITD AML patients led to the identification of missense and nonsense mutations of MLL3 (Garg et al. 2015). Prognostic analyses revealed that patients with MLL3 mutations had worse overall survival and relapse-free survival than those without MLL3 mutations (Garg et al. 2015). RNAi-knockdown of MLL3 in FLT3-ITD AML cell lines promoted cell proliferation and clonogenic growth and induced tumorigenicity in xenograft models (Garg et al. 2015). The investigators' additional analyses of The Cancer Genome Atlas (TCGA) data further supported that MLL3 is inactivated in AML (including FLT3-ITD AML) (Garg et al. 2015).

Two independent studies delved into understanding the mechanism by which mutated MLL4 contributes to diffuse large B-cell lymphoma (DLBCL) and follicular lymphoma (FL), the two most common types of NHL derived from germinal center B cells. In one study, MLL4 knockdown in hematopoietic progenitor cells (HPCs) with overexpression of the Bcl2 oncogene markedly accelerated lymphomagenesis in mice, which also developed splenomegaly (Ortega-Molina et al. 2015). Closer examination revealed a delayed germinal center involution and greater expansion of undifferentiated B cells, which were also unable to undergo proper class switch recombination and had reduced antibody production (Ortega-Molina et al. 2015). Gene expression analysis identified similar changes in transcription in mouse and human FL tumors, with the set of genes involved in early antigen immune signaling and B-cell differentiation being perturbed on MLL4 deficiency (Ortega-Molina et al. 2015). ChIP-seq analysis showed significant loss of H3K4 mono- and dimethylation at putative enhancers of lymphoid tumor suppressor genes that regulate B-cell signaling and that were downregulated in MLL4-deficient cells (Ortega-Molina et al. 2015). In a separate study, Pasqualucci and colleagues established conditional knockout mouse models mimicking two key stages of B-cell development, and found that conditional deletion of MLL4 early in B-cell development before germinal center initiation resulted in germinal center expansion with high B-cell proliferation, whereas MLL4 deficiency after germinal center initiation did not perturb B-cell development (Zhang et al. 2015). Loss of MLL4 before germinal center induction resulted in significant transcriptional changes, specifically up-regulation of anti-apoptotic genes and down-regulation of genes promoting B-cell differentiation, indicating that MLL4 depletion confers a survival and proliferative advantage for undifferentiated B cells (Zhang et al. 2015). Taken together, the two parallel studies illustrate specific mechanisms by which MLL4 mutation subverts B-cell identity to that of driving lymphoid malignancies.

MODELS FOR MALFUNCTION OF MLL3/MLL4/COMPASS FAMILY ON ENHANCERS IN CANCER

Because mutations of the enhancer-associated H3K4 monomethylases MLL3/MLL4 occur in cancer, combined with recent findings illustrating their tumor suppressor roles, disruption of MLL3/MLL4-regulated enhancer activity could contribute to cancer by a variety of mechanisms (Herz et al. 2014). A nonsense or missense mutation could render MLL3/MLL4 truncated or catalytically inactive, resulting in loss of function of the monomethylase. A truncating mutation could promote destabilization of this COMPASS family or decrease the affinity of MLL3/MLL4 with transcription factors and other cofactors, attenuating their binding at specific enhancers of key tumor suppressor genes that consequently diminishes activation of gene expression (Herz et al. 2014). In addition, these loss-of-function mutations of MLL3/MLL4 could have broader effects by reducing enhancer activity across the genome. In an alternative scenario, MLL3/MLL4 may drive tumorigenesis via a potential gain-of-function role, in which such gain-of-function mutations may stabilize the MLL3/MLL4 COMPASS family, heighten the catalytic activity, or strengthen the interaction of MLL3/MLL4 with other transcription factors, and subsequently increase their binding to enhancers at specific loci or across the genome (Herz et al. 2014). This gain-of-function role has been implicated in the recent study that discovered the up-regulation of the MLL3/MLL4 chromatin regulators in the context of oncogenic gain-of-function p53 and their potential contribution to gain-of-function p53-promoting cancer (Zhu et al. 2015). Although Zhu et al. (2015) did not report specific gain-of-function mutations in MLL4, we cannot exclude the possibility of MLL3/MLL4 gain-of-function mutations existing in breast cancer and other tumor types. With the two opposing model scenarios in mind, we still have key outstanding questions that should help further advance our understanding of the role of MLL3/MLL4 in cancer pathogenesis. For instance, how exactly do MLL3/MLL4 mutations aberrantly elicit gene expression to ultimately subvert cellular identity and promote tumorigenesis, and how do these underlying mechanisms differ across tumor types? Could mutations at particular domains have distinct molecular ramifications? Mutations within the PHD finger might disrupt recognition of the appropriate marks necessary for recruitment to chromatin, whereas mutations in the SET domain could alter the enzyme's catalytic activity. To date, specific downstream molecular effects of different regulatory domain interactions are still unclear (Henikoff and Shilatifard 2011). Furthermore, finding the key players involved in mobilizing MLL3/MLL4 to enhancer regions, and determining if these factors are themselves mutated to dysregulate MLL3/MLL4 recruitment and consequently their function will be important to understand the full spectrum of MLL3/4 contributions to cancer and help provide key therapeutic strategies for targeting MLL3/MLL4-mediated cancers.

UTX SUBUNIT OF MLL3/MLL4/COMPASS FAMILY AND CANCER

As shown in Figure 1, MLL3 and MLL4 each exist in a multimeric composition within COMPASS with additional components exclusive to the two methyltransferases. Interestingly, one such subunit, ubiquitously transcribed tetratricopeptide repeat on chromosome X (UTX), has also been documented to be frequently mutated in cancer (Kandoth et al. 2013; Lawrence et al. 2014). UTX (also known as KDM6A) is a histone lysine demethylase that specifically removes methyl groups from di- and trimethylated H3K27, the latter a histone mark typically associated with gene repression and chromatin compaction (Agger et al. 2007; Margueron et al. 2008; Smith et al. 2008; Herz et al. 2010). Methylation of H3K27 is deposited by Polycomb repressive complex 2 (PRC2) at both enhancers and promoters and subsequently recognized by PRC1 to silence transcription (Margueron and Reinberg 2011; Piunti and Shilatifard 2016). Thus, the WT enzymatic activity of UTX could serve to antagonize Polycomb-mediated repression. In *Dro-*

sophila, ChIP-seq studies found that UTX colocalized with Trr at transcriptional start sites and putative enhancers, with Trr-RNAi-induced loss of UTX resulting in increased trimethylated H3K27 and decreased H3K2ac at enhancer regions (Herz et al. 2012). Therefore, the presence of UTX in the Trr/MLL3/MLL4 branch of the COMPASS family suggests that UTX may act as an enhancer-specific H3K27 demethylase to facilitate the transition from inactive to active enhancers (Herz et al. 2012).

Somatic mutations of UTX are found in various tumor types, including bladder cancer, pancreatic cancer, renal carcinoma, and T-cell acute lymphoblastic leukemia (T-ALL) (van Haaften et al. 2009; Dalgliesh et al. 2010; Gui et al. 2011; Mar et al. 2012; Ntziachristos et al. 2014; Van der Meulen et al. 2015; Waddell et al. 2015). Thus far, the highest frequency of mutations documented to date is in bladder carcinoma (as high as 40%), in which UTX alterations have been found to occur more frequently in early stages and grades of bladder cancer (see cancer.sanger.ac.uk/cosmic; Gui et al. 2011; Gao et al. 2013; Kandoth et al. 2013; Kim et al. 2015). Although mutations are distributed

throughout the UTX protein, there appears to be a higher density of mutations localized around the Jumonji C (JmjC) demethylase domain (Fig. 4) (see cancer.sanger.ac.uk/cosmic). Accordingly, loss of UTX expression has been correlated with poor patient prognosis (Wang et al. 2010). UTX is expressed from the X chromosome and escapes X inactivation in females, while males have a paralog on the Y chromosome named UTY (Greenfield et al. 1998). In several female cancer cell lines, UTX mutations have been reported to be homozygous, while in males there is a tendency for mutation of the single UTX gene to be associated with genomic loss of UTY (van Haaften et al. 2009). This suggests that there is an allelic function of UTY for UTX, although studies have shown the lack of H3K27 demethylase activity of purified UTY (Hong et al. 2007; Lan et al. 2007). Nonetheless, the tendency for biallelic UTX inactivation in females and the two-hit UTX-UTY loss in males contributing to oncogenesis further indicates the potential tumor suppressor function of UTX (van Haaften et al. 2009).

The role of UTX as a cell growth regulator and tumor suppressor is being increasingly ex-

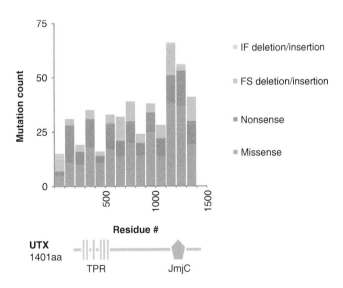

Figure 4. Identified mutations of UTX in cancer. Mutations were also obtained from the COSMIC database (refer to Fig. 3 legend for additional details). The plot was generated similarly as in Figure 3, with domain schematic of UTX to serve as reference. Domain annotation of UTX follows SMART using protein sequence obtained from NCBI (accessed January 8, 2016). TPR, Tetratricopeptide repeat; JmjC, Jumonji C.

plored. We and others showed in *Drosophila* that mutant clones of UTX show increased H3K27 trimethylation and are associated with a major growth advantage compared with the surrounding cells (Herz et al. 2010; Kanda et al. 2013). UTX was also found to control cell cycle and growth by antagonizing Notch signaling and modulating the activity of the known tumor suppressor retinoblastoma (Rb) protein, such that UTX mutations led to Notch overactivation and consequently to tumor-like growth in an Rb-dependent manner (Herz et al. 2010). A separate study using human fibroblasts found that UTX demethylated H3K27 trimethylation at key Rb-binding proteins, thereby maintaining their expression to facilitate cell-cycle arrest (Wang et al. 2010). To test the functional consequence of missense mutations identified in adenoid cystic carcinomas, WT or mutant UTX was overexpressed in HEK293T cells. Overexpression of mutant UTX led to increased cell growth and increased H3K27 trimethylation (Ho et al. 2013). Together, these studies support a tumor suppressor role for UTX from flies to mammals underlying the frequent mutation of UTX in cancer. Importantly, in *Drosophila*, loss of UTX also resulted in a significant reduction of global H3K4 monomethylation indicating that UTX could function at enhancer regions to mediate its tumor suppressor activity (Herz et al. 2010). This is further supported by the co-occurrence in bladder cancer of mutations of UTX and EP300 (p300) or CREBBP (CBP), histone acetyltransferases responsible for H3K27ac at active enhancers (Pasini et al. 2010; Gui et al. 2011).

Two independent teams recently showed that UTX is a tumor suppressor in T-ALL using a *NOTCH1*-induced T-ALL in vivo model. In one study, knockout UTX mice succumbed to the disease more quickly than those with WT UTX (Ntziachristos et al. 2014). The second study found that UTX knockdown significantly accelerated leukemic onset (Van der Meulen et al. 2015). T-ALL is an aggressive form of leukemia that is diagnosed more often in males than females, which may at least be partly attributed to the fact that UTX is expressed from the X chromosome. Indeed, sequencing of patient samples identified inactivating/truncating UTX mutations primarily in samples of male origin (Ntziachristos et al. 2014; Van der Meulen et al. 2015). Furthermore, gene expression profiling of UTX-deficient T-ALL mouse tumors showed that these tumors have downregulated expression of key tumor suppressor genes, strongly pinpointing UTX as a critical tumor suppressor in T-ALL (Ntziachristos et al. 2014; Van der Meulen et al. 2015). In fact, UTX overexpression via a doxycycline-inducible lentiviral system in a T-ALL cell line significantly promoted apoptosis (Ntziachristos et al. 2014).

The recent findings discussed above provide an initial understanding of the tumor suppressive role of UTX in human cancers; however, emerging evidence indicates that this role is cancer subtype- and tissue-specific. In a specific subtype of T-ALL driven by oncogenic transcription factor TAL1, UTX was identified as a coactivator of TAL1 that gets recruited to TAL1-targets to remove H3K27 trimethylation, thereby facilitating the expression of proproliferative and antiapoptotic genes (Benyoucef et al. 2016). Through depletion and overexpression experiments, the investigators indicated a selective oncogenic role of UTX in TAL1-positive T-ALL, but a tumor suppressive role in TAL1-negative T-ALL (Benyoucef et al. 2016). This functional distinction of UTX between molecular subtypes of T-ALL was further supported in patient-derived xenografts, in which treatment with H3K27 demethylase inhibitor GSK-J4 dramatically reduced percentage of human leukemic blasts and splenomegaly in TAL1-positive models, with no effect on TAL1-negative models (Benyoucef et al. 2016). In a separate study using breast cancer cells, aberrant UTX overexpression contributed to cell proliferation, anchorage-independent growth, and invasiveness, and UTX and MLL4 were intriguingly found to coregulate a set of genes linked with proliferation and invasiveness (Kim et al. 2014). Through knockdown and quantitative ChIP experiments, the investigators connected UTX demethylation of H3K27 trimethylation with increased MLL4-dependent H3K4 trimethylation at the promoters of cotarget genes, suggesting that UTX and MLL4 cooccupy the same loci to coordinately activate expression of oncogenes in breast cancer (Kim et al. 2014). Given findings from other studies, it is also possible that UTX and MLL4 may cooperatively control tumorigenic transcriptional programs in breast cancer by regulating enhancers of oncogenes, which in turn activates their associated promoters. Nevertheless, the ability of MLL3/MLL4 and UTX with the COMPASS family to act as tumor suppressors or oncogenes depends on the molecular and cellular circumstances, and reveals a critical need to define the precise mechanism through which MLL3/MLL4 and UTX act in each specific context and how their abnormal changes impact individual tumors.

CONCLUDING REMARKS

MLL3/MLL4 and UTX of the COMPASS family play a key tumor suppressor role across the breadth of cancers, although there are cases in which these chromatin modifiers may have an alternative tumorigenic role that is dependent on the cellular context. At this time, the mechanistic relationship between MLL3/MLL4 and UTX and their regulation of enhancer activity, whether global or local, in development and disease is still unclear. Given the complexity underlying the biological role of these chromatin proteins in cancer pathogenesis, additional research in various developmental processes is necessary to understand the fundamental mechanism and precise functional impact of MLL3/MLL4 and UTX alterations. With the rapid development and advancement of next-generation sequencing methods, determining novel mutations, and distinguishing those that are cooccurring and mutually exclusive in cancer will also be important to elucidate the signaling pathways likely deregulated by alterations of these chromatin-modifying enzymes. These technologies will also enable genome-wide identification of MLL3/MLL4/UTX-dependent enhancers in normal and cancerous conditions, with the potential to characterize enhancers of tumor suppressors and/or oncogenes. Use of groundbreaking tools such as CRISPR-Cas9 will permit further investigation

of these MLL3/MLL4/UTX-dependent enhancers, giving us the precision to genetically edit endogenous loci to examine enhancer function. Together, these studies will ultimately provide critical insights to facilitate the identification of therapeutic opportunities for cancer.

ACKNOWLEDGMENTS

We are grateful to Dr. Edwin Smith for critical reading and comments on this manuscript. C.S.S. is supported, in part, by National Institutes of Health/National Cancer Institute (NIH/NCI) training Grant T32CA09560. Studies in A.S.'s laboratory regarding the role of the COMPASS family in development and cancer are supported by NCI Grant R35CA197569.

REFERENCES

Agger K, Cloos PA, Christensen J, Pasini D, Rose S, Rappsilber J, Issaeva I, Canaani E, Salcini AE, Helin K. 2007. UTX and JMJD3 are histone H3K27 demethylases involved in *HOX* gene regulation and development. *Nature* **449:** 731–734.

Akhtar-Zaidi B, Cowper-Sal-lari R, Corradin O, Saiakhova A, Bartels CF, Balasubramanian D, Myeroff L, Lutterbaugh J, Jarrar A, Kalady MF, et al. 2012. Epigenomic enhancer profiling defines a signature of colon cancer. *Science* **336:** 736–739.

Allis CD, Berger SL, Cote J, Dent S, Jenuwein T, Kouzarides T, Pillus L, Reinberg D, Shi Y, Shiekhattar R, et al. 2007. New nomenclature for chromatin-modifying enzymes. *Cell* **131:** 633–636.

Ang SY, Uebersohn A, Spencer CI, Huang Y, Lee JE, Ge K, Bruneau BG. 2016. KMT2D regulates specific programs in heart development via histone H3 lysine 4 di-methylation. *Development* **143:** 810–821.

Aran D, Sabato S, Hellman A. 2013. DNA methylation of distal regulatory sites characterizes dysregulation of cancer genes. *Genome Biol* **14:** R21.

Ardehali MB, Mei A, Zobeck KL, Caron M, Lis JT, Kusch T. 2011. *Drosophila* Set1 is the major histone H3 lysine 4 trimethyltransferase with role in transcription. *EMBO J* **30:** 2817–2828.

Ashktorab H, Schaffer AA, Daremipouran M, Smoot DT, Lee E, Brim H. 2010. Distinct genetic alterations in colorectal cancer. *PLoS ONE* **5:** e8879.

Banerji J, Rusconi S, Schaffner W. 1981. Expression of a β-globin gene is enhanced by remote SV40 DNA sequences. *Cell* **27:** 299–308.

Benyoucef A, Palii CG, Wang C, Porter CJ, Chu A, Dai F, Tremblay V, Rakopoulos P, Singh K, Huang S, et al. 2016. UTX inhibition as selective epigenetic therapy against TAL1-driven T-cell acute lymphoblastic leukemia. *Genes Dev* **30:** 508–521.

Blackwood EM, Kadonaga JT. 1998. Going the distance: A current view of enhancer action. *Science* **281**: 60–63.

Breen TR, Harte PJ. 1991. Molecular characterization of the trithorax gene, a positive regulator of homeotic gene expression in *Drosophila*. *Mech Dev* **35**: 113–127.

Bulger M, Groudine M. 2011. Functional and mechanistic diversity of distal transcription enhancers. *Cell* **144**: 327–339.

Chen C, Liu Y, Rappaport AR, Kitzing T, Schultz N, Zhao Z, Shroff AS, Dickins RA, Vakoc CR, Bradner JE, et al. 2014. MLL3 is a haploinsufficient 7q tumor suppressor in acute myeloid leukemia. *Cancer cell* **25**: 652–665.

Creyghton MP, Cheng AW, Welstead GG, Kooistra T, Carey BW, Steine EJ, Hanna J, Lodato MA, Frampton GM, Sharp PA, et al. 2010. Histone H3K27ac separates active from poised enhancers and predicts developmental state. *Proc Natl Acad Sci* **107**: 21931–21936.

da Silva Almeida AC, Abate F, Khiabanian H, Martinez-Escala E, Guitart J, Tensen CP, Vermeer MH, Rabadan R, Ferrando A, Palomero T. 2015. The mutational landscape of cutaneous T cell lymphoma and Sézary syndrome. *Nat Genet* **47**: 1465–1470.

Dalgliesh GL, Furge K, Greenman C, Chen L, Bignell G, Butler A, Davies H, Edkins S, Hardy C, Latimer C, et al. 2010. Systematic sequencing of renal carcinoma reveals inactivation of histone modifying genes. *Nature* **463**: 360–363.

Djabali M, Selleri L, Parry P, Bower M, Young BD, Evans GA. 1992. A trithorax-like gene is interrupted by chromosome 11q23 translocations in acute leukaemias. *Nat Genet* **2**: 113–118.

Dorsett D. 1999. Distant liaisons: Long-range enhancer–promoter interactions in *Drosophila*. *Curr Opin Genet Dev* **9**: 505–514.

Dorsett D, Merkenschlager M. 2013. Cohesin at active genes: A unifying theme for cohesin and gene expression from model organisms to humans. *Curr Opin Cell Biol* **25**: 327–333.

Eissenberg JC, Shilatifard A. 2010. Histone H3 lysine 4 (H3K4) methylation in development and differentiation. *Dev Biol* **339**: 240–249.

Ellis MJ, Ding L, Shen D, Luo J, Suman VJ, Wallis JW, Van Tine BA, Hoog J, Goiffon RJ, Goldstein TC, et al. 2012. Whole-genome analysis informs breast cancer response to aromatase inhibition. *Nature* **486**: 353–360.

Forbes SA, Beare D, Gunasekaran P, Leung K, Bindal N, Boutselakis H, Ding M, Bamford S, Cole C, Ward S, et al. 2015. COSMIC: Exploring the world's knowledge of somatic mutations in human cancer. *Nucleic Acids Res* **43**: D805–811.

Gao J, Aksoy BA, Dogrusoz U, Dresdner G, Gross B, Sumer SO, Sun Y, Jacobsen A, Sinha R, Larsson E, et al. 2013. Integrative analysis of complex cancer genomics and clinical profiles using the cBioPortal. *Sci Signal* **6**: pl1.

Gao YB, Chen ZL, Li JG, Hu XD, Shi XJ, Sun ZM, Zhang F, Zhao ZR, Li ZT, Liu ZY, et al. 2014. Genetic landscape of esophageal squamous cell carcinoma. *Nat Genet* **46**: 1097–1102.

Garg M, Nagata Y, Kanojia D, Mayakonda A, Yoshida K, Haridas Keloth S, Zang ZJ, Okuno Y, Shiraishi Y, Chiba K, et al. 2015. Profiling of somatic mutations in acute myeloid leukemia with FLT3-ITD at diagnosis and relapse. *Blood* **126**: 2491–2501.

Grasso CS, Wu YM, Robinson DR, Cao X, Dhanasekaran SM, Khan AP, Quist MJ, Jing X, Lonigro RJ, Brenner JC, et al. 2012. The mutational landscape of lethal castration-resistant prostate cancer. *Nature* **487**: 239–243.

Greenfield A, Carrel L, Pennisi D, Philippe C, Quaderi N, Siggers P, Steiner K, Tam PP, Monaco AP, Willard HF, et al. 1998. The *UTX* gene escapes X inactivation in mice and humans. *Hum Mol Genet* **7**: 737–742.

Gu Y, Nakamura T, Alder H, Prasad R, Canaani O, Cimino G, Croce CM, Canaani E. 1992. The t(4;11) chromosome translocation of human acute leukemias fuses the *ALL-1* gene, related to *Drosophila trithorax*, to the *AF-4* gene. *Cell* **71**: 701–708.

Gui Y, Guo G, Huang Y, Hu X, Tang A, Gao S, Wu R, Chen C, Li X, Zhou L, et al. 2011. Frequent mutations of chromatin remodeling genes in transitional cell carcinoma of the bladder. *Nat Genet* **43**: 875–878.

Hallson G, Hollebakken RE, Li T, Syrzycka M, Kim I, Cotsworth S, Fitzpatrick KA, Sinclair DA, Honda BM. 2012. dSet1 is the main H3K4 di- and tri-methyltransferase throughout *Drosophila* development. *Genetics* **190**: 91–100.

Hanahan D, Weinberg RA. 2011. Hallmarks of cancer: The next generation. *Cell* **144**: 646–674.

Heintzman ND, Stuart RK, Hon G, Fu Y, Ching CW, Hawkins RD, Barrera LO, Van Calcar S, Qu C, Ching KA, et al. 2007. Distinct and predictive chromatin signatures of transcriptional promoters and enhancers in the human genome. *Nat Genet* **39**: 311–318.

Heintzman ND, Hon GC, Hawkins RD, Kheradpour P, Stark A, Harp LF, Ye Z, Lee LK, Stuart RK, Ching CW, et al. 2009. Histone modifications at human enhancers reflect global cell-type-specific gene expression. *Nature* **459**: 108–112.

Henikoff S, Shilatifard A. 2011. Histone modification: Cause or cog? *Trends Genet* **27**: 389–396.

Herz HM, Madden LD, Chen Z, Bolduc C, Buff E, Gupta R, Davuluri R, Shilatifard A, Hariharan IK, Bergmann A. 2010. The H3K27me3 demethylase dUTX is a suppressor of Notch- and Rb-dependent tumors in *Drosophila*. *Mol Cell Biol* **30**: 2485–2497.

Herz HM, Mohan M, Garruss AS, Liang K, Takahashi YH, Mickey K, Voets O, Verrijzer CP, Shilatifard A. 2012. Enhancer-associated H3K4 monomethylation by trithorax-related, the *Drosophila* homolog of mammalian Mll3/Mll4. *Genes Dev* **26**: 2604–2620.

Herz HM, Garruss A, Shilatifard A. 2013. SET for life: Biochemical activities and biological functions of SET domain-containing proteins. *Trends Biochem Sci* **38**: 621–639.

Herz HM, Hu D, Shilatifard A. 2014. Enhancer malfunction in cancer. *Mol Cell* **53**: 859–866.

Ho AS, Kannan K, Roy DM, Morris LG, Ganly I, Katabi N, Ramaswami D, Walsh LA, Eng S, Huse JT, et al. 2013. The mutational landscape of adenoid cystic carcinoma. *Nat Genet* **45**: 791–798.

Cite this article as *Cold Spring Harb Perspect Med* doi: 10.1101/cshperspect.a026427

Hong S, Cho YW, Yu LR, Yu H, Veenstra TD, Ge K. 2007. Identification of JmjC domain-containing UTX and JMJD3 as histone H3 lysine 27 demethylases. *Proc Natl Acad Sci* **104:** 18439–18444.

Hu D, Gao X, Morgan MA, Herz HM, Smith ER, Shilatifard A. 2013a. The MLL3/MLL4 branches of the COMPASS family function as major histone H3K4 monomethylases at enhancers. *Mol Cell Biol* **33:** 4745–4754.

Hu D, Garruss AS, Gao X, Morgan MA, Cook M, Smith ER, Shilatifard A. 2013b. The Mll2 branch of the COMPASS family regulates bivalent promoters in mouse embryonic stem cells. *Nat Struct Mol Biol* **20: 488:** 100–105.

Jones DT, Jager N, Kool M, Zichner T, Hutter B, Sultan M, Cho YJ, Pugh TJ, Hovestadt V, Stutz AM, et al. 2012. Dissecting the genomic complexity underlying medulloblastoma. *Nature* **488:** 100–105.

Kagey MH, Newman JJ, Bilodeau S, Zhan Y, Orlando DA, van Berkum NL, Ebmeier CC, Goossens J, Rahl PB, Levine SS, et al. 2010. Mediator and cohesin connect gene expression and chromatin architecture. *Nature* **467:** 430–435.

Kaikkonen MU, Spann NJ, Heinz S, Romanoski CE, Allison KA, Stender JD, Chun HB, Tough DF, Prinjha RK, Benner C, et al. 2013. Remodeling of the enhancer landscape during macrophage activation is coupled to enhancer transcription. *Mol Cell* **51:** 310–325.

Kanda H, Nguyen A, Chen L, Okano H, Hariharan IK. 2013. The *Drosophila* ortholog of *MLL3* and *MLL4*, *trithorax related*, functions as a negative regulator of tissue growth. *Mol Cell Biol* **33:** 1702–1710.

Kandoth C, McLellan MD, Vandin F, Ye K, Niu B, Lu C, Xie M, Zhang Q, McMichael JF, Wyczalkowski MA, et al. 2013. Mutational landscape and significance across 12 major cancer types. *Nature* **502:** 333–339.

Kantidakis T, Saponaro M, Mitter R, Horswell S, Kranz A, Boeing S, Aygun O, Kelly GP, Matthews N, Stewart A, et al. 2016. Mutation of cancer driver MLL2 results in transcription stress and genome instability. *Genes Dev* **30:** 408–420.

Kim JH, Sharma A, Dhar SS, Lee SH, Gu B, Chan CH, Lin HK, Lee MG. 2014. UTX and MLL4 coordinately regulate transcriptional programs for cell proliferation and invasiveness in breast cancer cells. *Cancer Res* **74:** 1705–1717.

Kim PH, Cha EK, Sfakianos JP, Iyer G, Zabor EC, Scott SN, Ostrovnaya I, Ramirez R, Sun A, Shah R, et al. 2015. Genomic predictors of survival in patients with high-grade urothelial carcinoma of the bladder. *Eur Urol* **67:** 198–201.

Klymenko T, Muller J. 2004. The histone methyltransferases Trithorax and Ash1 prevent transcriptional silencing by Polycomb group proteins. *EMBO Rep* **5:** 373–377.

Krogan NJ, Dover J, Khorrami S, Greenblatt JF, Schneider J, Johnston M, Shilatifard A. 2002. COMPASS, a histone H3 (Lysine 4) methyltransferase required for telomeric silencing of gene expression. *J Biol Chem* **277:** 10753–10755.

Kurdistani SK. 2012. Enhancer dysfunction: How the main regulators of gene expression contribute to cancer. *Genome Biol* **13:** 156.

Lan F, Bayliss PE, Rinn JL, Whetstine JR, Wang JK, Chen S, Iwase S, Alpatov R, Issaeva I, Canaani E, et al. 2007. A

histone H3 lysine 27 demethylase regulates animal posterior development. *Nature* **449:** 689–694.

Lawrence MS, Stojanov P, Mermel CH, Robinson JT, Garraway LA, Golub TR, Meyerson M, Gabriel SB, Lander ES, Getz G. 2014. Discovery and saturation analysis of cancer genes across 21 tumour types. *Nature* **505:** 495–501.

Lee J, Kim DH, Lee S, Yang QH, Lee DK, Lee SK, Roeder RG, Lee JW. 2009. A tumor suppressive coactivator complex of p53 containing ASC-2 and histone H3-lysine-4 methyltransferase MLL3 or its paralogue MLL4. *Proc Natl Acad Sci* **106:** 8513–8518.

Lee JE, Wang C, Xu S, Cho YW, Wang L, Feng X, Baldridge A, Sartorelli V, Zhuang L, Peng W, et al. 2013. H3K4 mono- and di-methyltransferase MLL4 is required for enhancer activation during cell differentiation. *eLife* **2:** e01503.

Letunic I, Doerks T, Bork P. 2015. SMART: Recent updates, new developments and status in 2015. *Nucleic Acids Res* **43:** D257–260.

Levine M, Cattoglio C, Tjian R. 2014. Looping back to leap forward: Transcription enters a new era. *Cell* **157:** 13–25.

Lin DC, Hao JJ, Nagata Y, Xu L, Shang L, Meng X, Sato Y, Okuno Y, Varela AM, Ding LW, et al. 2014. Genomic and molecular characterization of esophageal squamous cell carcinoma. *Nat Genet* **46:** 467–473.

Loven J, Hoke HA, Lin CY, Lau A, Orlando DA, Vakoc CR, Bradner JE, Lee TI, Young RA. 2013. Selective inhibition of tumor oncogenes by disruption of super-enhancers. *Cell* **153:** 320–334.

Mahmoudi T, Verrijzer CP. 2001. Chromatin silencing and activation by Polycomb and trithorax group proteins. *Oncogene* **20:** 3055–3066.

Maniatis T, Goodbourn S, Fischer JA. 1987. Regulation of inducible and tissue-specific gene expression. *Science* **236:** 1237–1245.

Mar BG, Bullinger L, Basu E, Schlis K, Silverman LB, Dohner K, Armstrong SA. 2012. Sequencing histone-modifying enzymes identifies UTX mutations in acute lymphoblastic leukemia. *Leukemia* **26:** 1881–1883.

Margueron R, Reinberg D. 2011. The Polycomb complex PRC2 and its mark in life. *Nature* **469:** 343–349.

Margueron R, Li G, Sarma K, Blais A, Zavadil J, Woodcock CL, Dynlacht BD, Reinberg D. 2008. Ezh1 and Ezh2 maintain repressive chromatin through different mechanisms. *Mol Cell* **32:** 503–518.

Miller T, Krogan NJ, Dover J, Erdjument-Bromage H, Tempst P, Johnston M, Greenblatt JF, Shilatifard A. 2001. COMPASS: A complex of proteins associated with a trithorax-related SET domain protein. *Proc Natl Acad Sci* **98:** 12902–12907.

Mohan M, Herz HM, Smith ER, Zhang Y, Jackson J, Washburn MP, Florens L, Eissenberg JC, Shilatifard A. 2011. The COMPASS family of H3K4 methylases in *Drosophila*. *Mol Cell Biol* **31:** 4310–4318.

Morgan MA, Shilatifard A. 2015. Chromatin signatures of cancer. *Genes Dev* **29:** 238–249.

Morin RD, Mendez-Lago M, Mungall AJ, Goya R, Mungall KL, Corbett RD, Johnson NA, Severson TM, Chiu R, Field M, et al. 2011. Frequent mutation of histone-modifying genes in non-Hodgkin lymphoma. *Nature* **476:** 298–303.

Ntziachristos P, Tsirigos A, Welstead GG, Trimarchi T, Bakogianni S, Xu L, Loizou E, Holmfeldt L, Strikoudis A, King B, et al. 2014. Contrasting roles of histone 3 lysine 27 demethylases in acute lymphoblastic leukaemia. *Nature* **514:** 513–517.

Ortega-Molina A, Boss IW, Canela A, Pan H, Jiang Y, Zhao C, Jiang M, Hu D, Agirre X, Niesvizky I, et al. 2015. The histone lysine methyltransferase KMT2D sustains a gene expression program that represses B cell lymphoma development. *Nat Med* **21:** 1199–1208.

Parsons DW, Li M, Zhang X, Jones S, Leary RJ, Lin JC, Boca SM, Carter H, Samayoa J, Bettegowda C, et al. 2011. The genetic landscape of the childhood cancer medulloblastoma. *Science* **331:** 435–439.

Pasini D, Malatesta M, Jung HR, Walfridsson J, Willer A, Olsson L, Skotte J, Wutz A, Porse B, Jensen ON, et al. 2010. Characterization of an antagonistic switch between histone H3 lysine 27 methylation and acetylation in the transcriptional regulation of Polycomb group target genes. *Nucleic Acids Res* **38:** 4958–4969.

Pasqualucci L, Trifonov V, Fabbri G, Ma J, Rossi D, Chiarenza A, Wells VA, Grunn A, Messina M, Elliot O, et al. 2011. Analysis of the coding genome of diffuse large B-cell lymphoma. *Nat Genet* **43:** 830–837.

Pirrotta V. 1998. Polycombing the genome: PcG, trxG, and chromatin silencing. *Cell* **93:** 333–336.

Piunti A, Shilatifard A. 2016. Epigenetic balance of gene expression by Polycomb and COMPASS families. *Science* **352:** aad9780.

Poux S, Horard B, Sigrist CJ, Pirrotta V. 2002. The *Drosophila* trithorax protein is a coactivator required to prevent re-establishment of polycomb silencing. *Development* **129:** 2483–2493.

Pugh TJ, Weeraratne SD, Archer TC, Pomeranz Krummel DA, Auclair D, Bochicchio J, Carneiro MO, Carter SL, Cibulskis K, Erlich RL, et al. 2012. Medulloblastoma exome sequencing uncovers subtype-specific somatic mutations. *Nature* **488:** 106–110.

Rada-Iglesias A, Bajpai R, Swigut T, Brugmann SA, Flynn RA, Wysocka J. 2011. A unique chromatin signature uncovers early developmental enhancers in humans. *Nature* **470:** 279–283.

Roguev A, Schaft D, Shevchenko A, Pijnappel WW, Wilm M, Aasland R, Stewart AF. 2001. The *Saccharomyces cerevisiae* Set1 complex includes an Ash2 homologue and methylates histone 3 lysine 4. *EMBO J* **20:** 7137–7148.

Ruault M, Brun ME, Ventura M, Roizes G, De Sario A. 2002. *MLL3*, a new human member of the *TRX/MLL* gene family, maps to 7q36, a chromosome region frequently deleted in myeloid leukaemia. *Gene* **284:** 73–81.

Schneider J, Wood A, Lee JS, Schuster R, Dueker J, Maguire C, Swanson SK, Florens L, Washburn MP, Shilatifard A. 2005. Molecular regulation of histone H3 trimethylation by COMPASS and the regulation of gene expression. *Mol Cell* **19:** 849–856.

Schultz J, Milpetz F, Bork P, Ponting CP. 1998. SMART, a simple modular architecture research tool: Identification of signaling domains. *Proc Natl Acad Sci* **95:** 5857–5864.

Sedkov Y, Benes JJ, Berger JR, Riker KM, Tillib S, Jones RS, Mazo A. 1999. Molecular genetic analysis of the *Drosophila trithorax*–related gene which encodes a novel SET domain protein. *Mech Dev* **82:** 171–179.

Shilatifard A. 2012. The COMPASS family of histone H3K4 methylases: Mechanisms of regulation in development and disease pathogenesis. *Annu Rev Biochem* **81:** 65–95.

Smith E, Shilatifard A. 2014. Enhancer biology and enhanceropathies. *Nat Struct Mol Biol* **21:** 210–219.

Smith ER, Lee MG, Winter B, Droz NM, Eissenberg JC, Shiekhattar R, Shilatifard A. 2008. *Drosophila* UTX is a histone H3 Lys27 demethylase that colocalizes with the elongating form of RNA polymerase II. *Mol Cell Biol* **28:** 1041–1046.

Stassen MJ, Bailey D, Nelson S, Chinwalla V, Harte PJ. 1995. The *Drosophila* trithorax proteins contain a novel variant of the nuclear receptor type DNA binding domain and an ancient conserved motif found in other chromosomal proteins. *Mech Dev* **52:** 209–223.

Sur IK, Hallikas O, Vaharautio A, Yan J, Turunen M, Enge M, Taipale M, Karhu A, Aaltonen LA, Taipale J. 2012. Mice lacking a *Myc* enhancer that includes human SNP rs6983267 are resistant to intestinal tumors. *Science* **338:** 1360–1363.

Tan J, Ong CK, Lim WK, Ng CC, Thike AA, Ng LM, Rajasegaran V, Myint SS, Nagarajan S, Thangaraju S, et al. 2015. Genomic landscapes of breast fibroepithelial tumors. *Nat Genet* **47:** 1341–1345.

Tkachuk DC, Kohler S, Cleary ML. 1992. Involvement of a homolog of *Drosophila* trithorax by 11q23 chromosomal translocations in acute leukemias. *Cell* **71:** 691–700.

Tschiersch B, Hofmann A, Krauss V, Dorn R, Korge G, Reuter G. 1994. The protein encoded by the *Drosophila* position-effect variegation suppressor gene Su(var)3-9 combines domains of antagonistic regulators of homeotic gene complexes. *EMBO J* **13:** 3822–3831.

Van der Meulen J, Sanghvi V, Mavrakis K, Durinck K, Fang F, Matthijssens F, Rondou P, Rosen M, Pieters T, Vandenberghe P, et al. 2015. The H3K27me3 demethylase UTX is a gender-specific tumor suppressor in T-cell acute lymphoblastic leukemia. *Blood* **125:** 13–21.

van Haaften G, Dalgliesh GL, Davies H, Chen L, Bignell G, Greenman C, Edkins S, Hardy C, O'Meara S, Teague J, et al. 2009. Somatic mutations of the histone H3K27 demethylase gene *UTX* in human cancer. *Nat Genet* **41:** 521–523.

van Nuland R, Smits AH, Pallaki P, Jansen PW, Vermeulen M, Timmers HT. 2013. Quantitative dissection and stoichiometry determination of the human SET1/MLL histone methyltransferase complexes. *Mol Cell Biol* **33:** 2067–2077.

Waddell N, Pajic M, Patch AM, Chang DK, Kassahn KS, Bailey P, Johns AL, Miller D, Nones K, Quek K, et al. 2015. Whole genomes redefine the mutational landscape of pancreatic cancer. *Nature* **518:** 495–501.

Wang P, Lin C, Smith ER, Guo H, Sanderson BW, Wu M, Gogol M, Alexander T, Seidel C, Wiedemann LM, et al. 2009. Global analysis of H3K4 methylation defines MLL family member targets and points to a role for MLL1-mediated H3K4 methylation in the regulation of transcriptional initiation by RNA polymerase II. *Mol Cell Biol* **29:** 6074–6085.

Wang JK, Tsai MC, Poulin G, Adler AS, Chen S, Liu H, Shi Y, Chang HY. 2010. The histone demethylase UTX enables RB-dependent cell fate control. *Genes Dev* **24:** 327–332.

Wu M, Wang PF, Lee JS, Martin-Brown S, Florens L, Washburn M, Shilatifard A. 2008. Molecular regulation of H3K4 trimethylation by Wdr82, a component of human Set1/COMPASS. *Mol Cell Biol* **28:** 7337–7344.

Zentner GE, Tesar PJ, Scacheri PC. 2011. Epigenetic signatures distinguish multiple classes of enhancers with distinct cellular functions. *Genome Res* **21:** 1273–1283.

Zhang J, Dominguez-Sola D, Hussein S, Lee JE, Holmes AB, Bansal M, Vlasevska S, Mo T, Tang H, Basso K, et al. 2015. Disruption of *KMT2D* perturbs germinal center B cell development and promotes lymphomagenesis. *Nat Med* **21:** 1190–1198.

Zhu J, Sammons MA, Donahue G, Dou Z, Vedadi M, Getlik M, Barsyte-Lovejoy D, Al-awar R, Katona BW, Shilatifard A, et al. 2015. Gain-of-function p53 mutants co-opt chromatin pathways to drive cancer growth. *Nature* **525:** 206–211.

Ziemin-van der Poel S, McCabe NR, Gill HJ, Espinosa R 3rd, Patel Y, Harden A, Rubinelli P, Smith SD, LeBeau MM, Rowley JD, et al. 1991. Identification of a gene, *MLL*, that spans the breakpoint in 11q23 translocations associated with human leukemias. *Proc Natl Acad Sci* **88:** 10735–10739.

The Many Roles of BAF (mSWI/SNF) and PBAF Complexes in Cancer

Courtney Hodges, Jacob G. Kirkland, and Gerald R. Crabtree

Departments of Pathology, Developmental Biology, and Genetics, Howard Hughes Medical Institute, Stanford University School of Medicine, Stanford, California 94305

Correspondence: crabtree@stanford.edu

During the last decade, a host of epigenetic mechanisms were found to contribute to cancer and other human diseases. Several genomic studies have revealed that ~20% of malignancies have alterations of the subunits of polymorphic BRG-/BRM-associated factor (BAF) and Polybromo-associated BAF (PBAF) complexes, making them among the most frequently mutated complexes in cancer. Recurrent mutations arise in genes encoding several BAF/PBAF subunits, including *ARID1A, ARID2, PBRM1, SMARCA4,* and *SMARCB1*. These subunits share some degree of conservation with subunits from related adenosine triphosphate (ATP)-dependent chromatin remodeling complexes in model organisms, in which a large body of work provides insight into their roles in cancer. Here, we review the roles of BAF- and PBAF-like complexes in these organisms, and relate these findings to recent discoveries in cancer epigenomics. We review several roles of BAF and PBAF complexes in cancer, including transcriptional regulation, DNA repair, and regulation of chromatin architecture and topology. More recent results highlight the need for new techniques to study these complexes.

EPIGENOMICS IN CANCER

Broadly defined, epigenetic factors contribute to the expression state of the genome by regulating heritable changes in gene expression independently of the DNA sequence. Chromatin-based epigenetic regulation occurs through a wide variety of mechanisms, including physical compaction and exclusion, recruitment of transcription machinery, or covalent modification of DNA and histones. Such features constitute the heritable physicochemical state of the genetic material, and are jointly referred to as the epigenetic landscape. The combinatorial regulation of these features represents the full spectrum of achievable cell-type diversity for the organism. Because epigenetic regulation contributes to cell-type functional specialization, it is essential for multicellular life.

An important component of the epigenetic state is the regulation provided by adenosine triphosphate (ATP)-dependent chromatin remodelers, which use ATP to physically remodel histones and other factors on chromatin. As we review below, genomic studies of primary tumors and cancer cell lines have revealed that ATP-dependent chromatin remodelers are among the most frequently disrupted genes in cancer. Because several important components of ATP-dependent chromatin remodelers are conserved between yeast, flies, and humans (Fig. 1), the basic research on chromatin remodeling performed in model organisms is taking on new relevance for disease biology. In many cases, fundamental observations from yeast and flies directly support our understanding of the role of chromatin remodelers in cancer; in other cases, the differences between these organisms and humans highlight important gaps in our knowledge of disease mechanisms.

In this review, we focus on the family of BRG-/BRM-associated factor ([BAF] or mSWI/SNF) and Polybromo-associated BAF (PBAF) complexes, whose subunits have been identified as major tumor suppressors in several malignancies (Davoli et al. 2013; Kadoch et al. 2013; Shain and Pollack 2013). Here, we relate the fundamental biology revealed by genetics, structural biology, and microscopy to the fast-moving field of cancer epigenomics. As we discuss below, the mechanisms revealed by fundamental studies inform our understanding of how epigenetic dysfunction contributes to cancer.

SWI/SNF AND RSC IN *Saccharomyces cerevisiae*

The SWI/SNF Complex

ATP-dependent chromatin remodelers were independently discovered in yeast, by screening for mutations that disrupt the ability of yeast to switch mating type (Stern et al. 1984) or activate sucrose fermentation pathways (Carlson et al. 1981; Neigeborn and Carlson 1984, 1987), both in response to extrinsic cues. Later work showed that many of these genes act in concert through a common complex that regulated transcription, termed SWI/SNF to honor both discoveries (Peterson and Herskowitz 1992; Winston and Carlson 1992; Cairns et al. 1994; Peterson et al. 1994). The observation that histone mutants were able to reverse the phenotypic defects associated with SWI/SNF mutation (Sternberg et al. 1987; Kruger et al. 1995) indicated that regulation of chromatin structure was the central function of the SWI/SNF complex. In vitro, ATP-dependent remodeling activity induces changes of position, phasing,

stability, or histone content of nucleosomes, and is well described in other reviews (Becker and Horz 2002; Narlikar et al. 2013). However, as we discuss below, SWI/SNF-like complexes have rich and biologically diverse regulatory roles in vivo that arise through mechanisms that are not entirely clear.

In yeast, SWI/SNF is a ~1.15-MDa protein complex (Smith et al. 2003) composed of Swi1, Snf2, Swi3, Snf5, Snf6, along with Swp- and actin-related proteins (ARPs) (Fig. 1). Most subunits of the complex, including the ATPase Snf2, are present as single copies in the complex, whereas several others integrate in multiple copies (two copies of Swi3, Swp82, Snf6, and Snf11, and three copies of Swp29) (Smith et al. 2003). Many of these subunits are required for the complex's biological activity and, in some cases, its biochemical stability (Estruch and Carlson 1990; Richmond and Peterson 1996). The complex's direct interaction with nucleosomal DNA is mediated by the catalytic subunit Snf2, whereas other subunits, such as Snf5, do not interact with nucleosomal DNA but instead contact the histone octamer (Dechassa et al. 2008).

Cells with SWI/SNF subunit mutations have disrupted chromatin structures, and fail to express many genes, leading to diverse phenotypic defects. As a result, several aspects of the complex's biological activity are illustrated by genetic deletion of its subunits.

Failure to activate gene expression affects several downstream processes. As an example, cells lacking the central ATPase Snf2 are viable but have impaired mating-type switching because of the inability to express the HO endonuclease needed for the process. Proper expression of HO depends not only on Snf2, but also Snf1 and Swi3 (Stern et al. 1984). In addition to the effects on sucrose metabolism, *snf2Δ* cells also have impaired sporulation. However, SWI/SNF activity is not uniformly activating. Although Snf2 plays a role in activation of many genes, it also is required for silencing of genes at rDNA and telomeric loci, either by direct or indirect means (Dror and Winston 2004; Manning and Peterson 2014).

SWI/SNF subunits also have important functions in maintaining proper chromatin-

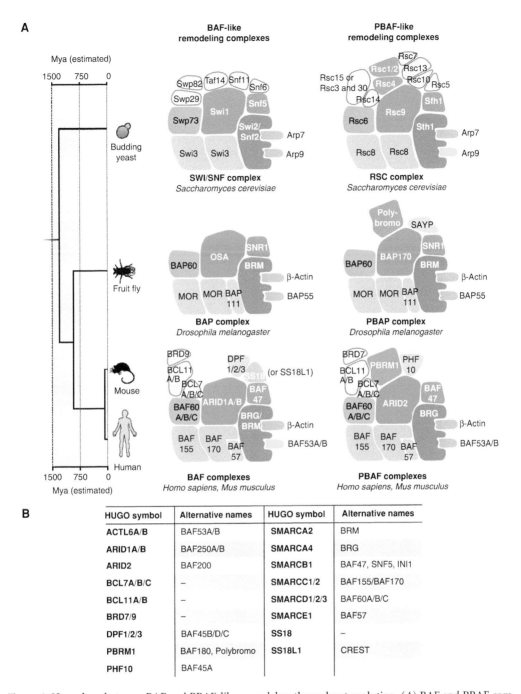

Figure 1. Homology between BAF and PBAF-like remodelers throughout evolution. (*A*) BAF and PBAF complexes in mammals share several features with Brahma-associated proteins (BAPs) and Polybromo-associated BAP (PBAP) complexes (*Drosophila melanogaster*), and SWI/SNF and RSC complexes (*Saccharomyces cerevisiae*), respectively. The similarities and differences between these complexes throughout evolution provide insight into their biological regulation and their roles in cancer. BAF/PBAF subunits labeled in a boldface white font have important roles in malignancy. Time since species divergence was estimated using TimeTree (Hedges et al. 2006), and plotted as a function of time, millions of years ago (Mya). (*B*) Summary of BAF and PBAF subunits and alternative names used in the text. In some cases, abbreviated names rather than the official human genome organization (HUGO) symbols are used in the text because of space constraints.

modification domains. In a reporter strain that lacks a native transfer RNA (tRNA) insulator element, nucleosomes occupy a region from which they are normally depleted, leading to loss of insulator function (Oki et al. 2004). In this system, artificial recruitment of Snf5 or Snf6 restores the nucleosome-depleted region over which heterochromatin marks cannot spread, thereby rescuing barrier function.

SWI/SNF mutations also cause increased sensitivity to DNA-damaging agents, including hydroxyurea, cisplatin, methyl methanesulfonate, and ultraviolet (UV) light (Birrell et al. 2001; Chai et al. 2005; Xia et al. 2007). These defects may arise because SWI/SNF subunits have roles in nucleotide excision repair (Gong et al. 2006) and DNA double-strand break (DSB) repair through the homologous recombination (HR) pathway (Chai et al. 2005). SWI/SNF defective cells also show increased sensitivity to the topoisomerase II inhibitor daunorubicin (Xia et al. 2007).

Because transcription, DNA repair, and chromatin modification domains are each influenced by the availability of accessible DNA, disruption of the nucleosome mobilization activity of SWI/SNF has distinct and pleiotropic effects.

The RSC Complex

In budding yeast, *STH1* codes for the ATPase subunit of the RSC complex, which also has the capacity to remodel the structure of chromatin. In addition to Sth1, whose ATPase domain is functionally interchangeable with Snf2 (Laurent et al. 1993), the RSC complex also has Rsc4, Rsc6, Rsc8, Rsc9, Sfh1, and several other dedicated subunits (Fig. 1). In budding yeast, there are two distinct RSC complexes, containing either of the two paralogs Rsc1 or Rsc2 that arose from gene duplication. Both subunits along with Rsc4 contain bromodomains, which interact with acetylated lysines. RSC complexes are ~10-fold more abundant that SWI/SNF (Cairns et al. 1996, 1999), which may explain why RSC complexes are essential but SWI/SNF is not. Electron microscopic (EM) reconstructions of the yeast RSC complex show a lobed

~1.3-MDa structure of similar scale to the SWI/SNF complex (Asturias et al. 2002; Leschziner et al. 2007; Chaban et al. 2008).

RSC remodels nucleosomes throughout the genome, regulating the positions and densities of histones near the promoters of genes transcribed by RNA polymerase II (Pol II), as well as genes transcribed by Pol III (Parnell et al. 2008; Hartley and Madhani 2009), and its activity affects the transcription state of both classes of genes.

Rsc2 is required for insulator boundary function at the *HMR* locus, and its mutation leads to a loss of the nucleosome-depleted region encompassed by the insulator (Dhillon et al. 2009). Additionally, loss of Rsc2 impairs HR and nonhomologous end joining (NHEJ), the two DSB-repair pathways, along with repair of DNA damaged by UV light (Chai et al. 2005; Shim et al. 2005; Srivas et al. 2013). Rsc2 is present at kinetochores and is required for proper sister chromosome cohesion and chromosome segregation (Hsu et al. 2003; Baetz et al. 2004), as well as maintenance of telomeres (Askree et al. 2004).

Brahma-Associated Protein (BAP)
AND Polybromo-Associated BAP (PBAP)
IN *Drosophila melanogaster*

The BAP Complex

Complexes similar to yeast SWI/SNF were discovered in *Drosophila* based on their ability to oppose *Polycomb* repressive activity (Kennison and Tamkun 1988; Tamkun et al. 1992). The central ATPase of this complex is the gene product of *brahma* (*brm*), giving rise to the name BAP complex. The BAP complex is defined as containing OSA and lacking BAP170, SAYP, and POLYBROMO (Fig. 1) (Mohrmann et al. 2004; Bouazoune and Brehm 2006; Chalkley et al. 2008).

In flies, the activating Trithorax group genes generally oppose the repressive activity of Polycomb group genes. Misregulation of developmental genes results in aberrant morphologies and ectopic locations of body parts. Proteins encoded by *brahma*, *osa*, and *moira* (*mor*) are members of the Trithorax group, a set of factors that oppose the repressive activity of Polycomb-group proteins (Kennison and Tamkun 1988; Tamkun et al. 1992; Papoulas et al. 1998; Collins et al. 1999; Crosby et al. 1999; Kal et al. 2000; Simon and Tamkun 2002; Kingston and Tamkun 2014).

Although Polycomb genes are conserved in animals, plants, and some fungi (Shaver et al. 2010), unicellular model yeasts lack Polycomb, suggesting that multicellular organisms have greater needs for repressive factors (as well as their regulators) for lineage-specific functions during development. As discussed below, failure to oppose Polycomb repressive activity in mammals plays an important role in malignancy. Nevertheless, despite this important regulatory role (likely present in the last common eukaryotic ancestor), the precise mechanisms of Polycomb opposition remain murky.

The PBAP Complex

Drosophila have a second BRM-containing complex, named PBAP. The subunit compositions of the BAP and PBAP complexes bear similarities to the functional specialization between SWI/SNF and RSC in yeast. PBAP complexes lack OSA and instead contain BAP170, SAYP, and POLYBROMO (Mohrmann et al. 2004; Bouazoune and Brehm 2006; Chalkley et al. 2008). Interestingly, BAP and PBAP subunits both genetically oppose Polycomb-mediated silencing, without regard to whether they are common to BAP and PBAP, or exclusive to one of the complexes (Kennison and Tamkun 1988; Tamkun et al. 1992; Papoulas et al. 1998; Collins et al. 1999; Crosby et al. 1999; Kal et al. 2000; Simon and Tamkun 2002). On the other hand, the complexes can also have distinct or even opposing functional roles. For example, PBAP but not BAP is required for germinal stem-cell maintenance (He et al. 2014). Moreover, BAP and PBAP have opposing roles in *Egfr* expression in wing development; although BAP positively regulates *Egfr* expression (Molnar et al. 2006; Terriente-Felix and de Celis 2009), PBAP instead negatively regulates *Egfr* (Rendina et al. 2010). Together, these observations

suggest that mutations in different subunits or complexes may result in distinct or even opposing changes to the genomic landscape, a fact that complicates straightforward predictions of their effects.

One clue regarding the distinct functions of BAP and PBAP complexes comes from microscopic examination of polytene chromosomes. Polytene chromosomes arise from successive rounds of replication without cell division, resulting in many copies of aligned condensed sister chromatids (Balbiani 1881). Visualization of chromatin domains using immunofluorescence shows that BAP and PBAP complexes have both overlapping and mutually exclusive domains. Polycomb domains are largely not found at both the overlapping and mutually exclusive BAP/PBAP domains (Armstrong et al. 2002; Mohrmann et al. 2004; Moshkin et al. 2007). This pattern suggests that BAP and PBAP work both cooperatively and independently at distinct sites to oppose Polycomb silencing.

BAF AND PBAF IN MAMMALS

BAF Complexes

In mammals, highly polymorphic BAF complexes (Wang et al. 1996a,b) are composed of a single central ATPase, either BRG (SMARCA4) or BRM (SMARCA2), and several BRG-/BRM-associated factors (BAF subunits) (Khavari et al. 1993). In addition to the subunits homologous to those in *Drosophila* or yeast, several other subunits appear to be dedicated to vertebrate or mammalian complexes, including SS18/SS18L1, BCL7A/B/C, BCL11/A/B, and BRD9 (see Fig. 1). Scanning force microscopy of BAF complexes isolated from HeLa cells show objects with similar appearances and dimensions as yeast SWI/SNF or RSC (Schnitzler et al. 2001); however, BAF subunits are frequently inactivated in long-term cell lines, and so caution is warranted when inferring the complex's characteristics based on immortalized cancer lines.

ChIP-seq studies in many cell types show that BAF complexes bind 20,000–40,000 sites genome-wide, with broad binding sites sometimes spanning 2–5 kbp, suggesting that more than one complex may operate at a given site (Ho et al. 2009a; Euskirchen et al. 2011). BAF complexes have many roles in development (Ho and Crabtree 2010); the presence of BAF complexes on chromatin correlates with enhancers (Rada-Iglesias et al. 2011), and its activity regulates a variety of important biological processes ranging from self-renewal and pluripotency in embryonic stem cells (Ho et al. 2009b), to cardiac development (Lickert et al. 2004), and neural differentiation (Yoo et al. 2009). BAF activity and transcription factor (TF) binding appear to be coupled idiosyncratically, as examples can be found in which TF binding requires BAF activity (Ho et al. 2011; Bao et al. 2015) or, alternatively, in which recruitment of BAF requires existing TF binding (Liu et al. 2001). Some of the complexes' subunits are tissue-specific; for example, BAF53B (ACTL6B), BAF45B (DPF1), and SS18L1 (CREST, a Ca^{2+}-responsive regulator), are found only in BAF complexes of mature, postmitotic neurons (Olave et al. 2002; Aizawa et al. 2004; Lessard et al. 2007; Staahl et al. 2013). BAF subunit composition is subject to tight regulation, as miRNA-based repression of *BAF53A* occurs either before or coincident with the last mitotic division of neurons (Yoo et al. 2009), and failure to express neural-specific subunits like BAF53B leads to defects in synaptogenesis and dendritic outgrowth (Lessard et al. 2007; Vogel-Ciernia et al. 2013). BAF subunit composition also contributes substantially to cell reprogramming (Singhal et al. 2010; Yoo et al. 2011), an instructive effect seemingly incompatible with simple mechanisms of nucleosome mobilization.

Before the modern tumor-sequencing era, the frequent absence of core BAF subunits in immortalized cell lines prompted early speculation that BAF subunits were tumor suppressors (Dunaief et al. 1994). Screening for *BRG* mutations revealed widespread defects in a number of different cancer cell lines, and ectopic expression of *BRG* in these lines often results in altered morphology (Wong et al. 2000). Moreover, many cell lines down-regulate both BRG and BRM ATPases (Reisman et al. 2002). In cultured cells, BAF complexes missing the core ATPase fail to bind the tumor suppressor RB1 and sup-

press E2F1 (Dunaief et al. 1994; Trouche et al. 1997), although it remains uncertain whether this feature reflects its central role in primary tumors. Moreover, as described below, several specific malignancies are driven entirely by BAF subunit dysfunction. Abundant evidence now shows that BAF complexes act as major tumor suppressors.

PBAF Complexes

PBAF complexes were first discovered by Tjian and colleagues in a search for factors that activated ligand-mediated transcription on nucleosomal templates (Lemon et al. 2001). PBAF complexes contain PBRM1 and ARID2 but lack ARID1A/B. PBAF complexes also contain BRD7 in place of BRD9 (Kaeser et al. 2008), BAF45A (PHF10) instead of BAF45B/C/D (DPF1/3/2), and lack SS18 (see Fig. 1) (Middeljans et al. 2012). EM reconstruction of PBAF complexes from HeLa cells show heterogeneous >1-MDa structures with similarities to RSC from yeast (Leschziner et al. 2005); however, because BAF and PBAF complexes are combinatorially assembled, the origin of the observed heterogeneity of these structures remains uncertain. Several subunits share some homology with subunits of the yeast RSC complex, and like RSC, PBAF complexes show significant occupancy at the kinetochores of mitotic chromosomes (Xue et al. 2000), suggesting an important conserved role in cell division.

PBAF subunits regulate cell differentiation and may be an important regulator of cell-type identity (Bajpai et al. 2010; Xu et al. 2012). Additionally, a large body of evidence shows that PBAF complexes have important roles in the maintenance of genomic integrity during mitosis, described in more detail below (see section on Nontranscriptional Roles of BAF/PBAF Complexes in Cancer).

BAF AND PBAF SUBUNITS ARE FREQUENTLY DISRUPTED IN CANCER

In mammals, 28 genes have been discovered to date with close sequence homology with the yeast Snf2 ATPase (Fig. 2A). Despite sharing

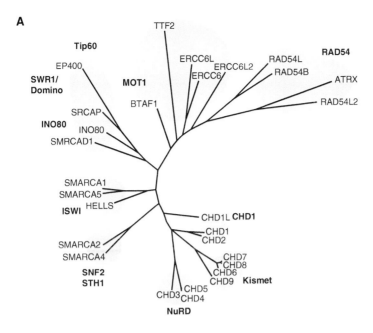

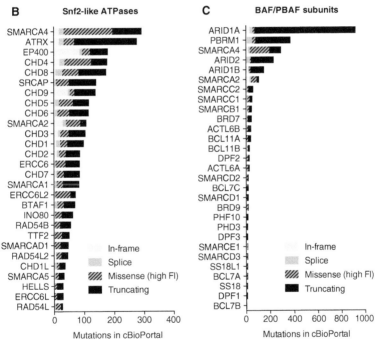

Figure 2. The family of human Snf2-like ATPases and their differing roles in cancer. (*A*) Human Snf2-like ATPases cluster into groups based on degree of sequence similarity. The chromatin remodelers from model organisms are shown near these groups in bold. Radial dendrogram constructed using TreeDyn (Chevenet et al. 2006). (*B*) Human Snf2-like ATPases are mutated at different frequencies across all cancer types. The total number of mutations appearing in cBioPortal (including public datasets from The Cancer Genome Atlas (TCGA), Cancer Cell Line Encyclopedia (CCLE), and others cited in the text) is summed for each gene and presented by the type of mutation. Missense mutations predicted to have neutral, low, or medium functional impact are not shown because of the unknown nature of their effects and increased likelihood to be background mutations. (*C*) The number of mutations of each BAF/PBAF subunit is presented as in *B*. *ARID1A* and *PBRM1* frequently undergo truncating mutation, but *BRG* (*SMARCA4*) frequently has missense mutations with high functional impact.

highly conserved Snf2-like ATPase domains, these ATPases play distinct biological roles and are nearly all functionally nonredundant between the different remodeling families, as reviewed elsewhere (Clapier and Cairns 2009). Based on their distinct biological activities, disruption of each of these remodelers is under a different selection pressure in cancer, resulting in a wide range of mutation frequencies (Fig. 2B). Across all cancer types, ~20% of human malignancies have defects in BAF-related complexes (Kadoch et al. 2013; Shain and Pollack 2013) making them among the most frequently mutated chromatin regulatory complexes in malignancy.

In addition to their ATPases BRG and BRM (Khavari et al. 1993), BAF and PBAF complexes contain a number of noncatalytic subunits that contribute to targeting of the complex to cognate loci, or have other unknown functions. BAF and PBAF complexes collectively contain eight bromodomains (six on PBRM1, one on either BRG or BRM, and one on BRD7 or BRD9), a region homologous with chromodomains (BAF155/170), two PHD finger proteins (BAF45 subunits), a large number of zinc finger and other DNA-binding domains that bind distinct architectural features such as AT-rich sequences or HMG recognition features (Wang et al. 1996a,b, 1998; Lessard et al. 2007). Some of these subunits are among the most frequently mutated genes in cancer, and highly subunit-specific mutation patterns contribute to different cancer types (Figs. 2C and 3). Below, we summarize the role of these subunits in the complex, and discuss their contribution to malignancy.

BRG (*SMARCA4*) Is Mutated in Many Different Malignancies

As defined by the overall number of truncating and high-functional-impact mutations, *BRG* is the most frequently mutated Snf2-like chromatin remodeling ATPase in cancer (Fig. 2C). Unlike many other tumor suppressors, hypermethylation and silencing of *BRG* is reported to be relatively uncommon (Medina et al. 2004; Ramos et al. 2014). However, heterozygous

and biallelic inactivation of *BRG* occurs in tumors of the breast (The Cancer Genome Atlas 2012b), lung (The Cancer Genome Atlas 2014c), stomach (The Cancer Genome Atlas 2014a), bladder (The Cancer Genome Atlas 2014b), colon (The Cancer Genome Atlas 2012a), and in several other tumor types and cell lines (Wong et al. 2000). Disruption of *BRG* is especially common in small cell ovarian cancer (90%–100%) (Jelinic et al. 2014; Ramos et al. 2014), cancers of the skin (up to 27%) (Hodis et al. 2012; Li et al. 2015; Shain et al. 2015; The Cancer Genome Atlas 2015), diffuse large B-cell lymphoma (10%), and non-small-cell lung cancers (~11%) (Imielinski et al. 2012; Rizvi et al. 2015), in which it has been reported as the fifth most frequently mutated gene (Medina et al. 2004). In some specific malignancies, such as certain thoracic sarcomas (Le Loarer et al. 2015), biallelic inactivation of *BRG* occurs at elevated frequencies. Although it was initially thought that in most cancers *BRG* mutations were generally homozygous (Medina and Sanchez-Cespedes 2008; Medina et al. 2008), it has since been determined that, in many cancer types, a large number of mutations of *BRG* are heterozygous, with many mutations clustering at conserved motifs of the ATPase domain. Accordingly, CRISPR-Cas9 tiling experiments have shown that the ATPase domain contains the most functionally important domain of BRG (Shi et al. 2015).

The ATPase domain of Snf2-like remodelers is composed of two conserved subdomains. The amino-terminal ATPase subdomain of BRG contains several residues highly conserved within the SF2 helicase superfamily (Jankowsky and Fairman 2007; Fairman-Williams et al. 2010). Based on crystal structures of homologous Snf2-like proteins (Durr et al. 2005; Thoma et al. 2005; Wollmann et al. 2011), many of these residues are predicted to contact ATP (Walker et al. 1982) or communicate the strain of ATP binding and hydrolysis to the site of DNA binding to exert mechanical force (Banroques et al. 2008), and are frequently mutated in a number of malignancies (Figs. 3 and 4). Some effects of these mutations have been characterized. For example, K785R and T910M, respectively,

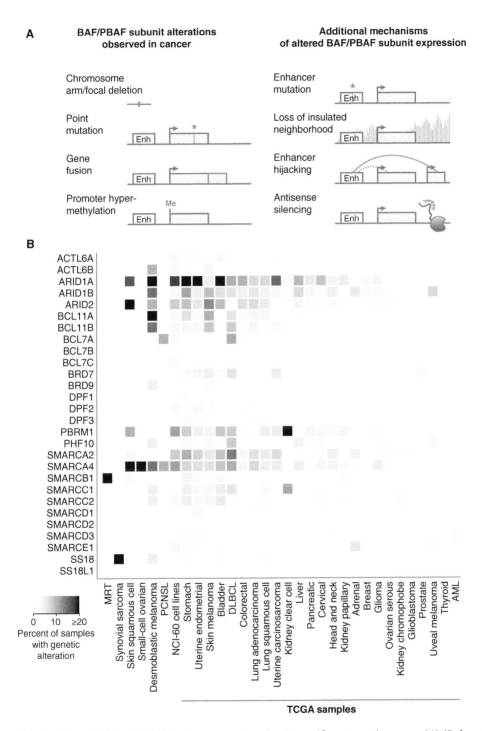

Figure 3. Mutations of BAF and PBAF subunits occur in subunit-specific patterns in cancer. (*A*) (*Left* panel) Illustration of the different types of genetic and epigenetic disruptions that affect BAF/PBAF subunits in cancer. Deletion of chromosome arms or foci leads to loss of a subunit allele, point mutations alter coding sequence, gene fusions lead to altered function, and hypermethylation of promoters associated with loss of expression. (*Legend continues on following page.*)

observed in melanoma (Hodis et al. 2012; The Cancer Genome Atlas 2015), medulloblastoma (Pugh et al. 2012), and several cancer cell lines, have severely reduced ATPase activity, leading to anaphase bridges and failure of topoisomerase IIa to bind DNA (Dykhuizen et al. 2013) (discussed in greater detail below). Adjacent residues are mutated in a number of different cancer types, suggesting that functional inactivation contributes to diverse malignancies. In other ATP-dependent remodelers, dominant-negative mutations result in phenotypes distinct from subunit deletion (Corona et al. 2004; Skene et al. 2014), suggesting that the particular mechanisms of inactivation may lead to different downstream effects.

Although the details remain murky, the carboxy-terminal subdomain of Snf2-like ATPases appears to cooperate with the amino-terminal subdomain to exert large-scale motions needed to translocate along DNA (Durr et al. 2005) but may also carry out some other uncharacterized activity. In the carboxy-terminal ATPase subdomain of BRG, R1192 is recurrently mutated in cancer of the stomach, liver, lung, melanoma, esophagus, and breast, as well as in gliomas (Figs. 3 and 4). Moreover, the homologous position is also mutated in *BTAF1*, *CHD1*, and *ATRX* in several different malignancies, suggesting this well-conserved position may be an Achilles' heel of Snf2-like remodelers. Other nearby mutations at conserved residues in Motif V of the carboxy-terminal subdomain severely compromise ATPase activity in the yeast SWI/SNF complex (Richmond and Peterson 1996; Smith and Peterson 2005). Although it is clear that commonly observed point mutations disrupt or completely abolish ATPase activity, a complete accounting of the downstream

effects of these mutations in malignancy has not yet been performed.

BRM, the paralog of BRG that is not a subunit of the PBAF complex, also shows similar clustering of mutations at the amino- and carboxy-terminal helicase-like subdomains, but is much less frequently mutated in cancers (Figs. 3 and 4). Interestingly, several in-frame deletions in the QLQ domain of BRM have been observed in primary tumors and several cancer cell lines (Reinhold et al. 2012). BRM and homologs are regulators of splicing (Batsche et al. 2006; Tyagi et al. 2009; Waldholm et al. 2011; Patrick et al. 2015); however, the effects of its mutation on alternative splicing remain unknown.

ARID1A in Uterine, Colorectal, Stomach, Bladder, and Other Cancers

By far the most frequently disrupted BAF subunit is *ARID1A* (*BAF250A*; Fig. 2B). Large regions of both ARID1A and its paralog ARID1B are low-complexity sequences with unknown function. ARID1A and ARID1B both contain an ARID DNA-binding domain as well as a homologous domain of unknown function (currently designated DUF3518 in Pfam; Fig. 4). Although the function of this carboxy-terminal domain has not yet been described, it has been speculated to have ubiquitin ligase activity (Li et al. 2010).

Among the earliest reports of the complex's tumor-suppressor role was the discovery that ~50% of ovarian clear cell carcinomas and endometriosis-associated ovarian carcinomas contain inactivating *ARID1A* mutations (Jones et al. 2010; Wiegand et al. 2010). Mutations of *ARID1A* have since been observed at high frequency in a number of studies, including uterine

---

Figure 3. (*Continued*) (*Right* panel) Mechanisms leading to altered BAF/PBAF subunit expression in cancer may also include mutations in enhancers, loss of insulated neighborhoods leading to spreading of heterochromatin over BAF/PBAF genes, enhancer hijacking, and antisense silencing. (*B*) Heat map of the frequency of subunit alterations across cancer types (frequency includes all nonsilent mutations, biallelic deletions, and gene fusions). Mutation frequencies for malignant rhabdoid tumor (MRT) and synovial sarcoma are inferred from available cytogenetic and mutation data, as described by works cited in the main text. All other data obtained from studies cited in the main text. PCNSL, Primary central nervous system lymphoma; DLBCL, diffuse large B-cell lymphoma; AML, acute myeloid leukemia.

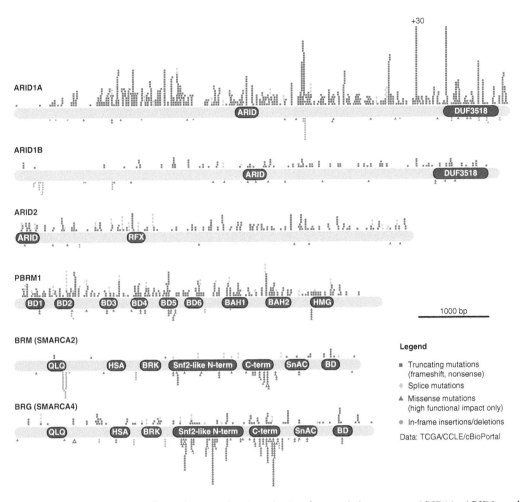

Figure 4. Cancer mutations of BAF/PBAF subunits arise in characteristic patterns. *ARID1A*, *ARID2*, and *PBRM1* are primarily affected by truncating mutations. The ATPases *BRG* and *BRM* show a high tendency for missense mutations at the conserved Snf2-like ATPase domains. Missense mutations predicted to have neutral, low, or medium functional impact are not shown because of the unknown nature of their effects and increased likelihood to be background mutations. N-term, Amino terminal; C-term, carboxy terminal.

endometrial carcinoma (34%) (Kandoth et al. 2013), colorectal cancers (10%) (The Cancer Genome Atlas 2012a), as well as cancers of the bladder (29%) (Gui et al. 2011), stomach (34%) (Wang et al. 2011a, 2014), cholangiocarcinomas (27%) (Jiao et al. 2013), neuroblastomas (11%) (Sausen et al. 2013), and pancreas (~5%) (Biankin et al. 2012). These recurrent loss-of-function mutations make ARID1A the premier tumor-suppressor subunit of the BAF complex; however, very little is known about the mechanisms of how this subunit contributes to malignancy.

Mutations of *ARID1A* are most frequently truncating mutations (frameshifts and non-sense mutations (Figs. 2C and 4), which may be degraded by nonsense-mediated decay. Although the ARID domain in mice is critical for the function of the protein, mutations in human cancer are not especially localized to the ARID domain; indeed no missense mutations expected to be of high functional impact have currently been reported in the ARID domain (Fig. 4). The few missense mutations present are generally predicted to have low or medium functional impact and are distributed uniformly over the gene. The hotspots of truncating mutations that occur are explained in part by frequent *ARID1A* mutations arising in tumors with mutated DNA polymerase ε (*POLE*). As a result of failed leading-strand proofreading during replication, *POLE*-mutated tumors often have huge numbers of C>T transitions, many of which convert arginine codons (CGA) to stop codons (TGA) (Alexandrov et al. 2013). Biallelic inactivation of *ARID1A* does occur, but in many cases (particularly in gastric and endometrial cancer) mutations occur in only a single allele (Kandoth et al. 2013). However, hypermethylation of the *ARID1A* promoter has been observed in many breast cancers; hence, epigenetic silencing mechanisms are also common (Zhang et al. 2013).

In ovarian cancer, mutation of *ARID1A* frequently co-occurs with activating mutations of phosphatidylinositol 3-kinase (*PI3K*). Interestingly, BRG binds the PI3K substrate PIP_2 (phosphatidylinositol 4,5-bisphosphate), a phospholipid with several roles in signaling and a regulator of actin-related proteins. Binding of PIP_2 by *Brg* regulates association of the complex with actin (Rando et al. 2002); therefore, activating mutations of PI3K may deplete PIP_2, leading to altered BAF localization or function (Zhao et al. 1998). Mice with *ARID1A/PI3K* double mutations, but not mice with only a single *ARID1A* or *PI3K* mutation, develop ovarian tumors with features similar to ovarian clear cell carcinoma (Chandler et al. 2015), suggesting the effects of PIP_2 may be mediated through ARID1A-containing complexes and providing new insight into the cooperation of these two genes in cancer.

ARID1B is mutated in several malignancies and, like *ARID1A*, these mutations are also mostly truncating (Fig. 4). However, *ARID1B* mutations are not as frequent as those of *ARID1A* (Fig. 2C). This discrepancy may reflect important functional differences between these two genes, or may instead reflect *ARID1B* expression in fewer cell types. Interestingly, *ARID1B* has been identified as one of the most important genes involved in neurodevelopmental disorders (Santen et al. 2012; Tsurusaki et al. 2012; Deciphering Developmental Disorders Study 2015), further illustrating the relevance of combinatorial subunit assembly of BAF complexes to development and human diseases.

PBRM1 in Clear Cell Renal Carcinoma

PBRM1 (BAF180, Polybromo) is named for the presence of six bromodomains in the protein, and is a defining subunit of the PBAF complex. In renal clear cell carcinoma (ccRCC), mutation or loss of *PBRM1* occurs in ∼41% of cases (Varela et al. 2011), making it the second-most frequently mutated gene in ccRCC. Like *ARID1A* in other cancers, the majority of mutations of *PBRM1* in ccRCC are truncating mutations (Figs. 2C and 4), which may not result in protein expression because of nonsense-mediated decay. However, many ccRCC cases have biallelic inactivation of *PBRM1*, through loss of one allele via focal/chromosomal deletion at chromosome arm 3p, and an inactivating mutation on the remaining allele. Furthermore, hypermethylation of the *PBRM1* promoter is generally absent in ccRCC (Ibragimova et al. 2013), indicating that inactivation occurs primarily through mutation or deletion. Some tumors do contain missense mutations, and although their functional impacts remain uncertain, their presence suggests a degree of nonredundancy between these domains. Although most bromodomains bind acetylated lysines from histones, the role of PBRM1's bromodomains toward targeting of PBAF complexes remains uncertain.

In ccRCC, *PBRM1* inactivation frequently coincides with mutation of the *VHL* (von Hippel–Lindau) tumor suppressor. Because of their close proximity on chromosome arm 3p, focal and chromosome arm-level deletions frequently affect both of these genes simultaneously. However, the striking frequency of inactivating point mutations of *PBRM1* alongside *VHL* and *BAP1* mutations suggests that joint inactivation of these genes may potentiate the oncogenic nature of these defects (Gerlinger et al. 2014).

Like *PBRM1*, *ARID2* (*BAF200*) encodes another subunit dedicated to PBAF complexes. *ARID2* is not a homolog of *ARID1A/B*, but is instead mutually exclusive with *ARID1A/B*, although the shared presence of an AT-rich interaction domain (ARID) in these three subunits suggests some common structural similarities. *ARID2* is frequently mutated in other malignancies but is apparently not targeted in ccRCC as frequently as *PBRM1*, suggesting that PBRM1 has an important and distinct functional role as a member of the PBAF complex in kidney cells. ARID2 has been reported to contribute to repression (Raab et al. 2015), and is frequently mutated in melanoma (Hodis et al. 2012; Ding et al. 2014; Lee et al. 2015), non-small-cell lung cancer (Manceau et al. 2013), as well as in ~18% of hepatitis-associated hepatocellular carcinomas (Li et al. 2011). Moreover, the discovery of frequent joint inactivation of *PBRM1*, *ARID2*, and *BAP1* in biliary-phenotype-displaying subtype of hepatic carcinomas (Fujimoto et al. 2015) suggests that in some contexts they may contribute to malignancy in a cooperative manner.

SMARCB1 in Malignant Rhabdoid Tumors

Perhaps the best-characterized example of a tumor-suppressor role for ATP-dependent chromatin remodeling comes from malignant rhabdoid tumors (MRTs). MRTs are rare but highly lethal childhood cancers that are caused by biallelic inactivation of *SMARCB1* (*BAF47*, *SNF5*, or *INI1*), which occurs in nearly all cases (Versteege et al. 1998; Roberts et al. 2000). The classic loss of heterozygosity observed for *SMARCB1* leads to aberrant activation of Hedgehog-Gli and Wnt/β-catenin pathways (Jagani et al. 2010; Mora-Blanco et al. 2014), and importantly, impairs the ability of BAF/PBAF complexes to regulate the placement and function of Polycomb repressive complexes. As a result of failure to oppose Polycomb, the repressive mark H3K27me3 accumulates at the tumor suppressor *p16/INK4A* (*CDKN2A*) locus (Wilson et al. 2010). The impaired opposition to Polycomb repression plays an important role in cancer, reminiscent of BAP/PBAP complexes throughout development in flies (Tamkun et al. 1992).

MRTs have remarkably stable diploid genomes except for deletions and mutations at chromosome *22q*, where *SMARCB1* is located (McKenna et al. 2008; McKenna and Roberts 2009; Lee et al. 2012). Exome sequencing also shows that these tumors have among the lowest mutational loads of any human tumor sequenced to date (Lawrence et al. 2014). Finally, ectopic expression of *SMARCB1* reverses Polycomb silencing at the tumor suppressor *p16/INK4A* locus, leading to cellular senescence (Oruetxebarria et al. 2004; Kia et al. 2008), indicating that these tumors are driven exclusively by epigenetic regulation (except for the original genetic inactivation of *SMARCB1*). In addition to MRTs, *SMARCB1* appears to play a role in a number of cancers and other neoplastic disorders, including prostate cancer, epithelioid sarcomas, familial schwannomatosis, and renal medullary carcinomas (Roberts and Biegel 2009; Prensner et al. 2013). Biallelic inactivation of *SMARCB1* has also been reported in 7%–10% of Ewing sarcomas (Jahromi et al. 2012).

In mouse models, conditional deletion of *SMARCB1* leads to T-cell lymphomas with short latency and 100% penetrance (Wang et al. 2011b), suggesting that *SMARCB1* inactivation can cause fast transformation alone without other genetic changes, as observed in MRTs. Interestingly, rhabdoid tumors are not seen in mice with *SMARCB1* inactivation, attesting to the tissue-specific and species-specific function of the complexes. Importantly, the pathogenesis of most human malignancies with BAF/PBAF mutations may be different from that of MRTs. With the exception of some noteworthy examples described below, the majority of cancers bearing BAF/PBAF subunit mutations are found in older age groups, in which tumors have long latencies, are highly mutated, and are genomically unstable. Thus, the low mutation rates observed in MRTs may be because of the extremely short latency between biallelic inactivation and transformation, which may not allow accumulation of mutations resulting from impairment of the complex's other functions.

SS18 in Synovial Sarcoma

Synovial sarcoma is an aggressive, poorly differentiated, stem-cell-like soft-tissue malignancy that typically arises in the extremities of young adults. The hallmark of synovial sarcoma is a highly characteristic translocation of chromosomes 18 and X, which fuses the dedicated BAF subunit SS18 (Middeljans et al. 2012) to the SSX fusion partner on the X chromosome (Crew et al. 1995; Naka et al. 2010). This fusion occurs in nearly all cases and in some cases is the only known cytogenetic abnormality. Despite the continued existence of the remaining wild-type SS18 allele and unaltered BAF47 alleles, the SS18–SSX fusion is preferentially assembled into the BAF complex concomitant with complete loss of BAF47 from the complex. BAF complexes containing the SS18–SSX fusion are retargeted to oncogenic loci such as *SOX2* and *PAX6*, where removal of the repressive histone mark H3K27me3 results in transformation (Kadoch and Crabtree 2013).

Forced overexpression of wild-type SS18, or shRNA-mediated knockdown of the SS18–SSX fusion, is sufficient to reverse oncogenic BAF subunit composition. Reversion leads to increased levels of H3K27me3 at *SOX2* and other oncogenic loci, and loss of proliferation (Kadoch and Crabtree 2013), indicating that transformation is maintained through epigenetic mechanisms. The reversible and remarkably specific pathogenesis of synovial sarcoma suggests that this tumor may be an attractive candidate for development of therapeutics.

A number of similarities exist between synovial sarcoma and MRTs. They are both childhood malignancies driven by a defining alteration of a single BAF subunit, and senescence can be achieved by repair of the affected subunit. Moreover, SMARCB1 activity is abolished in both cancers, albeit through different mechanisms. However, in contrast to MRTs, which transform by failing to oppose Polycomb activity at *p16/INK4A*, the genetic dominance of the SS18–SSX fusion in synovial sarcoma arises from its preferential assembly into BAF complexes, and its apparent ability to retarget the complex and oppose Polycomb at oncogenic loci (Kadoch and Crabtree 2013).

BAF53A and the Role of Nuclear Actin/ARPs

BAF53A (*ACTL6A*) is an ARP and a subunit of BAF/PBAF complexes that is rarely mutated in cancer. Instead, *BAF53A* frequently undergoes amplification in squamous cell malignancies from many different tissues of origin. BAF53A is required for maintenance of hematopoietic stem cell identity (Krasteva et al. 2012), and also maintains a progenitor state in epidermal cells by repressing *KLF4*, an activator of differentiation (Bao et al. 2013). Recent work has also shown that *BAF53A* is a target of miR-206, a microRNA missing in rhabdomyosarcomas (RMS). The resulting up-regulation of *BAF53A* in RMS cells contributes to the failure of myogenic cells to properly differentiate (Taulli et al. 2014), whereas its silencing inhibits proliferation of RMS cells, suggesting that BAF53A promotes proliferation and interferes with differentiation. Therefore, it is appealing to speculate that BAF53A may generally have oncogenic or mitogenic role in many cell types, perhaps based on interaction with the BRG/BRM helicase-SANT-associated (HSA) domain (Zhao et al. 1998; Rando et al. 2002; Szerlong et al. 2008). Ablation of the HSA domain from Sth1, the ATPase of RSC in yeast, causes the specific loss of ARPs from the complex, and a reduction in the activity of the ATPase.

ARPs are genetically essential subunits of SWI/SNF-like remodelers (Shen et al. 2003; Wu et al. 2007), and the interaction of actin and ARPs with the HSA domain from Snf2-like ATPases has long been thought to regulate their activity. Although complexes reconstituted without actin or ARPs can achieve remodeling activity comparable to intact complexes in vitro (Phelan et al. 1999), BAF53A/B is required for BAF function in vivo. The crystal structure of the Snf2 HSA domain with Arp7 and Arp9 shows that actin filament formation is unlikely because of the incompatible position adopted by the ARPs (Schubert et al. 2013). Thus, rather than binding filamentous actin, the ARPs in the complex may instead modulate ATPase activity or the coupling of ATP hydrolysis to remodeling activity.

Important structural differences exist between the yeast complexes and human complexes (Zhao et al. 1998). In yeast, both the SWI/SNF and RSC complexes contain Arp7 and Arp9 as obligate heterodimers (Szerlong et al.

2003), but lack actin itself (Cairns et al. 1998; Peterson et al. 1998). In contrast, BAF and PBAF complexes contain BAF53A/B and actin (Zhao et al. 1998). Therefore, it remains unclear whether the structures from yeast also apply directly to the mammalian complexes. As a result, it remains unknown how excess BAF53A may affect the activity of BAF and PBAF complexes.

NONTRANSCRIPTIONAL ROLES OF BAF/PBAF COMPLEXES IN CANCER

Involvement in DNA Repair and Chromosome Stability

In addition to their well-established roles as epigenetic regulators described above, BAF/PBAF complexes have several nontranscriptional roles that also contribute to malignancy. One nontranscriptional role in cancer is found in DNA-repair pathways. Various mechanisms have been proposed for recruitment of BAF/PBAF complexes to sites of DNA damage, including ATM-/ATR-dependent phosphorylation of BAF170 (Peng et al. 2009), and a direct interaction between γ-H2A.X and the BRG bromodomain (Lee et al. 2010). In addition, evidence now points to roles for BAF and PBAF in both NHEJ and HR pathways (Ogiwara et al. 2011; Watanabe et al. 2014; Brownlee et al. 2015; Qi et al. 2015). Therefore, in humans, BAF and PBAF complexes may help protect genomic integrity similar to the INO80 complex in budding yeast (Gerhold et al. 2015), which interestingly is not frequently mutated in cancer.

In addition to well-established pathways of DNA repair, PBAF complexes have other important roles for maintaining genomic stability. PBRM1 plays a critical role in sister chromatid cohesion, in which misregulation leads to genome instability, anaphase bridges, and aneuploidy (Brownlee et al. 2014). PBRM1 also plays a role in repriming stalled replication forks similar to yeast RSC complexes (Askree et al. 2004). Stalled replication forks are common sites of DNA damage, providing another important mechanism for ensuring genome integrity (Niimi et al. 2012). Recently, roles for PBAF but

not BAF have been identified in DNA-damage-induced transcriptional repression that involves PRC1/2 subunits (Kakarougkas et al. 2014), and ubiquitination of PCNA following DNA damage (Niimi et al. 2015).

BAF and PBAF subunits occupy regions that are critical for chromosome organization, such as the binding sites of CTCF (CCCTC-binding factor), cohesins, lamin, and replication origins (Euskirchen et al. 2011). In addition to the loop anchor sites formed by CTCF and cohesins, which appear to be master regulators of topological domains in stem cells and cancer cell lines (Kagey et al. 2010; Dixon et al. 2012; Yan et al. 2013; Dowen et al. 2014; Ji et al. 2015), several other chromatin organizational elements have been identified in eukaryotes, such as tRNA genes (Kirkland et al. 2013), repetitive elements (Lunyak et al. 2007), transposons (Lippman et al. 2004), and PRC1-binding sites (Bantignies et al. 2011; Schoenfelder et al. 2015; Wani et al. 2016). Chromatin architectural sites are often subject to epigenetic regulation (Bell and Felsenfeld 2000; Wang et al. 2012) and show significant BAF and PBAF enrichment (Euskirchen et al. 2011), suggesting that BAF and PBAF may play important roles in regulating overall chromatin architecture.

Synergy between BAF and Topoisomerase Function

Topoisomerases require nucleosome-free DNA (Sperling et al. 2011), and mutants of BRG that impair ATPase activity induce loss of topoisomerase IIa (TOP2A) binding to DNA, leading to topological defects, anaphase bridges, and partial arrest at the relatively uncharacterized decatenation checkpoint (Dykhuizen et al. 2013). Lung cancer cell lines with *BRG* mutations show increased sensitivity to topoisomerase II inhibitors when EZH2 is also inhibited (Fillmore et al. 2015), suggesting interplay between BAF, TOP2A, and PRC2 in the maintenance of chromatin topology. The importance of BAF's activity toward TOP2A function was recently underlined by the observation that mutations in BAF subunits predict responses to treatment with TOP2A inhibitors (Pang et al. 2015; Wijdeven et al. 2015). Importantly, the mechanism of opposition to EZH2 consists of more than opposing its methyltransferase activity (Kim et al. 2015), and may underlie an aspect of BAF's tumor-suppressor function with significant clinical importance.

TARGETING TUMORS WITH BAF/PBAF DEFICIENCIES

Recent reports suggest new approaches for targeting tumors with altered BAF/PBAF complexes based on synthetic lethality. In several tumor types, inactivation of one BAF/PBAF subunit induces dependency on the continued expression of that subunit's paralog. For example, tumors with *BRG* mutations frequently depend on the expression of *BRM* (Aguirre et al. 2014; Wilson et al. 2014), whereas tumors with *ARID1A* mutations often depend on *ARID1B* (Helming et al. 2014). Targeting these genetic dependencies represents a novel strategy to attack these tumors. Additionally, loss-of-function of BAF/PBAF subunits may lead to

increased Polycomb activity; therefore, inhibition of Polycomb silencing may be beneficial for patients with tumors bearing BAF/PBAF deficiencies. However, the effectiveness of these approaches may depend greatly on the downstream consequences of BAF/PBAF dysfunction within each cell type. For example, PRC2 inhibition may be more beneficial for MRTs than for synovial sarcoma, based on the molecular mechanisms of transformation described above, illustrating the continuing need to examine the epigenetic mechanisms within each tumor type.

PERSPECTIVE AND CLOSING REMARKS

Although abundant evidence indicates that BAF and PBAF defects contribute to malignancy by altering the epigenetic landscape to regulate transcription, many lines of evidence indicate that these defects have pleiotropic effects, because complexes participate in a number of other important chromatin regulatory processes. In addition to their roles in regulating transcrip-

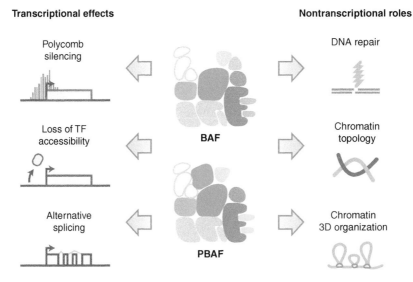

Figure 5. The effects of BAF and PBAF dysfunction in cancer. Dysfunctional BAF/PBAF complexes have been shown to deregulate Polycomb silencing of key tumor suppressors and oncogenes. In model systems, disruption of BAF- and PBAF-like complexes also affects DNA accessibility for transcription and other regulatory factors, and impacts splicing patterns. Given the conserved regulatory roles for BAF- and PBAF-like remodelers in DNA repair, maintenance of chromatin topology and 3D architecture, we anticipate that whole-genome sequencing and new techniques to examine 3D-chromatin architecture may reveal new roles for the complex in addition to its well-defined role as a transcriptional regulator.

tion, a body of work from cancer cell lines and model organisms indicates that BAF- and PBAF-like complexes contribute to several other processes, ranging from DNA recombination and repair to maintenance of 3D chromatin architecture and topology (Fig. 5).

Their numerous roles underscore the fact that the epigenetic state is more than simply a regulatory framework for transcription, but instead represents the sum physicochemical state of the genetic material, which impacts a large number of processes. As a result, the fundamental role of any given BAF and PBAF alteration in cancer is likely to be unique to each cancer type, and may reflect the idiosyncratic processes that drive each malignancy (whether oncogene addiction, autocrine signaling, mutagen exposure, chromosomal instability, etc.). Whole-genome sequencing and various new techniques to examine 3D-chromatin architecture may offer substantial insight into the full breadth of the effects of BAF and PBAF dysfunction, and reveal their diverse contributions toward oncogenesis and tumor biology.

ACKNOWLEDGMENTS

Figures and results reviewed here are based in part on data generated by the The Cancer Genome Atlas Research Network and hosted on cBioPortal (Cerami et al. 2012; Gao et al. 2013). This work is supported by a National Cancer Institute (NCI) career transition award K99CA187565 (C.H.), NCI Grant T32CA009151 (J.G.K.), National Institutes of Health (NIH) Grants R37NS046789 and R01CA163915 (G.R.C.), and the Howard Hughes Medical Institute (G.R.C.).

REFERENCES

Aguirre AJ, Jagani Z, Wang Z, Garraway LA, Hahn WC, Roberts CW, Hoffman GR, Rahal R, Buxton F, Xiang K, et al. 2014. Functional epigenetics approach identifies BRM/SMARCA2 as a critical synthetic lethal target in BRG1-deficient cancers. *Nat Med* **111**: 3128–3133.

Aizawa H, Hu SC, Bobb K, Balakrishnan K, Ince G, Gurevich I, Cowan M, Ghosh A. 2004. Dendrite development regulated by CREST, a calcium-regulated transcriptional activator. *Science* **303**: 197–202.

Alexandrov LB, Nik-Zainal S, Wedge DC, Aparicio SA, Behjati S, Biankin AV, Bignell GR, Bolli N, Borg A, Borresen-Dale AL, et al. 2013. Signatures of mutational processes in human cancer. *Nature* **500**: 415–421.

Armstrong JA, Papoulas O, Daubresse G, Sperling AS, Lis JT, Scott MP, Tamkun JW. 2002. The *Drosophila* BRM complex facilitates global transcription by RNA polymerase II. *EMBO J* **21**: 5245–5254.

Askree SH, Yehuda T, Smolikov S, Gurevich R, Hawk J, Coker C, Krauskopf A, Kupiec M, McEachern MJ. 2004. A genome-wide screen for *Saccharomyces cerevisiae* deletion mutants that affect telomere length. *Proc Natl Acad Sci* **101**: 8658–8663.

Asturias FJ, Chung WH, Kornberg RD, Lorch Y. 2002. Structural analysis of the RSC chromatin-remodeling complex. *Proc Natl Acad Sci* **99**: 13477–13480.

Baetz KK, Krogan NJ, Emili A, Greenblatt J, Hieter P. 2004. The *ctf13-30/CTF13* genomic haploinsufficiency modifier screen identifies the yeast chromatin remodeling complex RSC, which is required for the establishment of sister chromatid cohesion. *Mol Cell Biol* **24**: 1232–1244.

Bajpai R, Chen DA, Rada-Iglesias A, Zhang J, Xiong Y, Helms J, Chang CP, Zhao Y, Swigut T, Wysocka J. 2010. CHD7 cooperates with PBAF to control multipotent neural crest formation. *Nature* **463**: 958–962.

Balbiani E. 1881. Sur la structure du noyau des cellules salivaires chez les larves de *Chironomus* [On the structure of salivary cell nuclei in *Chironomus* larvae]. *Zool Anz* **4**: 662–667.

Banroques J, Cordin O, Doere M, Linder P, Tanner NK. 2008. A conserved phenylalanine of motif IV in superfamily 2 helicases is required for cooperative, ATP-dependent binding of RNA substrates in DEAD-box proteins. *Mol Cell Biol* **28**: 3359–3371.

Bantignies F, Roure V, Comet I, Leblanc B, Schuettengruber B, Bonnet J, Tixier V, Mas A, Cavalli G. 2011. Polycomb-dependent regulatory contacts between distant Hox loci in *Drosophila*. *Cell* **144**: 214–226.

Bao X, Tang J, Lopez-Pajares V, Tao S, Qu K, Crabtree GR, Khavari PA. 2013. ACTL6a enforces the epidermal progenitor state by suppressing SWI/SNF-dependent induction of KLF4. *Cell Stem Cell* **12**: 193–203.

Bao X, Rubin AJ, Qu K, Zhang J, Giresi PG, Chang HY, Khavari PA. 2015. A novel ATAC-seq approach reveals lineage-specific reinforcement of the open chromatin landscape via cooperation between BAF and p63. *Genome Biol* **16**: 284.

Batsche E, Yaniv M, Muchardt C. 2006. The human SWI/SNF subunit Brm is a regulator of alternative splicing. *Nat Struct Mol Biol* **13**: 22–29.

Becker PB, Horz W. 2002. ATP-dependent nucleosome remodeling. *Annu Rev Biochem* **71**: 247–273.

Bell AC, Felsenfeld G. 2000. Methylation of a CTCF-dependent boundary controls imprinted expression of the *Igf2* gene. *Nature* **405**: 482–485.

Biankin AV, Waddell N, Kassahn KS, Gingras MC, Muthuswamy LB, Johns AL, Miller DK, Wilson PJ, Patch AM, Wu J, et al. 2012. Pancreatic cancer genomes reveal aberrations in axon guidance pathway genes. *Nature* **491**: 399–405.

Birrell GW, Giaever G, Chu AM, Davis RW, Brown JM. 2001. A genome-wide screen in *Saccharomyces cerevisiae* for genes affecting UV radiation sensitivity. *Proc Natl Acad Sci* **98:** 12608–12613.

Bouazoune K, Brehm A. 2006. ATP-dependent chromatin remodeling complexes in *Drosophila*. *Chromosome Res* **14:** 433–449.

Brownlee PM, Chambers AL, Cloney R, Bianchi A, Downs JA. 2014. BAF180 promotes cohesion and prevents genome instability and aneuploidy. *Cell Rep* **6:** 973–981.

Brownlee PM, Meisenberg C, Downs JA. 2015. The SWI/SNF chromatin remodelling complex: Its role in maintaining genome stability and preventing tumourigenesis. *DNA Repair (Amst)* **32:** 127–133.

Cairns BR, Kim YJ, Sayre MH, Laurent BC, Kornberg RD. 1994. A multisubunit complex containing the *SWI1/ADR6, SWI2/SNF2, SWI3, SNF5*, and *SNF6* gene products isolated from yeast. *Proc Natl Acad Sci* **91:** 1950–1954.

Cairns BR, Lorch Y, Li Y, Zhang M, Lacomis L, Erdjument-Bromage H, Tempst P, Du J, Laurent B, Kornberg RD. 1996. RSC, an essential, abundant chromatin-remodeling complex. *Cell* **87:** 1249–1260.

Cairns BR, Erdjument-Bromage H, Tempst P, Winston F, Kornberg RD. 1998. Two actin-related proteins are shared functional components of the chromatin-remodeling complexes RSC and SWI/SNF. *Mol Cell* **2:** 639–651.

Cairns BR, Schlichter A, Erdjument-Bromage H, Tempst P, Kornberg RD, Winston F. 1999. Two functionally distinct forms of the RSC nucleosome-remodeling complex, containing essential AT hook, BAH, and bromodomains. *Mol Cell* **4:** 715–723.

Carlson M, Osmond BC, Botstein D. 1981. Mutants of yeast defective in sucrose utilization. *Genetics* **98:** 25–40.

Cerami E, Gao J, Dogrusoz U, Gross BE, Sumer SO, Aksoy BA, Jacobsen A, Byrne CJ, Heuer ML, Larsson E, et al. 2012. The cBio cancer genomics portal: An open platform for exploring multidimensional cancer genomics data. *Cancer Discov* **2:** 401–404.

Chaban Y, Ezeokonkwo C, Chung WH, Zhang F, Kornberg RD, Maier-Davis B, Lorch Y, Asturias FJ. 2008. Structure of a RSC-nucleosome complex and insights into chromatin remodeling. *Nat Struct Mol Biol* **15:** 1272–1277.

Chai B, Huang J, Cairns BR, Laurent BC. 2005. Distinct roles for the RSC and Swi/Snf ATP-dependent chromatin remodelers in DNA double-strand break repair. *Genes Dev* **19:** 1656–1661.

Chalkley GE, Moshkin YM, Langenberg K, Bezstarosti K, Blastyak A, Gyurkovics H, Demmers JA, Verrijzer CP. 2008. The transcriptional coactivator SAYP is a trithorax group signature subunit of the PBAP chromatin remodeling complex. *Mol Cell Biol* **28:** 2920–2929.

Chandler RL, Damrauer JS, Raab JR, Schisler JC, Wilkerson MD, Didion JP, Starmer J, Serber D, Yee D, Xiong J, et al. 2015. Coexistent ARID1A–PIK3CA mutations promote ovarian clear-cell tumorigenesis through pro-tumorigenic inflammatory cytokine signalling. *Nat Commun* **6:** 6118.

Chevenet F, Brun C, Banuls AL, Jacq B, Christen R. 2006. TreeDyn: Towards dynamic graphics and annotations for analyses of trees. *BMC Bioinformatics* **7:** 439.

Clapier CR, Cairns BR. 2009. The biology of chromatin remodeling complexes. *Annu Rev Biochem* **78:** 273–304.

Collins RT, Furukawa T, Tanese N, Treisman JE. 1999. Osa associates with the Brahma chromatin remodeling complex and promotes the activation of some target genes. *EMBO J* **18:** 7029–7040.

Corona DF, Armstrong JA, Tamkun JW. 2004. Genetic and cytological analysis of *Drosophila* chromatin-remodeling factors. *Methods Enzymol* **377:** 70–85.

Crew AJ, Clark J, Fisher C, Gill S, Grimer R, Chand A, Shipley J, Gusterson BA, Cooper CS. 1995. Fusion of *SYT* to two genes, *SSX1* and *SSX2*, encoding proteins with homology to the Kruppel-associated box in human synovial sarcoma. *EMBO J* **14:** 2333–2340.

Crosby MA, Miller C, Alon T, Watson KL, Verrijzer CP, Goldman-Levi R, Zak NB. 1999. The *trithorax* group gene *moira* encodes a Brahma-associated putative chromatin-remodeling factor in *Drosophila melanogaster*. *Mol Cell Biol* **19:** 1159–1170.

Davoli T, Xu AW, Mengwasser KE, Sack LM, Yoon JC, Park PJ, Elledge SJ. 2013. Cumulative haploinsufficiency and triplosensitivity drive aneuploidy patterns and shape the cancer genome. *Cell* **155:** 948–962.

Dechassa ML, Zhang B, Horowitz-Scherer R, Persinger J, Woodcock CL, Peterson CL, Bartholomew B. 2008. Architecture of the SWI/SNF-nucleosome complex. *Mol Cell Biol* **28:** 6010–6021.

Deciphering Developmental Disorders Study. 2015. Large-scale discovery of novel genetic causes of developmental disorders. *Nature* **519:** 223–228.

Dhillon N, Raab J, Guzzo J, Szyjka SJ, Gangadharan S, Aparicio OM, Andrews B, Kamakaka RT. 2009. DNA polymerase epsilon, acetylases and remodellers cooperate to form a specialized chromatin structure at a tRNA insulator. *EMBO J* **28:** 2583–2600.

Ding L, Kim M, Kanchi KL, Dees ND, Lu C, Griffith M, Fenstermacher D, Sung H, Miller CA, Goetz B, et al. 2014. Clonal architectures and driver mutations in metastatic melanomas. *PLoS ONE* **9:** e111153.

Dixon JR, Selvaraj S, Yue F, Kim A, Li Y, Shen Y, Hu M, Liu JS, Ren B. 2012. Topological domains in mammalian genomes identified by analysis of chromatin interactions. *Nature* **485:** 376–380.

Dowen JM, Fan ZP, Hnisz D, Ren G, Abraham BJ, Zhang LN, Weintraub AS, Schuijers J, Lee TI, Zhao K, et al. 2014. Control of cell identity genes occurs in insulated neighborhoods in mammalian chromosomes. *Cell* **159:** 374–387.

Dror V, Winston F. 2004. The Swi/Snf chromatin remodeling complex is required for ribosomal DNA and telomeric silencing in *Saccharomyces cerevisiae*. *Mol Cell Biol* **24:** 8227–8235.

Dunaief JL, Strober BE, Guha S, Khavari PA, Alin K, Luban J, Begemann M, Crabtree GR, Goff SP. 1994. The retinoblastoma protein and BRG1 form a complex and cooperate to induce cell cycle arrest. *Cell* **79:** 119–130.

Durr H, Korner C, Muller M, Hickmann V, Hopfner KP. 2005. X-ray structures of the *Sulfolobus solfataricus* SWI2/SNF2 ATPase core and its complex with DNA. *Cell* **121:** 363–373.

Dykhuizen EC, Hargreaves DC, Miller EL, Cui K, Korshunov A, Kool M, Pfister S, Cho YJ, Zhao K, Crabtree GR. 2013. BAF complexes facilitate decatenation of DNA by topoisomerase IIα. *Nature* **497:** 624–627.

Estruch F, Carlson M. 1990. *SNF6* encodes a nuclear protein that is required for expression of many genes in *Saccharomyces cerevisiae*. *Mol Cell Biol* **10:** 2544–2553.

Euskirchen GM, Auerbach RK, Davidov E, Gianoulis TA, Zhong G, Rozowsky J, Bhardwaj N, Gerstein MB, Snyder M. 2011. Diverse roles and interactions of the SWI/SNF chromatin remodeling complex revealed using global approaches. *PLoS Genet* **7:** e1002008.

Fairman-Williams ME, Guenther UP, Jankowsky E. 2010. SF1 and SF2 helicases: Family matters. *Curr Opin Struct Biol* **20:** 313–324.

Fillmore CM, Xu C, Desai PT, Berry JM, Rowbotham SP, Lin YJ, Zhang H, Marquez VE, Hammerman PS, Wong KK, et al. 2015. EZH2 inhibition sensitizes *BRG1* and *EGFR* mutant lung tumours to TopoII inhibitors. *Nature* **520:** 239–242.

Fujimoto A, Furuta M, Shiraishi Y, Gotoh K, Kawakami Y, Arihiro K, Nakamura T, Ueno M, Ariizumi S, Nguyen HH, et al. 2015. Whole-genome mutational landscape of liver cancers displaying biliary phenotype reveals hepatitis impact and molecular diversity. *Nat Commun* **6:** 6120.

Gao J, Aksoy BA, Dogrusoz U, Dresdner G, Gross B, Sumer SO, Sun Y, Jacobsen A, Sinha R, Larsson E, et al. 2013. Integrative analysis of complex cancer genomics and clinical profiles using the cBioPortal. *Sci Signal* **6:** pl1.

Gerhold CB, Hauer MH, Gasser SM. 2015. INO80-C and SWR-C: Guardians of the genome. *J Mol Biol* **427:** 637–651.

Gerlinger M, Horswell S, Larkin J, Rowan AJ, Salm MP, Varela I, Fisher R, McGranahan N, Matthews N, Santos CR, et al. 2014. Genomic architecture and evolution of clear cell renal cell carcinoma defined by multiregion sequencing. *Nat Genet* **46:** 225–233.

Gong F, Fahy D, Smerdon MJ. 2006. Rad4-Rad23 interaction with SWI/SNF links ATP-dependent chromatin remodeling with nucleotide excision repair. *Nat Struct Mol Biol* **13:** 902–907.

Gui Y, Guo G, Huang Y, Hu X, Tang A, Gao S, Wu R, Chen C, Li X, Zhou L, et al. 2011. Frequent mutations of chromatin remodeling genes in transitional cell carcinoma of the bladder. *Nat Genet* **43:** 875–878.

Hartley PD, Madhani HD. 2009. Mechanisms that specify promoter nucleosome location and identity. *Cell* **137:** 445–458.

He J, Xuan T, Xin T, An H, Wang J, Zhao G, Li M. 2014. Evidence for chromatin-remodeling complex PBAP-controlled maintenance of the *Drosophila* ovarian germline stem cells. *PLoS ONE* **9:** e103473.

Hedges SB, Dudley J, Kumar S. 2006. TimeTree: A public knowledge-base of divergence times among organisms. *Bioinformatics* **22:** 2971–2972.

Helming KC, Wang X, Wilson BG, Vazquez F, Haswell JR, Manchester HE, Kim Y, Kryukov GV, Ghandi M. 2014. ARID1B is a specific vulnerability in *ARID1A*-mutant cancers. *Nat Med* **20:** 251–254.

Ho L, Crabtree GR. 2010. Chromatin remodelling during development. *Nature* **463:** 474–484.

Ho L, Jothi R, Ronan JL, Cui K, Zhao K, Crabtree GR. 2009a. An embryonic stem cell chromatin remodeling complex, esBAF, is an essential component of the core pluripotency transcriptional network. *Proc Natl Acad Sci* **106:** 5187–5191.

Ho L, Ronan JL, Wu J, Staahl BT, Chen L, Kuo A, Lessard J, Nesvizhskii AI, Ranish J, Crabtree GR. 2009b. An embryonic stem cell chromatin remodeling complex, esBAF, is essential for embryonic stem cell self-renewal and pluripotency. *Proc Natl Acad Sci* **106:** 5181–5186.

Ho L, Miller EL, Ronan JL, Ho WQ, Jothi R, Crabtree GR. 2011. esBAF facilitates pluripotency by conditioning the genome for LIF/STAT3 signalling and by regulating polycomb function. *Nat Cell Biol* **13:** 903–913.

Hodis E, Watson IR, Kryukov GV, Arold ST, Imielinski M, Theurillat JP, Nickerson E, Auclair D, Li L, Place C, et al. 2012. A landscape of driver mutations in melanoma. *Cell* **150:** 251–263.

Hsu JM, Huang J, Meluh PB, Laurent BC. 2003. The yeast RSC chromatin-remodeling complex is required for kinetochore function in chromosome segregation. *Mol Cell Biol* **23:** 3202–3215.

Ibragimova I, Maradeo ME, Dulaimi E, Cairns P. 2013. Aberrant promoter hypermethylation of *PBRM1*, *BAP1*, *SETD2*, *KDM6A* and other chromatin-modifying genes is absent or rare in clear cell RCC. *Epigenetics* **8:** 486–493.

Imielinski M, Berger AH, Hammerman PS, Hernandez B, Pugh TJ, Hodis E, Cho J, Suh J, Capelletti M, Sivachenko A, et al. 2012. Mapping the hallmarks of lung adenocarcinoma with massively parallel sequencing. *Cell* **150:** 1107–1120.

Jagani Z, Mora-Blanco EL, Sansam CG, McKenna ES, Wilson B, Chen D, Klekota J, Tamayo P, Nguyen PT, Tolstorukov M, et al. 2010. Loss of the tumor suppressor Snf5 leads to aberrant activation of the Hedgehog–Gli pathway. *Nat Med* **16:** 1429–1433.

Jahromi MS, Putnam AR, Druzgal C, Wright J, Spraker-Perlman H, Kinsey M, Zhou H, Boucher KM, Randall RL, Jones KB, et al. 2012. Molecular inversion probe analysis detects novel copy number alterations in Ewing sarcoma. *Cancer Genet* **205:** 391–404.

Jankowsky E, Fairman ME. 2007. RNA helicases—One fold for many functions. *Curr Opin Struct Biol* **17:** 316–324.

Jelinic P, Mueller JJ, Olvera N, Dao F, Scott SN, Shah R, Gao J, Schultz N, Gonen M, Soslow RA, et al. 2014. Recurrent *SMARCA4* mutations in small cell carcinoma of the ovary. *Nat Genet* **46:** 424–426.

Ji X, Dadon DB, Powell BE, Fan ZP, Borges-Rivera D, Shachar S, Weintraub AS, Hnisz D, Pegoraro G, Lee TI, et al. 2015. 3D chromosome regulatory landscape of human pluripotent cells. *Cell Stem Cell* **18:** 262–275.

Jiao Y, Pawlik TM, Anders RA, Selaru FM, Streppel MM, Lucas DJ, Niknafs N, Guthrie VB, Maitra A, Argani P, et al. 2013. Exome sequencing identifies frequent inactivating mutations in *BAP1*, *ARID1A* and *PBRM1* in intrahepatic cholangiocarcinomas. *Nat Genet* **45:** 1470–1473.

Jones S, Wang TL, Shih Ie M, Mao TL, Nakayama K, Roden R, Glas R, Slamon D, Diaz LA Jr, Vogelstein B, et al. 2010. Frequent mutations of chromatin remodeling gene

ARID1A in ovarian clear cell carcinoma. *Science* **330:** 228–231.

Kadoch C, Crabtree GR. 2013. Reversible disruption of mSWI/SNF (BAF) complexes by the SS18-SSX oncogenic fusion in synovial sarcoma. *Cell* **153:** 71–85.

Kadoch C, Hargreaves DC, Hodges C, Elias L, Ho L, Ranish J, Crabtree GR. 2013. Proteomic and bioinformatic analysis of mammalian SWI/SNF complexes identifies extensive roles in human malignancy. *Nat Genet* **45:** 592–601.

Kaeser MD, Aslanian A, Dong MQ, Yates JR III, Emerson BM. 2008. BRD7, a novel PBAF-specific SWI/SNF subunit, is required for target gene activation and repression in embryonic stem cells. *J Biol Chem* **283:** 32254–32263.

Kagey MH, Newman JJ, Bilodeau S, Zhan Y, Orlando DA, van Berkum NL, Ebmeier CC, Goossens J, Rahl PB, Levine SS, et al. 2010. Mediator and cohesin connect gene expression and chromatin architecture. *Nature* **467:** 430–435.

Kakarougkas A, Ismail A, Chambers AL, Riballo E, Herbert AD, Kunzel J, Lobrich M, Jeggo PA, Downs JA. 2014. Requirement for PBAF in transcriptional repression and repair at DNA breaks in actively transcribed regions of chromatin. *Mol Cell* **55:** 723–732.

Kal AJ, Mahmoudi T, Zak NB, Verrijzer CP. 2000. The *Drosophila* Brahma complex is an essential coactivator for the *trithorax* group protein zeste. *Genes Dev* **14:** 1058–1071.

Kandoth C, Schultz N, Cherniack AD, Akbani R, Liu Y, Shen H, Robertson AG, Pashtan I, Shen R, Benz CC, et al. 2013. Integrated genomic characterization of endometrial carcinoma. *Nature* **497:** 67–73.

Kennison JA, Tamkun JW. 1988. Dosage-dependent modifiers of polycomb and antennapedia mutations in *Drosophila*. *Proc Natl Acad Sci* **85:** 8136–8140.

Khavari PA, Peterson CL, Tamkun JW, Mendel DB, Crabtree GR. 1993. BRG1 contains a conserved domain of the *SWI2/SNF2* family necessary for normal mitotic growth and transcription. *Nature* **366:** 170–174.

Kia SK, Gorski MM, Giannakopoulos S, Verrijzer CP. 2008. SWI/SNF mediates polycomb eviction and epigenetic reprogramming of the *INK4b-ARF-INK4a* locus. *Mol Cell Biol* **28:** 3457–3464.

Kim KH, Kim W, Howard TP, Vazquez F, Tsherniak A, Wu JN, Wang W, Haswell JR, Walensky LD, Hahn WC, et al. 2015. SWI/SNF-mutant cancers depend on catalytic and non-catalytic activity of EZH2. *Nat Med* **21:** 1491–1496.

Kingston RE, Tamkun JW. 2014. Transcriptional regulation by trithorax-group proteins. *Cold Spring Harb Perspect Biol* **6:** a019349.

Kirkland JG, Raab JR, Kamakaka RT. 2013. TFIIIC bound DNA elements in nuclear organization and insulation. *Biochim Biophys Acta* **1829:** 418–424.

Krasteva V, Buscarlet M, Diaz-Tellez A, Bernard MA, Crabtree GR, Lessard JA. 2012. The BAF53a subunit of SWI/SNF-like BAF complexes is essential for hemopoietic stem cell function. *Blood* **120:** 4720–4732.

Kruger W, Peterson CL, Sil A, Coburn C, Arents G, Moudrianakis EN, Herskowitz I. 1995. Amino acid substitutions in the structured domains of histones H3 and H4 partially relieve the requirement of the yeast SWI/SNF complex for transcription. *Genes Dev* **9:** 2770–2779.

Laurent BC, Treich I, Carlson M. 1993. The yeast SNF2/SWI2 protein has DNA-stimulated ATPase activity required for transcriptional activation. *Genes Dev* **7:** 583–591.

Lawrence MS, Stojanov P, Mermel CH, Robinson JT, Garraway LA, Golub TR, Meyerson M, Gabriel SB, Lander ES, Getz G. 2014. Discovery and saturation analysis of cancer genes across 21 tumour types. *Nature* **505:** 495–501.

Lee HS, Park JH, Kim SJ, Kwon SJ, Kwon J. 2010. A cooperative activation loop among SWI/SNF, γ-H2AX and H3 acetylation for DNA double-strand break repair. *EMBO J* **29:** 1434–1445.

Lee RS, Stewart C, Carter SL, Ambrogio L, Cibulskis K, Sougnez C, Lawrence MS, Auclair D, Mora J, Golub TR, et al. 2012. A remarkably simple genome underlies highly malignant pediatric rhabdoid cancers. *J Clin Invest* **122:** 2983–2988.

Lee JJ, Sholl LM, Lindeman NI, Granter SR, Laga AC, Shivdasani P, Chin G, Luke JJ, Ott PA, Hodi FS, et al. 2015. Targeted next-generation sequencing reveals high frequency of mutations in epigenetic regulators across treatment-naive patient melanomas. *Clin Epigenetics* **7:** 59.

Le Loarer F, Watson S, Pierron G, de Montpreville VT, Ballet S, Firmin N, Auguste A, Pissaloux D, Boyault S, Paindavoine S, et al. 2015. *SMARCA4* inactivation defines a group of undifferentiated thoracic malignancies transcriptionally related to BAF-deficient sarcomas. *Nat Genet* **47:** 1200–1205.

Lemon B, Inouye C, King DS, Tjian R. 2001. Selectivity of chromatin-remodelling cofactors for ligand-activated transcription. *Nature* **414:** 924–928.

Leschziner AE, Lemon B, Tjian R, Nogales E. 2005. Structural studies of the human PBAF chromatin-remodeling complex. *Structure* **13:** 267–275.

Leschziner AE, Saha A, Wittmeyer J, Zhang Y, Bustamante C, Cairns BR, Nogales E. 2007. Conformational flexibility in the chromatin remodeler RSC observed by electron microscopy and the orthogonal tilt reconstruction method. *Proc Natl Acad Sci* **104:** 4913–4918.

Lessard J, Wu JI, Ranish JA, Wan M, Winslow MM, Staahl BT, Wu H, Aebersold R, Graef IA, Crabtree GR. 2007. An essential switch in subunit composition of a chromatin remodeling complex during neural development. *Neuron* **55:** 201–215.

Li XS, Trojer P, Matsumura T, Treisman JE, Tanese N. 2010. Mammalian SWI/SNF—A subunit BAF250/ARID1 is an E3 ubiquitin ligase that targets histone H2B. *Mol Cell Biol* **30:** 1673–1688.

Li M, Zhao H, Zhang X, Wood LD, Anders RA, Choti MA, Pawlik TM, Daniel HD, Kannangai R, Offerhaus GJ, et al. 2011. Inactivating mutations of the chromatin remodeling gene *ARID2* in hepatocellular carcinoma. *Nat Genet* **43:** 828–829.

Li YY, Hanna GJ, Laga AC, Haddad RI, Lorch JH, Hammerman PS. 2015. Genomic analysis of metastatic cutaneous squamous cell carcinoma. *Clin Cancer Res* **21:** 1447–1456.

Lickert H, Takeuchi JK, Von Both I, Walls JR, McAuliffe F, Adamson SL, Henkelman RM, Wrana JL, Rossant J, Bruneau BG. 2004. Baf60c is essential for function of BAF chromatin remodelling complexes in heart development. *Nature* **432:** 107–112.

Lippman Z, Gendrel AV, Black M, Vaughn MW, Dedhia N, McCombie WR, Lavine K, Mittal V, May B, Kasschau KD, et al. 2004. Role of transposable elements in heterochromatin and epigenetic control. *Nature* **430:** 471–476.

Liu R, Liu H, Chen X, Kirby M, Brown PO, Zhao K. 2001. Regulation of CSF1 promoter by the SWI/SNF-like BAF complex. *Cell* **106:** 309–318.

Lunyak VV, Prefontaine GG, Nunez E, Cramer T, Ju BG, Ohgi KA, Hutt K, Roy R, Garcia-Diaz A, Zhu X, et al. 2007. Developmentally regulated activation of a SINE B2 repeat as a domain boundary in organogenesis. *Science* **317:** 248–251.

Manceau G, Letouze E, Guichard C, Didelot A, Cazes A, Corte H, Fabre E, Pallier K, Imbeaud S, Le Pimpec-Barthes F, et al. 2013. Recurrent inactivating mutations of *ARID2* in non-small cell lung carcinoma. *Int J Cancer* **132:** 2217–2221.

Manning BJ, Peterson CL. 2014. Direct interactions promote eviction of the Sir3 heterochromatin protein by the SWI/SNF chromatin remodeling enzyme. *Proc Natl Acad Sci* **111:** 17827–17832.

McKenna ES, Roberts CW. 2009. Epigenetics and cancer without genomic instability. *Cell Cycle* **8:** 23–26.

McKenna ES, Sansam CG, Cho YJ, Greulich H, Evans JA, Thom CS, Moreau LA, Biegel JA, Pomeroy SL, Roberts CW. 2008. Loss of the epigenetic tumor suppressor SNF5 leads to cancer without genomic instability. *Mol Cell Biol* **28:** 6223–6233.

Medina PP, Sanchez-Cespedes M. 2008. Involvement of the chromatin-remodeling factor *BRG1/SMARCA4* in human cancer. *Epigenetics* **3:** 64–68.

Medina PP, Carretero J, Fraga MF, Esteller M, Sidransky D, Sanchez-Cespedes M. 2004. Genetic and epigenetic screening for gene alterations of the chromatin-remodeling factor, *SMARCA4/BRG1*, in lung tumors. *Genes Chromosomes Cancer* **41:** 170–177.

Medina PP, Romero OA, Kohno T, Montuenga LM, Pio R, Yokota J, Sanchez-Cespedes M. 2008. Frequent *BRG1/SMARCA4*-inactivating mutations in human lung cancer cell lines. *Hum Mutat* **29:** 617–622.

Middeljans E, Wan X, Jansen PW, Sharma V, Stunnenberg HG, Logie C. 2012. SS18 together with animal-specific factors defines human BAF-type SWI/SNF complexes. *PLoS ONE* **7:** e33834.

Mohrmann L, Langenberg K, Krijgsveld J, Kal AJ, Heck AJ, Verrijzer CP. 2004. Differential targeting of two distinct SWI/SNF-related *Drosophila* chromatin-remodeling complexes. *Mol Cell Biol* **24:** 3077–3088.

Molnar C, Lopez-Varea A, Hernandez R, de Celis JF. 2006. A gain-of-function screen identifying genes required for vein formation in the *Drosophila melanogaster* wing. *Genetics* **174:** 1635–1659.

Mora-Blanco EL, Mishina Y, Tillman EJ, Cho YJ, Thom CS, Pomeroy SL, Shao W, Roberts CW. 2014. Activation of β-catenin/TCF targets following loss of the tumor suppressor SNF5. *Oncogene* **33:** 933–938.

Moshkin YM, Mohrmann L, van Ijcken WF, Verrijzer CP. 2007. Functional differentiation of SWI/SNF remodelers in transcription and cell cycle control. *Mol Cell Biol* **27:** 651–661.

Naka N, Takenaka S, Araki N, Miwa T, Hashimoto N, Yoshioka K, Joyama S, Hamada K, Tsukamoto Y, Tomita Y, et al. 2010. Synovial sarcoma is a stem cell malignancy. *Stem Cells* **28:** 1119–1131.

Narlikar GJ, Sundaramoorthy R, Owen-Hughes T. 2013. Mechanisms and functions of ATP-dependent chromatin-remodeling enzymes. *Cell* **154:** 490–503.

Neigeborn L, Carlson M. 1984. Genes affecting the regulation of *SUC2* gene expression by glucose repression in *Saccharomyces cerevisiae*. *Genetics* **108:** 845–858.

Neigeborn L, Carlson M. 1987. Mutations causing constitutive invertase synthesis in yeast: Genetic interactions with *snf* mutations. *Genetics* **115:** 247–253.

Niimi A, Chambers AL, Downs JA, Lehmann AR. 2012. A role for chromatin remodellers in replication of damaged DNA. *Nucleic Acids Res* **40:** 7393–7403.

Niimi A, Hopkins SR, Downs JA, Masutani C. 2015. The BAH domain of BAF180 is required for PCNA ubiquitination. *Mutat Res* **779:** 16–23.

Ogiwara H, Ui A, Otsuka A, Satoh H, Yokomi I, Nakajima S, Yasui A, Yokota J, Kohno T. 2011. Histone acetylation by CBP and p300 at double-strand break sites facilitates SWI/SNF chromatin remodeling and the recruitment of non-homologous end joining factors. *Oncogene* **30:** 2135–2146.

Oki M, Valenzuela L, Chiba T, Ito T, Kamakaka RT. 2004. Barrier proteins remodel and modify chromatin to restrict silenced domains. *Mol Cell Biol* **24:** 1956–1967.

Olave I, Wang W, Xue Y, Kuo A, Crabtree GR. 2002. Identification of a polymorphic, neuron-specific chromatin remodeling complex. *Genes Dev* **16:** 2509–2517.

Oruetxebarria I, Venturini F, Kekarainen T, Houweling A, Zuijderduijn LM, Mohd-Sarip A, Vries RG, Hoeben RC, Verrijzer CP. 2004. P16^{INK4a} is required for hSNF5 chromatin remodeler-induced cellular senescence in malignant rhabdoid tumor cells. *J Biol Chem* **279:** 3807–3816.

Pang B, de Jong J, Qiao X, Wessels LF, Neefjes J. 2015. Chemical profiling of the genome with anti-cancer drugs defines target specificities. *Nat Chem Biol* **11:** 472–480.

Papoulas O, Beek SJ, Moseley SL, McCallum CM, Sarte M, Shearn A, Tamkun JW. 1998. The *Drosophila* trithorax group proteins BRM, ASH1 and ASH2 are subunits of distinct protein complexes. *Development* **125:** 3955–3966.

Parnell TJ, Huff JT, Cairns BR. 2008. RSC regulates nucleosome positioning at Pol II genes and density at Pol III genes. *EMBO J* **27:** 100–110.

Patrick KL, Ryan CJ, Xu J, Lipp JJ, Nissen KE, Roguev A, Shales M, Krogan NJ, Guthrie C. 2015. Genetic interaction mapping reveals a role for the SWI/SNF nucleosome remodeler in spliceosome activation in fission yeast. *PLoS Genet* **11:** e1005074.

Peng G, Yim EK, Dai H, Jackson AP, Burgt I, Pan MR, Hu R, Li K, Lin SY. 2009. BRIT1/MCPH1 links chromatin remodelling to DNA damage response. *Nat Cell Biol* **11:** 865–872.

Peterson CL, Herskowitz I. 1992. Characterization of the yeast *SWI1*, *SWI2*, and *SWI3* genes, which encode a global activator of transcription. *Cell* **68:** 573–583.

Peterson CL, Dingwall A, Scott MP. 1994. Five *SWI/SNF* gene products are components of a large multisubunit complex required for transcriptional enhancement. *Proc Natl Acad Sci* **91:** 2905–2908.

Peterson CL, Zhao Y, Chait BT. 1998. Subunits of the yeast SWI/SNF complex are members of the actin-related protein (ARP) family. *J Biol Chem* **273:** 23641–23644.

Phelan ML, Sif S, Narlikar GJ, Kingston RE. 1999. Reconstitution of a core chromatin remodeling complex from SWI/SNF subunits. *Mol Cell* **3:** 247–253.

Prensner JR, Iyer MK, Sahu A, Asangani IA, Cao Q, Patel L, Vergara IA, Davicioni E, Erho N, Ghadessi M, et al. 2013. The long noncoding RNA *SChLAP1* promotes aggressive prostate cancer and antagonizes the SWI/SNF complex. *Nat Genet* **45:** 1392–1398.

Pugh TJ, Weeraratne SD, Archer TC, Pomeranz Krummel DA, Auclair D, Bochicchio J, Carneiro MO, Carter SL, Cibulskis K, Erlich RL, et al. 2012. Medulloblastoma exome sequencing uncovers subtype-specific somatic mutations. *Nature* **488:** 106–110.

Qi W, Wang R, Chen H, Wang X, Xiao T, Boldogh I, Ba X, Han L, Zeng X. 2015. BRG1 promotes the repair of DNA double-strand breaks by facilitating the replacement of RPA with RAD51. *J Cell Sci* **128:** 317–330.

Raab JR, Resnick S, Magnuson T. 2015. Genome-wide transcriptional regulation mediated by biochemically distinct SWI/SNF complexes. *PLoS Genet* **11:** e1005748.

Rada-Iglesias A, Bajpai R, Swigut T, Brugmann SA, Flynn RA, Wysocka J. 2011. A unique chromatin signature uncovers early developmental enhancers in humans. *Nature* **470:** 279–283.

Ramos P, Karnezis AN, Craig DW, Sekulic A, Russell ML, Hendricks WP, Corneveaux JJ, Barrett MT, Shumansky K, Yang Y, et al. 2014. Small cell carcinoma of the ovary, hypercalcemic type, displays frequent inactivating germline and somatic mutations in *SMARCA4*. *Nat Genet* **46:** 427–429.

Rando OJ, Zhao K, Janmey P, Crabtree GR. 2002. Phosphatidylinositol-dependent actin filament binding by the SWI/SNF-like BAF chromatin remodeling complex. *Proc Natl Acad Sci* **99:** 2824–2829.

Reinhold WC, Sunshine M, Liu H, Varma S, Kohn KW, Morris J, Doroshow J, Pommier Y. 2012. CellMiner: A web-based suite of genomic and pharmacologic tools to explore transcript and drug patterns in the NCI-60 cell line set. *Cancer Res* **72:** 3499–3511.

Reisman DN, Strobeck MW, Betz BL, Sciarrotta J, Funkhouser W Jr, Murchardt C, Yaniv M, Sherman LS, Knudsen ES, Weissman BE. 2002. Concomitant down-regulation of BRM and BRG1 in human tumor cell lines: Differential effects on RB-mediated growth arrest vs CD44 expression. *Oncogene* **21:** 1196–1207.

Rendina R, Strangi A, Avallone B, Giordano E. 2010. Bap170, a subunit of the *Drosophila* PBAP chromatin remodeling complex, negatively regulates the EGFR signaling. *Genetics* **186:** 167–181.

Richmond E, Peterson CL. 1996. Functional analysis of the DNA-stimulated ATPase domain of yeast SWI2/SNF2. *Nucleic Acids Res* **24:** 3685–3692.

Rizvi NA, Hellmann MD, Snyder A, Kvistborg P, Makarov V, Havel JJ, Lee W, Yuan J, Wong P, Ho TS, et al. 2015. Cancer immunology. Mutational landscape determines sensitivity to PD-1 blockade in non-small cell lung cancer. *Science* **348:** 124–128.

Roberts CW, Biegel JA. 2009. The role of SMARCB1/INI1 in development of rhabdoid tumor. *Cancer Biol Ther* **8:** 412–416.

Roberts CW, Galusha SA, McMenamin ME, Fletcher CD, Orkin SH. 2000. Haploinsufficiency of Snf5 (integrase interactor 1) predisposes to malignant rhabdoid tumors in mice. *Proc Natl Acad Sci* **97:** 13796–13800.

Santen GW, Aten E, Sun Y, Almomani R, Gilissen C, Nielsen M, Kant SG, Snoeck IN, Peeters EA, Hilhorst-Hofstee Y, et al. 2012. Mutations in SWI/SNF chromatin remodeling complex gene *ARID1B* cause Coffin–Siris syndrome. *Nat Genet* **44:** 379–380.

Sausen M, Leary RJ, Jones S, Wu J, Reynolds CP, Liu X, Blackford A, Parmigiani G, Diaz LA Jr, Papadopoulos N, et al. 2013. Integrated genomic analyses identify *ARID1A* and *ARID1B* alterations in the childhood cancer neuroblastoma. *Nat Genet* **45:** 12–17.

Schnitzler GR, Cheung CL, Hafner JH, Saurin AJ, Kingston RE, Lieber CM. 2001. Direct imaging of human SWI/SNF-remodeled mono- and polynucleosomes by atomic force microscopy employing carbon nanotube tips. *Mol Cell Biol* **21:** 8504–8511.

Schoenfelder S, Sugar R, Dimond A, Javierre BM, Armstrong H, Mifsud B, Dimitrova E, Matheson L, Tavares-Cadete F, Furlan-Magaril M, et al. 2015. Polycomb repressive complex PRC1 spatially constrains the mouse embryonic stem cell genome. *Nat Genet* **47:** 1179–1186.

Schubert HL, Wittmeyer J, Kasten MM, Hinata K, Rawling DC, Heroux A, Cairns BR, Hill CP. 2013. Structure of an actin-related subcomplex of the SWI/SNF chromatin remodeler. *Proc Natl Acad Sci* **110:** 3345–3350.

Shain AH, Pollack JR. 2013. The spectrum of SWI/SNF mutations, ubiquitous in human cancers. *PLoS ONE* **8:** e55119.

Shain AH, Garrido M, Botton T, Talevich E, Yeh I, Sanborn JZ, Chung J, Wang NJ, Kakavand H, Mann GJ, et al. 2015. Exome sequencing of desmoplastic melanoma identifies recurrent *NFKBIE* promoter mutations and diverse activating mutations in the MAPK pathway. *Nat Genet* **47:** 1194–1199.

Shaver S, Casas-Mollano JA, Cerny RL, Cerutti H. 2010. Origin of the polycomb repressive complex 2 and gene silencing by an E(z) homolog in the unicellular alga *Chlamydomonas*. *Epigenetics* **5:** 301–312.

Shen X, Ranallo R, Choi E, Wu C. 2003. Involvement of actin-related proteins in ATP-dependent chromatin remodeling. *Mol Cell* **12:** 147–155.

Shi J, Wang E, Milazzo JP, Wang Z, Kinney JB, Vakoc CR. 2015. Discovery of cancer drug targets by CRISPR-Cas9 screening of protein domains. *Nat Biotechnol* **33:** 661–667.

Shim EY, Ma JL, Oum JH, Yanez Y, Lee SE. 2005. The yeast chromatin remodeler RSC complex facilitates end joining repair of DNA double-strand breaks. *Mol Cell Biol* **25:** 3934–3944.

Simon JA, Tamkun JW. 2002. Programming off and on states in chromatin: Mechanisms of polycomb and trithorax group complexes. *Curr Opin Genet Dev* **12:** 210–218.

Cite this article as *Cold Spring Harb Perspect Med* doi: 10.1101/cshperspect.a026930

Singhal N, Graumann J, Wu G, Arauzo-Bravo MJ, Han DW, Greber B, Gentile L, Mann M, Scholer HR. 2010. Chromatin-remodeling components of the BAF complex facilitate reprogramming. *Cell* **141**: 943–955.

Skene PJ, Hernandez AE, Groudine M, Henikoff S. 2014. The nucleosomal barrier to promoter escape by RNA polymerase II is overcome by the chromatin remodeler Chd1. *eLife* **3**: e02042.

Smith CL, Peterson CL. 2005. A conserved Swi2/Snf2 ATPase motif couples ATP hydrolysis to chromatin remodeling. *Mol Cell Biol* **25**: 5880–5892.

Smith CL, Horowitz-Scherer R, Flanagan JF, Woodcock CL, Peterson CL. 2003. Structural analysis of the yeast SWI/SNF chromatin remodeling complex. *Nat Struct Biol* **10**: 141–145.

Sperling AS, Jeong KS, Kitada T, Grunstein M. 2011. Topoisomerase II binds nucleosome-free DNA and acts redundantly with topoisomerase I to enhance recruitment of RNA Pol II in budding yeast. *Proc Natl Acad Sci* **108**: 12693–12698.

Srivas R, Costelloe T, Carvunis AR, Sarkar S, Malta E, Sun SM, Pool M, Licon K, van Welsem T, van Leeuwen F, et al. 2013. A UV-induced genetic network links the RSC complex to nucleotide excision repair and shows dose-dependent rewiring. *Cell Rep* **5**: 1714–1724.

Staahl BT, Tang J, Wu W, Sun A, Gitler AD, Yoo AS, Crabtree GR. 2013. Kinetic analysis of npBAF to nBAF switching reveals exchange of SS18 with CREST and integration with neural developmental pathways. *J Neurosci* **33**: 10348–10361.

Stern M, Jensen R, Herskowitz I. 1984. Five *SWI* genes are required for expression of the *HO* gene in yeast. *J Mol Biol* **178**: 853–868.

Sternberg PW, Stern MJ, Clark I, Herskowitz I. 1987. Activation of the yeast *HO* gene by release from multiple negative controls. *Cell* **48**: 567–577.

Szerlong H, Saha A, Cairns BR. 2003. The nuclear actin-related proteins Arp7 and Arp9: A dimeric module that cooperates with architectural proteins for chromatin remodeling. *EMBO J* **22**: 3175–3187.

Szerlong H, Hinata K, Viswanathan R, Erdjument-Bromage H, Tempst P, Cairns BR. 2008. The HSA domain binds nuclear actin-related proteins to regulate chromatin-remodeling ATPases. *Nat Struct Mol Biol* **15**: 469–476.

Tamkun JW, Deuring R, Scott MP, Kissinger M, Pattatucci AM, Kaufman TC, Kennison JA. 1992. *brahma*: A regulator of *Drosophila* homeotic genes structurally related to the yeast transcriptional activator SNF2/SWI2. *Cell* **68**: 561–572.

Taulli R, Foglizzo V, Morena D, Coda DM, Ala U, Bersani F, Maestro N, Ponzetto C. 2014. Failure to downregulate the BAF53a subunit of the SWI/SNF chromatin remodeling complex contributes to the differentiation block in rhabdomyosarcoma. *Oncogene* **33**: 2354–2362.

Terriente-Felix A, de Celis JF. 2009. Osa, a subunit of the BAP chromatin-remodelling complex, participates in the regulation of gene expression in response to EGFR signalling in the *Drosophila* wing. *Dev Biol* **329**: 350–361.

The Cancer Genome Atlas. 2012a. Comprehensive molecular characterization of human colon and rectal cancer. *Nature* **487**: 330–337.

The Cancer Genome Atlas. 2012b. Comprehensive molecular portraits of human breast tumours. *Nature* **490**: 61–70.

The Cancer Genome Atlas. 2014a. Comprehensive molecular characterization of gastric adenocarcinoma. *Nature* **513**: 202–209.

The Cancer Genome Atlas. 2014b. Comprehensive molecular characterization of urothelial bladder carcinoma. *Nature* **507**: 315–322.

The Cancer Genome Atlas. 2014c. Comprehensive molecular profiling of lung adenocarcinoma. *Nature* **511**: 543–550.

The Cancer Genome Atlas. 2015. Genomic classification of cutaneous melanoma. *Cell* **161**: 1681–1696.

Thoma NH, Czyzewski BK, Alexeev AA, Mazin AV, Kowalczykowski SC, Pavletich NP. 2005. Structure of the SWI2/SNF2 chromatin-remodeling domain of eukaryotic Rad54. *Nat Struct Mol Biol* **12**: 350–356.

Trouche D, Le Chalony C, Muchardt C, Yaniv M, Kouzarides T. 1997. RB and hbrm cooperate to repress the activation functions of E2F1. *Proc Natl Acad Sci* **94**: 11268–11273.

Tsurusaki Y, Okamoto N, Ohashi H, Kosho T, Imai Y, Hibi-Ko Y, Kaname T, Naritomi K, Kawame H, Wakui K, et al. 2012. Mutations affecting components of the SWI/SNF complex cause Coffin–Siris syndrome. *Nat Genet* **44**: 376–378.

Tyagi A, Ryme J, Brodin D, Ostlund Farrants AK, Visa N. 2009. SWI/SNF associates with nascent pre-mRNPs and regulates alternative pre-mRNA processing. *PLoS Genet* **5**: e1000470.

Varela I, Tarpey P, Raine K, Huang D, Ong CK, Stephens P, Davies H, Jones D, Lin ML, Teague J, et al. 2011. Exome sequencing identifies frequent mutation of the SWI/SNF complex gene *PBRM1* in renal carcinoma. *Nature* **469**: 539–542.

Versteege I, Sevenet N, Lange J, Rousseau-Merck MF, Ambros P, Handgretinger R, Aurias A, Delattre O. 1998. Truncating mutations of hSNF5/INI1 in aggressive paediatric cancer. *Nature* **394**: 203–206.

Vogel-Ciernia A, Matheos DP, Barrett RM, Kramar EA, Azzawi S, Chen Y, Magnan CN, Zeller M, Sylvain A, Haettig J, et al. 2013. The neuron-specific chromatin regulatory subunit BAF53b is necessary for synaptic plasticity and memory. *Nat Neurosci* **16**: 552–561.

Waldholm J, Wang Z, Brodin D, Tyagi A, Yu S, Theopold U, Farrants AK, Visa N. 2011. SWI/SNF regulates the alternative processing of a specific subset of pre-mRNAs in *Drosophila melanogaster*. *BMC Mol Biol* **12**: 46.

Walker JE, Saraste M, Runswick MJ, Gay NJ. 1982. Distantly related sequences in the α- and β-subunits of ATP synthase, myosin, kinases and other ATP-requiring enzymes and a common nucleotide binding fold. *EMBO J* **1**: 945–951.

Wang W, Cote J, Xue Y, Zhou S, Khavari PA, Biggar SR, Muchardt C, Kalpana GV, Goff SP, Yaniv M, et al. 1996a. Purification and biochemical heterogeneity of the mammalian SWI-SNF complex. *EMBO J* **15**: 5370–5382.

Wang W, Xue Y, Zhou S, Kuo A, Cairns BR, Crabtree GR. 1996b. Diversity and specialization of mammalian SWI/SNF complexes. *Genes Dev* **10**: 2117–2130.

Wang W, Chi T, Xue Y, Zhou S, Kuo A, Crabtree GR. 1998. Architectural DNA binding by a high-mobility-group/kinesin-like subunit in mammalian SWI/SNF-related complexes. *Proc Natl Acad Sci* **95:** 492–498.

Wang K, Kan J, Yuen ST, Shi ST, Chu KM, Law S, Chan TL, Kan Z, Chan AS, Tsui WY, et al. 2011a. Exome sequencing identifies frequent mutation of *ARID1A* in molecular subtypes of gastric cancer. *Nat Genet* **43:** 1219–1223.

Wang X, Werneck MB, Wilson BG, Kim HJ, Kluk MJ, Thom CS, Wischhusen JW, Evans JA, Jesneck JL, Nguyen P, et al. 2011b. TCR-dependent transformation of mature memory phenotype T cells in mice. *J Clin Invest* **121:** 3834–3845.

Wang H, Maurano MT, Qu H, Varley KE, Gertz J, Pauli F, Lee K, Canfield T, Weaver M, Sandstrom R, et al. 2012. Widespread plasticity in CTCF occupancy linked to DNA methylation. *Genome Res* **22:** 1680–1688.

Wang K, Yuen ST, Xu J, Lee SP, Yan HH, Shi ST, Siu HC, Deng S, Chu KM, Law S, et al. 2014. Whole-genome sequencing and comprehensive molecular profiling identify new driver mutations in gastric cancer. *Nat Genet* **46:** 573–582.

Wani AH, Boettiger AN, Schordet P, Ergun A, Munger C, Sadreyev RI, Zhuang X, Kingston RE, Francis NJ. 2016. Chromatin topology is coupled to Polycomb group protein subnuclear organization. *Nat Commun* **7:** 10291.

Watanabe R, Ui A, Kanno S, Ogiwara H, Nagase T, Kohno T, Yasui A. 2014. SWI/SNF factors required for cellular resistance to DNA damage include ARID1A and ARID1B and show interdependent protein stability. *Cancer Res* **74:** 2465–2475.

Wiegand KC, Shah SP, Al-Agha OM, Zhao Y, Tse K, Zeng T, Senz J, McConechy MK, Anglesio MS, Kalloger SE, et al. 2010. *ARID1A* mutations in endometriosis-associated ovarian carcinomas. *N Engl J Med* **363:** 1532–1543.

Wijdeven RH, Pang B, van der Zanden SY, Qiao X, Blomen V, Hoogstraat M, Lips EH, Janssen L, Wessels L, Brummelkamp TR, et al. 2015. Genome-wide identification and characterization of novel factors conferring resistance to topoisomerase II poisons in cancer. *Cancer Res* **75:** 4176–4187.

Wilson BG, Wang X, Shen X, McKenna ES, Lemieux ME, Cho YJ, Koellhoffer EC, Pomeroy SL, Orkin SH, Roberts CW. 2010. Epigenetic antagonism between polycomb and SWI/SNF complexes during oncogenic transformation. *Cancer Cell* **18:** 316–328.

Wilson BG, Helming KC, Wang X, Kim Y, Vazquez F, Jagani Z, Hahn WC, Roberts CW. 2014. Residual complexes containing SMARCA2 (BRM) underlie the oncogenic drive of *SMARCA4* (*BRG1*) mutation. *Mol Cell Biol* **34:** 1136–1144.

Winston F, Carlson M. 1992. Yeast SNF/SWI transcriptional activators and the SPT/SIN chromatin connection. *Trends Genet* **8:** 387–391.

Wollmann P, Cui S, Viswanathan R, Berninghausen O, Wells MN, Moldt M, Witte G, Butryn A, Wendler P, Beckmann R, et al. 2011. Structure and mechanism of the Swi2/Snf2 remodeller Mot1 in complex with its substrate TBP. *Nature* **475:** 403–407.

Wong AK, Shanahan F, Chen Y, Lian L, Ha P, Hendricks K, Ghaffari S, Iliev D, Penn B, Woodland AM, et al. 2000. *BRG1*, a component of the SWI-SNF complex, is mutated in multiple human tumor cell lines. *Cancer Res* **60:** 6171–6177.

Wu JI, Lessard J, Olave IA, Qiu Z, Ghosh A, Graef IA, Crabtree GR. 2007. Regulation of dendritic development by neuron-specific chromatin remodeling complexes. *Neuron* **56:** 94–108.

Xia L, Jaafar L, Cashikar A, Flores-Rozas H. 2007. Identification of genes required for protection from doxorubicin by a genome-wide screen in *Saccharomyces cerevisiae*. *Cancer Res* **67:** 11411–11418.

Xu F, Flowers S, Moran E. 2012. Essential role of ARID2 protein-containing SWI/SNF complex in tissue-specific gene expression. *J Biol Chem* **287:** 5033–5041.

Xue Y, Canman JC, Lee CS, Nie Z, Yang D, Moreno GT, Young MK, Salmon ED, Wang W. 2000. The human SWI/SNF-B chromatin-remodeling complex is related to yeast rsc and localizes at kinetochores of mitotic chromosomes. *Proc Natl Acad Sci* **97:** 13015–13020.

Yan J, Enge M, Whitington T, Dave K, Liu J, Sur I, Schmierer B, Jolma A, Kivioja T, Taipale M, et al. 2013. Transcription factor binding in human cells occurs in dense clusters formed around cohesin anchor sites. *Cell* **154:** 801–813.

Yoo AS, Staahl BT, Chen L, Crabtree GR. 2009. MicroRNA-mediated switching of chromatin-remodelling complexes in neural development. *Nature* **460:** 642–646.

Yoo AS, Sun AX, Li L, Shcheglovitov A, Portmann T, Li Y, Lee-Messer C, Dolmetsch RE, Tsien RW, Crabtree GR. 2011. MicroRNA-mediated conversion of human fibroblasts to neurons. *Nature* **476:** 228–231.

Zhang X, Sun Q, Shan M, Niu M, Liu T, Xia B, Liang X, Wei W, Sun S, Zhang Y, et al. 2013. Promoter hypermethylation of *ARID1A* gene is responsible for its low mRNA expression in many invasive breast cancers. *PLoS ONE* **8:** e53931.

Zhao K, Wang W, Rando OJ, Xue Y, Swiderek K, Kuo A, Crabtree GR. 1998. Rapid and phosphoinositol-dependent binding of the SWI/SNF-like BAF complex to chromatin after T lymphocyte receptor signaling. *Cell* **95:** 625–636.

DNMT3A in Leukemia

Lorenzo Brunetti,[1,2,3,6] Michael C. Gundry,[1,2,4,6] and Margaret A. Goodell[1,2,3,5]

[1]Stem Cells and Regenerative Medicine Center, Baylor College of Medicine, Houston, Texas 77030

[2]Center for Cell and Gene Therapy, Baylor College of Medicine, Houston, Texas 77030

[3]CREO, University of Perugia, 06123 Perugia, Italy

[4]Department of Molecular & Human Genetics, Baylor College of Medicine, Houston, Texas 77030

[5]Texas Children's Hospital, and Houston Methodist Hospital, Houston, Texas 77030

Correspondence: goodell@bcm.edu

DNA methylation is an epigenetic process involved in development, aging, and cancer. Although the advent of new molecular techniques has enhanced our knowledge of how DNA methylation alters chromatin and subsequently affects gene expression, a direct link between epigenetic marks and tumorigenesis has not been established. DNMT3A is a de novo DNA methyltransferase that has recently gained relevance because of its frequent mutation in a large variety of immature and mature hematologic neoplasms. *DNMT3A* mutations are early events during cancer development and seem to confer poor prognosis to acute myeloid leukemia (AML) patients making this gene an attractive target for new therapies. Here, we discuss the biology of DNMT3A and its role in controlling hematopoietic stem cell fate decisions. In addition, we review how mutant DNMT3A may contribute to leukemogenesis and the clinical relevance of *DNMT3A* mutations in hematologic cancers.

D NA methylation is an epigenetic modification involved in key cellular processes such as transcriptional repression, genomic imprinting, and the suppression of repetitive elements. The first suggestion of a link between DNA methylation and cancer was the observation that human cancers tend to display global hypomethylation compared with normal controls (Feinberg and Vogelstein 1983). Subsequently, the field switched attention to focally hypermethylated regions with the hypothesis that epigenetic silencing of tumor suppressor genes through promoter hypermethylation would drive gene silencing, obviating the need for genetic inactivation of these pathways (Jones

and Laird 1999). Investigators then began looking for possible genes/pathways responsible for the observed methylation changes.

The deposition and maintenance of DNA methyl marks is orchestrated by DNA methyltransferases. In mammals, three genes encoding proteins with DNA methyltransferase activity have been identified: *DNMT1*, *DNMT3A*, and *DNMT3B* (Okano et al. 1998). DNMT3A and DNMT3B proteins are responsible for establishing the patterns of DNA methylation early in embryogenesis through de novo methylation of unmethylated CpG sites (Okano et al. 1999), and DNMT1 maintains such patterns throughout cell division by targeting hemimethylated

---

[6]These authors contributed equally to this work.

Cite this article as *Cold Spring Harb Perspect Med* doi: 10.1101/cshperspect.a030320

DNA and copying the methyl mark onto the nascent DNA strand. All three proteins are essential for mammalian development (Okano et al. 1999).

Since the discovery of altered DNA methylation patterns in human cancers, an abundance of data has been collected on the expression levels of the DNMT genes in tumors, with overexpression of DNMT1 the most frequently observed change (Issa et al. 1993; Robertson et al. 1999). In addition, overexpression of the de novo methyltransferases (DNMT3A/DNMT3B) has been broadly reported in cancers (Mizuno et al. 2001; Girault et al. 2003; Oh et al. 2007; Rahman et al. 2010), supporting an oncogenic role for the DNMT family proteins. The success of DNA hypomethylating agents, including 5-azacytidine and decitabine, in the treatment of a multiple hematologic malignancies further supported this hypothesis (Santini 2012).

It, therefore, came as a surprise when genomic studies uncovered loss-of-function mutations in the DNMT family across multiple cancers (Gao et al. 2011; Yang et al. 2015). Most strikingly, *DNMT3A* mutations are frequently detected in a variety of adult hematologic malignancies, often occurring as early events during leukemogenesis (Roller et al. 2013; Shlush et al. 2014; Yang et al. 2015). Since this discovery, the development of mouse models and biochemical studies on DNMT3A have improved our understanding of the function of DNMT3A and its role in normal and malignant hematopoiesis. However, significant questions remain unanswered regarding both molecular mechanisms and clinical prognostic and therapeutic value. Herein, we review the known biology of DNMT3A and discuss recent clinical studies with therapeutic implications for *DNMT3A* mutations in cancer.

DNMT3A STRUCTURE

DNMT3A is a 130-kDa protein that is highly conserved in vertebrates, with 98% homology between humans and mice (Okano et al. 1998; Xie et al. 1999). The gene is encoded by 23 exons on human chromosome 2p23 and is expressed in two major forms: a long isoform,

DNMT3A1, and a short isoform, DNMT3A2 (Fig. 1), which has been the subject of most studies because of its predominant expression in mouse embryonic stem (mES) cells (Chen et al. 2002). DNTM3A2 has been shown to interact with DNMT3L, a catalytically inactive member of the methyltransferase family whose expression is restricted to germ cells and early embryogenesis (Hata et al. 2002). Disruption of this interaction through loss of either protein leads to defects in imprinting and spermatogenesis (Kaneda et al. 2004; Nimura et al. 2006).

DNMT3A contains three well-studied domains. The Pro-Trp-Trp-Pro (PWWP) domain synergistically binds DNA and specific histone marks in gene bodies (see more below) (Wang et al. 2014). The ATRX-DNMT3-DNMT3L (ADD) domain binds H3 tail peptides (Otani et al. 2009). The *S*-adenosyl methionine (SAM)-dependent methyltransferase C5-type domain encodes the highly conserved methyltransferase (MTase) domain (Fig. 1), which recognizes and binds DNA, and after binding the cofactor SAM, attaches a methyl group to the C5 position of cytosine. The long isoform also contains 219 extra amino acids at the amino terminus, which have been shown to enhance the DNA-binding affinity and methylation activity of the protein in vitro (Suetake et al. 2011). De novo germline mutations in each of the three major domains occur in the recently described Tatton-Brown–Rahman syndrome (or *DNMT3A* overgrowth syndrome), which is characterized by tall stature, a distinctive facial appearance, and intellectual disability (Tatton-Brown et al. 2014).

In the nucleus, DNMT3A can exist in oligomeric form as dimers, tetramers, and larger structures through two distinct binding interfaces in the MTase domain. The oligomers are composed of homodimeric DNMT3A molecules or heterodimeric DNMT3A–DNMT3L molecules; however, the presence of a single binding interface in DNMT3L prevents further oligomerization past a 3L-3A-3A-3L tetramer state (Jurkowska et al. 2011). The structure of the tetramer allows for coordinated methylation of two independent CpG sites at an average distance of ∼9 bp (Zhang et al. 2009). Importantly, both the DNA and SAM binding affinity of

Cite this article as *Cold Spring Harb Perspect Med* doi: 10.1101/cshperspect.a030320

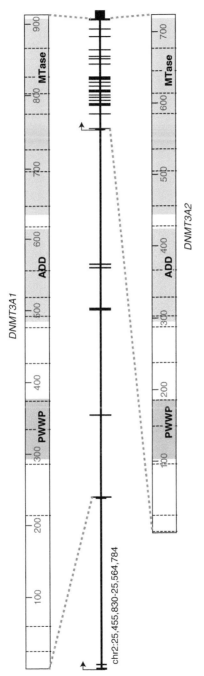

Figure 1. DNMT3A genomic locus. The *DNMT3A* gene is composed of 23 exons distributed across ~110,000 bp on chromosome 2. The two major isoforms, *DNMT3A1* and *DNMT3A2*, are depicted along with the promoters and exons from which they are transcribed. The vertical dotted lines drawn on the different isoforms represent exon–exon junctions.

DNMT3A, as well as the processivity of the enzyme, are tied to the oligomeric state of DNMT3A (Holz-Schietinger et al. 2011). In addition to programmed changes in oligomerization, such as those driven by developmental changes in DNMT3L expression and differential DNMT3A/3B isoform usage, a number of pathologic alterations including mutations at DNMT3A binding interfaces (Holz-Schietinger et al. 2012) have been shown to affect oligomerization and alter cell behavior. So far, oligomerization in adult somatic tissues, which lack DNMT3L, has not been assessed. It is possible that complex oligomers composed of different isoforms of DNMT3A and DNMT3B exist and that the dosages of these complexes affect regional de novo methylation. Further work is needed to better understand the regulation and importance of oligomerization on cellular phenotypes including cancer.

DNMT3A AND DNA METHYLATION

The distribution of mCpGs in the mammalian genome is nonrandom and the mechanism responsible for targeting DNMT3A and DNMT3B to specific genomic regions remains largely unknown. However, functional characterization of DNMT3A domains and studies on DNMT3A binding patterns and de novo methylation activity in embryonic stem cells (ESCs) have shed some light on the intrinsic and extrinsic factors controlling regional specificity.

Each of the three major domains in DNMT3A has been shown to associate with specific epigenetic marks. Crystal structure analysis recently showed an elegant autoinhibitory mechanism for DNMT3A, wherein the ADD domain physically blocks the MTase domain (Guo et al. 2015). When histone H3 with an unmodified lysine 4 tail is encountered (H3K4me0), the ADD–MTase interaction is disrupted, resulting in movement of the ADD domain enabling enzymatic activity. However, in the presence of amino-terminal H3 tail modifications, especially H3 lysine 4 trimethylation (H3K4me3), release from autoinhibition is significantly less efficient, providing a mechanism to ensure that actively expressed genes (marked by H3K4me3) are not subject to DNA methylation (Ooi et al. 2007; Otani et al. 2009; Zhang et al. 2010). Mutations in the ADD domain are found in some hematologic cancers. Although most cancer-associated *DNMT3A* mutations are presumed to represent a loss of function, it is conceivable that ADD domain mutations could result in loss of autoinhibition, leading to a gain of function and aberrant hypermethylation of actively expressed genes marked by histone H3K4me3, ultimately leading to their suppression; this scenario remains to be tested.

Similarly, the PWWP domain binds DNA and trimethylated histone H3 lysine 36 (H3K36me3) (Dhayalan et al. 2010). H3K36me3 marks active chromatin and is highly enriched in gene bodies and repetitive elements, including pericentromeric DNA (Chantalat et al. 2011). Indeed, DNMT3A colocalizes with pericentromic heterochromatic foci, and disruption of the PWWP domain in ES cells leads to a diffuse nuclear patterning of DNMT3A and a loss of methylation activity at major satellite repeats in pericentric heterochromatin (Chen et al. 2004). During gametogenesis and early embryogenesis, protection of the germline through suppression of repetitive elements is critical; this suppression may represent a major function of the PWWP domain during embryonic development. Furthermore, both the H3K36 methyltransferase SETD2 and the PWWP domain of DNMT3B have been shown to be required for targeting de novo methylation to transcribed genes (Baubec et al. 2015); thus, it is possible that the PWWP domain of DNMT3A acts in a similar manner.

Finally, the MTase domain of DNMT3A has been shown to preferentially bind unmethylated DNA (Yokochi and Robertson 2002). In vitro experiments have shown that some CpG flanking bases, specifically AT-rich sequences, can enhance methylation activity by several orders of magnitude (Lin et al. 2002), consistent with genome-wide data on enriched sequence motifs at methylated CpGs (Handa and Jeltsch 2005). As discussed previously, the MTase domain also cooperatively oligomerizes on DNA, forming nucleoprotein filaments that enhance the activity of the enzyme and confer a processive-like property to the protein (Emperle et al. 2014). The majority of somatic mutations in DNMT3A are heterozygous and act to disrupt this oligomerization, suggesting a key role for oligomerization in the normal function or stability of the protein.

Other chromatin-modifying enzymes have also been shown to associate with DNMT3A and may impact DNA methylation patterns. The majority of these binding partners associate either directly or indirectly with the ADD domain, including heterochromatin protein 1 (HP1), histone deacetylase 1 (HDAC1), UHRF1, and the histone lysine *N*-methyltransferases (EZH2, SUV39H1/2, SETDB1, and G9a) (Ayyanathan et al. 2003). The interplay between histone modifications and DNA methylation is complex, and the order of acquisition of histone marks and DNA methylation differs according to the specific histone mark and chromatin enzyme involved (Cedar and Bergman 2009). More data is needed on the complex relationships between histone marks and DNA methylation to trace the origin of epigenetic patterning during embryogenesis. Additionally, the relationship between these epigenetic marks in adult somatic tissues has not been studied and may be key to understanding the role of DNMT3A in cancer.

DNMT3A IN HEMATOPOIESIS

Although the majority of studies on DNMT3A have been performed in ESCs and germ cells, recent work indicates that DNMT3A may also regulate somatic stem cell differentiation (Challen et al. 2012; Hu et al. 2012; Wu et al. 2012; Dhawan et al. 2015). In the mouse hematopoietic system, deletion of *Dnmt3a* leads to a preference for self-renewal over differentiation and gradual expansion of the long-term hematopoietic stem cell (LT-HSC) compartment on serial transplantation (Fig. 2). A similar alteration in stem-cell fate is observed on loss of DNMT3A in neural stem cells (Hu et al. 2012) and on loss of DNMT3A/3B in mouse and human ES cells (Chen et al. 2003; Challen et al. 2012; Liao et al. 2015). Deletion of *Dnmt3b* does not appear to have a major effect on the LT-HSC

compartment, which is not surprising as the predominant DNMT3B isoform expressed in the hematopoietic system is enzymatically inactive (Challen et al. 2014). However, codeletion of *Dnmt3a/3b* enhanced LT-HSC expansion and virtually blocked differentiation, signifying residual DNMT3B activity.

Dnmt3a-null HSCs are phenotypically indistinguishable from wild-type cells, with similar levels of proliferation and apoptosis and an identical cell surface marker profile. *Dnmt3a*-null HSCs retain the ability to differentiate into all hematopoietic lineages, albeit with a lower differentiation quotient or reduced level of output per LT-HSC (Fig. 2). From a molecular standpoint, *Dnmt3a*-null HSCs display significant genome-wide hypomethylation with focal areas of hypermethylation that can be attributed to residual DNMT3B activity. Differentially methylated regions (DMRs) identified in *Dnmt3a*-null HSCs are distributed across all genomic features, but are significantly enriched at the edges of large hypomethylated regions known as methylation canyons (Jeong et al. 2014). These features were recently identified in multiple cell types (Xie et al. 2013), and often occur near developmental regulatory genes such as homeobox-containing genes. Analysis of gene expression changes after loss of DNMT3A revealed that many canyon-associated genes involved in HSC self-renewal are overexpressed in *Dnmt3a*-null HSCs and remain inappropriately turned on in differentiated progeny cells (Challen et al. 2012; Jeong et al. 2014). However, no clear relationship between canyon hypomethylation, altered gene expression, and phenotype has been established.

The expansion of *Dnmt3a*-null cells observed on competitive transplantation represents a potential mechanism for early leukemic transformation. In this model, cells that spontaneously acquire a mutation in *DNMT3A* would be more likely to self-renew than differentiate, and would therefore be protected from stem-cell exhaustion. Under conditions of stress, in which a larger fraction of the HSC pool is forced to cycle, selection and subsequent expansion of the *DNMT3A* mutant HSC pool could be even more pronounced. Indeed, trans-

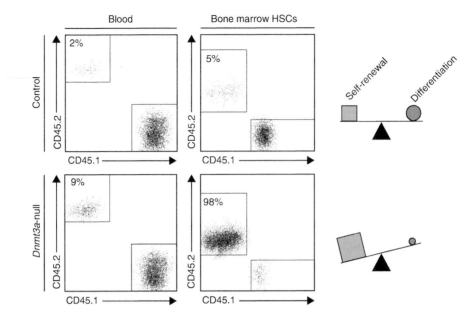

Figure 2. DNMT3A controls hematopoietic stem cell fate. In mice, serial competitive transplantation of wild-type (WT) hematopoietic stem cells (HSCs) (CD45.2) eventually leads to less efficient engraftment. The contribution of engrafted HSCs to hematopoiesis is proportional to their level of engraftment in the bone marrow (*top*). In contrast, *Dnmt3a*-null HSCs do not exhaust and instead begin to outcompete CD45.1 WT cells, resulting in an expanded contribution to the HSC compartment. The contribution of these *Dnmt3a*-null HSCs to blood production is minimal, reflecting an imbalance between self-renewal and differentiation driven by loss of DNMT3A (*bottom*).

planted *Dnmt3a*-null HSCs led to development of an array of myeloid and lymphoid neoplasms that recapitulate the full spectrum of human malignancies harboring *DNMT3A* mutations (Celik et al. 2015; Mayle et al. 2015).

DNMT3A MUTATIONS AND CANCER

Since the first description of *DNMT3A* mutations in acute myeloid leukemia (AML) (Ley et al. 2010), multiple exome and targeted resequencing studies have identified *DNMT3A* mutations in AML (Shlush et al. 2014; Klco et al. 2015; Ivey et al. 2016) as well as in a variety of other adult myeloid and lymphoid neoplasms (Roller et al. 2013). Interestingly, *DNMT3A* mutations are very rare in pediatric blood cancers, with the few identified mutations found in adolescents (Ho et al. 2011; Thol et al. 2011b; Shiba et al. 2012).

The majority of *DNMT3A* mutations found in hematopoietic disorders occur within the methyltransferase domain (MTD), with a significant enrichment for mutations at codon R882 (Fig. 3) (Cancer Genome Atlas Research 2013; Roller et al. 2013). However, non-R882 missense and truncating mutations are found in each of the major domains (Cancer Genome Atlas Research 2013; Roller et al. 2013). In AML and other myeloid neoplasms, R882 mutations are usually heterozygous (Ley et al. 2010; Cancer Genome Atlas Research 2013; Gaidzik et al. 2013; Gale et al. 2015). However, in T-cell acute lymphoblastic leukemia (T-ALL), a high frequency of non-R882 biallelic mutations has been reported (Grossmann et al. 2013; Roller et al. 2013), suggesting different selective pressures on *DNMT3A* mutations in myeloid and lymphoid disorders. Data from *Dnmt3a/Flt3-ITD* double-mutant mice are consistent with an influence of DNMT3A dosage. $Dnmt3a^{-/-}$ cells can generate both myeloid and lymphoid malignancies (Yang et al. 2016), whereas $Dnmt3a^{+/-}$ cells are more likely to develop

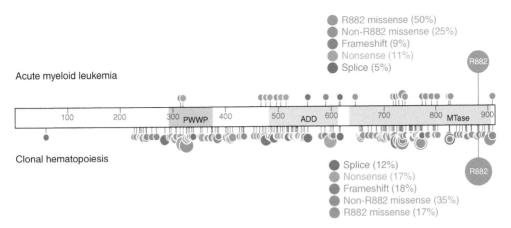

Figure 3. *DNMT3A* mutations in acute myeloid leukemia (AML) and clonal hematopoiesis. The spectrum of *DNMT3A* mutations in AML (TCGA data) and clonal hematopoiesis (Jaiswal et al. 2014) are depicted on a figure of the long isoform. AML TCGA data include 57 *DNMT3A* mutations from 51 individuals (cohort of 200 patients), whereas clonal hematopoiesis data include 403 *DNMT3A* mutations from 398 individuals (cohort of 17,182 healthy patients). Each lollipop represents a nonsynonymous mutation, with the size of the lollipop correlating to the mutation count (within the respective groups) and the color indicating the type of mutation.

myeloid cancers (Meyer et al. 2016; Yang et al. 2016). Taken together, these data strongly indicate that *DNMT3A* is a haploinsufficient tumor suppressor gene in myeloid leukemias, when cooperating mutations are present.

In AML, ~65% of *DNMT3A* mutations are heterozygous missense mutations affecting codon R882 (Ley et al. 2010; Gaidzik et al. 2013; Gale et al. 2015). R882H mutation (or mouse R878H mutation) has been shown to exert a dominant-negative effect over the wild-type protein (Holz-Schietinger et al. 2012; Kim et al. 2013; Russler-Germain et al. 2014). The mutant protein can dimerize with WT DNMT3A, but tetramers, which comprise a substantially more active form of the protein, are not able to form (Holz-Schietinger et al. 2012; Russler-Germain et al. 2014). The resulting low levels of DNMT3A homotetramers result in a significant reduction of methyltransferase activity, providing an explanation for the genome-wide hypomethylation observed in patients with R882 mutations (Cancer Genome Atlas Research 2013; Qu et al. 2014; Meyer et al. 2016; Yang et al. 2016). In addition, the R882 mutant may acquire other deleterious interactions. Recent data suggest that it may interact with the Polycomb repressive complex 1

(PRC1), unlike WT DNMT3A, resulting in down-regulation of genes involved in promoting hematopoietic differentiation such as *PU.1* and *Cebpa* (Koya et al. 2016).

In contrast to R882 mutations, there is still a complete lack of data related to the biological impact of non-R882 missense mutations. It is possible that mutations in different DNMT3A domains lead to different unique neomorphic functions that result in specific underlying mechanisms of pathogenesis. Characterizing the epigenetic landscapes and phenotypic effects of non-R882 missense mutations in more detail will be vital to our understanding of *DNMT3A*-mutated cancer biology.

DNMT3A MUTATIONS IN AML

DNMT3A mutations are detectable in ~20% of patients with de novo AML, as well as in secondary AML (Table 1) evolving from previous myelodysplastic syndromes or myeloproliferative neoplasms (sAML) (Ley et al. 2010; Thol et al. 2011a; Fried et al. 2012; Gaidzik et al. 2013; Roller et al. 2013). Some clinical and pathologic features characterize *DNMT3A*^{mut} AML patients at diagnosis. Compared with wild-type *DNMT3A* (*DNMT3A*^{wt}) patients, they tend to

Table 1. Prevalence of *DNMT3A* mutations in hematologic malignancies

Disease	Characteristics	Frequency (%)	References
Adult AML	De novo AML	62/281 (22.1)	Ley et al. 2010
	De novo CN-AML	36/123 (29.3)	Renneville et al. 2012
	Intermediate risk AML	272/914 (29.8)	Gale et al. 2015
	CN-AML	142/415 (34.2)	Marcucci et al. 2012
	89.5% de novo AML, 10.5% s-AML and t-AML	370/1770 (20.9)	Gaidzik et al. 2013
	t-AML and s-AML	t-AML 10/61 (16.4); s-AML 13/37 (35.1)	Fried et al. 2012
Pediatric AML	De novo AML	2/206 (1)	Liang et al. 2013
Adult T-ALL	T-ALL	Entire cohort 16/83 (19.3); early T-ALL 10/38 (26.3)	Grossmann et al. 2013
	ETP ALL	11/68 (16.2)	Neumann et al. 2013
MPAL	Adult T/myeloid MPAL	10/18 (55.6)	Wolach and Stone 2015
MDS	De novo MDS	127/944 (13.5)	Haferlach et al. 2014
MPN	All MPNs	10/155 (6.5)	Stegelmann et al. 2011
MDS/MPN	52 CMML, 20 s-AML evolved from CMML	CMML 2/52 (3.8); s-AML 6/20 (30)	Jankowska et al. 2011
	65 CMML-1, 38 CMML-2	CMML-1 2/65 (3.1); CMML-2 5/38 (13.2)	Roller et al. 2013
	JMML	3/100 (3)	Stieglitz et al. 2015
T-cell lymphomas	T-cell lymphomas	11/96 (11.4)	Couronné et al. 2012
	AITL and PTCL-NOS	21/79 (26.6)	Sakata-Yanagimoto et al. 2014
	CTCL	17/40 (42.5)	Choi et al. 2015

The large range of frequencies reported for AML depends on the subset of patients considered and whether the entire DNMT3A or only hotspots were sequenced.

AML, acute myeloid leukemia; CN-AML, caryotype normal-AML; t-AML, therapy-related AML; s-AML, secondary AML; T-ALL, T-cell acute lymphoblastic leukemia; ETP ALL, early T-cell precursor ALL; MPAL, mixed phenotype acute leukemia; MDS, myelodysplastic syndromes; MPN, myeloproliferative neoplasms; MDS/MPN, myelodysplastic syndrome/myelo-proliferative neoplasms; CMML, chronic myelomonocytic leukemia; JMML, juvenile myelomonocytic leukemia; AITL, angioimmunoblastic lymphoma; PTCL-NOS, peripheral T-cell lymphomas not otherwise specified; CTCL, cutaneous T-cell lymphoma.

be older, have higher white blood cell counts, and be frequently diagnosed with cytogenetically normal AML (CN-AML) with myelomonocytic or monocytic (French–American–British M4 and M5) blast morphology (Thol et al. 2011a; Marcucci et al. 2012; Gaidzik et al. 2013; Gale et al. 2015).

DNMT3A^{mut} AMLs frequently harbor *NPM1* (*NPM1*$^{c+}$) and *FLT3* mutations, with 60%–80% of *DNMT3A*^{mut} cases having *NPM1*$^{c+}$ and ~30% displaying both *NPM1* and *FLT3* mutations (Cancer Genome Atlas Research 2013; Gaidzik et al. 2013; Gale et al. 2015). Although there is evidence that *Npm1*$^{c+}$/*Flt3*-ITD mice (Mupo et al. 2013) and *Dnmt3a*^{mut}/*Flt3*-ITD mice (Meyer et al. 2016) develop leukemia with 100% penetrance,

no data have been published on *Dnmt3a*^{mut}/*Npm1*$^{c+}$ mice. It therefore remains unclear whether any of these mutations plays a central role in determining such specific clinical phenotype.

Prognostic Impact of *DNMT3A* Mutations in AML

Although a large number of studies reported that *DNMT3A*^{mut} patients fare worse compared with *DNMT3A*^{wt} patients (Ley et al. 2010; Shen et al. 2011; Thol et al. 2011a; Hou et al. 2012; Marcucci et al. 2012; Renneville et al. 2012; Ribeiro et al. 2012; Tie et al. 2014; Gale et al. 2015), the impact of *DNMT3A* mutations in clinical decision-making is still a matter of debate (Patel et al. 2012; Gaidzik et al. 2013). There is a lack of evidence that the increased risk of relapse is sufficient to indicate more aggressive strategies such as bone marrow transplantation during first complete remission for these patients. Interestingly, two groups reported that high-dose anthracyclines might overcome the negative impact of *DNMT3A* mutations (Patel et al. 2012; Sehgal et al. 2015), but larger prospective trials are needed to confirm this finding. Along with their classic topoisomerase II inhibitory properties, anthracyclines can cause nucleosome eviction from open chromatin, especially when delivered at high doses (Pang et al. 2013). Some data suggest that hypomethylated regions in *DNMT3A*^{mut} AMLs have a more open chromatin (Spencer et al. 2014), and it is possible that these regions are targets for nucleosome eviction induced by high-dose anthracycline.

Whether different types of mutations (e.g., missense vs. nonsense) in different amino acids (e.g., R882 vs. non-R882) have specific clinical consequences is unknown. The largest study reported worse outcomes for patients harboring R882 mutations, but not for patients with non-R882 mutations. Gale et al. (2015) investigated the impact of *DNMT3A* mutations in younger patients according to *NPM1* mutational status. R882 mutated patients had worse overall survival, in both *NPM1*$^{c+}$ and *NPM1*^{wt} subgroups, whereas the impact of non-R882 missense mutations was evident only in the *NPM1*$^{c+}$ patients, suggesting a less pronounced clinical impact for the non-R882 missense mutations. Although limited by small sample size, a subanalysis revealed that patients with *DNMT3A* truncating mutations had the same prognosis as *DNMT3A*^{wt} patients. Ahn et al. reported different outcomes for patients with R882 and non-R882 mutations treated with allogeneic bone marrow transplantation. R882 mutations conferred significant adverse prognostic impact, especially when associated with *FLT3*-ITD, whereas patients harboring non-R882 mutations had a similar prognosis to *DNMT3A* wild-type patients (Gaidzik et al. 2013; Ahn et al. 2016).

Although no definitive conclusion can be drawn, merging clinical and biological data, it seems likely that R882 mutations confer a neg-ative prognosis to *DNMT3A*^{mut} patients. On the other hand, non-R882 mutations are extremely heterogeneous, encompassing a variety of mutations, and therefore understanding the impact of such mutations may be difficult. Until results from larger prospective trials become available, published data suggest that patients with *DNMT3A*^{mut} AML should be stratified according to more validated risk parameters such as age, karyotype, the persistence of residual disease after treatment (i.e., minimal residual disease; MRD), and other molecular abnormalities (e.g., *NPM1*$^{c+}$, *FLT3*-ITD).

DNMT3A Mutations, Minimal Residual Disease, and Relapse

DNMT3A^{mut} cells are frequently detectable in patients with AML with long-lasting complete remission, raising the question of whether *DNMT3A* should be used to monitor MRD. Pløen et al. initially reported residual R882 *DNMT3A* mutations in a significant fraction of cells (mutant allele frequencies between <1% and 50%) in samples from 14 AML patients in complete remission. Five out of 14 patients did not relapse after a median follow-up of 53 months (Pløen et al. 2014). Recently, more studies have reported no clear association between the persistence of *DNMT3A* mutations in complete remission and worse clinical outcomes (Jeziskova et al. 2015; Ivey et al. 2016). The presence of *DNMT3A*^{mut} cells in patients with no evidence of residual leukemia is consistent with *DNMT3A* mutations being preleukemic events (Corces-Zimmerman et al. 2014; Shlush et al. 2014). The detection of *DNMT3A* mutations after successful treatment likely reflects the persistence of preleukemic clones instead of true leukemic cells, arguing against using *DNMT3A* as a marker for monitoring MRD.

The prevalence of residual *DNMT3A* mutations after treatment makes it reasonable to consider the hypothesis of AML relapse being, at least in some cases, the result of newly acquired mutations arising in a preleukemic *DNMT3A*-mutated clone. New prospective trials with large cohorts combined with genome-wide sequencing studies of samples acquired over time (e.g., diagnosis, remission, and relapse) will help to answer this question over the next few years.

DNMT3A MUTATIONS IN MYELODYSPLASTIC SYNDROMES (MDSs)

As in AML, *DNMT3A* mutations are early events in MDSs (Walter et al. 2011; Papaemmanuil et al. 2013); however, their prevalence is significantly lower than in AML (Table 1) and their prognostic value unclear. Although some studies suggest reduced overall survival and higher probability of evolution to AML for *DNMT3A*^mut MDS patients (Thol et al. 2011c; Walter et al. 2011), others have shown no prognostic significance (Bejar et al. 2012; Haferlach et al. 2014).

MDS patients are often treated with hypomethylating agents such as 5-azacytidine or decitabine (Santini 2012), which are thought to inhibit DNMT1 and restore normal methylation levels at promoter regions of tumor suppressor genes (Paul et al. 2010). One study, although limited by small numbers, suggested that *DNMT3A* mutations are predictive of better response to hypomethylating agents (Traina et al. 2014), but the biological explanation for this behavior is unknown. Data from *Dnmt3a*-null HSCs and *DNMT3A*^mut AML patients (Yan et al. 2011; Challen et al. 2012; Qu et al. 2014) revealed that, along with global DNA hypomethylation, some regions of the genome instead gain methylation. It may be that hypomethylating agents are acting at these sites to restore normal methylation levels. Additionally, in the context of lower DNA methylation levels, 5-azacytidine and decitabine may further reduce global DNA methylation leading to a synthetically lethal phenotype and targeted cell death. More data are needed to assess the real clinical impact of *DNMT3A* mutations in MDS and the implications for therapeutic approaches.

DNMT3A MUTATIONS IN MYELOPROLIFERATIVE NEOPLASMS AND MYELODYSPLASTIC/ MYELOPROLIFERATIVE NEOPLASMS

Myoproliferative neoplasms (MPNs) and myelodysplastic/myeloproliferative neoplasms (MDS/MPNs) are chronic myeloid neoplasms classically characterized by the expansion of mature myeloid cells in the bone marrow and extramedullary sites (e.g., spleen), with or without dysplastic features. *DNMT3A* mutations are less frequent genetic events in MPNs and MDS/ MPNs than in AML and MDSs (Table 1). *JAK2* is the most frequently mutated gene in MPNs and *JAK2*-V617F (*JAK2*^V617F) accounts for >95% of mutations of this gene (Kiladjian 2012). When co-occurring with *DNMT3A* mutations, the order of acquisition of *DNMT3A* and *JAK2* mutations may be relevant in defining different MPN disease phenotypes (Nangalia et al. 2015).

DNMT3A mutations are infrequent, but recurrent in MDS/MPNs. Indeed, they can be found in ∼2% of chronic myelomonocytic leukemia (CMML) (Jankowska et al. 2011) and 3% of juvenile myelomonocytic leukemia (Table 1) (Stieglitz et al. 2015). Jankowska et al. reported that 25% of patients with sAML arising from previous CMML harbored *DNMT3A* mutations, possibly suggesting a higher probability of evolution to AML for CMML patients with *DNTM3A* mutations; however, this study investigated a small cohort (Jankowska et al. 2011) and larger studies are needed. As described for AML, *DNMT3A*^mut CMML is reported to display significantly reduced DNA methylation, mainly at intergenic and intronic regions (Meldi et al. 2015) and low-DNA methylation levels are also reported in a *Dnmt3a*-null MDS/ MPN mouse model (Guryanova et al. 2015). However, whether this methylation pattern impacts the disease phenotype and eventually the prognosis of MDS/MPN patients is still unknown, and more studies are required to answer these important questions.

DNMT3A MUTATIONS IN LYMPHOID MALIGNANCIES

As reported for myeloid neoplasms, *DNMT3A* mutations in lymphoid malignances are recurrent (Table 1) and thought to be early events (see below). In T-ALL, the prevalence of biallelic hits is noticeably higher than in myeloid cancers (Grossmann et al. 2013; Roller et al. 2013), supporting the role of *DNMT3A* as a tumor suppressor in this context.

Adult early T-cell precursor ALL (ETP-ALL), a subset of T-ALL thought to arise from early thymocyte precursors that retain myeloid differentiating potential (Zhang et al. 2012), is particularly enriched for mutations in *DNMT3A* (Neumann et al. 2013). This finding creates a link between early immature T-ALL and myeloid malignancies, suggesting that some myeloid and lymphoid leukemias may arise from a shared $DNMT3A^{mut}$ precursor that maintains both myeloid and lymphoid potential. More studies are needed to enhance our understanding of the mechanisms that drive neoplastic differentiation toward the myeloid or lymphoid lineages.

Although $DNMT3A^{mut}$ T-ALL patients globally seem to fare worse compared with $DNMT3A^{wt}$ leukemias, it should be noted that this difference might reflect the early-T phenotype, which has been linked to a more aggressive disease evolution, and not the *DNMT3A* mutational status itself (Grossmann et al. 2013; Roller et al. 2013; Van Vlierberghe et al. 2013). However, the limited data available so far suggest an independent impact of *DNMT3A* mutations and further studies will verify its value for risk stratification.

DNMT3A is also recurrently mutated in peripheral T-cell lymphomas (Table 1) (Couronné et al. 2012; Sakata-Yanagimoto et al. 2014; Choi et al. 2015). As in myeloid neoplasms, $DNMT3A^{mut}$ tumor cells arise in an early progenitor that retains mutilineage potential, with the mutation detectable in both the neoplastic T-cells and normal B-cells of these lymphoma patients (Couronné et al. 2012). Strikingly, ∼70% of $DNMT3A^{mut}$ cases also harbor *TET2* mutations (Couronné et al. 2012) and mouse models of *Dnmt3a* mutation and *Tet2* knockout develop both AML and T-cell malignancies (Scourzic et al. 2016; Zhang et al. 2016). The interaction between $DNMT3A^{R882H}$ (thought to decrease global DNA methylation levels) and loss of TET2 (thought to increase global DNA methylation levels) generated a complex methylation landscape with increased DNA methylation and reduced expression of several tumor suppressors and hypomethylation and overexpression of T-cell key genes such as *Notch1* (Scourzic et al. 2016; Zhang et al. 2016). T-cell malignancies are still challenging to treat, thus a deeper understanding of these mechanisms is needed for the development of new therapeutic strategies.

DNMT3A MUTATIONS LINK AGE-RELATED CLONAL HEMATOPOIESIS, PRELEUKEMIC CLONES, AND AML

High-throughput sequencing studies showed that some somatic mutations in known cancer driver genes could be present in the blood of elderly individuals before any hematopoietic neoplasm appears (Busque et al. 2012; Wong et al. 2015). These data imply the possibility that predisposing mutations can be acquired early on during cancer development and that only a fraction of subjects with such mutations would eventually develop cancer. Recently, four independent groups reported the presence of subclonal somatic mutations in the peripheral blood of individuals with no history of hematologic diseases (Genovese et al. 2014; Jaiswal et al. 2014; Xie et al. 2014; McKerrell et al. 2015). The prevalence of this phenomenon increased with age, with >10% of subjects over 70 having at least one detectable somatic mutation. *DNMT3A* was by far the most frequently mutated gene in clonal hematopoiesis (Genovese et al. 2014; Jaiswal et al. 2014; Xie et al. 2014).

Individuals with clonal hematopoiesis have approximately a 10-fold higher risk of developing hematologic cancers compared with individuals without clonal hematopoiesis (Genovese et al. 2014; Jaiswal et al. 2014). Although the association with an increased risk of blood cancers may seem intuitive, strikingly clonal hematopoiesis is also found to significantly increase the risk of all-cause and cardiovascular-related mortality and these risks are higher in subjects with higher levels of clonality (Genovese et al. 2014; Jaiswal et al. 2014). Because several cells types that play key roles in the pathogenesis of cardiovascular diseases (e.g., macrophages) arise from hematopoietic progenitors, mutant cells could display altered functions that lead to an increased risk of coronary heart disease and ischemic stroke. Yet, it is unclear whether different somatic mutations associated with clonal hematopoiesis have different impacts on the risk of developing different diseases.

Organismal aging is known to be associated with decreased HSPCs fitness and alterations in the bone marrow microenvironment. Somatic mutations in some leukemia driver genes may allow for clonal expansion through increasing cellular fitness and promoting HSC self-renewal independent of age-related environmental changes. Alternatively, positive selection for cells with mutations in some genes may require the altered fitness landscape of aged bone marrow (McKerrell and Vassiliou 2015). *DNMT3A* mutations may confer a selective advantage to the cells in which they arise through both mechanisms. Although it has been shown that *Dnmt3a*-null HSCs have an enhanced serial engraftment capacity (Challen et al. 2012), less is known about the interaction of *DNMT3A* mutant cells and the surrounding environment. The aging microenvironment may select for the expansion of *DNMT3A* mutant cells at the expense of wild-type cells. Indeed, it has recently been reported that patients with autoimmune bone marrow failures such as aplastic anemia frequently harbor *DNMT3A* mutated clones (Yoshizato et al. 2015). These clones may expand as a result of stem cell exhaustion of wild-type HSCs forced to enter the cell cycle because of the profound cytopenia induced by marrow failure.

Whichever the mechanism leading to clonal expansion of hematopoietic progenitors, it seems likely that clonal hematopoiesis represents a primitive stage in the development of hematologic neoplasms. The vast majority of the genes reported as mutated in individuals with clonal hematopoiesis are well-known drivers of leukemia or MDS and many of them have been found in preleukemic clones in patients with AML (Corces-Zimmerman et al. 2014; Shlush et al. 2014). Preleukemic HSPCs may represent progenitors of patients with a history of clonal hematopoiesis that, after acquiring secondary genetic events, developed AML. It remains to be established why only a small frac-

tion of patients with clonal hematopoiesis develop hematologic cancers and what are the molecular and biological mechanisms that regulate the transition from clonal hematopoiesis to preleukemia and eventually to cancer. Addressing these issues would significantly enhance the development of new, targeted therapies to treat patients with MDS and leukemias.

THERAPEUTIC PERSPECTIVES AND FUTURE DIRECTIONS

DNMT3A mutations are one of the earliest genetic events during hematopoietic cancer development, making this gene an attractive target for new therapeutic approaches. Moreover, the negative overall worse prognosis for $DNMT3A^{mut}$ patients necessitates more efficient therapies. However, we still need to better understand how DNA methylation is regulated and how it impacts gene expression and phenotype. Although loss of DNMT3A may promote expression of self-renewal genes (Challen et al. 2012), whether these are viable candidates for therapeutic targets is unknown. The development of genome-wide CRISPR-Cas9 screens may facilitate the rapid detection of synthetically lethal phenotypes in normal and malignant hematopoietic cells (Koike-Yusa et al. 2014; Shi et al. 2015). The application of these screens to mouse and cell-based models will allow us to unravel the possible downstream effects of $DNMT3A^{mut}$ HSCs and discover potential targets for DNMT3A mutated cancers.

Beside high-dose anthracycline, there is a lack of data related to new strategies for $DNMT3A^{mut}$ patients with AML or other hematologic cancers. DNMT3A mutants may play a role in up-regulating *HOX* genes in AML. Novel H3-methyltransferase inhibitors such as DOT1L inhibitors (McLean et al. 2014), MLL–menin interaction inhibitors (Borkin et al. 2015), and MLL inhibitors (Cao et al. 2014) showed efficient knockdown of *HOX* cluster genes and leukemic cell death, both in vitro and in vivo. Although these compounds were mainly tested in the context of *MLL*-rearranged leukemia, a wider spectrum of leukemias, including $DNMT3A^{mut}$ AML, may be sensitive to this treatment. The bone marrow niche plays a relevant role in both hematopoietic and leukemic stem cell survival and function and it has been already proposed as a possible target for novel therapies (Krause et al. 2013). Both mouse and human data support the idea of a $DNMT3A^{mut}$ HSC having a competitive advantage over the wild-type counterpart (Fig. 2), but no study assessed the interplay between $DNMT3A^{mut}$ cells and the microenvironment so far. Specific connections between $DNMT3A^{mut}$ HSCs and the niche are likely to occur and they may promote expansion of $DNMT3A^{mut}$ clones, supporting age-related clonal hematopoiesis and leukemogenesis. A better understanding of these interactions will provide new insight for the development of novel therapeutic approaches.

CONCLUDING REMARKS

DNMT3A has recently been identified as one of the most commonly mutated genes in adult hematologic malignancies. The development of $Dnmt3a^{mut}$ mouse models has increased our understanding of how DNMT3A controls differentiation and self-renewal in the hematopoietic system and confirmed its role as a driver of leukemogenesis. The observation that *Dnmt3a*-null HSCs have a selective advantage over wild-type cells is consistent with the high prevalence of *DNMT3A* mutations in patients with clonal hematopoiesis, as well as with the detection of mutant *DNMT3A* in preleukemic clones, and with the persistence of *DNMT3A*-mutant cells in patients with AML in complete remission. However, whether these mutant cells are a long-term risk factor for leukemia relapse and need to be cleared with more aggressive therapies, such as allogeneic bone marrow transplantation, remains unclear. To develop new therapeutic strategies for the treatment of patients with *DNMT3A* mutations, it is crucial to better characterize the role of mutant DNMT3A in initiating and maintaining tumorigenesis.

ACKNOWLEDGMENTS

We thank C. Gillespie for critical reading of the manuscript. L.B. is funded by the University of Perugia Ph.D. program. M.C.G. is supported by Baylor Research Advocates for Student Scientists and Cancer Prevention and Research Institute of Texas Training Grant RP160283. We also acknowledge support from the National Institutes of Health (NIH) (DK092883, CA183252), Cancer Prevention & Research Institute of Texas (CPRIT) (RP140001), the Samuel Waxman Cancer Research Foundation, and the Edward P. Evans Foundation.

REFERENCES

Ahn JSS, Kim HJJ, Kim YKK, Lee SSS, Jung SHH, Yang DHH, Lee JJJ, Kim NY, Choi SH, Jung CW, et al. 2016. *DNMT3A* R882 mutation with *FLT3*-ITD positivity is an extremely poor prognostic factor in patients with normal-karyotype acute myeloid leukemia after allogeneic hematopoietic cell transplantation. *Biol Blood Marrow Transplant* **22**: 61–70.

Ayyanathan K, Lechner MS, Bell P, Maul GG, Schultz DC, Yamada Y, Tanaka K, Torigoe K, Rauscher FJ III. 2003. Regulated recruitment of HP1 to a euchromatic gene induces mitotically heritable, epigenetic gene silencing: A mammalian cell culture model of gene variegation. *Genes Dev* **17**: 1855–1869.

Baubec T, Colombo DF, Wirbelauer C, Schmidt J, Burger L, Krebs AR, Akalin A, Schubeler D. 2015. Genomic profiling of DNA methyltransferases reveals a role for DNMT3B in genic methylation. *Nature* **520**: 243–247.

Bejar R, Stevenson KE, Caughey BA, Abdel-Wahab O, Steensma DP, Galili N, Raza A, Kantarjian H, Levine RL, Neuberg D, et al. 2012. Validation of a prognostic model and the impact of mutations in patients with lower-risk myelodysplastic syndromes. *J Clin Oncol* **30**: 3376–3382.

Borkin D, He S, Miao H, Kempinska K, Pollock J, Chase J, Purohit T, Malik B, Zhao T, Wang J, et al. 2015. Pharmacologic inhibition of the Menin-MLL interaction blocks progression of MLL leukemia in vivo. *Cancer Cell* **27**: 589–602.

Busque L, Patel JP, Figueroa ME, Vasanthakumar A, Provost S, Hamilou Z, Mollica L, Li J, Viale A, Heguy A, et al. 2012. Recurrent somatic *TET2* mutations in normal elderly individuals with clonal hematopoiesis. *Nat Genet* **44**: 1179–1181.

Cancer Genome Atlas Research N. 2013. Genomic and epigenomic landscapes of adult de novo acute myeloid leukemia. *N Engl J Med* **368**: 2059–2074.

Cao F, Townsend EC, Karatas H, Xu J, Li L, Lee S, Liu L, Chen Y, Ouillette P, Zhu J, et al. 2014. Targeting MLL1 H3K4 methyltransferase activity in mixed-lineage leukemia. *Mol Cell* **53**: 247–261.

Cedar H, Bergman Y. 2009. Linking DNA methylation and histone modification: Patterns and paradigms. *Nat Rev Genet* **10**: 295–304.

Celik H, Mallaney C, Kothari A, Ostrander EL, Eultgen E, Martens A, Miller CA, Hundal J, Klco JM, Challen GA.

2015. Enforced differentiation of Dnmt3a-null bone marrow leads to failure with c-Kit mutations driving leukemic transformation. *Blood* **125:** 619–628.

Challen GA, Sun D, Jeong M, Luo M, Jelinek J, Berg JS, Bock C, Vasanthakumar A, Gu H, Xi Y, et al. 2012. Dnmt3a is essential for hematopoietic stem cell differentiation. *Nat Genet* **44:** 23–31.

Challen GA, Sun D, Mayle A, Jeong M, Luo M, Rodriguez B, Mallaney C, Celik H, Yang L, Xia Z, et al. 2014. Dnmt3a and Dnmt3b have overlapping and distinct functions in hematopoietic stem cells. *Cell Stem Cell* **15:** 350–364.

Chantalat S, Depaux A, Hery P, Barral S, Thuret JY, Dimitrov S, Gerard M. 2011. Histone H3 trimethylation at lysine 36 is associated with constitutive and facultative heterochromatin. *Genome Res* **21:** 1426–1437.

Chen T, Ueda Y, Xie S, Li E. 2002. A novel Dnmt3a isoform produced from an alternative promoter localizes to euchromatin and its expression correlates with active de novo methylation. *J Biol Chem* **277:** 38746–38754.

Chen T, Ueda Y, Dodge JE, Wang Z, Li E. 2003. Establishment and maintenance of genomic methylation patterns in mouse embryonic stem cells by Dnmt3a and Dnmt3b. *Mol Cell Biol* **23:** 5594–5605.

Chen T, Tsujimoto N, Li E. 2004. The PWWP domain of Dnmt3a and Dnmt3b is required for directing DNA methylation to the major satellite repeats at pericentric heterochromatin. *Mol Cell Biol* **24:** 9048–9058.

Choi J, Goh G, Walradt T, Hong BS, Bunick CG, Chen K, Bjornson RD, Maman Y, Wang T, Tordoff J, et al. 2015. Genomic landscape of cutaneous T cell lymphoma. *Nat Genet* **47:** 1011–1019.

Corces-Zimmerman RM, Hong WJ, Weissman IL, Medeiros BC, Majeti R. 2014. Preleukemic mutations in human acute myeloid leukemia affect epigenetic regulators and persist in remission. *Proc Natl Acad Sci* **111:** 2548–2553.

Couronné L, Bastard C, Bernard OA. 2012. *TET2* and *DNMT3A* mutations in human T-cell lymphoma. *N Engl J Med* **366:** 95–96.

Dhawan S, Tschen SI, Zeng C, Guo T, Hebrok M, Matveyenko A, Bhushan A. 2015. DNA methylation directs functional maturation of pancreatic beta cells. *J Clin Invest* **125:** 2851–2860.

Dhayalan A, Rajavelu A, Rathert P, Tamas R, Jurkowska RZ, Ragozin S, Jeltsch A. 2010. The Dnmt3a PWWP domain reads histone 3 lysine 36 trimethylation and guides DNA methylation. *J Biol Chem* **285:** 26114–26120.

Emperle M, Rajavelu A, Reinhardt R, Jurkowska RZ, Jeltsch A. 2014. Cooperative DNA binding and protein/DNA fiber formation increases the activity of the Dnmt3a DNA methyltransferase. *J Biol Chem* **289:** 29602–29613.

Feinberg AP, Vogelstein B. 1983. Hypomethylation distinguishes genes of some human cancers from their normal counterparts. *Nature* **301:** 89–92.

Fried I, Bodner C, Pichler MM, Lind K, Beham-Schmid C, Quehenberger F, Sperr WR, Linkesch W, Sill H, Wölfler A. 2012. Frequency, onset and clinical impact of somatic *DNMT3A* mutations in therapy-related and secondary acute myeloid leukemia. *Haematologica* **97:** 246–250.

Gaidzik VI, Schlenk RF, Paschka P, Stolzle A, Spath D, Kuendgen A, von Lilienfeld-Toal M, Brugger W, Derigs HG, Kremers S, et al. 2013. Clinical impact of *DNMT3A* mutations in younger adult patients with acute myeloid leukemia: Results of the AML Study Group (AMLSG). *Blood* **121:** 4769–4777.

Gale RE, Lamb K, Allen C, El-Sharkawi D, Stowe C, Jenkinson S, Tinsley S, Dickson G, Burnett AK, Hills RK, et al. 2015. Simpson's Paradox and the impact of different *DNMT3A* mutations on outcome in younger adults with acute myeloid leukemia. *J Clin Oncol* **33:** 2072–2083.

Gao Q, Steine EJ, Barrasa MI, Hockemeyer D, Pawlak M, Fu D, Reddy S, Bell GW, Jaenisch R. 2011. Deletion of the de novo DNA methyltransferase Dnmt3a promotes lung tumor progression. *Proc Natl Acad Sci* **108:** 18061–18066.

Genovese G, Kähler AK, Handsaker RE, Lindberg J, Rose SA, Bakhoum SF, Chambert K, Mick E, Neale BM, Fromer M, et al. 2014. Clonal hematopoiesis and blood-cancer risk inferred from blood DNA sequence. *N Engl J Med* **371:** 2477–2487.

Girault I, Tozlu S, Lidereau R, Bieche I. 2003. Expression analysis of DNA methyltransferases 1, 3A, and 3B in sporadic breast carcinomas. *Clin Cancer Res* **9:** 4415–4422.

Grossmann V, Haferlach C, Weissmann S, Roller A, Schindela S, Poetzinger F, Stadler K, Bellos F, Kern W, Haferlach T, et al. 2013. The molecular profile of adult T-cell acute lymphoblastic leukemia: Mutations in *RUNX1* and *DNMT3A* are associated with poor prognosis in T-ALL. *Genes Chromosomes Cancer* **52:** 410–422.

Guo X, Wang L, Li J, Ding Z, Xiao J, Yin X, He S, Shi P, Dong L, Li G, et al. 2015. Structural insight into autoinhibition and histone H3-induced activation of DNMT3A. *Nature* **517:** 640–644.

Guryanova OA, Lieu YK, Garrett-Bakelman FE, Spitzer B, Glass JL, Shank K, Martinez AB, Rivera SA, Durham BH, Rapaport F, et al. 2015. Dnmt3a regulates myeloproliferation and liver-specific expansion of hematopoietic stem and progenitor cells. *Leukemia* **30:** 1133–1142.

Haferlach T, Nagata Y, Grossmann V, Okuno Y, Bacher U, Nagae G, Schnittger S, Sanada M, Kon A, Alpermann T, et al. 2014. Landscape of genetic lesions in 944 patients with myelodysplastic syndromes. *Leukemia* **28:** 241–247.

Handa V, Jeltsch A. 2005. Profound flanking sequence preference of Dnmt3a and Dnmt3b mammalian DNA methyltransferases shape the human epigenome. *J Mol Biol* **348:** 1103–1112.

Hata K, Okano M, Lei H, Li E. 2002. Dnmt3L cooperates with the Dnmt3 family of de novo DNA methyltransferases to establish maternal imprints in mice. *Development* **129:** 1983–1993.

Ho PA, Kutny MA, Alonzo TA, Gerbing RB, Joaquin J, Raimondi SC, Gamis AS, Meshinchi S. 2011. Leukemic mutations in the methylation-associated genes *DNMT3A* and *IDH2* are rare events in pediatric AML: A report from the Children's Oncology Group. *Pediatr Blood Cancer* **57:** 204–209.

Holz-Schietinger C, Matje DM, Harrison MF, Reich NO. 2011. Oligomerization of DNMT3A controls the mechanism of de novo DNA methylation. *J Biol Chem* **286:** 41479–41488.

Holz-Schietinger C, Matje DM, Reich NO. 2012. Mutations in DNA methyltransferase (DNMT3A) observed in acute myeloid leukemia patients disrupt processive methylation. *J Biol Chem* 287: 30941–30951.

Hou HA, Kuo YY, Liu CY, Chou WC, Lee MC, Chen CY, Lin LI, Tseng MH, Huang CF, Chiang YC, et al. 2012. *DNMT3A* mutations in acute myeloid leukemia: Stability during disease evolution and clinical implications. *Blood* 119: 559–568.

Hu N, Strobl-Mazzulla P, Sauka-Spengler T, Bronner ME. 2012. DNA methyltransferase3A as a molecular switch mediating the neural tube-to-neural crest fate transition. *Genes Dev* 26: 2380–2385.

Issa JP, Vertino PM, Wu J, Sazawal S, Celano P, Nelkin BD, Hamilton SR, Baylin SB. 1993. Increased cytosine DNA-methyltransferase activity during colon cancer progression. *J Natl Cancer Inst* 85: 1235–1240.

Ivey A, Hills RK, Simpson MA, Jovanovic JV, Gilkes A, Grech A, Patel Y, Bhudia N, Farah H, Mason J, et al. 2016. Assessment of minimal residual disease in standard-risk AML. *N Engl J Med* 374: 422–433.

Jaiswal S, Fontanillas P, Flannick J, Manning A, Grauman PV, Mar BG, Lindsley RC, Mermel CH, Burtt N, Chavez A, et al. 2014. Age-related clonal hematopoiesis associated with adverse outcomes. *N Engl J Med* 371: 2488–2498.

Jankowska AM, Makishima H, Tiu RV, Szpurka H, Huang Y, Traina F, Visconte V, Sugimoto Y, Prince C, O'Keefe C, et al. 2011. Mutational spectrum analysis of chronic myelomonocytic leukemia includes genes associated with epigenetic regulation: *UTX, EZH2,* and *DNMT3A. Blood* 118: 3932–3941.

Jeong M, Sun D, Luo M, Huang Y, Challen GA, Rodriguez B, Zhang X, Chavez L, Wang H, Hannah R, et al. 2014. Large conserved domains of low DNA methylation maintained by Dnmt3a. *Nat Genet* 46: 17–23.

Jeziskova I, Musilova M, Culen M, Foltankova V, Dvorakova D, Mayer J, Racil Z. 2015. Distribution of mutations in *DNMT3A* gene and the suitability of mutations in R882 codon for MRD monitoring in patients with AML. *Int J Hematol* 102: 553–557.

Jones PA, Laird PW. 1999. Cancer epigenetics comes of age. *Nat Genet* 21: 163–167.

Jurkowska RZ, Rajavelu A, Anspach N, Urbanke C, Jankevicius G, Ragozin S, Nellen W, Jeltsch A. 2011. Oligomerization and binding of the Dnmt3a DNA methyltransferase to parallel DNA molecules: Heterochromatic localization and role of Dnmt3L. *J Biol Chem* 286: 24200–24207.

Kaneda M, Okano M, Hata K, Sado T, Tsujimoto N, Li E, Sasaki H. 2004. Essential role for de novo DNA methyltransferase Dnmt3a in paternal and maternal imprinting. *Nature* 429: 900–903.

Kiladjian JJ. 2012. The spectrum of JAK2-positive myeloproliferative neoplasms. *Hematol Am Soc Hematol Educ Program* 2012: 561–566.

Kim SJ, Zhao H, Hardikar S, Singh AK, Goodell MA, Chen T. 2013. A *DNMT3A* mutation common in AML exhibits dominant-negative effects in murine ES cells. *Blood* 122: 4086–4089.

Klco JM, Miller CA, Griffith M, Petti A, Spencer DH, Ketkar-Kulkarni S, Wartman LD, Christopher M, Lamprecht TL, Helton NM, et al. 2015. Association between mutation clearance after induction therapy and outcomes in acute myeloid leukemia. *JAMA* 314: 811–822.

Koike-Yusa H, Li Y, Tan EPP, Velasco-Herrera MDCdC, Yusa K. 2014. Genome-wide recessive genetic screening in mammalian cells with a lentiviral CRISPR-guide RNA library. *Nat Biotechnol* 32: 267–273.

Koya J, Kataoka K, Sato T, Bando M, Kato Y, Tsuruta-Kishino T, Kobayashi H, Narukawa K, Miyoshi H, Shirahige K, et al. 2016. *DNMT3A* R882 mutants interact with polycomb proteins to block haematopoietic stem and leukaemic cell differentiation. *Nat Commun* 7: 10924.

Krause DS, Scadden DT, Preffer FI. 2013. The hematopoietic stem cell niche—Home for friend and foe? *Cytometry B Clin Cytom* 84: 7–20.

Ley TJ, Ding L, Walter MJ, McLellan MD, Lamprecht T, Larson DE, Kandoth C, Payton JE, Baty J, Welch J, et al. 2010. *DNMT3A* mutations in acute myeloid leukemia. *N Engl J Med* 363: 2424–2433.

Liang DCC, Liu HCC, Yang CPP, Jaing THH, Hung IJJ, Yeh TCC, Chen SHH, Hou JYY, Huang YJJ, Shih YSS, et al. 2013. Cooperating gene mutations in childhood acute myeloid leukemia with special reference on mutations of *ASXL1, TET2, IDH1, IDH2,* and *DNMT3A. Blood* 121: 2988–2995.

Liao J, Karnik R, Gu H, Ziller MJ, Clement K, Tsankov AM, Akopian V, Gifford CA, Donaghey J, Galonska C, et al. 2015. Targeted disruption of *DNMT1, DNMT3A* and *DNMT3B* in human embryonic stem cells. *Nat Genet* 47: 469–478.

Lin IG, Han L, Taghva A, O'Brien LE, Hsieh CL. 2002. Murine de novo methyltransferase Dnmt3a demonstrates strand asymmetry and site preference in the methylation of DNA in vitro. *Mol Cell Biol* 22: 704–723.

Marcucci G, Metzeler KH, Schwind S, Becker H, Maharry K, Mrózek K, Radmacher MD, Kohlschmidt J, Nicolet D, Whitman SP, et al. 2012. Age-related prognostic impact of different types of *DNMT3A* mutations in adults with primary cytogenetically normal acute myeloid leukemia. *J Clin Oncol* 30: 742–750.

Mayle A, Yang L, Rodriguez B, Zhou T, Chang E, Curry CV, Challen GA, Li W, Wheeler D, Rebel VI, et al. 2015. *Dnmt3a* loss predisposes murine hematopoietic stem cells to malignant transformation. *Blood* 125: 629–638.

McKerrell T, Vassiliou GS. 2015. Aging as a driver of leukemogenesis. *Sci Transl Med* 7: 306fs338.

McKerrell T, Park N, Moreno T, Grove CS, Ponstingl H, Stephens J, Group U, Crawley C, Craig J, Scott MA, et al. 2015. Leukemia-associated somatic mutations drive distinct patterns of age-related clonal hemopoiesis. *Cell Rep* 10: 1239–1245.

McLean CM, Karemaker ID, van Leeuwen F. 2014. The emerging roles of DOT1L in leukemia and normal development. *Leukemia* 28: 2131–2138.

Meldi K, Qin T, Buchi F, Droin N, Sotzen J, Micol J-BB, Selimoglu-Buet D, Masala E, Allione B, Gioia D, et al. 2015. Specific molecular signatures predict decitabine response in chronic myelomonocytic leukemia. *J Clin Invest* 125: 1857–1872.

Meyer SE, Qin T, Muench DE, Masuda K, Venkatasubramanian M, Orr E, Suarez L, Gore SD, Delwel R, Paietta E, et al. 2016. Dnmt3a haploinsufficiency transforms $Flt3^{ITD}$ myeloproliferative disease into a rapid, spontaneous, and fully penetrant acute myeloid leukemia. *Cancer Discov* 6: 501–515.

Mizuno S, Chijiwa T, Okamura T, Akashi K, Fukumaki Y, Niho Y, Sasaki H. 2001. Expression of DNA methyltransferases *DNMT1*, *3A*, and *3B* in normal hematopoiesis and in acute and chronic myelogenous leukemia. *Blood* **97:** 1172–1179.

Mupo A, Celani L, Dovey O, Cooper JL, Grove C, Rad R, Sportoletti P, Falini B, Bradley A, Vassiliou GS. 2013. A powerful molecular synergy between mutant Nucleophosmin and *Flt3*[ITD] drives acute myeloid leukemia in mice. *Leukemia* **27:** 1917–1920.

Nangalia J, Nice FL, Wedge DC, Godfrey AL, Grinfeld J, Thakker C, Massie CE, Baxter J, Sewell D, Silber Y, et al. 2015. *DNMT3A* mutations occur early or late in patients with myeloproliferative neoplasms and mutation order influences phenotype. *Haematologica* **100:** e438–442.

Neumann M, Heesch S, Schlee C, Schwartz S, Gökbuget N, Hoelzer D, Konstandin NP, Ksienzyk B, Vosberg S, Graf A, et al. 2013. Whole-exome sequencing in adult ETP-ALL reveals a high rate of *DNMT3A* mutations. *Blood* **121:** 4749–4752.

Nimura K, Ishida C, Koriyama H, Hata K, Yamanaka S, Li E, Ura K, Kaneda Y. 2006. Dnmt3a2 targets endogenous Dnmt3L to ES cell chromatin and induces regional DNA methylation. *Genes Cells* **11:** 1225–1237.

Oh BK, Kim H, Park HJ, Shim YH, Choi J, Park C, Park YN. 2007. DNA methyltransferase expression and DNA methylation in human hepatocellular carcinoma and their clinicopathological correlation. *Int J Mol Med* **20:** 65–73.

Okano M, Xie S, Li E. 1998. Cloning and characterization of a family of novel mammalian DNA (cytosine-5) methyltransferases. *Nat Genet* **19:** 219–220.

Okano M, Bell DW, Haber DA, Li E. 1999. DNA methyltransferases Dnmt3a and Dnmt3b are essential for de novo methylation and mammalian development. *Cell* **99:** 247–257.

Ooi SK, Qiu C, Bernstein E, Li K, Jia D, Yang Z, Erdjument-Bromage H, Tempst P, Lin SP, Allis CD, et al. 2007. DNMT3L connects unmethylated lysine 4 of histone H3 to de novo methylation of DNA. *Nature* **448:** 714–717.

Otani J, Nankumo T, Arita K, Inamoto S, Ariyoshi M, Shirakawa M. 2009. Structural basis for recognition of H3K4 methylation status by the DNA methyltransferase 3A ATRX–DNMT3–DNMT3L domain. *EMBO Rep* **10:** 1235–1241.

Pang B, Qiao X, Janssen L, Velds A, Groothuis T, Kerkhoven R, Nieuwland M, Ovaa H, Rottenberg S, van Tellingen O, et al. 2013. Drug-induced histone eviction from open chromatin contributes to the chemotherapeutic effects of doxorubicin. *Nat Commun* **4:** 1908.

Papaemmanuil E, Gerstung M, Malcovati L, Tauro S, Gundem G, Van Loo P, Yoon CJ, Ellis P, Wedge DC, Pellagatti A, et al. 2013. Clinical and biological implications of driver mutations in myelodysplastic syndromes. *Blood* **122:** 3616.

Patel JP, Gönen M, Figueroa ME, Fernandez H, Sun Z, Racevskis J, Van Vlierberghe P, Dolgalev I, Thomas S, Aminova O, et al. 2012. Prognostic relevance of integrated genetic profiling in acute myeloid leukemia. *N Engl J Med* **366:** 1079–1089.

Paul TA, Bies J, Small D, Wolff L. 2010. Signatures of polycomb repression and reduced H3K4 trimethylation are associated with *p15INK4b* DNA methylation in AML. *Blood* **115:** 3098–3108.

Pløen GG, Nederby L, Guldberg P, Hansen M, Ebbesen LH, Jensen UB, Hokland P, Aggerholm A. 2014. Persistence of *DNMT3A* mutations at long-term remission in adult patients with AML. *Br J Haematol* **167:** 478–486.

Qu Y, Lennartsson A, Gaidzik VI, Deneberg S, Karimi M, Bengtzén S, Höglund M, Bullinger L, Döhner K, Lehmann S. 2014. Differential methylation in CN-AML preferentially targets non-CGI regions and is dictated by *DNMT3A* mutational status and associated with predominant hypomethylation of HOX genes. *Epigenetics* **9:** 1108–1119.

Rahman MM, Qian ZR, Wang EL, Yoshimoto K, Nakasono M, Sultana R, Yoshida T, Hayashi T, Haba R, Ishida M, et al. 2010. DNA methyltransferases 1, 3a, and 3b overexpression and clinical significance in gastroenteropancreatic neuroendocrine tumors. *Hum Pathol* **41:** 1069–1078.

Renneville A, Boissel N, Nibourel O, Berthon C, Helevaut N, Gardin C, Cayuela JMM, Hayette S, Reman O, Contentin N, et al. 2012. Prognostic significance of DNA methyltransferase 3A mutations in cytogenetically normal acute myeloid leukemia: A study by the Acute Leukemia French Association. *Leukemia* **26:** 1247–1254.

Ribeiro AF, Pratcorona M, Erpelinck-Verschueren C, Rockova V, Sanders M, Abbas S, Figueroa ME, Zeilemaker A, Melnick A, Löwenberg B, et al. 2012. Mutant *DNMT3A*: A marker of poor prognosis in acute myeloid leukemia. *Blood* **119:** 5824–5831.

Robertson KD, Uzvolgyi E, Liang G, Talmadge C, Sumegi J, Gonzales FA, Jones PA. 1999. The human DNA methyltransferases (DNMTs) 1, 3a and 3b: Coordinate mRNA expression in normal tissues and overexpression in tumors. *Nucleic Acids Res* **27:** 2291–2298.

Roller A, Grossmann V, Bacher U, Poetzinger F, Weissmann S, Nadarajah N, Boeck L, Kern W, Haferlach C, Schnittger S, et al. 2013. Landmark analysis of *DNMT3A* mutations in hematological malignancies. *Leukemia* **27:** 1573–1578.

Russler-Germain DA, Spencer DH, Young MA, Lamprecht TL, Miller CA, Fulton R, Meyer MR, Erdmann-Gilmore P, Townsend RR, Wilson RK, et al. 2014. The R882H DNMT3A mutation associated with AML dominantly inhibits wild-type DNMT3A by blocking its ability to form active tetramers. *Cancer Cell* **25:** 442–454.

Sakata-Yanagimoto M, Enami T, Yoshida K, Shiraishi Y, Ishii R, Miyake Y, Muto H, Tsuyama N, Sato-Otsubo A, Okuno Y, et al. 2014. Somatic *RHOA* mutation in angioimmunoblastic T cell lymphoma. *Nat Genet* **46:** 171–175.

Santini V. 2012. Novel therapeutic strategies: Hypomethylating agents and beyond. *Hematology Am Soc Hematol Educ Program* **2012:** 65–73.

Scourzic L, Couronne L, Pedersen MT, Della Valle V, Diop M, Mylonas E, Calvo J, Mouly E, Lopez CK, Martin N, et al. 2016. DNMT3A[R882H] mutant and Tet2 inactivation cooperate in the deregulation of DNA methylation control to induce lymphoid malignancies in mice. *Leukemia* **30:** 1388–1398.

Sehgal AR, Gimotty PA, Zhao J, Hsu JMM, Daber R, Morrissette JD, Luger S, Loren AW, Carroll M. 2015. *DNMT3A* mutational status affects the results of dose-escalated induction therapy in acute myelogenous leukemia. *Clin Cancer Res* **21:** 1614–1620.

Shen Y, Zhu YM, Fan X, Shi JY, Wang QR, Yan XJ. 2011. Gene mutation patterns and their prognostic impact in a cohort of 1185 patients with acute myeloid leukemia. *Blood* **118:** 5593–5603.

Shi J, Wang E, Milazzo JP, Wang Z, Kinney JB, Vakoc CR. 2015. Discovery of cancer drug targets by CRISPR-Cas9 screening of protein domains. *Nat Biotechnol* **33:** 661–667.

Shiba N, Taki T, Park MJJ, Shimada A, Sotomatsu M, Adachi S, Tawa A, Horibe K, Tsuchida M, Hanada R, et al. 2012. *DNMT3A* mutations are rare in childhood acute myeloid leukaemia, myelodysplastic syndromes and juvenile myelomonocytic leukaemia. *Br J Haematol* **156:** 413–414.

Shlush LI, Zandi S, Mitchell A, Chen WC, Brandwein JM, Gupta V, Kennedy JA, Schimmer AD, Schuh AC, Yee KW, et al. 2014. Identification of pre-leukaemic haematopoietic stem cells in acute leukaemia. *Nature* **506:** 328–333.

Spencer DH, Al-Khalil B, Russler-Germain D, Lamprecht T, Havey N, Fulton RS, O'Laughlin M, Fronick C, Wilson RK, Ley TJ. 2014. Whole-genome bisulfite sequencing of primary AML cells with the *DNMT3A* R882H mutation identifies regions of focal hypomethylation that are associated with open chromatin. *Blood* **124:** 608–608.

Stegelmann F, Bullinger L, Schlenk RF, Paschka P, Griesshammer M, Blersch C, Kuhn S, Schauer S, Döhner H, Döhner K. 2011. *DNMT3A* mutations in myeloproliferative neoplasms. *Leukemia* **25:** 1217–1219.

Stieglitz E, Taylor-Weiner AN, Chang TY, Gelston LC, Wang Y-DD, Mazor T, Esquivel E, Yu A, Seepo S, Olsen SR, et al. 2015. The genomic landscape of juvenile myelomonocytic leukemia. *Nat Genet* **47:** 1326–1333.

Suetake I, Mishima Y, Kimura H, Lee YH, Goto Y, Takeshima H, Ikegami T, Tajima S. 2011. Characterization of DNA-binding activity in the N-terminal domain of the DNA methyltransferase Dnmt3a. *Biochem J* **437:** 141–148.

Tatton-Brown K, Seal S, Ruark E, Harmer J, Ramsay E, Del Vecchio Duarte S, Zachariou A, Hanks S, O'Brien E, Aksglaede L, et al. 2014. Mutations in the DNA methyltransferase gene *DNMT3A* cause an overgrowth syndrome with intellectual disability. *Nat Genet* **46:** 385–388.

Thol F, Damm F, Lüdeking A, Winschel C, Wagner K, Morgan M, Yun H, Göhring G, Schlegelberger B, Hoelzer D, et al. 2011a. Incidence and prognostic influence of *DNMT3A* mutations in acute myeloid leukemia. *J Clin Oncol* **29:** 2889–2896.

Thol F, Heuser M, Damm F, Klusmann J-HH, Reinhardt K, Reinhardt D. 2011b. *DNMT3A* mutations are rare in childhood acute myeloid leukemia. *Haematologica* **96:** 1238–1240.

Thol F, Winschel C, Lüdeking A, Yun H, Friesen I, Damm F, Wagner K, Krauter J, Heuser M, Ganser A. 2011c. Rare occurrence of *DNMT3A* mutations in myelodysplastic syndromes. *Haematologica* **96:** 1870–1873.

Tie R, Zhang T, Fu H, Wang L, Wang Y, He Y, Wang B, Zhu N, Fu S, Lai X, et al. 2014. Association between *DNMT3A* mutations and prognosis of adults with de novo acute

myeloid leukemia: A systematic review and meta-analysis. *PLoS ONE* **9:** e96653.

Traina F, Visconte V, Elson P, Tabarroki A, Jankowska AM, Hasrouni E, Sugimoto Y, Szpurka H, Makishima H, O'Keefe CL, et al. 2014. Impact of molecular mutations on treatment response to DNMT inhibitors in myelodysplasia and related neoplasms. *Leukemia* **28:** 78–87.

Van Vlierberghe P, Ambesi-Impiombato A, De Keersmaecker K, Hadler M, Paietta E, Tallman MS, Rowe JM, Forne C, Rue M, Ferrando AA. 2013. Prognostic relevance of integrated genetic profiling in adult T-cell acute lymphoblastic leukemia. *Blood* **122:** 74–82.

Walter MJ, Ding L, Shen D, Shao J, Grillot M, McLellan M, Fulton R, Schmidt H, Kalicki-Veizer J, O'Laughlin M, et al. 2011. Recurrent *DNMT3A* mutations in patients with myelodysplastic syndromes. *Leukemia* **25:** 1153–1158.

Wang J, Qin S, Li F, Li S, Zhang W, Peng J, Zhang Z, Gong Q, Wu J, Shi Y. 2014. Crystal structure of human BS69 Bromo-ZnF-PWWP reveals its role in H3K36me3 nucleosome binding. *Cell Res* **24:** 890–893.

Wolach O, Stone RM. 2015. How I treat mixed-phenotype acute leukemia. *Blood* **125:** 2477–2485.

Wong TN, Ramsingh G, Young AL, Miller CA, Touma W, Welch JS, Lamprecht TL, Shen D, Hundal J, Fulton RS, et al. 2015. Role of *TP53* mutations in the origin and evolution of therapy-related acute myeloid leukaemia. *Nature* **518:** 552–555.

Wu Z, Huang K, Yu J, Le T, Namihira M, Liu Y, Zhang J, Xue Z, Cheng L, Fan G. 2012. Dnmt3a regulates both proliferation and differentiation of mouse neural stem cells. *J Neurosci Res* **90:** 1883–1891.

Xie S, Wang Z, Okano M, Nogami M, Li Y, He WW, Okumura K, Li E. 1999. Cloning, expression and chromosome locations of the human *DNMT3* gene family. *Gene* **236:** 87–95.

Xie W, Schultz MD, Lister R, Hou Z, Rajagopal N, Ray P, Whitaker JW, Tian S, Hawkins RD, Leung D, et al. 2013. Epigenomic analysis of multilineage differentiation of human embryonic stem cells. *Cell* **153:** 1134–1148.

Xie M, Lu C, Wang J, McLellan MD, Johnson KJ, Wendl MC, McMichael JF, Schmidt HK, Yellapantula V, Miller CA, et al. 2014. Age-related mutations associated with clonal hematopoietic expansion and malignancies. *Nat Med* **20:** 1472–1478.

Yan XJJ, Xu J, Gu ZHH, Pan CMM, Lu G, Shen Y, Shi JYY, Zhu YMM, Tang L, Zhang XWW, et al. 2011. Exome sequencing identifies somatic mutations of DNA methyltransferase gene *DNMT3A* in acute monocytic leukemia. *Nat Genet* **43:** 309–315.

Yang L, Rau R, Goodell MA. 2015. DNMT3A in haematological malignancies. *Nat Rev Cancer* **15:** 152–165.

Yang L, Rodriguez B, Mayle A, Park H, Lin X, Luo M, Jeong M, Curry CV, Kim SB, Ruau D, et al. 2016. DNMT3A loss drives enhancer hypomethylation in FLT3-ITD-associated leukemias. *Cancer Cell* **29:** 922–934.

Yokochi T, Robertson KD. 2002. Preferential methylation of unmethylated DNA by mammalian de novo DNA methyltransferase Dnmt3a. *J Biol Chem* **277:** 11735–11745.

Yoshizato T, Dumitriu B, Hosokawa K, Makishima H, Yoshida K, Townsley D, Sato-Otsubo A, Sato Y, Liu D, Suzuki H, et al. 2015. Somatic mutations and clonal hematopoiesis in aplastic anemia. *N Engl J Med* **373:** 35–47.

Zhang Y, Rohde C, Tierling S, Jurkowski TP, Bock C, Santacruz D, Ragozin S, Reinhardt R, Groth M, Walter J, et al. 2009. DNA methylation analysis of chromosome 21 gene promoters at single base pair and single allele resolution. *PLoS Genet* **5:** e1000438.

Zhang Y, Jurkowska R, Soeroes S, Rajavelu A, Dhayalan A, Bock I, Rathert P, Brandt O, Reinhardt R, Fischle W, et al. 2010. Chromatin methylation activity of Dnmt3a and Dnmt3a/3L is guided by interaction of the ADD domain with the histone H3 tail. *Nucleic Acids Res* **38:** 4246–4253.

Zhang J, Ding L, Holmfeldt L, Wu G, Heatley SL, Payne-Turner D, Easton J, Chen X, Wang J, Rusch M, et al. 2012. The genetic basis of early T-cell precursor acute lymphoblastic leukaemia. *Nature* **481:** 157–163.

Zhang X, Su J, Jeong M, Ko M, Huang Y, Park HJ, Guzman A, Lei Y, Huang YH, Rao A, et al. 2016. DNMT3A and TET2 compete and cooperate to repress lineage-specific transcription factors in hematopoietic stem cells. *Nat Genet* doi: 10.1038/ng.3610.

TET2 in Normal and Malignant Hematopoiesis

Robert L. Bowman[1] and Ross L. Levine[1,2]

[1]Human Oncology and Pathogenesis Program, Memorial Sloan Kettering Cancer Center, New York, New York 10021

[2]Leukemia Service, Department of Medicine, Memorial Sloan Kettering Cancer Center, New York, New York 10021

Correspondence: leviner@mskcc.org

The ten-eleven translocation (TET) family of enzymes were originally cloned from the translocation breakpoint of t(10;11) in infant acute myeloid leukemia (AML) with subsequent genomic analyses revealing somatic mutations and suppressed expression of TET family members across a range of malignancies, particularly enriched in hematological neoplasms. The TET family of enzymes is responsible for the hydroxylation of 5-methylcytosines (5-mC) to 5-hydroxymethylcytosine (5-hmC), followed by active and passive mechanisms leading to DNA demethylation. Given the complexity and importance of DNA methylation events in cellular proliferation and differentiation, it comes as no surprise that the TET family of enzymes is intricately regulated by both small molecules and regulatory cooperating proteins. Here, we review the structure and function of TET2, its interactions with cooperating mutations and small molecules, and its role in aberrant hematopoiesis.

Although often thought to be a stable epigenetic mark, recent research has revealed DNA methylation to be dynamic modification capable of regulating critical features of cellular proliferation, differentiation, and gene expression. Integral to this regulatory function are the enzymes necessary for both addition of the DNA-methyl mark and subsequent removal. Amongst these enzymes, the ten-eleven translocation (TET) family of proteins has emerged as critical regulators of the oxidation of 5-methylcytosine (5-mC) to 5-hydroxymethylcytosine (5-hmC). Since its recent identification in 2009, an explosion of studies has interrogated the roles of TET2 in malignancies of the blood and brain, developmental processes, and roles in inflammation. Genetic and biochemical studies in both human tumor specimen and animal models of disease have revealed TET2 as a critical node linking alterations in tumor metabolism to alterations in DNA methylation and modified chromatin. These features require a refined understanding of how to classify a cancer-associated gene that fits neither rigid definitions of an oncogene nor a tumor suppressor.

In this review, we will review the initial studies in hematologic malignancy that led to the discovery of TET2, the function and structure of the enzyme, its interactions with cooperating mutations and small molecules, and a perspective into other diseases in which TET2 mutations have been identified.

MUTATIONS IN HEMATOLOGIC DISEASE

In 2009, a series of papers identified somatic mutations in *TET2* in multiple hematologic malignancies (Delhommeau et al. 2009; Jankowska et al. 2009; Langemeijer et al. 2009; Tefferi et al. 2009a,b). Mapping of minimal regions of deletion in the 4q24 cytoband revealed *TET2* loss of heterozygosity (LOH) and somatic mutations in as many as 30%–50% of myelodysplastic syndrome (MDS) and myeloproliferative neoplasia (MPN) patients, whereas 32% of secondary acute myeloid leukemia (AML) patients harbored *TET2* mutations (Jankowska et al. 2009). Further genomic studies in MPN patients revealed the presence of *TET2* mutations in both *JAK2-V617F*-positive and -negative patients, with relatively equal distribution across essential thrombocytosis (ET), polycythemia vera (PV), and myelofibrosis (MF) (Tefferi et al. 2009b). In each of these studies, deletions as well as nonsense and missense mutations were found across multiple exons. Interestingly, most AML patients with *TET2* mutations retain expression of the wild-type allele with only 10% of patients possessing biallelic mutations (Delhommeau et al. 2009). Although the function of TET2 was not known at the time, these data were suggestive of a tumor suppressor role and potentially haploinsufficient loss-of-function role in *TET2* mutants.

TET2 mutations are present in multiple lymphoid and myeloid lineages, as well as CD34[+] progenitor cells, suggestive of an early clonal mutation in the stem cell compartment (Smith et al. 2010). In line with this early mutation designation, *TET2* mutants have continually been found at high allele frequency indicating that they are often the "first hit" in the multihit model of leukemogenesis (Smith et al. 2010; Papaemmanuil et al. 2016). These findings are reinforced by genetic studies identifying somatic *TET2* mutations in asymptomatic, healthy adults with clonal hematopoiesis (Smith et al. 2010; Busque et al. 2012). This, however, does not appear to always be the case, as two studies have shown that in MPN patients with *JAK2-V617F* mutations *TET2* can either present as the first hit or the second

hit based on mutant allele frequency (Abdel-Wahab et al. 2010; Ortmann et al. 2015). Interestingly, "JAK2-first" patients presented with significantly worse overall survival compared with "TET2-first" patients (Ortmann et al. 2015). In addition to the co-occurrence with *JAK2* mutations in MPN mentioned above, mutations in *TET2* have been shown to co-occur with mutations in *ASXL1, SRSF2, SF3B1, U2AF1,* and *CALR* (Rampal et al. 2014a). In AML, *TET2* shows comutational patterns with *NPM1, FLT3* and *DNMT3a* (Papaemmanuil et al. 2016). How these mutations cooperate in leukemogenesis remains an area of intense investigation and will be discussed later in this review.

Prognosis

TET2 mutational status has been found to be a variable prognostic indicator. One of the earliest studies of a cohort of 48 patients with systemic mastocytosis found no prognostic association with *TET2* mutational status (Tefferi et al. 2009a). Similarly, no survival association was found in a cohort of 63 patients with AML, chronic myelomonocytic leukemia (CMML), or MPN/MDS (Jankowska et al. 2009). Meanwhile, other studies showed a significant association with poor prognosis in AML (Abdel-Wahab et al. 2009) and a favorable prognostic association in MDS (Kosmider et al. 2009). In the largest cohorts to date by The Cancer Genome Atlas (TCGA) (Cancer Genome Atlas Research Network 2013) and by a group at the Sanger Institute (Papaemmanuil et al. 2016), there was no independent association with survival for *TET2* mutations. Interestingly, in a cohort of 211 MDS patients, there was an association in response to the hypomethylating agents decitabine and azacitidine, in which patients with *TET2* mutant AML were more likely to respond to therapy than those without the mutation, a finding that was more pronounced when the comutational partner *ASXL1* was not mutated (Bejar et al. 2014). Further studies into the functional role of *TET2* in disease progression and response to different therapeutic regiments may help clarify the prognostic value of *TET2* mutations in various hematologic malignancies.

TET2 FUNCTION AND STRUCTURE

When mutations in *TET2* were first discovered through mapping of the 4q24 region of loss/ LOH, the functions of this protein remained unknown. Shortly after, homology searches for the trypanosome proteins JBP1 and JBP2, enzymes known to oxidize methyl-thymine, identified the mammalian TET family as 2-oxoglutarate (2-OG) and Fe(II)-dependent enzymes (Tahiliani et al. 2009). These studies revealed that TET1 possessed enzymatic activity for con-

verting 5-mC to 5-hmC, and follow-up studies soon confirmed similar enzymatic activity for TET2 and TET3 (Ito et al. 2010; Ko et al. 2010). Subsequent studies would reveal that TET proteins are capable of generating iterative cytosine alterations leading to the formation of 5-formylcytosine (5-fC) and 5-carboxylcytosine (5-caC) (Fig. 1) (Ito et al. 2011). These intermediates were further shown to be substrates for thymine-DNA glycosylase (TDG)-mediated base excision repair (BER), converting the modified cytosine residue back to the unmethylated cytosine base (He et al. 2011; Maiti and Drohat 2011). Alternative demethylating mechanisms involve the APOBEC family mem-

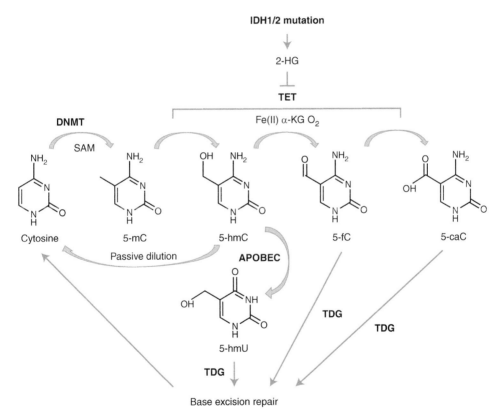

Figure 1. Reactions involved in TET-mediated oxidation of 5-methylcytosine (5-mC). Depicted here is cytosine-mediated methylation by the family of DNA methyltransferases (DNMT) with the substrate *S*-adenosyl methionine (SAM) leading to the formation of 5-mC. TET family members are then capable of mediating the iterative oxidation of 5-mC to 5-hydroxymethylcytosine (5-hmC), 5-formylcytosine (5-fC), and 5-carboxylcytosine (5-caC) in an Fe(II), O_2, and α-ketoglutarate (α-KG)-dependent reaction. These α-KG-dependent reactions can be inhibited by the oncometabolite 2-hydroxyglutarate (2-HG), which is a neomorphic by-product of mutant *IDH1* and *IDH2*. Each downstream product (5-hmC, 5-fC, and 5-caC) can serve as substrates for thymine DNA glycosylase (TDG) leading to base excision repair (BER) and eventual return to unmodified cytosine.

bers deaminating 5-hmC into 5-hydroxymeth-yluracil (5-hmU), which presents as a target for TDG and selective monofunctional uracil-DNA glycosylase 1 (SMUG1)-mediated BER (Bhutani et al. 2011). In addition to these active processes of DNA demehtylation, TET2 has been implicated in passive DNA demethylation, as 5-hmC serves as a poor substrate for the cell cycle–regulated DNMT1 leading to dilution of the 5-mC mark with each round of DNA replication and cell division (Hashimoto et al. 2012). It is important to note that the relative role of these different pathways in the ultimate removal of DNA modifications back to unmethylated cytosine remains to be fully delineated.

Recent biochemical studies have identified the structural components of TET2 that mediate these catalytic functions (Hu et al. 2013). Human TET2 encodes a 2002–amino acid, 223-kDa protein with a carboxy-terminal catalytic domain and poorly conserved amino-terminal domain. Biochemical studies on truncation variants were capable of reconstituting the enzymatic functions of TET2 within an 807–amino acid stretch (1129–1936) containing cysteine rich regions and a double-stranded β helix (DSBH) separated by an unstructured linker. This fragment was subsequently crystalized in complex with methylated DNA revealing coordination of a catalytic core by two zinc finger domains. These studies further revealed a cavity allowing for recognition of various modifications on the 5-mC base in the catalytic core. Subsequent studies showed that 5-hmC and 5-fC substrates possessed enzymatically unfavorable coordination of hydrogen bonds in the catalytic cavity, offering a potential explanation for the preferred substrate specificity of TET2 for 5-mC over that of 5-hmC and 5-fC (Hu et al. 2015), as well as the stability of 5-hmC in vivo (Ito et al. 2011).

REGULATION OF TET2 FUNCTION

Inhibition by IDH1/2 Mutant-Derived 2-Hydroxyglutarate

In addition to the genetic and biochemical studies above, one of the most important clues to understanding TET2 biology was the identification of mutually exclusive mutations in the metabolic enzymes isocitrate dehydrogenase 1 (IDH1) and IDH2, placing these enzymes in a putative genetic pathway (Abdel-Wahab et al. 2010). In a broader biological context, the identification of mutation in IDH1 and IDH2 was of fundamental importance in linking of altered cellular metabolism to the genomic age of cancer research. Although the altered glycolysis was a long appreciated hallmark of tumorigenesis (Hanahan and Weinberg 2011), it was not immediately clear how mutations in enzymes canonically involved in the citric acid cycle might impact tumorigenesis. The mechanistic role of these metabolic mutations began to take shape on the discovery that R132H mutant IDH1 was capable of producing 2-HG through an NADPH-dependent reduction of α-ketoglutarate (α-KG) (Dang et al. 2009). Soon after, these findings would be extended to the more common leukemic mutations of IDH2-R172K and IDH2-R140K (Ward et al. 2010). As IDH1/2 mutations were enriched in diseases with a relatively undifferentiated phenotype, low-grade glioma and leukemia, it was hypothesized that 2-HG might block differentiation albeit through unknown mechanisms. Definitive evidence would come later that year with the discovery that IDH1/2 mutant-derived 2-HG was capable of blocking differentiation and inhibiting the α-KG dependent enzyme TET2 (Figueroa et al. 2010). Critically, these studies revealed that IDH1/2 mutant AML patients displayed a hypermethylated phenotype (Figueroa et al. 2010), linking TET2 inhibition by 2-HG with the demethylating functions of TET proteins in development (Ito et al. 2011; Ko et al. 2011). The inhibitory capacity of 2-HG would later be extended to most α-KG-dependent enzymes (Xu et al. 2011), suggesting that IDH1/2 mutations might possess TET2-independent functions. Indeed, IDH1 mutant mice have been shown to down-regulate the DNA damage sensor ATM through altered histone methylation (Inoue et al. 2016). Further work will aim to identify therapeutic vulnerabilities that are shared between TET2 and IDH1/2 mutant AML, as well as those that are specifically relevant to the pleiotropic features of IDH1/2 mutant disease.

Although 2-HG has received much attention as a mutant IDH1/2 neometabolite, its production is not limited to these mutations. Importantly, 2-HG is a chiral molecule with the D-enantiomer being produced by mutant IDH1/2. Recent work has identified that under hypoxic conditions the L enantiomer of 2-HG is produced as a promiscuous bioproduct of lactate dehydrogenase A (LDHA)-mediated and malate dehydrogenase 1 (MDH1)-mediated reduction of α-KG (Intlekofer et al. 2015; Oldham et al. 2015). L2-HG was shown to function as a competitive inhibitor of the EGLN prolyl hydroxylase promoting hypoxia-inducible factor 1-α (HIF1-α) stability, whereas D2-HG served a substrate leading to HIF1-α degradation (Intlekofer et al. 2015). Meanwhile, both enantiomers are capable of inhibiting TET2 (Figueroa et al. 2010; Shim et al. 2014). These studies lead to the possibility that physiological production of L2-HG might play a role in modulating TET2 function in homeostasis, especially in the context of the hypoxic hematopoietic stem cell niche (Spencer et al. 2014).

TET2 Binding Proteins

The amino terminus of TET1 and TET3 contain a well-conserved CXXC domain that has been shown to mediate binding to unmethylated CpG residues (Xu et al. 2012); however, no such domain is present in TET2. Interestingly, a CXXC domain–containing protein, IDAX (CXXC4), is encoded 5′ of the TET2 genomic locus, suggestive of evolutionary splitting of the original TET2-CXXC gene into two separate genes (Ko et al. 2013). Biochemical studies revealed that IDAX is capable of binding both the amino terminus and catalytic domain of TET2, in which binding was associated with caspase-mediated TET2 degradation. This negative regulation of TET2 may present an additional mechanism for affecting 5-mC levels in malignancy, independent of genomic alterations to TET2 or mutations in IDH1/2. Indeed, IDAX has found to be overexpressed in villous adenomas in the colon (Nguyen

et al. 2010). In addition to its interaction with TET2, IDAX is a known inhibitor of WNT signaling (Hino et al. 2001), suggesting a potential source of cross talk between these pathways. Another interacting partner with WNT signaling, WT1, has also been shown to bind TET2 and TET3 (Rampal et al. 2014b), acting as a guide for TET2 to specific genomic loci associated with proliferation (Wang et al. 2015). In support of this observation, WT1 is mutated in AML, in a mutually exclusive pattern with TET2, and WT1 loss was further shown to phenocopy TET2 loss in hematopoiesis (Rampal et al. 2014b). In addition to these factors, the CRL4-VprBP complex has been shown to stabilize TET family members through mono-ubiquitination, increasing TET family members binding to DNA (Yu et al. 2013). Mutation at, or around, the TET2 monoubiquitination site at residue K1299 have been identified in several leukemia cell lines, offering another plausible mechanism for TET2 dysfunction in cancer (Nakagawa et al. 2015).

Vitamin C

Vitamin C has been shown to induce TET activity in embryonic stem (ES) cells, and to induce a global increase in 5-hmC content (Blaschke et al. 2013). Although this activity appeared to be specific to vitamin C and no other reducing agents, vitamin C affected both TET1 and TET2, the only TET family members expressed in ES cells. Interestingly, in this study, the investigators found that not all methylation marks were equally sensitive to subsequent demethylation. Indeed 5-hmC levels were most robustly affected at the promoters of genes, whereas methylation of retro elements remained unchanged. Vitamin C has previously been shown to regulate the activity of several iron-dependent dioxygenases; however, in this setting its effects did not appear to depend on either iron availability or α-KG concentration. In contrast, Hore et al. (2016) found in ES cells that vitamin C increased iron recycling and did not function as a cofactor. Although the details of the regulation may diverge, the capacity for vitamin C to induce TET2 activity is robust, with consistent effects in ES cells, mouse embryonic fibroblasts, T regulatory cells (Nair et al. 2016; Yue et al. 2016), and melanoma cells (Gustafson et al. 2015). Interestingly, in 2009, a single-arm clinical trial on 16 AML patients revealed a subset of patients that showed a clinical response following vitamin C deprivation (Park et al. 2009), highlighting the clinical relevance of vitamin C to leukemia biology. In addition to vitamin C, vitamin A has also been shown to play a role in inducing TET activity through the direct transcriptional regulation of both *TET2* and *TET3* (Hore et al. 2016). Understanding the mechanisms of both vitamin A and vitamin C activities on TET function could provide key insights into therapeutic options for both IDH and DNMT mutant cancers.

MECHANISMS OF CONTRIBUTION TO LEUKEMOGENESIS

The cancer genetics and biochemical studies discussed above provided substantial insight into the function of *TET2* in DNA methylation, yet understanding the cellular manifestations of these activities was made possible through the development of genetic mouse models. Conditional loss of TET2 activity by Vav:Cre-mediated removal of exon 3 led to an expansion of the lineage negative Sca.1$^+$ cKit$^+$ (LSK) cells in vivo and an increase in replating potential in a colony forming unit assay in vitro (Moran-Crusio et al. 2011). These studies further showed that *TET2*$^{KO/KO}$ bone marrow was capable of outcompeting *TET2*$^{WT/WT}$ bone marrow in competitive transplant assays and showed increased stem cell function and self-renewal. Finally, aged *TET2*$^{KO/KO}$ mice developed a CMML-like syndrome with expansion of the monocytes, increased spleen weight, and proliferative growth in the bone marrow, spleen, liver, and lung. Multiple studies published in the same year confirmed these findings (Ko et al. 2011; Li et al. 2011; Quivoron et al. 2011; Shide et al. 2012), with many studies revealing decreased 5-hmC levels in the LSK population. The expansion of the LSK and hematopoietic stem cell (HSC) populations in these mice mirror the findings in patient samples in which clonal *TET2* muta-

tions were found in healthy individuals with clonal hematopoiesis (Busque et al. 2012).

In AML and myeloproliferative disease, *TET2* mutations are typically present in concert with other mutations. Mutations in the fms related tyrosine kinase 3 (*FLT3*) are among the most common events in AML with point mutations in the tyrosine kinase domain, and internal tandem duplications (ITDs) near the juxtamembrane domain, leading to autoactivation of the kinase (Levis and Small 2003). Interestingly, when *TET2* loss was combined with a *FLT3-ITD* mutation there was a distinct set of genomic loci that underwent hypermethylation compared with either mutation alone (Shih et al. 2015). Among these loci, hypermethylation of the *GATA2* promoter led to a reduction in expression, blockade in differentiation and the development of a transplantable leukemia derived from the LSK progenitor compartment. Interestingly, in addition to the hypermethylated regions, there were more than 500 hypomethylated regions in the combined TET2-FLT3-ITD mutants that were not present in either mutant alone. These, at first paradoxical, findings may be partially explained by FLT3's role in the commitment to the myeloid lineage and thus hypomethylation of genes necessary for that engagement. Future studies on DNA methylation and 5-hmC will be of interest to determine if the loci-specific effects are indeed specific to this model or representative of a more general leukemic transformation phenotype. Additional models of mutational cooperation with TET2 loss include expression of a mutant c-KIT in mast cells (Soucie et al. 2012), expression of *AML-ETO* (Hatlen et al. 2016), and loss of Notch signaling (Lobry et al. 2013).

MUTATIONS IN OTHER MALIGNANCIES

T-Cell Lymphoma

In addition to myeloid malignancies detailed above, *TET2* mutations have also been identified in patients with T-cell lymphoma (Quivoron et al. 2011). One study found *TET2* mutations in 47% of angioimmunoblastic T-cell lymphomas (ATLs) and in 38% of peripheral T-cell lymphomas not otherwise specified (PCTL-NOS) (Lemonnier et al. 2012). Subsequent studies have identified *TET2* mutations in upward of 75% of ATL patients (Odejide et al. 2014). Interestingly, this study found multiple subclonal mutations in *TET2* within individual patients, all of which resulted in truncation or disruption of the final gene product. Unlike the myeloid leukemia setting, few mutations were found in CD34$^+$ progenitors or subsequent myeloid colonies derived from this population (Odejide et al. 2014). Collectively, these results place TET2 loss as a recurrent driver in ATL with acquisition of the mutation in a lineage committed stage, contrasting sharply with the myeloid malignancies. A related contrast was identified when patients presented with both *TET2* and *IDH1* mutations (Odejide et al. 2014), events that are largely mutually exclusive in myeloid malignancy. This may reflect a different role for subclonal TET2 loss of function in ATL versus the presumed expansion of a preleukemic clone in AML.

Melanoma

TET2 mutations have been predominantly associated with hematologic malignancies; however, whole-genome analyses through TCGA have identified additional mutations in melanoma and cutaneous squamous cell carcinoma (Cancer Genome Atlas Network 2015). Consistent with these genomic findings, loss of 5-hmC has been proposed to be a prevalent, epigenetic hallmark, of melanoma (Lian et al. 2012). Indeed, epigenetic silencing of *TET2* and *TET3* has been shown to drive TGF-β-dependent invasion and acquisition of EMT-like features (Gong et al. 2016). Subsequent in vivo studies showed that overexpression of TET2 blunted tumor growth and metastasis. Given the prominent role of dedifferentiation in metastatic melanoma, it will be interesting to investigate the potentially parallel roles of TET2 in hematopoietic and melanocyte differentiation.

Glioma

CpG-island hypermethylator phenotypes (CIMPs) have also been identified in low-grade gliomas, as well as some glioblastoma patients. Although *TET2* promoter methylation has been identified in glioma (Kim et al. 2011), the predominant mechanism appears to be driven by mutations in *IDH1*, with few loss-of-function mutations present in *TET2* (Kraus et al. 2015). This is in stark contrast to MPN and AML studies in which TET2, IDH1, and IDH2 mutations are all present. These studies support a growing literature showing mutant IDH1 and IDH2 elicit functions outside of 2-HG-mediated TET2 inhibition, including inhibition of histone demethylases (Lu et al. 2012), alteration of DNA damage repair (Inoue et al. 2016), and alterations in branched chain amino acid metabolism (Tonjes et al. 2013). This apparent tissue-specific distinction in mutational patterns may also be the result of tissue-specific gene expression levels of *TET2*, which is substantially more highly expressed in AML than in the gliomas (Fig. 2).

Other Diseases

In addition to melanoma, TET2 has been shown to be down-regulated in an androgen-dependent manner in prostate cancer, with lower expression conferring worse prognosis for patients (Nickerson et al. 2016). In colorectal cancer, TET2 has been shown to be excluded from the nucleus (Huang et al. 2016), with similar findings for TET1 in glioma (Waha et al. 2012).

CONCLUDING REMARKS

In sum, TET2 is a critical regulator of DNA methylation in development and malignancy. Genomic alterations of *TET2*, in addition to modulation of binding partners, lead to alterations in 5-hmC levels and downstream outputs on proliferation and maintenance of stem cells. Although mutations are enriched in hematologic neoplasms, TET2 loss of function has been observed in solid tumors as well. Collectively, these studies have provided insight into how TET2 contributes to disease and may provide clues for identifying specific therapeutic avenues for patients harboring these mutations.

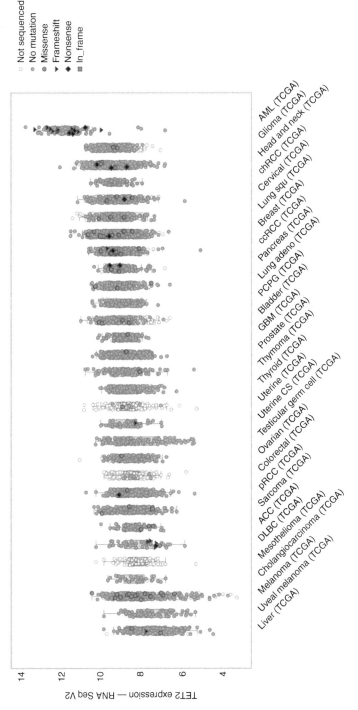

Figure 2. *TET2* expression across malignancy. Normalized RNA-Seq counts (log2) are shown for the indicated malignancies ranked from lowest to highest mean expression of *TET2*. Samples with frameshift mutations are denoted with an inverted triangle, nonsense mutations are denoted as a diamond, and in-frame mutations are shown with a square. Data was collected and graphed using the cBioPortal (see cbioportal.org/) (Gao et al. 2013).

ACKNOWLEDGMENTS

We thank members of the Levine laboratory for their discussion and critical insights.

REFERENCES

Abdel-Wahab O, Mullally A, Hedvat C, Garcia-Manero G, Patel J, Wadleigh M, Malinge S, Yao J, Kilpivaara O, Bhat R, et al. 2009. Genetic characterization of *TET1*, *TET2*, and *TET3* alterations in myeloid malignancies. *Blood* 114: 144–147.

Abdel-Wahab O, Manshouri T, Patel J, Harris K, Yao J, Hedvat C, Heguy A, Bueso-Ramos C, Kantarjian H, Levine RL, et al. 2010. Genetic analysis of transforming events that convert chronic myeloproliferative neoplasms to leukemias. *Cancer Res* 70: 447–452.

Bejar R, Lord A, Stevenson K, Bar-Natan M, Pérez-Ladaga A, Zaneveld J, Wang H, Caughey B, Stojanov P, Getz G, et al. 2014. *TET2* mutations predict response to hypomethylating agents in myelodysplastic syndrome patients. *Blood* 124: 2705–2712.

Bhutani N, Burns DM, Blau HM. 2011. DNA demethylation dynamics. *Cell* 146: 866–872.

Blaschke K, Ebata KT, Karimi MM, Zepeda-Martínez JA, Goyal P, Mahapatra S, Tam A, Laird DJ, Hirst M, Rao A, et al. 2013. Vitamin C induces Tet-dependent DNA demethylation and a blastocyst-like state in ES cells. *Nature* 500: 222–226.

Busque L, Patel JP, Figueroa ME, Vasanthakumar A, Provost S, Hamilou Z, Mollica L, Li J, Viale A, Heguy A, et al. 2012. Recurrent somatic *TET2* mutations in normal elderly individuals with clonal hematopoiesis. *Nat Genet* 44: 1179–1181.

Cancer Genome Atlas Network. 2015. Genomic classification of cutaneous melanoma. *Cell* 161: 1681–1696.

Cancer Genome Atlas Research Network. 2013. Genomic and epigenomic landscapes of adult de novo acute myeloid leukemia. *N Engl J Med* 368: 2059–2074.

Dang L, White DW, Gross S, Bennett BD, Bittinger MA, Driggers EM, Fantin VR, Jang HG, Jin S, Keenan MC, et al. 2009. Cancer-associated IDH1 mutations produce 2-hydroxyglutarate. *Nature* 462: 739–744.

Delhommeau F, Dupont S, Della Valle V, James C, Trannoy S, Masse A, Kosmider O, Le Couedic JP, Robert F, Alberdi A, et al. 2009. Mutation in *TET2* in myeloid cancers. *N Engl J Med* 360: 2289–2301.

Figueroa ME, Abdel-Wahab O, Lu C, Ward PS, Patel J, Shih A, Li Y, Bhagwat N, Vasanthakumar A, Fernandez HF, et al. 2010. Leukemic IDH1 and IDH2 mutations result in a hypermethylation phenotype, disrupt TET2 function, and impair hematopoietic differentiation. *Cancer Cell* 18: 553–567.

Gao J, Aksoy BA, Dogrusoz U, Dresdner G, Gross B, Sumer SO, Sun Y, Jacobsen A, Sinha R, Larsson E, et al. 2013. Integrative analysis of complex cancer genomics and clinical profiles using the cBioPortal. *Sci Signal* 6: pl1.

Gong F, Guo Y, Niu Y, Jin J, Zhang X, Shi X, Zhang L, Li R, Chen L, Ma RZ. 2016. Epigenetic silencing of TET2 and

TET3 induces an EMT-like process in melanoma. *Oncotarget*. doi: 10.18632/oncotarget.

Gustafson CB, Yang C, Dickson KM, Shao H, Van Booven D, Harbour JW, Liu ZJ, Wang G. 2015. Epigenetic reprogramming of melanoma cells by vitamin C treatment. *Clin Epigenetics* 7: 51.

Hanahan D, Weinberg RA. 2011. Hallmarks of cancer: The next generation. *Cell* 144: 646–674.

Hashimoto H, Liu Y, Upadhyay AK, Chang Y, Howerton SB, Vertino PM, Zhang X, Cheng X. 2012. Recognition and potential mechanisms for replication and erasure of cytosine hydroxymethylation. *Nucleic Acids Res* 40: 4841–4849.

Hatlen MA, Arora K, Vacic V, Grabowska EA, Liao W, Riley-Gillis B, Oschwald DM, Wang L, Joergens JE, Shih AH, et al. 2016. Integrative genetic analysis of mouse and human AML identifies cooperating disease alleles. *J Exp Med* 213: 25–34.

He YF, Li BZ, Li Z, Liu P, Wang Y, Tang Q, Ding J, Jia Y, Chen Z, Li L, et al. 2011. Tet-mediated formation of 5-carboxylcytosine and its excision by TDG in mammalian DNA. *Science* 333: 1303–1307.

Hino S, Kishida S, Michiue T, Fukui A, Sakamoto I, Takada S, Asashima M, Kikuchi A. 2001. Inhibition of the Wnt signaling pathway by Idax, a novel Dvl-binding protein. *Mol Cell Biol* 21: 330–342.

Hore TA, Von Meyenn F, Ravichandran M, Bachman M, Ficz G, Oxley D, Santos F, Balasubramanian S, Jurkowski TP, Reik W. 2016. Retinol and ascorbate drive erasure of epigenetic memory and enhance reprogramming to naïve pluripotency by complementary mechanisms. *Proc Natl Acad Sci* 113: 12202–12207.

Hu L, Li Z, Cheng J, Rao Q, Gong W, Liu M, Shi YG, Zhu J, Wang P, Xu Y. 2013. Crystal structure of TET2–DNA complex: Insight into TET-mediated 5mC oxidation. *Cell* 155: 1545–1555.

Hu L, Lu J, Cheng J, Rao Q, Li Z, Hou H, Lou Z, Zhang L, Li W, Gong W, et al. 2015. Structural insight into substrate preference for TET-mediated oxidation. *Nature* 527: 118–122.

Huang Y, Wang G, Liang Z, Yang Y, Cui L, Liu CY. 2016. Loss of nuclear localization of TET2 in colorectal cancer. *Clin Epigenetics* 8: 9.

Inoue S, Li WY, Tseng A, Beerman I, Elia AJ, Bendall SC, Lemonnier F, Kron KJ, Cescon DW, Hao Z, et al. 2016. Mutant *IDH1* downregulates ATM and alters DNA repair and sensitivity to DNA damage independent of TET2. *Cancer Cell* 30: 337–348.

Intlekofer AM, Dematteo RG, Venneti S, Finley LW, Lu C, Judkins AR, Rustenburg AS, Grinaway PB, Chodera JD, Cross JR, et al. 2015. Hypoxia induces production of L-2-hydroxyglutarate. *Cell Metab* 22: 304–311.

Ito S, Dalessio AC, Taranova OV, Hong K, Sowers LC, Zhang Y. 2010. Role of Tet proteins in 5mC to 5hmC conversion, ES-cell self-renewal and inner cell mass specification. *Nature* 466: 1129–1133.

Ito S, Shen L, Dai Q, Wu SC, Collins LB, Swenberg JA, He C, Zhang Y. 2011. Tet proteins can convert 5-methylcytosine to 5-formylcytosine and 5-carboxylcytosine. *Science* 333: 1300–1303.

Jankowska AM, Szpurka H, Tiu RV, Makishima H, Afable M, Huh J, O'Keefe CL, Ganetzky R, McDevitt MA, Maciejewski JP. 2009. Loss of heterozygosity 4q24 and *TET2* mutations associated with myelodysplastic/myeloproliferative neoplasms. *Blood* **113:** 6403–6410.

Kim YH, Pierscianek D, Mittelbronn M, Vital A, Mariani L, Hasselblatt M, Ohgaki H. 2011. *TET2* promoter methylation in low-grade diffuse gliomas lacking *IDH1/2* mutations. *J Clin Pathol* **64:** 850–852.

Ko M, Huang Y, Jankowska AM, Pape UJ, Tahiliani M, Bandukwala HS, An J, Lamperti ED, Koh KP, Ganetzky R, et al. 2010. Impaired hydroxylation of 5-methylcytosine in myeloid cancers with mutant TET2. *Nature* **468:** 839–843.

Ko M, Bandukwala HS, An J, Lamperti ED, Thompson EC, Hastie R, Tsangaratou A, Rajewsky K, Koralov SB, Rao A. 2011. Ten-eleven-translocation 2 (TET2) negatively regulates homeostasis and differentiation of hematopoietic stem cells in mice. *Proc Natl Acad Sci* **108:** 14566–14571.

Ko M, An J, Bandukwala HS, Chavez L, Äijö T, Pastor WA, Segal MF, Li H, Koh KP, Lähdesmäki H, et al. 2013. Modulation of TET2 expression and 5-methylcytosine oxidation by the CXXC domain protein IDAX. *Nature* **497:** 122–126.

Kosmider O, Gelsi-Boyer V, Cheok M, Grabar S, Della-Valle V, Picard F, Viguie F, Quesnel B, Beyne-Rauzy O, Solary E, et al. 2009. *TET2* mutation is an independent favorable prognostic factor in myelodysplastic syndromes (MDSs). *Blood* **114:** 3285–3291.

Kraus TF, Greiner A, Steinmaurer M, Dietinger V, Guibourt V, Kretzschmar HA. 2015. Genetic characterization of ten-eleven-translocation methylcytosine dioxygenase alterations in human glioma. *J Cancer* **6:** 832–842.

Langemeijer SM, Kuiper RP, Berends M, Knops R, Aslanyan MG, Massop M, Stevens-Linders E, van Hoogen P, van Kessel AG, Raymakers RA, et al. 2009. Acquired mutations in *TET2* are common in myelodysplastic syndromes. *Nat Genet* **41:** 838–842.

Lemonnier F, Couronné L, Parrens M, Jaïs JP, Travert M, Lamant L, Tournillac O, Rousset T, Fabiani B, Cairns RA, et al. 2012. Recurrent *TET2* mutations in peripheral T-cell lymphomas correlate with T_{FH}-like features and adverse clinical parameters. *Blood* **120:** 1466–1469.

Levis M, Small D. 2003. FLT3: It does matter in leukemia. *Leukemia* **17:** 1738–1752.

Li Z, Cai X, Cai CL, Wang J, Zhang W, Petersen BE, Yang FC, Xu M. 2011. Deletion of *Tet2* in mice leads to dysregulated hematopoietic stem cells and subsequent development of myeloid malignancies. *Blood* **118:** 4509–4518.

Lian CG, Xu Y, Ceol C, Wu F, Larson A, Dresser K, Xu W, Tan L, Hu Y, Zhan Q, et al. 2012. Loss of 5-hydroxymethylcytosine is an epigenetic hallmark of melanoma. *Cell* **150:** 1135–1146.

Lobry C, Ntziachristos P, Ndiaye-Lobry D, Oh P, Cimmino L, Zhu N, Araldi E, Hu W, Freund J, Abdel-Wahab O, et al. 2013. Notch pathway activation targets AML-initiating cell homeostasis and differentiation. *J Exp Med* **210:** 301–319.

Lu C, Ward PS, Kapoor GS, Rohle D, Turcan S, Abdel-Wahab O, Edwards CR, Khanin R, Figueroa ME, Melnick A, et al. 2012. IDH mutation impairs histone demethylation and

results in a block to cell differentiation. *Nature* **483:** 474–478.

Maiti A, Drohat AC. 2011. Thymine DNA glycosylase can rapidly excise 5-formylcytosine and 5-carboxylcytosine: Potential implications for active demethylation of CpG sites. *J Biol Chem* **286:** 35334–35338.

Moran-Crusio K, Reavie L, Shih A, Abdel-Wahab O, Ndiaye-Lobry D, Lobry C, Figueroa ME, Vasanthakumar A, Patel J, Zhao X, et al. 2011. *Tet2* loss leads to increased hematopoietic stem cell self-renewal and myeloid transformation. *Cancer Cell* **20:** 11–24.

Nair VS, Song MH, Oh KI. 2016. Vitamin C facilitates demethylation of the *Foxp3* enhancer in a Tet-dependent manner. *J Immunol* **196:** 2119–2131.

Nakagawa T, Lv L, Nakagawa M, Yu Y, Yu C, D'Alessio AC, Nakayama K, Fan HY, Chen X, Xiong Y. 2015. CRL4^{VprBP} E3 ligase promotes monoubiquitylation and chromatin binding of TET dioxygenases. *Mol Cell* **57:** 247–260.

Nguyen AV, Albers CG, Holcombe RF. 2010. Differentiation of tubular and villous adenomas based on Wnt pathway-related gene expression profiles. *Int J Mol Med* **26:** 121–125.

Nickerson ML, Das S, Im KM, Turan S, Berndt SI, Li H, Lou H, Brodie SA, Billaud JN, Zhang T, et al. 2016. TET2 binds the androgen receptor and loss is associated with prostate cancer. *Oncogene*. doi: 10.1038/onc.2016.376.

Odejide O, Weigert O, Lane AA, Toscano D, Lunning MA, Kopp N, Kim S, Van Bodegom D, Bolla S, Schatz JH, et al. 2014. A targeted mutational landscape of angioimmunoblastic T-cell lymphoma. *Blood* **123:** 1293–1296.

Oldham WM, Clish CB, Yang Y, Loscalzo J. 2015. Hypoxia-mediated increases in L-2-hydroxyglutarate coordinate the metabolic response to reductive stress. *Cell Metab* **22:** 291–303.

Ortmann CA, Kent DG, Nangalia J, Silber Y, Wedge DC, Grinfeld J, Baxter EJ, Massie CE, Papaemmanuil E, Menon S, et al. 2015. Effect of mutation order on myeloproliferative neoplasms. *N Engl J Med* **372:** 601–612.

Papaemmanuil E, Gerstung M, Bullinger L, Gaidzik VI, Paschka P, Roberts ND, Potter NE, Heuser M, Thol F, Bolli N, et al. 2016. Genomic classification and prognosis in acute myeloid leukemia. *N Engl J Med* **374:** 2209–2221.

Park CH, Kimler BF, Yi SY, Park SH, Kim K, Jung CW, Kim SH, Lee ER, Rha M, Kim S, et al. 2009. Depletion of L-ascorbic acid alternating with its supplementation in the treatment of patients with acute myeloid leukemia or myelodysplastic syndromes. *Eur J Haematol* **83:** 108–118.

Quivoron C, Couronne L, Della Valle V, Lopez CK, Plo I, Wagner-Ballon O, Do Cruzeiro M, Delhommeau F, Arnulf B, Stern MH, et al. 2011. TET2 inactivation results in pleiotropic hematopoietic abnormalities in mouse and is a recurrent event during human lymphomagenesis. *Cancer Cell* **20:** 25–38.

Rampal R, Ahn J, Abdel-Wahab O, Nahas M, Wang K, Lipson D, Otto GA, Yelensky R, Hricik T, McKenney AS, et al. 2014a. Genomic and functional analysis of leukemic transformation of myeloproliferative neoplasms. *Proc Natl Acad Sci* **111:** E5401–5410.

Rampal R, Alkalin A, Madzo J, Vasanthakumar A, Pronier E, Patel J, Li Y, Ahn J, Abdel-Wahab O, Shih A, et al. 2014b. DNA hydroxymethylation profiling reveals that *WT1* mutations result in loss of TET2 function in acute myeloid leukemia. *Cell Rep* **9:** 1841–1855.

Shide K, Kameda T, Shimoda H, Yamaji T, Abe H, Kamiunten A, Sekine M, Hidaka T, Katayose K, Kubuki Y, et al. 2012. TET2 is essential for survival and hematopoietic stem cell homeostasis. *Leukemia* **26:** 2216–2223.

Shih AH, Jiang Y, Meydan C, Shank K, Pandey S, Barreyro L, Antony-Debre I, Viale A, Socci N, Sun Y, et al. 2015. Mutational cooperativity linked to combinatorial epigenetic gain of function in acute myeloid leukemia. *Cancer Cell* **27:** 502–515.

Shim EH, Livi CB, Rakheja D, Tan J, Benson D, Parekh V, Kho EY, Ghosh AP, Kirkman R, Velu S, et al. 2014. L-2-Hydroxyglutarate: An epigenetic modifier and putative oncometabolite in renal cancer. *Cancer Discov* **4:** 1290–1298.

Smith AE, Mohamedali AM, Kulasekararaj A, Lim Z, Gaken J, Lea NC, Przychodzen B, Mian SA, Nasser EE, Shooter C, et al. 2010. Next-generation sequencing of the *TET2* gene in 355 MDS and CMML patients reveals low-abundance mutant clones with early origins, but indicates no definite prognostic value. *Blood* **116:** 3923–3932.

Soucie E, Hanssens K, Mercher T, Georgin-Lavialle S, Damaj G, Livideanu C, Chandesris MO, Acin Y, Letard S, de Sepulveda P, et al. 2012. In aggressive forms of mastocytosis, TET2 loss cooperates with c-KITD816V to transform mast cells. *Blood* **120:** 4846–4849.

Spencer JA, Ferraro F, Roussakis E, Klein A, Wu J, Runnels JM, Zaher W, Mortensen LJ, Alt C, Turcotte R, et al. 2014. Direct measurement of local oxygen concentration in the bone marrow of live animals. *Nature* **508:** 269–273.

Tahiliani M, Koh KP, Shen YH, Pastor WA, Bandukwala H, Brudno Y, Agarwal S, Iyer LM, Liu DR, Aravind L, et al. 2009. Conversion of 5-methylcytosine to 5-hydroxymethylcytosine in mammalian DNA by MLL partner TET1. *Science* **324:** 930–935.

Tefferi A, Levine RL, Lim KH, Abdel-Wahab O, Lasho TL, Patel J, Finke CM, Mullally A, Li CY, Pardanani A, et al. 2009a. Frequent *TET2* mutations in systemic mastocytosis: Clinical, *KITD816V* and *FIP1L1-PDGFRA* correlates. *Leukemia* **23:** 900–904.

Tefferi A, Pardanani A, Lim KH, Abdel-Wahab O, Lasho TL, Patel J, Gangat N, Finke CM, Schwager S, Mullally A, et al. 2009b. *TET2* mutations and their clinical correlates in polycythemia vera, essential thrombocythemia and myelofibrosis. *Leukemia* **23:** 905–911.

Tonjes M, Barbus S, Park YJ, Wang W, Schlotter M, Lindroth AM, Pleier SV, Bai AH, Karra D, Piro RM, et al. 2013. BCAT1 promotes cell proliferation through amino acid catabolism in gliomas carrying wild-type IDH1. *Nat Med* **19:** 901–908.

Waha A, Müller T, Gessi M, Waha A, Isselstein LJ, Luxen D, Freihoff D, Freihoff J, Becker A, Simon M, et al. 2012. Nuclear exclusion of TET1 is associated with loss of 5-hydroxymethylcytosine in *IDH1* wild-type gliomas. *Am J Pathol* **181:** 675–683.

Wang Y, Xiao M, Chen X, Chen L, Xu Y, Lv L, Wang P, Yang H, Ma S, Lin H, et al. 2015. WT1 recruits TET2 to regulate its target gene expression and suppress leukemia cell proliferation. *Mol Cell* **57:** 662–673.

Ward PS, Patel J, Wise DR, Abdel-Wahab O, Bennett BD, Coller HA, Cross JR, Fantin VR, Hedvat CV, Perl AE, et al. 2010. The common feature of leukemia-associated IDH1 and IDH2 mutations is a neomorphic enzyme activity converting α-ketoglutarate to 2-hydroxyglutarate. *Cancer Cell* **17:** 225–234.

Xu W, Yang H, Liu Y, Yang Y, Wang P, Kim SH, Ito S, Yang C, Wang P, Xiao MT, et al. 2011. Oncometabolite 2-hydroxyglutarate is a competitive inhibitor of α-ketoglutarate-dependent dioxygenases. *Cancer Cell* **19:** 17–30.

Xu Y, Xu C, Kato A, Tempel W, Abreu JG, Bian C, Hu Y, Hu D, Zhao B, Cerovina T, et al. 2012. Tet3 CXXC domain and dioxygenase activity cooperatively regulate key genes for *Xenopus* eye and neural development. *Cell* **151:** 1200–1213.

Yu C, Zhang YL, Pan WW, Li XM, Wang ZW, Ge ZJ, Zhou JJ, Cang Y, Tong C, Sun QY, et al. 2013. CRL4 complex regulates mammalian oocyte survival and reprogramming by activation of TET proteins. *Science* **342:** 1518–1521.

Yue X, Trifari S, Äijö T, Tsagaratou A, Pastor WA, Zepeda-Martínez JA, Lio CWJ, Li X, Huang Y, Vijayanand P, et al. 2016. Control of Foxp3 stability through modulation of TET activity. *J Exp Med* **213:** 377–397.

Exploitation of EP300 and CREBBP Lysine Acetyltransferases by Cancer

Narsis Attar[1,2] and Siavash K. Kurdistani[1,2,3,4]

[1]Department of Biological Chemistry, David Geffen School of Medicine, University of California, Los Angeles, California 90095

[2]Molecular Biology Institute, David Geffen School of Medicine, University of California, Los Angeles, California 90095

[3]Department of Pathology and Laboratory Medicine, David Geffen School of Medicine, University of California, Los Angeles, California 90095

[4]Eli and Edythe Broad Center of Regenerative Medicine and Stem Cell Research, David Geffen School of Medicine, University of California, Los Angeles, California 90095

Correspondence: skurdistani@mednet.ucla.edu

p300 and CREB-binding protein (CBP), two homologous lysine acetyltransferases in metazoans, have a myriad of cellular functions. They exert their influence mainly through their roles as transcriptional regulators but also via nontranscriptional effects inside and outside of the nucleus on processes such as DNA replication and metabolism. The versatility of p300/CBP as molecular tools has led to their exploitation by viral oncogenes for cellular transformation and by cancer cells to achieve and maintain an oncogenic phenotype. How cancer cells use p300/CBP in their favor varies depending on the cellular context and is evident by the growing list of loss- and gain-of-function genetic alterations in p300 and CBP in solid tumors and hematological malignancies. Here, we discuss the biological functions of p300/CBP and how disruption of these functions by mutations and alterations in expression or subcellular localization contributes to the cancer phenotype.

EP300 (hereafter referred to as p300) and its closely related paralog CREB-binding protein (CREBBP, hereafter CBP) are ubiquitously expressed transcriptional coactivators and major lysine acetyltransferases (KATs) in metazoans. They regulate transcription by serving as scaffolds that bridge sequence-specific DNA-binding factors and the basal transcriptional machinery (Chan and La Thangue 2001), and facilitate transcription through acetylation of histones, transcription factors, and autoacetylation (Sterner and Berger 2000; Black et al. 2006; Pugh 2006; Das et al. 2014). p300 and CBP are large proteins with multiple functional domains accommodating diverse protein–protein interactions. This has enabled a large number of disparate transcription factors to use p300/CBP as cofactors in regulating the expression of thousands of genes in essentially all cell types (Chan and La Thangue 2001). The large number of proteins that interact with p300 and CBP underscore the widespread influence of these

Cite this article as *Cold Spring Harb Perspect Med* doi: 10.1101/cshperspect.a026534

coactivators on essential cellular functions. p300 and CBP regulate several fundamental biological processes including proliferation, cell cycle, cell differentiation, and the DNA damage response (Shi and Mello 1998; Goodman and Smolik 2000; Grossman 2001; Polesskaya et al. 2001). But the versatility of these proteins has also made it difficult to discern their specific involvements in distinct biological processes and pathophysiological states. Another impediment to understanding the roles of p300 and CBP is the overlapping contribution of these proteins to the same molecular processes such as gene regulation, hence, the commonly used designation "p300/CBP."

Although p300/CBP have been implicated in cancer development, the specific contributions of each acetyltransferase to the cancer phenotype have been less precisely defined. This is in part attributable to the participation of p300/CBP in diverse and, at times, antagonistic cellular pathways such as tumor-suppressive and pro-oncogenic processes. The challenge is even greater when attempting to understand the consequences of mutations in p300 and CBP, which have been identified in numerous cancer genome studies. To fully understand how genetic or epigenetic alterations of p300/CBP contribute to the cancer phenotype, it is important to determine which cellular pathways are specifically affected by mutations in p300 or CBP. One important consideration is that p300 and CBP, despite significant sequence homology, also perform nonoverlapping cellular functions and can cooperate with distinct binding partners. In this review, we discuss the major functions of p300/CBP in the cell and how cancer cells exploit these functions to their advantage.

p300/CBP ORCHESTRATE THE CELL CYCLE AND REGULATE PROLIFERATION

p300 was initially identified through its physical association with the adenovirus transforming protein E1A and determined to be essential for adenovirus-mediated oncogenic transformation (Whyte et al. 1989; Sawada et al. 1997). Soon after, the E1A–p300 interaction

was shown to be critical for the G_1–S phase transition in adenovirus-infected cells (Howe et al. 1990). A number of other oncogenic viral proteins (e.g., SV40/polyoma LT, HPV E7) were subsequently shown to also target p300 or CBP to promote cellular transformation (Eckner et al. 1996; Zimmermann et al. 1999; Bernat et al. 2003). The frequent exploitations of p300/CBP as cofactors for viral oncoproteins highlighted the fundamental role of these proteins in regulating cellular proliferation, and raised the possibility that alterations of p300/CBP may also contribute to nonviral mechanisms of tumorigenesis.

E1A was initially reported to inhibit p300/CBP-mediated KAT activity and transcriptional activation, which led to a hypothesis that p300 may normally function as a negative regulator of S phase entry (Arany et al. 1995; Yang et al. 1996; Chakravarti et al. 1999). However, many studies have since revealed a major role for p300/CBP in promoting growth and cell-cycle progression. In fact, on adenoviral infection p300 is recruited to genes with functions in cell cycle and proliferation to promote their full activation and S phase entry in otherwise nondividing cells (Ferrari et al. 2008, 2014).

The functions of p300 and CBP in regulating cell-cycle progression are partly mediated through their influence on transcription by being recruited to gene regulatory regions, such as enhancers and promoters, via sequence-specific DNA-binding transcription factors. Once bound, they facilitate subsequent regulatory events to ultimately direct RNA polymerase II activation. p300 in particular contributes to the formation of the transcription pre-initiation complex, a large multiprotein complex required for expression of genes. p300 does this partly through dynamic association with and dissociation from the transcriptional machinery, which is facilitated by p300 auto-acetylation activity (Black et al. 2006). The pervasive participation of p300 and CBP in transcriptional regulation is evident in their binding to >16,000 genes in human cells (Smith et al. 2004; Ramos et al. 2010). Not all binding events lead to transcriptional activation and a growing body of evidence indicates a gene-repressive role for p300/CBP in certain contexts (Santoso and Kadonaga 2006; Sankar et al. 2008; Ferrari et al. 2014). p300 and CBP also regulate the cell cycle through interactions with or acetylation of proteins involved in cell-cycle progression, such as the DNA replication machinery and histones for the purpose of DNA replication through chromatin.

Transcriptional Coactivation and the Cell Cycle

One of the earliest cell-based models showing the role of p300/CBP in cell-cycle progression involved depletion of p300 and CBP, through microinjection of an antibody against both proteins, which was found to limit S phase entry (Ait-Si-Ali et al. 2000). This defect was reversed by overexpression of exogenous CBP, indicating a direct function of CBP in promoting cell-cycle progression. p300 and CBP serve as transcriptional coactivators for the E2F transcription factor family, which are central for expression of genes required for G_1/S transition (Trouche and Kouzarides 1996; Trouche et al. 1996; Wang et al. 2007). In addition, p300/CBP acetylate the E2F proteins themselves (e.g., E2F1), leading to enhanced DNA-binding and gene activation (Martínez-Balbás et al. 2000; Marzio et al. 2000). The acetyltransferase activity of CBP is regulated in a cell-cycle-dependent manner and peaks at the G_1/S boundary possibly as a consequence of cyclin/Cdk-mediated phosphorylation of CBP before initiation of S phase (Ait-Si-Ali et al. 1998). Cell-cycle-dependent transcription of the major histone genes for DNA replication is also dependent on p300/CBP, which are recruited by NPAT, the general histone expression regulator (He et al. 2011). Therefore, the transcriptional coactivator functions of p300/CBP mediate S phase entry through proper expression of DNA replication and cellular growth genes.

Nontranscriptional Effects of p300/CBP on Cell Cycle

p300/CBP may regulate DNA replication through modifying the histones surrounding the DNA replication origins. These two acetyltransferases are responsible for the bulk of histone H3 lysine 18 acetylation (H3K18ac) and H3K27ac, modifications associated with active promoters and enhancers (Horwitz et al. 2008; Jin et al. 2011). H3K18ac is also associated with active DNA replication in certain cell types (Li et al. 2014). p300/CBP may also directly regulate the DNA replication machinery by acetylating two major endonucleases involved in Okazaki fragment processing, FEN1 and Dna2, inhibiting and stimulating their activities, respectively (Hasan et al. 2001; Balakrishnan et al. 2010). This differential regulation is suggested to lead to increased accuracy of DNA replication (Balakrishnan et al. 2010). Complementing these results, pharmacological inhibition of p300 KAT activity prolongs S phase because of reduced replication fork velocity and defects in timing of replication origin firing and synchronization (Prieur et al. 2011). Altogether, p300/CBP regulate various aspects of the DNA replication process, including the choice and timing of origin firing and the assembly of the newly synthesized DNA into chromatin.

p300/CBP function in other phases of the cell cycle as well. Depletion of CBP leads to a delay in mitosis and accumulation of cells in G_2/M because of the aberrant activity of the anaphase-promoting complex (APC/C), an E3 ubiquitin ligase required for progression through mitosis (Turnell et al. 2005). Taken together with the above, these findings place p300 and CBP at multiple positions along the cell cycle and emphasize the functions of these proteins in promoting progression through the entire cell cycle (Fig. 1).

Consistent with the critical roles of p300/CBP in cell-cycle regulation, significant growth defects are observed when these proteins are depleted in cells or organisms. Mouse models null for p300 or CBP are embryonic lethal and, although p300-null cells obtained from these embryos are viable, they show reduced proliferation (Yao et al. 1998). This also occurs when p300 is transiently or stably depleted (Yuan et al. 1999; Iyer et al. 2007). Therefore, loss of p300/CBP in most contexts leads to decreased proliferation.

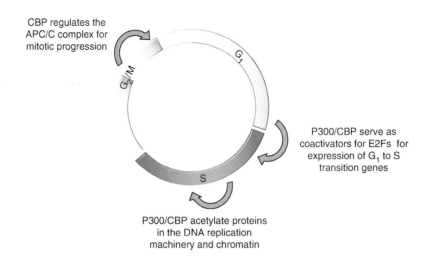

CBP regulates the APC/C complex for mitotic progression

G_2/M

G_1

P300/CBP serve as coactivators for E2Fs for expression of G_1 to S transition genes

S

P300/CBP acetylate proteins in the DNA replication machinery and chromatin

Figure 1. p300/CREB-binding protein (CBP) regulate the cell cycle at multiple points. The diagram summarizes the known functions of p300/CBP at different points along the cell cycle. The E2F family of transcription factors uses p300/CBP as transcriptional coactivators to facilitate expression of E2F target genes that orchestrate the transition from G_1 to S. p300 also facilitates S phase progression by acetylating the DNA replication machinery (e.g., FEN1 and Dna2) and the histones surrounding the replication origins (e.g., H3K18). CBP promotes progression through mitosis by regulating the function of the anaphase-promoting complex (APC)/C complex.

p300 AND CBP IN CANCER

p300/CBP as Classic Tumor Suppressors

Early indications of tumor suppression by p300/CBP came from findings in a rare congenital developmental disorder, Rubinstein–Taybi syndrome (RTS). Germline heterozygous mutations in CBP and less frequently in p300 are observed in RTS and may play a role in the pathogenesis of this disease. RTS patients have an increased incidence of cancer with ∼5% diagnosed with childhood tumors of neural crest origin (Miller and Rubinstein 1995). p300/CBP mutations in RTS are variable and encompass microdeletions, truncating mutations as well as point mutations in different domains (Petrij et al. 1995; Roelfsema and Peters 2007). A number of these genetic lesions reduce acetyltransferase and/or transcriptional activities of p300/CBP implicating the reduction of these functions in the etiology of RTS-associated malignancies (Roelfsema and Peters 2007), a contention that is supported by studies in mice (Tanaka et al. 1997; Rebel et al. 2002; Alarcón et al. 2004).

Investigations of primary tumor samples helped to strengthen the tumor-suppressive functions of p300/CBP in humans. Work by Gayther et al. identified the first cancer-associated inactivating genetic lesions in p300 in breast and colorectal primary tumors and cell lines (Gayther et al. 2000). The majority of cases harbored inactivation or deletion of the second allele of p300. Studies of larger cohorts of solid tumors including colorectal, gastric, ovarian, and hepatocellular carcinomas also detected loss of heterozygosity (LOH) at the p300 or CBP loci at frequencies ranging from 1% to 50% (Bryan et al. 2002; Tillinghast et al. 2003; Koshiishi et al. 2004; Dancy and Cole 2015). A small fraction of p300/CBP LOH events in these studies were accompanied by somatic mutations in the second allele confirming earlier findings. Tumors showing LOH indicate that haploinsufficiency of p300/CBP may be a factor in the pathogenesis of cancer. This is consistent with the idea that a limiting cellular pool of p300/CBP may be a biological determinant of their effects on the cell. In fact there is evidence that these proteins are haploinsufficient because p300/CBP heterozygote null embryos have reduced survival (Yao et al. 1998). Therefore, different molecular pathways have to compete for a limited pool of p300/CBP to regulate their target genes (Kamei et al. 1996; Huang et al. 2007). So the reduced availability of p300/CBP through LOH may contribute to cancer development or progression by altering the equilibrium between the various p300/CBP-dependent pathways. Additional evidence suggesting a tumor-suppressive function for p300/CBP came from oral and cervical carcinoma cell lines. These cell lines, which harbor either a homozygous mutation in p300 or a heterozygous truncation of p300 with inactivation of the normal allele, show reduced proliferation on introduction of a normal copy of p300 (Suganuma et al. 2002).

p300/CBP may exert tumor-suppressive effects through promoting the functions of other bona fide tumor suppressors, such as p53, RB1, BRCA1, or through inducing transforming growth factor β (TGF-β)-responsive genes (Nishihara et al. 1998; Pao et al. 2000; Chan et al. 2001; Grossman 2001). The involvement of p300/CBP in p53-mediated functions is extensively studied and occurs at multiple levels. In response to DNA damage, p300/CBP augment p53-dependent transcriptional activation of genes required for cell-cycle arrest and DNA repair (Grossman 2001). In addition, p300 promotes the nuclear accumulation and stability of p53 in response to genotoxic stress. Interestingly, in unstressed conditions and during recovery from DNA damage, p300 is thought to ensure degradation of p53 for resumption of the cell cycle after DNA repair (Grossman et al. 1998; Grossman 2001; Kawai et al. 2001). BRCA1, which is frequently mutated in familial breast and ovarian cancers, plays a role in cell-cycle checkpoint, DNA damage repair, and transcriptional regulation (Monteiro et al. 1996; Wu et al. 2010). The latter role has linked BRCA1 to p300/CBP, which enhance BRCA1-mediated transcriptional activation (Pao et al. 2000; Mullan et al. 2006). Similarly p300/CBP mediate the effects of TGF-β signaling by serving as transcriptional coactivators for Smad3, a downstream effector of this tumor-suppressive pathway (Feng et al. 1998; Derynck et al. 2001).

p300/CBP as Drivers of Cancer Growth

Despite the tumor-suppressive roles of p300/CBP, several lines of evidence suggest that these KATs can also participate in promoting cancer. Although inactivating mutations in p300/CBP are found in certain cancers (Kalkhoven 2004; Pasqualucci et al. 2011), some cancer-linked point mutations are in fact gain-of-function alterations in p300/CBP that could contribute to cancer development (Ringel and Wolberger 2013). In addition to the acetyltransferase domains, important structural features of p300/CBP include three cysteine/histidine-rich zinc-binding domains (CH1-3), a bromodomain, and a recently identified RING (Really Interesting New Gene) domain within the larger CH2 region. The RING domain contacts the active site of the KAT domain blocking substrate binding and decreasing acetyltransferase activity in vitro. Disruption of the RING domain enhances p300 KAT activity (Delvecchio et al. 2013). Mutations in the p300 RING domain are found in malignancies including melanoma, endometrial and colorectal carcinoma (Forbes et al. 2015), as well as in RTS, and may boost p300 KAT activity in these settings (Delvecchio et al. 2013). How increased KAT activity of p300 or CBP promotes malignancy is not clear. In addition to acetylation of H3K18 and H3K27, p300/CBP also mediate the acetylation of histone H3 lysine 56, a modification associated with nucleosome assembly in yeast and DNA replication and repair in mammals (Li et al. 2008; Yuan et al. 2009; Vempati et al. 2010). Increased cellular levels of histone H3K56ac are observed in a number of epithelial tumors and relate to tumor stage and an undifferentiated phenotype (Das et al. 2009). Enhanced KAT activity of p300 or CBP may lead to increased acetylation of H3K56 in certain cancers.

Another mode of p300/CBP acetyltransferase gain-of-function involves translocation events in hematological malignancies such as myelodysplastic syndrome and acute myeloid leukemia. These translocations occur between p300 or CBP and monocytic leukemia zing-finger (MOZ), MOZ-related factor (MORF), or myeloid/lymphoid or mixed-lineage leukemia (MLL) (Kitabayashi et al. 2001; Panagopoulos et al. 2001). The translocation events more commonly generate a fusion protein containing the carboxy-terminal region of CBP with or without its KAT domain. The MOZ/MORF-CBP as well as MOZ-p300 fusion proteins maintain the KAT domains from both parent proteins potentially resulting in highly active lysine acetyltransferases (Yang and Ullah 2007). In essentially all translocation events, the amino-terminal region of CBP is excluded from the fusion protein. The novel functions that are gained by the fusion protein can contribute to the oncogenic nature of these translocations. Mutations in the KAT-inhibitory RING domain of p300 are also detected in myelodysplastic syndrome (Forbes et al. 2015). This further suggests a key role for increased p300/CBP KAT activity in the pathogenesis of these hematological malignancies.

Mutations in p300/CBP Are Nonrandom

The widespread application of next generation sequencing has revealed an abundance of somatic genetic mutations with frequencies of up to 30% in p300 and CBP in various types of cancer. Although earlier studies of p300/CBP had uncovered gross or partial gene deletions, only a relatively small fraction of all p300/CBP genetic lesions in cancer are of this nature. The majority of alterations are, in fact, missense point mutations (Fig. 2) that occur essentially throughout the p300 and CBP proteins with a higher frequency in the KAT domains, suggesting a selective pressure in cancers for alteration of this activity (Fig. 3A). Certain residues in the KAT domain of p300 and CBP (e.g., D1399 and

Y1467 for p300 and Y1450 and Y1503 for CBP) that are known to reduce or abolish the KAT activity when mutated are among the most frequent mutations in cancer (Delvecchio et al. 2013; Forbes et al. 2015). It is unclear, however, whether and how the other frequently mutated residues in this domain are important for the KAT activity. The top four and five most frequent missense mutations in p300 and CBP, respectively, are highlighted in Fig. 3B. These mutations in p300 including those thought to reduce KAT activity are clustered near the site of acetyl-CoA binding (Fig. 3B) (Liu et al. 2008; Maksimoska et al. 2014). The location of the other uncharacterized mutations near this site suggests these may also have an effect on the KAT activity.

Tables 1 and 2 list the cancers with p300 and CBP mutation frequencies, respectively, of 5% or higher as reported in the COSMIC (Catalogue of Somatic Mutations in Cancer) database (Forbes et al. 2015). Both p300 and CBP are frequently mutated in skin squamous cell carcinoma followed by certain types of lymphomas. Alterations of p300/CBP may therefore be key contributory milestones to the development or progression of these cancers. Alternatively, these specific cellular contexts may provide a more permissive background for accumulation of genetic lesions in p300/CBP with no selective pressure to avoid them. Tables 3 and 4 include lists of cancers with ≤1% frequency of p300 or CBP mutations, respectively. Of interest in these

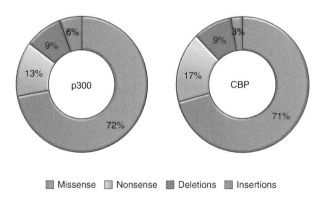

Missense Nonsense Deletions Insertions

Figure 2. Frequency of the different types of mutations in p300 and CBP in cancer. The majority of genetic lesions to p300 or CBP in cancer are missense mutations followed by nonsense mutations, deletions, and a small number of insertions.

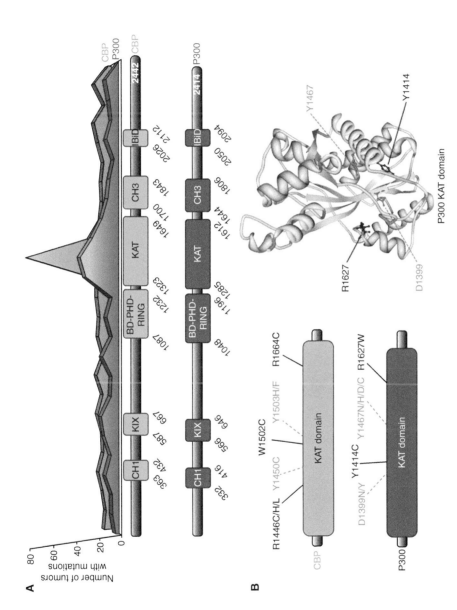

Figure 3. The acetyltransferase domains of p300 and CBP are hotspots for mutations in cancer. (*A*) The number of tumors of different origins with missense mutations along the p300 and CBP proteins domain structures is shown (Forbes et al. 2015). CH1, Cysteine/histidine-rich region 1; BR, bromodomain; PHD, plant homeodomain finger; RING, Really Interesting New Gene finger domain; KAT, lysine acetyltransferase domain; CH3, cysteine/histidine-rich region 3 (also referred to as TAZ2); IBiD, IRF3-binding domain. (*B*) The residues in the KAT domains of p300 and CBP that are frequently mutated in cancer are indicated. Residues in red are important for the KAT activity. The four most common residues mutated in p300 reside close to the acetyl-CoA-binding site as indicated in the crystal structure (Liu et al. 2008; Maksimoska et al. 2014).

Table 1. Cancer subtypes with higher frequency of p300 mutations

Cancer subtype	N	Samples mutated (%)
Skin squamous cell carcinoma	82	26.8
Marginal zone B-cell lymphoma	15	13.3
Bladder carcinoma	425	9.4
Follicular lymphoma	68	8.8
Lung small-cell carcinoma	48	8.3
Endometrial carcinoma	387	7.2
Esophageal squamous cell carcinoma	511	5.9
Cervical squamous cell carcinoma	193	5.2
Breast carcinoma (estrogen-receptor-positive)	80	5.0

Data were obtained from the publicly available COSMIC database (cancer.sanger.ac.uk).

groups are prostate and pancreas carcinomas, which may be under selective pressure to preserve the normal functions of p300/CBP for growth or may alter the functions of p300 or CBP through nongenetic means.

Beyond Genetic Alterations

In the absence of genetic defects, mechanisms such as changes in expression or subcellular localization can alter p300/CBP-associated functions in cancer. Analysis of The Cancer Genome Atlas (TCGA) data indicates differences in expression levels of p300 and CBP in multiple types of carcinomas and hematological malignancies. The effects of these changes in mediating the neoplastic phenotype cannot be described under one umbrella as both higher and lower levels of p300/CBP are found. These expression differences in some cases are accompanied by changes in gene copy number as a result of gross gene amplifications or deletions. Several studies of primary tumors of varying origins have also revealed changes in p300/CBP protein levels, which in either direction are prognostic in many cases. A study of 95 prostate cancer lesions revealed that increased p300 protein levels, as compared with adjacent

normal tissue, correlate with increased proliferation, tumor volume, and extraprostatic involvement (Debes et al. 2003). A function of p300 in the progression of prostate cancer has also been proposed as it mediates androgen-dependent as well as independent transactivation of the androgen receptor (Debes et al. 2002). Additionally, increased p300 expression correlates with poor survival and aggressive phenotypes in breast, hepatocellular, esophageal, and cutaneous squamous cell carcinoma (Li et al. 2011a,b; Xiao et al. 2011; Chen et al. 2014). Consistently, pharmacological inhibition of p300/CBP KAT activities in a panel of primary melanoma cell lines sensitizes cells to DNA-damaging chemotherapeutic agents (Yan et al. 2013). Higher nuclear CBP protein levels have also been detected in precancerous hyperplastic and dysplastic laryngeal lesions, suggesting overexpression of p300/CBP may contribute to different steps of cancer development and growth (Karamouzis et al. 2002). Conversely, reduced levels of p300/CBP have been detected in certain cancers. Pasqualucci et al. (2011) detected loss of p300 and/or CBP expression in 8% of diffuse large B-cell lymphoma with no genetic lesion in these genes. The significance of these changes has been underscored by the prognostic value of this information. A study of

Table 2. Cancer subtypes with higher frequency of CBP mutations

Cancer subtype	N	Samples mutated (%)
Follicular lymphoma	66	33.3
Skin squamous cell carcinoma	77	28.6
Marginal zone B-cell lymphoma	15	13.3
Diffuse large B-cell lymphoma	242	12.0
Salivary gland carcinoma	63	9.5
Bladder carcinoma	438	8.9
Endometrial carcinoma	337	8.0
Lung small-cell carcinoma	52	7.7
Breast carcinoma (estrogen-receptor-positive)	80	7.5

Data were obtained from the publicly available COSMIC database (cancer.sanger.ac.uk).

Table 3. Cancer subtypes with low frequency of p300 mutations

Cancer subtype	N	Samples mutated (%)
Acute lymphoblastic leukemia	438	0.9
Clear cell renal cell carcinoma	692	0.9
Hepatocellular carcinoma	628	0.8
Chronic lymphocytic leukemia	798	0.5
Ovarian carcinoma	693	0.3
Prostate adenocarcinoma	755	0.3
Pancreatic carcinoma	593	0.2

Data were obtained from the publicly available COSMIC database (cancer.sanger.ac.uk).

327 melanoma samples found that decreased nuclear levels of p300 associated with disease progression and poor overall survival (Rotte et al. 2013).

Interestingly, an increase in cytoplasmic levels of p300 was observed in melanoma and correlated significantly with tumor size and disease progression in early stages (Rotte et al. 2013; Bhandaru et al. 2014). These findings suggest that a shift in the subcellular localization of p300 may be involved in the progression of this cancer and highlight the importance of p300/CBP cytoplasmic functions. However, few investigations addressing cytoplasmic functions of p300/CBP have been conducted. Among these are studies that provide evidence for a p53-directed E4 ligase activity associated with cytoplasmic p300/CBP leading to polyubiquitination and degradation of cytoplasmic p53 (Grossman et al. 2007; Shi et al. 2009). The compartmentalized regulation of p53 by p300/CBP provides an explanation for previously reported opposing effects of these KATs on p53, namely, destabilization of the protein and promoting its nuclear functions. Hence, the physical separation of p300/CBP's nuclear transcriptional function from the cytosolic E3/E4 ligase activity is an important aspect of p53 regulation by p300/CBP. Extranuclear functions of p53 involve triggering apoptosis through interactions with mitochondrial outer membrane proteins, which contribute to its tu-

mor-suppressive effects (O'Brate and Giannakakou 2003). Inappropriate accumulation of cytosolic p300/CBP can, therefore, suppress p53-mediated apoptosis in response to stress signals.

The Effects of p300/CBP on Chromatin in Cancer

Deregulated chromatin targeting by p300/CBP can have profound effects in cancer. For instance, alterations of H3K27ac, a major in vivo target of p300/CBP at enhancer loci, is observed in numerous cancers indicating disrupted activities of p300/CBP at specific enhancer elements (Akhtar-Zaidi et al. 2012). These "variant enhancer loci" (VELs) as they were termed, correlate with aberrant expression of their putative target genes. The enhancers with inappropriately acquired H3K27ac associate frequently with genes that have known contributions to many phenotypic hallmarks of cancer (Hnisz et al. 2013). Thus, alterations in the distribution of H3K27ac (and likely H3K18ac, the other major target of p300/CBP) and, hence, enhancer activity in cancer can promote tumorigenesis through promoting an oncogenic gene expression program. Possible mechanisms of deregulated chromatin targeting may involve mutant or inappropriately expressed p300/CBP as well as changes in transcription factors that recruit these coactivators to gene regulatory elements. For example, genetic lesions in the acetyllysine-binding bromodomain of p300/CBP can lead

Table 4. Cancer subtypes with low frequency of CBP mutations

Cancer subtype	N	Samples mutated (%)
Pancreatic carcinoma	593	1.0
Breast carcinoma (triple-negative)	121	0.8
Clear cell renal cell carcinoma	692	0.6
Chronic lymphocytic leukemia	798	0.5
Prostate adenocarcinoma	762	0.4
Acute myeloid leukemia	785	0.4

Data were obtained from the publicly available COSMIC database (cancer.sanger.ac.uk).

to alterations in specificity or strength of chromatin binding and to redistribution of H3K18/27ac across the genome. Such a scenario could also occur through gene amplifications or translocations, giving rise to aberrant formation of enhancer loci in some tumors (Hnisz et al. 2013).

Functions of p300/CBP on chromatin can also be co-opted by viral oncogenes leading to cellular transformation. Adenovirus small E1A (e1a), a splice variant of E1A that is responsible for reprogramming the expression of thousands of host genes, relies on interactions with p300/CBP to coerce normal, primary cell-cycle-arrested fibroblasts into S phase (Howe et al. 1990). The e1a protein causes the recruitment of p300/CBP to and increased H3K18ac (but not H3K27ac) at promoter regions of cell-cycle genes for full transcriptional activation. In parallel, e1a also actively represses cell-type-specific genes by opposing the functions of p300/CBP at promoters and enhancers of these genes as evident by substantial deacetylation of H3K18 and H3K27 at these sites on e1a expression (Ferrari et al. 2010, 2014). Furthermore, e1a represses cellular defense response genes by forming a trimeric complex between e1a, RB1, and, surprisingly, p300 itself, which acetylates RB1 to prevent its normal inactivation by phosphorylation (Ferrari et al. 2014). The repressive RB1-e1a-p300 complex binds to the promoter and gene body regions of defense response genes, and in some cases fully coating entire gene loci and preventing their activation by the host cell. This repression is accompanied by condensation of the local chromatin environment (Ferrari et al. 2014). The overall effect of e1a in the 24 h after entry into a cell is to turn off cell identity and the antiviral cellular defense genes, and to turn on genes that are required for entry into S phase and DNA replication. The bulk of this oncogenic reprogramming depends on interactions of e1a with p300/CBP and RB1. These findings provide a blueprint for understanding how nonviral oncogenesis may also depend on precise exploitations of p300/CBP to achieve similar cellular outcomes.

HISTONE ACETYLATION BALANCE AND ITS IMPLICATIONS FOR CANCER

In addition to targeted recruitment and acetylation of specific genomic loci, histone acetyltransferases function globally throughout the genome in a seemingly nontargeted manner by mechanisms that are not yet clear (Vogelauer et al. 2000). When coupled to the global actions of lysine deacetylases (KDACs), the opposing but continual functions of KATs and KDACs result in fast turnover of histone acetylation (Waterborg 2001), which consumes acetyl coenzyme A and generates acetate anions. In primary tumor tissues, cancer cells show marked differences in the global levels of histone modifications including acetylation, which are prognostic of clinical outcome in many types of solid tumors (Kurdistani 2007). Specifically, lower global level of H3K18ac is associated with cancer-related mortality and/or morbidity in prostate, kidney, lung, pancreatic, and breast cancers (Seligson et al. 2005, 2009; Manuyakorn et al. 2010; Mosashvilli et al. 2010; Kurdistani 2011). Cancer-associated genetic lesions in p300/CBP resulting in reduced KAT activity can certainly lead to the global loss of H3K18/27ac. However, recent work from our laboratory has revealed an unanticipated function for global histone acetylation in regulating intracellular pH (McBrian et al. 2013). We found that in multiple cancer or normal cell lines, the balance of KAT and KDAC activities is shifted toward the latter in response to acidic cellular environment, resulting in histones that are globally and continuously deacetylated. This leads to liberated acetate anions that are in turn used by the membrane-bound monocarboxylate transporters to export protons out of the cell, thus buffering the intracellular pH. Proliferating cells including cancer cells need to maintain an alkaline intracellular pH relative to the outside for cell growth and division (Webb et al. 2011). Because cancer tissues commonly show low pH in vivo, it is possible that enhanced global histone deacetylation serves to maintain a viable intracellular pH in these tumors, providing a growth advantage (Parks et al. 2011). This chromatin response to acidity is an active process resulting in the continuous generation of free acetate molecules through enhanced deacetylation and thus depends on intact or even enhanced histone acetylation. Therefore, the function of KATs in maintaining global histone acetylation is imperative to this pH-regulatory function of chromatin. In this regard, loss of KAT function may in fact reduce fitness in tumors exposed to an acidic environment.

LINKING CELLULAR ENERGETICS AND THE EPIGENOME

p300 and CBP target a significant number of nonhistone proteins for acetylation, including cytosolic proteins involved in essential metabolic processes. This can potentially coordinate cytoplasmic and chromatin-related functions of p300/CBP. The involvement of p300 in regulating metabolism via targeting the M2 isoform of pyruvate kinase, PKM2, is one such example (Lv et al. 2013). A majority of cancers express PKM2, which unlike the constitutively active PKM1 isoform, shows lower activity and is allosterically activated by an upstream glycolytic intermediate (Christofk et al. 2008; Wong et al. 2015). The slower enzymatic rate of PKM2 is thought to essentially serve as a road blockage that causes a logjam in upstream glycolytic reactions, forcing glycolysis intermediates into branching pathways, the products of which, such as nucleotide precursors, are required for general cellular biosynthesis. p300 acetylates a lysine residue (K433) unique to PKM2, which abolishes allosteric activation and enhances nuclear localization of this PK isoform. The switch between cytoplasmic metabolic function and nuclear protein kinase activity of PKM2 regulated by p300 occurs in response to mitogens and oncogenic signals and may be involved in tumorigenesis (Lv et al. 2013). In the nucleus PKM2 phosphorylates histone H3 at threonine 11, a modification shown to be required for cell-cycle progression and tumorigenesis (Yang et al. 2012). These findings indicate a role of p300 in mediating the proliferative program in cancer cells through switching a metabolic enzyme to a nuclear kinase to create a chromatin state conducive for cell replication. The regula-

tion of PKM2 localization and activity is just one recent indication of the broader influence of p300/CBP beyond their nuclear functions as transcriptional coactivators.

FUTURE DIRECTIONS AND THERAPEUTIC APPLICATIONS

The central roles of p300/CBP in regulating cell proliferation have spurred efforts to develop specific inhibitors of the enzymatic activities as well as protein–protein interactions of p300/CBP. KAT inhibitors with higher specificity toward p300/CBP show antiproliferative effects in preclinical studies of cancer (Santer et al. 2011; Yang et al. 2013). Small molecules that inhibit p300/CBP interactions with other proteins also show promising clinical use. ICG-001, which specifically inhibits CBP binding to β-catenin, a component of the Wnt signaling pathway, reduces tumorigenic phenotypes and enhances drug sensitivity in both acute lymphoblastic leukemia (ALL) and nasopharyngeal carcinoma (Emami et al. 2004; Gang et al. 2014; Chan et al. 2015). This approach is thought to take advantage of differential co-activator usage by β-catenin. β-catenin may mediate the opposing outcomes of Wnt signaling by using CBP or p300 to either stimulate proliferation or initiate differentiation, respectively (Ma et al. 2005; Teo and Kahn 2010). Interestingly, the effect of ICG-001 is independent of CBP mutational status in ALL (Gang et al. 2014). Most CBP mutations in ALL are found carboxy terminal to the β-catenin binding site where ICG-001 binds (Mulligan et al. 2011). These findings suggest that the preservation of CBP-β-catenin interaction and not the mutations in other regions of CBP may underlie progression of ALL.

CONCLUDING REMARKS

The many functions of p300/CBP can be differentially exploited in cancer depending on the context, cellular identity, and perhaps environmental cues to confer a growth advantage. The paradigm of cancer as an evolutionary system suggests the sequential acquisition of somatic mutations in a fluctuating microenvironment to gain fitness. Considering such a system, the order and nature of other oncogenic events can dictate the selection for or against p300/CBP alterations that are most advantageous for survival and growth, therefore branding these proteins as tumor suppressors or oncogenes.

ACKNOWLEDGMENTS

We thank Michael Carey and Trent Su for valuable input and discussions in preparing this manuscript. N.A. is supported by a Ruth L. Kirschstein National Research Service Award (CA186619-02) and S.K.K. by a National Institutes of Health Grant (CA178415).

REFERENCES

Ait-Si-Ali S, Ramirez S, Barre FX, Dkhissi F, Magnaghi-Jaulin L, Girault JA, Robin P, Knibiehler M, Pritchard LL, Ducommun B, et al. 1998. Histone acetyltransferase activity of CBP is controlled by cycle-dependent kinases and oncoprotein E1A. *Nature* **396:** 184–186.

Ait-Si-Ali S, Polesskaya A, Filleur S, Ferreira R, Duquet A, Robin P, Vervish A, Trouche D, Cabon F, Harel-Bellan A. 2000. CBP/p300 histone acetyl-transferase activity is important for the G_1/S transition. *Oncogene* **19:** 2430–2437.

Akhtar-Zaidi B, Cowper-Sal-lari R, Corradin O, Saiakhova A, Bartels CF, Balasubramanian D, Myeroff L, Lutterbaugh J, Jarrar A, Kalady MF, et al. 2012. Epigenomic enhancer profiling defines a signature of colon cancer. *Science* **336:** 736–739.

Alarcón JM, Malleret G, Touzani K, Vronskaya S, Ishii S, Kandel ER, Barco A. 2004. Chromatin acetylation, memory, and LTP are impaired in CBP[+/−] mice: A model for the cognitive deficit in Rubinstein–Taybi syndrome and its amelioration. *Neuron* **42:** 947–959.

Arany Z, Newsome D, Oldread E, Livingston DM, Eckner R. 1995. A family of transcriptional adaptor proteins targeted by the E1A oncoprotein. *Nature* **374:** 81–84.

Balakrishnan L, Stewart J, Polaczek P, Campbell JL, Bambara RA. 2010. Acetylation of Dna2 endonuclease/helicase and flap endonuclease 1 by p300 promotes DNA stability by creating long flap intermediates. *J Biol Chem* **285:** 4398–4404.

Bernat A, Avvakumov N, Mymryk JS, Banks L. 2003. Interaction between the HPV E7 oncoprotein and the transcriptional coactivator p300. *Oncogene* **22:** 7871–7881.

Bhandaru M, Ardekani GS, Zhang G, Martinka M, McElwee KJ, Li G, Rotte A. 2014. A combination of p300 and Braf expression in the diagnosis and prognosis of melanoma. *BMC Cancer* **14:** 398.

Black JC, Choi JE, Lombardo SR, Carey M. 2006. A mechanism for coordinating chromatin modification and pre-initiation complex assembly. *Mol Cell* **23:** 809–818.

Bryan EJ, Jokubaitis VJ, Chamberlain NL, Baxter SW, Dawson E, Choong DYH, Campbell IG. 2002. Mutation analysis of EP300 in colon, breast and ovarian carcinomas. *Int J Cancer* **102:** 137–141.

Chakravarti D, Ogryzko V, Kao HY, Nash A, Chen H, Nakatani Y, Evans RM. 1999. A viral mechanism for inhibition of p300 and PCAF acetyltransferase activity. *Cell* **96:** 393–403.

Chan HM, La Thangue NB. 2001. p300/CBP proteins: HATs for transcriptional bridges and scaffolds. *J Cell Sci* **114:** 2363–2373.

Chan HM, Krstic-Demonacos M, Smith L, Demonacos C, La Thangue NB. 2001. Acetylation control of the retinoblastoma tumour-suppressor protein. *Nat Cell Biol* **3:** 667–674.

Chan KC, Chan LS, Ip JCY, Lo C, Yip TTC, Ngan RKC, Wong RNS, Lo KW, Ng WT, Lee AWM, et al. 2015. Therapeutic targeting of CBP/β-catenin signaling reduces cancer stem-like population and synergistically suppresses growth of EBV-positive nasopharyngeal carcinoma cells with cisplatin. *Sci Rep* **5:** 9979.

Chen MK, Cai MY, Luo RZ, Tian X, Liao QM, Zhang XY, Han JD. 2014. Overexpression of p300 correlates with poor prognosis in patients with cutaneous squamous cell carcinoma. *Br J Dermatol* **172:** 111–119.

Christofk HR, Vander Heiden MG, Harris MH, Ramanathan A, Gerszten RE, Wei R, Fleming MD, Schreiber SL, Cantley LC. 2008. The M2 splice isoform of pyruvate kinase is important for cancer metabolism and tumour growth. *Nature* **452:** 230–233.

Dancy BM, Cole PA. 2015. Protein lysine acetylation by p300/CBP. *Chem Rev* **115:** 2419–2452.

Das C, Lucia MS, Hansen KC, Tyler JK. 2009. CBP/p300-mediated acetylation of histone H3 on lysine 56. *Nature* **459:** 113–117.

Das C, Roy S, Namjoshi S, Malarkey CS, Jones DNM, Kutateladze TG, Churchill ME, Tyler JK. 2014. Binding of the histone chaperone ASF1 to the CBP bromodomain promotes histone acetylation. *Proc Natl Acad Sci* **111:** E1072–E1081.

Debes JD, Schmidt LJ, Huang H, Tindall DJ. 2002. P300 mediates androgen-independent transactivation of the androgen receptor by interleukin 6. *Cancer Res* **62:** 5632–5636.

Debes JD, Sebo TJ, Lohse CM, Murphy LM, Haugen DAL, Tindall DJ. 2003. p300 in prostate cancer proliferation and progression. *Cancer Res* **63:** 7638–7640.

Delvecchio M, Gaucher J, Aguilar-Gurrieri C, Ortega E, Panne D. 2013. Structure of the p300 catalytic core and implications for chromatin targeting and HAT regulation. *Nat Struct Mol Biol* **20:** 1040–1046.

Derynck R, Akhurst RJ, Balmain A. 2001. TGF-β signaling in tumor suppression and cancer progression. *Nat Genet* **29:** 117–129.

Eckner R, Ludlow JW, Lill NL, Oldread E, Arany Z, Modjtahedi N, DeCaprio JA, Livingston DM, Morgan JA. 1996. Association of p300 and CBP with simian virus 40 large T antigen. *Mol Cell Biol* **16:** 3454–3464.

Emami KH, Nguyen C, Ma H, Kim DH, Jeong KW, Eguchi M, Moon RT, Teo J, Oh SW, Kim HY, et al. 2004. A small molecule inhibitor of β-catenin/CREB-binding protein transcription. *Proc Natl Acad Sci* **101**: 12682–12687.

Feng XH, Zhang Y, Wu RY, Derynck R. 1998. The tumor suppressor Smad4/DPC4 and transcriptional adaptor CBP/p300 are coactivators for Smad3 in TGF-β-induced transcriptional activation. *Genes Dev* **12**: 2153–2163.

Ferrari R, Pellegrini M, Horwitz GA, Xie W, Berk AJ, Kurdistani SK. 2008. Epigenetic reprogramming by adenovirus e1a. *Science* **321**: 1086–1088.

Ferrari R, Berk AJ, Kurdistani SK. 2010. Viral manipulation of the host epigenome for oncogenic transformation. *Nat Rev Genet* **10**: 290–294.

Ferrari R, Gou D, Jawdekar G, Johnson SA, Nava M, Su T, Yousef AF, Zemke NR, Pellegrini M, Kurdistani SK, et al. 2014. Adenovirus small E1A employs the lysine acetylases p300/CBP and tumor suppressor Rb to repress select host genes and promote productive virus infection. *Cell Host Microbe* **16**: 663–676.

Forbes SA, Beare D, Gunasekaran P, Leung K, Bindal N, Boutselakis H, Ding M, Bamford S, Cole C, Ward S, et al. 2015. COSMIC: Exploring the world's knowledge of somatic mutations in human cancer. *Nucleic Acids Res* **43**: D805–D811.

Gang EJ, Hsieh YT, Pham J, Zhao Y, Nguyen C, Huantes S, Park E, Naing K, Klemm L, Swaminathan S, et al. 2014. Small-molecule inhibition of CBP/catenin interactions eliminates drug-resistant clones in acute lymphoblastic leukemia. *Oncogene* **33**: 2169–2178.

Gayther SA, Batley SJ, Linger L, Bannister A, Thorpe K, Chin SF, Daigo Y, Russell P, Wilson A, Sowter HM, et al. 2000. Mutations truncating the EP300 acetylase in human cancers. *Nat Genet* **24**: 300–303.

Goodman RH, Smolik S. 2000. CBP/p300 in cell growth, transformation, and development. *Genes Dev* **14**: 1553–1577.

Grossman SR. 2001. p300/CBP/p53 interaction and regulation of the p53 response. *Eur J Biochem* **268**: 2773–2778.

Grossman SR, Perez M, Kung AL, Joseph M, Mansur C, Xiao ZX, Kumar S, Howley PM, Livingston DM. 1998. p300/MDM2 complexes participate in MDM2-mediated p53 degradation. *Mol Cell* **2**: 405–415.

Grossman SR, Grossman SR, Deato ME, Tagami H, Nakatani Y, Livingston DM. 2007. Polyubiquitination of p53 by a ubiquitin ligase activity of p300. *Science* **342**: 342–345.

Hasan S, Stucki M, Hassa PO, Imhof R, Gehrig P, Hunziker P, Hübscher U, Hottiger MO. 2001. Regulation of human flap endonuclease-1 activity by acetylation through the transcriptional coactivator p300. *Mol Cell* **7**: 1221–1231.

He H, Yu FX, Sun C, Luo Y. 2011. CBP/p300 and SIRT1 are involved in transcriptional regulation of S phase specific histone genes. *PLoS ONE* **6**: e22088.

Hnisz D, Abraham BJ, Lee TI, Lau A, Saint-André V, Sigova AA, Hoke HA, Young RA. 2013. Super-enhancers in the control of cell identity and disease. *Cell* **155**: 934–947.

Horwitz GA, Zhang K, McBrian MA, Grunstein M, Kurdistani SK, Berk AJ. 2008. Adenovirus small e1a alters global patterns of histone modification. *Science* **321**: 1084–1085.

Howe JA, Mymryk JS, Egan C, Branton PE, Bayley ST. 1990. Retinoblastoma growth suppressor and a 300-kDa protein appear to regulate cellular DNA synthesis. *Proc Natl Acad Sci* **87**: 5883–5887.

Huang WC, Ju TK, Hung MC, Chen CC. 2007. Phosphorylation of CBP by IKKα promotes cell growth by switching the binding preference of CBP from p53 to NF-κB. *Mol Cell* **26**: 75–87.

Iyer NG, Xian J, Chin SF, Bannister AJ, Daigo Y, Aparicio S, Kouzarides T, Caldas C. 2007. p300 is required for orderly G_1/S transition in human cancer cells. *Oncogene* **26**: 21–29.

Jin Q, Yu LR, Wang L, Zhang Z, Kasper LH, Lee JE, Wang C, Brindle PK, Dent SYR, Ge K. 2011. Distinct roles of GCN5/PCAF-mediated H3K9ac and CBP/p300-mediated H3K18/27ac in nuclear receptor transactivation. *EMBO J* **30**: 249–262.

Kalkhoven E. 2004. CBP and p300: HATs for different occasions. *Biochem Pharmacol* **68**: 1145–1155.

Kamei Y, Xu L, Heinzel T, Torchia J, Kurokawa R, Gloss B, Lin SC, Heyman RA, Rose DW, Glass CK, et al. 1996. A CBP-integrator complex mediates transcriptional activation and AP-1 inhibition by nuclear receptors. *Cell* **85**: 403–414.

Karamouzis MV, Papadas T, Varakis I, Sotiropoulou-Bonikou G, Papavassiliou AG. 2002. Induction of the CBP transcriptional co-activator early during laryngeal carcinogenesis. *J Cancer Res Clin Oncol* **128**: 135–140.

Kawai H, Nie L, Wiederschain D, Yuan ZM. 2001. Dual role of p300 in the regulation of p53 stability. *J Biol Chem* **276**: 45928–45932.

Kitabayashi I, Aikawa Y, Yokoyama A, Hosoda F, Nagai M, Kakazu N, Abe T, Ohki M. 2001. Fusion of MOZ and p300 histone acetyltransferases in acute monocytic leukemia with a t(8;22)(p11;q13) chromosome translocation. *Leukemia* **15**: 89–94.

Koshiishi N, Chong JM, Fukasawa T, Ikeno R, Hayashi Y, Funata N, Nagai H, Miyaki M, Matsumoto Y, Fukayama M. 2004. P300 gene alterations in intestinal and diffuse types of gastric carcinoma. *Gastric Cancer* **7**: 85–90.

Kurdistani SK. 2007. Histone modifications as markers of cancer prognosis: A cellular view. *Br J Cancer* **97**: 1–5.

Kurdistani S. 2011. CRCnetBASE—Histone modifications in cancer biology and prognosis. *Prog Drug Res* **67**: 91–106.

Li Q, Zhou H, Wurtele H, Davies B, Horazdovsky B, Verreault A, Zhang Z. 2008. Acetylation of histone H3 lysine 56 regulates replication-coupled nucleosome assembly. *Cell* **134**: 244–255.

Li M, Luo RZ, Chen JW, Cao Y, Lu JB, He JH, Wu QL, Cai MY. 2011a. High expression of transcriptional coactivator p300 correlates with aggressive features and poor prognosis of hepatocellular carcinoma. *J Transl Med* **9**: 5.

Li Y, Yang HX, Luo RZ, Zhang Y, Li M, Wang X, Jia WH. 2011b. High expression of p300 has an unfavorable impact on survival in resectable esophageal squamous cell carcinoma. *Ann Thorac Surg* **91**: 1531–1538.

Li B, Su T, Ferrari R, Li JY, Kurdistani SK. 2014. A unique epigenetic signature is associated with active DNA replication loci in human embryonic stem cells. *Epigenetics* **9**: 257–267.

Liu X, Wang L, Zhao K, Thompson PR, Hwang Y, Marmorstein R, Cole PA. 2008. The structural basis of protein acetylation by the p300/CBP transcriptional coactivator. *Nature* **451:** 846–850.

Lv L, Xu YP, Zhao D, Li FL, Wang W, Sasaki N, Jiang Y, Zhou X, Li TT, Guan KL, et al. 2013. Mitogenic and oncogenic stimulation of K433 acetylation promotes PKM2 protein kinase activity and nuclear localization. *Mol Cell* **52:** 340–352.

Ma H, Nguyen C, Lee KS, Kahn M. 2005. Differential roles for the coactivators CBP and p300 on TCF/β-catenin-mediated survivin gene expression. *Oncogene* **24:** 3619–3631.

Maksimoska J, Segura-Peña D, Cole PA, Marmorstein R. 2014. Structure of the p300 histone acetyltransferase bound to acetyl-coenzyme A and its analogues. *Biochemistry* **53:** 3415–3422.

Manuyakorn A, Paulus R, Farrell J, Dawson NA, Tze S, Cheung-Lau G, Hines OJ, Reber H, Seligson DB, Horvath S, et al. 2010. Cellular histone modification patterns predict prognosis and treatment response in resectable pancreatic adenocarcinoma: Results from RTOG 9704. *J Clin Oncol* **28:** 1358–1365.

Martínez-Balbás MA, Bauer UM, Nielsen SJ, Brehm A, Kouzarides T. 2000. Regulation of E2F1 activity by acetylation. *EMBO J* **19:** 662–671.

Marzio G, Wagener C, Gutierrez MI, Cartwright P, Helin K, Giacca M. 2000. E2F family members are differentially regulated by reversible acetylation. *J Biol Chem* **275:** 10887–10892.

McBrian MA, Behbahan IS, Ferrari R, Su T, Huang TW, Li K, Hong CS, Christofk HR, Vogelauer M, Seligson DB, et al. 2013. Histone acetylation regulates intracellular pH. *Mol Cell* **49:** 310–321.

Miller RW, Rubinstein JH. 1995. Tumors in Rubinstein–Taybi syndrome. *Am J Med Genet* **56:** 112–115.

Monteiro AN, August A, Hanafusa H. 1996. Evidence for a transcriptional activation function of BRCA1 C-terminal region. *Proc Natl Acad Sci* **93:** 13595–13599.

Mosashvilli D, Kahl P, Mertens C, Holzapfel S, Rogenhofer S, Hauser S, Büttner R, Von Ruecker A, Müller SC, Ellinger J. 2010. Global histone acetylation levels: Prognostic relevance in patients with renal cell carcinoma. *Cancer Sci* **101:** 2664–2669.

Mullan PB, Quinn JE, Harkin DP. 2006. The role of BRCA1 in transcriptional regulation and cell cycle control. *Oncogene* **25:** 5854–5863.

Mullighan CG, Zhang J, Kasper LH, Lerach S, Payne-Turner D, Phillips LA, Heatley SL, Holmfeldt L, Collins-Underwood JR, Ma J, et al. 2011. CREBBP mutations in relapsed acute lymphoblastic leukaemia. *Nature* **471:** 235–239.

Nishihara A, Hanai JI, Okamoto N, Yanagisawa J, Kato S, Miyazono K, Kawabata M. 1998. Role of p300, a transcriptional coactivator, in signalling of TGF-β. *Genes Cells* **3:** 613–623.

O'Brate A, Giannakakou P. 2003. The importance of p53 location: Nuclear or cytoplasmic zip code? *Drug Resist Updat* **6:** 313–322.

Panagopoulos I, Fioretos T, Isaksson M, Samuelsson U, Billström R, Strömbeck B, Mitelman F, Johansson B. 2001.

Fusion of the MORF and CBP genes in acute myeloid leukemia with the t(10;16)(q22;p13). *Hum Mol Genet* **10:** 395–404.

Pao GM, Janknecht R, Ruffner H, Hunter T, Verma IM. 2000. CBP/p300 interact with and function as transcriptional coactivators of BRCA1. *Proc Natl Acad Sci* **97:** 1020–1025.

Parks SK, Chiche J, Pouyssegur J. 2011. pH control mechanisms of tumor survival and growth. *J Cell Physiol* **226:** 299–308.

Pasqualucci L, Dominguez-Sola D, Chiarenza A, Fabbri G, Grunn A, Trifonov V, Kasper LH, Lerach S, Tang H, Ma J, et al. 2011. Inactivating mutations of acetyltransferase genes in B-cell lymphoma. *Nature* **471:** 189–195.

Petrij F, Giles RH, Dauwerse HG, Saris JJ, Hennekam RC, Masuno M, Tommerup N, van Ommen GJ, Goodman RH, Peters DJ, et al. 1995. Rubinstein–Taybi syndrome caused by mutations in the transcriptional co-activator CBP. *Nature* **376:** 348–351.

Polesskaya A, Naguibneva I, Fritsch L, Duquet A, Ait-Si-Ali S, Robin P, Vervisch A, Pritchard LL, Cole P, Harel-Bellan A. 2001. CBP/p300 and muscle differentiation: No HAT, no muscle. *EMBO J* **20:** 6816–6825.

Prieur A, Besnard E, Babled A, Lemaitre JM. 2011. p53 and p16(INK4A) independent induction of senescence by chromatin-dependent alteration of S phase progression. *Nat Commun* **2:** 473.

Pugh BF. 2006. HATs off to PIC assembly. *Mol Cell* **23:** 776–777.

Ramos YFM, Hestand MS, Verlaan M, Krabbendam E, Ariyurek Y, van Galen M, van Dam H, van Ommen GJB, den Dunnen JT, Zantema A, et al. 2010. Genome-wide assessment of differential roles for p300 and CBP in transcription regulation. *Nucleic Acids Res* **38:** 5396–5408.

Rebel VI, Kung AL, Tanner EA, Yang H, Bronson RT, Livingston DM. 2002. Distinct roles for CREB-binding protein and p300 in hematopoietic stem cell self-renewal. *Proc Natl Acad Sci* **99:** 14789–14794.

Ringel AE, Wolberger C. 2013. A new RING tossed into an old HAT. *Structure* **72:** 181–204.

Roelfsema JH, Peters DJM. 2007. Rubinstein–Taybi syndrome: Clinical and molecular overview. *Expert Rev Mol Med* **9:** 1–16.

Rotte A, Bhandaru M, Cheng Y, Sjoestroem C, Martinka M, Li G. 2013. Decreased expression of nuclear p300 is associated with disease progression and worse prognosis of melanoma patients. *PLoS ONE* **8:** 1–12.

Sankar N, Baluchamy S, Kadeppagari RK, Singhal G, Weitzman S, Thimmapaya B. 2008. p300 provides a corepressor function by cooperating with YY1 and HDAC3 to repress c-Myc. *Oncogene* **27:** 5717–5728.

Santer FR, Höschele PPS, Oh SJ, Erb HHH, Bouchal J, Cavarretta IT, Parson W, Meyers DJ, Cole PA, Culig Z. 2011. Inhibition of the acetyltransferases p300 and CBP reveals a targetable function for p300 in the survival and invasion pathways of prostate cancer cell lines. *Mol Cancer Ther* **10:** 1644–1655.

Santoso B, Kadonaga JT. 2006. Reconstitution of chromatin transcription with purified components reveals a chromatin-specific repressive activity of p300. *Nat Struct Mol Biol* **13:** 131–139.

Sawada Y, Ishino M, Miura K, Ohtsuka E, Fujinaga K. 1997. Identification of specific amino acid residues of adenovirus 12 E1A involved in transformation and p300 binding. *Virus Genes* **15:** 161–170.

Seligson DB, Horvath S, Shi T, Yu H, Tze S, Grunstein M, Kurdistani SK. 2005. Global histone modification patterns predict risk of prostate cancer recurrence. *Nature* **435:** 1262–1266.

Seligson DB, Horvath S, McBrian MA, Mah V, Yu H, Tze S, Wang Q, Chia D, Goodglick L, Kurdistani SK. 2009. Global levels of histone modifications predict prognosis in different cancers. *Am J Pathol* **174:** 1619–1628.

Shi Y, Mello C. 1998. A CBP/p300 homolog specifies multiple differentiation pathways in *Caenorhabditis elegans*. *Genes Dev* **12:** 943–955.

Shi D, Pop MS, Kulikov R, Love IM, Kung AL, Grossman SR. 2009. CBP and p300 are cytoplasmic E4 polyubiquitin ligases for p53. *Proc Natl Acad Sci* **106:** 16275–16280.

Smith JL, Freebern WJ, Collins I, De Siervi A, Montano I, Haggerty CM, McNutt MC, Butscher WG, Dzekunova I, Petersen DW, et al. 2004. Kinetic profiles of p300 occupancy in vivo predict common features of promoter structure and coactivator recruitment. *Proc Natl Acad Sci* **101:** 11554–11559.

Sterner DE, Berger SL. 2000. Acetylation of histones and transcription-related factors. *Microbiol Mol Biol Rev* **64:** 435–459.

Suganuma T, Kawabata M, Ohshima T, Ikeda MA. 2002. Growth suppression of human carcinoma cells by reintroduction of the p300 coactivator. *Proc Natl Acad Sci* **99:** 13073–13078.

Tanaka Y, Naruse I, Maekawa T, Masuya H, Shiroishi T, Ishii S. 1997. Abnormal skeletal patterning in embryos lacking a single Cbp allele: A partial similarity with Rubinstein–Taybi syndrome. *Proc Natl Acad Sci* **94:** 10215–10220.

Teo JL, Kahn M. 2010. The Wnt signaling pathway in cellular proliferation and differentiation: A tale of two coactivators. *Adv Drug Deliv Rev* **62:** 1149–1155.

Tillinghast GW, Partee J, Albert P, Kelley JM, Burtow KH, Kelly K. 2003. Analysis of genetic stability at the EP300 and CREBBP loci in a panel of cancer cell lines. *Genes Chromosom Cancer* **37:** 121–131.

Trouche D, Kouzarides T. 1996. E2F1 and E1A(12S) have a homologous activation domain regulated by RB and CBP. *Proc Natl Acad Sci* **93:** 1439–1442.

Trouche D, Cook A, Kouzarides T. 1996. The CBP co-activator stimulates E2F1/DP1 activity. *Nucleic Acids Res* **24:** 4139–4145.

Turnell AS, Stewart GS, Grand RJA, Rookes SM, Martin A, Yamano H, Elledge SJ, Gallimore PH. 2005. The APC/C and CBP/p300 cooperate to regulate transcription and cell-cycle progression. *Nature* **438:** 690–695.

Vempati RK, Jayani RS, Notani D, Sengupta A, Galande S, Haldar D. 2010. p300–mediated acetylation of histone H3 lysine 56 functions in DNA damage response in mammals. *J Biol Chem* **285:** 28553–28564.

Vogelauer M, Wu J, Suka N, Grunstein M. 2000. Global histone acetylation and deacetylation in yeast. *Nature* **408:** 495–498.

Wang H, Larris B, Peiris TH, Zhang L, Le Lay J, Gao Y, Greenbaum LE. 2007. C/EBP β activates E2F-regulated genes in vivo via recruitment of the coactivator CREB-binding protein/P300. *J Biol Chem* **282:** 24679–24688.

Waterborg JH. 2001. Dynamics of histone acetylation in *Saccharomyces cerevisiae*. *Biochemistry* **40:** 2599–2605.

Webb BA, Chimenti M, Jacobson MP, Barber DL. 2011. Dysregulated pH: A perfect storm for cancer progression. *Nat Rev Cancer* **11:** 671–677.

Whyte P, Williamson NM, Harlow E. 1989. Cellular targets for transformation by the adenovirus E1A proteins. *Cell* **56:** 67–75.

Wong N, Ojo D, Yan J, Tang D. 2015. PKM2 contributes to cancer metabolism. *Cancer Lett* **356:** 184–191.

Wu J, Lu LY, Yu X. 2010. The role of BRCA1 in DNA damage response. *Protein Cell* **1:** 117–123.

Xiao XS, Cai MY, Chen JW, Guan XY, Kung HF, Zeng YX, Xie D. 2011. High expression of p300 in human breast cancer correlates with tumor recurrence and predicts adverse prognosis. *Chinese J Cancer Res* **23:** 201–207.

Yan G, Eller MS, Elm C, Larocca CA, Ryu B, Panova IP, Dancy BM, Bowers EM, Meyers D, Lareau L, et al. 2013. Selective inhibition of p300 HAT blocks cell-cycle progression, induces cellular senescence, and inhibits the DNA damage response in melanoma cells. *J Invest Dermatol* **133:** 2444–2452.

Yang XJ, Ullah M. 2007. MOZ and MORF, two large MYSTic HATs in normal and cancer stem cells. *Oncogene* **26:** 5408–5419.

Yang XJ, Ogryzko VV, Nishikawa J, Howard BH, Nakatani Y. 1996. A p300/CBP-associated factor that competes with the adenoviral oncoprotein E1A. *Nature* **382:** 319–324.

Yang W, Xia Y, Hawke D, Li X, Liang J, Xing D, Aldape K, Hunter T, Alfred Yung WK, Lu Z. 2012. PKM2 phosphorylates histone H3 and promotes gene transcription and tumorigenesis. *Cell* **150:** 685–696.

Yang H, Pinello CE, Luo J, Li D, Wang Y, Zhao LY, Jahn SC, Saldanha SA, Planck J, Geary KR, et al. 2013. Small-molecule inhibitors of acetyltransferase p300 identified by high-throughput screening are potent anticancer agents. *Mol Cancer Ther* **12:** 610–620.

Yao TP, Oh SP, Fuchs M, Zhou ND, Ch'ng LE, Newsome D, Bronson RT, Li E, Livingston DM, Eckner R. 1998. Gene dosage-dependent embryonic development and proliferation defects in mice lacking the transcriptional integrator p300. *Cell* **93:** 361–372.

Yuan ZM, Huang Y, Ishiko T, Nakada S, Utsugisawa T, Shioya H, Utsugisawa Y, Yokoyama K, Weichselbaum R, Shi Y, et al. 1999. Role for p300 in stabilization of p53 in the response to DNA damage. *J Biol Chem* **274:** 1883–1886.

Yuan J, Pu M, Zhang Z, Lou Z. 2009. Histone H3-K56 acetylation is important for genomic stability in mammals. *Cell Cycle* **8:** 1747–1753.

Zimmermann H, Degenkolbe R, Bernard HU, O'Connor MJ. 1999. The human papillomavirus type 16 E6 oncoprotein can down-regulate p53 activity by targeting the transcriptional coactivator CBP/p300. *J Virol* **73:** 6209–6219.

The Role of Additional Sex Combs-Like Proteins in Cancer

Jean-Baptiste Micol[1,2,3] and Omar Abdel-Wahab[3]

[1]Hematology Department, INSERM UMR1170, Gustave Roussy Cancer Campus Grand Paris, Villejuif, France

[2]Université Paris-Sud, Faculté de Médecine, Le Kremlin-Bicêtre, Paris, France

[3]Human Oncology and Pathogenesis Program and Leukemia Service, Department of Medicine, Memorial Sloan Kettering Cancer Center, New York, New York 10065

Correspondence: abdelwao@mskcc.org

Additional sex combs-like (ASXL) proteins are mammalian homologs of Addition of sex combs (Asx), a protein that regulates the balance of trithorax and Polycomb function in *Drosophila*. All three ASXL family members (ASXL1, ASXL2, and ASXL3) are affected by somatic or de novo germline mutations in cancer or rare developmental syndromes, respectively. Although Asx is characterized as a catalytic partner for the deubiquitinase Calypso (or BAP1), there are domains of ASXL proteins that are distinct from Asx and the roles and redundancies of ASXL members are not yet well understood. Moreover, it is not yet fully clarified if commonly encountered *ASXL1* mutations result in a loss of protein or stable expression of a truncated protein with dominant-negative or gain-of-function properties. This review summarizes our current knowledge of the biological and functional roles of ASXL members in development, cancer, and transcription.

Additional sex combs-like (ASXL) genes encode three proteins (ASXL1, ASXL2, and ASXL3), that share conserved domains and are orthologs of the *Drosophila Additional sex combs* (*Asx*) gene. Although the exact functions of ASXL family members are not well understood, cancer genomic studies identified highly recurrent somatic mutations in *ASXL1* in 2009 and a series of studies have since revealed that these mutations are consistent predictors of adverse outcome (Bejar et al. 2011; Thol et al. 2011; Patel et al. 2012; Itzykson et al. 2013; Haferlach et al. 2014). Since then, recurrent mutations in *ASXL2* and *ASXL3* have been identified in specific subsets of cancer and *ASXL1/3* germline mutations have been found to underlie rare developmental syndromes. These findings have resulted in a great interest in understanding the function of each ASXL protein, the protein complexes they exist in, and their roles and redundancies in the development of normal and malignant cell types. In this review, we describe the discovery of mammalian *ASXL* genes and the mutations affecting each member. We also present a review and interpretation of the biochemical, epigenomic, and biological studies that have been performed to date to dissect their function.

Cite this article as *Cold Spring Harb Perspect Med* doi: 10.1101/cshperspect.a026526

DISCOVERY OF THE *DROSOPHILA* ADDITIONAL SEX COMBS (*Asx*) GENE

As mentioned above, the mammalian *ASXL* genes derive their name from their *Drosophila* homolog *Asx. Asx* was originally discovered based on a genetic screen in *Drosophila* whereby mutations of *Asx* were found to enhance the phenotype of both Trithorax (TrxG) as well as Polycomb group (PcG) gene mutants (Fig. 1) (Sinclair et al. 1998; Milne et al. 1999). These phenotypes suggested that *Asx* is required to both maintain repression as well as activate expression of *Hox* genes in *Drosophila*. This dual function of *Asx* placed *Asx* as a member of the enhancer of Trithorax and Polycomb group (ETP) of genes, which are thought to mediate the balance between PcG and TrxG function.

Despite the fact that *Asx* was first uncovered 18 years ago, the molecular basis for its dual function in both repression as well as activation of *Hox* gene expression is not well understood. *Asx* encodes a chromatin-associated protein whose binding pattern to polytene chromosomes overlaps with PcG proteins (Sinclair et al. 1998). However, other than a well-described physical interaction with Calypso (termed BAP1 [BRCA1-associated protein 1] in mammalian cells) to oppose the function of the Polycomb repressive complex 1 (PRC1) (Scheuermann et al. 2010), the complexes that Asx participates in are not well defined (Fig. 2).

DISCOVERY OF THE MAMMALIAN *ASXL* GENE FAMILY

Efforts to understand the molecular and structural basis for *Asx*'s function as an ETP gene led to the discovery of mammalian homologs of *Asx*. Unlike *Drosophila*, mammals have 3 *Asx* homologs: *ASXL1* (Fisher et al. 2003), *ASXL2* (Katoh and Katoh 2003), and *ASXL3* (Fig. 1) (Katoh and Katoh 2003). *ASXL1, ASXL2*, and *ASXL3* are located on human chromosomes 20q11.21, 2p23.3, and 18q12.1, respectively. Phylogenetic analyses of the coding sequences of *ASXL* family members reveal that *ASXL1* and *ASXL2* orthologs are more closely related to one another than to *ASXL3* (Katoh 2013). Although

ASXL1 and *ASXL2* are widely expressed in mammalian tissues, *ASXL3* expression is restricted to mostly the brain and the eye and is not appreciably expressed in hematopoietic cells (Fisher et al. 2006; Bainbridge et al. 2013; LaFave et al. 2015). There is ~40% amino acid homology between ASXL1, ASXL2, and ASXL3 overall, but this homology increases to ~70% in the conserved domains of ASXL proteins including the ASXN, ASX homology (ASXH), ASXM1, and ASXM2 domains and the carboxy-terminal cysteine cluster plant homedomain (PHD) (Fig. 3).

Mammalian *ASXL* family members each contain an ASXN domain located at the amino terminus, which is predicted to be a DNA-binding domain (Sanchez-Pulido et al. 2012). Interestingly, this domain is not present in *Asx*, suggesting an important divergence in function between *Asx* and mammalian ASXL genes. The ASXH domain, in contrast, is highly homologous to the *Drosophila* DEUBAD (deubiquitinase adaptor) domain and corresponds to the BAP1 interaction site (Scheuermann et al. 2010). The ASXH domain is present in all three mammalian ASXL family members (Figs. 3 and 4), each of which have been recently shown to bind to BAP1 (Sahtoe et al. 2016). The ASXM1 and ASXM2 domains of ASXL proteins have been reported to bind to several nuclear hormone receptors (NHRs), including the androgen receptor (Grasso et al. 2012), estrogen receptor (Park et al. 2015), and PPARγ (Park et al. 2011).

Currently, our knowledge of the function of ASXL1-3 is limited by a lack of systematic functional investigation of the conserved protein domains of ASXL family members. For example, although phylogenetic analyses suggest that the PHD domain of ASXL proteins may bind histone H3 lysine 4 trimethyl (H3K4me3) (Aravind and Iyer 2012; Katoh 2015), the function of the carboxy-terminal PHD domain of ASXL proteins is in need of definition. One recent report revealed that the PHD domain of ASXL2 binds H3K4me1 and H3K4me2, but validation of this finding with quantitative measurement of binding affinity and other confirmatory assays was not performed nor were the

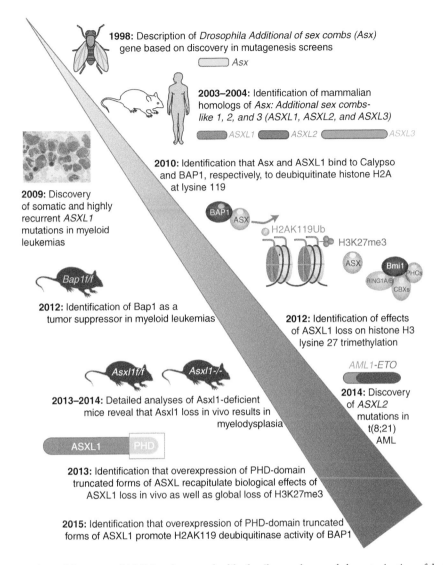

Figure 1. Timeline of discovery of Additional sex combs-like family members and characterization of their effects on the epigenome, development, and cancer.

functions of ASXL1 or ASXL3 PHD domains studied (Park et al. 2015). Moreover, the possibility that the ASXL PHD domains may bind to nonhistone proteins remains to be tested.

In addition to further analysis of conserved domains of ASXL proteins, systematic exploration of the function of the domains that are not conserved between ASXL members needs to be performed. It is hypothesized that differing functions of these regions might endow each ASXL member with specialized functions in mammalian cells. Proteomic analysis of binding partners for each ASXL family member may be very helpful in identifying shared and unique binding partners for each ASXL family member in mammalian cells.

DISCOVERY OF THE H2A DEUBIQUITINASE COMPLEX

As noted above, ASXL members have been shown to interact with BAP1 through physical

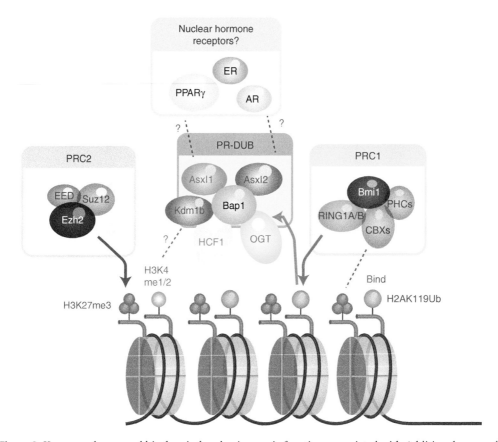

Figure 2. Known and proposed biochemical and epigenomic functions associated with Additional sex combs-like (ASXL) family members. ASXL1 and ASXL2 have been copurified with BAP1 in numerous reports and serve as a deubiquitinase for histone H2A lysine 119 with BAP1 termed the Polycomb repressive deubiquitinase (PR-DUB). At least one report has suggested that ASXL1 and ASXL2 form mutually exclusive complexes with BAP1 in the PR-DUB. In addition, HCF1, OGT, and KDM1B all appear to be in complex with BAP1 as well. Whether the association of KDM1B with BAP1 has any influence on histone H3 lysine 4 mono- or dimethylation (H3K4me1/2) is currently unclear. In addition, to opposing the function of the Polycomb repressive complex 1 (PRC1)-mediated H2AK119 ubiquitination, ASXL1 loss has also been repeatedly associated with global loss of histone H3 lysine 27 trimethylation (H3K27me3). Whether the H3K27me3 loss associated with ASXL1 loss is secondary to loss of H2AK119Ub and/or because of impaired Polycomb repressive complex 2 (PRC2) remains to be further clarified. Finally, ASXL1 and ASXL2 have been proposed to physically interact and/or functionally affect the function of a number of nuclear hormone receptors (NHRs) including PPARγ and the estrogen and androgen receptors (ER and AR, respectively).

interaction at the ASXH domain of ASXL proteins. BAP1 owes its name to its discovery as a protein that interacts with the RING finger domain of BRCA1 (Jensen et al. 1998). BAP1 contains as amino-terminal ubiquitin carboxy-terminal hydrolase domain and was initially thought to be important in the deubiquitination of BRCA1 (a finding that has subsequently been disproven [Mallery et al. 2002]). Interest-

ingly, in 2007, a forward genetic screen for mutations that cause loss of PcG repression in *Drosophila* identified *Calypso*, the ortholog of BAP1, as a novel PcG gene involved in repression of *Hox* genes (Gaytan de Ayala Alonso et al. 2007). As noted earlier, PcG proteins consist of several multiprotein complexes (the PRC1 and PRC2 complexes) that repress gene expression. PRC1 harbors an E3 ligase activity that

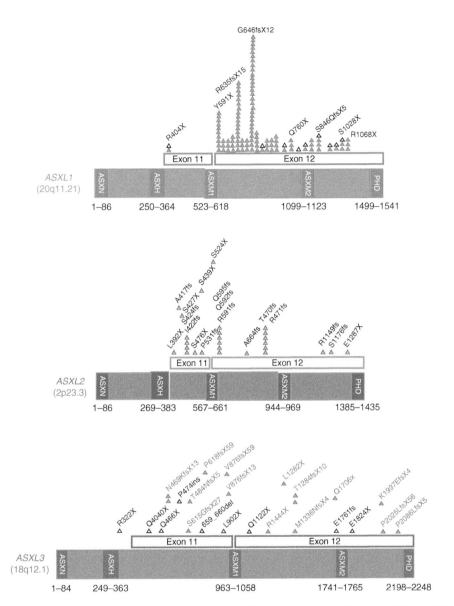

Figure 3. Protein domain structure and location of amino acids affected by mutations in ASXL1, ASXL2, and ASXL3. Only nonsense or frameshift mutations are shown for each protein here. Mutations shown in black triangles are previously reported as germline mutations in Bohring–Opitz syndrome (ASXL1) or Bohring–Opitz-like syndrome (ASXL3), whereas mutations in red, blue, or green triangles have been reported as somatic mutations in *ASXL1*, *ASXL2*, or *ASXL3*, respectively. For ASXL1, only somatic mutations that have been described in more than one patient or as a germline mutation are shown.

monoubiquitinates histone 2A at lysine 119 (H2AK119). Biochemical characterization of BAP1 by Scheuermann et al. (2010) identified that BAP1 and ASXL1 coexist in a complex that they termed the Polycomb repressive deubiquitinase (PR-DUB) complex. They subsequently showed that BAP1 requires ASXL1 as a cofactor to deubiquitinate H2AK119Ub (Scheuermann et al. 2010). Consistent with this, *Drosophila* mutants lacking either *Calypso* or *Asx* had an increase in global H2AK119Ub (Scheuermann et al. 2010).

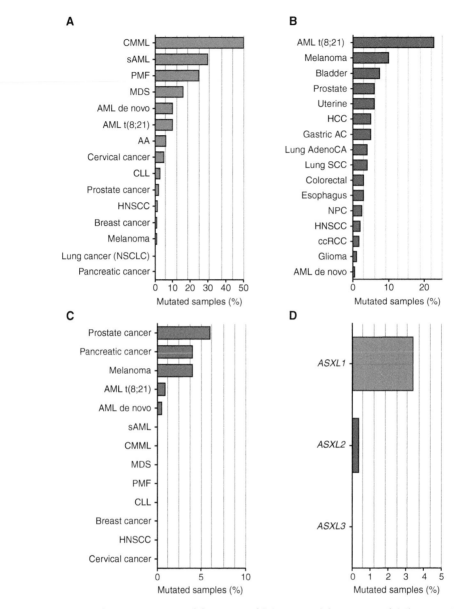

Figure 4. Histogram of somatic mutational frequency of (*A*) *ASXL1*, (*B*) *ASXL2*, and (*C*) *ASXL3* in cancer, as well as in (*D*) clonal hematopoiesis of indeterminate potential. AA, Aplastic anemia; AML, acute myeloid leukemia; ccRCC, clear cell renal cell carcinoma; CMML, chronic myelomonocytic leukemia; CLL, chronic lymphocytic leukemia; HCC, hepatocellular carcinoma; HNSCC, head and neck squamous cell carcinoma; MDS, myelodysplastic syndromes; PMF, primary myelofibrosis; sAML, secondary AML.

The above data by Scheuermann et al. provided the first molecular function for *Asx* and suggested that mammalian ASXL members might similarly function as cofactors for BAP1. However, unlike *Drosophila*, mammals have three ASXL members, and it is unclear how much redundancy in function occurs between each ASXL member in their interaction with BAP1. As mentioned earlier, the interaction between BAP1 and ASXL1 occurs through the ASXH domain. ASXH homology between ASXL family members is ∼60%, and all three

Cite this article as *Cold Spring Harb Perspect Med* doi: 10.1101/cshperspect.a026526

ASXL paralogs bind BAP1 and stimulate H2AK119 deubiquitination in vitro (Sahtoe et al. 2016).

In parallel to the discovery of BAP1's enzymatic function on H2AK119Ub, a tumor suppressor role of BAP1 became increasingly clear with genomic and functional investigation in cancer (reviewed recently by Carbone et al. 2013). *BAP1* is located on 3p21.1, a region deleted in multiple malignancies, including 30%–60% of mesothelioma and 40–65% of uveal melanoma patients. Subsequently, *BAP1* was shown to be affected by germline mutations in mesothelioma (Testa et al. 2011) and melanoma (Harbour et al. 2010). More recently, the concept of BAP1 as a tumor suppressor was further extended after the description of *BAP1* somatic mutations in a variety of additional cancer types (Carbone et al. 2013). This includes somatic *BAP1* mutations in 40% of uveal melanoma (Harbour et al. 2010; Wiesner et al. 2011), 20% of pleural mesothelioma (Bott et al. 2011; Testa et al. 2011), and 15% of clear cell renal carcinoma (Bott et al. 2011; Guo et al. 2012) patients, respectively.

To clarify the role of *BAP1* in mammalian development and cancer further, Dey et al. generated *Bap1* conditional knockout (cKO) mice as well as transgenic knockin mice expressing a 3xFlag tag (Dey et al. 2012). Proteomic analysis of BAP1 interacting partners using this system revealed that BAP1 interacts with ASXL1 and ASXL2 in vivo as well as host cell factor 1 (HCF-1), O-GlcNAc transferase (OGT), and Forkhead box protein K1/2 (FOXK1/2) (Dey et al. 2012). Similar results were seen with proteomic analysis of FLAG-HA tandem purification of tagged BAP1 and ASXL1 expressed in 293T cells more recently (Wu et al. 2015). In addition, the mass spectrometric analysis by Dey et al. (2012) also identified a physical interaction between BAP1 and the H3K4 demethylase KDM1B (also known as LSD1B), an intriguing interaction that might implicate BAP1/ASXL in KDM1B function and will need to be validated further.

Despite the above analyses of interaction partners between BAP1 and ASXL1/2 in mammalian cells, it is not clear how much of the function of ASXL1 or ASXL2 relates to the function of BAP1 and other members of the BAP–ASXL complexes, which have been identified. Unlike *Drosophila*, lack of ASXL1 in mammalian cells has not been clearly linked to changes in H2AK119Ub (Abdel-Wahab et al. 2012), possibly because of redundancies of multiple ASXL members in mammalian cells. Moreover, murine models with cKO of *Bap1* versus *Asxl1* (described below) show distinct phenotypes. Finally, there is a lack of obvious clinical overlap of diseases involving mutations in *BAP1* versus *ASXL1*. Each of these findings suggests some potential independent roles of BAP1 and ASXL1.

DISCOVERY OF MUTATIONS IN *ASXL* GENE FAMILY MEMBERS IN CANCER

Investigations into the mechanistic function and biological roles of ASXL proteins increased tremendously with the discovery of frequent somatic mutations in *ASXL1* in patients with myeloid leukemias (Gelsi-Boyer et al. 2009). Further sequencing efforts have since identified that *ASXL1* mutations are highly recurrent throughout all myeloid leukemias (Figs. 3 and 4). Mutations affecting *ASXL1* most frequently occur as heterozygous frameshift or nonsense mutations, usually in exons 11 or 12, just before the PHD domain. In addition, the locus of *ASXL1*, 20q11.21, is occasionally affected by somatic deletions in myeloid leukemias. Moreover, clinical correlative analyses have repeatedly identified that *ASXL1* mutations predominate in older patients and are linked to an adverse outcome independent of other known clinical prognosticators (Bejar et al. 2011; Thol et al. 2011; Patel et al. 2012; Itzykson et al. 2013; Haferlach et al. 2014).

The human disorder in which *ASXL1* mutations appear to be most frequent is chronic myelomonocytic leukemia (CMML), a disease with overlapping features of myeloproliferative neoplasms (MPNs) and myelodysplastic syndromes (MDSs). *ASXL1* mutations are present in 40%–50% of CMML patients and appear to be a unique predictor of adverse outcome in this disease (Itzykson et al. 2013).

Similar to CMML, *ASXL1* mutations have been shown to be important biomarkers of adverse outcome in the MPNs, MDSs, and acute myeloid leukemia (AML). Although *ASXL1* mutations are infrequent in the chronic and relatively benign MPNs essential thrombocytosis (ET) and polycythemia vera (PV), they are enriched in those PV/ET patients who experience transformation to a more aggressive clinical state known as myelofibrosis (MF) (Stein et al. 2011).

In MDS, *ASXL1* mutations represent one of the most frequent gene mutations and are present in 10%–25% of patients (Figs. 3 and 4) (Bejar et al. 2011; Thol et al. 2011; Haferlach et al. 2014). Interestingly there appears to be an enrichment of *ASXL1* mutations in AML patients with a history of preceding MDS (Fernandez-Mercado et al. 2012) or in AML with MDS-related changes (Devillier et al. 2015) compared with de novo AML. In AML, *ASXL1* mutations are 4–5 times more likely in older patients, explaining the discrepancy in *ASXL1* mutational frequencies reported across AML sequencing studies (Metzeler et al. 2011). Consistent with this, *ASXL1* mutations have been repeatedly identified in individuals with clonal hematopoiesis of indeterminate potential, a condition whose frequency increases with advanced age (Genovese et al. 2014; Jaiswal et al. 2014; Xie et al. 2014).

Very recently, several studies have identified that *ASXL1* and *ASXL2* mutations are enriched in a very specific subset of AML, AML with t(8;21) (Huether et al. 2014; Duployez et al. 2015b), which represent ∼10% of all AML. ASXL family member mutations occur in ∼30% of pediatric and adult t(8;21) AML patients (Figs. 3 and 4) (Micol et al. 2014). This high frequency is mainly a result of *ASXL2* mutations (23%), which are otherwise almost never observed in other forms of leukemia. Interestingly, *ASXL1* and *ASXL2* mutations are mutually exclusive in t(8;21) AML suggesting that the mutations may have convergent downstream and/or are synthetic lethal effects with one another. A recent targeted mutational analysis of t(8;21) AML suggests that these mutations are also mutually exclusive with mutations in other epigenetic modifiers, such as *EZH2*, and when associated with tyrosine kinase mutations are associated with an increased risk of relapse (Duployez et al. 2016). In sequencing analysis of t(8;21) AML, a single patient with an *ASXL3* mutation has been described but otherwise *ASXL3* mutations have been very infrequently reported in hematological malignancies (Duployez et al. 2015a).

ASXL1 mutations have also been described recently in additional rare myeloid malignancies including aplastic anemia (Yoshizato et al. 2015), chronic neutrophilic leukemia (Elliott et al. 2015), and systemic mastocytosis (Damaj et al. 2014). They have also been described as secondary somatic events in patients with hereditary hematological malignancies because of congenital *GATA2* deficiency (Micol and Abdel-Wahab 2014; West et al. 2014) and congenital neutropenia (Beekman et al. 2012). In contrast to their frequency in myeloid disorders, *ASXL* mutations have only been sporadically observed in lymphoid malignancies, being noted in very rare cases of chronic lymphocytic leukemia (Quesada et al. 2012) and acute lymphoblastic leukemia (Prebet et al. 2013).

DISCOVERY OF MUTATIONS IN *ASXL* GENE FAMILY MEMBERS IN DEVELOPMENTAL DISORDERS

In addition to somatic mutations of *ASXL* family members in cancer, de novo germline *ASXL* mutations have been recently discovered to underlie developmental disorders as well. In particular, *ASXL1* mutations have been described in ∼50% of patients with Bohring–Opitz syndrome (BOS) (Hoischen et al. 2011; Magini et al. 2012; Avila et al. 2013; Dangiolo et al. 2015), a rare disease defined by distinctive craniofacial features, intellectual disability, and severe feedings problems. *ASXL1*-mutant BOS patients appear to have specific clinical characteristics, including myopia and hypertrichosis, compared with *ASXL1* wild-type patients (Magini et al. 2012). *ASXL3* mutations have also been discovered in patients with a BOS-like syndrome, a close but distinct congenital disorder (Bainbridge et al. 2013; Dinwiddie et al. 2013; Srivastava et al. 2015). Moreover, no *ASXL3* mutations have been found in BOS patients wild-type for *ASXL1* (Russell and Graham 2013), suggesting true genetic differences between BOS and BOS-like conditions.

Interestingly, heterozygous nonsense and out-of-frame frameshift germline mutations in *ASXL1* and *ASXL3* occur exclusively in the two last exons of these two genes, exactly as seen as with somatic mutations affecting these genes in patients with cancer (Fig. 3). Despite this, and in contrast to patients with germline *BAP1* mutations, patients with BOS and BOS-like conditions have not been identified to have consistently increased risk of cancer. Given that only ∼50 patients with BOS have been described and that BOS patients frequently die as infants, cancer risk associated with this syndrome is not well defined. However, BOS syndrome patients have been reported to have nephroblastomatosis, meduloblastoma, and Wilms tumor, suggesting that germline *ASXL1/3* mutations could increase the risk for cancer predisposition (Russell et al. 2015).

PHENOTYPIC ANALYSES OF *ASXL* FUNCTION IN VIVO

To understand the role of *ASXL* family members in mammalian development and cancer further, a number of murine models targeting the expression of *Asxl* family member have been described. Currently, gene-trap constitutive KO models as well as cKO models have been described for *Asxl1* and *Asxl2* and no murine models targeting *Asxl3* have been described (Table 1). These mice have been helpful in elucidating the effects of Asxl1 versus Asxl2 on tissue development. Constitutive homozygous loss of *Asxl1* is associated with partial (Fisher et al. 2010a,b; McGinley et al. 2014; Wang et al. 2014) to complete embryonic lethality (Abdel-Wahab et al. 2013) depending on the model. Constitutive *Asxl1*-deficient mice display developmental abnormalities including dwarfism and anophthalmia, features not shared by *Asxl2* KO mice (Baskind et al. 2009). *Asxl2* KO mice, in contrast, die in the first few days following birth because of failure to close the ductus

arteriosus inducing fatal pulmonary hypertension (McGinley et al. 2014). Both *Asxl1* and *Asxl2* KO mice have alterations of the axial skeleton with posterior (PcG phenotype) and anterior (TrxG phenotype) transformations, analogous to the effects of *Asx* loss in *Drosophila* described earlier (Sinclair et al. 1998; Milne et al. 1999). At least two prior studies have noted that ASXL1 and ASXL2 are important mediators of the function of the nuclear receptor PPARγ in adipogenesis (Park et al. 2011; Izawa et al. 2015). Consistent with this, *Asxl2* KO mice appear to develop osteopetrosis, lipodystrophy, and insulin resistance because of a failure of PPARγ activation (Table 1) (Izawa et al. 2015). Given data that ASXL1 and ASXL2 may have opposing roles in adipogenesis based on in vitro fat differentiation assays (Park et al. 2011), it would be important to determine whether *Asxl1* KO mice have alterations in metabolism and bone metabolism as well. Finally, *Asxl1* KO mice have also been reported to have defects in kidney size and glomerular podocyte formation because of impaired WT1 signaling in early kidney development (Moon et al. 2015).

Shortly after the discovery of somatic *ASXL1* mutations in myeloid leukemias, the in vivo effects of *Asxl1* alterations on hematopoiesis were studied using *Asxl1* gene-trap mice (Table 1) (Fisher et al. 2010b). In an initial report, postnatal hematopoietic cells as well as fetal liver hematopoietic cells from surviving *Asxl1* KO mice were found to have defects in B- and T-lymphopoiesis as well as myeloid skewing but were not found to develop overt leukemia or lymphoma. At the same time, results of competitive or serial transplantation assays were not reported. Thus, comprehensive assessment of the effects of *Asxl1* loss on hematopoietic stem cell function and self-renewal were not clear from this study (Fisher et al. 2010b). In contrast to these data, longitudinal analysis of another *Asxl1* KO model identified that mice deficient for *Asxl1* developed pancytopenia and morphologic dysplasia (Table 1) (Wang et al. 2014). Moreover, these features were also present in mice with heterozygous deletion of *Asxl1* suggesting that *Asxl1* may serve as a haploinsuffi-

Table 1. Germline genetically engineered murine models investigating the function of ASXL family members in cancer and development

Mouse model	Biological phenotype	Epigenetic alterations	References
$Asxl1^{tm1Bc}$ mutant	Partial embryonic lethality, alterations of the axial skeleton, early death, reduction in body weight and thymus, splenomegaly; defects in frequency of differentiation of lymphoid and myeloid progenitors but no hematological phenotype	$Hoxa4$ and $Hoxa7$ repression	Fisher et al. 2010a,b
EIIa-cre $Asxl1^{fl/fl}$	100% embryonic lethality, craniofacial abnormalities	Not described	Abdel-Wahab et al. 2013
Mx1-cre $Asxl1^{fl/fl}$	MDS-like phenotype; aggressive MDS	Reduced global H3K27me3, increase expression of $HoxA$ genes	
$Asxl1^{-/-}$ (Asxl1:nlacZ/nGFP)	80% embryonic lethality, 78% of newborn died after 1 day, developmental abnormalities; MDS-like disease	Not described	Wang et al. 2014
$Asxl1^{+/-}$ (Asxl1:nlacZ/nGFP)	Haploinsufficiency sufficient for the development of MDS-like and MDS/MPN-like disease	Increased expression of $HoxA$ genes ($HoxA5$, 7, 9, and 10); decrease of global levels of H3K27me3 and H3K4me3	
$Asxl1^{tm1a}$	Partial embryonic lethality, reduced body weight and display cleft palate, anophthalmia, ventricular septal defects, lung defects	Not described	McGinley et al. 2014
$Asxl1$-null mice	Defects in kidney size and glomerular podocyte morphology	Not described	Moon et al. 2015
Germline $Bap1^{-/-}$ Cre-ERT2 $Bap1^{fl/fl}$	Embryonic lethality MDS/CMML-like disease	Not described	Dey et al. 2012
$Asxl2$ gene trap mutant mice ($Asxl2Gt(AQ0356)$)	Partially embryonic lethal, transformation in the axial skeleton, reduced body weight, congenital heart malformations, low bone mineral density, osteopetrosis, lipodystrophy, insulin resistance	Reduction in the bulk level of H3K27me3 in cardiac tissue	Baskind et al. 2009; Farber et al. 2011; Lai et al. 2012; Lai and Wang 2013; Izawa et al. 2015
Mx1-cre $Asxl2^{fl/fl}$	Cytopenias (leukopenia, thrombocytopenia), defect in hematopoietic self-renewal	Reduction in H3K4me1	Micol et al. 2015

CMML, Chronic myelomonocytic leukemia; MDS, myelodysplastic syndrome; MPN, myeloproliferative neoplasm.

cient tumor suppressor in the hematopoietic system (Wang et al. 2014).

Given the early lethality associated with constitutive *Asxl1* loss and the need to study the effects of *Asxl1* loss in a hematopoietic-specific manner following postnatal development, we developed a cKO *Asxl1* model (Abdel-Wahab et al. 2013). Mice with hematopoietic-specific deletion of *Asxl1* (*Vav*-cre *Asxl1*$^{fl/fl}$ mice) as well as postnatal, inducible deletion of *Asxl1* (*Mx1*-cre *Asxl1*$^{fl/fl}$ mice) were observed to develop progressive multilineage cytopenias and dysplasia with increased numbers of hematopoietic stem and progenitor cells (HSPCs) with impaired mature cell differentiation—features consistent with the human disease MDS (Table 1). These features were further apparent in mice with hematopoietic-specific deletion of both *Tet2* and *Asxl1* (Abdel-Wahab et al. 2013), a combined genetic alteration present in ∼30% of patients with CMML (Itzykson et al. 2013) and 15% of patients with high-risk MDS (Bejar et al. 2011).

The above data suggest that *Asxl1* functions as a haploinsufficient tumor suppressor in the hematopoietic system. Consistent with these germline genetically engineered murine models of *Asxl1* loss, we had also previously observed that RNAi-mediated depletion of *Asxl1* in a murine retroviral bone marrow transplant (BMT) assay with concomitant overexpression of NRasG12D resulted in a disorder with features of MDS/MPN in vivo (Abdel-Wahab et al. 2012). Despite these data, the nature of *ASXL1* mutations as heterozygous mutations in a restricted domain have suggested that *ASXL1* mutations might confer a gain-of-function, rather than loss-of-function, by generating a stable truncated protein lacking the carboxy-terminal PHD domain that either serves as a dominant-negative function or generates a new function. To this end, Inoue et al. performed a series of in vitro and in vivo experiments with retroviral overexpression of cDNA bearing ASXL1 truncated forms (Inoue et al. 2013). Overexpression of truncated forms of ASXL1 in a murine BMT model along with NRASG12D was found to result in a phenotype very similar to that seen with Asxl1 loss as described earlier (Abdel-Wahab et al. 2012). These results suggest that ASXL1 mutations might confer a dominant-negative function mimicking the biological effects of complete ASXL1 loss. However, further efforts to define the actual function of the ASXL1 PHD domain and to directly compare complete loss of *ASXL1* versus expression of mutant *ASXL1* from its endogenous locus are still needed to clarify the functional impact of *ASXL1* mutations.

In contrast to the time-dependent bone marrow failure identified with hematopoietic-specific deletion of *Asxl1* in vivo, hematopoietic-specific deletion of *Bap1* in vivo has been found to result in an abrupt and aggressive disease with features of MDS and MPN (Dey et al. 2012). The discrepant phenotype between loss of *Bap1* versus *Asxl1* in vivo suggests that loss of *Asxl1* function may be compensated for by its paralog *Asxl2* and/or that *Bap1* and *Asxl1* affect hematopoiesis through divergent downstream effects. To address the potential role of Asxl2 in hematopoiesis, we recently generated *Asxl2* cKO mice to study the effects of postnatal deletion of *Asxl2* on hematopoiesis (Micol et al. 2015). Interestingly, *Asxl2* cKO mice (*Mx1*-cre *Asxl2*$^{fl/fl}$) develop cytopenias and impairments in HSPC self-renewal, which are far more apparent than seen with *Asxl1* cKO (*Mx1*-cre *Asxl1*$^{fl/fl}$) studied in parallel (Micol et al. 2015). These data suggest that ASXL1 and ASXL2 have nonoverlapping effects in hematopoiesis.

EFFECT OF *ASXL* MUTATIONS ON ASXL FUNCTION

Despite 7 years of research on *ASXL1* mutations in cancer now, it is still not entirely clear whether the recurrent nonsense and frameshift mutations affecting *ASXL1* result in a loss of ASXL1 protein expression or expression of a stable, truncated form of ASXL1 lacking the carboxy-terminal PHD domain. We previously failed to detect any ASXL1 protein expressed in cell lines bearing homozygous truncating mutations in *ASXL1* (Abdel-Wahab et al. 2012). Moreover, we noted reduced full-length ASXL1 protein expression in cell lines bearing heterozygous *ASXL1* mutations without any evidence of truncated ASXL1 protein forms. These data suggested that *ASXL1* mutations are associated with nonsense-mediated decay and loss-of-function. Consistent with this, the expression of *ASXL1* mutants in the presence of cyclohexamide results in a rapid degradation of the proteins suggesting reduced stability of the mutant proteins (Abdel-Wahab et al. 2012).

In contrast to the above, Inoue et al. (2015) used new amino-terminal anti-ASXL1 antibodies and mTRAQ-based mass spectrometric analysis in an effort to detect potential truncated ASXL1 proteins. Interestingly, truncated forms of ASXL1 were detected in two cell lines bearing homozygous truncating *ASXL1* mutations that had not previously been studied. Nonetheless, it is not clear whether such truncated proteins are recurrently found in cells bearing *ASXL1* mutations or how stable these truncated proteins are. To this end, recent work by Inoue et al. has also identified that ASXL1 protein stability appears to be mediated in part by ubiquitination at residue lysine 351 (K351). Interestingly, K351 is deubiquitinated, not by BAP1, by the deubiquitinase USP7 (Inoue et al. 2015). Further efforts to manipulate the expression of wild-type versus mutant ASXL1 by mutating the K351 residue of ASXL1 may help elucidate the function of endogenous mutations in *ASXL1*. In addition, efforts to introduce an easily detectable epitope tag into the amino terminus of the *ASXL1* locus in cells with endogenous *ASXL1* mutations may likewise be extremely helpful in understanding the effect of *ASXL1* mutations on expression of ASXL1.

Although both ASXL1 and ASXL2 have been identified to bind to BAP1, recent biochemical characterization of BAP1 complexes by Daou et al. (2015) suggest that BAP1 forms two mutually exclusive complexes with ASXL1 and ASXL2. In this study, BAP1 was identified to stabilize ASXL2 via deubiquitination of ASXL2. Consistent with this, the protein level of ASXL2, but not ASXL1, was clearly dependent on BAP1 abundance (Daou et al. 2015). These data suggest that loss of ASXL2 function may consistently accompany *BAP1* mutations and deletions, adding further importance to understanding the role of ASXL2 loss on cancer pathogenesis.

ROLE OF ASXL PROTEINS IN H2A DEUBIQUITINATION

As noted earlier, phenotypic analyses in vitro and in vivo have suggested that overexpression of stable truncated forms of ASXL1 appear to result in similar biological phenotypes as seen with ASXL1 loss (Inoue et al. 2013). To understand the potential mechanistic effects of expression of PHD-domain truncated forms of ASXL1, Balasubramani et al. recently studied the effects of overexpression of ASXL1 truncated forms on H2AK119Ub. Overexpression of several truncated forms of ASXL1, but not full-length ASXL1, in combination with overexpression of BAP1 resulted in clear depletion of global H2AK119Ub as well as a striking reduction of histone H3 lysine 27 trimethylation (H3K27me3) (Balasubramani et al. 2015). Despite these results suggesting that ASXL1 truncation mutants might serve a gain-of-function mutations by hyperactivating PR-DUB activity, it is important to note that these experiments were performed using overexpression of both the enzymatic deubiquitinase as well as truncated forms of ASXL1—neither of which are clearly overexpressed in cells bearing these mutations. Moreover, the effects of loss of ASXL1 were not assessed in parallel with expression of the truncated forms of ASXL1 and, thus, it is unclear whether these mutations might actually confer a dominant negative activity. Future efforts to study the epigenomic and biological effects of isogenic cells with engineered mutations in the endogenous locus of *ASXL1* will hopefully clarify the effects of physiologic expression of these mutations further.

ROLE OF ASXL PROTEINS AND BAP1 IN PRC2 FUNCTION

In addition to potential effects on H2A deubiquitination, *ASXL1* mutations have also been linked to alterations in PRC2 function as a result of the repeated observation of changes in H3K27 methylation in the setting of *ASXL1* loss or mutation (Abdel-Wahab et al. 2012, 2013; Inoue et al. 2013; Wang et al. 2014; Balasubramani et al. 2015). Although work from our group (Abdel-Wahab et al. 2012) and others (Inoue et al. 2013) suggested that ASXL1 loss was associated with loss of H3K27me because of altered EZH2 recruitment at key loci, more recent work has suggested that this alteration in H3K27 methylation with ASXL1 loss is not because of direct interaction between ASXL1 and PRC2 components (Wu et al. 2015). In contrast, Balasubramani et al. suggest that the depletion of H3K27me3 seen in the presence of *ASXL1* alterations may be secondary to depletion of H2AK119Ub and subsequent failure of PRC2 recruitment (Balasubramani et al. 2015). It is also important to note recent work identifying that EZH2 activity is also required for malignant transformation mediated by BAP1 loss in several tissue types (LaFave et al. 2015). These data underlie a relationship between PR-DUB and PRC2 function. Further efforts to analyze the genome-wide binding of ASXL1 as well as the binding of PRC2 components in the presence and absence of ASXL1 will hopefully help to elucidate the effects of ASXL1 alterations on PRC2 localization.

CONCLUSIONS

The majority of our understanding of ASXL family members to date emanates from analyses of ASXL1 and BAP1 function and mutations. However, it is not clear exactly how much of the function of ASXL1 in vivo relates to serving in an enzymatic complex with BAP1 versus other functions in regulating PRC2 activity and/or the function of other nuclear receptors. Further effort to understand the protein complexes that each ASXL family member participates in as well as the biological functions of each ASXL family member in vivo will be critical in understanding the roles and potential redundancies of each protein. In addition, efforts to study cells with engineered mutations in the endogenous locus of *ASXL1* to model human-disease-associated alterations will be critical. Given the clinical importance of *ASXL1* mutations in myeloid leukemias, knowledge of the precise effect of these mutations may be important in efforts to therapeutically target the downstream effects of *ASXL1* mutations and/or the mutant ASXL1 protein itself (if such a protein is actually stably expressed in cells). Indeed, recent work using genome editing to correct a homozygous *ASXL1* nonsense mutation in vivo suggests that restoring normal expression of ASXL1 may have important therapeutic implications for leukemia (Valletta et al. 2015). Along these same lines, it will be critically important to define the function of each ASXL member's

PHD domain, given the hypothesis that removal of the ASXL1/2 PHD domain may alone be pathogenic.

ACKNOWLEDGMENTS

J.-B.M. is supported by a grant from the Fondation de France. O.A.-W. is supported by grants from the National Institutes of Health (NIH) (R01 HL128239), the Edward P. Evans Foundation, and the United States Department of Defense Bone Marrow Failure Research Program (W81XWH-12-1-0041 and BM150092), the NIH (1K08CA160647-01), the Josie Robertson Investigator Program, the Damon Runyon Foundation, the Starr Foundation, and the Center for Experimental Therapeutics at Memorial Sloan Kettering Cancer Center.

REFERENCES

Abdel-Wahab O, Adli M, LaFave LM, Gao J, Hricik T, Shih AH, Pandey S, Patel JP, Chung YR, Koche R, et al. 2012. *ASXL1* mutations promote myeloid transformation through loss of PRC2-mediated gene repression. *Cancer Cell* 22: 180–193.

Abdel-Wahab O, Gao J, Adli M, Dey A, Trimarchi T, Chung YR, Kuscu C, Hricik T, Ndiaye-Lobry D, Lafave LM, et al. 2013. Deletion of Asxl1 results in myelodysplasia and severe developmental defects in vivo. *J Exp Med* 210: 2641–2659.

Aravind L, Iyer LM. 2012. The HARE-HTH and associated domains: Novel modules in the coordination of epigenetic DNA and protein modifications. *Cell Cycle* 11: 119–131.

Avila M, Kirchhoff M, Marle N, Hove HD, Chouchane M, Thauvin-Robinet C, Masurel A, Mosca-Boidron AL, Callier P, Mugneret F, et al. 2013. Delineation of a new chromosome 20q11.2 duplication syndrome including the *ASXL1* gene. *Am J Med Genet A* 161: 1594–1598.

Bainbridge MN, Hu H, Muzny DM, Musante L, Lupski JR, Graham BH, Chen W, Gripp KW, Jenny K, Wienker TF, et al. 2013. De novo truncating mutations in *ASXL3* are associated with a novel clinical phenotype with similarities to Bohring–Opitz syndrome. *Genome Med* 5: 11.

Balasubramani A, Larjo A, Bassein JA, Chang X, Hastie RB, Togher SM, Lahdesmaki H, Rao A. 2015. Cancer-associated *ASXL1* mutations may act as gain-of-function mutations of the ASXL1–BAP1 complex. *Nat Commun* 6: 7307.

Baskind HA, Na L, Ma Q, Patel MP, Geenen DL, Wang QT. 2009. Functional conservation of *Asxl2*, a murine homolog for the *Drosophila* enhancer of trithorax and polycomb group gene *Asx. PLoS ONE* 4: e4750.

Beekman R, Valkhof MG, Sanders MA, van Strien PM, Haanstra JR, Broeders L, Geertsma-Kleinekoort WM,

Veerman AJ, Valk PJ, Verhaak RG, et al. 2012. Sequential gain of mutations in severe congenital neutropenia progressing to acute myeloid leukemia. *Blood* **119:** 5071–5077.

Bejar R, Stevenson K, Abdel-Wahab O, Galili N, Nilsson B, Garcia-Manero G, Kantarjian H, Raza A, Levine RL, Neuberg D, et al. 2011. Clinical effect of point mutations in myelodysplastic syndromes. *N Engl J Med* **364:** 2496–2506.

Bott M, Brevet M, Taylor BS, Shimizu S, Ito T, Wang L, Creaney J, Lake RA, Zakowski MF, Reva B, et al. 2011. The nuclear deubiquitinase BAP1 is commonly inactivated by somatic mutations and 3p21.1 losses in malignant pleural mesothelioma. *Nat Genet* **43:** 668–672.

Carbone M, Yang H, Pass HI, Krausz T, Testa JR, Gaudino G. 2013. BAP1 and cancer. *Nat Rev Cancer* **13:** 153–159.

Damaj G, Joris M, Chandesris O, Hanssens K, Soucie E, Canioni D, Kolb B, Durieu I, Gyan E, Livideanu C, et al. 2014. *ASXL1* but not *TET2* mutations adversely impact overall survival of patients suffering systemic mastocytosis with associated clonal hematologic non-mast-cell diseases. *PLoS ONE* **9:** e85362.

Dangiolo SB, Wilson A, Jobanputra V, Anyane-Yeboa K. 2015. Bohring–Opitz syndrome (BOS) with a new *ASXL1* pathogenic variant: Review of the most prevalent molecular and phenotypic features of the syndrome. *Am J Med Genet A* **167:** 3161–3166.

Daou S, Hammond-Martel I, Mashtalir N, Barbour H, Gagnon J, Iannantuono NV, Nkwe NS, Motorina A, Pak H, Yu H, et al. 2015. The BAP1/ASXL2 histone H2A deubiquitinase complex regulates cell proliferation and is disrupted in cancer. *J Biol Chem* **290:** 28643–28663.

Devillier R, Mansat-De Mas V, Gelsi-Boyer V, Demur C, Murati A, Corre J, Prebet T, Bertoli S, Brecqueville M, Arnoulet C, et al. 2015. Role of *ASXL1* and *TP53* mutations in the molecular classification and prognosis of acute myeloid leukemias with myelodysplasia-related changes. *Oncotarget* **6:** 8388–8396.

Dey A, Seshasayee D, Noubade R, French DM, Liu J, Chaurushiya MS, Kirkpatrick DS, Pham VC, Lill JR, Bakalarski CE, et al. 2012. Loss of the tumor suppressor BAP1 causes myeloid transformation. *Science* **337:** 1541–1546.

Dinwiddie DL, Soden SE, Saunders CJ, Miller NA, Farrow EG, Smith LD, Kingsmore SF. 2013. De novo frameshift mutation in ASXL3 in a patient with global developmental delay, microcephaly, and craniofacial anomalies. *BMC Med Genomics* **6:** 32.

Duployez N, Micol JB, Boissel N, Petit A, Geffroy S, Bucci M, Lapillonne H, Renneville A, Leverger G, Ifrah N, et al. 2015a. Unlike *ASXL1* and *ASXL2* mutations, *ASXL3* mutations are rare events in acute myeloid leukemia with t(8;21). *Leuk Lymphoma* **57:** 199–200.

Duployez N, Willekens C, Marceau-Renaut A, Boudry-Labis E, Preudhomme C. 2015b. Prognosis and monitoring of core-binding factor acute myeloid leukemia: Current and emerging factors. *Exp Rev Hematol* **8:** 43–56.

Duployez N, Marceau-Renaut A, Boissel N, Petit A, Bucci M, Geffroy S, Lapillonne H, Renneville A, Ragu C, Figeac M, et al. 2016. Comprehensive mutational profiling of core binding factor acute myeloid leukemia. *Blood* **19:** 2451–2459.

Elliott MA, Pardanani A, Hanson CA, Lasho TL, Finke CM, Belachew AA, Tefferi A. 2015. *ASXL1* mutations are frequent and prognostically detrimental in *CSF3R*-mutated chronic neutrophilic leukemia. *Am J Hematol* **90:** 653–656.

Farber CR, Bennett BJ, Orozco L, Zou W, Lira A, Kostem E, Kang HM, Furlotte N, Berberyan A, Ghazalpour A, et al. 2011. Mouse genome-wide association and systems genetics identify *Asxl2* as a regulator of bone mineral density and osteoclastogenesis. *PLoS Genet* **7:** e1002038.

Fernandez-Mercado M, Yip BH, Pellagatti A, Davies C, Larrayoz MJ, Kondo T, Perez C, Killick S, McDonald EJ, Odero MD, et al. 2012. Mutation patterns of 16 genes in primary and secondary acute myeloid leukemia (AML) with normal cytogenetics. *PLoS ONE* **7:** e42334.

Fisher CL, Berger J, Randazzo F, Brock HW. 2003. A human homolog of *Additional sex combs*, *ADDITIONAL SEX COMBS-LIKE 1*, maps to chromosome 20q11. *Gene* **306:** 115–126.

Fisher CL, Randazzo F, Humphries RK, Brock HW. 2006. Characterization of *Asxl1*, a murine homolog of *Additional sex combs*, and analysis of the *Asx-like* gene family. *Gene* **369:** 109–118.

Fisher CL, Lee I, Bloyer S, Bozza S, Chevalier J, Dahl A, Bodner C, Helgason CD, Hess JL, Humphries RK, et al. 2010a. *Additional sex combs-like 1* belongs to the enhancer of trithorax and polycomb group and genetically interacts with *Cbx2* in mice. *Dev Biol* **337:** 9–15.

Fisher CL, Pineault N, Brookes C, Helgason CD, Ohta H, Bodner C, Hess JL, Humphries RK, Brock HW. 2010b. Loss-of-function *Additional sex combs like 1* mutations disrupt hematopoiesis but do not cause severe myelodysplasia or leukemia. *Blood* **115:** 38–46.

Gaytan de Ayala Alonso A, Gutierrez L, Fritsch C, Papp B, Beuchle D, Muller J. 2007. A genetic screen identifies novel polycomb group genes in *Drosophila*. *Genetics* **176:** 2099–2108.

Gelsi-Boyer V, Trouplin V, Adelaide J, Bonansea J, Cervera N, Carbuccia N, Lagarde A, Prebet T, Nezri M, Sainty D, et al. 2009. Mutations of polycomb-associated gene *ASXL1* in myelodysplastic syndromes and chronic myelomonocytic leukaemia. *Br J Haematol* **145:** 788–800.

Genovese G, Kahler AK, Handsaker RE, Lindberg J, Rose SA, Bakhoum SF, Chambert K, Mick E, Neale BM, Fromer M, et al. 2014. Clonal hematopoiesis and blood-cancer risk inferred from blood DNA sequence. *N Engl J Med* **371:** 2477–2487.

Grasso CS, Wu YM, Robinson DR, Cao X, Dhanasekaran SM, Khan AP, Quist MJ, Jing X, Lonigro RJ, Brenner JC, et al. 2012. The mutational landscape of lethal castration-resistant prostate cancer. *Nature* **487:** 239–243.

Guo G, Gui Y, Gao S, Tang A, Hu X, Huang Y, Jia W, Li Z, He M, Sun L, et al. 2012. Frequent mutations of genes encoding ubiquitin-mediated proteolysis pathway components in clear cell renal cell carcinoma. *Nat Genet* **44:** 17–19.

Haferlach T, Nagata Y, Grossmann V, Okuno Y, Bacher U, Nagae G, Schnittger S, Sanada M, Kon A, Alpermann T, et al. 2014. Landscape of genetic lesions in 944 patients with myelodysplastic syndromes. *Leukemia* **28:** 241–247.

Harbour JW, Onken MD, Roberson ED, Duan S, Cao L, Worley LA, Council ML, Matatall KA, Helms C, Bowcock

AM. 2010. Frequent mutation of *BAP1* in metastasizing uveal melanomas. *Science* **330:** 1410–1413.

Hoischen A, van Bon BW, Rodriguez-Santiago B, Gilissen C, Vissers LE, de Vries P, Janssen I, van Lier B, Hastings R, Smithson SF, et al. 2011. De novo nonsense mutations in *ASXL1* cause Bohring–Opitz syndrome. *Nat Genet* **43:** 729–731.

Huether R, Dong L, Chen X, Wu G, Parker M, Wei L, Ma J, Edmonson MN, Hedlund EK, Rusch MC, et al. 2014. The landscape of somatic mutations in epigenetic regulators across 1,000 paediatric cancer genomes. *Nat Commun* **5:** 3630.

Inoue D, Kitaura J, Togami K, Nishimura K, Enomoto Y, Uchida T, Kagiyama Y, Kawabata KC, Nakahara F, Izawa K, et al. 2013. Myelodysplastic syndromes are induced by histone methylation-altering *ASXL1* mutations. *J Clin Invest* **123:** 4627–4640.

Inoue D, Nishimura K, Kozuka-Hata H, Oyama M, Kitamura T. 2015. The stability of epigenetic factor ASXL1 is regulated through ubiquitination and USP7-mediated deubiquitination. *Leukemia* **29:** 2257–2260.

Itzykson R, Kosmider O, Renneville A, Gelsi-Boyer V, Meggendorfer M, Morabito M, Berthon C, Ades L, Fenaux P, Beyne-Rauzy O, et al. 2013. Prognostic score including gene mutations in chronic myelomonocytic leukemia. *J Clin Oncol* **31:** 2428–2436.

Izawa T, Rohatgi N, Fukunaga T, Wang QT, Silva MJ, Gardner MJ, McDaniel ML, Abumrad NA, Semenkovich CF, Teitelbaum SL, et al. 2015. ASXL2 regulates glucose, lipid, and skeletal homeostasis. *Cell Rep* **11:** 1625–1637.

Jaiswal S, Fontanillas P, Flannick J, Manning A, Grauman PV, Mar BG, Lindsley RC, Mermel CH, Burtt N, Chavez A, et al. 2014. Age-related clonal hematopoiesis associated with adverse outcomes. *N Engl J Med* **371:** 2488–2498.

Jensen DE, Proctor M, Marquis ST, Gardner HP, Ha SI, Chodosh LA, Ishov AM, Tommerup N, Vissing H, Sekido Y, et al. 1998. BAP1: A novel ubiquitin hydrolase which binds to the BRCA1 RING finger and enhances BRCA1-mediated cell growth suppression. *Oncogene* **16:** 1097–1112.

Katoh M. 2013. Functional and cancer genomics of *ASXL* family members. *Br J Cancer* **109:** 299–306.

Katoh M. 2015. Functional proteomics of the epigenetic regulators ASXL1, ASXL2 and ASXL3: A convergence of proteomics and epigenetics for translational medicine. *Expert Rev Proteomics* **12:** 317–328.

Katoh M, Katoh M. 2003. Identification and characterization of *ASXL2* gene in silico. *Int J Oncol* **23:** 845–850.

LaFave LM, Beguelin W, Koche R, Teater M, Spitzer B, Chramiec A, Papalexi E, Keller MD, Hricik T, Konstantinoff K, et al. 2015. Loss of BAP1 function leads to EZH2-dependent transformation. *Nat Med* **21:** 1344–1349.

Lai HL, Wang QT. 2013. *Additional sex combs-like 2* is required for polycomb repressive complex 2 binding at select targets. *PLoS ONE* **8:** e73983.

Lai HL, Grachoff M, McGinley AL, Khan FF, Warren CM, Chowdhury SA, Wolska BM, Solaro RJ, Geenen DL, Wang QT. 2012. Maintenance of adult cardiac function requires the chromatin factor Asxl2. *J Mol Cell Cardiol* **53:** 734–741.

Magini P, Della Monica M, Uzielli ML, Mongelli P, Scarselli G, Gambineri E, Scarano G, Seri M. 2012. Two novel patients with Bohring–Opitz syndrome caused by de novo *ASXL1* mutations. *Am J Med Genet A* **158:** 917–921.

Mallery DL, Vandenberg CJ, Hiom K. 2002. Activation of the E3 ligase function of the BRCA1/BARD1 complex by polyubiquitin chains. *EMBO J* **21:** 6755–6762.

McGinley AL, Li Y, Deliu Z, Wang QT. 2014. Additional sex combs-like family genes are required for normal cardiovascular development. *Genesis* **52:** 671–686.

Metzeler KH, Becker H, Maharry K, Radmacher MD, Kohlschmidt J, Mrozek K, Nicolet D, Whitman SP, Wu YZ, Schwind S, et al. 2011. *ASXL1* mutations identify a high-risk subgroup of older patients with primary cytogenetically normal AML within the ELN Favorable genetic category. *Blood* **118:** 6920–6929.

Micol JB, Abdel-Wahab O. 2014. Collaborating constitutive and somatic genetic events in myeloid malignancies: *ASXL1* mutations in patients with germline *GATA2* mutations. *Haematologica* **99:** 201–203.

Micol JB, Duployez N, Boissel N, Petit A, Geffroy S, Nibourel O, Lacombe C, Lapillonne H, Etancelin P, Figeac M, et al. 2014. Frequent *ASXL2* mutations in acute myeloid leukemia patients with t(8;21)/*RUNX1-RUNX1T1* chromosomal translocations. *Blood* **124:** 1445–1449.

Micol JB, Duployez N, Pastore A, Williams R, Kim E, Lee S, Durham B, Chung YR, Cho H, Preudhomme C, Abdel-Wahab O. 2015. ASXL2 is a novel mediator of RUNX1-ETO transcriptional function and collaborates with RUNX1-ETO to promote leukemogenesis. *57th ASH Annual Meeting and Exposition*, Abstract 302. Orlando, FL, December 5–8.

Milne TA, Sinclair DA, Brock HW. 1999. The *Additional sex combs* gene of *Drosophila* is required for activation and repression of homeotic loci, and interacts specifically with *Polycomb* and *super sex combs*. *Mol Gen Genet* **261:** 753–761.

Moon S, Um SJ, Kim EJ. 2015. Role of Asxl1 in kidney podocyte development via its interaction with Wtip. *Biochem Biophys Res Commun* **466:** 560–566.

Park UH, Yoon SK, Park T, Kim EJ, Um SJ. 2011. Additional sex comb-like (ASXL) proteins 1 and 2 play opposite roles in adipogenesis via reciprocal regulation of peroxisome proliferator-activated receptor γ. *J Biol Chem* **286:** 1354–1363.

Park UH, Kang MR, Kim EJ, Kwon YS, Hur W, Yoon SK, Song BJ, Park JH, Hwang JT, Jeong JC, et al. 2015. ASXL2 promotes proliferation of breast cancer cells by linking ERα to histone methylation. *Oncogene* doi: 10.1038/onc.2015.443.

Patel JP, Gonen M, Figueroa ME, Fernandez H, Sun Z, Racevskis J, Van Vlierberghe P, Dolgalev I, Thomas S, Aminova O, et al. 2012. Prognostic relevance of integrated genetic profiling in acute myeloid leukemia. *N Engl J Med* **366:** 1079–1089.

Prebet T, Carbuccia N, Raslova H, Favier R, Rey J, Arnoulet C, Vey N, Vainchenker W, Birnbaum D, Mozziconacci MJ. 2013. Concomitant germ-line *RUNX1* and acquired *ASXL1* mutations in a T-cell acute lymphoblastic leukemia. *Eur J Haematol* **91:** 277–279.

Quesada V, Conde L, Villamor N, Ordonez GR, Jares P, Bassaganyas L, Ramsay AJ, Bea S, Pinyol M, Martinez-Trillos A, et al. 2012. Exome sequencing identifies recurrent mutations of the splicing factor *SF3B1* gene in chronic lymphocytic leukemia. *Nat Genet* **44:** 47–52.

Russell B, Graham JM Jr. 2013. Expanding our knowledge of conditions associated with the *ASXL* gene family. *Genome Med* **5:** 16.

Russell B, Johnston JJ, Biesecker LG, Kramer N, Pickart A, Rhead W, Tan WH, Brownstein CA, Kate Clarkson L, Dobson A, et al. 2015. Clinical management of patients with *ASXL1* mutations and Bohring–Opitz syndrome, emphasizing the need for Wilms tumor surveillance. *Am J Med Genet A* **167:** 2122–2131.

Sahtoe DD, van Dijk WJ, Ekkebus R, Ovaa H, Sixma TK. 2016. BAP1/ASXL1 recruitment and activation for H2A deubiquitination. *Nat Commun* **7:** 10292.

Sanchez-Pulido L, Kong L, Ponting CP. 2012. A common ancestry for BAP1 and Uch37 regulators. *Bioinformatics* **28:** 1953–1956.

Scheuermann JC, de Ayala Alonso AG, Oktaba K, Ly-Hartig N, McGinty RK, Fraterman S, Wilm M, Muir TW, Muller J. 2010. Histone H2A deubiquitinase activity of the Polycomb repressive complex PR-DUB. *Nature* **465:** 243–247.

Sinclair DA, Milne TA, Hodgson JW, Shellard J, Salinas CA, Kyba M, Randazzo F, Brock HW. 1998. The additional sex combs gene of *Drosophila* encodes a chromatin protein that binds to shared and unique Polycomb group sites on polytene chromosomes. *Development* **125:** 1207–1216.

Srivastava A, Ritesh KC, Tsan YC, Liao R, Su F, Cao X, Hannibal MC, Keegan CE, Chinnaiyan AM, Martin DM, et al. 2015. De novo dominant *ASXL3* mutations alter H2A deubiquitination and transcription in Bainbridge–Ropers syndrome. *Hum Mol Genet* **25:** 597–608.

Stein BL, Williams DM, O'Keefe C, Rogers O, Ingersoll RG, Spivak JL, Verma A, Maciejewski JP, McDevitt MA, Moliterno AR. 2011. Disruption of the *ASXL1* gene is frequent in primary, post-essential thrombocytosis and post-polycythemia vera myelofibrosis, but not essential thrombocytosis or polycythemia vera: Analysis of molecular genetics and clinical phenotypes. *Haematologica* **96:** 1462–1469.

Testa JR, Cheung M, Pei J, Below JE, Tan Y, Sementino E, Cox NJ, Dogan AU, Pass HI, Trusa S, et al. 2011. Germline *BAP1* mutations predispose to malignant mesothelioma. *Nat Genet* **43:** 1022–1025.

Thol F, Friesen I, Damm F, Yun H, Weissinger EM, Krauter J, Wagner K, Chaturvedi A, Sharma A, Wichmann M, et al. 2011. Prognostic significance of *ASXL1* mutations in patients with myelodysplastic syndromes. *J Clin Oncol* **29:** 2499–2506.

Valletta S, Dolatshad H, Bartenstein M, Yip BH, Bello E, Gordon S, Yu Y, Shaw J, Roy S, Scifo L, et al. 2015. *ASXL1* mutation correction by CRISPR/Cas9 restores gene function in leukemia cells and increases survival in mouse xenografts. *Oncotarget* **6:** 44061–44071.

Wang J, Li Z, He Y, Pan F, Chen S, Rhodes S, Nguyen L, Yuan J, Jiang L, Yang X, et al. 2014. Loss of *Asxl1* leads to myelodysplastic syndrome-like disease in mice. *Blood* **123:** 541–553.

West RR, Hsu AP, Holland SM, Cuellar-Rodriguez J, Hickstein DD. 2014. Acquired *ASXL1* mutations are common in patients with inherited *GATA2* mutations and correlate with myeloid transformation. *Haematologica* **99:** 276–281.

Wiesner T, Obenauf AC, Murali R, Fried I, Griewank KG, Ulz P, Windpassinger C, Wackernagel W, Loy S, Wolf I, et al. 2011. Germline mutations in *BAP1* predispose to melanocytic tumors. *Nat Genet* **43:** 1018–1021.

Wu X, Bekker-Jensen IH, Christensen J, Rasmussen KD, Sidoli S, Qi Y, Kong Y, Wang X, Cui Y, Xiao Z, et al. 2015. Tumor suppressor ASXL1 is essential for the activation of INK4B expression in response to oncogene activity and anti-proliferative signals. *Cell Res* **25:** 1205–1218.

Xie M, Lu C, Wang J, McLellan MD, Johnson KJ, Wendl MC, McMichael JF, Schmidt HK, Yellapantula V, Miller CA, et al. 2014. Age-related mutations associated with clonal hematopoietic expansion and malignancies. *Nat Med* **20:** 1472–1478.

Yoshizato T, Dumitriu B, Hosokawa K, Makishima H, Yoshida K, Townsley D, Sato-Otsubo A, Sato Y, Liu D, Suzuki H, et al. 2015. Somatic mutations and clonal hematopoiesis in aplastic anemia. *N Engl J Med* **373:** 35–47.

Cite this article as *Cold Spring Harb Perspect Med* doi: 10.1101/cshperspect.a026526

SETting the Stage for Cancer Development: SETD2 and the Consequences of Lost Methylation

Catherine C. Fahey[1] and Ian J. Davis[1,2]

[1]Lineberger Comprehensive Cancer Center, University of North Carolina at Chapel Hill, Chapel Hill, North Carolina 27599-7295

[2]Departments of Genetics and Pediatrics, University of North Carolina at Chapel Hill, Chapel Hill, North Carolina 27599-7295

Correspondence: ian_davis@med.unc.edu

The H3 lysine 36 histone methyltransferase *SETD2* is mutated across a range of human cancers. Although other enzymes can mediate mono- and dimethylation, SETD2 is the exclusive trimethylase. SETD2 associates with the phosphorylated carboxy-terminal domain of RNA polymerase and modifies histones at actively transcribed genes. The functions associated with SETD2 are mediated through multiple effector proteins that bind trimethylated H3K36. These effectors directly mediate multiple chromatin-regulated processes, including RNA splicing, DNA damage repair, and DNA methylation. Although alterations in each of these processes have been associated with SETD2 loss, the relative role of each in the development of cancer is not fully understood. Critical vulnerabilities resulting from SETD2 loss may offer a strategy for potential therapeutics.

Large-scale sequencing efforts of human cancers have identified recurrent mutations and deletions in a variety of chromatin regulatory proteins. SETD2, a methyltransferase that trimethylates histone H3 at lysine 36 (H3K36me3), has been found to be both mutated and deleted in select range of cancers suggesting a role in tumor suppression. In this review, we will discuss the range of SETD2 mutations and how disruption of this gene could foster the development of cancer.

ENZYMATIC FUNCTION AND STRUCTURE OF SETD2

SETD2 (also known as HYPB or KMT3a) is the sole methyltransferase that mediates trimethyl-ation of histone H3.1 and H3 variants (Fig. 1) (Edmunds et al. 2008). Although SETD2 shows biochemical evidence of mono- and dimethyla-tion activity, in cells it seems to exclusively mediate trimethylation because SETD2 silencing results in a near complete loss of H3K36 trimethylation without decreasing mono- or dimethylation levels (Edmunds et al. 2008; Yuan et al. 2009). In higher eukaryotes, other enzymes, including NSD1, WHSC1 (NSD2), and SETMAR, are able to mono- and dimethylate H3K36 (reviewed in Wagner and Carpenter 2012). In contrast, the homolog of SETD2 in *Saccharomyces cerevisiae*, Set2, exclusively mediates all H3K36 methylation states (Strahl et al. 2002). Although extensive genetic experimentation with yeast Set2 has informed our under-

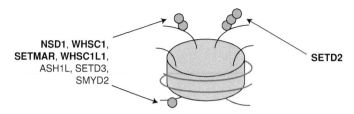

Figure 1. H3K36 methyltransferases. Methyltransferases shown to mono-, di-, and trimethylate H3K36. Those shown in bold have been shown in cell-based assays and/or in vivo. (Figure adapted from data in Wagner and Carpenter 2012.)

standing of the biochemical properties of critical SETD2 domains as well as the possible roles, the redundancy of mono- and dimethylation activities in humans offers an important caveat when extrapolating results from *S. cerevisiae*.

The functions of SETD2 have been attributed to several domains in the protein (Fig. 2). These domains share sequence homology as well as functional similarity with those in yeast Set2. The methyltransferase activity is mediated by the conserved SET domain (Strahl et al. 2002; Li et al. 2005). SETD2 contains two known protein-binding domains: SRI (Set2 Rpd1 Interacting) and WW. The SRI domain mediates association with the hyperphosphorylated carboxy-terminal domain (CTD) of RNA polymerase II (RNAPII) (Li et al. 2005; Sun et al. 2005; Xiao et al. 2003). In yeast, this interaction is required for H3K36 activity, as deletion of the RNAPII CTD decreases H3K36 methylation levels (Li et al. 2003; Xiao et al. 2003). The WW domain, which precedes the SRI domain, may mediate intramolecular interaction (Gao et al. 2014). Based on the property of WW domain interaction with phosphorylated proteins, this domain could also mediate other protein interactions (Lu et al. 1999). About half of SETD2 consists of a large amino terminal domain that is not shared by yeast Set2 and is of unknown function.

SETD2 MUTATION IN CANCER

Hundreds of distinct SETD2 mutations have been identified across a wide range of human tumors, including epithelial, central nervous system (CNS), and hematopoietic (Fig. 2; Table 1) (Gao et al. 2013; Cerami et al. 2012). *SETD2*

mutation was first described in clear cell renal cell carcinoma (ccRCC). In a cohort of 407 ccRCC tumors, truncating mutations were observed in 12 samples (Dalgliesh et al. 2010). *SETD2* mutation was also found in ccRCC cell lines (Duns et al. 2012). In The Cancer Genome Atlas (TCGA) study of ccRCC, *SETD2* was the third-most-commonly mutated gene with a prevalence of 15.6%. SETD2 is located on chromosome 3p, which shows near universal loss of heterozygosity in ccRCC (Zbar et al. 1987). Chromosome 3p is also the location of the well-known tumor suppressor von Hippel–Lindau (VHL). VHL expression is lost in most cases of sporadic ccRCC, and germline mutation is associated with a high penetrance of ccRCC (Stolle et al. 1998). Mutations of *SETD2* affect the remaining allele, and frequently have a significant impact on gene function. Most mutations tend to be inactivating frameshift or nonsense mutations, although missense mutations in critical domains have been detected (Gerlinger et al. 2012; The Cancer Genome Atlas 2013a; Gossage et al. 2014; Simon et al. 2014). In a study of 128 sporadic ccRCC tumors that specifically examined genes known to be mutated in ccRCC, SETD2 was mutated in ∼16% (Gossage et al. 2014). Five frameshift, 10 nonsense, and two splice site mutations were observed. Of three nonsynonymous missense mutations, one altered the SET domain (Gossage et al. 2014). Studies of intratumoral genetic heterogeneity also support a key role for SETD2 loss. Sequencing multiple sites in a single kidney tumor together with metastatic sites identified multiple distinct SETD2 mutations each likely to disrupt function. This convergent tumor evolution suggests that SETD2 mutation

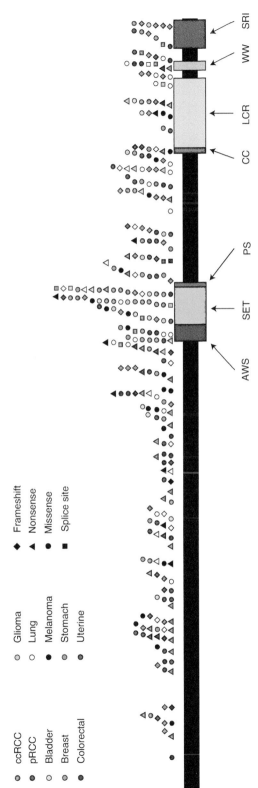

Figure 2. Schematic representation of SETD2 and cancer-associated mutations. SETD2 domains: AWS (associated with SET), SET (Su(var)3-9, enhancer-of-zeste, trithorax), PS (post-SET), CC (coiled coil), LCR (low charge region), WW, SRI (Set2 Rpb1 Interacting). Mutation lists were obtained from CBioPortal on February 5, 2016, and separated into cancer types. Duplicates were removed. Mutations are plotted by color (cancer type) and shape (mutation type). ccRCC, Clear cell renal cell carcinoma; pRCC, papillary renal cell carcinoma.

Table 1. Cancers associated with SETD2 mutation

Cancer type	Mutation % (total samples)	References
Clear cell renal cell carcinoma	15.6% (418 samples)	The Cancer Genome Atlas 2013a
High-grade glioma	15% (543 samples)	Fontebasso et al. 2013
Uterine carcinosarcoma	13.6% (22 samples)	Jones et al. 2014
Uterine corpus endometrioid carcinoma	9% (240 samples)	The Cancer Genome Atlas et al. 2013
Acute lymphocytic leukemia	12% (125 samples) 10% (94 samples)	Zhang et al. 2012; Mar et al. 2014
Bladder urothelial carcinoma	10.2% (107 samples) 6% (50 samples)	The Cancer Genome Atlas 2014b; Van Allen et al. 2014
Desmoplastic melanoma	10% (20 samples)	Berger et al. 2012
Melanoma	8% (25 samples)	Shain et al. 2015
Cutaneous melanoma	5.5% (91 samples)	Berger et al. 2012; Hodis et al. 2012
Lung adenocarcinoma	9% (230 samples) 5.5% (182 samples)	Imielinski et al. 2012; The Cancer Genome Atlas 2014c
Colorectal adenocarcinoma	8.3% (72 samples) 6.1% (212 samples)	The Cancer Genome Atlas 2012; Seshagiri et al. 2012
Pancreatic adenocarcinoma	8.3% (109 samples)	Witkiewicz et al. 2015
Stomach adenocarcinoma	7% (287 samples)	The Cancer Genome Atlas 2014a
Papillary renal cell carcinoma	7.6% (157 samples)	The Cancer Genome Atlas et al. 2016
Cutaneous squamous cell carcinoma	6.9% (29 samples)	Li et al. 2015

Cancers are selected for which the mutation rate in CBioPortal (Cerami et al. 2012; Gao et al. 2013) exceeded 5% and a publication was available. Indicated mutation rate reflects published results. Additional cancers discussed in the text are also included.

is a critical event for a subset of ccRCCs (Gerlinger et al. 2012). Suggestive of a link with aggressive disease, a lower level of H3K36 trimethylation was observed in tumors in metastases compared with primary tumors (Ho et al. 2015).

SETD2 mutations have also been identified in multiple other cancers. High-severity *SETD2* mutation was observed in 15%–28% of pediatric and 8% of adult high-grade gliomas (HGGs) (Fontebasso et al. 2013). In contrast, *SETD2* mutations were not identified in low-grade gliomas. In addition, all tumors with *SETD2* mutation were located in the cerebral hemispheres (Fontebasso et al. 2013). However, mutation of *SETD2* was detected in a single diffuse intrinsic pontine glioma, although it co-occurred with an H3.1 K27M mutation, a common feature of these tumors (Wu et al. 2014). Intriguingly, high-grade hemispheric gliomas of children

and young adults are also commonly associated with mutation of histone H3.3 (*H3F3A*) at glycine 34 (Schwartzentruber et al. 2012; Wu et al. 2012). These mutations were nonoverlapping with SETD2 and were associated with reduced H3K36 methylation (Schwartzentruber et al. 2012; Sturm et al. 2012; Fontebasso et al. 2013). Gliomas also commonly harbor IDH1 mutations, which result in the generation of the 2-hydroxyglutarate (2-HG) oncometabolite (Dang et al. 2009). These mutations are not mutually exclusive with SETD2 mutation. Although 2-HG inhibits histone demethylases, including those that can act on H3K36, it is not clear whether IDH1 mutation directly affects H3K36me3 (Chowdhury et al. 2011; Xu et al. 2011; Lu et al. 2012; Fontebasso et al. 2013). Overall, these findings indicate that dysregulation of H3K36me3 is a common event in glioma. The finding linking H3.3 mutation with

Cite this article as *Cold Spring Harb Perspect Med* doi: 10.1101/cshperspect.a026468

reduced H3K36 methylation will be discussed in more detail below.

SETD2 mutations have also been identified in acute leukemias. In a study of early T-cell precursor acute lymphoblastic leukemia (ETP-ALL), approximately 10% showed deletion or high-severity mutation of *SETD2* (Zhang et al. 2012). In a separate study, SETD2 mutation was detected in approximately 6% of ALL and AML samples (Zhu et al. 2014). Mutation of SETD2 was more common in both acute lymphoid and myeloid leukemias with mixed lineage leukemia (MLL)-rearrangement compared to acute lymphoblastic leukemia (ALL) and acute myeloid leukemia (AML) with an intact MLL gene. Further supporting *SETD2* loss as a critical event in leukemia development, *SETD2* mutations were commonly nonsense or frameshift, and approximately a quarter of samples carried biallelic *SETD2* mutations (Zhu et al. 2014). A comparison of matched primary and relapsed ALL samples suggested that mutations in epigenetic regulators as a class were more common at relapse, and this included mutations in *SETD2* (Mar et al. 2014). In this study, SETD2 showed a mutation rate of 5% in a pilot cohort and 12% in a larger validation set. The validation set contained a higher fraction of MLL-rearranged samples possibly explaining the discrepancy in mutation frequency. Suggesting that the importance of SETD2 mutation is greater in acute leukemias in children, the study of AML by TCGA identified only a single SETD2 mutation among 191 adult samples (The Cancer Genome Atlas 2013b). The link between SETD2 loss and MLL rearrangement is provocative since, like SETD2, MLL fusion proteins can associate with components of the transcriptional complex, and the combined alterations in these proteins may lead to transcriptional dysregulation (Milne et al. 2010; Yokoyama et al. 2010).

SETD2 mutations have also been observed at a low frequency in a range of other tumors. 6% of melanoma and chronic lymphocytic leukemia showed SETD2 alteration (Berger et al. 2012; Lee et al. 2015; Parker et al. 2016). *SETD2* alterations were observed in high-risk, but not low-risk, gastrointestinal stromal tumors (Huang et al. 2015). SETD2 mutation has also

been described in phyllodes tumors of the breast, but not in breast fibroadenoma (Tan et al. 2015; Liu et al. 2016). Among other genitourinary tumors, *SETD2* mutation is found in 10% of bladder tumors and the papillary subtype of renal cell carcinoma (The Cancer Genome Atlas 2014b; The Cancer Genome Atlas et al. 2016). Although many of these mutations are monoallelic and consequently predicted to lead to haploinsufficiency, mutations are not the exclusive mechanism for modulating SETD2 activity.

Decreased H3K36me3 has also been observed in the context of nonmutant SETD2 in ccRCC (Simon et al. 2014). miR-106b-5p, a micro RNA known to regulate SETD2, was elevated in a cohort of 40 ccRCC tumor samples; levels of this miRNA inversely correlated with SETD2 mRNA and protein levels (Xiang et al. 2015). SETD2 mRNA levels were also decreased in a subset of patients with AML and lymphoma (Zhu et al. 2014).

In addition to alterations in SETD2, H3K36me3 can be lost by mutation of the methyl acceptor site in histones or by mutations in neighboring amino acids. Virtually all chondroblastomas harbor a lysine 36 to methionine variant in histone H3.3 (H3.3K36M) (Behjati et al. 2013). Expression of the H3.3K36M mutant in cells led to depletion of all H3K36 trimethylation (Lewis et al. 2013). H3.3K36M binds and directly inhibits the activity of SETD2 and the dimethylation activity of MMSET (Fang et al. 2016). Mutations in H3.3 at the neighboring G34 residue have been described in almost all giant-cell bone tumors (Behjati et al. 2013) and, as previously mentioned, in high-grade gliomas (Schwartzentruber et al. 2012; Wu et al. 2014). Overall, the common finding of loss or inhibition of SETD2 across a wide range of cancers suggests the importance of disrupted H3K36 methylation in cancer development.

TRANSCRIPTION AND RNA SPLICING

Chromatin influences many cellular processes including transcription, replication and DNA damage repair. Many studies have linked SETD2 and transcription. In yeast, Set2 and H3K36 tri-methylation are associated with gene bodies, and H3K36me3 levels correlate both with the level of transcription and the position in the gene (Krogan et al. 2003; Bannister et al. 2005). H3K36me3 signals are enriched at exons, although the higher levels of nucleosome occupancy at exons may explain this difference (Schwartz et al. 2009). Treatment with a transcription inhibitor causes a decrease in H3K36me3 levels (de Almeida et al. 2011), suggesting that active transcription is necessary for H3K36me3 placement. These data are consistent with the model that Set2 and SETD2 are targeted to elongating RNAPII.

Our understanding of the role of SETD2 in transcription is largely based on studies of Set2 in yeast. In yeast, Set2 partially functions to prevent cryptic initiation, aberrant transcription from internal sites. H3K36 methylation recruits the Rpd3C(S) complex, which includes a histone deacetylase (Carrozza et al. 2005; Keogh et al. 2005), leading to the deacetylation of histones in gene bodies. H3K36 methylation also suppresses of the interaction between histone H3 and the histone chaperone Asf1 (Venkatesh et al. 2012). Preventing the incorporation of new acetylated histones maintains a hypoacetylated state, thereby stabilizing nucleosomes that decreases the chance of a segment of gene being aberrantly recognized by the transcriptional initiation complex as a promoter (Carrozza et al. 2005; Li et al. 2007; Lickwar et al. 2009).

The role of SETD2 in transcription in higher eukaryotes is more complicated. In addition to the separation of methylation activities across multiple enzymes, genes in higher eukaryotes contain introns and are regulated by alternative splicing, and DNA itself can be methylated. H3K36 methylation levels differ based on exon utilization (Kolasinska-Zwierz et al. 2009), with alternatively spliced exons having lower levels of H3K36me3 than those that are constitutively included. Altering SETD2 levels influenced the inclusion of exons in genes known to be alternatively spliced (Luco et al. 2010). Deletion of the splice acceptor site in the β-globin intron leads to shifts in H3K36me3 signal (Kim et al. 2011), and intronless genes have lower levels of

H3K36 trimethylation (de Almeida et al. 2011). Chemical or RNAi-mediated inhibition of splicing decreased H3K36me3 levels. Together these data suggest a close relationship between trimethylation of H3K36 and RNA splicing. Perhaps reflecting aberrant transcription or RNA processing, silencing SETD2 results in mRNA accumulation in the nucleus (Yoh et al. 2008).

Several studies have explored the impact of SETD2 loss on transcription in kidney cancer. Examining primary ccRCC, H3K36me3-deficient tumors show alterations in splicing and evidence of intron retention (Simon et al. 2014). This association was also detected in transcriptomic ccRCC data from TCGA. Similarly, differential splicing and altered exon utilization was observed in *SETD2* knockout cells (Ho et al. 2015). However, no difference in exon usage or intron retention was observed in other studies in which SETD2 was silenced using RNAi (Kanu et al. 2015). SETD2 has also been associated with aberrant transcriptional termination (Grosso et al. 2015). In the absence of appropriate termination, RNAPII complexes can read through into neighboring genes yielding chimeric RNAs.

Several proteins that bind H3K36 methylation offer a link between SETD2 and RNA splicing (Fig. 3). LEDGF (PSIP1) exists as two isoforms, p52 and p75, based on the inclusion of six additional 3′ exons (Singh et al. 2000). Both forms contain a PWWP domain, which interacts with di- and trimethylated H3K36. The short form, LEDGF/p52, interacts with proteins involved in alternative splicing (Pradeepa et al. 2012). ZMYND11, also a PWWP domain–containing protein, binds H3.3K36 and associates with regulators of RNA splicing (Guo et al. 2014). MRG15 is a chromodomain-containing protein that binds di- and trimethylated H3K36 (Zhang et al. 2006) and recruits polypyrimidine tract-binding protein (PTB) to alternatively spliced exons (Luco et al. 2010). PTB then binds to silencing elements causing repression of specific exons.

In embryonic stem cells, MRG15 also recruits the lysine demethylase KDM5B to H3K36me3 marked chromatin (Xie et al. 2011). Silencing KDM5B resulted in recruit-

Cite this article as *Cold Spring Harb Perspect Med* doi: 10.1101/cshperspect.a026468

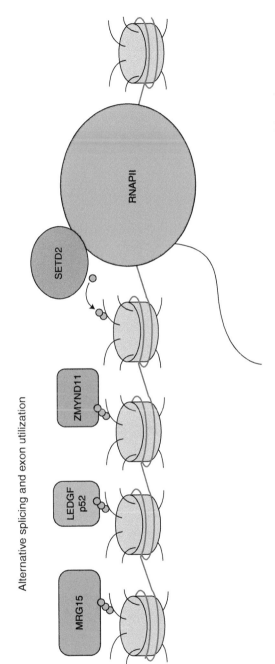

Figure 3. Transcription and RNA processing. SETD2 associates with RNAPII to posttranslationally modify nucleosomes. H3K36me3 is directly bound by readers, which mediate RNA processing through downstream effectors.

ment of unphosphorylated RNAPII to intragenic regions marked by H3K4me3, potential sites of cryptic initiation. Knockdown of KDM5B increased levels of unspliced transcripts, possibly reflecting aberrant transcription. Downregulation of SETD2 was also found to be associated with increased RNA abundance at noninitiating exons, potentially indicating transcriptional initiation from these sites (Carvalho et al. 2013). Taken together, the extent to which cryptic initiation in yeast is a function of dimethylation (rather than trimethylation) or that RNA alteration in higher eukaryotic cells results from aberrant splicing (rather than cryptic initiation) are unknown.

Overall, SETD2-mediated histone modification and its interaction with specific binding partners offers an explanation for the variation in transcription and aberrant RNA detected after SETD2 loss. Although intron retention has been shown to be a mechanism of tumor-suppressor inactivation (Jung et al. 2015), how SETD2 loss facilitates tumor development remains unknown.

CHROMATIN STRUCTURE

Several studies have shown that H3K36me3 loss results in alterations in chromatin architecture. Silencing SETD2 impairs the recruitment of the Facilitates Chromatin Transcription (FACT) complex to transcribed chromatin, which results in increased sensitivity to MNase digestion, suggesting alteration in nucleosomal interaction with DNA (Carvalho et al. 2013). This effect was particularly evident at internal exons. Chromatin alterations in response to SETD2 silencing was also observed in a kidney cancer cell line (Kanu et al. 2015). A similar observation was made by examining chromatin accessibility in H3K36me3-deficient primary renal cell carcinomas using formaldehyde-assisted isolation of regulatory elements (FAIRE) (Simon et al. 2014). Overall, enhanced accessibility corresponded to regions typically marked by H3K36me3. By examining individual genic features, signal increases were most striking immediately preceding exons, suggestive of a specific effect at splice acceptor sites. Together,

these studies support a link between SETD2, chromatin accessibility and splicing.

DNA REPLICATION AND DAMAGE REPAIR

Substantial evidence supports the involvement of SETD2 and H3K36 methylation in DNA damage repair by homologous recombination (HR) (Fig. 4). SETD2 silencing resulted in a loss of HR at experimentally induced sites of double-strand breaks (DSBs) (Aymard et al. 2014). In particular, SETD2 loss decreased levels of ATM phosphorylation and consequently p53 phosphorylation (Carvalho et al. 2014) with decreased levels of p53 transcriptional targets (Xie et al. 2008; Carvalho et al. 2014). SETD2 has been shown to associate with and potentially stabilize p53, which may partially account for the difference in transcriptional targets (Xie et al. 2008). SETD2 loss was also associated with decreased recruitment of the HR proteins 53BP1, RPA, and RAD51 to chromatin (Aymard et al. 2014; Pfister et al. 2014; Kanu et al. 2015). This effect seems to be mediated by the histone methylation activity of SETD2 since reintroduction of a catalytically dead SETD2 mutant failed to rescue RPA or RAD51 foci formation, and depletion of H3K36me3 by overexpression of the KDM4 demethylase or H3.3K36M also delayed RAD51 foci formation (Pfister et al. 2014).

LEDGF may bridge H3K36 methylation and the DNA damage response mechanism. In contrast to the role of the p52 isoform in transcription, the LEDGF/p75 isoform recruits carboxy-terminal binding protein and interacting protein (CtIP) to sites of DNA damage (Daugaard et al. 2012). CtIP processes DNA ends to enable binding of RAD51 (reviewed in Symington 2010). Depletion of SETD2 results in decreased LEDGF bound to chromatin, reduced CtIP recruitment to DSB and levels of single-stranded DNA near the DSB, suggesting impaired resection (Pfister et al. 2014). This suggests a model in which regions marked by H3K36me3 by SETD2 are bound by LEDGF following DSB, leading to recruitment of CtIP and RAD51 and, ultimately, repair by HR. Whether SETD2 is recruited to sites of DNA damage that will

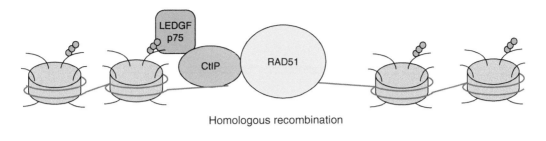

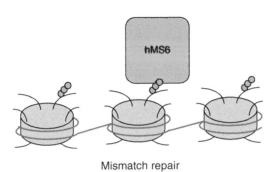

Figure 4. DNA damage repair. Specific H3K36me3 readers direct either homologous recombination (*top*) or mismatch repair (*bottom*).

go on to repair by homologous recombination, or conversely, that homologous recombination is more likely at regions already marked by H3K36me3 remains unclear. In contrast to H3K36me3, H3K36me2 is rapidly induced after irradiation (Fnu et al. 2011). Increased levels of early nonhomologous end joining (NHEJ) factors were detected by immunoprecipitation of H3K36me2 following radiation.

In addition to HR, SETD2 is involved in mismatch repair (MMR) (Fig. 4). hMSH6, a component of the hMutSα complex that recognizes mismatches in the genome, also contains a PWWP domain that mediates interaction with methylated H3K36 (Li et al. 2013). SETD2 silencing or overexpression of KDM4 decreased MSH6 foci formation associated with increased microsatellite instability (MSI) (Li et al. 2013; Awwad and Ayoub 2015). However, ccRCC tumors samples with biallelic SETD2 loss did not show classic findings of MSI, increased breakpoints or a substantially increased mutation load, compared with tumor cells with monoallelic loss (Kanu et al. 2015). DNA breaks identified in tumors with monoalleic SETD2 loss showed significantly lower levels of H3K36me3. These data are consistent with the model that H3K36me3 marked sites are protected from breakage. Taken together, these studies suggest that H3K36 methylation functions in DNA damage repair with methylation status biasing toward different repair pathways, with dimethylation favoring NHEJ and trimethylation favoring HR and MMR. Although SETD2 deficient kidney cancers are not characterized by increased mutational level, it remains possible that intratumoral heterogeneity limits our ability to detect this feature (Gerlinger et al. 2012).

DNA METHYLATION AND REPLICATION

Alterations in DNA methylation have been linked to SETD2 and H3K36me3 loss. The DNA methyltransferases DNMT3A and DNMT3B contain a PWWP domain enabling binding to methylated H3K36 (Dhayalan et al. 2010; Baubec et al. 2015). DNMT3B was enriched at H3K36me3 marked gene bodies, and *SETD2* knockout reduced DNMT3B binding (Baubec et al. 2015). In this study, SETD2 loss was associated with decreased de novo DNA methylation, although a separate study did not observe this association (Hahn et al. 2011). Alterations in DNA methylation correlated with SETD2 loss have been shown in ccRCC. In the TCGA analysis, *SETD2* mutation was associated with decreased DNA methylation at regions that are normally marked by H3K36me3 in kidney (The Cancer Genome Atlas 2013a). Increased chromatin accessibility was also associated with regions of DNA hypomethylation in *SETD2* mutant tumors (Simon et al. 2014). SETD2 loss has also been associated with increased DNA methylation at intergenic regions (Tiedemann et al. 2016). Overall, these data suggest that H3K36me3 directs DNA methyltransferases to gene bodies but in the absence of this histone modification, methylation increases elsewhere.

H3K36me3 levels are cell-cycle regulated with a peak in early S phase then declining to low levels that persist during G_2/M (Li et al. 2013). This pattern suggests that SETD2 is most active during DNA replication. In support of a role during replication, depletion of SETD2 in kidney cancer cells slowed replication fork progression and led to an accumulation of cells in S phase (Kanu et al. 2015). However, in isogenic SETD2 knockout cell lines, cell-cycle differences were not observed (Pfister et al. 2015).

SETD2 IN CANCER DEVELOPMENT AND THERAPEUTICS

How SETD2 loss results in cancer development remains unknown. However, several studies have linked H3K36 methylation to aberrant differentiation or proliferation. SETD2 loss disrupts murine embryonic stem-cell differentiation, possibly by altering intracellular signaling (Zhang et al. 2014). Expression of H3.3 mutants that inhibit H3K36 methylation in chondrocytes and mesenchymal progenitor cells disrupted differentiation (Fang et al. 2016; Lu et al. 2016). Mouse mesenchymal progenitor cells (MPCs) that stably express either wildtype or K36M mutant H3.3 formed tumors after subcutaneous injection in immunocompro-

mised mice (Lu et al. 2016). In renal primary epithelial tubule cells, cells considered to be the progenitor for ccRCC, SETD2 knockdown resulted in continued proliferation well past the point at which these cells typically senesce (Li et al. 2016). In models of MLL-rearranged leukemia, SETD2 loss is associated with increased colony formation, proliferation, and accelerated leukemia development after transplantation (Zhu et al. 2014). Taken together, these studies support a role of SETD2 in facilitating faithful differentiation. Interestingly, germline mutations in SETD2 as well as NSD1 (the H3K36 dimethylase; see Fig. 1) have been associated with Sotos and Sotos-like overgrowth syndromes (Kurotaki et al. 2002; Luscan et al. 2014; Tlemsani et al. 2016). Sotos syndrome has been associated with an increased frequency of malignancy, particularly acute leukemias and lymphomas, Wilms tumor, and neuroblastoma (reviewed in Lapunzina and Cohen 2005).

The clinical implications of *SETD2* loss in cancer have primarily focused on ccRCC. *SETD2* mutation is associated with worse cancer-specific survival in the TCGA dataset (Hakimi et al. 2013b). Additionally, *SETD2* mutation was a univariate predicator of time to recurrence, and was found at higher percentages in late stage tumors. Tumors with *BAP1*, *SETD2*, or *KDM5C* mutation were more likely to present with advanced stage (Hakimi et al. 2013a). In metastatic RCC, low SETD2 expression was associated with reduced overall and progression-free survival, and was an independent prognostic marker for these endpoints (Wang et al. 2016). SETD2 expression was lower in breast tumors, compared with matched normal tissue (Al Sarakbi et al. 2009; Newbold and Mokbel 2010), with expression inversely correlated with increasing tumor stage (Al Sarakbi et al. 2009). In these patients, SETD2 mRNA levels were lower in patients with poor outcomes, such as metastasis, local recurrence, and cancer-specific death.

Several studies have explored whether SETD2 loss sensitizes tumor cells to targeted agents. TGX221, a selective PI3Kβ inhibitor, was selectively toxic to RCC cells that were mutant for both *VHL* and *SETD2*, whereas cells lacking either mutation were not sensitive (Feng et al. 2015). Treatment with this compound resulted in decreased migration and invasion of mutant cell lines. Using a synthetic lethality screening strategy, H3K36me3-deficient cell lines were found to be sensitive to WEE1 inhibition (Pfister et al. 2015). The proposed target of this synthetic lethal interaction is RRM2, a ribonucleotide reductase subunit. SETD2 deficiency and WEE1 inhibition each decreased RRM2 levels, and the combination resulted in further depletion. WEE1 inhibition in the context of SETD2 deficiency critically reduces the dNTP pool causing cells to accumulate in non-replicating S phase, replication stress, and cell death.

CONCLUDING REMARKS

Large sequencing studies have increasingly implicated mutations in epigenetic modifiers as critical events in cancer development and have identified *SETD2* loss as a key feature of multiple types of cancer. SETD2 has been implicated in many chromatin-directed nuclear processes, including transcriptional regulation, DNA damage repair, DNA methylation, and replication. These effects are likely mediated by H3K36me3-binding reader proteins. Consequently, shifts in H3K36 di- and trimethylation are expected to lead to loss of appropriate reader targeting or redistribution. The relative importance of the H3K36-associated functions to cancer development remains unclear. Although the focus of SETD2 research has been primarily on histone regulation, it is possible that SETD2 may have important nonhistone targets.

As SETD2 mutation is associated with more aggressive cancer, it is important to fully understand the effect of SETD2 loss on oncogenesis. Vulnerabilities created by SETD2 through deregulated transcription and DNA replication may offer therapeutic strategies.

ACKNOWLEDGMENTS

We thank Brian D. Strahl for critical reading and feedback. This project is supported by University of North Carolina (UNC) Medical Scientist Training Program T32 GM008719-12 and National Research Service Award (NRSA) award F30 CA192643-02 (C.C.F.) and R01 CA198482 and R01 CA166447 (I.J.D.).

REFERENCES

Al Sarakbi W, Sasi W, Jiang WG, Roberts T, Newbold RF, Mokbel K. 2009. The mRNA expression of SETD2 in human breast cancer: Correlation with clinico-pathological parameters. *BMC Cancer* **9:** 290.

Awwad SW, Ayoub N. 2015. Overexpression of KDM4 lysine demethylases disrupts the integrity of the DNA mismatch repair pathway. *Biol Open* **4:** 498–504.

Aymard F, Bugler B, Schmidt CK, Guillou E, Caron P, Briois S, Iacovoni JS, Daburon V, Miller KM, Jackson SP, et al. 2014. Transcriptionally active chromatin recruits homologous recombination at DNA double-strand breaks. *Nat Struct Mol Biol* **21:** 366–374.

Bannister AJ, Schneider R, Myers F a, Thorne AW, Crane-Robinson C, Kouzarides T. 2005. Spatial distribution of di- and tri-methyl lysine 36 of histone H3 at active genes. *J Biol Chem* **280:** 17732–17736.

Baubec T, Colombo DF, Wirbelauer C, Schmidt J, Burger L, Krebs AR, Akalin A, Schübeler D. 2015. Genomic profiling of DNA methyltransferases reveals a role for DNMT3B in genic methylation. *Nature* **520:** 243–247.

Behjati S, Tarpey PS, Presneau N, Scheipl S, Pillay N, Van Loo P, Wedge DC, Cooke SL, Gundem G, Davies H, et al. 2013. Distinct H3F3A and H3F3B driver mutations define chondroblastoma and giant cell tumor of bone. *Nat Genet* **45:** 1479–1482.

Berger MF, Hodis E, Heffernan TP, Deribe YL, Lawrence MS, Protopopov A, Ivanova E, Watson IR, Nickerson E, Ghosh P, et al. 2012. Melanoma genome sequencing reveals frequent PREX2 mutations. *Nature* **485:** 502–506.

Carrozza MJ, Li B, Florens L, Suganuma T, Swanson SK, Lee KK, Shia W-J, Anderson S, Yates J, Washburn MP, et al. 2005. Histone H3 methylation by Set2 directs deacetylation of coding regions by Rpd3S to suppress spurious intragenic transcription. *Cell* **123:** 581–592.

Carvalho S, Raposo AC, Martins FB, Grosso AR, Sridhara SC, Rino J, Carmo-Fonseca M, de Almeida SF. 2013. Histone methyltransferase SETD2 coordinates FACT recruitment with nucleosome dynamics during transcription. *Nucleic Acids Res* **41:** 2881–2893.

Carvalho S, Vítor AACA, Sridhara SCS, Martins FB, Raposo AC, Desterro JMP, Ferreira J, de Almeida SF, Filipa BM, Ana CR, et al. 2014. SETD2 is required for DNA double-strand break repair and activation of the p53-mediated checkpoint. *eLife* **3:** e02482.

Cerami E, Gao J, Dogrusoz U, Gross BE, Sumer SO, Aksoy BA, Jacobsen A, Byrne CJ, Heuer ML, Larsson E, et al. 2012. The cBio Cancer Genomics Portal: An open platform for exploring multidimensional cancer genomics data. *Cancer Discov* **2:** 401–404.

Chowdhury R, Yeoh KK, Tian Y, Hillringhaus L, Bagg EA, Rose NR, Leung IKH, Li XS, Woon ECY, Yang M, et al. 2011. Scientific report. *Nat Publ Gr* **12:** 463–469.

Dalgliesh GL, Furge K, Greenman C, Chen L, Bignell G, Butler A, Davies H, Edkins S, Hardy C, Latimer C, et al. 2010. Systematic sequencing of renal carcinoma reveals inactivation of histone modifying genes. *Nature* **463:** 360–363.

Dang L, White DW, Gross S, Bennett BD, Bittinger MA, Driggers EM, Fantin VR, Jang HG, Jin S, Keenan MC, et al. 2009. Cancer-associated IDH1 mutations produce 2-hydroxyglutarate. *Nature* **462:** 739–744.

Daugaard M, Baude A, Fugger K, Povlsen LK, Beck H, Sørensen CS, Petersen NHT, Sorensen PHB, Lukas C, Bartek J, et al. 2012. LEDGF (p75) promotes DNA-end resection and homologous recombination. *Nat Struct Mol Biol* **19:** 803–810.

de Almeida SF, Grosso AR, Koch F, Fenouil R, Carvalho S, Andrade J, Levezinho H, Gut M, Eick D, Gut I, et al. 2011. Splicing enhances recruitment of methyltransferase HYPB/Setd2 and methylation of histone H3 Lys36. *Nat Struct Mol Biol* **18:** 977–983.

Dhayalan A, Rajavelu A, Rathert P, Tamas R, Jurkowska RZ, Ragozin S, Jeltsch A. 2010. The Dnmt3a PWWP domain reads histone 3 lysine 36 trimethylation and guides DNA methylation. *J Biol Chem* **285:** 26114–26120.

Duns G, Hofstra RMW, Sietzema JG, Hollema H, van Duivenbode I, Kuik A, Giezen C, Jan O, Bergsma J, Bijnen H, et al. 2012. Targeted exome sequencing in clear cell renal cell carcinoma tumors suggests aberrant chromatin regulation as a crucial step in ccRCC development. *Hum Mutat* **33:** 1059–1062.

Edmunds JW, Mahadevan LC, Clayton AL. 2008. Dynamic histone H3 methylation during gene induction: HYPB/Setd2 mediates all H3K36 trimethylation. *EMBO J* **27:** 406–420.

Fang D, Gan H, Lee J, Han J, Wang Z, Riester SM, Jin L, Chen J, Zhou H, Wang J, et al. 2016. The histone H3.3K36M mutation reprograms the epigenome of chondroblastomas. *Science* **352:** 1344–1348.

Feng C, Ding G, Jiang H, Ding Q, Wen H. 2015. Loss of MLH1 confers resistance to PI3Kβ inhibitors in renal clear cell carcinoma with SETD2 mutation. *Tumour Biol* **36:** 3457–3464.

Fnu S, Williamson EA, De Haro LP, Brenneman M, Wray J, Shaheen M, Radhakrishnan K, Lee SH, Nickoloff JA, Hromas R. 2011. Methylation of histone H3 lysine 36 enhances DNA repair by nonhomologous end-joining. *Proc Natl Acad Sci* **108:** 540–545.

Fontebasso AM, Schwartzentruber J, Khuong-Quang DA, Liu XY, Sturm D, Korshunov A, Jones DTW, Witt H, Kool M, Albrecht S, et al. 2013. Mutations in SETD2 and genes affecting histone H3K36 methylation target hemispheric high-grade gliomas. *Acta Neuropathol* **125:** 659–669.

Gao J, Aksoy BA, Dogrusoz U, Dresdner G, Gross B, Sumer SO, Sun Y, Jacobsen A, Sinha R, Larsson E, et al. 2013. Integrative analysis of complex cancer genomics and clinical profiles using the cBioPortal. *Sci Signal* **6:** pl1.

Gao YG, Yang H, Zhao J, Jiang YJ, Hu HY. 2014. Autoinhibitory structure of the WW domain of HYPB/SETD2 regulates its interaction with the proline-rich region of huntingtin. *Structure* **22:** 378–386.

Gerlinger M, Rowan AJ, Horswell S, Larkin J, Endesfelder D, Gronroos E, Martinez P, Matthews N, Stewart A, Tarpey P, et al. 2012. Intratumor heterogeneity and branched evolution revealed by multiregion sequencing. *N Engl J Med* **366:** 883–892.

Gossage L, Murtaza M, Slatter AF, Lichtenstein CP, Warren A, Haynes B, Marass F, Roberts I, Shanahan SJ, Claas A, et al. 2014. Clinical and pathological impact of VHL, PBRM1, BAP1, SETD2, KDM6A, and JARID1c in clear cell renal cell carcinoma. *Genes Chromosomes Cancer* **53:** 38–51.

Grosso AR, Leite AP, Carvalho S, Matos MR, Martins FB, Vítor AC, Desterro JMP, Carmo-Fonseca M, de Almeida SF. 2015. Pervasive transcription read-through promotes aberrant expression of oncogenes and RNA chimeras in renal carcinoma. *eLife* **4:** 1–16.

Guo R, Zheng L, Park JW, Lv R, Chen H, Jiao F, Xu W, Mu S, Wen H, Qiu J, et al. 2014. BS69/ZMYND11 reads and connects histone H3.3 Lysine 36 trimethylation-decorated chromatin to regulated pre-mRNA processing. *Mol Cell* **56:** 298–310.

Hahn MA, Wu X, Li AX, Hahn T, Pfeifer GP. 2011. Relationship between gene body DNA methylation and intragenic H3K9me3 and H3K36me3 chromatin marks. *PLoS ONE* **6:** e18844.

Hakimi AA, Chen Y-B, Wren J, Gonen M, Abdel-Wahab O, Heguy A, Liu H, Takeda S, Tickoo SK, Reuter VE, et al. 2013a. Clinical and pathologic impact of select chromatin-modulating tumor suppressors in clear cell renal cell carcinoma. *Eur Urol* **63:** 848–854.

Hakimi AA, Ostrovnaya I, Reva B, Schultz N, Chen Y-B, Gonen M, Liu H, Takeda S, Voss MH, Tickoo SK, et al. 2013b. Adverse outcomes in clear cell renal cell carcinoma with mutations of 3p21 epigenetic regulators BAP1 and SETD2: A report by MSKCC and the KIRC TCGA research network. *Clin Cancer Res* **19:** 3259–3267.

Ho TH, Park IY, Zhao H, Tong P, Champion MD, Yan H, Monzon F, Hoang, Tamboli P, Parker S, et al. 2015. High-resolution profiling of histone h3 lysine 36 trimethylation in metastatic renal cell carcinoma. *Oncogene* **35:** 1565–1574.

Hodis E, Watson IR, Kryukov G V, Arold ST, Imielinski M, Theurillat JP, Nickerson E, Auclair D, Li L, Place C, et al. 2012. A landscape of driver mutations in melanoma. *Cell* **150:** 251–263.

Huang KK, McPherson JR, Tay ST, Das K, Tan IB, Ng CCY, Chia N-Y, Zhang SL, Myint SS, Hu L, et al. 2015. SETD2 histone modifier loss in aggressive GI stromal tumours. *Gut* doi: 10.1136/gutjnl-2015-309482.

Imielinski M, Berger AH, Hammerman PS, Hernandez B, Pugh TJ, Hodis E, Cho J, Suh J, Capelletti M, Sivachenko A, et al. 2012. Mapping the hallmarks of lung adenocarcinoma with massively parallel sequencing. *Cell* **150:** 1107–1120.

Jones S, Stransky N, McCord CL, Cerami E, Lagowski J, Kelly D, Angiuoli S V, Sausen M, Kann L, Shukla A, et al. 2014. Genomic analyses of gynaecologic carcinosarcomas reveal frequent mutations in chromatin remodelling genes. *Nat Commun* **5:** 5006.

Jung H, Lee D, Lee J, Park D, Kim YJ, Park WY, Hong D, Park PJ, Lee E. 2015. Intron retention is a widespread mechanism of tumor-suppressor inactivation. *Nat Genet* **47:** 1242–1248.

Cite this article as *Cold Spring Harb Perspect Med* doi: 10.1101/cshperspect.a026468

Kanu N, Grönroos E, Martinez P, Burrell Ra, Yi Goh X, Bartkova J, Maya-Mendoza A, Mistrík M, Rowan J, Patel H, et al. 2015. SETD2 loss-of-function promotes renal cancer branched evolution through replication stress and impaired DNA repair. *Oncogene* **34:** 5699–5708.

Keogh MC, Kurdistani SK, Morris SA, Ahn SH, Podolny V, Collins SR, Schuldiner M, Chin K, Punna T, Thompson NJ, et al. 2005. Cotranscriptional set2 methylation of histone H3 lysine 36 recruits a repressive Rpd3 complex. *Cell* **123:** 593–605.

Kim S, Kim H, Fong N, Erickson B, Bentley DL. 2011. Pre-mRNA splicing is a determinant of histone H3K36 methylation. *Proc Natl Acad Sci* **108:** 13564–13569.

Kolasinska-Zwierz P, Down T, Latorre I, Liu T, Liu XS, Ahringer J. 2009. Differential chromatin marking of introns and expressed exons by H3K36me3. *Nat Genet* **41:** 376–381.

Krogan NJ, Kim M, Tong A, Golshani A, Cagney G, Canadien V, Richards DP, Beattie BK, Emili A, Boone C, et al. 2003. Methylation of histone H3 by Set2 in Saccharomyces cerevisiae is linked to transcriptional elongation by RNA polymerase II. *Mol Cell Biol* **23:** 4207–4218.

Kurotaki N, Imaizumi K, Harada N, Masuno M, Kondoh T, Nagai T, Ohashi H, Naritomi K, Tsukahara M, Makita Y, et al. 2002. Haploinsufficiency of NSD1 causes Sotos syndrome. *Nat Genet* **30:** 365–366.

Lapunzina P, Cohen MM. 2005. Risk of tumorigenesis in overgrowth syndromes: A comprehensive review. *Am J Med Genet-Semin Med Genet* **137C:** 53–71.

Lee JJ, Sholl LM, Lindeman NI, Granter SR, Laga AC, Shivdasani P, Chin G, Luke JJ, Ott PA, Hodi FS, et al. 2015. Targeted next-generation sequencing reveals high frequency of mutations in epigenetic regulators across treatment-naïve patient melanomas. *Clin Epigenetics* **7:** 59.

Lewis PW, Müller MM, Koletsky MS, Cordero F, Lin S, Banaszynski L a, Garcia B a, Muir TW, Becher OJ, Allis CD. 2013. Inhibition of PRC2 activity by a gain-of-function H3 mutation found in pediatric glioblastoma. *Science* **340:** 857–861.

Li B, Howe L, Anderson S, Yates JR, Workman JL. 2003. The Set2 histone methyltransferase functions through the phosphorylated carboxyl-terminal domain of RNA polymerase II. *J Biol Chem* **278:** 8897–8903.

Li M, Phatnani HP, Guan Z, Sage H, Greenleaf AL, Zhou P. 2005. Solution structure of the Set2-Rpb1 interacting domain of human Set2 and its interaction with the hyperphosphorylated C-terminal domain of Rpb1. *Proc Natl Acad Sci* **102:** 17636–17641.

Li B, Gogol M, Carey M, Pattenden SG, Seidel C, Workman JL. 2007. Infrequently transcribed long genes depend on the Set2/Rpd3S pathway for accurate transcription. *Genes Dev* **21:** 1422–1430.

Li F, Mao G, Tong D, Huang J, Gu L, Yang W, Li G-M. 2013. The histone mark H3K36me3 regulates human DNA mismatch repair through its interaction with MutSα. *Cell* **153:** 590–600.

Li YY, Hanna GJ, Laga AC, Haddad RI, Lorch JH, Hammerman PS. 2015. Genomic analysis of metastatic cutaneous squamous cell carcinoma. *Clin Cancer Res* **21:** 1447–1456.

Li J, Kluiver J, Osinga J, Westers H, van Werkhoven MB, Seelen MA, Sijmons RH, van den Berg A, Kok K. 2016.

Functional studies on primary tubular epithelial cells indicate a tumor suppressor role of SETD2 in clear cell renal cell carcinoma. *Neoplasia* **18:** 339–346.

Lickwar CR, Rao B, Shabalin Aa, Nobel AB, Strahl BD, Lieb JD. 2009. The set2/Rpd3S pathway suppresses cryptic transcription without regard to gene length or transcription frequency. *PLoS ONE* **4:** e4886.

Liu S-Y, Joseph NM, Ravindranathan A, Stohr BA, Greenland NY, Vohra P, Hosfield E, Yeh I, Talevich E, Onodera C, et al. 2016. Genomic profiling of malignant phyllodes tumors reveals aberrations in FGFR1 and PI-3 kinase/RAS signaling pathways and provides insights into intratumoral heterogeneity. *Mod Pathol* **29:** 1012–1027.

Lu PJ, Zhou XZ, Shen M, Lu KP. 1999. Function of WW domains as phosphoserine- or phosphothreonine-binding modules. *Science* **283:** 1325–1328.

Lu C, Ward P, Kapoor G, Rohle D. 2012. IDH mutation impairs histone demethylation and results in a block to cell differentiation. *Nature* **483:** 474–478.

Lu C, Jain SU, Hoelper D, Bechet D, Molden RC, Ran L, Murphy D, Venneti S, Hameed M, Pawel BR, et al. 2016. Histone H3K36 mutations promote sarcomagenesis through altered histone methylation landscape. *Science* **352:** 844–849.

Luco RF, Pan Q, Tominaga K, Blencowe BJ, Pereira-Smith OM, Misteli T. 2010. Regulation of alternative splicing by histone modifications. *Science* **327:** 996–1000.

Luscan A, Laurendeau I, Malan V, Francannet C, Odent S, Giuliano F, Lacombe D, Touraine R, Vidaud M, Pasmant E, et al. 2014. Mutations in SETD2 cause a novel overgrowth condition. *J Med Genet* **51:** 512–517.

Mar BG, Bullinger LB, McLean KM, Grauman P V, Harris MH, Stevenson K, Neuberg DS, Sinha AU, Sallan SE, Silverman LB, et al. 2014. Mutations in epigenetic regulators including SETD2 are gained during relapse in paediatric acute lymphoblastic leukaemia. *Nat Commun* **5:** 3469.

Milne TA, Kim J, Wang GG, Stadler SC, Basrur V, Whitcomb SJ, Wang Z, Ruthenburg AJ, Elenitoba-Johnson KSJ, Roeder RG, et al. 2010. Multiple interactions recruit MLL1 and MLL1 fusion proteins to the HOXA9 locus in leukemogenesis. *Mol Cell* **38:** 853–863.

Newbold RF, Mokbel K. 2010. Evidence for a tumour suppressor function of SETD2 in human breast cancer: A new hypothesis. *Anticancer Res* **30:** 3309–3311.

Parker H, Rose-Zerilli MJJ, Larrayoz M, Clifford R, Edelmann J, Blakemore S, Gibson J, Wang J, Ljungström V, Wojdacz TK, et al. 2016. Genomic disruption of the histone methyltransferase SETD2 in chronic lymphocytic leukaemia. *Leukemia* doi: 10.1038/leu.2016.134.

Pfister SX, Ahrabi S, Zalmas LP, Sarkar S, Aymard F, Bachrati CZ, Helleday T, Legube G, LaThangue NB, Porter ACG, et al. 2014. SETD2-dependent histone H3K36 trimethylation is required for homologous recombination repair and genome stability. *Cell Rep* **7:** 2006–2018.

Pfister SX, Markkanen E, Jiang Y, Sarkar S, Woodcock M, Orlando G, Mavrommati I, Pai C-C, Zalmas L-P, Drobnitzky N, et al. 2015. Inhibiting WEE1 selectively kills histone H3K36me3-deficient cancers by dNTP starvation. *Cancer Cell* **28:** 557–568.

Pradeepa MM, Sutherland HG, Ule J, Grimes GR, Bickmore W a. 2012. Psip1/Ledgf p52 binds methylated histone

H3K36 and splicing factors and contributes to the regulation of alternative splicing. *PLoS Genet* **8:** e1002717.

Schwartz S, Meshorer E, Ast G. 2009. Chromatin organization marks exon–intron structure. *Nat Struct Mol Biol* **16:** 990–995.

Schwartzentruber J, Korshunov A, Liu XY, Jones DTW, Pfaff E, Jacob K, Sturm D, Fontebasso AM, Quang DAK, Tönjes M, et al. 2012. Driver mutations in histone H3.3 and chromatin remodelling genes in paediatric glioblastoma. *Nature* **482:** 226–231.

Seshagiri S, Stawiski EW, Durinck S, Modrusan Z, Storm EE, Conboy CB, Chaudhuri S, Guan Y, Janakiraman V, Jaiswal BS, et al. 2012. Recurrent R-spondin fusions in colon cancer. *Nature* **488:** 660–4.

Shain AH, Garrido M, Botton T, Talevich E, Yeh I, Sanborn JZ, Chung J, Wang NJ, Kakavand H, Mann GJ, et al. 2015. Exome sequencing of desmoplastic melanoma identifies recurrent NFKBIE promoter mutations and diverse activating mutations in the MAPK pathway. *Nat Genet* **47:** 1194–1199.

Simon JM, Hacker KE, Singh D, Brannon a R, Parker JS, Weiser M, Ho TH, Kuan PF, Jonasch E, Furey TS, et al. 2014. Variation in chromatin accessibility in human kidney cancer links H3K36 methyltransferase loss with widespread RNA processing defects. *Genome Res* **24:** 241–250.

Singh DP, Kimura A, Chylack LT, Shinohara T. 2000. Lens epithelium-derived growth factor (LEDGF/p75) and p52 are derived from a single gene by alternative splicing. *Gene* **242:** 265–273.

Stolle C, Glenn G, Zbar B, Humphrey JS, Choyke P, Walther M, Pack S, Hurley K, Andrey C, Klausner R, et al. 1998. Improved detection of germline mutations in the von Hippel–Lindau disease tumor suppressor gene. *Hum Mutat* **12:** 417–23.

Strahl BD, Grant PA, Briggs SD, Sun ZW, Bone JR, Caldwell JA, Mollah S, Cook RG, Shabanowitz J, Hunt DF, et al. 2002. Set2 is a nucleosomal histone H3-selective methyltransferase that mediates transcriptional repression. *Mol Cell Biol* **22:** 1298–1306.

Sturm D, Witt H, Hovestadt V, Khuong-Quang DA, Jones DTW, Konermann C, Pfaff E, Tönjes M, Sill M, Bender S, et al. 2012. Hotspot mutations in H3F3A and IDH1 define distinct epigenetic and biological subgroups of glioblastoma. *Cancer Cell* **22:** 425–437.

Sun XJ, Wei J, Wu XY, Hu M, Wang L, Wang HH, Zhang QH, Chen SJ, Huang QH, Chen Z. 2005. Identification and characterization of a novel human histone H3 lysine 36-specific methyltransferase. *J Biol Chem* **280:** 35261–35271.

Symington LS. 2010. Mechanism and regulation of DNA end resection in eukaryotes. *Crit Rev Biochem Mol Biol* **51:** 195–212.

Tan J, Ong CK, Lim WK, Ng CCY, Thike AA, Ng LM, Rajasegaran V, Myint SS, Nagarajan S, Thangaraju S, et al. 2015. Genomic landscapes of breast fibroepithelial tumors. *Nat Genet* **47:** 1341–1345.

The Cancer Genome Atlas. 2012. Comprehensive molecular characterization of human colon and rectal cancer. *Nature* **487:** 330–337.

The Cancer Genome Atlas. 2013a. Comprehensive molecular characterization of clear cell renal cell carcinoma. *Nature* **499:** 43–49.

The Cancer Genome Atlas. 2013b. Genomic and epigenomic landscapes of adult de novo acute myeloid leukemia. *N Engl J Med* **368:** 2059–2074.

The Cancer Genome Atlas. 2014a. Comprehensive molecular characterization of gastric adenocarcinoma. *Nature* **513:** 202–209.

The Cancer Genome Atlas. 2014b. Comprehensive molecular characterization of urothelial bladder carcinoma. *Nature* **507:** 315–322.

The Cancer Genome Atlas. 2014c. Comprehensive molecular profiling of lung adenocarcinoma. *Nature* **511:** 543–550.

The Cancer Genome Atlas; Kandoth C, Schultz N, Cherniack AD, Akbani R, Liu Y, Shen H, Robertson AG, Pashtan I, Shen R, et al. 2013. Integrated genomic characterization of endometrial carcinoma. *Nature* **497:** 67–73.

The Cancer Genome Atlas; Linehan WM, Spellman PT, Ricketts CJ, Creighton CJ, Fei SS, Davis C, Wheeler DA, Murray BA, Schmidt L, et al. 2016. Comprehensive molecular characterization of papillary renal-cell carcinoma. *N Engl J Med* **374:** 135–145.

Tiedemann RL, Hlady RA, Hanavan PD, Lake DF, Tibes R, Lee J-H, Choi J, Ho TH, Robertson KD. 2016. Dynamic reprogramming of DNA methylation in SETD2-deregulated renal cell carcinoma. *Oncotarget* **7:** 1927–1946.

Tlemsani C, Luscan A, Leulliot N, Bieth E, Afenjar A, Baujat G, Doco-Fenzy M, Goldenberg A, Lacombe D, Lambert L, et al. 2016. *SETD2* and *DNMT3A* screen in the Sotos-like syndrome French cohort. *J Med Genet* doi: 10.1136/jmedgenet-2015-103638.

Van Allen EM, Mouw KW, Kim P, Iyer G, Wagle N, Al-Ahmadie H, Zhu C, Ostrovnaya I, Kryukov GV, O'Connor KW, et al. 2014. Somatic ERCC2 mutations correlate with cisplatin sensitivity in muscle-invasive urothelial carcinoma. *Cancer Discov* **4:** 1140–1153.

Venkatesh S, Smolle M, Li H, Gogol MM, Saint M, Kumar S, Natarajan K, Workman JL. 2012. Set2 methylation of histone H3 lysine 36 suppresses histone exchange on transcribed genes. *Nature* **489:** 452–455.

Wagner EJ, Carpenter PB. 2012. Understanding the language of Lys36 methylation at histone H3. *Nat Rev Mol Cell Biol* **13:** 115–126.

Wang J, Liu L, Qu Y, Xi W, Xia Y, Bai Q, Xiong Y, Long Q, Xu J, Guo J. 2016. Prognostic value of SETD2 expression in patients with metastatic renal cell carcinoma treated with tyrosine kinase inhibitors. *J Urol* **195:** 1363–1370.

Witkiewicz AK, McMillan EA, Balaji U, Baek G, Lin WC, Mansour J, Mollaee M, Wagner KU, Koduru P, Yopp A, et al. 2015. Whole-exome sequencing of pancreatic cancer defines genetic diversity and therapeutic targets. *Nat Commun* **6:** 6744.

Wu G, Broniscer A, McEachron TA, Lu C, Paugh BS, Becksfort J, Qu C, Ding L, Huether R, Parker M, et al. 2012. Somatic histone H3 alterations in pediatric diffuse intrinsic pontine gliomas and non-brainstem glioblastomas. *Nat Genet* **44:** 251–253.

Wu G, Diaz AK, Paugh BS, Rankin SL, Ju B, Li Y, Zhu X, Qu C, Chen X, Zhang JJ, et al. 2014. The genomic landscape of diffuse intrinsic pontine glioma and pediatric non-brainstem high-grade glioma. *Nat Genet* **46:** 444–450.

Xiang W, He J, Huang C, Chen L, Tao D, Wu X. 2015. miR-106b-5p targets tumor suppressor gene SETD2 to inactive its function in clear cell renal cell carcinoma. *Oncotarget* **6:** 4066–4079.

Xiao T, Hall H, Kizer KO, Shibata Y, Hall MC, Borchers CH, Strahl BD. 2003. Phosphorylation of RNA polymerase II CTD regulates H3 methylation in yeast. *Genes Dev* **17:** 654–663.

Xie P, Tian C, An L, Nie J, Lu K, Xing G, Zhang L, He F. 2008. Histone methyltransferase protein SETD2 interacts with p53 and selectively regulates its downstream genes. *Cell Signal* **20:** 1671–1678.

Xie L, Pelz C, Wang W, Bashar A, Varlamova O, Shadle S, Impey S. 2011. KDM5B regulates embryonic stem cell self-renewal and represses cryptic intragenic transcription. *EMBO J* **30:** 1473–1484.

Xu W, Yang H, Liu Y, Yang Y, Wang P, Kim S, Ito S, Yang C, Wang P. 2011. Oncometabolite 2-hydroxyglutarate is a competitive inhibitor of α-ketoglutarate-dependent dioxygenases. *Cancer Cell* **19:** 17–30.

Yoh SM, Lucas JS, Jones KA. 2008. The Iws1:Spt6:CTD complex controls cotranscriptional mRNA biosynthesis and HYPB/Setd2-mediated histone H3K36 methylation. *Genes Dev* **22:** 3422–3434.

Yokoyama A, Lin M, Naresh A, Kitabayashi I, Cleary ML. 2010. A higher-order complex containing AF4 and ENL family proteins with P-TEFb facilitates oncogenic and physiologic MLL-dependent transcription. *Cancer Cell* **17:** 198–212.

Yuan W, Xie J, Long C, Erdjument-Bromage H, Ding X, Zheng Y, Tempst P, Chen S, Zhu B, Reinberg D. 2009. Heterogeneous nuclear ribonucleoprotein L is a subunit of human KMT3a/set2 complex required for H3 Lys-36 trimethylation activity in vivo. *J Biol Chem* **284:** 15701–15707.

Zbar B, Brauch H, Talmadge C, Linehan M. 1987. Loss of alleles of loci on the short arm of chromosome 3 in renal cell carcinoma. *Nature* **327:** 721–724.

Zhang P, Du J, Sun B, Dong X, Xu G, Zhou J, Huang Q, Liu Q, Hao Q, Ding J. 2006. Structure of human MRG15 chromo domain and its binding to Lys36-methylated histone H3. *Nucleic Acids Res* **34:** 6621–6628.

Zhang J, Ding L, Holmfeldt L, Wu G, Heatley SL, Payne-Turner D, Easton J, Chen X, Wang J, Rusch M, et al. 2012. The genetic basis of early T-cell precursor acute lymphoblastic leukaemia. *Nature* **481:** 157–163.

Zhang Y, Xie S, Zhou Y, Xie Y, Liu P, Sun M, Xiao H, Jin Y, Sun X, Chen Z, et al. 2014. H3K36 histone methyltransferase Setd2 is required for murine embryonic stem cell differentiation toward endoderm. *Cell Rep* **8:** 1989–2002.

Zhu X, He F, Zeng H, Ling S, Chen A, Wang Y, Yan X, Wei W, Pang Y, Cheng H, et al. 2014. Identification of functional cooperative mutations of SETD2 in human acute leukemia. *Nat Genet* **46:** 287–293.

ATRX and DAXX: Mechanisms and Mutations

Michael A. Dyer,[1] Zulekha A. Qadeer,[2,3] David Valle-Garcia,[2] and Emily Bernstein[2,3]

[1]Department of Developmental Neurobiology, St. Jude Children's Research Hospital, Memphis, Tennessee 38105

[2]Departments of Oncological Sciences and Dermatology, Icahn School of Medicine at Mount Sinai, New York, New York 10029

[3]Graduate School of Biomedical Sciences, Icahn School of Medicine at Mount Sinai, New York, New York 10029

Correspondence: michael.dyer@stjude.org; emily.bernstein@mssm.edu

Recent genome sequencing efforts in a variety of cancers have revealed mutations and/or structural alterations in *ATRX* and *DAXX*, which together encode a complex that deposits histone variant H3.3 into repetitive heterochromatin. These regions include retrotransposons, pericentric heterochromatin, and telomeres, the latter of which show deregulation in *ATRX/DAXX*-mutant tumors. Interestingly, *ATRX* and *DAXX* mutations are often found in pediatric tumors, suggesting a particular developmental context in which these mutations drive disease. Here we review the functions of ATRX and DAXX in chromatin regulation as well as their potential contributions to tumorigenesis. We place emphasis on the chromatin remodeler ATRX, which is mutated in the developmental disorder for which it is named, α-thalassemia, mental retardation, X-linked syndrome, and at high frequency in a number of adult and pediatric tumors.

Recent whole-genome and exome sequencing efforts across a wide spectrum of cancers have unexpectedly revealed chromatin remodeling-encoding genes as being frequently altered (Jiao et al. 2011; Schwartzentruber et al. 2012; Cheung and Dyer 2013; Helming et al. 2014). The *ATRX* gene, which encodes a SWI/SNF-like chromatin remodeling protein, is frequently mutated in a variety of tumors, including adult lower-grade gliomas, pediatric glioblastoma multiforme, pediatric adrenocortical carcinoma, osteosarcoma, and neuroblastoma (Table 1). Some, but not all, of these cancers also harbor *DAXX* and/or *H3F3A* (H3.3) mutations (Schwartzentruber et al. 2012; Wu et al. 2012).

Given that ATRX and DAXX form a histone chaperone complex that deposits histone variant H3.3 into specific genomic regions, it is intriguing that mutations in the histone as well as its chaperone complex have been identified (Schwartzentruber et al. 2012). Although certain H3.3 mutations appear to be associated with altered Polycomb repressive complex 2 (PRC2) activity (e.g., H3.3K27M) (Bender et al. 2013; Chan et al. 2013; Lewis et al. 2013; Funato et al. 2014), the mechanisms by which *ATRX* and *DAXX* mutations promote oncogenesis may be distinct from those elicited by H3.3 mutation.

Given the identification of *ATRX* and *DAXX* mutations in cancer, these factors have received

Cite this article as *Cold Spring Harb Perspect Med* doi: 10.1101/cshperspect.a026567

Table 1. ATRX alterations identified in human cancers

Disease	Frequency (%)	Type of alteration	Age group	ALT association	Co-occurrence of DAXX/H3.3 mutations	References
Whole-genome and exome studies						
PanNets	19	Deletions, point	Adult	Yes	Yes	Heaphy et al. 2011; Jiao et al. 2011
LGG	42	Indels, point	Adult	Yes	No	Kannan et al. 2012; Johnson et al. 2014; TCGA 2015
GBM	29/16	Indels, point	Pediatric/adult	Yes	Yes/no	Liu et al. 2012b; Schwartzentruber et al. 2012
DIPG	8	Point	Pediatric	N/A	Yes/no	Khuong-Quang et al. 2012; Wu et al. 2014
NB	8/28	Indels, point, in-frame fusions	Pediatric/adolescent	Yes	No	Cheung et al. 2012; Molenaar et al. 2012; Pugh et al. 2013; Peifer et al. 2015; Valentijn et al. 2015
OS	29	Indels, point, in-frame fusions	Pediatric/adolescent	Yes	No	Chen et al. 2014
Adrenocortical tumors	32	Indels, point	Pediatric/adolescent	Yes	Yes	Assié et al. 2014; Pinto et al. 2015

Disease	Frequency (%)	Type of alteration	Age group	ALT association	References
mRNA/protein studies					
Melanoma	N/A	Transcriptional down-regulation, protein loss	Adult	N/A	Liau et al. 2015b
Leiomyosarcoma	33	Protein loss	Adult	Yes	Liau et al. 2015c
Angiosarcoma	18	Protein loss	Adult	Yes	Liau et al. 2015a
Liposarcoma	29	Protein loss	Adult	Yes	Lee et al. 2015
Soft tissue sarcoma	13	Protein loss	Adult	Yes	Liau et al. 2015a,b,c

PanNets, Pancreatic neuroendocrine tumors; LGG, low-grade glioma; GBM, glioblastoma multiforme; DIPG, diffuse intrinsic pontine glioma; NB, neuroblastoma; OS, osteosarcoma; ALT, alternative lengthening of telomeres; N/A, not applicable.

Cite this article as *Cold Spring Harb Perspect Med* doi: 10.1101/cshperspect.a026567

renewed attention, particularly in the field of pediatric oncology. To date, however, their role in promoting pathogenesis in both adult and pediatric tumors remains unclear. Here we review the functions of the chromatin remodeler ATRX and the histone chaperone DAXX in the context of cellular biology, development, and disease.

ATRX SYNDROME

ATRX was first discovered through efforts to identify the genetic lesion that contributes to the α-thalassemia, mental retardation, X-linked (ATRX) syndrome (Gibbons et al. 2003). Patients are characterized during early childhood by distinctive craniofacial features, significant developmental delays, genital anomalies and sterility, microcephaly, and severe intellectual disabilities (Gibbons et al. 2012). Coupled to these features, patients with ATRX syndrome also present with varying degrees of anemia caused by α-thalassemia, a condition caused by deficient α-globin expression (Gibbons et al. 2008). Consistent with a recessive X-linked inheritance pattern, ATRX syndrome is predominant in males. However, there are rare reports of ATRX syndrome in 46XX heterozygous females, which show an uneven pattern of X-inactivation that favors the expression of the mutant allele (Badens et al. 2006b).

Importantly, mutations in *ATRX* are the only genetic lesions that cause ATRX syndrome (Gibbons et al. 2008). This suggests that ATRX is essential for diverse developmental processes across the ectodermal (neural crest and central nervous system [CNS]), mesodermal (bone and erythroid lineage), and endodermal (genitalia) lineages. Interestingly, mutations in genes that encode proteins that interact with ATRX, such as *DAXX*, have not been identified in patients with ATRX syndrome. Moreover, ATRX patients do not have an increased incidence of cancer. Taken together, ATRX plays a unique role in development and raises questions about tissue specificity, as well as its role in promoting cancer.

Considering the monogenic origins of ATRX syndrome, it is important to identify the type of mutations involved and their clinical severity. For example, mutations in the amino-terminal ADD (ATRX-DNTM3-DNMT3L) domain produce more severe psychomotor phenotypes than do mutations in the carboxy-terminal helicase domain (Badens et al. 2006a). Interestingly, ATRX syndrome mutations almost exclusively lie in either of these two domains (Fig. 1A). ATRX's helicase domain has DNA translocase activity via ATP-dependent hydrolysis (Xue et al. 2003; Mitson et al. 2011). In fact, inactivating mutations in this domain results in translocase defects, and one particular disease-causing mutation even uncouples ATP hydrolysis from DNA binding (Mitson et al. 2011). The ADD domain contains a GATA-like domain and a PHD (plant homeodomain) finger, which together recognize H3K9me3 when unmethylated at H3K4 (Dhayalan et al. 2011; Eustermann et al. 2011; Iwase et al. 2011). Interestingly, a nonsense mutation at residue 37 of ATRX is associated with a milder phenotype than that associated with missense mutations in the ADD and helicase domains (Guerrini et al. 2000). The nonsense mutation at residue 37 is spliced out of a subset of transcripts, which partially restores ATRX protein function. Indeed, it has been proposed that virtually all mutations in patients with ATRX syndrome represent hypomorphs caused by protein destabilizing effects of the mutations (Gibbons et al. 2008). Because mutations in numerous cancers occur outside of the ADD and helicase domains and span widely across the *ATRX* coding region (Fig. 1), they have been proposed to be loss-of-function mutations (Watson et al. 2015). Consequently, ATRX syndrome mutations may shed light on our understanding of somatic *ATRX* mutations in cancer.

MOUSE MODELS OF ATRX DEFICIENCY

Consistent with the hypothesis that ATRX is required for embryonic development, *Atrx*-null mice die at E9.5 (Garrick et al. 2006). Moreover, conditional inactivation of *Atrx* in the developing forebrain results in perinatal lethality. The frontal cortex, subiculum, and hippocampus are reduced in size, and the dentate

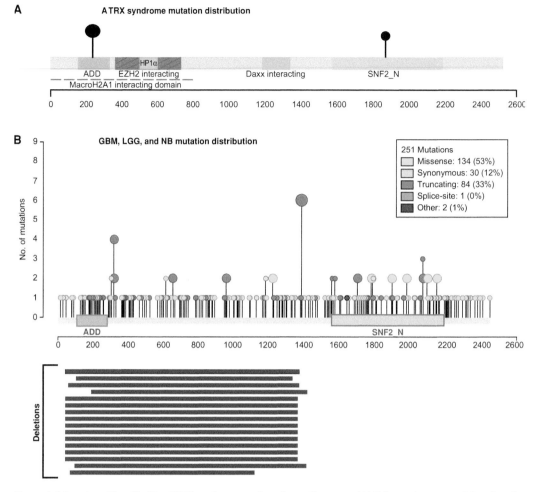

Figure 1. Mutations identified in ATRX syndrome and a subset of cancers. (A) Schematic summarizing that the mutations identified in patients with ATRX syndrome are concentrated in the ADD and SNF2_N helicase domains. Mutations are represented by a black needle; larger needle head size and height indicate larger mutation frequency. (B) Needle plot depicting the distribution of mutations identified in GBM (glioblastoma multiforme), LGG (low-grade glioma), and NB (neuroblastoma) along the ATRX protein product (Gonzalez-Perez et al. 2013). The needles' head size and height indicate mutational recurrence. Note that these mutations are present throughout the gene locus and not concentrated at one particular domain. Blue bars below protein product represent in-frame fusions identified in NB (Cheung et al. 2012).

gyrus is completely absent in *Atrx*^{Lox}*;Foxg1-Cre* mice (Bérubé et al. 2005). There is no defect in neural progenitor cell proliferation, but cells undergo apoptosis as they exit the cell cycle and initiate their differentiation programs. This effect appears to be cell autonomous. Thus, the phenotype in mice is much more severe than that in humans, because patients with *ATRX* mutations can survive to adulthood.

These findings in genetically engineered mouse models are consistent with the hypothesis that almost all *ATRX* alleles in patients with ATRX syndrome represent hypomorphs (Gibbons et al. 2008). To date, hypomorphic alleles of *Atrx* have not been modeled in mice.

There are additional cell-type-specific defects in development upon *Atrx* inactivation. For example, the amacrine and horizontal interneurons are lost in *Atrx*-deficient retinae (Medina et al. 2009). Up to 23% of patients with ATRX syndrome have visual anomalies (Medina et al. 2009), but whether these are attributable to defects in interneuron differentiation and/or survival remains unknown. The consequence of *Atrx* inactivation in skeletal muscle results in defective regenerative capacity of satellite cells (Huh et al. 2012). Skeletal muscle homeostasis requires continuous production of myoblasts from satellite cells, which then differentiate and form myotubes. Myoblasts derived from *Atrx*-deficient satellite cells are defective in cell cycle progression and show evidence of DNA damage. The proliferating myoblasts have elevated levels of both phosphorylated ATM and Chk1, which is indicative of replicative stress, and this effect is cell autonomous. In addition, the telomeres are unstable in *Atrx*-deficient myoblasts and show increased rates of merging, duplication, and bridging (Huh et al. 2012). These findings are consistent with the reported roles of ATRX in human cells (see below).

Collectively, these studies highlight the major mechanisms underlying developmental defects that result from *Atrx* inactivation during development. Specifically, they underscore the lineage-specific defects in differentiation (CNS) and in progenitor cell proliferation as a result of replicative stress-induced DNA damage (muscle). Similar defects have been reported in other cell types and tissues, such as embryonic stem cells, macrophages, Sertoli cells, and limb mesenchyme (Bagheri-Fam et al. 2011; Conte et al. 2012; Solomon et al. 2013; Clynes et al. 2014). It is possible that in lineages such as muscle wherein there is massive proliferative expansion through a stem-cell-related mechanism (i.e., satellite cells), the DNA replication phenotype is predominant. However, in the developing CNS wherein proliferation during development is primarily limited to multipotent progenitor cells, replicative stress might not be as important as lineage−specific defects during differentiation. Thus, the varied roles of Atrx in differentiation can be a starting point to explain the phenotypes across mesodermal, endodermal, and ectodermal lineages. Moreover, it can provide a context for understanding the role of ATRX in cancer.

DEATH DOMAIN−ASSOCIATED PROTEIN DAXX

DAXX was originally identified as a Fas death receptor binding protein that induced apoptosis via JNK pathway activation (Yang et al. 1997). Thus, it was coined the death domain−associated protein, DAXX. However, its role in apoptosis remains unclear because of conflicting reports, including the fact that the knockout mouse resulted in extensive apoptosis and embryonic lethality, rather than the expected antiapoptotic phenotype (Michaelson et al. 1999). Moreover, DAXX was found to be primarily a nuclear protein in which it localizes to promyelocytic leukemia (PML) bodies and was suggested to promote sensitivity to the Fas receptor from its nuclear location (Torii et al. 1999; Zhong et al. 2000). DAXX was also reported to interact with a series of transcription factors (TFs) and histone deacetylase (HDAC) complexes (Hollenbach et al. 1999, 2002; Li et al. 2000). Thus, by the early 2000s, the primary role proposed for DAXX function was in regulating transcription, although the mechanism by which it did so was unclear.

Although ATRX function was studied in the context of ATRX syndrome and DAXX function in the context of apoptosis, two independent biochemical studies showed that ATRX and DAXX are in fact components of a distinct nuclear complex (Xue et al. 2003; Tang et al. 2004). It was not until the purification of histone variant chaperone complexes some years later by the groups of Hamiche and Allis that the function of the ATRX/DAXX complex would be assigned as an H3.3-specfic deposition complex (see below) (Goldberg et al. 2010; Drane et al. 2010; Lewis et al. 2010). Here, we outline the molecular mechanisms of ATRX and DAXX function as a complex as well as their independent functions in chromatin.

ATRX AND DAXX REGULATE HISTONE VARIANT DEPOSITION

Biochemical investigation of H3.3 chaperone complexes identified ATRX and DAXX, and found this complex to be necessary for H3.3

deposition at telomeres (Goldberg et al. 2010; Lewis et al. 2010) and pericentric heterochromatin (Drane et al. 2010) (Fig. 2A). Interestingly, these studies identified a second histone chaperone complex for H3.3, as the HIRA complex had been previously identified to deposit H3.3 at euchromatic regions of the genome, such as transcriptional start sites and gene bodies (Ahmad and Henikoff 2002; Tagami et al. 2004). The studies that identified the role of the ATRX/DAXX complex in H3.3 deposition not only revealed that distinct factors are responsible for H3.3 incorporation at specific genomic locations but also uncovered an unexpected role for H3.3 at heterochromatic regions of the genome.

Within the ATRX/DAXX complex, DAXX was shown as the component containing H3.3 histone chaperone activity (Drane et al. 2010; Lewis et al. 2010). For example, DAXX binds directly to H3.3 and contributes to the deposition of H3.3-H4 tetramers onto naked DNA (Drane et al. 2010; Lewis et al. 2010). Elegant structural studies resolved DAXX in complex with an H3.3/H4 dimer and showed that residue G90 (unique to H3.3 vs. canonical H3), is critical for this interaction (Elsässer et al. 2012; Liu et al. 2012a). Consistent with its role as a chromatin remodeler, ATRX binds to DAXX to incorporate H3.3 into telomeric, pericentromeric, and other repetitive DNA (Fig. 2A) (Drane et al. 2010; Goldberg et al. 2010; Wong 2010). Although the precise role of ATRX in this process remains unclear, it is well established that ATRX can recognize H3K9me3 through its ADD domain and is likely responsible for the recruitment of its binding partner DAXX to such regions.

Before studies implicating ATRX as a component of an H3.3 histone chaperone complex, its role in transcription was studied primarily at the α-globin gene cluster in erythroid cells because of the α-thalassemia phenotype observed in ATRX syndrome patients. In this particular case, ATRX promotes expression of the *HBA* genes at the α-globin locus on the subtelomere of chromosome 16 (Law et al. 2010). Our group identified a role for the transcriptionally repressive H2A variant macroH2A at the α-globin lo-cus and telomeres (Fig. 2A). In the absence of ATRX, macroH2A is found enriched at these regions, concomitant with reduced α-globin expression—a prevalent feature of ATRX syndrome patients. Interestingly, this function for ATRX in negatively regulating macroH2A deposition at the α-globin locus is DAXX-independent (Ratnakumar et al. 2012).

ATRX BINDS TO REPETITIVE REGIONS OF THE GENOME

Initial immunofluorescence studies of ATRX found it to be localized to condensed DAPI-dense regions, suggesting that ATRX is primarily associated with heterochromatin (McDowell et al. 1999; Bérubé et al. 2000; Xue et al. 2003). More recently, ChIP-seq studies identified ATRX enrichment at telomeric, subtelomeric, and pericentric repeats, consistent with the functional studies described above (Law et al. 2010). Additional ChIP-seq analyses found that ATRX is also enriched at the silenced allele of imprinted regions, particular families of retrotransposons, a subset of G-rich intragenic regions, and the 3′ exons of genes belonging to the C2H2 Zinc Finger (ZNFs) family of transcription factors (Elsässer et al. 2015; He et al. 2015; Levy et al. 2015; Sadic et al. 2015; Voon et al. 2015; Valle-Garcia et al. 2016). Some of these recent findings are outlined below.

Telomeres

The ATRX/DAXX complex is enriched at telomeric repeats where it has been shown to regulate H3.3 deposition (Goldberg et al. 2010; Law et al. 2010; Lewis et al. 2010; Wong 2010) (Fig. 2A). Although the precise function of H3.3 deposition at telomeres remains unclear, ATRX depletion leads to telomere dysfunction (Wong 2010; Watson et al. 2013; Ramamoorthy and Smith 2015; Watson et al. 2015). For example, ATRX loss leads to increased DNA damage, increased rates of telomeric end fusions, persistent telomere cohesion, and general genomic instability (Wong 2010; Watson et al. 2013, 2015; Ramamoorthy

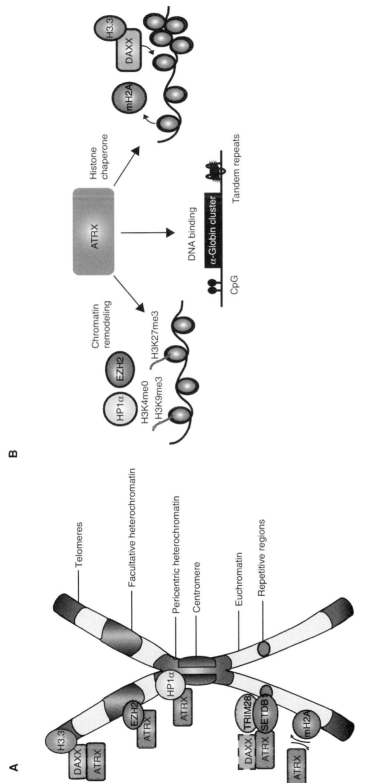

Figure 2. ATRX has multiple functions in the chromatin landscape. (*A*) The SWI/SNF-like chromatin remodeler ATRX is found in various complexes throughout the chromosome to maintain distinct chromatin states. (*B*) Model summarizing ATRX's roles in the epigenetic landscape. (*Left*) ATRX has chromatin remodeling function to bind and regulate genomic regions enriched for H3K9me3 via its ADD domain and HP1α. It also recruits EZH2 to deposit H3K27me3 to regulate transcription. (*Right*) ATRX functions to negatively regulate macroH2A (mH2a) deposition at the α-globin locus and telomeres and is critical in H3.3 deposition along with DAXX, particularly at repetitive regions (e.g., telomeres). (*Center*) ATRX also binds DNA through its SNF2_N helicase domain to resolve tandem repeats to promote transcription at the α-globin locus.

and Smith 2015). Intriguingly, alternative lengthening of telomeres (ALT) is a frequent feature of *ATRX*-mutated cancers (see below), suggesting that ATRX indeed has an important role in the regulation of telomeres.

G4 Structures

ATRX syndrome patients display varying degrees of severity of α-thalassemia (Gibbons 2006). Chromatin immunoprecipitation followed by high-throughput sequencing (ChIP-seq), revealed that ATRX directly binds to G-rich tandem repeats upstream of the α-globin genes (Fig. 2B) (Law et al. 2010). These repeats tend to form G-quadruplexes (G4) structures—secondary DNA structures that can lead to replicative stress (Huh et al. 2012, Leung et al. 2013, Clynes et al. 2014) or block transcriptional processes (Levy et al. 2015). Such structures are typically found in tandem repeats at telomeres and pericentromeric repeats, but not restricted to them (e.g., intragenic CpGs, differentially methylated regions, etc.) (Huppert and Balasubramanian 2005). Using recombinant ATRX protein, in vitro assays showed binding to preformed G4 oligonucleotides, suggesting that this property of ATRX is in fact direct (Law et al. 2010).

As ATRX binds G4 DNA, it has been suggested to resolve these structures (Law et al. 2010; Clynes and Gibbons 2013). Consistent with this, the severity of the α-thalassemia phenotype in ATRX syndrome patients directly correlates with VNTR (variable number tandem repeat) size at the α-globin locus (Law et al. 2010). In the absence of ATRX, larger VNTR regions form more G4 structures with increased impact on the transcription of the α-globin genes. This is corroborated by findings that show the phenotypes associated with ATRX loss are enhanced with compounds that stabilize G4 DNA (Watson et al. 2013; Clynes et al. 2014, 2015). Further, it has been suggested that one of the mechanisms by which ATRX suppresses recombination at telomeric repeats is by resolving G4 structures through the deposition of H3.3 (Clynes and Gibbons 2013; Clynes et al. 2015).

Transposable Elements

The ATRX/DAXX complex also mediates the deposition of H3.3 at retrotransposons, particularly those belonging to the endogenous retrovirus (ERV) family (Elsässer et al. 2015; He et al. 2015; Sadic et al. 2015). The corepressor TRIM28 (also known as KAP1) and the H3K9 histone methyltransferase SETDB1 (also known as ESET), colocalize with ATRX at ERVs (Fig. 2A). ATRX-bound retrotransposons have high levels of H3K9me3 that are dependent on the ATRX/DAXX complex. Knockdown of either component, or H3.3 itself, reduces the H3K9me3 levels at these repetitive regions and induces aberrant transcription (Elsässer et al. 2015; Sadic et al. 2015). Moreover, ATRX/DAXX-mediated silencing of ERVs is critical for preimplantation embryos during the wave of global DNA demethylation (He et al. 2015). Collectively, these studies highlight the critical role of ATRX in silencing repetitive elements through regulation of H3K9me3.

ZNF Genes

In line with these findings, we identified the 3′ exons of ZNF genes as ATRX targets in human cells (Valle-Garcia et al. 2016). ZNFs are the largest family of TFs in the human genome (Nowick et al. 2011). Although these regions are not considered repetitive per se, ZNFs share high levels of DNA similarity at their 3′ exons, which contain motifs encoding for their zinc finger domains. In fact, ZNF genes can encode up to 40 copies of the zinc finger motif. Unlike other ATRX-regulated regions (i.e., telomeric repeats, retrotransposons, etc.), ZNFs are actively transcribed and are highly enriched in both H3K36me3 and H3K9me3 at their 3′ exons. Interestingly, depletion of ATRX decreases H3K9me3 levels at ZNF 3′ exons and other regions presenting an H3K9me3/H3K36me3 atypical chromatin signature. As a consequence, cells with reduced levels of H3K9me3 show increased levels of DNA damage, suggesting that ATRX binds to the 3′ exons of ZNFs to maintain their genomic stability through preservation of H3K9me3.

Cite this article as *Cold Spring Harb Perspect Med* doi: 10.1101/cshperspect.a026567

Overall, a common feature of ATRX targets is the presence of H3K9me3. The loss of H3K9me3 at repetitive regions, via ATRX loss-of-function, appears to promote genomic instability. However, it remains unclear how the loss of ATRX results in reduced H3K9me3 levels. Is ATRX required to maintain H3K9me3 through binding of its ADD domain? Does such docking of ATRX at H3K9me3-modified chromatin facilitate DAXX-mediated H3.3 deposition? What is the role of H3.3 in regulating heterochromatin dynamics? Although all evidence to date suggests that increased genomic instability is a driving feature of *ATRX*-mutated cancers, this important area of research requires further investigation.

ADDITIONAL FUNCTIONS OF ATRX

Before the discovery that ATRX cooperates with DAXX to deposit H3.3 at H3K9me3-containing repetitive regions, ATRX was reported to interact with various heterochromatin-related proteins. These protein interaction partners include HP1α and EZH2, among others (Fig. 2) (Cardoso et al. 1998; Lechner et al. 2005). HP1α was originally identified to interact with ATRX through a yeast two-hybrid screen (Le Douarin et al. 1996) and later shown to colocalize at pericentric heterochromatin (McDowell et al. 1999). Additional recombinant studies identified a variant PxVxL motif within ATRX that binds to the chromoshadow domain of HP1α (Lechner et al. 2005). Curiously, this study and others were unable to copurify these two proteins from cellular extracts (Lechner et al. 2005; Rosnoblet et al. 2011). Thus, although ATRX and HP1α co-occupy similar genomic regions and cooperate in heterochromatin regulation, their interaction may be transient, such as during the recruitment of ATRX to H3K9me3-enriched regions (Eustermann et al. 2011).

A yeast two-hybrid analysis coupled to in vitro binding assays showed that ATRX also binds to EZH2 (Cardoso et al. 1998); however, until recently, this interaction had not been characterized in cells. Sarma et al. (2014) reported that ATRX plays a role in targeting PRC2 (Polycomb repressive complex 2) in

mouse embryonic stem cells, particularly to the inactive X chromosome (Xi). In this case, the ATRX-EZH2 interaction occurs via the *Xist* RNA, which is necessary to recruit ATRX to the Xi and facilitate deposition of H3K27me3. As described above, female ATRX syndrome patients can display skewed X-chromosome inactivation (De la Fuente et al. 2011) and, therefore, H3K27me3 deposition at the Xi (and other genomic regions) may be defective in these patients.

How DAXX plays a role in the above processes remains unclear. ATRX may have particular DAXX-independent roles in the cell. As noted above, *DAXX* mutations have not been identified in patients with ATRX syndrome. Furthermore, there are certain cancers that harbor *ATRX* mutations without evidence of *DAXX* mutations (see below). We previously showed that ATRX regulates macroH2A deposition in a DAXX-independent manner as well (Ratnakumar et al. 2012). Therefore, such DAXX-independent roles of ATRX could mediate critical functions that are perturbed in *ATRX*-mutant cancers.

ATRX/DAXX MUTATIONS IN CANCER

Recent years have witnessed a myriad of whole genome/exome sequencing studies identifying mutations in *ATRX*, and to a lesser extent *DAXX*, across a spectrum of tumor types (Table 1). The first observations came from pancreatic neuroendocrine tumors (PanNETs), in which DNA sequencing studies uncovered frequent mutations in the chromatin factors MEN1, DAXX, and ATRX (Jiao et al. 2011). Notably, *DAXX* and *ATRX* mutations were found to be mutually exclusive, yet were altered in a strikingly high percentage of tumors (43%). These mutations ranged from point mutations to insertion or deletions of bases (indels) and were not localized to any particular domains of *ATRX*, suggesting loss-of-function. In accordance, both *ATRX* and *DAXX* mutations correlated with loss of protein expression by IHC analysis. Intriguingly, these mutations associated with ALT as evidenced by fluorescence in situ hybridization (Heaphy et al. 2011; Jiao et al. 2011).

Subsequently, *ATRX* and *DAXX* mutations were identified in pediatric glioblasoma multiforme (GBM) (Fig. 1B) (Schwartzentruber et al. 2012). Remarkably, point mutations in *H3F3A* leading to critical amino acid substitutions at residues K27 (K27M) or G34 (G34R/V) of the histone tail were also identified in pediatric GBM and diffuse pontine glioma (DIPG), a rare pediatric glioma of the brainstem (Khuong-Quang et al. 2012; Wu et al. 2012). In contrast to pediatric GBM, DIPGs harbor only H3.3K27M mutations (not G34R/V) and are extremely prevalent (70%–80%) (Khuong-Quang et al. 2012; Wu et al. 2012). *ATRX* mutations are found far less frequently (~10%) in DIPG and tend to occur in older children (Khuong-Quang et al. 2012). In the context of pediatric GBM, mutations in the ATRX-DAXX-H3.3 axis occur in ~45% of patients (Schwartzentruber et al. 2012) and show a strong correlation between *ATRX* mutations and ALT (Heaphy et al. 2011; Schwartzentruber et al. 2012).

H3F3A mutations in pediatric cancers represent the first high-frequency mutations in a histone variant to be identified (Khuong-Quang et al. 2012; Schwartzentruber et al. 2012; Wu et al. 2012). Intriguingly, H3.3K27M mutations were not found to frequently co-occur with *ATRX* mutations. Several groups showed that the H3.3K27M mutation induces a dramatic global decrease in H3K27me3 levels. This may be because of sequestration and/or inhibition of the PRC2 subunit, EZH2 (Bender et al. 2013; Chan et al. 2013; Lewis et al. 2013; Funato et al. 2014). This is compelling given that ATRX has also been shown to interact with EZH2 (Cardoso et al. 1998; Sarma et al. 2014) and suggests that both *ATRX* and H3.3K27M mutations could be independently driving Polycomb dysfunction in cancer. In contrast, H3.3G34R/V mutations do co-occur with *ATRX* mutations, suggesting distinct functions of each in tumors harboring these mutations (Schwartzentruber et al. 2012; Wu et al. 2014). Given the young age of patients diagnosed with pediatric GBM and DIPG, *ATRX* mutations may be promoting defects in lineage differentiation in neural stem cells during brain development that culminate in aberrant cell growth and tumor formation. This idea is corroborated by severe defects in the developing brain in the *Atrx*-deficient mouse (Medina et al. 2009) and in ATRX syndrome in which patients present with brain abnormalities and intellectual disabilities (Gibbons et al. 2008).

Studies in adult glioma patient cohorts showed a strong correlation of *ATRX* mutations with *IDH1* and p53 mutations in lower grade gliomas (LGGs) (astrocytomas and oligodendrogliomas) and secondary GBM (Kannan et al. 2012; Liu et al. 2012b). These mutations are mutually exclusive from (1) *FUPB1/CIC* mutations, genes that regulate cell growth and are frequently mutated in oligodendroglioma, and (2) the chromosome 1p/19q co-deletion subtype (Olar and Sulman 2015). Moreover, no *H3F3A* or *DAXX* mutations have been identified in adult GBM (Table 1). Similar to pediatric GBM, there is a strong correlation between *ATRX* mutations and ALT status and such mutations appear to be exclusive from TERT promoter mutations (Liu et al. 2012b; Ceccarelli et al. 2016). These collective findings classify ATRX status to a molecular subtype of adult glioma that is highly recurrent with intermediate prognosis (Olar and Sulman 2015). *IDH1* mutations promote DNA and histone hypermethylation through production of an oncometabolite 2-hydroxygluterate (2-HG) that inhibits DNA and histone demethylases (Lu et al. 2012; Turcan et al. 2012). In adult recurrent gliomas, *ATRX* mutations are sustained in the presence of p53 loss to cooperate with mutant *IDH1* to remodel the epigenetic landscape. In contrast, *Atrx* ablation in adult mouse neurons leads to increased DNA damage and p53-mediated apoptosis (Bérubé et al. 2005), suggesting that ATRX loss in cancer requires additional alterations for survival.

Large-scale efforts to identify mutations in additional pediatric tumors have identified *ATRX* as frequently mutated in neuroblastoma (NB), osteosarcoma (OS), and adrenocortical tumors (Table 1) (Cheung et al. 2012; Assié et al. 2014; Chen et al. 2014; Pinto et al. 2015). Notably, no *H3F3A* and few *DAXX* mutations were detected in these patient populations. In addition to point mutations and indels, these studies also identified large deletions near the amino-terminal region of *ATRX* leading to in-frame fusions of the gene product (Fig. 1B). Notably, such in-frame fusions remove the chromatin binding modules of ATRX including the ADD domain, DAXX binding domain, and the macroH2A1-, HP1α-, and EZH2-interacting regions (Fig. 1B). *ATRX* mutations and in-frame fusions in NB associate with older age at diagnosis (28% found in adolescent patients), with poor prognosis (Table 1) (Cheung et al. 2012).

ATRX syndrome patients show craniofacial abnormalities and developmental delays (Gibbons et al. 2008), supporting the notion that *ATRX* mutations alter proper neural crest cell lineage specification during development. Thus, *ATRX* in-frame fusions and mutations in tumors in cells of this lineage may occur early in multipotent progenitors. The selective pressure to produce in-frame fusions in NB and OS would indicate a possible gain-of-function for these new *ATRX* gene products to promote a cellular growth advantage in such progenitors, although this remains to be tested.

Along with tumor sequencing efforts, several recent studies highlight the loss of *ATRX* mRNA and protein levels in multiple tumor types, including melanoma and various sarcomas (Table 1). These studies showed that as tumors become more advanced or invasive, there is a loss of *ATRX* expression by qPCR and/or IHC analysis (Qadeer et al. 2014; Liau et al. 2015a,b,c; Lee et al. 2015). Many of these studies uncovered a correlation between ATRX loss and ALT and proposed this co-occurrence as a prognostic factor for poor outcome (Lee et al. 2015; Liau et al. 2015a,b,c).

ATRX/DAXX LOSS AND ALT

The ALT pathway in cancer was first described in 1997 (Bryan et al. 1997). A subsequent somatic cell hybridization study showed that when cells displaying ALT were fused to normal cells, the ALT phenotype was reversed (Perrem et al. 1999). This finding suggests that ALT is caused by depletion of a factor(s) that could be rescued by normal cells. As described above,

diverse cancer types with mutations in *ATRX* and *DAXX* are associated with ALT. Approximately 80% of tumors with ALT harbor mutations in *ATRX* or *DAXX*, and conversely 70%–80% of cancers with *ATRX* mutations show ALT (Lovejoy et al. 2012).

The G-rich telomeric repeats can adopt G4 structures, and H3.3 may prevent G4 formation. Consistent with this model, ATRX-mediated suppression of ALT in U-2 OS cells is dependent on DAXX and H3.3 deposition at telomeres (Clynes et al. 2015). Moreover, this suppression is associated with a reduction in replication fork stalling, likely because of limited formation of G4 structures, which are a substrate for the homologous recombination (HR) at telomeres that results in ALT. An independent study suggested that ALT is dependent on sensing of replicative stress by ATR, and that the role of ATRX in recognizing G4 structures and preventing replicative stress is integral to the ALT phenotype in cancer cells lacking ATRX (O'Sullivan et al. 2014). Clynes et al. (2015) have proposed that in the absence of ATRX, DAXX cannot deposit H3.3 at the G-rich repeats found in telomeres, leading to formation of G4 structures, replication fork stalling, and HR of telomeres through the MRE11–RAD50–NBS1 (MRN) complex, which ultimately leads to ALT. The authors also propose that as a secondary level of regulation, ATRX can sequester the MRN complex to prevent it from mediating HR at telomeres (Clynes et al. 2015). In the absence of ATRX, the MRN complex is released to facilitate HR at telomeres. This model is supported by the findings that patients with ATRX syndrome have aberrant chromatin at telomeres and pericentromeric DNA (Clynes and Gibbons 2013; Clynes et al. 2013).

Ramamoorthy and colleagues have proposed an alternative mechanism to explain the contribution of *ATRX* mutations to the ALT phenotype. The telomeric poly(ADP-ribose) polymerase tankyrase 1 is required for the resolution of sister telomere cohesion during mitosis (Dynek and Smith 2004). In the absence of ATRX, the histone variant macroH2A1.1 negatively regulates tankyrase, thus inhibiting its ability to resolve sister telomere cohesion.

This, in turn, allows for aberrant telomere recombination between sister chromatids (Ramamoorthy and Smith 2015).

Although evidence points toward ATRX loss of function driving tumorigenicity through aberrant DNA recombination at repetitive elements, there may be additional functional roles for *ATRX* mutations in cancer. This is supported by the fact that loss of ATRX alone is not sufficient to drive to ALT (Clynes et al. 2015). ATRX loss may promote cancer via transcriptional changes and/or deregulation of the DNA replication machinery, which remains to be tested.

CONCLUDING REMARKS

Given the high frequency of *ATRX* and *DAXX* mutations in cancer, these chromatin regulators likely play a key role in pathogenesis. However, it remains unclear how. Is it through promoting ALT? If loss of ATRX or DAXX alone is insufficient to promote ALT, what other key factors or mutations are required? It also remains to be determined whether histone variant deposition by the ATRX/DAXX complex is related to its tumor suppressive function. Further, we question why some pediatric cancers show mutations in *ATRX*, *DAXX*, and *H3F3A* (e.g., pediatric GBM), whereas others predominantly harbor mutations only in *ATRX* (e.g., NB and OS) Are these mutations patterns as clear-cut as reported? Or does it relate to their cell of origin? In turn, how does ATRX regulate cell fate decisions during development that go awry in cancer?

Given the diversity of mutations identified in *ATRX*, we also question if they are all loss-of-function. For example, in-frame deletions of *ATRX*, such as those found in NB and OS, may have the potential to generate new protein products devoid of critical protein interaction domains. Moreover, as complete absence of ATRX is not well tolerated in development (i.e., CNS), we question whether ATRX alterations are truly loss-of-function or hypomorphic? We also query whether ATRX controls recombination of other repetitive sequences in the genome besides telomeres (e.g., ZNFs), and whether such recombination events play a role

in tumorigenesis. Finally, can our understanding of ATRX and DAXX function be modeled in development and disease to provide clues of their roles in promoting cancer (i.e., mouse models)? To our knowledge, there are currently no in vivo models mimicking *ATRX* or *DAXX* mutations that address these important questions.

Given that ATRX has critical roles in maintenance of repetitive regions (particularly telomeres), as well as cell cycle regulation (Wong 2010; Ramamoorthy and Smith 2015; Watson et al. 2015; Huh et al. 2016), targeting these perturbed pathways could serve as therapeutic strategies for *ATRX*-mutant cancers. For example, Flynn et al. (2015) recently reported that ATRX null, ALT positive cells are sensitive to inhibition of the ATR kinase, a key regulator of homologous recombination. Corroborating these findings, another study showed that ATRX ablation caused increased stalled fork replication, DNA damage and subsequent hyperactivation of PARP-1 and ATM kinase. Thus, it was proposed that PARP-1 inhibitors might further exacerbate DNA damage and promote cell death in ATRX null cells (Huh et al. 2016). We posit that other strategies might include the targeting of epigenetic factors that work together with ATRX, which may also be deregulated, such as SETDB1 or EZH2. Collectively, we anticipate that addressing the important questions raised in this review will provide critical information for deciphering and treating these devastating cancers.

ACKNOWLEDGMENTS

Funding is provided by a graduate fellowship from the National Council of Science and Technology (CONACyT) (239663, CVU 257385) to D.V.-G., the National Cancer Institute (NCI) T32-CA078207 to Z.A.Q., the National Institutes of Health (NIH) EY014867, EY018599, and CA168875, Cancer Center Support from the NCI (CA21765), support from the American Lebanese Syrian Associated Charities (ALSAC), and a grant from Alex's Lemonade Stand Foundation for Childhood Cancer to M.A.D., and St. Baldrick's Foundation to E.B.

REFERENCES

Ahmad K, Henikoff S. 2002. The histone variant H3.3 marks active chromatin by replication-independent nucleosome assembly. *Mol Cell* **9:** 1191–1200.

Assié G, Letouzé E, Fassnacht M, Jouinot A, Luscap W, Barreau O, Omeiri H, Rodriguez S, Perlemoine K, René-Corail F, et al. 2014. Integrated genomic characterization of adrenocortical carcinoma. *Nat Genet* **46:** 607–612.

Badens C, Lacoste C, Philip N, Martini N, Courrier S, Giuliano F, Verloes A, Munnich A, Leheup B, Burglen L, et al. 2006a. Mutations in PHD-like domain of the *ATRX* gene correlate with severe psychomotor impairment and severe urogenital abnormalities in patients with *ATRX* syndrome. *Clin Genet* **70:** 57–62.

Badens C, Martini N, Courrier S, DesPortes V, Touraine R, Levy N, Edery P. 2006b. ATRX syndrome in a girl with a heterozygous mutation in the ATRX Zn finger domain and a totally skewed X-inactivation pattern. *Am J Med Genet A* **140:** 2212–2215.

Bagheri-Fam S, Argentaro A, Svingen T, Combes AN, Sinclair AH, Koopman P, Harley VR. 2011. Defective survival of proliferating Sertoli cells and androgen receptor function in a mouse model of the ATR-X syndrome. *Hum Mol Genet* **20:** 2213–2224.

Bender S, Tang Y, Lindroth AM, Hovestadt V, Jones DTW, Kool M, Zapatka M, Northcott PA, Sturm D, Wang W, et al. 2013. Reduced H3K27me3 and DNA hypomethylation are major drivers of gene expression in K27M mutant pediatric high-grade gliomas. *Cancer Cell* **24:** 660–672.

Bérubé NG, Smeenk CA, Picketts DJ. 2000. Cell cycle-dependent phosphorylation of the ATRX protein correlates with changes in nuclear matrix and chromatin association. *Hum Mol Genet* **9:** 539–547.

Bérubé NG, Mangelsdorf M, Jagla M, Vanderluit J, Garrick D, Gibbons RJ, Higgs DR, Slack RS, Picketts DJ. 2005. The chromatin-remodeling protein ATRX is critical for neuronal survival during corticogenesis. *J Clin Invest* **115:** 258–267.

Bryan TM, Englezou A, Dalla-Pozza L, Dunham MA, Reddel RR. 1997. Evidence for an alternative mechanism for maintaining telomere length in human tumors and tumor-derived cell lines. *Nat Med* **3:** 1271–1274.

Cardoso C, Timsit S, Villard L, Khrestchatisky M, Fontès M, Colleaux L. 1998. Specific interaction between the *XNP/ATR-X* gene product and the SET domain of the human EZH2 protein. *Hum Mol Genet* **7:** 679–684.

Ceccarelli M, Barthel FP, Malta TM, Sabedot TS, Salama SR, Murray BA, Morozova O, Newton Y, Radenbaugh A, Pagnotta SM, et al. 2016. Molecular profiling reveals biologically discrete subsets and pathways of progression in diffuse glioma. *Cell* **164:** 550–563.

Chan K-M, Fang D, Gan H, Hashizume R, Yu C, Schroeder M, Gupta N, Mueller S, James CD, Jenkins R, et al. 2013. The histone H3.3K27M mutation in pediatric glioma reprograms H3K27 methylation and gene expression. *Genes Dev* **27:** 985–990.

Chen X, Bahrami A, Pappo A, Easton J, Dalton J, Hedlund E, Ellison D, Shurtleff S, Wu G, Wei L, et al. 2014. Recurrent somatic structural variations contribute to tumorigenesis in pediatric osteosarcoma. *Cell Rep* **7:** 104–112.

Cheung N-KV, Dyer MA. 2013. Neuroblastoma: Developmental biology, cancer genomics and immunotherapy. *Nat Rev Cancer* **13:** 397–411.

Cheung N-KV, Zhang J, Lu C, Parker M, Bahrami A, Tickoo SK, Heguy A, Pappo AS, Federico S, Dalton J, et al. 2012. Association of age at diagnosis and genetic mutations in patients with neuroblastoma. *JAMA* **307:** 1062–1071.

Clynes D, Gibbons RJ. 2013. ATRX and the replication of structured DNA. *Curr Opin Genet Dev* **23:** 289–294.

Clynes D, Higgs DR, Gibbons RJ. 2013. The chromatin remodeller ATRX: A repeat offender in human disease. *Trends Biochem Sci* **38:** 461–466.

Clynes D, Jelinska C, Xella B, Ayyub H, Taylor S, Mitson M, Bachrati CZ, Higgs DR, Gibbons RJ. 2014. ATRX dysfunction induces replication defects in primary mouse cells. *PLoS ONE* **9:** e92915.

Clynes D, Jelinska C, Xella B, Ayyub H, Scott C, Mitson M, Taylor S, Higgs DR, Gibbons RJ. 2015. Suppression of the alternative lengthening of telomere pathway by the chromatin remodelling factor ATRX. *Nat Commun* **6:** 7538.

Conte D, Huh M, Goodall E, Delorme M, Parks RJ, Picketts DJ. 2012. Loss of Atrx sensitizes cells to DNA damaging agents through p53-mediated death pathways. *PLoS ONE* **7:** e52167.

De La Fuente R, Baumann C, Viveiros MM. 2011. Role of ATRX in chromatin structure and function: Implications for chromosome instability and human disease. *Reproduction* **142:** 221–234.

Dhayalan A, Tamas R, Bock I, Tattermusch A, Dimitrova E, Kudithipudi S, Ragozin S, Jeltsch A. 2011. The ATRX-ADD domain binds to H3 tail peptides and reads the combined methylation state of K4 and K9. *Hum Mol Genet* **20:** 2195–2203.

Drane P, Ouararhni K, Depaux A, Shuaib M, Hamiche A. 2010. The death-associated protein DAXX is a novel histone chaperone involved in the replication-independent deposition of H3.3. *Genes Dev* **24:** 1253–1265.

Dynek JN, Smith S. 2004. Resolution of sister telomere association is required for progression through mitosis. *Science* **304:** 97–100.

Elsässer SJ, Huang H, Lewis PW, Chin JW, Allis CD, Patel DJ. 2012. DAXX envelops a histone H3.3-H4 dimer for H3.3-specific recognition. *Nature* **491:** 560–565.

Elsässer SJ, Noh KM, Diaz N, Allis CD, Banaszynski LA. 2015. Histone H3.3 is required for endogenous retroviral element silencing in embryonic stem cells. *Nature* **522:** 240–244.

Eustermann S, Yang J-C, Law MJ, Amos R, Chapman LM, Jelinska C, Garrick D, Clynes D, Gibbons RJ, Rhodes D, et al. 2011. Combinatorial readout of histone H3 modifications specifies localization of ATRX to heterochromatin. *Nat Struct Mol Biol* **18:** 777–782.

Flynn RL, Cox KE, Jeitany M, Wakimoto H, Bryll AR, Ganem NJ, Bersani F, Pineda JR, Suvà ML, Benes CH, et al. 2015. Alternative lengthening of telomeres renders cancer cells hypersensitive to ATR inhibitors. *Science* **347:** 273–277.

Funato K, Major T, Lewis PW, Allis CD, Tabar V. 2014. Use of human embryonic stem cells to model pediatric gliomas with H3.3K27M histone mutation. *Science* **346:** 1529–1533.

Garrick D, Sharpe JA, Arkell R, Dobbie L, Smith AJH, Wood WG, Higgs DR, Gibbons RJ. 2006. Loss of Atrx affects trophoblast development and the pattern of X-inactivation in extraembryonic tissues. *PLoS Genet* **2:** e58.

Gibbons R. 2006. α-Thalassaemia-mental retardation, X linked. *Orphanet J Rare Dis* **1:** 15.

Gibbons RJ. 2012. α-Thalassemia, mental retardation, and myelodysplastic syndrome. *Cold Spring Harb Perspect Med* **2:** a011759.

Gibbons RJ, Pellagatti A, Garrick D, Wood WG, Malik N, Ayyub H, Langford C, Boultwood J, Wainscoat JS, Higgs DR. 2003. Identification of acquired somatic mutations in the gene encoding chromatin-remodeling factor ATRX in the α-thalassemia myelodysplasia syndrome (ATMDS). *Nat Genet* **34:** 446–449.

Gibbons RJ, Wada T, Fisher CA, Malik N, Mitson MJ, Steensma DP, Fryer A, Goudie DR, Krantz ID, Traeger-Synodinos J. 2008. Mutations in the chromatin-associated protein ATRX. *Hum Mutat* **29:** 796–802.

Goldberg AD, Banaszynski LA, Noh KM, Lewis PW, Elsaesser SJ, Stadler S, Dewell S, Law M, Guo X, Li X, et al. 2010. Distinct factors control histone variant H3.3 localization at specific genomic regions. *Cell* **140:** 678–691.

Gonzalez-Perez A, Perez-Llamas C, Deu-Pons J, Tamborero D, Schroeder MP, Jene-Sanz A, Santos A, Lopez-Bigas N. 2013. IntOGen-mutations identifies cancer drivers across tumor types. *Nat Methods* **10:** 1081–1082.

Guerrini R, Shanahan JL, Carrozzo R, Bonanni P, Higgs DR, Gibbons RJ. 2000. A nonsense mutation of the *ATRX* gene causing mild mental retardation and epilepsy. *Ann Neurol* **47:** 117–121.

He Q, Kim H, Huang R, Lu W, Tang M, Shi F, Yang D, Zhang X, Huang J, Liu D, et al. 2015. The Daxx/Atrx complex protects tandem repetitive elements during DNA hypomethylation by promoting H3K9 trimethylation. *Cell Stem Cell* **17:** 273–286.

Heaphy CM, de Wilde RF, Jiao Y, Klein AP, Edil BH, Shi C, Bettegowda C, Rodriguez FJ, Eberhart CG, Hebbar S, et al. 2011. Altered telomeres in tumors with ATRX and DAXX mutations. *Science* **333:** 425.

Helming KC, Wang X, Roberts CWM. 2014. Vulnerabilities of mutant SWI/SNF complexes in cancer. *Cancer Cell* **26:** 309–317.

Hollenbach AD, Sublett JE, McPherson CJ, Grosveld G. 1999. The Pax3-FKHR oncoprotein is unresponsive to the Pax3-associated repressor hDaxx. *EMBO J* **18:** 3702–3711.

Hollenbach AD, McPherson CJ, Mientjes EJ, Iyengar R, Grosveld G. 2002. Daxx and histone deacetylase II associate with chromatin through an interaction with core histones and the chromatin-associated protein Dek. *J Cell Sci* **115:** 3319–3330.

Huh MS, Price O'Dea T, Ouazia D, McKay BC, Parise G, Parks RJ, Rudnicki MA, Picketts DJ. 2012. Compromised genomic integrity impedes muscle growth after Atrx inactivation. *J Clin Invest* **122:** 4412–4423.

Huh MS, Ivanochko D, Hashem LE, Curtin M, Delorme M, Goodall E, Yan K, Picketts DJ. 2016. Stalled replication

forks within heterochromatin require ATRX for protection. *Cell Death Dis* **7:** e2220.

Huppert JL, Balasubramanian S. 2005. Prevalence of quadruplexes in the human genome. *Nucleic Acids Res* **33:** 2908–2916.

Iwase S, Xiang B, Ghosh S, Ren T, Lewis PW, Cochrane JC, Allis CD, Picketts DJ, Patel DJ, Li H, et al. 2011. ATRX ADD domain links an atypical histone methylation recognition mechanism to human mental-retardation syndrome. *Nat Struct Mol Biol* **18:** 769–776.

Jiao Y, Shi C, Edil BH, de Wilde RF, Klimstra DS, Maitra A, Schulick RD, Tang LH, Wolfgang CL, Choti MA, et al. 2011. DAXX/ATRX, MEN1, and mTOR pathway genes are frequently altered in pancreatic neuroendocrine tumors. *Science* **331:** 1199–1203.

Johnson BE, Mazor T, Hong C, Barnes M, Aihara K, McLean CY, Fouse SD, Yamamoto S, Ueda H, Tatsuno K, et al. 2014. Mutational analysis reveals the origin and therapy-driven evolution of recurrent glioma. *Science* **343:** 189–193.

Kannan K, Inagaki A, Silber J, Gorovets D, Zhang J, Kastenhuber ER, Heguy A, Petrini JH, Chan TA, Huse JT. 2012. Whole-exome sequencing identifies ATRX mutation as a key molecular determinant in lower-grade glioma. *Oncotarget* **3:** 1194–1203.

Khuong-Quang DA, Buczkowicz P, Rakopoulos P, Liu XY, Fontebasso AM, Bouffet E, Bartels U, Albrecht S, Schwartzentruber J, Letourneau L, et al. 2012. K27M mutation in histone H3.3 defines clinically and biologically distinct subgroups of pediatric diffuse intrinsic pontine gliomas. *Acta Neuropathol* **124:** 439–447.

Law MJ, Lower KM, Voon HPJ, Hughes JR, Garrick D, Viprakasit V, Mitson M, De Gobbi M, Marra M, Morris A, et al. 2010. ATR-X syndrome protein targets tandem repeats and influences allele-specific expression in a size-dependent manner. *Cell* **143:** 367–378.

Lechner MS, Schultz DC, Negorev D, Maul GG, Rauscher FJ. 2005. The mammalian heterochromatin protein 1 binds diverse nuclear proteins through a common motif that targets the chromoshadow domain. *Biochem Biophys Res Commun* **331:** 929–937.

Le Douarin B, Nielsen AL, Garnier JM, Ichinose H, Jeanmougin F, Losson R, Chambon P. 1996. A possible involvement of TIF1 α and TIF1 β in the epigenetic control of transcription by nuclear receptors. *EMBO J* **15:** 6701–6715.

Lee JC, Jeng YM, Liau JY, Tsai JH, Hsu HH, Yang CY. 2015. Alternative lengthening of telomeres and loss of ATRX are frequent events in pleomorphic and dedifferentiated liposarcomas. *Mod Pathol* **28:** 1064–1073.

Leung JW-C, Ghosal G, Wang W, Shen X, Wang J, Li L, Chen J. 2013. α-Thalassemia/mental retardation syndrome X-linked gene product ATRX is required for proper replication restart and cellular resistance to replication stress. *J Biol Chem* **288:** 6342–6350.

Levy MA, Kernohan KD, Jiang Y, Bérubé NG. 2015. ATRX promotes gene expression by facilitating transcriptional elongation through guanine-rich coding regions. *Hum Mol Genet* **24:** 1824–1835.

Lewis PW, Elsaesser SJ, Noh KM, Stadler SC, Allis CD. 2010. Daxx is an H3.3-specific histone chaperone and cooperates with ATRX in replication-independent chromatin assembly at telomeres. *Proc Natl Acad Sci* **107**: 14075–14080.

Lewis PW, Müller MM, Koletsky MS, Cordero F, Lin S, Banaszynski LA, Garcia BA, Muir TW, Becher OJ, Allis CD. 2013. Inhibition of PRC2 activity by a gain-of-function H3 mutation found in pediatric glioblastoma. *Science* **340**: 857–861.

Li R, Pei H, Watson DK, Papas TS. 2000. EAP1/Daxx interacts with ETS1 and represses transcriptional activation of ETS1 target genes. *Oncogene* **19**: 745–753.

Liau JY, Lee JC, Tsai JH, Yang CY, Liu TL, Ke ZL, Hsu HH, Jeng YM. 2015a. Comprehensive screening of alternative lengthening of telomeres phenotype and loss of ATRX expression in sarcomas. *Mod Pathol* **28**: 1545–1554.

Liau JY, Tsai JH, Jeng YM, Lee JC, Hsu HH, Yang CY. 2015b. Leiomyosarcoma with alternative lengthening of telomeres is associated with aggressive histologic features, loss of ATRX expression, and poor clinical outcome. *Am J Surg Pathol* **39**: 236–244.

Liau JY, Tsai JH, Yang CY, Lee JC, Liang CW, Hsu HH, Jeng YM. 2015c. Alternative lengthening of telomeres phenotype in malignant vascular tumors is highly associated with loss of ATRX expression and is frequently observed in hepatic angiosarcomas. *Hum Pathol* **46**: 1360–1366.

Liu CP, Xiong C, Wang M, Yu Z, Yang N, Chen P, Zhang Z, Li G, Xu R-M. 2012a. Structure of the variant histone H3.3-H4 heterodimer in complex with its chaperone DAXX. *Nat Struct Mol Biol* **19**: 1287–1292.

Liu XY, Gerges N, Korshunov A, Sabha N, Khuong-Quang DA, Fontebasso AM, Fleming A, Hadjadj D, Schwartzentruber J, Majewski J, et al. 2012b. Frequent ATRX mutations and loss of expression in adult diffuse astrocytic tumors carrying IDH1/IDH2 and TP53 mutations. *Acta Neuropathol* **124**: 615–625.

Lovejoy CA, Li W, Reisenweber S, Thongthip S, Bruno J, de Lange T, De S, Petrini JHJ, Sung PA, Jasin M, et al. 2012. Loss of ATRX, genome instability, and an altered DNA damage response are hallmarks of the alternative lengthening of telomeres pathway. *PLoS Genet* **8**: e1002772.

Lu C, Ward PS, Kapoor GS, Rohle D, Turcan S, Abdel-Wahab O, Edwards CR, Khanin R, Figueroa ME, Melnick A, et al. 2012. IDH mutation impairs histone demethylation and results in a block to cell differentiation. *Nature* **483**: 474–478.

McDowell TL, Gibbons RJ, Sutherland H, O'Rourke DM, Bickmore WA, Pombo A, Turley H, Gatter K, Picketts DJ, Buckle VJ, et al. 1999. Localization of a putative transcriptional regulator (ATRX) at pericentromeric heterochromatin and the short arms of acrocentric chromosomes. *Proc Natl Acad Sci* **96**: 13983–13988.

Medina CF, Mazerolle C, Wang Y, Bérubé NG, Coupland S, Gibbons RJ, Wallace VA, Picketts DJ. 2009. Altered visual function and interneuron survival in *Atrx* knockout mice: Inference for the human syndrome. *Hum Mol Genet* **18**: 966–977.

Michaelson JS, Bader D, Frank K, Kozak C, Leder P. 1999. Loss of Daxx, a promiscuously interacting protein, results in extensive apoptosis in early mouse development. *Genes Dev* **13**: 1918–1923.

Mitson M, Kelley LA, Sternberg MJE, Higgs DR, Gibbons RJ. 2011. Functional significance of mutations in the Snf2 domain of ATRX. *Hum Mol Genet* **20**: 2603–2610.

Molenaar JJ, Koster J, Zwijnenburg DA, van Sluis P, Valentijn LJ, van der Ploeg I, Hamdi M, van Nes J, Westerman BA, van Arkel J, et al. 2012. Sequencing of neuroblastoma identifies chromothripsis and defects in neuritogenesis genes. *Nature* **483**: 589–593.

Nowick K, Fields C, Gernat T, Caetano-Anolles D, Kholina N, Stubbs L. 2011. Gain, loss and divergence in primate Zinc-Finger genes: A rich resource for evolution of gene regulatory differences between species. *PLoS ONE* **6**: e21553.

Olar A, Sulman EP. 2015. Molecular markers in low-grade glioma-toward tumor reclassification. *Semin Radiat Oncol* **25**: 155–163.

O'Sullivan RJ, Arnoult N, Lackner DH, Oganesian L, Haggblom C, Corpet A, Almouzni G, Karlseder J. 2014. Rapid induction of alternative lengthening of telomeres by depletion of the histone chaperone ASF1. *Nat Struct Mol Biol* **21**: 167–174.

Peifer M, Hertwig F, Roels F, Dreidax D, Gartlgruber M, Menon R, Krämer A, Roncaioli JL, Sand F, Heuckmann JM, et al. 2015. Telomerase activation by genomic rearrangements in high-risk neuroblastoma. *Nature* **526**: 700–704.

Perrem K, Bryan TM, Englezou A, Hackl T, Moy EL, Reddel RR. 1999. Repression of an alternative mechanism for lengthening of telomeres in somatic cell hybrids. *Oncogene* **18**: 3383–3390.

Pinto EM, Chen X, Easton J, Finkelstein D, Liu Z, Pounds S, Rodriguez-Galindo C, Lund TC, Mardis ER, Wilson RK, et al. 2015. Genomic landscape of paediatric adrenocortical tumours. *Nat Commun* **6**: 6302.

Pugh TJ, Morozova O, Attiyeh EF, Asgharzadeh S, Wei JS, Auclair D, Carter SL, Cibulskis K, Hanna M, Kiezun A, et al. 2013. The genetic landscape of high-risk neuroblastoma. *Nat Genet* **45**: 279–284.

Qadeer ZA, Harcharik S, Valle-Garcia D, Chen C, Birge MB, Vardabasso C, Duarte LF, Bernstein E. 2014. Decreased expression of the chromatin remodeler ATRX associates with melanoma progression. *J Invest Dermatol* **134**: 1768–1772.

Ramamoorthy M, Smith S. 2015. Loss of ATRX suppresses resolution of telomere cohesion to control recombination in ALT cancer cells. *Cancer Cell* **28**: 357–369.

Ratnakumar K, Duarte LF, LeRoy G, Hasson D, Smeets D, Vardabasso C, Bonisch C, Zeng T, Xiang B, Zhang DY, et al. 2012. ATRX-mediated chromatin association of histone variant macroH2A1 regulates α-globin expression. *Genes Dev* **26**: 433–438.

Rosnoblet C, Vandamme J, Völkel P, Angrand P-O. 2011. Analysis of the human HP1 interactome reveals novel binding partners. *Biochem Biophys Res Commun* **413**: 206–211.

Sadic D, Schmidt K, Groh S, Kondofersky I, Ellwart J, Fuchs C, Theis FJ, Schotta G. 2015. Atrx promotes heterochromatin formation at retrotransposons. *EMBO Rep* **16**: 836–850.

Sarma K, Cifuentes-Rojas C, Ergun A, Del Rosario A, Jeon Y, White F, Sadreyev R, Lee JT. 2014. ATRX directs binding of PRC2 to Xist RNA and Polycomb targets. *Cell* **159**: 869–883.

Schwartzentruber J, Korshunov A, Liu X-Y, Jones DTW, Pfaff E, Jacob K, Sturm D, Fontebasso AM, Quang D-

AK, Tönjes M, et al. 2012. Driver mutations in histone H3.3 and chromatin remodelling genes in paediatric glioblastoma. *Nature* **482:** 226–231.

Solomon LA, Russell BA, Watson LA, Beier F, Bérubé NG. 2013. Targeted loss of the ATR-X syndrome protein in the limb mesenchyme of mice causes brachydactyly. *Hum Mol Genet* **22:** 5015–5025.

Tagami H, Ray-Gallet D, Almouzni G, Nakatani Y. 2004. Histone H3.1 and H3.3 complexes mediate nucleosome assembly pathways dependent or independent of DNA synthesis. *Cell* **116:** 51–61.

Tang J, Wu S, Liu H, Stratt R, Barak OG, Shiekhattar R, Picketts DJ, Yang X. 2004. A novel transcription regulatory complex containing death domain-associated protein and the ATR-X syndrome protein. *J Biol Chem* **279:** 20369–20377.

TCGA. 2015. The Cancer Genome Atlas, cancergenome .nih.gov.

Torii S, Egan DA, Evans RA, Reed JC. 1999. Human Daxx regulates Fas-induced apoptosis from nuclear PML oncogenic domains (PODs). *EMBO J* **18:** 6037–6049.

Turcan S, Rohle D, Goenka A, Walsh LA, Fang F, Yilmaz E, Campos C, Fabius AWM, Lu C, Ward PS, et al. 2012. IDH1 mutation is sufficient to establish the glioma hypermethylator phenotype. *Nature* **483:** 479–483.

Valentijn LJ, Koster J, Zwijnenburg DA, Hasselt NE, van Sluis P, Volckmann R, van Noesel MM, George RE, Tytgat GAM, Molenaar JJ, et al. 2015. TERT rearrangements are frequent in neuroblastoma and identify aggressive tumors. *Nat Genet* **47:** 1411–1414.

Valle-Garcia D, Qadeer ZA, McHugh D, Ghiraldini FG, Chowdhury AH, Hasson D, Dyer MA, Recillas-Targa F, Bernstein E. 2016. ATRX binds to atypical chromatin domains at the 3′ exons of ZNF genes to preserve H3K9me3 enrichment. *Epigenetics* doi: 10.1080/ 15592294.2016.1169351.

Voon HPJ, Hughes JR, Rode C, DeLaRosa-Velazquez IA, Jenuwein T, Feil R, Higgs DR, Gibbons RJ. 2015. ATRX plays a key role in maintaining silencing at interstitial heterochromatic loci and imprinted genes. *Cell Rep* **11:** 405–418.

Watson LA, Solomon LA, Li JR, Jiang Y, Edwards M, Shin-Ya K, Beier F, Berube NG. 2013. *Atrx* deficiency induces telomere dysfunction, endocrine defects, and reduced life span. *J Clin Invest* **123:** 2049–2063.

Watson LA, Goldberg H, Bérubé NG. 2015. Emerging roles of ATRX in cancer. *Epigenomics* **7:** 1365–1378.

Wong LH. 2010. Epigenetic regulation of telomere chromatin integrity in pluripotent embryonic stem cells. *Epigenomics* **2:** 639–655.

Wu G, Broniscer A, McEachron TA, Lu C, Paugh BS, Becksfort J, Qu C, Ding L, Huether R, Parker M, et al. 2012. Somatic histone H3 alterations in pediatric diffuse intrinsic pontine gliomas and non-brainstem glioblastomas. *Nat Genet* **44:** 251–253.

Wu G, Diaz AK, Paugh BS, Rankin SL, Ju B, Li Y, Zhu X, Qu C, Chen X, Zhang J, et al. 2014. The genomic landscape of diffuse intrinsic pontine glioma and pediatric non-brainstem high-grade glioma. *Nat Genet* **46:** 444–450.

Xue Y, Gibbons R, Yan Z, Yang D, McDowell TL, Sechi S, Qin J, Zhou S, Higgs D, Wang W. 2003. The ATRX syndrome protein forms a chromatin-remodeling complex with Daxx and localizes in promyelocytic leukemia nuclear bodies. *Proc Natl Acad Sci* **100:** 10635–10640.

Yang X, Khosravi-Far R, Chang HY, Baltimore D. 1997. Daxx, a novel Fas-binding protein that activates JNK and apoptosis. *Cell* **89:** 1067–1076.

Zhong S, Salomoni P, Ronchetti S, Guo A, Ruggero D, Pandolfi PP. 2000. Promyelocytic leukemia protein (PML) and Daxx participate in a novel nuclear pathway for apoptosis. *J Exp Med* **191:** 631–640.

The Chromodomain Helicase DNA-Binding Chromatin Remodelers: Family Traits that Protect from and Promote Cancer

Alea A. Mills

Cold Spring Harbor Laboratory, Cold Spring Harbor, New York, 11724

Correspondence: mills@cshl.edu

A plethora of mutations in chromatin regulators in diverse human cancers is emerging, attesting to the pivotal role of chromatin dynamics in tumorigenesis. A recurrent theme is inactivation of the chromodomain helicase DNA-binding (CHD) family of proteins— ATP-dependent chromatin remodelers that govern the cellular machinery's access to DNA, thereby controlling fundamental processes, including transcription, proliferation, and DNA damage repair. This review highlights what is currently known about how genetic and epigenetic perturbation of CHD proteins and the pathways that they regulate set the stage for cancer, providing new insight for designing more effective anti-cancer therapies.

The advent of high-throughput sequencing technologies has made it increasingly straightforward to interrogate the genomes of human tumors in an effort to identify cancer-driving mutations, diagnose tumor subtypes, and implement regimens for personalized therapies. The picture that is emerging from these studies is that lesions in genes encoding chromatin regulators are among the most prevalent mutations, underscoring the importance of chromatin structure and function in tumorigenesis. Indeed, efforts of The Cancer Genome Atlas Consortium have established that chromatin regulators are some of the most frequent mutations in 12 major types of human cancer, including glioma and leukemia, as well as tumors of the breast, bladder, colon, kidney, and lung (Kandoth et al. 2013).

The chromodomain helicase DNA-binding (CHD) family of chromatin remodelers is one type of chromatin regulator that is frequently lost or inactivated in a diverse array of human cancers. The CHD family consists of nine members, CHD1–9 (Fig. 1). CHD5 was the first CHD protein shown to have a functional role in cancer (Bagchi et al. 2007). CHD family members share chromatin organizing (CHROMO) domains that bind specifically modified histones and an SNF2-like ATP-dependent helicase domain that facilitates nucleosome mobilization (Marfella and Imbalzano 2007). The family name reflects the fact that CHD1, the original CHD protein identified, tends to interact with AT-rich regions of DNA, implying that it has a DNA-binding domain (Delmas et al. 1993). In addition to core motifs characteristic of the

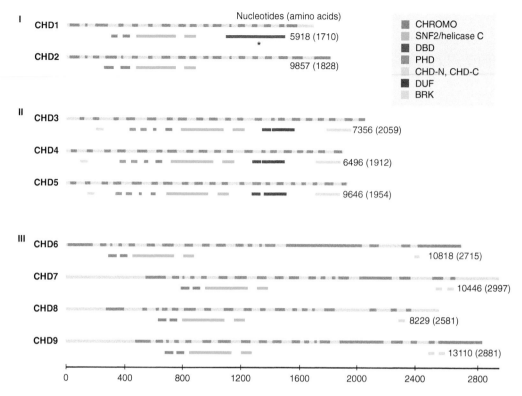

Figure 1. The chromodomain helicase DNA-binding (CHD) family of chromatin remodelers. CHD proteins are classified into three subfamilies (Roman numerals) based on their functional motifs (see legend). The human CHD family based on Ensembl is drawn to scale, with light and dark gray bars depicting alternating exons (*above*) and the functional motifs from PFAM (a database of protein families of multiple sequence alignments generated using hidden Markov models) shown in color (*below*) for each CHD member. The number of nucleotides and amino acid residues for the CHD transcript and protein, respectively, are shown. BRK, Brahma and Kismet domains; CHD, chromodomain helicase DNA binding; CHROMO, chromodomain; CHD-N, CHD-C: CHD_N and CHD_C are shown in upstream and downstream regions, respectively; DBD, DNA-binding domain (based on Delmas et al. 1993) rather than PFAM; DUF, domain of unknown function; PHD, plant homeodomain; SNF2/Helicase C, SNF2_N, and Helicase_C are shown in upstream and downstream regions, respectively. (From Li and Mills 2014; reproduced and modified, with express permission, from Future Medicine, 2014.)

family as a whole, individual CHD proteins have other domains that classify them into three subfamilies. Subfamily I, which includes human CHD1 and CHD2, is classified based on members having SNF2 domains homologous to CHD1 proteins of other organisms, with mouse Chd1 having an AT-rich DNA-binding domain (Delmas et al. 1993). Subfamily II, which includes CHD3, CHD4, and CHD5, has dual plant homeodomains (PHDs). Subfamily III, which includes CHD6, CHD7, CHD8, and CHD9, has Brahma and Kismet domains. CHD proteins affect chromatin compaction and therefore the

cellular machinery's access to DNA; thus, these enzymes control fundamental biological processes, including transcription, cellular proliferation, and DNA damage repair. Given their key role in these crucial cellular processes, it is perhaps not surprising that loss or inactivation of CHD proteins is pivotal in a range of developmental syndromes and cancers. This review focuses on the mechanisms by which perturbation of CHD-mediated chromatin dynamics regulates tumorigenesis. I discuss what is known about the biological roles of these chromatin remodelers, highlight recent evidence for the genetic and epigenetic factors that are upstream, in parallel, and downstream from CHD proteins, and discuss the current view on how perturbation of different members of each CHD subfamily modulates the tumorigenic process and affects cancer patient survival.

SUBFAMILY I: CHD1 and CHD2

CHD1—the founding member of the CHD family—was initially discovered as a DNA-binding protein that, based on its functional motifs, was proposed to regulate chromatin structure and gene expression (Delmas et al. 1993; Stokes and Perry 1995). Subfamily I CHD proteins have been implicated in transcriptional regulation, and this has recently been shown to impact proliferation, repair of DNA damage, and pluripotency. The first demonstration for any CHD protein playing a functional role in cancer was when the subfamily II member CHD5 was identified as a tumor suppressor mapping to human 1p36—a genomic interval frequently deleted in a variety of cancers (Bagchi et al. 2007; Bagchi and Mills 2008) (discussed below). But more recently, it has become apparent that like CHD5, subfamily I CHD proteins are also lost or inactivated in several cancers. However, their gain of function has also been shown to promote cancer. Deregulation of subfamily I proteins has a profound effect on invasion, metastasis, and patient survival. CHD1 is the CHD subfamily I member with the tightest link to cancer, although CHD2 has also been implicated in tumor suppression.

Factors upstream of CHD subfamily I proteins include environmental factors, hormones, chromatin regulators, and signaling pathways. Exposure to cigarette smoke correlates with hypermethylation of the *CHD1* promoter (Lyn-Cook et al. 2014). Promoter methylation is associated with RNA polymerase II (Pol II) stalling and compromised transcriptional activation, whereas demethylating agents abrogate this effect by inducing Pol II phosphorylation at serine 2 to promote transcriptional elongation (Tao et al. 2011). Estrogen signaling plays a key role in promoting proliferation of estrogen

receptor (ER)-positive breast cancer. Estrogen inhibits expression of miR26a and miR26b, microRNAs that target and degrade *CHD1* transcript and promote proliferation of breast cancer cells (Tan et al. 2014). c-MYC is required for both the inhibition of miR26 and the increase in CHD1 expression in response to estrogen. The investigators show that depletion of CHD1 abrogates the pro-proliferative effect of estrogen, indicating that CHD1 potentiates oncogenesis, at least in the context of ERα-positive breast cancer. Factors interacting with components of the transcriptional machinery and histone modifiers also converge upstream of *CHD1* to modulate its expression. For example, the Pol II–associated factor hPAF2/PD2 mediates MLL-mediated deposition of the H3K4me2/3 covalent modifications characteristic of transcriptionally active genes and facilitates *CHD1* expression in pancreatic cancer cells (Dey et al. 2011). Another example is the protein arginine methyltransferase Prmt6, which inhibits expression of *Chd1* (Lee et al. 2012). Prmt6 evokes H3R2me2—a covalent modification antagonistic to the H3K4me3 mark associated with transcriptional activation. CHD1 is also modified by SUMOylation in KRAS mutant colorectal cancer cells by the SUMO E2 ligase UBC9 (Yu et al. 2015a). While this study indicates that activation of the RAS/RAF pathway covalently modifies CHD1 and that UBC9 is required for KRAS-mediated transformation, the finding that CHD1 depletion inhibits the transformed phenotype suggests that SUMOylation endows CHD1 with pro-oncogenic activity, rather than it inhibiting its tumor-suppressive activity.

Factors working in parallel with CHD1 include hPAF1/PD2 that binds and facilitates nuclear import of CHD1 where it is then positioned to bind H3K4me2/3 via its dual CHROMO domains (Flanagan et al. 2005), promote nucleosome destabilization, and modulate transcription in pancreatic cancer cells (Dey et al. 2011). Because hPAF1/PD2 is proposed to have an oncogenic function and is aberrantly overexpressed in pancreatic cancer, and hPAF/PD2 facilitates CHD1 activity, this again suggests that CHD1 plays a pro-oncogenic role. There is some indication that CHD1 functionally interacts with MAP3K7, as co-deletion of *CHD1* and *MAP3K7* occurs in prostate cancer and co-suppression of Chd1 and Map3k7 in mouse prostate epithelial stem/progenitor cells inhibits differentiation and causes aggressive prostate tumors (Rodrigues et al. 2015). CHD1 also works in concert with the androgen receptor (AR), as it is required for AR-dependent transcriptional activation of androgen-responsive genes in prostate cancer (Burkhardt et al. 2013).

Factors downstream from CHD1 include the AR-responsive tumor suppressor genes *NKX3-1*, *FOXO1*, and *PPARγ*, consistent with CHD1 being a coactivator of AR-mediated transcriptional activation (Burkhardt et al. 2013). In mouse embryonic stem (ES) cells, Chd1 facilitates expression of the pluripotency genes *Oct4* and *Nanog* (Lee et al. 2012). The mechanism proposed is that increased expression of the histone arginine methyltransferase Prmt6 occurs upon differentiation thereby evoking H3R2me2, a covalent histone modification that counteracts the transcriptional activation mark H3K4me3. Simultaneously, Chd1 expression is compromised, and there is also less Chd1 bound at the promoters of *Oct4* and *Nanog* because these regions have less H3K4me3 compared with undifferentiated ES cells, and Chd1 is known to bind H3K4me3 via its CHROMO domains. A separate study found that H3K4me3 is not sufficient for recruiting CHD1 to promoters; instead, CHD1 is recruited to target loci through its interaction with components of the transcriptional machinery in an activation-dependent manner where it regulates H3/H3.3 occupancy and chromatin accessibility at transcriptional start sites of its target genes (Siggens et al. 2015).

Inactivating lesions affecting *CHD1* include promoter methylation in breast and other cancers (Lyn-Cook et al. 2014), and mutation in colorectal cancers with high levels of microsatellite instability (Kim et al. 2011). But the most striking evidence for CHD1 inactivation is in prostate cancer, where it is deleted or mutated (Grasso et al. 2012; Huang et al. 2012; Liu et al. 2012; Burkhardt et al. 2013; Martin et al. 2013; Blattner et al. 2014; Gao et al. 2014; Scott et al.

2014; Tereshchenko et al. 2014; Attard et al. 2015; Fisher et al. 2015; Sowalsky et al. 2015). Indeed, homozygous deletion of *CHD1* is the second most common genetic event in prostate cancer after *PTEN* deletion (Liu et al. 2012). Chromosome rearrangements that cause overexpression of ETS family members, most commonly translocations between the androgen-regulated gene *transmembrane protease serine 2* (*TMPRSS2*) and the *ERG* gene, are frequent in some types of prostate cancer (Clark and Cooper 2009). *CHD1* lesions occur in ETS fusion-negative prostate cancer (Grasso et al. 2012; Martin et al. 2013; Tereshchenko et al. 2014), indicating that the *CHD1* status defines a unique prostate cancer subtype (Attard et al. 2015; Fisher et al. 2015). Whereas CHD1 mutation and ETS fusions are mutually exclusive, CHD1 inactivation co-occurs with speckle-type PTB/POZ protein mutations (Blattner et al. 2014) and *MAP3K7* deletion (Rodrigues et al. 2015), suggesting that these lesions cooperate with CHD1 loss to drive tumorigenesis in the prostate. Inactivation of CHD1 has also been correlated with anchorage-independent growth (Yu et al. 2015a), enhanced invasiveness (Huang et al. 2012; Liu et al. 2012), and compromised differentiation and increased stemness in mouse prostate epithelial stem/progenitor cells (Rodrigues et al. 2015). These findings from human studies are in agreement with the observations made from work in the mouse. For example, Prmt6 inhibits Chd1 occupancy at the pluripotency genes *Oct4* and *Nanog*, and Chd1 inactivation augments expression of *Oct4* and *Nanog* and enhances stemness (Lee et al. 2012). Furthermore, Chd1 is essential for the open chromatin state and pluripotency of ES cells and is required for the reprogramming of somatic cells (Gaspar-Maia et al. 2009).

There is also some evidence that CHD2 plays a role in cancer, although this view is not nearly as clear as it is for CHD1. In this regard, there are some similarities between CHD1 and CHD2. For example, CHD2 may also be hormone responsive, as human chorionic gonadotropin that is released systemically during pregnancy causes transcriptional induction of CHD2, which has been proposed, along with other chromatin regulators, to prevent breast cancer (Russo and Russo 2012). CHD2 also regulates H3/H3.3 occupancy (Siggens et al. 2015). As is the case for CHD1 (and other CHDs, see below), colorectal tumors with high microsatellite instability have CHD2 mutations (Kim et al. 2011). There is some, although scant, evidence that CHD2 is inactivated in human cancers, including chronic lymphoblastic leukemia (CLL) and monoclonal B lymphocytosis, a B-cell expansion syndrome that can progress to CLL (Rodriguez et al. 2015). These *CHD2* mutations in CLL are associated with mutations in genes encoding immunoglobulin heavy chain variable regions. CHD2 is also down-regulated in colorectal cancer (Bandres et al. 2007). CHD2 has also been implicated in neurodevelopment, as haploinsufficiency of CHD2 is associated with neurological deficits, including developmental delay, intellectual disability, epilepsy, and behavioral anomalies (Chenier et al. 2014). Neurological symptoms are a characteristic feature of deregulation of CHD subfamily III (and to a lesser extent to subfamily II) members (discussed below).

Mouse models echo the theme that Chd2 functions more in development than in tumorigenesis, as $Chd2^{-/-}$ mice have compromised viability, growth delay (Marfella et al. 2006), and lordokyphosis (Kulkarni et al. 2008). Furthermore, a congenic mouse backcross study identified *Chd2* as a candidate obesity gene (Sarahan et al. 2011). Perhaps in line with the idea that Chd2 has some cancer-specific roles, genetic linkage analysis identified *Chd2* as one of three candidate genes for a genetic modifier of breast cancer, suggesting an explanation for why *p53* heterozygous mutant mice are uniquely susceptible to developing mammary gland tumors when established in the BALB/c genetic background (Koch et al. 2007). Heterozygosity for mutant *Chd2* alleles is associated with extramedullary hematopoiesis and susceptibility to lymphoma (Nagarajan et al. 2009). In a follow-up study, the investigators found that Chd2-deficient cells are sensitive to DNA-damaging agents and do not efficiently repair DNA damage induced by ultraviolet or ionizing radiation, leading them to conclude that Chd2 (like other

Chd proteins, discussed below) facilitates DNA repair and maintains genomic stability (Rajagopalan et al. 2012). Yet reports using a different Chd2 compromised mouse model concluded that heterozygotes succumb to non-neoplastic lesions in a number of organs, but are not susceptible to frank cancer (Marfella et al. 2006). Thus, while there is ample evidence that CHD1 functions as a tumor suppressor particularly in the prostate, and that in some cases it promotes oncogenesis, there is currently only tangential evidence that CHD2 shares these cancer-associated roles with its closest sibling.

SUBFAMILY II: CHD3, CHD4, and CHD5

Several of the biological processes ascribed to CHD subfamily II proteins are similar to those of subfamily I; for example, essentially all subfamily II proteins control transcription (Zhang et al. 1998; Srinivasan et al. 2006; Denslow and Wade 2007; Lee and Das 2010) and DNA damage repair (Stanley et al. 2013; Hall et al. 2014). But whereas subfamily I members function to maintain the pluripotent state in ES cells (Gaspar-Maia et al. 2009; Lee et al. 2012), the CHD II subfamily of proteins includes potent modulators of cellular proliferation, senescence, and apoptosis (Bagchi et al. 2007). The prototypical member of this family, CHD5, is a tumor suppressor whose inactivation is a predominant theme in a variety of human cancers. Lesions in CHD4, and to a lesser extent in CHD3, also occur in human cancer, with loss or inactivation of CHD subfamily II members being associated with chemoresistance, epithelial–mesenchymal transition (EMT), metastasis (Wang et al. 2011; Wu et al. 2012), and poor overall patient survival (Garcia et al. 2010; Wong et al. 2011; Wu et al. 2012; Du et al. 2013; Wang et al. 2013; Hall et al. 2014; Xie et al. 2015).

The best body of evidence for a CHD II subclass protein having a functional role in cancer exists for CHD5, likely due at least in part to the fact that CHD5 was the earliest CHD member defined to have a tumor-suppressive role (Bagchi et al. 2007). *CHD5* maps to 1p36—a region of the genome frequently deleted in human cancers; modeling these deletions in the mouse using chromosome engineering pinpointed a 4.3-Mb genomic interval that encodes a product with potent tumor-suppressive activity. Although heterozygous loss of this interval leads to immortalization, oncogenic transformation, and spontaneous tumorigenesis, gain of dosage of this interval induces cellular senescence, excessive apoptosis, and perinatal lethality. Because the gain of dosage phenotypes of compromised proliferation and enhanced senescence can be rescued by depleting Chd5, we conclude that Chd5 (and not any of the other genes in the duplicated region that were tested) is responsible for these phenotypes. Thus, Chd5 is a highly dosage-dependent tumor suppressor that must be diploid: having only one copy predisposes to cancer; having three copies causes death.

Factors upstream of CHD subfamily II members include environmental factors, DNA methylation, miRNAs, DNA tumor virus-encoded oncogenes, chromatin regulators, transcription factors, and signaling pathways. CHD5 responds to environmental cues; for example, genestein—a compound present in soybeans—enhances *CHD5* expression (Li et al. 2012). In neuroblastoma cells in which *CHD5* is silenced by promoter methylation, genestein inhibits expression of the DNA methyltransferase DNMT3B, thereby reducing *CHD5* methylation and activating its transcription. MicroRNAs (miRNAs), miR211 and miR454, target and degrade *CHD5* mRNA in colorectal cancer (Cai et al. 2012) and hepatocellular carcinoma (Yu et al. 2015b), respectively. The chromatin regulator JMJD2A (also known as KDM4A, a member of the jumonji domain containing two families of lysine demethylases) inhibits RAS-mediated senescence to drive transformation (Mallette 2012). Importantly, JMJD2A was shown to inhibit the ability of RAS to induce *CHD5* expression (Mallette 2012), thereby compromising p53-mediated pathways that are downstream (Bagchi 2007). The promoter of *CHD5* has binding sites for transcription factors, including LEF1/TCF, SP1, and AP2, suggesting that *CHD5* is transcriptionally regulated by components of the WNT/β-catenin pathway (Fatemi et al. 2014). Aberrant insulin-like

growth factor 1 (IGF-1) signaling promotes *CHD5* promoter hypermethylation, thereby inhibiting *CHD5* expression to drive hepatocellular carcinoma development (Fang et al. 2015). RAS normally induces CHD5 expression, but aberrant up-regulation of JMJD2A inhibits this activation (Mallette and Richard 2012). This finding is consistent with Chd5-compromised cells being exquisitely sensitive to oncogenic transformation (Bagchi et al. 2007).

CHD5 is a component of the nucleosome remodeling and deacetylase (NURD) complex (Quan and Yusufzai 2014; Quan et al. 2014; Kolla et al. 2015). Therefore, factors functioning in parallel with CHD5 include NURD complex components such as MTA, GATAD2A, HDAC1/2, RBBP4/7, MDB2/3. Like CHD1, CHD5 is a nucleosome remodeler—an ATP-dependent enzyme that repositions or exchanges nucleosomes (Quan and Yusufzai 2014). A unique nucleosome "unwrapping" activity was discovered for CHD5 that at least in vitro, appeared to be distinct from the subfamily II member CHD4; perhaps this capability provides CHD5 with specific roles. In addition, the carboxyl terminus of CHD5 is distinct from CHD3 and CHD4, which may equip it with exclusive function distinct from its closest subfamily members.

Factors downstream from CHD5 include its transcriptional targets. Chd5 inhibits proliferation by transcriptionally activating *Cdkn2a*, a locus that encodes multiple tumor suppressors, including p16Ink4a and p19Arf (Bagchi et al. 2007; Bagchi and Mills 2008). Decreased *Chd5* dosage cripples p16Ink4a/Rb and p19Arf/p53-tumor-suppressive pathways, setting the stage for cancer; on the other hand, enhanced dosage of the interval encoding *Chd5* exacerbates these tumor-suppressive pathways, causing over exuberant apoptosis that depletes stem cells and is incompatible with life. What distinguishes subfamily II from other CHD proteins is the presence of tandem PHD zinc-finger motifs (see Fig. 1). We and others reported that the PHDs of CHD5 bind the amino-terminal tail of unmodified histone H3 (Oliver et al. 2012; Paul et al. 2013). The dual nature of the juxtaposed PHDs may have functional importance, as

PHD1 and PHD2 of CHD5 simultaneously bind two H3 amino termini, which together enhance binding affinity four- to 11-fold (Oliver et al. 2012). PHD-mediated H3 binding is crucial for Chd5's ability to regulate transcription, inhibit proliferation, and function as a tumor suppressor (Paul et al. 2013). In addition to Chd5-inducing *Cdkn2a* expression, it regulates gene expression globally. We identified Chd5-bound loci across the genome and found that Chd5 regulates a cascade of cancer pathways and chromatin regulators such as the polycomb repressive group complex (PcG) oncoprotein Bmi1 (Paul et al. 2013). CHD5 is also linked to other PcG proteins, as CHD5 and EZH2 transcriptionally inhibit each other's expression (Xie et al. 2015). CHD5 can also induce expression of *WEE1* to engage cell-cycle arrest at the G_2/M checkpoint (Quan et al. 2014).

Mouse models were key for revealing that Chd5 plays a critical role in maintaining genomic integrity. Chd5 plays a dynamic role in remodeling the genome during maturation of the male germline (Li et al. 2014; Zhuang et al. 2014). The process of sperm maturation or "spermiogenesis," an intricate process that occurs in haploid spermatids following meiosis, is one of the most extensive examples of chromatin remodeling known. A consequence of Chd5 deficiency in the male germline is alterations in chromatin compaction because of inefficient removal of canonical histones, deregulated incorporation of transition proteins and protamines, unbridled DNA damage, and compromised fertility (Li and Mills 2014; Li et al. 2014). These findings are in agreement with the low *CHD5* expression levels found in testes of infertile men. CHD5's role in unpackaging the genome to remove canonical histones and repackaging it, first with transition proteins and ultimately with protamines in mature sperm, and by doing so to maintain genomic integrity, may be unique to this particular subfamily member, as to date neither CHD3 nor CHD4 have been implicated in infertility. Perhaps Chd5's role in the male germline is a result of its unique nucleosome unwrapping activity (Quan and Yusufzai 2014) or its distinctive carboxy-terminal region. Another possibility is that the three Chd II family members are expressed differently in testes. CHD5 also functions to maintain the genome in somatic cells, as compromised CHD5 enhances the DNA damage response in pancreatic adenocarcinoma cells, a finding that correlates with decreased patient survival (Hall et al. 2014).

CHD5 is frequently lost or inactivated in diverse human cancers. Loss-of-function *CHD5* lesions, including compromised expression, promoter hypermethylation, deletion, and/or mutation, have been reported in glioma (Bagchi et al. 2007; Mulero-Navarro and Esteller 2008; Wang et al. 2013), neuroblastoma (Fujita et al. 2008; Garcia et al. 2010; Koyama et al. 2012; Li et al. 2012), lung cancer (Zhao et al. 2012), prostate cancer (Robbins et al. 2011), breast cancer (Mulero-Navarro and Esteller 2008; Wu et al. 2012), pancreatic adenocarcinoma (Hall et al. 2014), gastric cancer (Wang et al. 2009; Qu et al. 2013), bladder cancer (Wu et al. 2015), ovarian cancer (Gorringe et al. 2008; Wong et al. 2011), gallbladder carcinoma (Du et al. 2013), colorectal cancer (Mulero-Navarro and Esteller 2008; Mokarram et al. 2009; Cai et al. 2012; Fatemi et al. 2014), hepatocellular carcinoma (Zhao et al. 2014; Fang et al. 2015; Xie et al. 2015), melanoma (Lang et al. 2011), leukemia (Zhao et al. 2014), and laryngeal squamous cell carcinoma (Wang et al. 2011).

CHD5 expression correlates directly with overall patient survival for several cancers, including glioma (Wang et al. 2013), neuroblastoma (Garcia et al. 2010), as well as for cancers of the ovary (Wong et al. 2011), breast (Wu et al. 2012), gallbladder (Du et al. 2013), pancreas (Hall et al. 2014), and liver (Xie et al. 2015). The fact that *CHD5* lesions tend to be heterozygous (Henrich et al. 2012) suggests that reactivation of the wild-type locus may be effective as a therapeutic strategy. Indeed, *CHD5* induction using demethylating agents and transcriptional up-regulation of CHD5 decreases proliferation and compromises invasion (Fatemi et al. 2014). Thus, there is ample experimental and clinical evidence for CHD5's potent tumor-suppressive role, suggesting that strategies to induce it could provide new avenues for treating diverse types of cancer.

The subfamily II CHD protein CHD4 has a number of similarities with CHD5. CHD4 is also a component of the NURD and is also a chromatin remodeler that regulates transcription, proliferation, and DNA damage repair. Like CHD5, CHD4 has strong ATPase activity (Quan and Yusufzai 2014). However, there are important differences, for instance, assays performed in vitro suggest that CHD4 does not unwrap nucleosomes nearly as efficiently as CHD5. The distinction between the different CHD II subfamily members in vivo, however, is not at present clear.

As found for CHD5, factors upstream of CHD4 include environmental insults such as tobacco smoke (Yamada et al. 2015). MYC enhances CHD4's interaction with MTA and NURD during transformation (Zhang et al. 2005). The HPV16 oncoprotein E7 binds CHD4, evokes histone deacetylase activity, and enhances proliferation (Brehm et al. 1999).

Factors working in parallel with CHD4 include the NURD components (Quan and Yusufzai 2014; Kolla et al. 2015). Interestingly, it has been proposed that NURD complexes containing CHD4 are mutually exclusive with those containing CHD5 (Quan and Yusufzai 2014). CHD4 interacts with BRD4/NSD3 (Rahman et al. 2011), ZFHX4 (Chudnovsky et al. 2014), p300 acetyltransferase (Qi et al. 2015), p300/GATA4 (Hosokawa et al. 2013), TWIST (Fu et al. 2011), and HSF (Khaleque et al. 2008).

Factors downstream from CHD4 include its transcriptional targets. CHD4 has been shown to couple histone deacetylase activity to promoter hypermethylation in colorectal cancer (Cai et al. 2014). CHD4 interacts with NAB2 to regulate expression of genes encoding early growth response (EGR) transactivators (Srinivasan et al. 2006). CHD4 is recruited to MBD2/p66α-bound methylated DNA, an interaction abrogated in breast cancer (Desai et al. 2015). CHD4 inhibits E-cadherin, thereby inhibiting EMT and metastasis of lung cancer cells (Fu et al. 2011).

CHD4 lesions include its mutation in endometrial cancer (Le Gallo et al. 2012) and uterine serous carcinoma (Zhao et al. 2013), perhaps analogous to the single-nucleotide polymor-phisms in *CHD5* that are associated with endometriosis (Falconer et al. 2012). CHD4 maintains tumor-initiating cells in glioblastoma (Chudnovsky et al. 2014). As is the case for CHD5, CHD4 functions in the DNA damage response (Stanley et al. 2013; Qi et al. 2015). CHD4 deficiency has been reported to contribute to chemoresistance in BRCA mutant cells (Guillemette et al. 2015), and targeting CHD4 is able to deplete EpCam[+] liver cancer cells (Nio et al. 2015). Thus, while there are some indications that CHD4 functions as a tumor suppressor, targeting it therapeutically has been proposed as a way to overcome chemoresistance (Nio et al. 2015), a finding that has also been shown for CHD subfamily I and III members (discussed below).

There is scant evidence for a role for CHD3 in cancer. Like its closest siblings, CHD3 is part of the NURD complex (Kolla et al. 2015), and the crystal and NMR structures were recently solved (Torchy et al. 2015). CHD3, like CHD4, interacts with the transcriptional corepressors NAB2 to inhibit expression of EGR target genes (Srinivasan et al. 2006). CHD3 appears to take on the family job of inducing the DNA damage repair response (Stanley et al. 2013; Klement et al. 2014). CHD3 regulates heterochromatin formation, thereby stimulating ATM-induced double-strand break repair, KAP-1 phosphorylation, and recruitment of ACF/SNF2 to sites of DNA damage (Klement et al. 2014). Thus, while there is substantial evidence that CHD5 functions as a tumor suppressor, and there is accumulating support for CHD4 playing a somewhat similar role, it is early days for CHD3. Time will tell whether this trait is conserved throughout the subfamily.

SUBFAMILY III: CHD6, CHD7, CHD8, and CHD9

Like the other subfamilies, subfamily III CHD proteins have been implicated in transcriptional regulation, cellular proliferation, and repair of DNA damage, and therefore have also been shown to be deregulated in cancer and to affect overall patient survival. But mutations in members of this subfamily have also been heavily

Cite this article as *Cold Spring Harb Perspect Med* doi: 10.1101/cshperspect.a026450

implicated in developmental and neurological syndromes that are not associated with frank malignancy (Ronan et al. 2013), including coloboma of the eye, heart defects, atresia of the choanae, retardation of growth and/or development, genital and/or urinary abnormalities, and ear abnormalities and deafness (CHARGE) syndrome, schizophrenia, and autism (Layman et al. 2010). Furthermore, similar to the subfamily I CHD proteins, loss or inactivation of members of subfamily III CHD proteins sometimes correlate with enhanced patient survival, suggesting that members of this subfamily have both oncogenic and tumor-suppressive functions. At present, CHD8 appears to have the strongest link with cancer, although CHD7 has also been reported to modulate cancer-related pathways and to impact patient survival.

Factors upstream of CHD subfamily III proteins include hormones, environmental factors, and DNA methylation. Estrogen's pro-proliferative effect in breast cancer cells effect occurs via cyclin D1-mediated activation of cyclin E2/CDK2, and CHD8 is required for efficient E2F1 recruitment to the promoter of cyclin E2 and its transcriptional activation (Caldon et al. 2009). In gastric cancers, CHD8 mutations are associated with the presence of *Fusibacterium*, a pathogen that is part of the gut microbiome (Tahara et al. 2014a), and CHD8 is silenced by promoter methylation in prostate cancer (Damaschke et al. 2014). At the parallel level, CHD8 interacts with c-MYC (Dingar et al. 2015). CHD8 also interacts with CCCTC-binding factor thereby affecting transcriptional output through modulation of chromatin insulation, DNA methylation, and histone acetylation (Ishihara et al. 2006). Indeed, CHD8 protein expression is reduced in prostate cancer and demethylating agents such as 5-aza-2′-deoxycytidine induces *CHD8* expression at the transcriptional level (Damaschke et al. 2014). CHD8 is a coregulator of AR, and transcriptional activation the AR-responsive genes, such as TMPRSS2, requires CHD8 (Menon et al. 2010). CHD8 transcriptionally regulates genes encoding components of the WNT/β-catenin pathway and cell-cycle regulators (Sawada et al. 2013) and also has an effect on expression of

genes implicated in cancer and neurogenesis (Sugathan et al. 2014). CHD8 inhibits β-catenin signaling by recruiting histone H1 to promoters of WNT target genes (Nishiyama et al. 2012). CHD8 inhibits p53-mediated apoptosis by its ability to recruit histone H1, a process that occurs during embryogenesis (Nishiyama et al. 2009) but apparently not in the context of malignancy—at least in the setting that was analyzed (Sawada et al. 2013).

CHD8 is mutated in breast cancer (Pongor et al. 2015), deleted in 36% of gastric cancers and 29% of colorectal cancers (Kim et al. 2011), and silenced by promoter methylation in prostate cancer (Damaschke et al. 2014). CHD8 is mutated in CpG island methylator phenotype 1 (CIMP1)-positive colorectal cancers (Tahara et al. 2014a) and CHD8 mutations correlate with *Fusibacterium* status, CIMP1 positivity, microsatellite instability, as well as mutations in BRAF, KRAS, and P53 (Tahara et al. 2014b). A mouse model of BCR-Abl-driven acute lymphoblastic leukemia found that *Chd8* knockdown causes apoptosis, suggesting that targeting CHD8 is an effective therapy for patients with B-lymphoid malignancies (Shingleton and Hemann 2015). A genetic screen in a mouse model of acute myelogenous leukemia (AML) revealed that CHD8 is required for the ability of BRD4 to maintain AML through the H3K36-specific methyltransferase NSD3-short (Shen et al. 2015). Seemingly at odds with these mouse studies, high CHD8 expression in gastric cancers correlates with favorable patient survival (Sawada et al. 2013). In fact, enhanced nuclear expression of CHD8 has been shown to correlate with decreased survival and increased metastasis in patients with prostate cancer (Damaschke et al. 2014). Thus, while some reports suggest that CHD8 has tumor-suppressive functions, others clearly indicate that it is endowed with more nefarious pro-oncogenic capabilities. This apparent dichotomy warrants further clarification.

Some of the factors upstream of CHD7 are similar to those that regulate CHD8, such as environmental insults by *Fusibacterium* and the correlation between *CHD7* mutation, CIMP1, and genomic instability (Kim et al. 2011; Tahara et al. 2014a,b). *CHD7* is also mutated in response to tobacco smoke in small-cell lung cancer, having an in-frame duplication of exons 3–7, or being expressed as a fusion with PVT-1 (Pleasance et al. 2010). The factors that work in parallel to CHD7 are not well understood, although one example is that studies in mouse show that Chd7 is recruited by Smad1, Smad5, and Smad8 to promoters of cardiogenic genes (Liu et al. 2014). Pathways downstream from Chd7 include induction of Bmp signaling in mice (Jiang et al. 2012; Liu et al. 2014) and inhibition of CHK1 phosphorylation-dependent DNA damage repair in response to gemcitabine in human pancreatic cancer cells (Colbert et al. 2014). CHD7 and ES cell genes are aberrantly up-regulated in cutaneous T-cell lymphoma, leading to stem-cell-like features (Litvinov et al. 2014). Consistent with these findings made using human cells, conditional deletion of Chd7 in mice reveals that Chd7 maintains quiescence, thereby preventing premature depletion of neural stem cells (Jones et al. 2015).

A number of mouse models show Chd7's pleiotropic roles in development. For example, the spontaneous heterozygous Chd7 mutations in "looper" and "whirligig" mice lead to CHARGE syndrome-like features (Ogier et al. 2014) and olfaction and reproductive defects (Bergman et al. 2010), respectively. Heterozygous disruption of Chd7 causes hearing loss (Hurd et al. 2011), ear defects (Adams et al. 2007; Tian et al. 2012), and defects in puberty and reproduction (Layman et al. 2011), whereas homozygous disruption of Chd7 causes embryonic lethality at 11 d of gestation (Sperry et al. 2014). While these studies highlight the critical role of Chd7 in development (reviewed in Layman et al. 2010), it is also clear that Chd7 modulates pathways central in tumorigenesis. Indeed, the CHARGE syndrome phenotypes of Chd7-compromised mice are at least partially a result of enhanced p53 activity (Van Nostrand and Attardi 2014). This is reminiscent of the finding that gain of Chd5 dosage enhances p53 activity, leading to developmental abnormalities and neonatal lethality caused by over exuberant apoptosis (Bagchi et al. 2007). But in contrast to Chd5, which promotes p53-mediated apoptosis, Chd7 appears to inhibit it (Bagchi et al. 2007).

Whereas there are loss-of-function mutations in CHD7 (like CHD8) in colorectal cancer, and these lesions correlate with mutations in BRAF, P53, and KRAS (Tahara et al. 2014b), there is enhanced expression of CHD7 in cutaneous T-cell lymphoma (Litvinov et al. 2014). Consistent with the concept that CHD7 promotes oncogenesis, low-level CHD7 protein expression correlates with enhanced survival of patients with pancreatic cancer (Colbert et al. 2014). This study showed that CHD7 depletion enhances the sensitivity of pancreatic cancer cells to gemcitabine by triggering DNA damage via ATR-mediated phosphorylation of CHK1. As has been suggested for CHD8, CHD7 depletion may render current therapies more effective by enhancing cell death, at least in the case of pancreatic cancer.

The roles of the remaining subfamily III proteins, CHD6 and CHD9, in cancer are at present much more obscure. CHD6 has been reported to map within a minimally common region of amplification in colorectal cancer (Ali Hassan et al. 2014), and CHD6 is mutated in both colorectal tumors (Mouradov et al. 2014) and transitional cell carcinoma of the bladder (Gui et al. 2011). In addition, it has been suggested that CHD6, like many other CHD proteins, regulates DNA damage repair (Stanley et al. 2013). Evidence for CHD9 playing a role in cancer is even less compelling, but CHD9 mutations have been reported in gastric and colorectal cancers (Kim et al. 2011). Whether these mutations are bone fide drivers or merely passengers of tumorigenesis remains to be evaluated. Thus, there is clear evidence that CHD subfamily III members, in particular CHD8 and to some extent CHD7, are critical cancer genes. However, in stark contrast to the tumor-suppressive roles ascribed to members of subfamily II, subfamily III members also have pro-oncogenic roles in some settings and their inhibition is proving to be an effective therapeutic strategy.

CONCLUDING REMARKS

The CHD family shares core motifs, with unique features equipping different members with highly variable functions. While subfamily II members have been defined as potent tumor suppressors, members of subfamily I and subfamily III have tumor-suppressive capabilities in some contexts but oncogenic capabilities in others. Lesions in members of each subfamily can define tumor subtype and predict patient survival. Whereas activation of subfamily II CHD members may hold promise as an effective therapeutic strategy, inactivation of subfamily I and III CHD members reveal vulnerabilities that conquer chemoresistance, which may be exploited in the oncology clinic.

ACKNOWLEDGMENTS

I thank members of my laboratory for helpful discussions. While every effort has been made to include a thorough and comprehensive review of the current literature, relevant publications may have been inadvertently omitted. This work was supported by the National Institutes of Health (R01CA190997 and R21OD018332) and the Stanley Family Foundation.

REFERENCES

Adams ME, Hurd EA, Beyer LA, Swiderski DL, Raphael Y, Martin DM. 2007. Defects in vestibular sensory epithelia and innervation in mice with loss of Chd7 function: Implications for human CHARGE syndrome. *J Comp Neurol* **504:** 519–532.

Ali Hassan NZ, Mokhtar NM, Kok Sin T, Mohamed Rose I, Sagap I, Harun R, Jamal R. 2014. Integrated analysis of copy number variation and genome-wide expression profiling in colorectal cancer tissues. *PLoS ONE* **9:** e92553.

Attard G, Parker C, Eeles RA, Schroder F, Tomlins SA, Tannock I, Drake CG, de Bono JS. 2016. Prostate cancer. *Lancet* **387:** 70–82.

Bagchi A, Mills AA. 2008. The quest for the 1p36 tumor suppressor. *Cancer Res* **68:** 2551–2556.

Bagchi A, Papazoglu C, Wu Y, Capurso D, Brodt M, Francis D, Bredel M, Vogel H, Mills AA. 2007. CHD5 is a tumor suppressor at human 1p36. *Cell* **128:** 459–475.

Bandres E, Malumbres R, Cubedo E, Honorato B, Zarate R, Labarga A, Gabisu U, Sola JJ, Garcia-Foncillas J. 2007. A gene signature of 8 genes could identify the risk of recurrence and progression in Dukes' B colon cancer patients. *Oncol Rep* **17:** 1089–1094.

Bergman JE, Bosman EA, van Ravenswaaij-Arts CM, Steel KP. 2010. Study of smell and reproductive organs in a mouse model for CHARGE syndrome. *Eur J Hum Genet* **18:** 171–177.

Blattner M, Lee DJ, O'Reilly C, Park K, MacDonald TY, Khani F, Turner KR, Chiu YL, Wild PJ, Dolgalev I, et al. 2014. SPOP mutations in prostate cancer across demographically diverse patient cohorts. *Neoplasia* **16:** 14–20.

Brehm A, Nielsen SJ, Miska EA, McCance DJ, Reid JL, Bannister AJ, Kouzarides T. 1999. The E7 oncoprotein associates with Mi2 and histone deacetylase activity to promote cell growth. *EMBO J* **18:** 2449–2458.

Burkhardt L, Fuchs S, Krohn A, Masser S, Mader M, Kluth M, Bachmann F, Huland H, Steuber T, Graefen M, et al. 2013. CHD1 is a 5q21 tumor suppressor required for ERG rearrangement in prostate cancer. *Cancer Res* **73:** 2795–2805.

Cai C, Ashktorab H, Pang X, Zhao Y, Sha W, Liu Y, Gu X. 2012. MicroRNA-211 expression promotes colorectal cancer cell growth in vitro and in vivo by targeting tumor suppressor CHD5. *PLoS ONE* **7:** e29750.

Cai Y, Geutjes EJ, de Lint K, Roepman P, Bruurs L, Yu LR, Wang W, van Blijswijk J, Mohammad H, de Rink I, et al. 2014. The NuRD complex cooperates with DNMTs to maintain silencing of key colorectal tumor suppressor genes. *Oncogene* **33:** 2157–2168.

Caldon CE, Sergio CM, Schutte J, Boersma MN, Sutherland RL, Carroll JS, Musgrove EA. 2009. Estrogen regulation of cyclin E2 requires cyclin D1 but not c-Myc. *Mol Cell Biol* **29:** 4623–4639.

Chenier S, Yoon G, Argiropoulos B, Lauzon J, Laframboise R, Ahn JW, Ogilvie CM, Lionel AC, Marshall CR, Vaags AK, et al. 2014. CHD2 haploinsufficiency is associated with developmental delay, intellectual disability, epilepsy and behavioural problems. *J Neurodev Disord* **6:** 9.

Chudnovsky Y, Kim D, Zheng S, Whyte WA, Bansal M, Bray MA, Gopal S, Theisen MA, Bilodeau S, Thiru P, et al. 2014. ZFHX4 interacts with the NuRD core member CHD4 and regulates the glioblastoma tumor-initiating cell state. *Cell Rep* **6:** 313–324.

Clark JP, Cooper CS. 2009. ETS gene fusions in prostate cancer. *Nat Rev Urol* **6:** 429–439.

Colbert LE, Petrova AV, Fisher SB, Pantazides BG, Madden MZ, Hardy CW, Warren MD, Pan Y, Nagaraju GP, Liu EA, et al. 2014. CHD7 expression predicts survival outcomes in patients with resected pancreatic cancer. *Cancer Res* **74:** 2677–2687.

Damaschke NA, Yang B, Blute ML Jr, Lin CP, Huang W, Jarrard DF. 2014. Frequent disruption of chromodomain helicase DNA-binding protein 8 (CHD8) and functionally associated chromatin regulators in prostate cancer. *Neoplasia* **16:** 1018–1027.

Delmas V, Stokes DG, Perry RP. 1993. A mammalian DNA-binding protein that contains a chromodomain and an SNF2/SWI2-like helicase domain. *Proc Natl Acad Sci* **90:** 2414–2418.

Denslow SA, Wade PA. 2007. The human Mi-2/NuRD complex and gene regulation. *Oncogene* **26:** 5433–5438.

Desai MA, Webb HD, Sinanan LM, Scarsdale JN, Walavalkar NM, Ginder GD, Williams DC Jr. 2015. An intrinsically disordered region of methyl-CpG binding domain protein 2 (MBD2) recruits the histone deacetylase core of the NuRD complex. *Nucleic Acids Res* **43:** 3100–3113.

Dey P, Ponnusamy MP, Deb S, Batra SK. 2011. Human RNA polymerase II–association factor 1 (hPaf1/PD2) regulates histone methylation and chromatin remodeling in pancreatic cancer. *PLoS ONE* **6:** e26926.

Dingar D, Kalkat M, Chan PK, Srikumar T, Bailey SD, Tu WB, Coyaud E, Ponzielli R, Kolyar M, Jurisica I, et al. 2015. BioID identifies novel c-MYC interacting partners in cultured cells and xenograft tumors. *J Proteomics* **118:** 95–111.

Du X, Wu T, Lu J, Zang L, Song N, Yang T, Zhao H, Wang S. 2013. Decreased expression of chromodomain helicase DNA-binding protein 5 is an unfavorable prognostic marker in patients with primary gallbladder carcinoma. *Clin Transl Oncol* **15:** 198–204.

Falconer H, Sundqvist J, Xu H, Vodolazkaia A, Fassbender A, Kyama C, Bokor A, D'Hooghe TM. 2012. Analysis of common variations in tumor-suppressor genes on chr1p36 among Caucasian women with endometriosis. *Gynecol Oncol* **127:** 398–402.

Fang QL, Yin YR, Xie CR, Zhang S, Zhao WX, Pan C, Wang XM, Yin ZY. 2015. Mechanistic and biological significance of DNA methyltransferase 1 upregulated by growth factors in human hepatocellular carcinoma. *Int J Oncol* **46:** 782–790.

Fatemi M, Paul TA, Brodeur GM, Shokrani B, Brim H, Ashktorab H. 2014. Epigenetic silencing of CHD5, a novel tumor-suppressor gene, occurs in early colorectal cancer stages. *Cancer* **120:** 172–180.

Fisher KW, Montironi R, Lopez Beltran A, Moch H, Wang L, Scarpelli M, Williamson SR, Koch MO, Cheng L. 2015. Molecular foundations for personalized therapy in prostate cancer. *Curr Drug Target* **16:** 103–114.

Flanagan JF, Mi LZ, Chruszcz M, Cymborowski M, Clines KL, Kim Y, Minor W, Rastinejad F, Khorasanizadeh S. 2005. Double chromodomains cooperate to recognize the methylated histone H3 tail. *Nature* **438:** 1181–1185.

Fu J, Qin L, He T, Qin J, Hong J, Wong J, Liao L, Xu J. 2011. The TWIST/Mi2/NuRD protein complex and its essential role in cancer metastasis. *Cell Res* **21:** 275–289.

Fujita T, Igarashi J, Okawa ER, Gotoh T, Manne J, Kolla V, Kim J, Zhao H, Pawel BR, London WB, et al. 2008. CHD5, a tumor suppressor gene deleted from 1p36.31 in neuroblastomas. *J Natl Cancer Inst* **100:** 940–949.

Gao D, Vela I, Sboner A, Iaquinta PJ, Karthaus WR, Gopalan A, Dowling C, Wanjala JN, Undvall EA, Arora VK, et al. 2014. Organoid cultures derived from patients with advanced prostate cancer. *Cell* **159:** 176–187.

Garcia I, Mayol G, Rodriguez E, Sunol M, Gershon TR, Rios J, Cheung NK, Kieran MW, George RE, Perez-Atayde AR, et al. 2010. Expression of the neuron-specific protein CHD5 is an independent marker of outcome in neuroblastoma. *Mol Cancer* **9:** 277.

Gaspar-Maia A, Alajem A, Polesso F, Sridharan R, Mason MJ, Heidersbach A, Ramalho-Santos J, McManus MT, Plath K, Meshorer E, et al. 2009. Chd1 regulates open chromatin and pluripotency of embryonic stem cells. *Nature* **460:** 863–868.

Gorringe KL, Choong DY, Williams LH, Ramakrishna M, Sridhar A, Qiu W, Bearfoot JL, Campbell IG. 2008. Mutation and methylation analysis of the chromodomain-helicase-DNA binding 5 gene in ovarian cancer. *Neoplasia* **10:** 1253–1258.

Grasso CS, Wu YM, Robinson DR, Cao X, Dhanasekaran SM, Khan AP, Quist MJ, Jing X, Lonigro RJ, Brenner JC, et al. 2012. The mutational landscape of lethal castration-resistant prostate cancer. *Nature* **487:** 239–243.

Gui Y, Guo G, Huang Y, Hu X, Tang A, Gao S, Wu R, Chen C, Li X, Zhou L, et al. 2011. Frequent mutations of chromatin remodeling genes in transitional cell carcinoma of the bladder. *Nat Genet* **43:** 875–878.

Guillemette S, Serra RW, Peng M, Hayes JA, Konstantinopoulos PA, Green MR, Cantor SB. 2015. Resistance to therapy in BRCA2 mutant cells due to loss of the nucleosome remodeling factor CHD4. *Genes Dev* **29:** 489–494.

Hall WA, Petrova AV, Colbert LE, Hardy CW, Fisher SB, Saka B, Shelton JW, Warren MD, Pantazides BG, Gandhi K, et al. 2014. Low CHD5 expression activates the DNA damage response and predicts poor outcome in patients undergoing adjuvant therapy for resected pancreatic cancer. *Oncogene* **33:** 5450–5456.

Henrich KO, Schwab M, Westermann F. 2012. 1p36 tumor suppression—A matter of dosage? *Cancer Res* **72:** 6079–6088.

Hosokawa H, Tanaka T, Suzuki Y, Iwamura C, Ohkubo S, Endoh K, Kato M, Endo Y, Onodera A, Tumes DJ, et al. 2013. Functionally distinct Gata3/Chd4 complexes coordinately establish T helper 2 (Th2) cell identity. *Proc Natl Acad Sci* **110:** 4691–4696.

Huang S, Gulzar ZG, Salari K, Lapointe J, Brooks JD, Pollack JR. 2012. Recurrent deletion of CHD1 in prostate cancer with relevance to cell invasiveness. *Oncogene* **31:** 4164–4170.

Hurd EA, Adams ME, Layman WS, Swiderski DL, Beyer LA, Halsey KE, Benson JM, Gong TW, Dolan DF, Raphael Y, et al. 2011. Mature middle and inner ears express Chd7 and exhibit distinctive pathologies in a mouse model of CHARGE syndrome. *Hear Res* **282:** 184–195.

Ishihara K, Oshimura M, Nakao M. 2006. CTCF-dependent chromatin insulator is linked to epigenetic remodeling. *Mol Cell* **23:** 733–742.

Jiang X, Zhou Y, Xian L, Chen W, Wu H, Gao X. 2012. The mutation in Chd7 causes misexpression of Bmp4 and developmental defects in telencephalic midline. *Am J Pathol* **181:** 626–641.

Jones KM, Saric N, Russell JP, Andoniadou CL, Scambler PJ, Basson MA. 2015. CHD7 maintains neural stem cell quiescence and prevents premature stem cell depletion in the adult hippocampus. *Stem Cells* **33:** 196–210.

Kandoth C, McLellan MD, Vandin F, Ye K, Niu B, Lu C, Xie M, Zhang Q, McMichael JF, Wyczalkowski MA, et al. 2013. Mutational landscape and significance across 12 major cancer types. *Nature* **502:** 333–339.

Khaleque MA, Bharti A, Gong J, Gray PJ, Sachdev V, Ciocca DR, Stati A, Fanelli M, Calderwood SK. 2008. Heat shock factor 1 represses estrogen-dependent transcription through association with MTA1. *Oncogene* **27:** 1886–1893.

Kim MS, Chung NG, Kang MR, Yoo NJ, Lee SH. 2011. Genetic and expressional alterations of CHD genes in gastric and colorectal cancers. *Histopathology* **58:** 660–668.

Klement K, Luijsterburg MS, Pinder JB, Cena CS, Del Nero V, Wintersinger CM, Dellaire G, van Attikum H, Goodarzi AA. 2014. Opposing ISWI- and CHD-class chromatin remodeling activities orchestrate heterochromatic DNA repair. *J Cell Biol* **207:** 717–733.

Koch JG, Gu X, Han Y, El-Naggar AK, Olson MV, Medina D, Jerry DJ, Blackburn AC, Peltz G, Amos CI, et al. 2007. Mammary tumor modifiers in BALB/cJ mice heterozygous for p53. *Mamm Genome* **18:** 300–309.

Kolla V, Naraparaju K, Zhuang T, Higashi M, Kolla S, Blobel GA, Brodeur GM. 2015. The tumour suppressor CHD5 forms a NuRD-type chromatin remodelling complex. *Biochem J* **468:** 345–352.

Koyama H, Zhuang T, Light JE, Kolla V, Higashi M, McGrady PW, London WB, Brodeur GM. 2012. Mechanisms of CHD5 inactivation in neuroblastomas. *Clin Cancer Res* **18:** 1588–1597.

Kulkarni S, Nagarajan P, Wall J, Donovan DJ, Donell RL, Ligon AH, Venkatachalam S, Quade BJ. 2008. Disruption of chromodomain helicase DNA binding protein 2 (CHD2) causes scoliosis. *Am J Med Genet A* **146A:** 1117–1127.

Lang J, Tobias ES, Mackie R. 2011. Preliminary evidence for involvement of the tumour suppressor gene CHD5 in a family with cutaneous melanoma. *Br J Dermatol* **164:** 1010–1016.

Layman WS, Hurd EA. 2011. Reproductive dysfunction and decreased GnRH neurogenesis in a mouse model of CHARGE syndrome. *Hum Mol Genet* **20:** 3138–3150.

Layman WS, Hurd EA, Martin DM. 2010. Chromodomain proteins in development: Lessons from CHARGE syndrome. *Clin Genet* **78:** 11–20.

Lee S, Das HK. 2010. Transcriptional regulation of the presenilin-1 gene controls γ-secretase activity. *Front Biosci* **2:** 22–35.

Lee YH, Ma H, Tan TZ, Ng SS, Soong R, Mori S, Fu XY, Zernicka-Goetz M, Wu Q. 2012. Protein arginine methyltransferase 6 regulates embryonic stem cell identity. *Stem Cells Dev* **21:** 2613–2622.

Le Gallo M, O'Hara AJ, Rudd ML, Urick ME, Hansen NF, O'Neil NJ, Price JC, Zhang S, England BM, Godwin AK, et al. 2012. Exome sequencing of serous endometrial tumors identifies recurrent somatic mutations in chromatin-remodeling and ubiquitin ligase complex genes. *Nat Genet* **44:** 1310–1315.

Li W, Mills AA. 2014. Packing for the journey: CHD5 remodels the genome. *Cell Cycle* **13:** 1833–1834.

Li H, Xu W, Huang Y, Huang X, Xu L, Lv Z. 2012. Genistein demethylates the promoter of CHD5 and inhibits neuroblastoma growth in vivo. *Int J Mol Med* **30:** 1081–1086.

Li W, Wu J, Kim SY, Zhao M, Hearn SA, Zhang MQ, Meistrich ML, Mills AA. 2014. Chd5 orchestrates chromatin remodelling during sperm development. *Nat Commun* **5:** 3812.

Litvinov IV, Netchiporouk E, Cordeiro B, Zargham H, Pehr K, Gilbert M, Zhou Y, Moreau L, Woetmann A, Odum N, et al. 2014. Ectopic expression of embryonic stem cell and other developmental genes in cutaneous T-cell lymphoma. *Oncoimmunology* **3:** e970025.

Liu W, Lindberg J, Sui G, Luo J, Egevad L, Li T, Xie C, Wan M, Kim ST, Wang Z, et al. 2012. Identification of novel CHD1-associated collaborative alterations of genomic structure and functional assessment of CHD1 in prostate cancer. *Oncogene* **31:** 3939–3948.

Liu Y, Harmelink C, Peng Y, Chen Y, Wang Q, Jiao K. 2014. CHD7 interacts with BMP R-SMADs to epigenetically regulate cardiogenesis in mice. *Hum Mol Genet* **23:** 2145–2156.

Lyn-Cook L, Word B, George N, Lyn-Cook B, Hammons G. 2014. Effect of cigarette smoke condensate on gene promoter methylation in human lung cells. *Tob Induc Dis* **12:** 15.

Mallette FA, Richard S. 2012. JMJD2A promotes cellular transformation by blocking cellular senescence through transcriptional repression of the tumor suppressor CHD5. *Cell Rep* **2:** 1233–1243.

Marfella CG, Imbalzano AN. 2007. The Chd family of chromatin remodelers. *Mutat Res* **618:** 30–40.

Marfella CG, Ohkawa Y, Coles AH, Garlick DS, Jones SN, Imbalzano AN. 2006. Mutation of the SNF2 family member Chd2 affects mouse development and survival. *J Cell Physiol* **209:** 162–171.

Martin TJ, Peer CJ, Figg WD. 2013. Uncovering the genetic landscape driving castration-resistant prostate cancer. *Cancer Biol Ther* **14:** 399–400.

Menon T, Yates JA, Bochar DA. 2010. Regulation of androgen-responsive transcription by the chromatin remodeling factor CHD8. *Mol Endocrinol* **24:** 1165–1174.

Mokarram P, Kumar K, Brim H, Naghibalhossaini F, Saberifiroozi M, Nouraie M, Green R, Lee E, Smoot DT, Ashktorab H. 2009. Distinct high-profile methylated genes in colorectal cancer. *PLoS ONE* **4:** e7012.

Mouradov D, Sloggett C, Jorissen RN, Love CG, Li S, Burgess AW, Arango D, Strausberg RL, Buchanan D, Wormald S, et al. 2014. Colorectal cancer cell lines are representative models of the main molecular subtypes of primary cancer. *Cancer Res* **74:** 3238–3247.

Mulero-Navarro S, Esteller M. 2008. Chromatin remodeling factor CHD5 is silenced by promoter CpG island hypermethylation in human cancer. *Epigenetics* **3:** 210–215.

Nagarajan P, Onami TM, Rajagopalan S, Kania S, Donnell R, Venkatachalam S. 2009. Role of chromodomain helicase DNA-binding protein 2 in DNA damage response signaling and tumorigenesis. *Oncogene* **28:** 1053–1062.

Nio K, Yamashita T, Okada H, Kondo M, Hayashi T, Hara Y, Nomura Y, Zeng SS, Yoshida M, Hayashi T, et al. 2015. Defeating EpCAM+ liver cancer stem cells by targeting chromatin remodeling enzyme CHD4 in human hepatocellular carcinoma. *J Hepatol* **63:** 1164–1172.

Nishiyama M, Oshikawa K, Tsukada Y, Nakagawa T, Iemura S, Natsume T, Fan Y, Kikuchi A, Skoultchi AI, Nakayama KI. 2009. CHD8 suppresses p53-mediated apoptosis through histone H1 recruitment during early embryogenesis. *Nat Cell Biol* **11:** 172–182.

Nishiyama M, Skoultchi AI, Nakayama KI. 2012. Histone H1 recruitment by CHD8 is essential for suppression of the Wnt-β-catenin signaling pathway. *Mol Cell Biol* **32:** 501–512.

Ogier JM, Carpinelli MR, Arhatari BD, Symons RC, Kile BT, Burt RA. 2014. CHD7 deficiency in "Looper," a new mouse model of CHARGE syndrome, results in ossicle malformation, otosclerosis and hearing impairment. *PLoS ONE* **9:** e97559.

Oliver SS, Musselman CA, Srinivasan R, Svaren JP, Kutateladze TG, Denu JM. 2012. Multivalent recognition of histone tails by the PHD fingers of CHD5. *Biochemistry* **51:** 6534–6544.

Paul S, Kuo A, Schalch T, Vogel H, Joshua-Tor L, McCombie WR, Gozani O, Hammell M, Mills AA. 2013. Chd5 requires PHD-mediated histone 3 binding for tumor suppression. *Cell Rep* **3:** 92–102.

Pleasance ED, Stephens PJ, O'Meara S, McBride DJ, Meynert A, Jones D, Lin ML, Beare D, Lau KW, Greenman C, et al. 2010. A small-cell lung cancer genome with complex signatures of tobacco exposure. *Nature* **463:** 184–190.

Pongor L, Kormos M, Hatzis C, Pusztai L, Szabo A, Gyorffy B. 2015. A genome-wide approach to link genotype to clinical outcome by utilizing next generation sequencing and gene chip data of 6,697 breast cancer patients. *Genome Med* **7:** 104.

Qi W, Chen H, Xiao T, Wang R, Li T, Han L, Zeng X. 2015. Acetyltransferase p300 collaborates with chromodomain helicase DNA-binding protein 4 (CHD4) to facilitate DNA double-strand break repair. *Mutagenesis* **31:** 193–203.

Qu Y, Dang S, Hou P. 2013. Gene methylation in gastric cancer. *Int J Clin Chem* **424:** 53–65.

Quan J, Yusufzai T. 2014. The tumor suppressor chromodomain helicase DNA-binding protein 5 (CHD5) remodels nucleosomes by unwrapping. *J Biol Chem* **289:** 20717–20726.

Quan J, Adelmant G, Marto JA, Look AT, Yusufzai T. 2014. The chromatin remodeling factor CHD5 is a transcriptional repressor of WEE1. *PLoS ONE* **9:** e108066.

Rahman S, Sowa ME, Ottinger M, Smith JA, Shi Y, Harper JW, Howley PM. 2011. The Brd4 extraterminal domain confers transcription activation independent of pTEFb by recruiting multiple proteins, including NSD3. *Mol Cell Biol* **31:** 2641–2652.

Rajagopalan S, Nepa J, Venkatachalam S. 2012. Chromodomain helicase DNA-binding protein 2 affects the repair of X-ray and UV-induced DNA damage. *Environ Mol Mutagen* **53:** 44–50.

Robbins CM, Tembe WA, Baker A, Sinari S, Moses TY, Beckstrom-Sternberg S, Beckstrom-Sternberg J, Barrett M, Long J, Chinnaiyan A, et al. 2011. Copy number and targeted mutational analysis reveals novel somatic events in metastatic prostate tumors. *Genome Res* **21:** 47–55.

Rodrigues LU, Rider L, Nieto C, Romero L, Karimpour-Fard A, Loda M, Lucia MS, Wu M, Shi L, Cimic A, et al. 2015. Coordinate loss of MAP3K7 and CHD1 promotes aggressive prostate cancer. *Cancer Res* **75:** 1021–1034.

Rodriguez D, Bretones G, Quesada V, Villamor N, Arango JR, Lopez-Guillermo A, Ramsay AJ, Baumann T, Quiros PM, Navarro A, et al. 2015. Mutations in CHD2 cause defective association with active chromatin in chronic lymphocytic leukemia. *Blood* **126:** 195–202.

Ronan JL, Wu W, Crabtree GR. 2013. From neural development to cognition: Unexpected roles for chromatin. *Nat Rev Genet* **14:** 347–359.

Russo J, Russo IH. 2012. Molecular basis of pregnancy-induced breast cancer prevention. *Horm Mol Biol Clin Investig* **9:** 3–10.

Sarahan KA, Fisler JS, Warden CH. 2011. Four out of eight genes in a mouse chromosome 7 congenic donor region are candidate obesity genes. *Physiol Genomics* **43:** 1049–1055.

Sawada G, Ueo H, Matsumura T, Uchi R, Ishibashi M, Mima K, Kurashige J, Takahashi Y, Akiyoshi S, Sudo T, et al. 2013. CHD8 is an independent prognostic indicator that regulates Wnt/β-catenin signaling and the cell cycle in gastric cancer. *Oncol Rep* **30:** 1137–1142.

Scott AF, Mohr DW, Ling H, Scharpf RB, Zhang P, Liptak GS. 2014. Characterization of the genomic architecture and mutational spectrum of a small cell prostate carcinoma. *Genes* **5:** 366–384.

Shen C, Ipsaro JJ, Shi J, Milazzo JP, Wang E, Roe JS, Suzuki Y, Pappin DJ, Joshua-Tor L, Vakoc CR. 2015. NSD3-short is an adaptor protein that couples BRD4 to the CHD8 chromatin remodeler. *Mol Cell* **60:** 847–859.

Shingleton JR, Hemann MT. 2015. The chromatin regulator CHD8 is a context-dependent mediator of cell survival in murine hematopoietic malignancies. *PLoS ONE* **10:** e0143275.

Siggens L, Corddeddu L, Ronnerblad M, Lennartsson A, Ekwall K. 2015. Transcription-coupled recruitment of human CHD1 and CHD2 influences chromatin accessibility and histone H3 and H3.3 occupancy at active chromatin regions. *Epigenetics Chromatin* **8:** 4.

Sowalsky AG, Xia Z, Wang L, Zhao H, Chen S, Bubley GJ, Balk SP, Li W. 2015. Whole transcriptome sequencing reveals extensive unspliced mRNA in metastatic castration-resistant prostate cancer. *Mol Cancer Res* **13:** 98–106.

Sperry ED, Hurd EA, Durham MA, Reamer EN, Stein AB, Martin DM. 2014. The chromatin remodeling protein CHD7, mutated in CHARGE syndrome, is necessary for proper craniofacial and tracheal development. *Dev Dyn* **243:** 1055–1066.

Srinivasan R, Mager GM, Ward RM, Mayer J, Svaren J. 2006. NAB2 represses transcription by interacting with the CHD4 subunit of the nucleosome remodeling and deacetylase (NuRD) complex. *J Biol Chem* **281:** 15129–15137.

Stanley FK, Moore S, Goodarzi AA. 2013. CHD chromatin remodelling enzymes and the DNA damage response. *Mutat Res* **750:** 31–44.

Stokes DG, Perry RP. 1995. DNA-binding and chromatin localization properties of CHD1. *Mol Cell Biol* **15:** 2745–2753.

Sugathan A, Biagioli M, Golzio C, Erdin S, Blumenthal I, Manavalan P, Ragavendran A, Brand H, Lucente D, Miles J, et al. 2014. CHD8 regulates neurodevelopmental pathways associated with autism spectrum disorder in neural progenitors. *Proc Natl Acad Sci* **111:** E4468–E4477.

Tahara T, Yamamoto E, Madireddi P, Suzuki H, Maruyama R, Chung W, Garriga J, Jelinek J, Yamano HO, Sugai T, et al. 2014a. Colorectal carcinomas with CpG island methylator phenotype 1 frequently contain mutations in chromatin regulators. *Gastroenterology* **146:** 530–538.

Tahara T, Yamamoto E, Suzuki H, Maruyama R, Chung W, Garriga J, Jelinek J, Yamano HO, Sugai T, An B, et al. 2014b. Fusobacterium in colonic flora and molecular features of colorectal carcinoma. *Cancer Res* **74:** 1311–1318.

Tan S, Ding K, Li R, Zhang W, Li G, Kong X, Qian P, Lobie PE, Zhu T. 2014. Identification of miR-26 as a key mediator of estrogen stimulated cell proliferation by targeting CHD1, GREB1 and KPNA2. *Breast Cancer Res* **16:** R40.

Tao Y, Liu S, Briones V, Geiman TM, Muegge K. 2011. Treatment of breast cancer cells with DNA demethylating agents leads to a release of Pol II stalling at genes with DNA-hypermethylated regions upstream of TSS. *Nucleic Acids Res* **39:** 9508–9520.

Tereshchenko IV, Zhong H, Chekmareva MA, Kane-Goldsmith N, Santanam U, Petrosky W, Stein MN, Ganesan S, Singer EA, Moore D, et al. 2014. ERG and CHD1 heterogeneity in prostate cancer: Use of confocal microscopy in assessment of microscopic foci. *Prostate* **74:** 1551–1559.

Tian C, Yu H, Yang B, Han F, Zheng Y, Bartels CF, Schelling D, Arnold JE, Scacheri PC, Zheng QY. 2012. Otitis media in a new mouse model for CHARGE syndrome with a deletion in the Chd7 gene. *PLoS ONE* **7:** e34944.

Torchy MP, Hamiche A, Klaholz BP. 2015. Structure and function insights into the NuRD chromatin remodeling complex. *Cell Mol Life Sci* **72:** 2491–2507.

Van Nostrand JL, Attardi LD. 2014. Guilty as CHARGED: p53's expanding role in disease. *Cell Cycle* **13:** 3798–3807.

Wang X, Lau KK, So LK, Lam YW. 2009. CHD5 is downregulated through promoter hypermethylation in gastric cancer. *J Biomed Sci* **16:** 95.

Wang J, Chen H, Fu S, Xu ZM, Sun KL, Fu WN. 2011. The involvement of CHD5 hypermethylation in laryngeal squamous cell carcinoma. *Oral Oncol* **47:** 601–608.

Wang L, He S, Tu Y, Ji P, Zong J, Zhang J, Feng F, Zhao J, Gao G, Zhang Y. 2013. Downregulation of chromatin remodeling factor CHD5 is associated with a poor prognosis in human glioma. *J Clin Neurosci* **20:** 958–963.

Wong RR, Chan LK, Tsang TP, Lee CW, Cheung TH, Yim SF, Siu NS, Lee SN, Yu MY, Chim SS, et al. 2011. CHD5 downregulation associated with poor prognosis in epithelial ovarian cancer. *Gynecol Obstet Invest* **72:** 203–207.

Wu X, Zhu Z, Li W, Fu X, Su D, Fu L, Zhang Z, Luo A, Sun X, Fu L, et al. 2012. Chromodomain helicase DNA binding protein 5 plays a tumor suppressor role in human breast cancer. *Breast Cancer Res* **14:** R73.

Wu S, Yang Z, Ye R, An D, Li C, Wang Y, Wang Y, Huang Y, Liu H, Li F, et al. 2015. Novel variants in MLL confer to bladder cancer recurrence identified by whole-exome sequencing. *Oncotarget* **7:** 2629–2645.

Xie CR, Li Z, Sun HG, Wang FQ, Sun Y, Zhao WX, Zhang S, Zhao WX, Wang XM, Yin ZY. 2015. Mutual regulation between CHD5 and EZH2 in hepatocellular carcinoma. *Oncotarget* **6:** 40940–40952.

Yamada M, Sato N, Ikeda S, Arai T, Sawabe M, Mori S, Yamada Y, Muramatsu M, Tanaka M. 2015. Association of the chromodomain helicase DNA-binding protein 4 (CHD4) missense variation p.D140E with cancer: Potential interaction with smoking. *Genes Chromosomes Cancer* **54:** 122–128.

Yu B, Swatkoski S, Holly A, Lee LC, Giroux V, Lee CS, Hsu D, Smith JL, Yuen G, Yue J, et al. 2015a. Oncogenesis driven by the Ras/Raf pathway requires the SUMO E2 ligase Ubc9. *Proc Natl Acad Sci* **112:** E1724–E1733.

Yu L, Gong X, Sun L, Yao H, Lu B, Zhu L. 2015b. miR-454 functions as an oncogene by inhibiting CHD5 in hepatocellular carcinoma. *Oncotarget* **6:** 39225–39234.

Zhang Y, LeRoy G, Seelig HP, Lane WS, Reinberg D. 1998. The dermatomyositis-specific autoantigen Mi2 is a component of a complex containing histone deacetylase and nucleosome remodeling activities. *Cell* **95:** 279–289.

Zhang XY, DeSalle LM, Patel JH, Capobianco AJ, Yu D, Thomas-Tikhonenko A, McMahon SB. 2005. Metastasis-associated protein 1 (MTA1) is an essential downstream effector of the c-MYC oncoprotein. *Proc Natl Acad Sci* **102:** 13968–13973.

Zhao R, Yan Q, Lv J, Huang H, Zheng W, Zhang B, Ma W. 2012. CHD5, a tumor suppressor that is epigenetically silenced in lung cancer. *Lung cancer* **76:** 324–331.

Zhao S, Choi M, Overton JD, Bellone S, Roque DM, Cocco E, Guzzo F, English DP, Varughese J, Gasparrini S, et al. 2013. Landscape of somatic single-nucleotide and copy-number mutations in uterine serous carcinoma. *Proc Natl Acad Sci* **110:** 2916–2921.

Zhao R, Meng F, Wang N, Ma W, Yan Q. 2014. Silencing of CHD5 gene by promoter methylation in leukemia. *PLoS ONE* **9:** e85172.

Zhuang T, Hess RA, Kolla V, Higashi M, Raabe TD, Brodeur GM. 2014. CHD5 is required for spermiogenesis and chromatin condensation. *Mech Dev* **131:** 35–46.

Cohesin Mutations in Cancer

Magali De Koninck and Ana Losada

Chromosome Dynamics Group, Molecular Oncology Programme, Spanish National Cancer Research Centre (CNIO), Madrid E-28029, Spain

Correspondence: alosada@cnio.es

Cohesin is a large ring-shaped protein complex, conserved from yeast to human, which participates in most DNA transactions that take place in the nucleus. It mediates sister chromatid cohesion, which is essential for chromosome segregation and homologous recombination (HR)-mediated DNA repair. Together with architectural proteins and transcriptional regulators, such as CTCF and Mediator, respectively, it contributes to genome organization at different scales and thereby affects transcription, DNA replication, and locus rearrangement. Although cohesin is essential for cell viability, partial loss of function can affect these processes differently in distinct cell types. Mutations in genes encoding cohesin subunits and regulators of the complex have been identified in several cancers. Understanding the functional significance of these alterations may have relevant implications for patient classification, risk prediction, and choice of treatment. Moreover, identification of vulnerabilities in cancer cells harboring cohesin mutations may provide new therapeutic opportunities and guide the design of personalized treatments.

Cohesin is one of the three structural maintenance of chromosomes (SMC) complexes that exist in eukaryotic cells. The other two are condensin and the Smc5/6 complex. They are all composed of an SMC heterodimer and additional non-SMC subunits arranged in a characteristic domain architecture (Haering and Gruber 2016). Remarkably, bacteria and archea also possess SMC complexes, although in this case the SMC proteins homodimerize. In all three kingdoms of life, the functions of SMC complexes are critical for genome organization, chromosome duplication, and segregation. In particular, cohesin was initially identified for its role in sister chromatid cohesion (Guacci et al. 1997; Michaelis et al. 1997; Losada et al. 1998), a requirement for proper chromosome segregation in mitosis and meiosis, as well as for homologous recombination (HR)-mediated DNA repair (Nasmyth and Haering 2009). In addition, cohesin is currently recognized as a major player in higher-order chromatin structure together with the CCCTC-binding factor (CTCF) (Phillips-Cremins et al. 2013; Mizuguchi et al. 2014). How the same complex can perform all of these different functions is far from understood. Germline mutations in cohesin and its regulators are at the origin of human developmental syndromes collectively known as cohesinopathies, the most prevalent of which is Cornelia de Lange syndrome (CdLS) (Horsfield et al. 2012). Recent sequencing efforts of cancer genomes have revealed the presence of somatic mutations in cohesin in several cancer types. Understanding how cohesin works and how it is regulated will likely help us recognize the contribution of these mutations to tumor initiation and progression.

THE BASIC BIOLOGY OF COHESIN

Composition and Architecture

Cohesin is a ring-shaped complex that consists of Smc1, Smc3, Rad21, and SA (see Table 1 for nomenclature). SMCs are 1000–1500 amino-acid-long proteins that contain two coiled-coil stretches separated by a flexible globular domain called "hinge." When folded at this domain, the amino and carboxyl termini of the protein are brought in proximity to create an ATPase head domain (hd; Fig. 1). Smc1 and Smc3 interact stably through their hinges and on the other end are bridged by the Rad21 subunit (Haering et al. 2002, 2004; Gligoris et al. 2014; Huis in 't Veld et al. 2014). The central region of Rad21 binds the fourth subunit of cohesin, SA (Haering et al. 2002; Orgil et al. 2015). SA is composed of many homologous huntingtin, elongation factor 3, A subunit, and TOR (HEAT) repeats and likely serves as an interaction platform for cohesin-interacting proteins (Hara et al. 2014). Among these are two regulatory subunits associated with chromatin-bound cohesin complexes throughout the cell cycle, Pds5 and Wapl (Sumara et al.

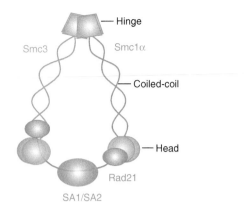

Figure 1. Cohesin composition and architecture. When the Smc1 and Smc3 proteins are folded at their flexible hinge domains, the NTP-binding motif and the DA box present at their amino- and carboxy-terminal globular domains come together to form a functional ATPase. Smc1 and Smc3 interact through their hinges, whereas the kleisin subunit Rad21 bridges their head domains and associates with SA. The outer diameter of the resulting ring-shaped complex is estimated at ~50 nm and could hold two 10-nm chromatin fibers.

2000; Losada et al. 2005; Gandhi et al. 2006; Kueng et al. 2006). There are also transient or position-specific interactors such as CTCF (Xiao et al. 2011) or the telomeric protein TRF1 (Canudas et al. 2007). In vertebrate somatic cells cohesin contains one of two SA sub-

Table 1. Nomenclature of cohesin subunits and regulators

Category	Yeast (*Saccharomyces cerevisiae*)	Mouse/human
Cohesin	Smc1	Smc1α (*SMC1A*) Smc1β (*SMC1B*)
	Smc3	Smc3 (*SMC3*)
	Scc1 Rec8	Rad21 (*RAD21*) Rad21L (*RAD21L1*) Rec8 (*REC8*)
	Scc3	SA1 (*STAG1*) SA2 (*STAG2*) SA3 (*STAG3*)
Associated factors	Pds5	Pds5A (*PDS5A*) Pds5B (*PDS5B*)
	Wapl/Rad61	Wapl (*WAPL*)
	−	Sororin (*CDCA5*)
Loader	Scc2	Nipbl (*NIPBL*)
	Scc4	Mau2 (*MAU2*)
CoAT	Eco1	Esco1 (*ESCO1*) Esco2 (*ESCO2*)
CoDAC	Hos1	Hdac8 (*HDAC8*)
Mitotic regulators	Sgo1	Sgo1 (*SGOL1*)
	Separase/Esp1	Separase (*ESPL1*)
	Securin/Pds1	Securin (*PTTG1*)

Meiosis-specific variants are shown in red. For mouse/human, the name of the gene appears in parentheses. CoAT, cohesin acetyl transferase; CoDAC, cohesin deacetylase.

units, SA1 or SA2 (Losada et al. 2000; Sumara et al. 2000; Remeseiro et al. 2012a). Additional meiosis-specific versions of all cohesin subunits exist except Smc3 (in red in Table 1).

Cohesin Interaction with DNA

Evidence from a number of in vivo and in vitro studies supports a model in which cohesin entraps the chromatin fiber within its ring structure (Gruber et al. 2003; Haering et al. 2008). Cohesin loading occurs in G_1 and requires ATP hydrolysis and a heterodimeric complex composed of Nipbl and Mau2 (Fig. 2) (Arumugam et al. 2003; Weitzer et al. 2003; Gillespie and Hirano 2004; Watrin et al. 2006). In vitro, loading can occur in the absence of the loader, albeit very inefficiently (Murayama and Uhlmann 2014). After loading, cohesin binding to chromatin is dynamic and unloading mediated by Wapl occurs throughout the cell cycle (Gerlich et al. 2006; Bernard et al. 2008). Some evidence

supports the idea that DNA enters and exits the cohesin ring through different interfaces or "gates" (Nasmyth 2011; Buheitel and Stemmann 2013; Eichinger et al. 2013). The entry gate requires dissociation of the Smc1 and Smc3 hinges (Gruber et al. 2006), whereas the exit gate would be located in the interface formed by the coiled coil emerging from the Smc3 hd and two α helices in the amino terminus of Rad21 (Gligoris et al. 2014; Huis in 't Veld et al. 2014). The opening/closure of this second gate is regulated by the ATPase activity of the SMCs, by DNA sensing through two lysines present in the Smc3 hd (K105 and K106 in human Smc3), and by Pds5-Wapl (Murayama and Uhlmann 2015).

Cohesin Distribution

Chromatin immunoprecipitation (ChIP) studies provide a genome-wide view of cohesin distribution. In yeast, cohesin accumulates in a

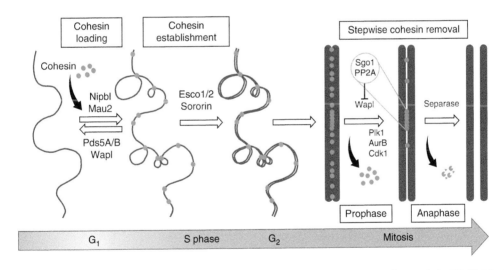

Figure 2. Cohesin and its regulators throughout the cell cycle. Cohesin is loaded on chromatin by Nipbl-Mau2 throughout the cell cycle, starting in early G_1. Pds5 and Wapl associate with chromatin-bound cohesin and promote its unloading. Cohesin complexes may be encircling a single chromatin fiber or two fibers at the base of a chromatin loop that brings distal regions in proximity. During S phase, acetylation of Smc3 by CoATs Esco1/2 and Sororin recruitment, both facilitated by Pds5A/B (not depicted), results in cohesion establishment. A fraction of cohesin remains dynamic even after DNA replication (not depicted). Whether cohesin at the base of chromatin loops is also involved in tethering sister chromatids is not known. In prophase, most cohesin dissociates from chromatin in a process that requires Wapl and phosphorylation of cohesin and Sororin. Sgo1 and its partner PP2A prevent the dissociation of a population of cohesin, enriched at centromeres. This population is removed at the onset of anaphase when Securin (not depicted) is destroyed and the active Separase cleaves Rad21.

50-kb region around centromeres and at sites of convergent transcription (Glynn et al. 2004; Lengronne et al. 2004). Because there is little colocalization of cohesin and its loader, it has been proposed that once topologically entrapping chromatin, the cohesin ring can slide away from the loading site (Hu et al. 2011; Ocampo-Hafalla and Uhlmann 2011). In contrast, cohesin and its loader do colocalize at sites of active transcription in *Drosophila* (Misulovin et al. 2008). Comparison of cohesin distribution in two different mouse tissues also reveals a correlation between active transcription and the presence of cohesin (Cuadrado et al. 2015). Importantly, cohesin accumulates at CTCF binding sites along the human and mouse genomes, but only when CTCF is present (Parelho et al. 2008; Rubio et al. 2008; Wendt et al. 2008; Remeseiro et al. 2012b). Whether cohesin is loaded at these CTCF sites or is loaded elsewhere and then slides to reach them is not known. The number of Nipbl positions identified by ChIP is 5–10 times lower that the number of cohesin or CTCF sites, and cohesin and its loader colocalize only at a subset of sites near active genes (Kagey et al. 2010) but not at CTCF sites (Zuin et al. 2014b). Considering the data from different model organisms, it is possible that cohesin is loaded mainly at sites of active transcription and then moves along the genome until it finds an obstacle, such as CTCF or another chromatin-binding protein with which it interacts. This ability of cohesin to slide along DNA may be consistent with "loop-extrusion" models recently proposed to explain how cohesin and CTCF contribute to the formation of chromatin loops (de Wit et al. 2015; Nichols and Corces 2015; Sanborn et al. 2015).

Cohesion Establishment and Dissolution

Cohesion establishment is coupled to DNA replication and requires acetylation of the two aforementioned lysine residues in the Smc3 hd by cohesin acetyltransferases (CoATs) and Sororin recruitment (Fig. 2) (Rolef Ben-Shahar et al. 2008; Unal et al. 2008; Zhang et al. 2008). The functional links between Smc3 acetylation, ATP hydrolysis, and DNA entrapment

are not clear yet (Heidinger-Pauli et al. 2010b; Ladurner et al. 2014; Camdere et al. 2015). There are two CoATs in mammalian cells, Esco1 and Esco2, with partially redundant functions (Hou and Zou 2005; Whelan et al. 2012; Minamino et al. 2015; Rahman et al. 2015). Pds5 proteins, which exist in two versions in vertebrate cells, Pds5A and Pds5B, are also required for cohesion establishment in yeast and mouse cells (Vaur et al. 2012; Carretero et al. 2013; Chan et al. 2013). As a result of establishment, Sororin displaces Wapl from its Pds5-interaction site and thereby counteracts its unloading activity (Nishiyama et al. 2010; Ouyang et al. 2016). In this way, a fraction of cohesin complexes tethering the two sister chromatids become stably bound to chromatin and maintain cohesion until mitosis (Gerlich et al. 2006; Schmitz et al. 2007). Whether a single complex embraces the two sister chromatids or two complexes are required, each one embracing a sister, is still a matter of debate (Eng et al. 2015).

At the time of chromosome segregation, cohesin dissociates from DNA in two steps, in prophase and anaphase (Fig. 2) (Losada et al. 1998; Waizenegger et al. 2000). The prophase pathway requires SA phosphorylation by Plk1 and release of Sororin, after phosphorylation by Cdk1 and Aurora B, to restore Wapl unloading activity (Losada et al. 2002; Sumara et al. 2002; Dreier et al. 2011; Nishiyama et al. 2013). Shugoshin (Sgo1) and its partner, the protein phosphatase 2A (PP2A), prevent cohesin release around centromeres (McGuinness et al. 2005). Sgo1 outcompetes the binding of Wapl to SA-Rad21 (Hara et al. 2014), whereas PP2A counteracts Sororin dissociation (Liu et al. 2013b). Pericentromeric cohesin remains on chromatin and is essential to hold the sister chromatids together until all the chromosomes establish proper attachments to opposite spindle poles (Toyoda and Yanagida 2006; Liu et al. 2013a). Once this task is completed, activation of the anaphase-promoting complex (APC/C) leads to degradation of Securin and activation of Separase (Shindo et al. 2012). The protease cleaves the kleisin subunit of chromatin-bound cohesin and sister chromatid separation ensues (Hauf et al. 2001). In yeast, all chromatin-bound cohesin is released from chromatin in anaphase by this cleavage pathway (Uhlmann et al. 2000). Acetylated cohesin removed from chromosomes in prophase and anaphase is deacetylated by a cohesin deacetylase (CoDAC), Hdac8 in human cells and Hos1 in yeast (Beckouet et al. 2010; Borges et al. 2010; Deardorff et al. 2012).

Cohesin Functions

Our knowledge of cohesin functions comes from studies in many different experimental systems, most notably yeast and human cells, but also *Drosophila*, zebrafish, and *Xenopus* egg extracts. Mouse models carrying knockout alleles for genes encoding cohesin subunits or their regulators have also been generated and characterized to different extents (Table 2). As mentioned above, cohesin was first recognized as a mediator of sister chromatid cohesion (Fig. 3, left). During mitosis, cohesion contributes to the proper orientation of sister kinetochores (Sakuno et al. 2009) and prevents the premature separation of sister chromatids under the pulling forces of spindle microtubules, whereas chromosomes try to align at the metaphase plate (Daum et al. 2011). In the absence of cohesin, chromosome missegregation is commonly observed (Sonoda et al. 2001; Vass et al. 2003; Toyoda and Yanagida 2006; Barber et al. 2008; Solomon et al. 2013; Covo et al. 2014). Cohesin is also important for HR-driven DNA repair (Sjogren and Nasmyth 2001; Schmitz et al. 2007; Heidinger-Pauli et al. 2010a; Xu et al. 2010; Wu et al. 2012). Cohesin promotes usage of the sister chromatid as a template for faithful repair while preventing both damage-induced recombination between homologs (Covo et al. 2010) and end joining of distal double-strand breaks (DSB) (Gelot et al. 2015). A most intriguing function of cohesin is cohesion between mother and daughter centrioles, and its regulation bears similarities with that of sister chromatid cohesion (Wang et al. 2008; Beauchene et al. 2010; Schockel et al. 2011; Mohr et al. 2015).

Cohesin-SA1 and cohesin-SA2 coexist in vertebrate cells and mediate cohesion at telomeres and centromeres, respectively (Canudas

and Smith 2009; Remeseiro et al. 2012a). Telomeres are repeated regions prone to fork stalling (Sfeir et al. 2009). Cohesin-SA1 likely stabilizes stalled forks and facilitates their restart by HR, consistent with results in budding yeast (Tittel-Elmer et al. 2012). In SA1-null mouse embryo fibroblasts (MEFs), faulty telomere replication leads to chromosome missegregation (Remeseiro et al. 2012a). Cohesin-SA2 is preferentially recruited to laser-induced DNA damage sites in postreplicative human cells (Kong et al. 2014) but both complexes are loaded at double-strand breaks (DSBs) generated by a restriction enzyme (Caron et al. 2012). HR-mediated DNA repair in cells exposed to replication stress is also facilitated by both complexes (Remeseiro et al. 2012a).

Cohesin performs additional functions that do not require cohesion establishment (Fig. 3, right). These functions could be related to the ability of cohesin to tether chromatin fibers at the base of a chromatin loop to facilitate long-range interactions. Cohesin contributes to the spatial organization of the genome, together with CTCF. Recently developed chromosome conformation capture (3C)-related technologies together with improved microscopy and computational modeling offer the picture of a genome partitioned in "topological" domains that are conserved among cell types and even in evolution (Dixon et al. 2012). Within these domains, more local contacts allow or prevent communication between enhancers and promoters (Kagey et al. 2010; Phillips-Cremins et al. 2013; Dowen et al. 2014; Tang et al. 2015). Down-regulation of cohesin leads to a loss of contacts and deregulation of gene expression (Hadjur et al. 2009; Mishiro et al. 2009; Nativio et al. 2009; Seitan et al. 2013; Sofueva et al. 2013; Zuin et al. 2014a). Importantly, different loci display very different sensitivities to loss of cohesin (Ing-Simmons et al. 2015; Viny et al. 2015). In the pancreata of SA1 heterozygous mice, for instance, a twofold decrease in SA1 protein levels is sufficient to alter the chromatin architecture and the expression of the *Reg* gene cluster. The resulting down-regulation of Reg proteins, involved in inflammation, may contribute to the increased incidence of pancreatic

Table 2. Mouse models of cohesin subunits and regulators

Targeted gene	References	Phenotype
SMC3	White et al. 2013; Viny et al. 2015	Embryonic lethality (prior to E14.5). Heterozygous animals have reduced body weight and higher mortality rates, and a subset showed a distinct craniofacial morphology (reminiscent of Cornelia de Lange syndrome [CdLS]). Deletion in hematopoietic compartment in adult mice results in rapid lethality.
RAD21	Xu et al. 2010; Seitan et al. 2011	Embryonic lethality (prior to E8.5). Heterozygous mouse embryo fibroblasts (MEFs) are defective in homologous recombination (HR)-mediated DNA repair. Heterozygous animals show increased sensitivity to irradiation, particularly in the gastrointestinal tract and the hematopoietic system. Deletion in thymocytes results in reduced differentiation efficiency and impairs TCRα locus rearrangement.
STAG1	Remeseiro et al. 2012a,b	Embryonic lethality (from E12.5). Null MEFs show telomere cohesion defects leading to faulty replication and chromosome missegregation, altered transcription, and decreased colocalization of cohesin at CTCF sites and promoters. Heterozygous animals show increased incidence and earlier onset of cancer but are protected against acute carcinogenesis.
WAPL	Tedeschi et al. 2013	Embryonic lethality. Heterozygous animals healthy. Conditional elimination in MEFs leads to aberrant retention of cohesin on chromatin, altered transcription, defects in cell-cycle progression, and chromosome segregation.
PDS5A	Zhang et al. 2009; Carretero et al. 2013	Late embryonic lethality (from E12.5) or death soon after birth (depending on the allele) with cleft palate, skeletal patterning defects, growth retardation, congenital heart defects. Null MEFs proliferate slowly but show no chromosome missegregation.
PDS5B	Zhang et al. 2007; Carretero et al. 2013	Late embryonic lethality (from E12.5) or death soon after birth (depending on the allele) with multiple congenital anomalies, including heart defects, cleft palate, fusion of the ribs, short limbs. Null MEFs show centromere cohesion defects and delocalization of the chromosomal passenger complex (CPC), chromosome missegregation, and aneuploidy.
NIPBL	Kawauchi et al. 2009; Remeseiro et al. 2013; Smith et al. 2014	Embryonic lethality (prior to E9.5). Up to 80% of heterozygous animals die during the first weeks of life and display CdLS-like defects such as small size, craniofacial anomalies, heart defects, delayed bone maturation, and behavioral disturbances. Heterozygous MEFs show gene expression alterations but no cohesion defects.
MAU2	Smith et al. 2014	Embryonic lethality (prior to E9.5). Heterozygous animals are normal.
ESCO2	Whelan et al. 2012	Embryonic lethality (prior to E8). Conditional elimination in MEFs leads to defects in centromere cohesion and chromosome segregation.
HDAC8	Haberland et al. 2009	Homozygous mice show perinatal lethality with dramatic skull abnormalities.
SGO1	Yamada et al. 2012	Embryonic lethality. Heterozygous MEFs show chromosome missegregation and aneuploidy. Heterozygous animals viable but display increased susceptibility to colon and liver cancer induced by treatment with azoxymethane.

Continued

Table 2. *Continued*

Targeted gene	References	Phenotype
ESPL1	Kumada et al. 2006; Wirth et al. 2006	Embryonic lethality (E3.5). Conditional elimination in MEFs leads to proliferation defects and polyploidy, with multiple chromosomes connected at their centromeric regions. Depletion in bone marrow causes aplasia.
PTTG1	Mei et al. 2001; Kumada et al. 2006	Homozygous mice are viable but display testicular and splenic hypoplasia, thymic hyperplasia, and thrombocytopenia. Null MEFs grow slowly in culture and accumulate in G_2.

cancer observed in SA1 heterozygous mice (Remeseiro et al. 2012a).

Cohesin depletion also increases RNA polymerase II pausing at cohesin binding genes in *Drosophila*, suggesting that it regulates its transition to elongation (Schaaf et al. 2013). In human cells, cohesin-SA1 is specifically involved in interactions with the super elongation complex (SEC) involved in mobilization of the paused polymerase (Izumi et al. 2015). Transcriptional

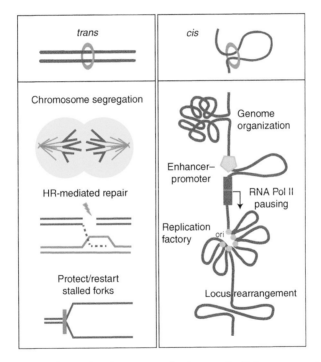

Figure 3. Cohesin functions. Cohesin plays important roles in several cellular processes involving DNA. These roles rely on the ability of cohesin to hold two DNA strands in *trans* (the sister chromatids) or in *cis* (e.g., at the base of a chromatin loop). Accurate chromosome segregation in mitosis and meiosis, HR-mediated DNA repair, and restart and/or protection of stalled replication forks require sister chromatid cohesion (*left*). DNA looping mediated by cohesin in collaboration with CTCF, Mediator, or transcription factors, among others, likely provides a major organizational principle for the genome (*right*). This organization regulates transcription both globally, through generation of active/silent domains, and locally, facilitating interactions between enhancer and promoters required for gene activation or RNA Pol II pause release. It also facilitates coordinated origin firing at replication factories and recombination at loci such as IgH or TCRα. For simplicity, a single cohesin ring embracing the two DNA fibers is drawn, but alternative configurations are possible (Eng et al. 2015).

control at the level of elongation is key for a number of developmental genes (Smith and Shilatifard 2013) and has been linked to pathogenesis in some leukemias (Lin et al. 2010).

In addition to transcription regulation, several lines of evidence support the idea that chromatin loops stabilized by cohesin organize DNA replication factories to promote efficient origin firing (Guillou et al. 2010), and facilitate V(D)J recombination (Degner et al. 2011) and T-cell receptor α locus rearrangement (Seitan et al. 2011).

COHESIN MUTATIONS IN CANCER

Recent pan-cancer studies have placed cohesin and its regulators among the networks most frequently mutated in cancer (Kandoth et al. 2013; Lawrence et al. 2014; Leiserson et al. 2015). Mutations in genes encoding cohesin subunits had been first reported in colorectal cancer after targeted sequencing of genes essential for chromosome segregation in yeast (Barber et al. 2008). A few years later, mutations in the gene encoding SA2, STAG2, were found in glioblastoma, Ewing sarcoma, and melanoma (Solomon et al. 2011). These two studies pointed to chromosome missegregation as the main contribution of cohesin dysfunction to tumorigenesis. However, sequencing of acute myeloid leukemia (AML) samples revealed the presence of recurrent mutations in STAG2, SMC3, RAD21, and SMC1A that were not associated with cytogenetic abnormalities, implying alternative pathological pathways (Welch et al. 2012). The correlation between STAG2 mutations and aneuploidy in bladder cancer was also unclear (Balbas-Martinez et al. 2013; Guo et al. 2013). In the next sections we will review these and other recent studies that provide evidence for the presence of cohesin mutations in cancer.

Cohesin Mutations in Myeloid Malignancies

Identification of Mutations in Cancer Cells

AML results from the aberrant proliferation and impaired differentiation of hematopoietic stem and progenitor cells. Reports using next-generation sequencing in AML samples appeared by 2012 and identified mutations in cohesin genes (Ding et al. 2012; Dolnik et al. 2012; Walter et al. 2012; Welch et al. 2012). According to The Cancer Genome Atlas (TCGA) Research Network (2013), AML genomes have fewer mutations than most adult cancers and, among them, 13% correspond to cohesin-related genes (Fig. 4A). Thol et al. (2014) performed targeted sequencing of genes encoding the five cohesin core subunits in samples from 389 AML patients and identified mutations in all of them (collectively in 6% of the cases). Most patients carrying cohesin mutations had a normal karyotype, supporting the hypothesis that they do not affect genome integrity. Importantly, cohesin mutations were found in myeloid malignancies other than AML (Kon et al. 2013). STAG2 and RAD21 were the most mutated cohesin genes (Fig. 4B). An even higher frequency of cohesin mutations (∼15%, most of them in STAG2) was found in MDS samples in another study (Haferlach et al. 2014). Even in the absence of cohesin mutations, low expression of cohesin components was detected in a significant fraction of myeloid malignancies (Thota et al. 2014). Mutations in additional components of the cohesin network including PDS5B, NIPBL, or ESCO2 were also identified in some studies (Fig. 4A,B).

The prognostic impact of cohesin mutations in myeloid disorders is unclear, with studies reporting a positive (Kihara et al. 2014), negative (Thota et al. 2014), or no significant effect (Thol et al. 2014) on survival. The presence of these mutations in the major tumor populations points to their early origin during the neoplastic process (Kon et al. 2013; Thol et al. 2014). Interestingly, two patients analyzed by Kon et al. harbored each two independent subclones with different STAG2 mutations, which suggests that loss of STAG2 could confer a strong advantage to preexisting leukemic cells during clonal evolution. Analysis of clonal dynamics in another report revealed that cohesin mutations were not commonly present in the founder clone but rather promoted clonal expansion and transformation to more aggressive disease

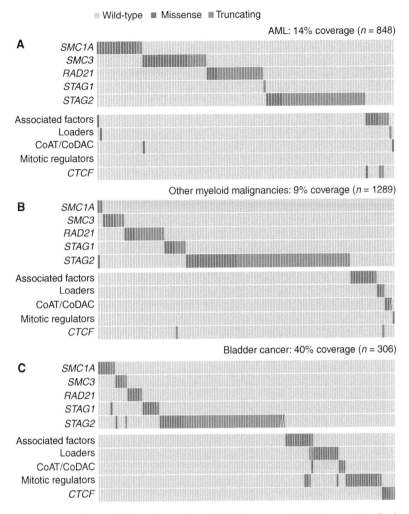

Figure 4. Cohesin mutations in myeloid malignancies and bladder cancer. Mutation matrix of cohesin subunits, associated factors (*PDS5A, PDS5B, WAPL, CDC5A*), loaders (*NIPBL, MAU2*), CoAT/CoDAC (*ESCO1, ESCO2, HDAC8*), mitotic regulators (*SGOL1, ESPL1, PTTG1*), and *CTCF* in AML (*A*), other myeloid malignancies (*B*), and bladder cancer (*C*). Nonsense, frameshift, and splice site mutations are grouped as truncating mutations (red), and missense and inframe indels are labeled as missense mutations (blue). (Data from Ding et al. 2012; Jan et al. 2012; Welch et al. 2012; Balbas-Martinez et al. 2013; Cancer Genome Atlas Research Network 2013, 2014; Guo et al. 2013; Kon et al. 2013; Yoshida et al. 2013; Lohr et al. 2014; Pellagatti et al. 2014; Thota et al. 2014.)

(Thota et al. 2014). Cohesin mutations often co-occurred with other mutations, most often in nucleophosmin (*NPM1*), epigenetic regulators such as *ASXL1, TET2,* or *DNMT3A,* or transcription factors like *RUNX1* (Cancer Genome Atlas Research Network 2013; Walter et al. 2013; Kihara et al. 2014; Thol et al. 2014). Further support for a key role of cohesin mutations in malignant transformation came from the analysis of Down syndrome–related acute

megakaryocitic leukemia (DS-AMKL) (Yoshida et al. 2013). Children with Down syndrome often suffer transient abnormal myelopoiesis (TAM) that, in some cases, evolves to DS-AMKL. Genomic profiling of TAM, DS-AMKL, and non-DS-AMKL samples revealed that TAM is caused by a *GATA1* mutation and progression to DS-AMKL requires additional mutations. More than half of them were found in cohesin genes. Here again, most cases with mutated cohesin had normal karyotypes, except for constitutive trisomy 21.

In summary, sequencing studies have revealed a high prevalence of mutations in the genes encoding cohesin components in AML and other myeloid cancers (Fig. 4A,B). As expected from proteins working as part of the same complex, cohesin mutations are mutually exclusive. Although mutation rates differ among these studies, they typically account collectively for less than 10% of the cases. Cohesin mutations are usually heterozygous, with the exception of those present in the X-linked genes *STAG2* and *SMC1A* in male patients. In female samples, mutations in *STAG2* and *SMC1A* often reside in the active chromosome. While *SMC1A*, *SMC3*, and *STAG1* mutations are often missense, those in *STAG2* and *RAD21* are usually truncating (i.e., frameshift, nonsense, or splice site mutations). No clear mutation hotspot has been identified in any of these genes, a characteristic of tumor suppressors. Importantly, no association of cohesin mutations and unstable karyotypes or aneuploidy has been reported. Thus, the contribution of cohesin dysfunction to development of myeloid malignancies is possibly not related to cohesion defects and genomic instability, and instead could be the result of altered transcription.

Functional Studies in Hematopoietic Cells

Cohesin mutations may reduce the levels of cohesin complexes in the cell or may alter their functionality. Understanding the contribution of these changes to tumorigenesis requires functional studies. Mazumdar et al. (2015) introduced a missense SMC1A mutant (SMC1A R711G) and a RAD21 truncation mutant (RAD21 Q592*), previously identified in AML (Cancer Genome Atlas Research Network 2013), in primary human hematopoietic stem and progenitor cells (HSPCs). No clear defects in proliferation or cell death were observed, but differentiation was impaired. The defects were restricted to the most immature populations of cord blood cells, consistent with the observation of cohesin mutations in the most immature forms of AML (Welch et al. 2012). Increased chromatin accessibility was observed in regions enriched for DNA-binding motifs of ERG, GATA2, and RUNX1, all transcription factors (TFs) involved in maintenance of the stem-cell program in HSPCs (Wilson et al. 2010). Knockdown of any of these factors reversed the differentiation block of cohesin mutants.

Two additional studies have explored the consequences of cohesin knockdown (kd) in hematopoiesis. Transgenic mice carrying inducible shRNAs against SA2, Smc1α, and Rad21 allowed ubiquitous and inducible cohesin kd in vivo in adult mice (Mullenders et al. 2015). Efficient reduction of cohesin levels, at least in the hematopoetic organs, was well tolerated, suggesting that a small fraction of cohesin complexes is sufficient to carry out its essential functions. Lineage skewing toward myeloid lineage commitment was observed in the spleen of cohesin-deficient mice, as well as in HSPCs, and gene expression changed accordingly. SA2 kd led to increased chromatin accessibility in regions enriched in the GATA motif. In the other study, mice carrying a conditional *KO* allele of *SMC3* were used (Viny et al. 2015). Complete ablation of *SMC3* in the hematopoietic compartment led to rapid lethality, whereas deletion of a single *SMC3* allele resulted in increased cell renewal capacity of HSPCs and reduced expression of transcription factors and other genes associated with lineage commitment. Hematopoietic progenitors of Scm3 heterozygous animals displayed increased accessibility in regions harboring binding sites for yet another transcription factor, STAT5.

Taken all together, these studies suggest that decreased cohesin levels may promote transformation of HSPCs through delaying or skewing differentiation and instead enforcing stem-cell programs. They appear to do so through modulation of chromatin accessibility of TFs involved in stem-cell maintenance. An alternative possibility is that the observed changes are a consequence rather that the cause of the differentiation block. Cohesin has been proposed to regulate the expression of cell identity genes together with CTCF (Dowen et al. 2014), or with tissue-specific TFs (Schmidt et al. 2010) through the control of chromosome structure. To extend the studies described here, it would be informative to actually compare cohesin binding to chromatin in cells carrying or not the cohesin mutants, or partially deficient in cohesin subunits. This comparison could be performed in bulk by using chromatin fractionation, at genome wide scale resolution by using ChIP-seq and at specific sites by ChIP-qPCR. It would also be of interest to look for changes in chromatin architecture near the promoters of genes required to promote or prevent terminal myeloid differentiation, including the abovementioned TFs. For instance, cohesin-mediated contacts between *cis*-regulatory elements modulate tissue-specific RUNX1 expression in zebrafish embryos, and probably also in human hematopoietic cells (Horsfield et al. 2007; Marsman et al. 2014).

Importantly, *SMC3* haploinsufficiency by itself did not result in AML, but it enhanced tumorigenesis when combined with A FLT3-internal tandem duplication (ITD) mutation often found in AML (Viny et al. 2015). Aged cohesin knockdown mice in the study by Mullenders et al. (2015) developed phenotypes resembling myeloid neoplasias, but did not develop frank AML. These observations are consistent with the idea that cohesin mutations cooperate with additional mutations to promote myeloid malignancies.

Cohesin Mutations in Bladder Cancer

Urothelial bladder cancer (UBC) is a heterogeneous disease. Tumors are classified according to the stage of invasion (Tis-T4), and graded based on their cellular characteristics. At diagnosis, around 60% of bladder cancers are non-muscle-invasive (NMIBC) papillary tumors of low grade. Stage T1 tumors, which have penetrated the epithelial basement membrane but have not invaded the muscle, are mostly of high grade. Also aggressive are the muscle-invasive bladder cancers (MIBCs). Sequencing of 99 low-grade tumors revealed mutations in *STAG2* (16%), *NIPBL* (4%), *SMC1A* (3%), and *SMC3* (2%), as well as in the gene-encoding separase,

ESPL1 (6%) (Guo et al. 2013). Individuals with *STAG2* mutations had worse prognosis and increased number of copy number variations (CNVs), indicative of increased genomic instability. Soon afterward, a discovery exome sequencing screen ($n = 17$), followed by a prevalence screen ($n = 60$), identified mutations in *STAG2* (16%) and some other cohesin subunits in UBC (Balbas-Martinez et al. 2013). *STAG2* was mutated mainly in tumors of low stage or grade, commonly genomically stable, and unlike the previous study, its loss was associated with improved outcome. Moreover, chromosome number changes were not associated with *STAG2* deficiency. Another analysis of aggressive MIBCs identified *STAG2* mutations in 14 out of 131 tumors (11%), often co-occurring with mutations in epigenetic regulators (Cancer Genome Atlas Research Network 2014). Tumors with mutations in other cohesin subunits (9%) and cohesin regulators (12%) were also identified (Fig. 4C).

Solomon et al. (2013) reported higher mutation frequencies after sequencing *STAG2* in 111 tumors of different stages/grades. Around 36% and 27% of *STAG2* mutations were found in pTa and pT1 NMIBCs, respectively, and 16% in MIBCs. In low-grade NMIBCs, loss of *STAG2* expression was significantly associated with increased disease-free survival, whereas the opposite was observed in MIBCs. Chromosomal copy number aberrations were found in many tumor samples, but even in the presence of wild-type *STAG2*. Another study sequencing *STAG2* in 307 bladder tumors confirmed higher mutation frequencies in NMIBC noninvasive tumors (33%) or superficially invasive tumors (21%), and lower in MIBCs (13%) (Taylor et al. 2013). No significant association was found with disease recurrence in either NMIBC or MIBCs. Whole chromosome copy number alterations measured by aCGH showed an inverse relationship to *STAG2* mutation. No association between *STAG2* mutation and outcome was found in another report analyzing 109 high grade UBCs (16% *STAG2* mutation rate) (Kim et al. 2014).

Taken all together, we can conclude that in UBC: (1) *STAG2* mutation frequencies are higher than for genes encoding other cohesin subunits, and also higher than in other cancers; (2) most mutations in *STAG2* are truncating, whereas other cohesin genes harbor missense mutations, similar to what was observed in myeloid syndromes; (3) cohesin mutations appear more frequently in lower grade/stage UBCs; and (4) there is no clear correlation with prognosis or aneuploidy.

Functional Studies in Bladder Cancer Cell Lines

Very limited functional experiments in bladder cells have been reported to date. SA2 kd in UBC cell lines with normal SA2 expression did not consistently alter chromosome number in one study (Balbas-Martinez et al. 2013) but it did in another (Solomon et al. 2013). Similarly, reintroduction of *STAG2* cDNA in UBC cell lines with truncating *STAG2* mutations led to a significant decrease in colony formation in one study (Balbas-Martinez et al. 2013) but did not affect proliferation in vitro or in xenografts in the other (Solomon et al. 2013). To interpret these conflicting observations, it would be important to quantify the functional cohesin complexes remaining in the cell under each experimental condition. Given the role of cohesin-SA2 in centromeric cohesion, one would expect chromosome segregation defects when *STAG2* expression is lost (Canudas and Smith 2009). Indeed, SA2 kd led to chromosome missegregation in some human cell lines (Barber et al. 2008; Solomon et al. 2011; Kleyman et al. 2014). In contrast, and similar to UBCs, premature sister chromatid separation was not observed after SA2 kd in mouse hematopoietic progenitor cells, and was detected in only a small percentage of cells after efficient kd of Smc1α, Smc3, or Rad21 (Mullenders et al. 2015). Arm cohesion mediated by cohesin-SA1 may compensate for the loss of centromeric cohesion in the absence of cohesin-SA2, and the relative levels of both complexes may differ among cell types. In addition, a low amount of cohesin may be sufficient to maintain cohesion in mitosis. In yeast, cohesin levels must be reduced below 13% to result in detectable cohesion and segregation defects (Heidinger-Pauli et al. 2010a).

Genes involved in chromatin regulation are more frequently mutated in urothelial carcinoma than in any other common cancer studied so far. Cohesin belongs to this category. These mutations likely modulate the activity levels of various TFs and pathways implicated in cancer (Cancer Genome Atlas Research Network 2014). Mutations in *STAG2* may be more frequent because the gene is located in the X chromosome and because in the absence of cohesin-SA2, cohesin-SA1 may be sufficient to perform essential cohesin functions. It is also possible that transcriptional dysregulation of key genes involved in tumorigenesis depends on cohesin-SA2, not on cohesin-SA1. So far the functional specificities of cohesin-SA1 and cohesin-SA2 in terms of chromatin regulation are poorly understood. In MEFs, genome-wide distribution of both complexes is similar and overlaps with the distribution of CTCF. Upon ablation of *STAG1*, however, cohesin could be detected at additional sites that showed less overlap with promoters and CTCF (Remeseiro et al. 2012b). It was then proposed that cohesin-SA1 could be more important than cohesin-SA2 for transcriptional regulation (Cuadrado et al. 2012). Consistent with this possibility, the transcriptomes of paired human gliobastoma cell lines with and without STAG2 expression did not differ significantly (Solomon et al. 2011). However, in one of the studies described in the previous sections, bone marrow cells treated with shRNAs against SA2 did display significant alterations in gene expression. Moreover, these alterations were similar to those observed on kd of Smc1α, suggesting that cohesin-mediated transcriptional regulation in HPSCs relies specifically on cohesin-SA2 (Mullenders et al. 2015). The reasons underlying this specificity are unknown. Even the relative abundance of cohesin-SA1 versus cohesin-SA2 in different cell types could be different. It will be of great interest to compare gene expression profiles before and after SA2 kd in bladder cell lines, as well as other cell types, to better understand how the cohesin variants contribute to cell proliferation and gene expression in a tissue-specific manner.

Cite this article as *Cold Spring Harb Perspect Med* doi: 10.1101/cshperspect.a026476

Cohesin Mutations in Other Cancers

A look at TCGA database reveals the presence of cohesin mutations in many additional cancers (Fig. 5). Bladder cancer is the one in which alteration of the cohesin network components is more common, followed by melanoma, colorectal, and lung cancers. Other than bladder cancer, *STAG2* mutations are most frequent in Ewing sarcoma (EWS). This a pediatric tumor of the bone and soft tissues characterized genetically by the presence of translocations involving ETS family transcription factors such as EWS-FLI (Brohl et al. 2014; Crompton et al. 2014; Tirode et al. 2014; Agelopoulos et al. 2015). In one study looking for secondary genetic lesions, somatic mutations were detected in *STAG2* (17%), *CDKN2A* (12%), and *TP53* (7%). Although mutations in *STAG2* and *CDKN2A* were mutually exclusive, *STAG2* and *TP53* mutations co-occurred particularly in aggressive tumors (Tirode et al. 2014). Mutation rates ranged from 8% to 21% in the other studies. In some cases, loss of expression was detected without mutation, suggesting another mechanism of *STAG2* inactivation (Crompton et al. 2014). EWS is among the most genetically stable cancers. No association with aneuploidy was detected although tumors without *STAG2* showed an increased number of somatic CNVs. However, it was not clear if this was because of the association with *TP53* mutations (Crompton et al. 2014; Tirode et al. 2014).

One intriguing observation is the high frequency of mutations in meiosis-specific cohesin genes (labeled in yellow in Fig. 5). Whether these mutations have actual consequences (e.g., the genes are expressed in the tumor cells and act as dominant negative mutant proteins), or are just passenger or silent mutations, remains to be addressed. Similarly, it is also unclear whether meiotic versions of cohesin subunits become expressed in tumors with mutations in their somatic counterparts to compensate for their loss. For instance, truncating mutations in the X-linked *SMC1A* gene have been described in tumor samples from male patients. Maybe Smc1β is expressed in these cells and forms functional complexes with Rad21 and SA1/2 (Mannini et al. 2015).

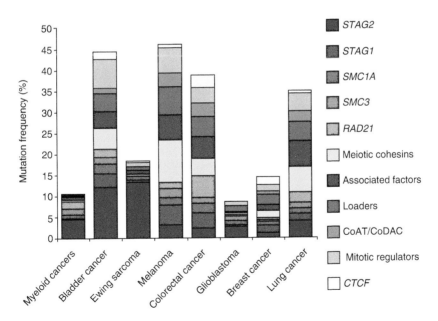

Figure 5. Mutations in cohesin complex and related genes in cancer. Bar graph showing mutations in the indicated cancer types for cohesin subunits and cohesin-related genes (grouped as in Fig. 4). (Data obtained from the cBioPortal for Cancer Genomics [Cerami et al. 2012; Gao et al. 2013] and from studies mentioned in Fig. 4.)

In addition to mutations, CNVs encompassing cohesin genes have also been observed in several cancer types. Of all cohesin genes, *RAD21* is the most frequently amplified (e.g., 20% in invasive breast cancer) (Ciriello et al. 2015). This could be because of its proximity to the *MYC* locus, also present in 8q24. Abnormal levels of cohesin may also contribute to tumorigenesis. Finally, whole-genome sequencing data from more than 200 samples of colorectal cancer (CRC) patients together with ChIP-seq analyses in a CRC cell line revealed a high incidence of mutations in cohesin/CTCF binding sites in the noncoding genome (Katainen et al. 2015). A fraction of these mutations are predicted to affect CTCF binding affinity to *cis*-regulatory elements and could therefore contribute to tumorigenesis through aberrant expression of their target genes. Another epigenetic mechanism recently described in gliomas involves disruption of boundary elements through hypermethylation of CTCF/cohesin binding sites leading to oncongene activation (Flavahan et al. 2016). Thus, there are multiple ways in which cohesin dysfunction may contribute to tumorigenesis.

CONCLUSIONS AND PERSPECTIVES

Mutations in genes encoding cohesin subunits and regulators have been now identified in many cancer genomes, but their relevance for tumor initiation and progression is unknown. It is also unclear how they affect the functionality of the complex, particularly in the case of missense mutations. Because most mutations are heterozygous, they may either reduce the amount of fully functional cohesin complexes in the cell or even have a dominant negative effect. The diverse tasks accomplished by cohesin require different amounts of the complex and may rely on a particular variant. In the majority of tumors or cancer cell lines analyzed, there is no clear correlation between the presence of cohesin mutations and aneuploidy. Thus, cohesion is unlikely to be the function impaired in these tumor cells. Moreover, the recent functional studies performed in hematopoietic cells point to changes in chromatin ac-

cessibility and transcription as the most striking consequences of cohesin dysfunction. Future studies will tell if the same is true in other cell types. Identification of vulnerabilities in cancer cells harboring cohesin mutations will be an important step toward targeted therapies. A study in yeast and *Caenorhabditis elegans* identified a strong synthetic lethality between mutations in cohesin genes and genes involved in replication fork progression and stability, including Poly (ADP-ribose) polymerase or PARP (McLellan et al. 2012). Consistent with this observation, SMC1 down-regulation sensitized triple-negative breast cancer cells to PARP inhibition (Yadav et al. 2013). *STAG2*-mutated glioblastoma cell lines also displayed increased sensitivity to PARP inhibitors, especially when used in combination with DNA-damaging agents (Bailey et al. 2014). A better understanding of how cohesin works and how it contributes to proliferation, cell-identity determination, and homeostasis will hopefully guide improvements in diagnosis and treatment of cancer and other diseases related with cohesin dysfunction.

ACKNOWLEDGMENTS

The graphs in Figures 4 and 5 contain data obtained by TCGA Research Network (cancergenome.nih.gov) that have not been published yet. We therefore thank all researchers within this network as well as those involved in making these data available through the cBioPortal for Cancer Genomics (www .cbioportal.org). Our own research on cohesin is funded by the Spanish Ministry of Economy (MINECO) and The European Regional Development Fund (FEDER) (Grant BFU2013-48481-R to A.L. and Fellowship BES-2014-069166 to M.D.K.).

REFERENCES

Agelopoulos K, Richter GH, Schmidt E, Dirksen U, von Heyking K, Moser B, Klein HU, Kontny U, Dugas M, Poos K, et al. 2015. Deep sequencing in conjunction with expression and functional analyses reveals activation of FGFR1 in Ewing sarcoma. *Clin Cancer Res* **21:** 4935–4946.

Arumugam P, Gruber S, Tanaka K, Haering CH, Mechtler K, Nasmyth K. 2003. ATP hydrolysis is required for cohesin's association with chromosomes. *Curr Biol* **13**: 1941–1953.

Bailey ML, O'Neil NJ, van Pel DM, Solomon DA, Waldman T, Hieter P. 2014. Glioblastoma cells containing mutations in the cohesin component *STAG2* are sensitive to PARP inhibition. *Mol Cancer Ther* **13**: 724–732.

Balbas-Martinez C, Sagrera A, Carrillo-de-Santa-Pau E, Earl J, Marquez M, Vazquez M, Lapi E, Castro-Giner F, Beltran S, Bayes M, et al. 2013. Recurrent inactivation of *STAG2* in bladder cancer is not associated with aneuploidy. *Nat Genet* **45**: 1464–1469.

Barber TD, McManus K, Yuen KW, Reis M, Parmigiani G, Shen D, Barrett I, Nouhi Y, Spencer F, Markowitz S, et al. 2008. Chromatid cohesin defects may underlie chromosome instability in human colorectal cancers. *Proc Natl Acad Sci* **105**: 3443–3448.

Beauchene NA, Diaz-Martinez LA, Furniss K, Hsu WS, Tsai HJ, Chamberlain C, Esponda P, Gimenez-Abian JF, Clarke DJ. 2010. Rad21 is required for centrosome integrity in human cells independently of its role in chromosome cohesion. *Cell Cycle* **9**: 1774–1780.

Beckouet F, Hu B, Roig MB, Sutani T, Komata M, Uluocak P, Katis VL, Shirahige K, Nasmyth K. 2010. An Smc3 acetylation cycle is essential for establishment of sister chromatid cohesion. *Mol Cell* **39**: 689–699.

Bernard P, Schmidt CK, Vaur S, Dheur S, Drogat J, Genier S, Ekwall K, Uhlmann F, Javerzat JP. 2008. Cell-cycle regulation of cohesin stability along fission yeast chromosomes. *EMBO J* **27**: 111–121.

Borges V, Lehane C, Lopez-Serra L, Flynn H, Skehel M, Rolef Ben-Shahar T, Uhlmann F. 2010. Hos1 deacetylates Smc3 to close the cohesin acetylation cycle. *Mol Cell* **39**: 677–688.

Brohl AS, Solomon DA, Chang W, Wang J, Song Y, Sindiri S, Patidar R, Hurd L, Chen L, Shern JF, et al. 2014. The genomic landscape of the Ewing sarcoma family of tumors reveals recurrent *STAG2* mutation. *PLoS Genet* **10**: e1004475.

Buheitel J, Stemmann O. 2013. Prophase pathway-dependent removal of cohesin from human chromosomes requires opening of the Smc3-Scc1 gate. *EMBO J* **32**: 666–676.

Camdere G, Guacci V, Stricklin J, Koshland D. 2015. The ATPases of cohesin interface with regulators to modulate cohesin-mediated DNA tethering. *eLife* **4**: e11315.

Cancer Genome Atlas Research Network. 2013. Genomic and epigenomic landscapes of adult de novo acute myeloid leukemia. *N Engl J Med* **368**: 2059–2074.

Cancer Genome Atlas Research Network. 2014. Comprehensive molecular characterization of urothelial bladder carcinoma. *Nature* **507**: 315–322.

Canudas S, Smith S. 2009. Differential regulation of telomere and centromere cohesion by the Scc3 homologues SA1 and SA2, respectively, in human cells. *J Cell Biol* **187**: 165–173.

Canudas S, Houghtaling BR, Kim JY, Dynek JN, Chang WG, Smith S. 2007. Protein requirements for sister telomere association in human cells. *EMBO J* **26**: 4867–4878.

Caron P, Aymard F, Iacovoni JS, Briois S, Canitrot Y, Bugler B, Massip L, Losada A, Legube G. 2012. Cohesin protects genes against γH2AX induced by DNA double-strand breaks. *PLoS Genet* **8**: e1002460.

Carretero M, Ruiz-Torres M, Rodriguez-Corsino M, Barthelemy I, Losada A. 2013. Pds5B is required for cohesion establishment and Aurora B accumulation at centromeres. *EMBO J* **32**: 2938–2949.

Cerami E, Gao J, Dogrusoz U, Gross BE, Sumer SO, Aksoy BA, Jacobsen A, Byrne CJ, Heuer ML, Larsson E, et al. 2012. The cBio cancer genomics portal: An open platform for exploring multidimensional cancer genomics data. *Cancer Discov* **2**: 401–404.

Chan KL, Gligoris T, Upcher W, Kato Y, Shirahige K, Nasmyth K, Beckouet F. 2013. Pds5 promotes and protects cohesin acetylation. *Proc Natl Acad Sci* **110**: 13020–13025.

Ciriello G, Gatza ML, Beck AH, Wilkerson MD, Rhie SK, Pastore A, Zhang H, McLellan M, Yau C, Kandoth C, et al. 2015. Comprehensive molecular portraits of invasive lobular breast cancer. *Cell* **163**: 506–519.

Covo S, Westmoreland JW, Gordenin DA, Resnick MA. 2010. Cohesin is limiting for the suppression of DNA damage-induced recombination between homologous chromosomes. *PLoS Genet* **6**: e1001006.

Covo S, Puccia CM, Argueso JL, Gordenin DA, Resnick MA. 2014. The sister chromatid cohesion pathway suppresses multiple chromosome gain and chromosome amplification. *Genetics* **196**: 373–384.

Crompton BD, Stewart C, Taylor-Weiner A, Alexe G, Kurek KC, Calicchio ML, Kiezun A, Carter SL, Shukla SA, Mehta SS, et al. 2014. The genomic landscape of pediatric Ewing sarcoma. *Cancer Discov* **4**: 1326–1341.

Cuadrado A, Remeseiro S, Gomez-Lopez G, Pisano DG, Losada A. 2012. The specific contributions of cohesin-SA1 to cohesion and gene expression: Implications for cancer and development. *Cell Cycle* **11**: 2233–2238.

Cuadrado A, Remeseiro S, Grana O, Pisano DG, Losada A. 2015. The contribution of cohesin-SA1 to gene expression and chromatin architecture in two murine tissues. *Nucleic Acids Res* **43**: 3056–3067.

Daum JR, Potapova TA, Sivakumar S, Daniel JJ, Flynn JN, Rankin S, Gorbsky GJ. 2011. Cohesion fatigue induces chromatid separation in cells delayed at metaphase. *Curr Biol* **21**: 1018–1024.

Deardorff MA, Bando M, Nakato R, Watrin E, Itoh T, Minamino M, Saitoh K, Komata M, Katou Y, Clark D, et al. 2012. *HDAC8* mutations in Cornelia de Lange Syndrome affect the cohesin acetylation cycle. *Nature* **489**: 313–317.

Degner SC, Verma-Gaur J, Wong TP, Bossen C, Iverson GM, Torkamani A, Vettermann C, Lin YC, Ju Z, Schulz D, et al. 2011. CCCTC-binding factor (CTCF) and cohesin influence the genomic architecture of the Igh locus and antisense transcription in pro-B cells. *Proc Natl Acad Sci* **108**: 9566–9571.

de Wit E, Vos ES, Holwerda SJ, Valdes-Quezada C, Verstegen MJ, Teunissen H, Splinter E, Wijchers PJ, Krijger PH, de Laat W. 2015. CTCF binding polarity determines chromatin looping. *Mol Cell* **60**: 676–684.

Ding L, Ley TJ, Larson DE, Miller CA, Koboldt DC, Welch JS, Ritchey JK, Young MA, Lamprecht T, McLellan MD, et al. 2012. Clonal evolution in relapsed acute myeloid leukaemia revealed by whole-genome sequencing. *Nature* **481**: 506–510.

Dixon JR, Selvaraj S, Yue F, Kim A, Li Y, Shen Y, Hu M, Liu JS, Ren B. 2012. Topological domains in mammalian genomes identified by analysis of chromatin interactions. *Nature* **485**: 376–380.

Dolnik A, Engelmann JC, Scharfenberger-Schmeer M, Mauch J, Kelkenberg-Schade S, Haldemann B, Fries T, Kronke J, Kuhn MW, Paschka P, et al. 2012. Commonly altered genomic regions in acute myeloid leukemia are enriched for somatic mutations involved in chromatin remodeling and splicing. *Blood* **120**: e83–92.

Dowen JM, Fan ZP, Hnisz D, Ren G, Abraham BJ, Zhang LN, Weintraub AS, Schuijers J, Lee TI, Zhao K, et al. 2014. Control of cell identity genes occurs in insulated neighborhoods in mammalian chromosomes. *Cell* **159**: 374–387.

Dreier MR, Bekier ME II, Taylor WR. 2011. Regulation of sororin by Cdk1-mediated phosphorylation. *J Cell Sci* **124**: 2976–2987.

Eichinger CS, Kurze A, Oliveira RA, Nasmyth K. 2013. Disengaging the Smc3/kleisin interface releases cohesin from *Drosophila* chromosomes during interphase and mitosis. *EMBO J* **32**: 656–665.

Eng T, Guacci V, Koshland D. 2015. Interallelic complementation provides functional evidence for cohesin–cohesin interactions on DNA. *Mol Biol Cell* **26**: 4224–4235.

Flavahan WA, Drier Y, Liau BB, Gillespie SM, Venteicher AS, Stemmer-Rachamimov AO, Suva ML, Bernstein BE. 2016. Insulator dysfunction and oncogene activation in IDH mutant gliomas. *Nature* **529**: 110–114.

Gandhi R, Gillespie PJ, Hirano T. 2006. Human Wapl is a cohesin-binding protein that promotes sister-chromatid resolution in mitotic prophase. *Curr Biol* **16**: 2406–2417.

Gao J, Aksoy BA, Dogrusoz U, Dresdner G, Gross B, Sumer SO, Sun Y, Jacobsen A, Sinha R, Larsson E, et al. 2013. Integrative analysis of complex cancer genomics and clinical profiles using the cBioPortal. *Sci Signal* **6**: pl1.

Gelot C, Guirouilh-Barbat J, Le Guen T, Dardillac E, Chailleux C, Canitrot Y, Lopez BS. 2015. The cohesin complex prevents the end joining of distant DNA double-strand ends. *Mol Cell* **61**: 15–26.

Gerlich D, Koch B, Dupeux F, Peters JM, Ellenberg J. 2006. Live-cell imaging reveals a stable cohesin-chromatin interaction after but not before DNA replication. *Curr Biol* **16**: 1571–1578.

Gillespie PJ, Hirano T. 2004. Scc2 couples replication licensing to sister chromatid cohesion in *Xenopus* egg extracts. *Curr Biol* **14**: 1598–1603.

Gligoris TG, Scheinost JC, Burmann F, Petela N, Chan KL, Uluocak P, Beckouet F, Gruber S, Nasmyth K, Lowe J. 2014. Closing the cohesin ring: Structure and function of its Smc3-kleisin interface. *Science* **346**: 963–967.

Glynn EF, Megee PC, Yu HG, Mistrot C, Unal E, Koshland DE, DeRisi JL, Gerton JL. 2004. Genome-wide mapping of the cohesin complex in the yeast *Saccharomyces cerevisiae*. *PLoS Biol* **2**: E259.

Gruber S, Haering CH, Nasmyth K. 2003. Chromosomal cohesin forms a ring. *Cell* **112**: 765–777.

Gruber S, Arumugam P, Katou Y, Kuglitsch D, Helmhart W, Shirahige K, Nasmyth K. 2006. Evidence that loading of cohesin onto chromosomes involves opening of its SMC hinge. *Cell* **127**: 523–537.

Guacci V, Hogan E, Koshland D. 1997. Centromere position in budding yeast: Evidence for anaphase A. *Mol Biol Cell* **8**: 957–972.

Guillou E, Ibarra A, Coulon V, Casado-Vela J, Rico D, Casal I, Schwob E, Losada A, Mendez J. 2010. Cohesin organizes chromatin loops at DNA replication factories. *Genes Dev* **24**: 2812–2822.

Guo G, Sun X, Chen C, Wu S, Huang P, Li Z, Dean M, Huang Y, Jia W, Zhou Q, et al. 2013. Whole-genome and whole-exome sequencing of bladder cancer identifies frequent alterations in genes involved in sister chromatid cohesion and segregation. *Nat Genet* **45**: 1459–1463.

Haberland M, Mokalled MH, Montgomery RL, Olson EN. 2009. Epigenetic control of skull morphogenesis by histone deacetylase 8. *Genes Dev* **23**: 1625–1630.

Hadjur S, Williams LM, Ryan NK, Cobb BS, Sexton T, Fraser P, Fisher AG, Merkenschlager M. 2009. Cohesins form chromosomal *cis*-interactions at the developmentally regulated IFNG locus. *Nature* **460**: 410–413.

Haering CH, Gruber S. 2016. SnapShot: SMC protein complexes part I. *Cell* **164**: 326–326.e321.

Haering CH, Lowe J, Hochwagen A, Nasmyth K. 2002. Molecular architecture of SMC proteins and the yeast cohesin complex. *Mol Cell* **9**: 773–788.

Haering CH, Schoffnegger D, Nishino T, Helmhart W, Nasmyth K, Lowe J. 2004. Structure and stability of cohesin's Smc1-kleisin interaction. *Mol Cell* **15**: 951–964.

Haering CH, Farcas AM, Arumugam P, Metson J, Nasmyth K. 2008. The cohesin ring concatenates sister DNA molecules. *Nature* **454**: 297–301.

Haferlach T, Nagata Y, Grossmann V, Okuno Y, Bacher U, Nagae G, Schnittger S, Sanada M, Kon A, Alpermann T, et al. 2014. Landscape of genetic lesions in 944 patients with myelodysplastic syndromes. *Leukemia* **28**: 241–247.

Hara K, Zheng G, Qu Q, Liu H, Ouyang Z, Chen Z, Tomchick DR, Yu H. 2014. Structure of cohesin subcomplex pinpoints direct shugoshin–Wapl antagonism in centromeric cohesion. *Nat Struct Mol Biol* **21**: 864–870.

Hauf S, Waizenegger IC, Peters JM. 2001. Cohesin cleavage by separase required for anaphase and cytokinesis in human cells. *Science* **293**: 1320–1323.

Heidinger-Pauli JM, Mert O, Davenport C, Guacci V, Koshland D. 2010a. Systematic reduction of cohesin differentially affects chromosome segregation, condensation, and DNA repair. *Curr Biol* **20**: 957–963.

Heidinger-Pauli JM, Onn I, Koshland D. 2010b. Genetic evidence that the acetylation of the Smc3p subunit of cohesin modulates its ATP-bound state to promote cohesion establishment in Saccharomyces cerevisiae. *Genetics* **185**: 1249–1256.

Horsfield JA, Anagnostou SH, Hu JK, Cho KH, Geisler R, Lieschke G, Crosier KE, Crosier PS. 2007. Cohesin-dependent regulation of Runx genes. *Development* **134**: 2639–2649.

Horsfield JA, Print CG, Monnich M. 2012. Diverse developmental disorders from the one ring: Distinct molecular pathways underlie the cohesinopathies. *Front Genet* **3**: 171.

Hou F, Zou H. 2005. Two human orthologues of Eco1/Ctf7 acetyltransferases are both required for proper sister-chromatid cohesion. *Mol Biol Cell* **16**: 3908–3918.

Hu B, Itoh T, Mishra A, Katoh Y, Chan KL, Upcher W, Godlee C, Roig MB, Shirahige K, Nasmyth K. 2011. ATP hydrolysis is required for relocating cohesin from sites occupied by its Scc2/4 loading complex. *Curr Biol* **21:** 12–24.

Huis in 't Veld PJ, Herzog F, Ladurner R, Davidson IF, Piric S, Kreidl E, Bhaskara V, Aebersold R, Peters JM. 2014. Characterization of a DNA exit gate in the human cohesin ring. *Science* **346:** 968–972.

Ing-Simmons E, Seitan VC, Faure AJ, Flicek P, Carroll T, Dekker J, Fisher AG, Lenhard B, Merkenschlager M. 2015. Spatial enhancer clustering and regulation of enhancer-proximal genes by cohesin. *Genome Res* **25:** 504–513.

Izumi K, Nakato R, Zhang Z, Edmondson AC, Noon S, Dulik MC, Rajagopalan R, Venditti CP, Gripp K, Samanich J, et al. 2015. Germline gain-of-function mutations in *AFF4* cause a developmental syndrome functionally linking the super elongation complex and cohesin. *Nat Genet* **47:** 338–344.

Jan M, Snyder TM, Corces-Zimmerman MR, Vyas P, Weissman IL, Quake SR, Majeti R. 2012. Clonal evolution of preleukemic hematopoietic stem cells precedes human acute myeloid leukemia. *Sci Transl Med* **4:** 149ra118.

Kagey MH, Newman JJ, Bilodeau S, Zhan Y, Orlando DA, van Berkum NL, Ebmeier CC, Goossens J, Rahl PB, Levine SS, et al. 2010. Mediator and cohesin connect gene expression and chromatin architecture. *Nature* **467:** 430–435.

Kandoth C, McLellan MD, Vandin F, Ye K, Niu B, Lu C, Xie M, Zhang Q, McMichael JF, Wyczalkowski MA, et al. 2013. Mutational landscape and significance across 12 major cancer types. *Nature* **502:** 333–339.

Katainen R, Dave K, Pitkanen E, Palin K, Kivioja T, Valimaki N, Gylfe AE, Ristolainen H, Hanninen UA, Cajuso T, et al. 2015. CTCF/cohesin-binding sites are frequently mutated in cancer. *Nat Genet* **47:** 818–821.

Kawauchi S, Calof AL, Santos R, Lopez-Burks ME, Young CM, Hoang MP, Chua A, Lao T, Lechner MS, Daniel JA, et al. 2009. Multiple organ system defects and transcriptional dysregulation in the *Nipbl*$^{+/-}$ mouse, a model of Cornelia de Lange Syndrome. *PLoS Genet* **5:** e1000650.

Kihara R, Nagata Y, Kiyoi H, Kato T, Yamamoto E, Suzuki K, Chen F, Asou N, Ohtake S, Miyawaki S, et al. 2014. Comprehensive analysis of genetic alterations and their prognostic impacts in adult acute myeloid leukemia patients. *Leukemia* **28:** 1586–1595.

Kim PH, Cha EK, Sfakianos JP, Iyer G, Zabor EC, Scott SN, Ostrovnaya I, Ramirez R, Sun A, Shah R, et al. 2014. Genomic predictors of survival in patients with high-grade urothelial carcinoma of the bladder. *Eur Urol* **67:** 198–201.

Kleyman M, Kabeche L, Compton DA. 2014. *STAG2* promotes error correction in mitosis by regulating kinetochore-microtubule attachments. *J Cell Sci* **127:** 4225–4233.

Kon A, Shih LY, Minamino M, Sanada M, Shiraishi Y, Nagata Y, Yoshida K, Okuno Y, Bando M, Nakato R, et al. 2013. Recurrent mutations in multiple components of the cohesin complex in myeloid neoplasms. *Nat Genet* **45:** 1232–1237.

Kong X, Ball AR Jr, Pham HX, Zeng W, Chen HY, Schmiesing JA, Kim JS, Berns M, Yokomori K. 2014. Distinct functions of human cohesin-SA1 and cohesin-SA2 in double-strand break repair. *Mol Cell Biol* **34:** 685–698.

Kueng S, Hegemann B, Peters BH, Lipp JJ, Schleiffer A, Mechtler K, Peters JM. 2006. Wapl controls the dynamic association of cohesin with chromatin. *Cell* **127:** 955–967.

Kumada K, Yao R, Kawaguchi T, Karasawa M, Hoshikawa Y, Ichikawa K, Sugitani Y, Imoto I, Inazawa J, Sugawara M, et al. 2006. The selective continued linkage of centromeres from mitosis to interphase in the absence of mammalian separase. *J Cell Biol* **172:** 835–846.

Ladurner R, Bhaskara V, Huis in 't Veld PJ, Davidson IF, Kreidl E, Petzold G, Peters JM. 2014. Cohesin's ATPase activity couples cohesin loading onto DNA with Smc3 acetylation. *Curr Biol* **24:** 2228–2237.

Lawrence MS, Stojanov P, Mermel CH, Robinson JT, Garraway LA, Golub TR, Meyerson M, Gabriel SB, Lander ES, Getz G. 2014. Discovery and saturation analysis of cancer genes across 21 tumour types. *Nature* **505:** 495–501.

Leiserson MD, Vandin F, Wu HT, Dobson JR, Eldridge JV, Thomas JL, Papoutsaki A, Kim Y, Niu B, McLellan M, et al. 2015. Pan-cancer network analysis identifies combinations of rare somatic mutations across pathways and protein complexes. *Nat Genet* **47:** 106–114.

Lengronne A, Katou Y, Mori S, Yokobayashi S, Kelly GP, Itoh T, Watanabe Y, Shirahige K, Uhlmann F. 2004. Cohesin relocation from sites of chromosomal loading to places of convergent transcription. *Nature* **430:** 573–578.

Lin C, Smith ER, Takahashi H, Lai KC, Martin-Brown S, Florens L, Washburn MP, Conaway JW, Conaway RC, Shilatifard A. 2010. AFF4, a component of the ELL/P-TEFb elongation complex and a shared subunit of MLL chimeras, can link transcription elongation to leukemia. *Mol Cell* **37:** 429–437.

Liu H, Jia L, Yu H. 2013a. Phospho-H2A and cohesin specify distinct tension-regulated Sgo1 pools at kinetochores and inner centromeres. *Curr Biol* **23:** 1927–1933.

Liu H, Rankin S, Yu H. 2013b. Phosphorylation-enabled binding of SGO1-PP2A to cohesin protects sororin and centromeric cohesion during mitosis. *Nat Cell Biol* **15:** 40–49.

Lohr JG, Stojanov P, Carter SL, Cruz-Gordillo P, Lawrence MS, Auclair D, Sougnez C, Knoechel B, Gould J, Saksena G, et al. 2014. Widespread genetic heterogeneity in multiple myeloma: Implications for targeted therapy. *Cancer Cell* **25:** 91–101.

Losada A, Hirano M, Hirano T. 1998. Identification of *Xenopus* SMC protein complexes required for sister chromatid cohesion. *Genes Dev* **12:** 1986–1997.

Losada A, Yokochi T, Kobayashi R, Hirano T. 2000. Identification and characterization of SA/Scc3p subunits in the *Xenopus* and human cohesin complexes. *J Cell Biol* **150:** 405–416.

Losada A, Hirano M, Hirano T. 2002. Cohesin release is required for sister chromatid resolution, but not for condensin-mediated compaction, at the onset of mitosis. *Genes Dev* **16:** 3004–3016.

Losada A, Yokochi T, Hirano T. 2005. Functional contribution of Pds5 to cohesin-mediated cohesion in human cells and *Xenopus* egg extracts. *J Cell Sci* **118:** 2133–2141.

Mannini L, Cucco F, Quarantotti V, Amato C, Tinti M, Tana L, Frattini A, Delia D, Krantz ID, Jessberger R, et al. 2015.

SMC1B is present in mammalian somatic cells and interacts with mitotic cohesin proteins. *Sci Rep* **5**: 18472.

Marsman J, O'Neill AC, Kao BR, Rhodes JM, Meier M, Antony J, Monnich M, Horsfield JA. 2014. Cohesin and CTCF differentially regulate spatiotemporal runx1 expression during zebrafish development. *Biochim Biophys Acta* **1839**: 50–61.

Mazumdar C, Shen Y, Xavy S, Zhao F, Reinisch A, Li R, Corces MR, Flynn RA, Buenrostro JD, Chan SM, et al. 2015. Leukemia-associated cohesin mutants dominantly enforce stem cell programs and impair human hematopoietic progenitor differentiation. *Cell Stem Cell* **17**: 675–688.

McGuinness BE, Hirota T, Kudo NR, Peters JM, Nasmyth K. 2005. Shugoshin prevents dissociation of cohesin from centromeres during mitosis in vertebrate cells. *PLoS Biol* **3**: e86.

McLellan JL, O'Neil NJ, Barrett I, Ferree E, van Pel DM, Ushey K, Sipahimalani P, Bryan J, Rose AM, Hieter P. 2012. Synthetic lethality of cohesins with PARPs and replication fork mediators. *PLoS Genet* **8**: e1002574.

Mei J, Huang X, Zhang P. 2001. Securin is not required for cellular viability, but is required for normal growth of mouse embryonic fibroblasts. *Curr Biol* **11**: 1197–120.

Michaelis C, Ciosk R, Nasmyth K. 1997. Cohesins: Chromosomal proteins that prevent premature separation of sister chromatids. *Cell* **91**: 35–45.

Minamino M, Ishibashi M, Nakato R, Akiyama K, Tanaka H, Kato Y, Negishi L, Hirota T, Sutani T, Bando M, et al. 2015. Esco1 acetylates cohesin via a mechanism different from that of Esco2. *Curr Biol* **25**: 1694–1706.

Mishiro T, Ishihara K, Hino S, Tsutsumi S, Aburatani H, Shirahige K, Kinoshita Y, Nakao M. 2009. Architectural roles of multiple chromatin insulators at the human apolipoprotein gene cluster. *EMBO J* **28**: 1234–1245.

Misulovin Z, Schwartz YB, Li XY, Kahn TG, Gause M, Macarthur S, Fay JC, Eisen MB, Pirrotta V, Biggin MD, et al. 2008. Association of cohesin and Nipped-B with transcriptionally active regions of the *Drosophila* melanogaster genome. *Chromosoma* **117**: 89–102.

Mizuguchi T, Fudenberg G, Mehta S, Belton JM, Taneja N, Folco HD, FitzGerald P, Dekker J, Mirny L, Barrowman J, et al. 2014. Cohesin-dependent globules and heterochromatin shape 3D genome architecture in *S. pombe*. *Nature* **516**: 432–435.

Mohr L, Buheitel J, Schockel L, Karalus D, Mayer B, Stemmann O. 2015. An Alternatively spliced bifunctional localization signal reprograms human shugoshin 1 to protect centrosomal instead of centromeric cohesin. *Cell Rep* **12**: 2156–2168.

Mullenders J, Aranda-Orgilles B, Lhoumaud P, Keller M, Pae J, Wang K, Kayembe C, Rocha PP, Raviram R, Gong Y, et al. 2015. Cohesin loss alters adult hematopoietic stem cell homeostasis, leading to myeloproliferative neoplasms. *J Exp Med* **212**: 1833–1850.

Murayama Y, Uhlmann F. 2014. Biochemical reconstitution of topological DNA binding by the cohesin ring. *Nature* **505**: 367–371.

Murayama Y, Uhlmann F. 2015. DNA entry into and exit out of the cohesin ring by an interlocking gate mechanism. *Cell* **163**: 1628–1640.

Nasmyth K. 2011. Cohesin: A catenase with separate entry and exit gates? *Nat Cell Biol* **13**: 1170–1177.

Nasmyth K, Haering CH. 2009. Cohesin: Its roles and mechanisms. *Annu Rev Genet* **43**: 525–558.

Nativio R, Wendt KS, Ito Y, Huddleston JE, Uribe-Lewis S, Woodfine K, Krueger C, Reik W, Peters JM, Murrell A. 2009. Cohesin is required for higher-order chromatin conformation at the imprinted IGF2-H19 locus. *PLoS Genet* **5**: e1000739.

Nichols MH, Corces VG. 2015. A CTCF code for 3D genome architecture. *Cell* **162**: 703–705.

Nishiyama T, Ladurner R, Schmitz J, Kreidl E, Schleiffer A, Bhaskara V, Bando M, Shirahige K, Hyman AA, Mechtler K, et al. 2010. Sororin mediates sister chromatid cohesion by antagonizing Wapl. *Cell* **143**: 737–749.

Nishiyama T, Sykora MM, Huis in 't Veld PJ, Mechtler K, Peters JM. 2013. Aurora B and Cdk1 mediate Wapl activation and release of acetylated cohesin from chromosomes by phosphorylating Sororin. *Proc Natl Acad Sci* **110**: 13404–13409.

Ocampo-Hafalla MT, Uhlmann F. 2011. Cohesin loading and sliding. *J Cell Sci* **124**: 685–691.

Orgil O, Matityahu A, Eng T, Guacci V, Koshland D, Onn I. 2015. A conserved domain in the scc3 subunit of cohesin mediates the interaction with both mcd1 and the cohesin loader complex. *PLoS Genet* **11**: e1005036.

Ouyang Z, Zheng G, Tomchick DR, Luo X, Yu H. 2016. Structural basis and IP requirement for Pds5-dependent cohesin dynamics. *Mol Cell* **62**: 248–259.

Parelho V, Hadjur S, Spivakov M, Leleu M, Sauer S, Gregson HC, Jarmuz A, Canzonetta C, Webster Z, Nesterova T, et al. 2008. Cohesins functionally associate with CTCF on mammalian chromosome arms. *Cell* **132**: 422–433.

Pellagatti A, Fernandez-Mercado M, Di Genua C, Larrayoz MJ, Killick S, Dolatshad H, Burns A, Calasanz MJ, Schuh A, Boultwood J. 2014. Whole-exome sequencing in del(5q) myelodysplastic syndromes in transformation to acute myeloid leukemia. *Leukemia* **28**: 1148–1151.

Phillips-Cremins JE, Sauria ME, Sanyal A, Gerasimova TI, Lajoie BR, Bell JS, Ong CT, Hookway TA, Guo C, Sun Y, et al. 2013. Architectural protein subclasses shape 3D organization of genomes during lineage commitment. *Cell* **153**: 1281–1295.

Rahman S, Jones MJ, Jallepalli PV. 2015. Cohesin recruits the Esco1 acetyltransferase genome wide to repress transcription and promote cohesion in somatic cells. *Proc Natl Acad Sci* **112**: 11270–11275.

Remeseiro S, Cuadrado A, Carretero M, Martínez P, Drosopoulos WC, Cañamero M, Schildkraut CL, Blasco MA, Losada A. 2012a. Cohesin-SA1 deficiency drives aneuploidy and tumourigenesis in mice due to impaired replication of telomeres. *EMBO J* **31**: 2076–2089.

Remeseiro S, Cuadrado A, Gómez-López G, Pisano DG, Losada A. 2012b. A unique role of cohesin-SA1 in gene regulation and development. *EMBO J* **31**: 2090–2102.

Remeseiro S, Cuadrado A, Kawauchi S, Calof AL, Lander AD, Losada A. 2013. Reduction of Nipbl impairs cohesin loading locally and affects transcription but not cohesion-dependent functions in a mouse model of Cornelia de Lange Syndrome. *Biochim Biophys Acta* **1832**: 2097–2102.

Rolef Ben-Shahar T, Heeger S, Lehane C, East P, Flynn H, Skehel M, Uhlmann F. 2008. Eco1-dependent cohesin acetylation during establishment of sister chromatid cohesion. *Science* **321**: 563–566.

Rubio ED, Reiss DJ, Welcsh PL, Disteche CM, Filippova GN, Baliga NS, Aebersold R, Ranish JA, Krumm A. 2008. CTCF physically links cohesin to chromatin. *Proc Natl Acad Sci* **105**: 8309–8314.

Sakuno T, Tada K, Watanabe Y. 2009. Kinetochore geometry defined by cohesion within the centromere. *Nature* **458**: 852–858.

Sanborn AL, Rao SS, Huang SC, Durand NC, Huntley MH, Jewett AI, Bochkov ID, Chinnappan D, Cutkosky A, Li J, et al. 2015. Chromatin extrusion explains key features of loop and domain formation in wild-type and engineered genomes. *Proc Natl Acad Sci* **112**: E6456–6465.

Schaaf CA, Kwak H, Koenig A, Misulovin Z, Gohara DW, Watson A, Zhou Y, Lis JT, Dorsett D. 2013. Genome-wide control of RNA polymerase II activity by cohesin. *PLoS Genet* **9**: e1003382.

Schmidt D, Schwalie PC, Ross-Innes CS, Hurtado A, Brown GD, Carroll JS, Flicek P, Odom DT. 2010. A CTCF-independent role for cohesin in tissue-specific transcription. *Genome Res* **20**: 578–588.

Schmitz J, Watrin E, Lenart P, Mechtler K, Peters JM. 2007. Sororin is required for stable binding of cohesin to chromatin and for sister chromatid cohesion in interphase. *Curr Biol* **17**: 630–636.

Schockel L, Mockel M, Mayer B, Boos D, Stemmann O. 2011. Cleavage of cohesin rings coordinates the separation of centrioles and chromatids. *Nat Cell Biol* **13**: 966–972.

Seitan VC, Hao B, Tachibana-Konwalski K, Lavagnolli T, Mira-Bontenbal H, Brown KE, Teng G, Carroll T, Terry A, Horan K, et al. 2011. A role for cohesin in T-cell-receptor rearrangement and thymocyte differentiation. *Nature* **476**: 467–471.

Seitan VC, Faure AJ, Zhan Y, McCord RP, Lajoie BR, Ing-Simmons E, Lenhard B, Giorgetti L, Heard E, Fisher AG, et al. 2013. Cohesin-based chromatin interactions enable regulated gene expression within preexisting architectural compartments. *Genome Res* **23**: 2066–2077.

Sfeir A, Kosiyatrakul ST, Hockemeyer D, MacRae SL, Karlseder J, Schildkraut CL, de Lange T. 2009. Mammalian telomeres resemble fragile sites and require TRF1 for efficient replication. *Cell* **138**: 90–103.

Shindo N, Kumada K, Hirota T. 2012. Separase sensor reveals dual roles for separase coordinating cohesin cleavage and cdk1 inhibition. *Dev Cell* **23**: 112–123.

Sjogren C, Nasmyth K. 2001. Sister chromatid cohesion is required for postreplicative double-strand break repair in *Saccharomyces cerevisiae*. *Curr Biol* **11**: 991–995.

Smith E, Shilatifard A. 2013. Transcriptional elongation checkpoint control in development and disease. *Genes Dev* **27**: 1079–1088.

Smith TG, Laval S, Chen F, Rock MJ, Strachan T, Peters H. 2014. Neural crest cell-specific inactivation of Nipbl or Mau2 during mouse development results in a late onset of craniofacial defects. *Genesis* **52**: 687–694.

Sofueva S, Yaffe E, Chan WC, Georgopoulou D, Vietri Rudan M, Mira-Bontenbal H, Pollard SM, Schroth GP, Tanay A, Hadjur S. 2013. Cohesin-mediated interactions organize chromosomal domain architecture. *EMBO J* **32**: 3119–3129.

Solomon DA, Kim T, Diaz-Martinez LA, Fair J, Elkahloun AG, Harris BT, Toretsky JA, Rosenberg SA, Shukla N, Ladanyi M, et al. 2011. Mutational inactivation of *STAG2* causes aneuploidy in human cancer. *Science* **333**: 1039–1043.

Solomon DA, Kim JS, Bondaruk J, Shariat SF, Wang ZF, Elkahloun AG, Ozawa T, Gerard J, Zhuang D, Zhang S, et al. 2013. Frequent truncating mutations of *STAG2* in bladder cancer. *Nat Genet* **45**: 1428–1430.

Sonoda E, Matsusaka T, Morrison C, Vagnarelli P, Hoshi O, Ushiki T, Nojima K, Fukagawa T, Waizenegger IC, Peters JM, et al. 2001. Scc1/Rad21/Mcd1 is required for sister chromatid cohesion and kinetochore function in vertebrate cells. *Dev Cell* **1**: 759–770.

Sumara I, Vorlaufer E, Gieffers C, Peters BH, Peters JM. 2000. Characterization of vertebrate cohesin complexes and their regulation in prophase. *J Cell Biol* **151**: 749–762.

Sumara I, Vorlaufer E, Stukenberg PT, Kelm O, Redemann N, Nigg EA, Peters JM. 2002. The dissociation of cohesin from chromosomes in prophase is regulated by Polo-like kinase. *Mol Cell* **9**: 515–525.

Tang Z, Luo OJ, Li X, Zheng M, Zhu JJ, Szalaj P, Trzaskoma P, Magalska A, Wlodarczyk J, Ruszczycki B, et al. 2015. CTCF-mediated human 3D genome architecture reveals chromatin topology for transcription. *Cell* **163**: 1611–1627.

Taylor CF, Platt FM, Hurst CD, Thygesen HH, Knowles MA. 2013. Frequent inactivating mutations of *STAG2* in bladder cancer are associated with low tumour grade and stage and inversely related to chromosomal copy number changes. *Hum Mol Genet* **23**: 1964–1974.

Tedeschi A, Wutz G, Huet S, Jaritz M, Wuensche A, Schirghuber E, Davidson IF, Tang W, Cisneros DA, Bhaskara V, et al. 2013. Wapl is an essential regulator of chromatin structure and chromosome segregation. *Nature* **501**: 564–568.

Thol F, Bollin R, Gehlhaar M, Walter C, Dugas M, Suchanek KJ, Kirchner A, Huang L, Chaturvedi A, Wichmann M, et al. 2014. Mutations in the cohesin complex in acute myeloid leukemia: Clinical and prognostic implications. *Blood* **123**: 914–920.

Thota S, Viny AD, Makishima H, Spitzer B, Radivoyevitch T, Przychodzen B, Sekeres MA, Levine RL, Maciejewski JP. 2014. Genetic alterations of the cohesin complex genes in myeloid malignancies. *Blood* **124**: 1790–1798.

Tirode F, Surdez D, Ma X, Parker M, Le Deley MC, Bahrami A, Zhang Z, Lapouble E, Grossetete-Lalami S, Rusch M, et al. 2014. Genomic landscape of Ewing sarcoma defines an aggressive subtype with co-association of *STAG2* and *TP53* mutations. *Cancer Discov* **4**: 1342–1353.

Tittel-Elmer M, Lengronne A, Davidson MB, Bacal J, Francois P, Hohl M, Petrini JH, Pasero P, Cobb JA. 2012. Cohesin association to replication sites depends on Rad50 and promotes fork restart. *Mol Cell* **48**: 98–108.

Toyoda Y, Yanagida M. 2006. Coordinated requirements of human topo II and cohesin for metaphase centromere alignment under Mad2-dependent spindle checkpoint surveillance. *Mol Biol Cell* **17**: 2287–2302.

Uhlmann F, Wernic D, Poupart MA, Koonin EV, Nasmyth K. 2000. Cleavage of cohesin by the CD clan protease separin triggers anaphase in yeast. *Cell* **103:** 375–386.

Unal E, Heidinger-Pauli JM, Kim W, Guacci V, Onn I, Gygi SP, Koshland DE. 2008. A molecular determinant for the establishment of sister chromatid cohesion. *Science* **321:** 566–569.

Vass S, Cotterill S, Valdeolmillos AM, Barbero JL, Lin E, Warren WD, Heck MM. 2003. Depletion of Drad21/ Scc1 in *Drosophila* cells leads to instability of the cohesin complex and disruption of mitotic progression. *Curr Biol* **13:** 208–218.

Vaur S, Feytout A, Vazquez S, Javerzat JP. 2012. Pds5 promotes cohesin acetylation and stable cohesin-chromosome interaction. *EMBO Rep* **13:** 645–652.

Viny AD, Ott CJ, Spitzer B, Rivas M, Meydan C, Papalexi E, Yelin D, Shank K, Reyes J, Chiu A, et al. 2015. Dose-dependent role of the cohesin complex in normal and malignant hematopoiesis. *J Exp Med* **212:** 1819–1832.

Waizenegger IC, Hauf S, Meinke A, Peters JM. 2000. Two distinct pathways remove mammalian cohesin from chromosome arms in prophase and from centromeres in anaphase. *Cell* **103:** 399–410.

Walter MJ, Shen D, Ding L, Shao J, Koboldt DC, Chen K, Larson DE, McLellan MD, Dooling D, Abbott R, et al. 2012. Clonal architecture of secondary acute myeloid leukemia. *N Engl J Med* **366:** 1090–1098.

Walter MJ, Shen D, Shao J, Ding L, White BS, Kandoth C, Miller CA, Niu B, McLellan MD, Dees ND, et al. 2013. Clonal diversity of recurrently mutated genes in myelodysplastic syndromes. *Leukemia* **27:** 1275–1282.

Wang X, Yang Y, Duan Q, Jiang N, Huang Y, Darzynkiewicz Z, Dai W. 2008. sSgo1, a major splice variant of Sgo1, functions in centriole cohesion where it is regulated by Plk1. *Dev Cell* **14:** 331–341.

Watrin E, Schleiffer A, Tanaka K, Eisenhaber F, Nasmyth K, Peters JM. 2006. Human Scc4 is required for cohesin binding to chromatin, sister-chromatid cohesion, and mitotic progression. *Curr Biol* **16:** 863–874.

Weitzer S, Lehane C, Uhlmann F. 2003. A model for ATP hydrolysis-dependent binding of cohesin to DNA. *Curr Biol* **13:** 1930–1940.

Welch JS, Ley TJ, Link DC, Miller CA, Larson DE, Koboldt DC, Wartman LD, Lamprecht TL, Liu F, Xia J, et al. 2012. The origin and evolution of mutations in acute myeloid leukemia. *Cell* **150:** 264–278.

Wendt KS, Yoshida K, Itoh T, Bando M, Koch B, Schirghuber E, Tsutsumi S, Nagae G, Ishihara K, Mishiro T, et al. 2008. Cohesin mediates transcriptional insulation by CCCTC-binding factor. *Nature* **451:** 796–801.

Whelan G, Kreidl E, Wutz G, Egner A, Peters JM, Eichele G. 2012. Cohesin acetyltransferase Esco2 is a cell viability factor and is required for cohesion in pericentric heterochromatin. *EMBO J* **31:** 71–82.

White JK, Gerdin AK, Karp NA, Ryder E, Buljan M, Bussell JN, Salisbury J, Clare S, Ingham NJ, Podrini C, et al. 2013. Genome-wide generation and systematic phenotyping of knockout mice reveals new roles for many genes. *Cell* **154:** 452–464.

Wilson NK, Foster SD, Wang X, Knezevic K, Schutte J, Kaimakis P, Chilarska PM, Kinston S, Ouwehand WH, Dzierzak E, et al. 2010. Combinatorial transcriptional control in blood stem/progenitor cells: Genome-wide analysis of ten major transcriptional regulators. *Cell Stem Cell* **7:** 532–544.

Wirth KG, Wutz G, Kudo NR, Desdouets C, Zetterberg A, Taghybeeglu S, Seznec J, Ducos GM, Ricci R, Firnberg N, et al. 2006. Separase: A universal trigger for sister chromatid disjunction but not chromosome cycle progression. *J Cell Biol* **172:** 847–860.

Wu N, Kong X, Ji Z, Zeng W, Potts PR, Yokomori K, Yu H. 2012. Scc1 sumoylation by Mms21 promotes sister chromatid recombination through counteracting Wapl. *Genes Dev* **26:** 1473–1485.

Xiao T, Wallace J, Felsenfeld G. 2011. Specific sites in the C terminus of CTCF interact with the SA2 subunit of the cohesin complex and are required for cohesin-dependent insulation activity. *Mol Cell Biol* **31:** 2174–2183.

Xu H, Balakrishnan K, Malaterre J, Beasley M, Yan Y, Essers J, Appeldoorn E, Tomaszewski JM, Vazquez M, Verschoor S, et al. 2010. Rad21-cohesin haploinsufficiency impedes DNA repair and enhances gastrointestinal radiosensitivity in mice. *PLoS ONE* **5:** e12112.

Yadav S, Sehrawat A, Eroglu Z, Somlo G, Hickey R, Liu X, Awasthi YC, Awasthi S. 2013. Role of SMC1 in overcoming drug resistance in triple negative breast cancer. *PLoS ONE* **8:** e64338.

Yamada HY, Yao Y, Wang X, Zhang Y, Huang Y, Dai W, Rao CV. 2012. Haploinsufficiency of SGO1 results in deregulated centrosome dynamics, enhanced chromosomal instability and colon tumorigenesis. *Cell Cycle* **11:** 479–488.

Yoshida K, Toki T, Okuno Y, Kanezaki R, Shiraishi Y, Sato-Otsubo A, Sanada M, Park MJ, Terui K, Suzuki H, et al. 2013. The landscape of somatic mutations in Down syndrome–related myeloid disorders. *Nat Genet* **45:** 1293–1299.

Zhang B, Jain S, Song H, Fu M, Heuckeroth RO, Erlich JM, Jay PY, Milbrandt J. 2007. Mice lacking sister chromatid cohesion protein PDS5B exhibit developmental abnormalities reminiscent of Cornelia de Lange syndrome. *Development* **134:** 3191–3201.

Zhang J, Shi X, Li Y, Kim BJ, Jia J, Huang Z, Yang T, Fu X, Jung SY, Wang Y, et al. 2008. Acetylation of Smc3 by Eco1 is required for S phase sister chromatid cohesion in both human and yeast. *Mol Cell* **31:** 143–151.

Zhang B, Chang J, Fu M, Huang J, Kashyap R, Salavaggione E, Jain S, Kulkarni S, Deardorff MA, Uzielli ML, et al. 2009. Dosage effects of cohesin regulatory factor PDS5 on mammalian development: Implications for cohesinopathies. *PLoS ONE* **4:** e5232.

Zuin J, Dixon JR, van der Reijden MI, Ye Z, Kolovos P, Brouwer RW, van de Corput MP, van der Werken HJ, Knoch TA, van Ijcken WF, et al. 2014a. Cohesin and CTCF differentially affect chromatin architecture and gene expression in human cells. *Proc Natl Acad Sci* **111:** 996–1001.

Zuin J, Franke V, van Ijcken WF, van der Sloot A, Krantz ID, van der Reijden MI, Nakato R, Lenhard B, Wendt KS. 2014b. A cohesin-independent role for NIPBL at promoters provides insights in CdLS. *PLoS Genet* **10:** e1004153.

HDACs and HDAC Inhibitors in Cancer Development and Therapy

Yixuan Li and Edward Seto

George Washington University Cancer Center, Department of Biochemistry and Molecular Medicine, George Washington University, Washington, DC 20037

Correspondence: seto@gwu.edu

Over the last several decades, it has become clear that epigenetic abnormalities may be one of the hallmarks of cancer. Posttranslational modifications of histones, for example, may play a crucial role in cancer development and progression by modulating gene transcription, chromatin remodeling, and nuclear architecture. Histone acetylation, a well-studied posttranslational histone modification, is controlled by the opposing activities of histone acetyltransferases (HATs) and histone deacetylases (HDACs). By removing acetyl groups, HDACs reverse chromatin acetylation and alter transcription of oncogenes and tumor suppressor genes. In addition, HDACs deacetylate numerous nonhistone cellular substrates that govern a wide array of biological processes including cancer initiation and progression. This review will discuss the role of HDACs in cancer and the therapeutic potential of HDAC inhibitors (HDACi) as emerging drugs in cancer treatment.

Histone function is modulated by multiple posttranslational modifications, including reversible acetylation of the amino-terminal ε-group of lysines on histones. Histone acetylation is tightly controlled by a balance between the opposing activities of histone acetyltransferases (HATs) and histone deacetylases (HDACs, also known as lysine deacetylases or KDACs). There are 18 potential human HDACs grouped into four classes. By removing the acetyl groups from the ε-amino lysine residues on histone tails, HDACs may play a critical role in transcription regulation (Seto and Yoshida 2014).

Given that histone modification modulates chromatin structure and gene expression, it is not surprising that abnormal alterations in his-

tone acetylation are associated with cancer development. For example, global loss of acetylation at lysine 16 and trimethylation at lysine 20 of histone H4 is reported to be a common abnormality in human cancer (Fraga et al. 2005), and a low level of histone H3 lysine 18 acetylation (H3K18ac) was found to be a predictor of poor survival in pancreatic, breast, prostate, and lung cancers. In parallel, research increasingly shows aberrant expression of HDACs is frequently observed in various human cancers. Although it is not known whether the changes in histone modification are related to specific alterations in HDACs expression (there are obviously many other mechanisms that can explain why cancer cells might exploit HDACs to support tumorigenesis), they do nevertheless contribute to the overall principle of targeting HDACs for cancer therapy.

Because approximately equal numbers of genes are activated and repressed by HDAC inhibition, other mechanisms besides histone modification are involved in HDAC-mediated gene regulation. In addition to histones, HDACs also deacetylate a large number of nonhistone proteins. This is consistent with the discovery of many acetylated nonhistone proteins by global analysis in human cells (Choudhary et al. 2009). In tumorigenesis, the finely tuned acetylation status at the whole proteome level is greatly impaired by dysregulated deacetylases (Parbin et al. 2014). Through hyperacetylation of histone and nonhistone targets, HDACi enable the reestablishment of cellular acetylation homeostasis and restore normal expression and function of numerous proteins that may reverse cancer initiation and progression. This article describes recent advances in our understanding of the role of HDACs in cancer and the implications of HDACi in the treatment of cancer.

DYSREGULATION AND MUTATION OF HDACs IN HUMAN CANCER

Based on sequence homology to yeast, 18 human HDACs are grouped into four classes. Class I Rpd3-like enzymes are comprised of HDAC1, 2, 3, and 8. Class II Hda1-like enzymes are further divided into two subclasses: IIa (HDAC4, 5, 6, 7, and 9) and IIb (HDAC6 and 10). Class III Sir2-like enzymes consist of seven sirtuins, which are NAD-dependent protein deacetylases and/or ADP ribosylases. Sirtuins have been shown to regulate many cellular processes including survival, aging, stress response, and metabolism. Class IV contains only HDAC11, which shares sequences similarity to both class I and II proteins.

HDACs are involved in multiple different stages of cancer (Fig. 1). Aberrant expression of classical (class I, II, IV) HDACs has been linked to a variety of malignancies, including solid and hematological tumors (Table 1). In most cases, a high level of HDACs is associated with advanced disease and poor outcomes in patients. For example, high expression of HDAC1, 2, and 3 are associated with poor outcomes in gastric and ovarian cancers (Weichert et al. 2008a,b; Sudo et al. 2011), and high expression of HDAC8 correlates with advanced-stage disease and poor survival in neuroblastoma (Oehme et al. 2009; Rettig et al. 2015). HDACs have also been found broadly dysregulated in multiple myeloma (MM). Overexpression of class I HDACs, particularly HDAC1, is associated with inferior patient outcomes (Mithraprabhu et al. 2014).

The mechanisms by which individual HDACs regulate tumorigenesis are quite diverse. Because HDACs induce a range of cellular and molecular effects through hyperacetylation of histone and nonhistone substrates, HDACs could either repress tumor suppressor gene expression or regulate the oncogenic cell-signaling pathway via modification of key molecules. However, the contribution of HDACs to cancer may not necessarily be related to the level of HDAC expression, because aberrant activity of HDACs is also common in cancer development (West and Johnstone 2014). Certain HDACs function as the catalytic subunits of large corepressor complexes and could be aberrantly recruited to target genes by oncogenic proteins to drive tumorigenesis. For example, the aberrant recruitment of HDAC1, 2, or 3 by oncogenic fusion proteins AML1-ETO and PML-RAR contributes to the pathogenesis of acute myeloid leukemia (AML) (Hug and Lazar 2004; Falkenberg and Johnstone 2014).

Although the broad anticancer effects of HDACi predict an oncogenic role of HDACs in tumor development, in some cancers it has been found that genetic inactivation of HDACs might have tumorigenic effects. *HDAC1* somatic mutations were detected in 8.3% of dedifferentiated liposarcoma and *HDAC4* homozygous deletions occurred in 4% of melanomas (Stark and Hayward 2007; Taylor et al. 2011). Truncating mutations of HDAC2 have been observed in human epithelial cancers with microsatellite instability, which causes a loss of HDAC2 protein expression and confers cells more resistant to HDACi (Ropero et al. 2006). Furthermore, the ectopic expression of

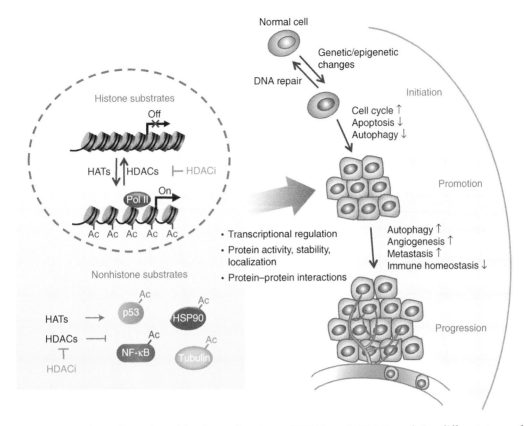

Figure 1. A simplistic illustration of the diverse functions of HDACs and HDACi regulating different stages of cancer through multiple different mechanisms and changing different biological processes. Far right, ↑ indicates promotion or up-regulation, ↓ indicates repression or down-regulation.

HDAC2 in mutant cancer cells induces a reduction of tumor cell growth in vitro and in vivo, suggesting a putative tumor-suppressor role for HDAC2 in this cellular setting. Class II HDACs may also function as tumor suppressor in certain circumstances. Low expression of HDAC10 is associated with poor prognosis in lung and gastric cancer patients (Osada et al. 2004; Jin et al. 2014). HDAC6 is down-regulated in human hepatocellular carcinoma (HCC) tissues, and low expression of HDAC6 is associated with poor prognosis in liver transplantation patients. Knockdown of HDAC6 promotes angiogenesis in HCC by HIF-1α/VEGFA axis (Lv et al. 2015). Also, a recent study revealed a dual role of HDAC1 in cancer initiation and maintenance. HDAC1 antagonizes the oncogenic activity of PML-RAR during the preleukemic stage of acute promyelocytic leukemia (APL), but favors the growth of APL cells at the leukemic stage (Santoro et al. 2013), indicating that elucidation of the role of HDACs at each step of tumorigenesis in different tumor cell types will provide a rationale for targeting HDACs in cancer therapy.

POSSIBLE MECHANISMS OF HDACs IN CANCER DEVELOPMENT

Although identification of substrate specificity and biological function for individual HDACs still requires more comprehensive investigations, it is well known that HDACs play crucial roles in cancer by deacetylating histone and nonhistone proteins, which are involved in the regulation of cell cycle, apoptosis, DNA-damage response, metastasis, angiogenesis, autophagy, and other cellular processes (Fig. 1).

Table 1. Dysregulation and mutation of HDACs in human cancer

Cancer types	HDACs	Prognostic relevance	Genetic evidence	Molecular mechanism	References
Solid tumors					
Neuroblastoma	HDAC8	High transcript level correlates with advanced-stage disease and poor survival in neuroblastoma	Knockdown and inhibition of HDAC8 promotes cell-cycle arrest and differentiation, delays cell growth, and induces cell death in vitro and in vivo	HDAC8 inhibition induces p21$^{WAF1/CIP1}$ and $NTRK1/TrkA$ gene expression and enhances retinoic acid-mediated differentiation by regulating CREB phosphorylation	Oehme et al. 2009; Rettig et al. 2015
	HDAC10	High expression correlates with poor overall patient survival in advanced INSS stage 4 neuroblastoma	Knockdown and inhibition of HDAC10 in neuroblastoma cells interrupted autophagic flux resulting in an increase of sensitization to cytotoxic drug treatment	HDAC10 controls autophagic processing and resistance to cytotoxic drugs via interaction with Hsp70 family proteins	Oehme et al. 2013
Medulloblastoma	HDAC2	Overexpressed in medulloblastoma subgroups with poor prognosis	HDAC2 depletion induces cell death and attenuates cell growth; MYC amplified and HDAC2 overexpressing cell lines are more sensitive to class I HDACi	N/A	Ecker et al. 2015
	HDAC5, 9	Up-regulated in high-risk medulloblastoma, and their expression is associated with poor survival	Depletion of either HDAC5 or HDAC9 in MB cells resulted in a reduction of cell proliferation and increase in cell death	N/A	Milde et al. 2010
Lung	HDAC1, 3	High expression correlates with a poor prognosis in patients with lung adenocarcinoma	N/A	N/A	Minamiya et al. 2010, 2011

	HDAC2	Abundant expression is observed in lung cancer tissues	HDAC2 inactivation represses tumor cell growth in vitro and in vivo	HDAC2 depletion activates apoptosis via p53 and Bax activation and Bcl2 suppression induces cell-cycle arrest by induction of p21 and suppression of cyclin E2, cyclin D1, and CDK2	Jung et al. 2012
	HDAC5, 10	Low expression is associated with poor prognosis in lung cancer patients	N/A	N/A	Osada et al. 2004
Gastric	HDAC1, 2, 3	High expression is associated with nodal tumor spread and decreased overall patient survival	N/A	N/A	Weichert et al. 2008b; Sudo et al. 2011
	HDAC4	Up-regulated in gastric tumor cells compared with adjacent normal tissues	HDAC4 inhibition has a synergistic effect with docetaxel treatment	HDAC4 inhibition increased the level of cleaved caspases 3 and 9	Colarossi et al. 2014
	HDAC10	Low expression is a poor prognosis marker for gastric cancer patients	N/A	N/A	Jin et al. 2014
Liver	HDAC1	Highly expressed in human HCCs and liver cancer cell lines	HDAC1 inactivation impairs G_1/S cell-cycle transition and causes autophagic cell death	Knockdown of HDAC1 induces p21 and p27 expression and suppresses cyclin D1 and CDK2 expression	Xie et al. 2012
	HDAC1, 2, 3	Up-regulated in human HCCs and liver cancer cell lines	The knockdown of HDAC1–3 leads to increased apoptosis and decreased proliferation	Knockdown and inhibition of HDAC1–3 up-regulates miR-449 and down-regulates c-MET expression, and reduced c-MET dephosphorylates ERK1/2 and inhibits tumor growth	Buurman et al. 2012
	HDAC1, 2, 3	High expression is associated with poor survival in low-grade and early-stage tumors	N/A	N/A	Quint et al. 2011
	HDAC3, 5	Up-regulation is correlated with DNA copy number gains	N/A	N/A	Lachenmayer et al. 2012

Continued

Table 1. *Continued*

Cancer types	HDACs	Prognostic relevance	Genetic evidence	Molecular mechanism	References
	HDAC5	Up-regulated in HCC tissues	Knockdown of HDAC5 promotes cell apoptosis and inhibits tumor cell growth in vitro and in vivo	HDAC5 knockdown promotes apoptosis by up-regulating cyto C, caspase 3, p53, and bax, and induces G_1 phase cell-cycle arrest by up-regulating p21 and down-regulating cyclin D1 and CDK2/4/6	Fan et al. 2014
	HDAC5	Up-regulated in human HCC tissues	Overexpression of HDAC5 promotes tumor cell proliferation, while knockdown of HDAC5 inhibits cell proliferation	HDAC5 promotes cell proliferation by up-regulation of Six1	Feng et al. 2014
	HDAC6	Low expression is associated with poor prognosis in liver transplantation patients	Knockdown of HDAC6 promotes HUVEC migration, proliferation, and tube formation in vitro, and suppresses HCC cell apoptosis and promotes HCC cell proliferation in hypoxia	HDAC6 knockdown promotes angiogenesis in HCC by HIF-1α/VEGFA axis	Lv et al. 2015
Pancreatic	HDAC2	Highly expressed in PDAC	HDAC2 confers resistance toward etoposide in PDAC cells	HDAC2 knockdown up-regulates NOXA expression, which sensitizes tumor cells toward etoposide-induced apoptosis	Fritsche et al. 2009
	HDAC6	Highly expressed in human pancreatic cancer tissues	Knockdown and inhibition of HDAC6 impairs the motility of cancer cells	HDAC6 interacts with CLIP-170 and stimulates the migration of pancreatic cancer cells	Li et al. 2014
	HDAC7	Overexpression is associated with poor prognosis	Knockdown of HDAC7 inhibits tumor cell growth	N/A	Ouaissi et al. 2008, 2014
Colorectal	HDAC2 mutation	Truncating mutations are found in microsatellite unstable sporadic colorectal cancers, which lead to a loss of expression of the protein	HDAC2 mutation renders cells more resistant to antiproliferative and proapoptotic effects of the HDAC inhibitor	N/A	Ropero et al. 2006

Cite this article as *Cold Spring Harb Perspect Med* doi: 10.1101/cshperspect.a026831

Cancer	HDAC	Description	Function		References
	HDAC1, 2, 3	Highly expressed in a subset of colorectal carcinomas; HDAC2 is an independent prognostic factor in colorectal carcinoma	Knockdown of HDAC1 and HDAC2 but not HDAC3 suppresses tumor cell growth	N/A	Weichert et al. 2008d
	HDAC1, 2, 3, 5, 7	Up-regulated in human colorectal cancer; HDAC2 is an early biomarker of colon carcinogenesis	N/A	N/A	Stypula-Cyrus et al. 2013
Breast	HDAC1, 2, 3	HDAC1 was highly expressed in hormone receptor-positive tumors; HDAC2 and 3 are highly expressed in poorly differentiated and hormone receptor negative tumors	N/A	N/A	Zhang et al. 2005; Muller et al. 2013
	HDAC1, 6	High expression is good prognostic factors for estrogen-receptor-positive invasive ductal carcinomas	N/A	N/A	Zhang et al. 2004; Seo et al. 2014
Ovarian	HDAC1, 2, 3	High expression is associated with a poor outcome	Knockdown of HDAC1 reduces cell proliferation via down-regulating cyclin A expression, and knockdown of HDAC3 reduces the cell migration with elevated E-cadherin		Hayashi et al. 2010
	HDAC1, 2, 3	High-level expression is associated with a poor prognosis in ovarian endometrioid carcinomas	N/A	N/A	Weichert et al. 2008a
Cervical	HDAC10	Low expression correlates with lymph node metastasis in cervical cancer	Knockdown of HDAC10 promotes cervical cancer cell migration and invasion	HDAC10 inhibits MMP2 and -9 expression	Song et al. 2013

Continued

Table 1. *Continued*

Cancer types	HDACs	Prognostic relevance	Genetic evidence	Molecular mechanism	References
Prostate	HDAC1, 2, 3	Highly expressed in prostate carcinomas HDAC2 is an independent prognostic marker in prostate cancer cohort	N/A	N/A	Weichert et al. 2008c
Renal	HDAC1, 2	Highly expressed in renal cell cancer, but none of them are associated with patient survival	N/A	N/A	Fritzsche et al. 2008
Bladder	HDAC1, 2, 3	High expression is associated with higher tumor grades; high HDAC1 is a poor prognostic factor in urothelial bladder cancer	N/A	N/A	Poyet et al. 2014
	HDAC2, 4, 5, 7, 8	Up-regulation of HDAC2, 8 and down-regulation of HDAC4, 5, and 7 mRNA are observed in urothelial cancer	N/A	N/A	Niegisch et al. 2013
Melanoma	HDAC3, 8	HDAC8 was increased in BRAF-mutated melanoma; HDAC8 and 3 overexpression are associated with improved survival of patients with stage IV metastatic melanoma	N/A	N/A	Wilmott et al. 2015
Hematological tumors					
ALL	HDAC1–9	HDAC2, 3, 6, 7, 8 are up-regulated in ALL samples; HDAC1 and 4 show high expression in T-ALL and HDAC6 and 9 are highly expressed in B-lineage ALL; higher expression of HDAC7 and 9 is associated with a poor prognosis in childhood ALL	N/A	N/A	Moreno et al. 2010

Cite this article as *Cold Spring Harb Perspect Med* doi: 10.1101/cshperspect.a026831

CLL	HDAC4	High expression is associated with high initial leukocyte count, T cell ALL and prednisone poor response	HDAC4 knockdown enhanced etoposide's cytotoxic activity	N/A	Gruhn et al. 2013
	HDAC1, 3, 6, 7, 9, 10, SIRT1 and 6	Higher expressions are associated with poor prognosis and more advanced disease stage	N/A	N/A	Wang et al. 2011
	HDAC6, 7, 10 and SIRT2, 3, 6	Overexpression of HDAC7 and 10 and underexpression of HDAC6 and SIRT3 are correlated with a poor prognosis	N/A	N/A	Van Damme et al. 2012
AML	HDAC5, 6, SIRT1 and 4	HDAC6 and SIRT1 are overexpressed, and HDAC5 and SIRT4 are underexpressed in AML samples	N/A	N/A	Bradbury et al. 2005
DLBCL	HDAC1	Highly expressed in cases of DLBCL and correlated with a poor survival	N/A	N/A	Min et al. 2012
	HDAC2	Highly expressed in nodal lymphomas, which is associated with shorter survival	N/A	N/A	Lee et al. 2014b
	HDAC1, 2, 6	The expression is higher in cases of DLBCL or PTCL; high HDAC6 level is associated with favorable outcome in DLBCL, but with a negative outcome in PTCL	N/A	N/A	Marquard et al. 2009
	HDAC3	Overexpression is observed in phospho STAT3-positive ABC-type DLBCL	HDAC3 knockdown inhibited survival of pSTAT3-positive DLBCL cells	HDAC3 knockdown unregulated STAT3Lys685 acetylation but prevented STAT3Tyr705 phosphorylation	Gupta et al. 2012

Continued

Table 1. *Continued*

Cancer types	HDACs	Prognostic relevance	Genetic evidence	Molecular mechanism	References
CTCL	HDAC2, 6	HDAC2 is highly expressed in aggressive rather than indolent CTCL; HDAC6 is associated with a favorable outcome independent of the subtype	N/A	N/A	Marquard et al. 2008
HL	HDAC1, 2, 3	Overexpressed in HL tissue samples; high HDAC1 expression is correlated with a worse outcome	N/A	N/A	Adams et al. 2010
Myeloma	HDAC1	Overexpression of class I HDAC, particularly HDAC1, is associated with poor prognosis in myeloma	N/A	N/A	Mithraprabhu et al. 2014

HCC, Human hepatocellular carcinoma; PDAC, pancreatic ductal adenocarcinoma; ALL, acute lymphoblastic leukemia; CLL, chronic lymphocytic leukemia; AML, acute myelogenous leukemia; DLBCL, diffuse large B-cell lymphoma; CTCL, cutaneous T-cell lymphomas; HL, Hodgkin's lymphoma.

Cite this article as *Cold Spring Harb Perspect Med* doi: 10.1101/cshperspect.a026831

Cell Cycle

HDAC inhibition has been shown to have antiproliferative effects by inducing cell-cycle arrest in G_1 via up-regulation of cyclin-dependent kinase (CDK) inhibitors or down-regulation of cyclins and CDKs (Chun 2015). HDAC1 and 2 directly bind to the promoters of the $p21^{WAF1/CIP1}$, $p27^{KIP1}$, and $p57^{KIP2}$ genes and negatively regulate their expression (Yamaguchi et al. 2010; Zupkovitz et al. 2010). Loss of HDAC1 and 2 induces expression of CDK inhibitors, leading to a cell-cycle block in G_1. Knockdown of HDAC5 leads to a significant up-regulation of p21 and down-regulation of cyclin D1 and CDK2/4/6, which results in G_1-phase cell-cycle arrest in human HCC cells (Fan et al. 2014). HDAC inhibition might block the cellular G_1/S transition by reactivating Rb function by dephosphorylation and subsequently inhibiting E2F activities in the transcription of genes for G_1 progression. Trichostatin A (TSA) suppressed retinal pigment epithelium (RPE) cell proliferation via a G_1 phase arrest, caused through inhibition of Rb phosphorylation, reduction of cyclinD1/CDK4/6 complexes, and induction of p21 and p27 (Xiao et al. 2014). However, TSA also induced G_1 phase arrest in malignant tumor cells with mutated Rb, indicating an Rb-independent G_1 arrest (Tomosugi et al. 2012).

In addition to controlling the G_1/S transition, HDAC1 knockdown in tumor cells impairs G_2/M transition and inhibits cell growth as evidenced by a reduction of mitotic cells and an increased percentage of apoptotic cells (Senese et al. 2007). HDAC10 regulates the G_2/M transition via modulation of cyclin A2 expression. The effect of HDAC10 on cyclin A2 transcription was dependent on let-7 and HMGA2 (Li et al. 2015b). Consistent with gene knockdown results, inhibition of HDACs by inhibitors including TSA, SAHA, and VPA cause cell-cycle arrest at the G_2/M boundary in a variety of tumor cell lines, supporting pleiotropic roles of HDAC throughout the cell cycle (Juengel et al. 2014).

Besides transcriptional repression of cell-cycle-related genes at the G_1/S and G_2/M cell-cycle checkpoints, HDACs might also regulate cell-cycle progression in a transcription-independent manner. During mitosis, A-kinase-anchoring proteins AKAP95 and HA95 recruit HDAC3 along with Aurora B. In this context, HDAC3 deacetylates histone H3, which in turn allows maximal phosphorylation of Ser10 by Aurora B, leading to HP1β dissociation from mitotic chromosomes. The HDAC3-AKAP95/HA95-Aurora B pathway is required for normal mitotic progression (Li et al. 2006). Indeed, the histone deacetylase inhibitor LBH589 can induce G_2-M cell-cycle arrest and apoptosis in renal cancer cells through degradation of Aurora A and B kinases by targeting HDAC3 and HDAC6 (Cha et al. 2009). Moreover, HDAC3 directly interacts with cyclin A and regulates cyclin A stability by modulating its acetylation status. An abrupt loss of HDAC3 at metaphase facilitates cyclin A acetylation by PCAF/GCN5, which targets cyclin A for degradation. Given that cyclin A is crucial for S phase progression and entry into mitosis, HDAC3 knockdown causes cell accumulation in S and G_2/M phases (Vidal-Laliena et al. 2013). Collectively, HDAC inhibition can arrest the cell cycle at either G_1/S or G_2/M phase, suggesting HDACs as therapeutic targets for abnormal cell growth and proliferation in cancer.

Apoptosis

HDACs have been shown to regulate apoptosis in a variety of cancer cells through changing expression of pro- and antiapoptotic proteins. Treatment of tumor cells with HDACi can either directly activate apoptosis through the extrinsic (death receptor)/intrinsic (mitochondria) pathway, or enhance the susceptibility of tumor cells to apoptosis. The extrinsic apoptosis pathway induced by HDACi is via diverse mechanisms including up-regulation of cell surface death receptors and/or ligands including FAS/APO1-FASL, TNF-TNF receptors and TRAIL-TRAIL receptors, reductions in the level of cytoplasmic FLICE-like inhibitory protein (c-FLIP), and enhanced recruitment of DISC formation (Zhang and Zhong 2014). Two nonselective HDACi, vorinostat and panobinostat, have been identified as modulators of FLIP expression in several preclinical cancer models (Bangert et al. 2012). Inhibition of HDAC1, 2, and/or 3, but not HDAC6, is necessary for efficient FLIP down-regulation and caspase-8 activation in NSCLC. HDACi sensitized cancer cells for TRAIL-induced apoptosis in a FLIP- and caspase-8-dependent manner (Riley et al. 2013). Depletion of HDAC2 also synergizes pancreatic cancer cells toward TRAIL-induced apoptosis with an increased expression of TRAIL receptor DR5 (TRAIL-R2) (Schuler et al. 2010).

The intrinsic cell death pathway involves the interplay of the pro- and antiapoptotic Bcl-2 superfamily of proteins, and HDAC inhibition could induce the intrinsic pathway by decreasing the expression of antiapoptotic proteins, and increasing the expression of proapoptotic proteins (Zhang and Zhong 2014). HDAC2 depletion results in regression of tumor cell growth and activation of apoptosis via p53 and Bax activation and Bcl2 suppression in human lung cancer cells (Jung et al. 2012). In gastric cancer cells, HDAC2 knockdown selectively induced the expression of proapoptotic factors Bax, AIF, and Apaf-1, but repressed the expression of antiapoptotic Bcl-2 (Kim et al. 2013). HDAC1 and HDAC8 cooperate to repress BMF (Zhang et al. 2006; Kang et al. 2014). Inhibition of HDAC8 by methylselenopyruvate (MSP), a competitive inhibitor of HDAC8, was sufficient to activate BMF transcription and promote BMF-mediated apoptosis in colon cancer cells (Kang et al. 2014). HDAC3 down-regulates PUMA expression in gastric cancer cells and HDACi, like TSA, promotes PUMA expression through enhancing the binding of p53 to the PUMA promoter (Feng et al. 2013). TSA treatment can also effectively overcome resistance to DNA-damage-induced cell death by reactivating PUMA expression in renal cell carcinoma cells (Zhou et al. 2014).

The relevance of p53 in HDACi-induced apoptosis is controversial. HDACi can activate p53, but does not necessarily require p53 for induction of anticancer action (Sonnemann et al. 2014). Most studies point to a p53-independent action of HDACi because the antican-

cer effect of HDACi is not influenced by the tumor's p53 status (Ellis et al. 2009). Other studies, however, suggest an essential role of p53 in the response of tumor cells to HDACi treatment (Bajbouj et al. 2012). Using isogenic HCT-116 colon cancer cell lines with different p53 status, the antitumor effects of vorinostat, apicidin, and VPA were largely independent of p53, whereas entinostat-induced cell death partially depends on p53 (Sonnemann et al. 2014), indicating that HDACi may regulate apoptotic processes via both p53-dependent and independent pathways.

DNA-Damage Response

Numerous studies have shown that HDACs have important roles in DNA-damage repair (DDR) responses because HDACs are critical in modulating chromatin remodeling and maintaining dynamic acetylation equilibrium of DNA-damage-related proteins (Li and Zhu 2014). HDAC1 and HDAC2 are recruited to DNA-damage sites to deacetylate histones H3K56 and H4K16, and facilitate nonhomologous end-joining (NHEJ) (Miller et al. 2010), suggesting a direct role for these two enzymes during DNA replication and double-strand break (DSB) repair. HDAC3 is also associated with DNA-damage control, although it is not localized to DSB DNA-damage sites (Miller et al. 2010). Inactivation of HDAC3 causes genomic instability, and deletion of HDAC3 in the liver leads to hepatocellular carcinoma (Bhaskara et al. 2010). Besides altering histone acetylation status, class I HDACs also regulate other proteins involved in the DNA-damage response, including ATR, ATM, BRCA1, and FUS (Thurn et al. 2013). HDAC inhibition can repress DSB repair and render cancer cells more susceptible to ionizing radiation (IR) and DNA-damaging-agents-induced cell death (Koprinarova et al. 2011).

Among the class II HDACs, HDAC4, HDAC6, HDAC9 and HDAC10, have each been implicated in DNA-damage-repair processes. HDAC4 colocalizes with 53BP to nuclear foci after DSB. Depletion of HDAC4 reduces 53BP1 expression and abrogates the DNA-damage-induced G_2 checkpoint (Kao et al. 2003).

Cite this article as *Cold Spring Harb Perspect Med* doi: 10.1101/cshperspect.a026831

HDAC9 and HDAC10 are reported to be required for homologous recombination (HR). Depletion of HDAC9 or HDAC10 inhibits HR and sensitizes cells to mitomycin C treatment (Kotian et al. 2011). The mismatch repair (MMR) system recognizes DNA mismatches that occur during DNA replication or recombination, and corrects these defects to maintain genomic integrity. MutS protein homolog 2 (MSH2), a key DNA mismatch repair protein, is regulated by class IIb HDACs. HDAC6 sequentially deacetylates and ubiquitinates MSH2, causing a cellular tolerance to DNA damage and decreased cellular DNA mismatch repair activities by down-regulation of MSH2 (Zhang et al. 2014). However, the deacetylation of MSH2 by HDAC10 might promote DNA mismatch repair activity (Radhakrishnan et al. 2015).

SIRT1 is a critical component of the DNA-damage response pathway that regulates multiple steps of DDR, including damage sensing, signal transduction, DNA repair, and apoptosis (Gorospe and de Cabo 2008). SIRT1 interacts with and deacetylates several DDR proteins, including Ku70, NBS1, APE1, XPA, PARP-1, TopBP1, and KAP1 (Luna et al. 2013; Li and Zhu 2014; Wang et al. 2014; Lin et al. 2015). SIRT1 antagonizes p53 acetylation and facilitates cancer cells survival after DNA damage (Luo et al. 2001; Vaziri et al. 2001), making SIRT1 a promising target in cancer therapy. However, SIRT1 also plays an essential role in maintaining genome integrity and stability (Wang et al. 2008; Palacios et al. 2010), so the challenge still remains to modulate SIRT1 function in such a manner that it will be beneficial for cancer therapy. Recent research demonstrated that the role of SIRT1 in response to DNA damage requires posttranslational modifications such as site-specific phosphorylation and ubiquitination. HIPK2 interacts and phosphorylates SIRT1 at Ser682 after DNA damage. Phosphorylation of SIRT1 inhibits SIRT1 deacetylase activity on p53, which in turn potentiates apoptotic p53 target gene expression and DNA-damage-induced apoptosis (Conrad et al. 2015). SIRT1 is also ubiquitinated by MDM2 during DDR, and ubiquitination of SIRT1 affects its function in cell death and survival in response to DNA damage (Peng et al. 2015).

SIRT6, also important in DNA repair, was first found to suppress genomic instability by regulating base excision DNA repair (BER) (Mostoslavsky et al. 2006). Recent studies have demonstrated that SIRT6 is involved in homologous recombination (HR) by deacetylating carboxy-terminal binding protein (CtBP) and interacting protein (CtIP) (Kaidi et al. 2010). SIRT6 is rapidly recruited to DNA-damage sites and stimulates DSB repair by mono-ADP-ribosylation of PARP1(Mao et al. 2011). SIRT6 also recruits the ISWI-chromatin remodeler SNF2H to DSBs, and deacetylates histone H3K56, preventing genomic instability through chromatin remodeling (Toiber et al. 2013). Together, results from these studies suggest that at least some of the class III sirtuins (Sir2 proteins) have an equally critical role in DNA-damage response compared to the classical HDACs.

Metastasis

Epithelial-to-mesenchymal transition (EMT) is a major process in cancer cell invasion and metastasis, and emerging studies have demonstrated the key role of HDACs in EMT regulation in a variety of cancer contexts. EMT is characterized by the loss of epithelial cell markers, namely, epithelial-cadherin (CDH1), and several transcriptional repressors of CDH1 have been identified, including Snail, Slug, Twist, ZEB1, and ZEB2. A mechanism of their action involves recruitment of HDACs to the CDH1 promoter resulting in deacetylation of H3 and H4 histones. Snail recruits HDAC1/2 and Sin3A complex to the CDH1 promoter for histone deacetylation, and the snail/HDAC1/HDAC2 repressor complex contributes to CDH1 silencing in the metastasis of pancreatic cancer (von Burstin et al. 2009). Moreover, the complex of snail/HDAC1/HDAC2 is required for EZH2-mediated CDH1 repression in nasopharyngeal carcinoma cells (Tong et al. 2012). Recruitment of HDACs to the CHD1 promoter is also regulated by ZEB1 in human pancreatic cancer cells (Aghdassi et al. 2012). Given that ZEB1 could also alter the splicing of CDH1 exon 11, a recent study revealed a dual effect of ZEB1 and its interacting class I HDACs; the decrease in CDH1 is the result of a combination of transcriptional inhibition and aberrant splicing (Liao et al. 2013). Treatment of cells with the HDACs inhibitor, LBH589 (panobinostat), induces CDH1 expression, and represses EMT and metastasis in triple-negative breast cancer (TNBC) cells. Besides preventing ZEB-mediated repression of CDH1 by inhibiting the class I HDACs corepressors, the effect of LBH589 is partially mediated by inhibition of ZEB expression (Rhodes et al. 2014). The expression of ZEB1 is also reduced in pancreatic cancer cells after mocetinostat treatment that sensitizes the undifferentiated, ZEB1-expressing cancer cells for chemotherapy (Meidhof et al. 2015). These findings indicate the therapeutic potential of inhibition of class I HDACs in targeting EMT and metastasis of cancer cells.

The role of SIRT1 in EMT regulation depends on the tumor type. In prostate cancer cells, SIRT1 induces cell migration in vitro and metastasis in vivo by cooperating with ZEB1 to suppress CDH1 transcription (Byles et al. 2012). A recent study reveals a vital role of MPP8-SIRT1 interaction in CDH1 silencing (Sun et al. 2015). The deacetylation of MPP8 at K439 by SIRT1 increases MPP8 protein stability, and MPP8 in turn facilitates SIRT1 recruitment to the CDH1 promoter for H4K16 deacetylation by regulating SIRT1-ZEB1 interaction. Thus, disruption of MPP8-SIRT1 interaction derepresses CDH1 expression and reduces cell motility and invasiveness in prostate cancer cells (Sun et al. 2015). In breast cancer, SIRT1 overexpression is associated with decreased miR-200a. miR-200a negatively regulated SIRT1 expression and reduced EMT (Eades et al. 2011). miR-204 also inhibits EMT in gastric cancer and osteosarcoma cells by directly targeting and repressing SIRT1 at the posttranscriptional level (Shi et al. 2015b). However, an opposite role of SIRT1 in cancer metastasis was indicated by the demonstration that SIRT1 reduces EMT in breast epithelial cells by deacetylating Smad4 and repressing the effect of TGF-β signaling on MMP7 (Simic et al. 2013). A similar mechanism is also found in oral squamous cell carcinoma (OSCC) cells, suggesting the key role of the SIRT1/Smad4/MMP7 pathway in EMT process (Chen et al. 2014).

Angiogenesis

Tumor growth and metastasis depend on angiogenesis. Angiogenesis is triggered by hypoxia or hypoxic microenvironment, and the cellular response to hypoxia is primarily regulated by the transcription factor hypoxia-inducible factors 1 α (HIF-1α). Many HDACs are associated with HIF-1α activity as cell treatment with HDACi causes HIF-1α degradation and functional repression. HDAC1 and HDAC4 directly deacetylate HIF-1α and block degradation of the protein (Yoo et al. 2006; Geng et al. 2011). Instead of regulating HIF-1α acetylation, HDAC5 and HDAC6 facilitate HIF-1α maturation and stabilization by deacetylating its chaperones, HSP70 and HSP90 (Kong et al. 2006; Chen et al. 2015). Inhibition of HDAC5 and 6 results in hyperacetylation of these chaperones, accumulation of the immature HIF-1α complex, and degradation of HIF-1α by the 20S proteasome. HDAC4, HDAC5, and HDAC7 increased transcriptional activity of HIF-1α by promoting its association with p300 (Kato et al. 2004; Seo et al. 2009). In contrast, SIRT1-mediated deacetylation of HIF-1α at Lys674 inhibits HIF-1α activity by blocking p300 recruitment. The suppression of SIRT1 under hypoxic conditions provides a positive feedback loop that maintains a high level of HIF-1 activity (Lim et al. 2010).

Although the antiangiogenic activity of HDAC inhibition has been demonstrated to be associated with decreased expression of proangiogenic genes, the specific effect of individual HDAC enzymes on angiogenic gene expression is controversial. KLF-4 recruits HDAC2 and HDAC3 at the VEGF promoter and represses its transcription. The up-regulation of VEGF in cancer is associated with loss of KLF-4-HDAC-mediated transcriptional repression (Ray et al. 2013). HDAC5 is another negative regulator of angiogenesis by repressing proangiogenic gene expression, such as FGF2 or Slit2, in endothelial cells (Urbich et al. 2009). Recent studies demonstrated a dual role of HDAC6 in angiogenesis. HDAC6 promotes angiogenesis by deacetylating cortactin in endothelial cells, thereby regulating endothelial cell migration and sprouting (Kaluza et al. 2011). Another study reported the antiangiogenic effect of HDAC6 in HCC, as depletion of HDAC6 facilitates angiogenesis by up-regulating the expression of HIF-1α and VEGFA (Lv et al. 2015). HDAC7 is crucial in maintaining vascular integrity and endothelial angiogenic functions, such as tube formation, migration and proliferation (Turtoi et al. 2012). HDAC9 positively regulates endothelial cell sprouting and vascular growth by the repression of the miR-17-92 cluster, which reduces the expression of proangiogenic proteins (Kaluza et al. 2013). Taken together, HDACs play important roles in angiogenesis by modulating a multitude of pro- and antiangiogenic factors, indicating that they are potential targets for antiangiogenesis in cancer therapy.

Autophagy

The role of autophagy in cancer is complex. The failure to properly modulate autophagy in response to oncogenic stresses has been implicated both positively and negatively in tumorigenesis. On the one hand, autophagy functions as a surveillance mechanism to remove damaged organelles and cellular components, which might prevent normal cells from transforming to tumor cells. So the loss of autophagy proteins appears to promote cancer development. On the other hand, for established tumors, autophagy can help cancer cell survival under conditions of metabolic stress, and it might also confer resistance to anticancer therapies (Zhi and Zhong 2015). Consistent with the dual role of autophagy in cancer, many HDAC family members show both pro- and antiautophagy activities (Koeneke et al. 2015). Depletion or inhibition of HDAC1 is reported to induce autophagy by promoting accumulation of the autophagosomal marker LC3-II (Xie et al. 2012). However, in mouse models, deletion of both HDAC1 and HDAC2 in skeletal muscle blocks autophagy flux (Moresi et al. 2012). Recent re-search indicates that the oncogenic role of class IIa HDAC4 and HDAC5 in cancer cells would be derived at least partially via decreasing autophagic flux, but the detailed mechanism needs further investigation.

The key role of HDAC6 in autophagy was first established through the observation that HDAC6 provide an essential link between autophagy and the ubiquitin proteasome system (UPS) in neurodegenerative diseases. When the UPS is impaired, autophagy is strongly activated and acts as a compensatory degradation system in an HDAC6-dependent manner (Pandey et al. 2007). In HDAC6 knockout mouse embryonic fibroblasts (MEFs), HDAC6 appears to be important for ubiquitin-selective quality control (QC) autophagy, but not starvation-induced autophagy (Lee et al. 2010a). A similar mechanism is observed in mitophagy, a selective degradation of mitochondria. The parkin-mediated mitochondrial ubiquitination recruits HDAC6 and p62, which assemble the autophagy machinery and lead to mitochondrial clearance (Lee et al. 2010b). Besides Ub-based selective autophagy, HDAC6 is associated with autophagic clearance of IFN-induced ISG15-conjugated proteins. HDAC6 and p62 independently bind ISG15 and facilitate the autophagosome/lysosome degradation of ISG15 conjugates (Nakashima et al. 2015). The role of HDAC6 in the fusion event is to control acetylation of salt-inducible kinase 2 (SIK2). HDAC6-mediated deacetylation activates SIK2 kinase activity and promotes autophagosome processing (Yang et al. 2013). Although HDAC6 is dispensable for starvation-induced autophagy in MEFs (Lee et al. 2010a), another study demonstrated that HDAC6 is involved in this nonselective degradation by deacetylating LC3-II in HeLa cells (Liu et al. 2013). The acetylation level of LC3B-II is decreased upon serum deprivation and HDAC6 is at least partially responsible for deacetylating LC3-II. In neuroblastoma, depletion and inhibition of HDAC10 disables efficient autophagosome/lysosome fusion and interrupts autophagic flux, resulting in an increase of sensitization to cytotoxic drug treatment (Oehme et al. 2013). The deacetylation of Hsp70 protein families by HDAC10 might contribute to autophagy-mediated cell survival (Oehme et al. 2013). Overall, class IIb HDACs seem to mainly regulate autophagic flux at the level of autophagosome–autolysosome fusion via deacetylation of cytoplasmic proteins (Koeneke et al. 2015).

Sirtuins also participate in regulating autophagy. Sirt1 activity is necessary for the induction of starvation-induced autophagy by directly deacetylating critical regulators of the autophagy machinery, including Atg5, Atg7, Atg8, and LC3 (Lee et al. 2008; Huang et al. 2015a). In embryonic stem cells (ESCs), SIRT1 mediates oxidative stress-induced autophagy at least in part by PI3K/Beclin 1 and mTOR pathways (Ou et al. 2014). SIRT1 and the PI3K/Akt/mTOR pathway are also found to be related to Plumbagin (PLB)-induced autophagy in prostate cancer cells (Zhou et al. 2015). SIRT3, a mitochondrial deacetylase, evokes mitophagy under oxidative stress or starvation conditions (Tseng et al. 2013; Webster et al. 2013). SIRT3 is found to confer cytoprotective effects by activating antioxidant defenses and mitophagy under mitochondrial proteotoxic stress and reestablishing homeostasis in breast cancer cells (Papa and Germain 2014). Two recent reports suggest a connection between autophagy with SIRT5 and SIRT6. SIRT5 reduces ammonia-induced autophagy and mitophagy by regulating glutamine metabolism (Papa and Germain 2014). SIRT6 is involved in autophagy activation during cigarette-smoke-induced cellular senescence via attenuation of IGF-Akt-mTOR signaling (Takasaka et al. 2014). Collectively, a better understanding of the context-dependent effects of individual HDACs enzymes on autophagic process will give us an advantage to treat cancers by exploiting this area in a specifically targeted manner.

ANTICANCER EFFECT OF HDAC INHIBITORS

The availability of HDACi has not only accelerated our understanding of HDAC functions and mechanism of actions, but also presented a promising new class of compounds for cancer treatment. To date, numerous synthetic or natural molecules that target classes I, II, and IV enzymes have been developed and characterized, although interest in the class III Sirtuin family is increasing. Here we only describe the potential role of classical HDACi in cancer therapy. Because classical HDACs display Zn^{2+}-dependent deacetylase activity, the binding of HDACi to the Zn^{2+} ion, which resides at the active site of HDACs, interferes with the activity of HDACs, thereby inhibiting their enzymatic function. On the bases of chemical structures, HDACi are classified into four groups, including hydroxamates, benzamides, short-chain fatty acids, and cyclic peptides (Table 2). Most of these molecules have been developed as anticancer agents with varying specificity and efficiency, pharmacokinetic properties, and toxicological characteristics.

The rationale for targeting HDACs in cancer therapy is that altered HDAC expression and/or function is frequently observed in a variety of cancer types. The disrupted acetylation homeostasis in cells might contribute to tumorigenesis, and the nature of reversible modulation by HDACs makes them attractive targets for cancer treatment. HDACs reversibly modify the acetylation status histones and nonhistones and cause widespread changes in genes expression without a change in DNA sequence. HDACi can counteract the abnormal acetylation status of proteins found in cancer cells and can reactivate the expression of tumor suppressors, resulting in induction of cell-cycle arrest, apoptosis, differentiation, and inhibition of angiogenesis and metastasis (Fig. 1). Furthermore, cancer cells are more sensitive to HDACi-induced apoptosis than normal cells (Ungerstedt et al. 2005), providing additional therapeutic potential of HDACi.

Currently, there are numerous HDACi under clinical development (Table 2), which can be divided into three groups based on their specificity: (1) nonselective HDACi, such as vorinostat, belinostat, and panobinostat; (2) selective HDACi, such as class I HDACi (romidepsin and entinostat) and HDAC6 inhibitor (ricolinostat); and (3) multipharmacological HDACi, such as CUDC-101 and CUDC-907.

Table 2. HDAC inhibitors currently under clinical investigations

HDACis	Specificity	Cancer types	Clinical trial	References
Hydroxamic acid				
Vorinostat (SAHA)	Classes I, II, and IV	CTCL	FDA approved in 2006	Mann et al. 2007
Belinostat (Beleodaq/ PXD101)	Classes I, II, and IV	PTCL	FDA approved in 2014	McDermott and Jimeno 2014
Panobinostat (LBH-589)	Classes I, II, and IV	MM	FDA approved in 2015	Richardson et al. 2015
Resminostat (4SC-201)	Classes I and II	Advanced colorectal and hepatocellular carcinoma; HL	Phase II trial	Brunetto et al. 2013; Zhao and Lawless 2015
Givinostat (ITF2357)	Classes I and II	CLL; MM; HL	Phase II trial	Galli et al. 2010; Locatelli et al. 2014
Pracinostat (SB939)	Classes I, II, and IV	AML	Phase II trial	Zorzi et al. 2013
Abexinostat (PCI-24781)	Classes I and II	Metastatic solid tumors; HL; non-HL; CLL	Phase I trial	Choy et al. 2015; Morschhauser et al. 2015
Quisinostat (JNJ-26481585)	Class I and II HDACs	Advanced solid tumor; lymphoma; CTCL	Phase I and II trial	Venugopal et al. 2013
MPT0E028	HDAC1, 2, 6	Advanced solid tumor	Phase I trial	Zwergel et al. 2015
CHR-3996	Class I	Solid tumor	Phase I trial	Banerji et al. 2012
CUDC-101	Classes I and II HDAC, EGFR, HER2	Solid tumor	Phase I trial	Shimizu et al. 2015
CUDC-907	Classes I and II HDAC, PI3K	MM; lymphoma; solid tumor	Phase I trial	Qian et al. 2012
Benzamides				
Entinostat (MS-275)	Class I	Solid and hematological malignancies	Phase I and II trial	Knipstein and Gore 2011
Mocetinostat (MGCD0103)	Class I and IV	Solid and hematological malignancies	Phase I and II trial	Younes et al. 2011
Tacedinaline (CI-994)	Class I	MM; lung and pancreatic cancer	Phase II and III trial	Pauer et al. 2004
Ricolinostat (ACY-1215)	HDAC6	MM; lymphoma	Phase I and II trial	Santo et al. 2012
Chidamide (CS055/ HBI-8000)	HDAC1, 2, 3, and 10	Breast cancer; NSCLC	Phase II and III trial	Dong et al. 2012; Shi et al. 2015a
Cyclic peptides				
Romidepsin (Depsipeptide/ FK228)	Class I	CTCL; PTCL	FDA approved in 2009 and 2011	Frye et al. 2012
Aliphatic fatty acids				
Valproic acid (VPA)	Class I and II	Solid and hematological malignancies	Phase I and II trial	Bilen et al. 2015
Phenylbutyrate	Classes I and II	Solid and hematological malignancies	Phase I and II trial	Iannitti and Palmieri 2011
AR-42	Class I and IIb	AML	Phase I trial	Guzman et al. 2014
Pivanex (AN-9)	Classes I and II	NSCLC; myeloma; CLL	Phase II trial	Reid et al. 2004

AML, Acute myeloid leukemia; CLL, chronic lymphocytic leukemia; CTCL, cutaneous T-cell lymphoma; HL, Hodgkin's lymphoma; MM, multiple myeloma; NSCLC, non-small-cell lung cancer; PTCL, peripheral T-cell lymphoma.

Nonselective Broad-Spectrum HDAC Inhibitors

The most extensively studied and commonly used HDACi are nonselective HDACi. For example, vorinostat (SAHA), a hydroxamate class agent, was the first histone deacetylase inhibitor approved by the Food and Drug Administration (FDA) to treat patients with cutaneous T-cell lymphoma (CTCL). Like TSA, SAHA inhibits all zinc-dependent HDACs in low nanomolar ranges, although more recent studies suggest that most hydroxamate-based HDACi have a weak effect on class IIa enzymes (Bradner et al. 2010). Preclinical studies have demonstrated that SAHA induces apoptosis and cell-cycle arrest, and reduces the proliferation and metastatic potential of tumor cells. SAHA also sensitized tumor cells to chemotherapy and/or radiotherapy (Shi et al. 2014; Xue et al. 2015). In addition to SAHA, two hydroxamate-based nonselective HDACi, belinostat (Beleodaq/PXD101) and panobinostat (LBH-589), were recently approved by the FDA to treat peripheral T-cell lymphomas (PTCL) and multiple myeloma, respectively. Both of these drugs are also under investigation in combination therapies for solid tumors.

Selective HDAC Inhibitors

To date, there are relatively few highly selective HDACi, but compounds with proposed selectivity for several class I and class II HDACs have been developed (Table 3).

Class I HDACi

HDAC1 and HDAC2. Romidepsin (FK-228), the second histone deacetylase inhibitor approved for the treatment of CTCL and PTCL, exhibits a stronger inhibition toward HDAC1 and HDAC2 enzymes at low nanomolar levels (Furumai et al. 2002). Its antitumor efficacy has been demonstrated in different cancer models (McGraw 2013; Karthik et al. 2014). BRD8430, compound 60 and MRLB-223 are three novel HDAC1 and HDAC2 inhibitors under preclinical studies. The selective inhibition of HDAC1 and HDAC2 by BRD8430 and compound 60

induced differentiation and cell death in neuroblastoma cells, and synergistically activated retinoic acid signaling in combination treatment with 13-*cis* retinoic acid (Frumm et al. 2013). MRLB-223 induced tumor cell death via the intrinsic apoptotic pathway in a p53-independent manner. However, MRLB-223 had less effect on induction of apoptosis and therapeutic efficacy as seen using the broad-spectrum histone deacetylase inhibitor vorinostat (Newbold et al. 2013).

HDAC3. RGFP966 is an *N*-(*o*-aminophenyl) carboxamide HDAC3-selective inhibitor (Malvaez et al. 2013). RGFP966 decreased growth and increased apoptosis of refractory CTCL cells by targeting DNA replication (Wells et al. 2013). Consistent with previous research demonstrating the contribution of HDAC3 to the effects of SAHA on DNA replication (Conti et al. 2010), HDAC3 inhibition by RGFP966 reduces DNA replication fork velocity and causes replication stress in CTCL cells (Wells et al. 2013).

A recent preclinical study demonstrated that HDAC3 represents a promising therapeutic target in multiple myeloma (MM) (Minami et al. 2014). HDAC3 inhibition by BG45, a HDAC3-selective inhibitor, induces significant apoptosis in MM cells, without affecting normal donor PBMCs. BG45-induced MM cell toxicity might be associated with hyperacetylation and hypophosphorylation of STAT3. HDAC3 inhibition, but not HDAC1 or HDAC2, significantly enhances bortezomib-induced cell death in vitro and in vivo, providing the preclinical rationale for combination treatment of MM with HDAC3 and proteasome inhibitors.

T247 and T326 are identified as HDAC3-selective inhibitors by screening a series of compounds assembled using "click chemistry" (Suzuki et al. 2013). In cell-based assays, T247 and T326 selectively enhance the acetylation of NF-κB, a substrate of HDAC3, but did not regulate HDAC1 and HDAC6 substrates, suggesting they are HDAC3-selective inhibitors. T247 and T326 inhibited the growth of colon and prostate cancer cells (Suzuki et al. 2013). In TMEM16A-expressing cancer cells, T247 also exerts a suppressive effect on cancer cell viability via downregulating TMEM16A (Matsuba et al. 2014).

Table 3. Specific HDAC inhibitors in cancer therapy

HDACis	Specificity	Cancer types	Stage	References
Class I				
Romidepsin (Depsipeptide/ FK228)	HDAC1, 2	CTCL and PTCL	FDA approved	Furumai et al. 2002; Frye et al. 2012
BRD8430	HDAC1, 2	Neuroblastoma	Preclinical	Frumm et al. 2013
Compound 60	HDAC1, 2	Neuroblastoma	Preclinical	Methot et al. 2008; Frumm et al. 2013; Schroeder et al. 2013
MRLB-223	HDAC1, 2	Lymphomas	Preclinical	Newbold et al. 2013
Entinostat (MS-275)	HDAC1, 2, 3	Multiple cancer cells	Clinical trial	Hu et al. 2003; Khan et al. 2008; Knipstein and Gore 2011
CHR-3996	HDAC1, 2, 3	Multiple cancer cells	Clinical trial	Moffat et al. 2010; Banerji et al. 2012
Tacedinaline (CI-994)	HDAC1, 2, 3	Multiple myeloma; lung and pancreatic cancer	Clinical trial	Kraker et al. 2003; Pauer et al. 2004
Apicidin	HDAC1, 2, 3	Multiple cancer cells	Preclinical	Olsen et al. 2012; Ahn et al. 2015
RGFP966	HDAC3	CTCL	Preclinical	Wells et al. 2013
BG45	HDAC3	Myeloma	Preclinical	Minami et al. 2014
T247 and T326	HDAC3	Colon and prostate cancer	Preclinical	Suzuki et al. 2013
PCI-34051	HDAC8	Lymphoma, neuroblastoma	Preclinical	Balasubramanian et al. 2008; Oehme et al. 2009; Rettig et al. 2015
Compound 2 (Cpd2)	HDAC8	Neuroblastoma	Preclinical	Krennhrubec et al. 2007; Oehme et al. 2009; Rettig et al. 2015
C149 (NCC149)	HDAC8	Multiple cancer cells	Preclinical	Suzuki et al. 2012; Suzuki et al. 2014
Compound 22 d	HDAC8	Lung cancer	Preclinical	Huang et al. 2012
Class IIa				
TMP269	HDAC4, 5, 7, 9	Multiple myeloma	Preclinical	Lobera et al. 2013; Kikuchi et al. 2015
MC1568	HDAC4, 5, 6, 7, 9	Gastric, colorectal, pancreatic and breast cancer	Preclinical	Mai et al. 2005; Duong et al. 2008; Wang et al. 2012; Colarossi et al. 2014; Ishikawa et al. 2014
LMK235	HDAC4, 5	Multiple cancer cells	Preclinical	Marek et al. 2013
Class IIb				
Ricolinostat (ACY-1215)	HDAC6	Multiple myeloma, lymphoma, glioblastoma	Clinical trial	Santo et al. 2012; Amengual et al. 2015; Li et al. 2015a; Mishima et al. 2015
Tubacin	HDAC6	Multiple cancer cells	Preclinical	Haggarty et al. 2003; Aldana-Masangkay et al. 2011
Tubastatin A	HDAC6	Multiple cancer cells	Preclinical	Butler et al. 2010
C1A	HDAC6	Multiple cancer cells	Preclinical	Kaliszczak et al. 2013
HPOB	HDAC6	Multiple cancer cells	Preclinical	Lee et al. 2013
Nexturastat A (Compound 5g)	HDAC6	Melanoma	Preclinical	Bergman et al. 2012
Compound 12	HDAC6	Colorectal cancer	Preclinical	Lee et al. 2014a
Befexamac	HDAC6, 10	Neuroblastoma, lung cancer	Preclinical	Bantscheff et al. 2011; Oehme et al. 2013; Li et al. 2015b; Scholz et al. 2015

CTCL, Cutaneous T-cell lymphoma; PTCL, peripheral T cell lymphoma.

HDAC8. HDAC8 has proven to be the most promising target to achieve selectivity. The unique features of HDAC8, such as conformational variability of the L1 and L2 loop segments (Dowling et al. 2008) and the presence of serine 39 phosphorylation near the active site (Lee et al. 2004), led to the design of higher selective inhibitors for HDAC8.

Modifications to the hydroxamic acid scaffold resulted in the discovery of PCI-34051, a potent HDAC8-specific inhibitor with a >200-fold selectivity over other HDACs. It induces caspase-dependent apoptosis in T-cell-derived malignant cells, but not in a panel of solid tumor cell lines or other hematopoietic cells. Mechanistically, PLCγ1-dependent calcium mobilization from the endoplasmic reticulum (ER) and, in turn, release of cytochrome *c* from mitochondria might contribute to PCI-34051-induced cell death (Balasubramanian et al. 2008). Besides T-cell leukemia and lymphoma, human and murine-derived malignant peripheral nerve sheath tumors (MPNST) cells also exhibited "sensitivity" to HDAC8 inhibitors: PCI-34051 and its variant PCI-48012 (Lopez et al. 2015). HDAC8 inhibition-induced S phase cell-cycle arrest and apoptosis in MPNST cells, but the underlying mechanism remains unclear. Given that high HDAC8 expression is significantly correlated with advanced stage and poor outcome in neuroblastoma (Oehme et al. 2009), HDAC8 inhibition by selective inhibitors, compound 2 (Cpd2) and PCI-34051, induced cell differentiation, cell-cycle arrest, and cell death in neuroblastoma cells, while untransformed cells were not affected (Rettig et al. 2015). PCI-48012, an in vivo stable variant of PCI-34051 with improved pharmacokinetic properties, displayed a significant antitumor activity without toxicity in xenograft mouse models. PCI-48012 in combination with retinoic acid further enhanced differentiation in neuroblastoma cells and delayed tumor growth in vivo (Rettig et al. 2015).

Class IIa HDACi

In contrast to class I HDACs, much less is known about the molecular mechanisms and therapeutic potential of targeting class IIa HDACs (HDAC4, 5, 7, and 9), and there is a lack of pharmacological tools to specifically probe class IIa HDAC activities (Lobera et al. 2013). A high-throughput screen (HTS) identified trifluoromethyloxadiazole (TMFO) derivatives as inhibitors selective for class IIa HDACs. Although TMP269, a compound in the TFMO series, has a modest growth inhibitory effect in multiple myeloma (MM) cell lines, it enhances protease inhibitor carfilzomib-induced apoptosis by activating ER stress signaling (Kikuchi et al. 2015), providing the combination of the inhibition of both proteasome and class IIa HDACs as a novel treatment strategy in MM.

MC1568 and MC1575 are derivatives of aroyl-pyrrolyl-hydroxyamides (APHAs), showing selectivity toward class IIa HDACs and HDAC6 (Mai et al. 2005; Fleming et al. 2014). Although class IIa HDACs are mainly involved in tissue-specific growth and differentiation, rather than in cell proliferation, MC1568 and MC1575 treatment still displayed antiproliferative effects in estrogen-receptor-positive breast cancer cells (Duong et al. 2008) as well as human melanoma cells (Venza et al. 2013). MC1568 significantly enhanced MGCD0103-induced apoptosis and G_2/M arrest in pancreatic cancer cells (Wang et al. 2012). The additional treatment with MC1568 to simvastatin led to further induction of p27 expression and displayed a considerable synergistic antiproliferative effect in colorectal cancer cells (Ishikawa et al. 2014). MC1568 also had a synergistic effect with docetaxel treatment to increase cytotoxicity in gastric cancer cells (Colarossi et al. 2014).

Another new hydroxamate-based histone deacetylase inhibitor, LMK235, showed high selectivity for HDAC4 and HDAC5 (Marek et al. 2013). Compared with SAHA, LMK235 is less toxic and more suitable for the treatment of some cancers. Consistent with a recent study where silencing of HDAC4 was able to sensitize ovarian cancer cells to cisplatin (Stronach et al. 2011), the combination of LMK235 with cisplatin-enhanced cisplatin sensitivity in resistant cells (Marek et al. 2013).

YK-4-272 and tasquinimod are two novel unconventional HDACi, which either target HDAC nuclear-cytoplasmic shuttling or alter the interaction of class IIa HDACs with their partners. YK-4-272 represses the growth of human prostate cancer cells in vitro and in vivo (Kong et al. 2012). YK-4-272 binds HDAC4, and the localization of YK-4-272 in the cytoplasm traps and sequestrates HDAC4 in cytoplasm, resulting in increased acetylation of tubulin and nuclear histones in prostate cancer cells. However SAHA treatment also causes an accumulation of HDAC4 in cytoplasm similar to YK-4-272, which suggests the possibility that cytoplasmic restriction of class II HDACs is an indirect effect of class I inhibition. So far, the cytoplasmic functions of class IIa HDACs is not well known and this uncertain function could be amplified by inhibition of HDAC nuclear transport, limiting the use of the HDAC shuttling inhibitor.

Class IIb HDACi

HDAC6 inhibition has been intensively studied and a number of HDAC6-selective inhibitors are developed, such as tubacin and tubastatin A; however, their poor pharmacokinetic properties prevented them from further clinical development. Among HDAC6-specific inhibitors available, ricolinostat (ACY-1215) was the first one entered in clinical studies of patients with relapsed/refractory multiple myeloma or lymphoma (Santo et al. 2012; Amengual et al. 2015). In multiple myeloma, the highly secretory antibody-producing cells are heavily reliant on protein handling pathways, including the unfolded protein response (UPR), proteasome, aggresome, and autophagy pathways. So targeting both of the proteasome and aggresome degradation pathways by proteasome and HDAC6 inhibitors, respectively, induces accumulation of polyubiquitinated proteins, followed by activation of apoptotic cascades and synergistic cytotoxicity. ACY-1215 in combination with bortezomib triggered synergistic anti-MM activity without significant adverse effects (Santo et al. 2012), and similar anti-MM effects were obtained by combination treatment of ACY-1215

with another proteasome inhibitor, carfilzomib (Mishima et al. 2015). Besides hematological tumors, recent research indicated that ACY-1215 also significantly inhibited glioblastoma multiforme (GBM) cell growth (Li et al. 2015a). C1A is another HDAC6-selective inhibitor, which modulates HDAC6 downstream targets (α-tubulin and HSP90) and induces growth inhibition of a panel of cancer cell lines. To date, HDAC10-specific inhibitors are not yet available. Like other selective HDAC inhibitors, development of HDAC10-selective inhibitors might help clarify the function and mechanism of action of HDAC10, and potentially provide additional anticancer drugs.

Multipharmacological HDAC Inhibitors

Tumor heterogeneity requires a comprehensive approach to target multiple pathways underlying the initiation and progression of cancers. To enhance the therapeutic efficacy of HDACi, the combination with other anticancer agents have been explored and evaluated in preclinical and clinical studies. Another promising approach is to generate a single chemical compound that acts on multiple targets. CUDC-101, with a potent inhibitory activity against EGFR, HER2, and HDACs, is currently being evaluated in clinical trials as a treatment for advanced solid tumors, such as head and neck, gastric, breast, liver, and non-small-cell lung cancer tumors (Cai et al. 2010; Galloway et al. 2015). Recent research also indicates the antitumor effect of CUDC-101 in EGFR-overexpressing glioblastoma and anaplastic thyroid cancer (Liffers et al. 2015; Zhang et al. 2015a). CUDC-907 is another dual-acting agent developed by the same research group to inhibit both HDACs and phosphoinositide 3-kinase (PI3K) (Qian et al. 2012) and its clinical trials are underway for the treatment of lymphoma and multiple myeloma as well as advanced/relapsed solid tumors. Romidepsin (FK228, depsipeptide) is a potent class I histone deacetylase inhibitor that has FDA approval for the treatment of cutaneous and peripheral T-cell lymphomas, and recent research demonstrated that FK228 and its analogs (FK-A5 and FK-A11) act as HDACs and

PI3K dual inhibitors (Saijo et al. 2012; Saijo et al. 2015).

Numerous chemical hybrid molecules containing both HDACi activities and an additional anticancer module are under development, dual targeting HDACs and estrogen receptor (Gryder et al. 2013; Tang et al. 2015), retinoid X receptor (RXR) (Wang et al. 2015), topoisomerase I/II (Guerrant et al. 2013), 1α, 25-vitamin D (Lamblin et al. 2010), receptor tyrosine kinases (RTKs) (Zhang et al. 2013), tubulin (Zhang et al. 2015b), or DNA methyltransferase (Shukla et al. 2015), potentially leading to a rational efficacy in cancer therapy. Additionally, the hybrid of a nitric oxide (NO) donor and a histone deacetylase inhibitor has been developed and displayed outstanding antiproliferative activity in tumor cells (Duan et al. 2015).

CLINICAL LANDSCAPE OF HDAC INHIBITORS IN CANCER THERAPY

After vorinostat (SAHA) was approved to treat CTCL in 2006, the other three HDACi, romidepsin, belinostat, and panobinostat, have since been approved by the FDA for the treatment of cancer. Currently, more than 20 different HDACi are in different phases of clinical trials as single agents or in combination with chemotherapy or radiation therapy in patients with hematologic or solid tumors.

The efficacy of HDACi tested in clinical trials has been largely restricted to hematological malignancies, with positive therapeutic responses in leukemias, lymphomas, and multiple myeloma; however, the clinical outcomes in solid tumors are disappointing when used as monotherapy. It is not entirely clear why HDACi are more effective in hematological malignancies. One reason might be the poor pharmacokinetic properties of some HDACi, such as a short drug half-life that restricts them to distribute to solid tumors. Selective and accurate drug delivery of HDACi may help to overcome the issues associated with inefficient bioavailability. For example, HDACi conjugated to folic and pteroic acids selectively targets folate receptor (FR)-overexpressed solid tumors (Sodji

et al. 2015). The other reason might be that HDACi do not target solid tumors. Identifying those cancers or patients where HDAC deregulation is important for tumor development might contribute to rational cancer therapy.

Another obstacle that limits the use of HDACi in patients is their side effects and toxicity displayed during early-phase clinical trials. The common toxicities related to vorinostat, romidepsin, and belinostat were nausea, vomiting, anorexia, and fatigue that are mostly manageable, but some may cause more serious adverse events. In general, acute toxicity of nonselective HDACi is mainly through HDAC1-3 inhibition, so these compounds from unrelated chemical classes have a similar toxicity profile. HDACi have a broad effect on chromatin and can reverse the aberrant epigenetic changes in cancers. However, although the inhibition of HDACs may reactivate some tumor suppressors, they can also affect numerous other genes (Guha 2015). Although the second-generation of HDACi have been developed with improved pharmacodynamic and pharmacokinetic values, given that these new agents possess similar specificity profiles as their parental compounds, it is unclear whether these newer agents will have improved and less toxic clinical outcomes. Currently, major efforts in therapeutic strategies are focused on developing selective inhibitors and studying combination therapies, with the aim of increasing potency against specific cancer types and overcoming drug toxicity and resistance.

HDACi are continuously explored for used in combination with other antitumor agents to optimize their efficacy and toxicity. Combining HDACi with primary chemotherapeutic agents that induce DNA damage or apoptosis has shown very promising results in preclinical research studies. HDAC inhibition might resensitize tumor cells to the primary agents and overcome therapy resistance. For example, hypoxia-induced cisplatin resistance in NSCLC can be overcome by combining cisplatin with panobinostat by increasing histone acetylation and destabilization of HIF-1α (Fischer et al. 2015). ERCC1 and p53 were reported to have a predictive role for the efficacy of combined panobinostat and cisplatin treatment (Fischer et al. 2015). HDAC inhibitions could also overcome resistance to mTOR inhibitors (e.g., everolimus, temsirolimus, sirolimus, and ridaforolimus) in advanced solid tumors or lymphoma (Dong et al. 2013; Beagle et al. 2015; Zibelman et al. 2015).

Given that cross talk exists between DNA methylation and histone deacetylation in gene expression, a combination of HDACi and DNA methyltransferases (DNMTs) have been shown to produce a synergistic effect on reactivation of tumor-suppressor genes and represent a promising future therapeutic approach. Large phase I and II trials are currently underway to assess the efficacy of two chromatin-modifying agents, azacitidine and entinostat, for the treatment of chronic myelomonocytic leukemia, acute myeloid leukemia, NSCLC, advanced breast cancer, and metastatic colorectal cancer (Juergens et al. 2011; Prebet et al. 2014). Recent research demonstrated that combined MS-275 and azacitidine treatment is more efficient and selectively targeted esophageal cancer cells by inducing DNA damage, cell viability loss, apoptosis, and decreasing cell migration (Ahrens et al. 2015).

Preclinical studies also indicate a synergistic antitumor effect of HDACi with other epigenetic-targeted drugs, such as lysine-specific histone demethylase inhibitors (Vasilatos et al. 2013; Fiskus et al. 2014). These observations are consistent with recent findings that broad-acting HDAC inhibitors have minimal effect on promoter acetylation, but rather they promote H3K27 trimethylation, a silencing-associated histone modification (Halsall et al. 2015). These and other studies on the basic mechanisms of HDACs, HDACi, and their relationships with other histone modifications will no doubt guide the choice of future combination therapies.

Similarly, preclinical evidence from studies of HDACi together with proteasome inhibitors (e.g., bortezomib, carfilzomib, and marizomib) provides a strong scientific rationale for combination therapy. Given that HDAC6 facilitates misfolded protein aggresome formation for proteosome-independent proteolysis, dual targeting of HDAC6 and proteasomes can produce synergistic effects in lymphoma and multiple myeloma (Amengual et al. 2015; Mishima et al. 2015). However, the combination of proteasome and class I HDAC-specific inhibitors in nasopharyngeal carcinoma cells induced a significant apoptosis through an ROS-dependent and ER stress-induced mechanism, independent of HDAC6 inhibition (Hui and Chiang 2014). Because vorinostat and bortezomib are both FDA-approved drugs for the treatment of CTCL and multiple myeloma, respectively, the combination of these two agents has been tested in a variety of preclinical models and in clinical trials. Recent research explored the synergistic effect of vorinostat and bortezomib on host immune response and found cotreatment of HPV-expressing cervical cancer cells with bortezomib and vorinostat led to a tumor-specific immunity by rendering tumor cells more susceptible to killing by antigen-specific CD8$^+$ T cells, suggesting that activated host immune surveillance contributes to antitumor effects (Huang et al. 2015b).

HDACi have also been evaluated in combination with a hormone antagonist for the treatment of patients whose tumors express hormone receptors. Three phase II clinical trials are currently carried out with vorinostat and tamoxifen for the treatment of breast cancer (Munster et al. 2011). Although histone deacetylation plays a key role in estrogen receptor gene silencing, it remains unclear whether the addition of HDACi actually reactivates functional estrogen receptor α expression (de Cremoux et al. 2015). A recent study demonstrates that Bcl-2 down-regulation and induction of proapoptotic proteins by combined estrogen receptor and HDAC inhibition leads to apoptotic cell death of tamoxifen-resistant cells (Raha et al. 2015).

HDACi have been shown to enhance the immunogenicity of cancer cells (Murakami et al. 2008; Christiansen et al. 2011; Jazirehi et al. 2014), and the antitumor efficacy of HDACi in vivo also relies on an intact immune system (West et al. 2013, 2014), so the combination of HDACi with immunotherapy is a promising strategy for the treatment of cancer. The efficacy of HDACi can be significantly enhanced by the concurrent administration of various immunotherapeutic approaches, such as cancer vaccines, adoptive T-cell transfer, and immune checkpoint inhibitors (Park et al. 2015). For example, coadministration of HDACi with antibodies against cytotoxic T-lymphocyte antigen 4 (CTLA4) could further enhance the infiltration of CD4$^+$ T cells and achieve a synergistic therapeutic effect on tumors by promoting antitumor immune responses (Cao et al. 2015). A recent preclinical study indicated that HDACi in combination with immunomodulatory drugs, such as lenalidomide and pomalidomide, showed a synergistic cytotoxicity in multiple myeloma by down-regulating c-Myc expression. A phase I trial is currently underway to assess the effect of ACY-241, a next-generation selective inhibitor of HDAC6, with and without pomalidomide and low-dose dexamethasone for treatment of multiple myeloma. Another phase II clinical trial evaluated the class I histone deacetylase inhibitor romidepsin in combination with lenalidomide in patients with peripheral T-cell lymphoma.

There is also substantial evidence that HDACi such as vorinostat enhance radiation sensitization by inhibiting DNA-damage repair, inducing apoptosis, inhibiting proliferation and angiogenesis, and enhancing immune surveillance for cancer (Son et al. 2014).

SUMMARY AND PERSPECTIVE

Studies over the past few decades have demonstrated that HDACs play a critical role in the development of cancer by reversibly modulating acetylation status of histone and nonhistone proteins. As an eraser of histone acetylation and a key regulator of epigenetics, HDACs have been found to dysregulate and/or function incorrectly in cancer, providing a crucial attractive target against cancer. However, the precise function of HDACs as a central mediator of proliferation and tumorigenic capacity still remains a conundrum. Although genetic knockdown or knockout of HDACs in a variety of cancer cells induces cell-cycle arrest and apoptosis, a putative tumor suppressor role of HDACs is also observed in certain circumstances. More studies

are needed to systematically dissect the role of individual HDACs in different cancer types at different stages of tumorigenesis. Clearly, the development of HDACi, in particular selective inhibitors, could help clarify the function of distinct HDACs, and a better comprehension of HDACs in cancer will give us a mechanistic-based rationale for the clinical use of HDACi as antitumor agents. So far, the most common HDACi under preclinical and clinical evaluation are broad spectrum nonselective HDACi. The effectiveness of nonselective HDACi for the treatment of cancer relies on its broad-spectrum inhibition against HDACs, which is also the major reason for toxicity of these agents. Therefore, current emphasis is placed on developing HDACi with higher target specificity that might be more efficacious with less toxicity. In parallel, research is increasingly showing that combination therapy might be another important direction to enhance the therapeutic efficacy of HDACi. Further elucidation of the mechanisms of action of HDACs and HDACi will provide a bright future for the use of HDACi as one of many tools in the fight against cancer.

ACKNOWLEDGMENTS

This work is supported in part by National Institutes of Health (NIH) Grants R01CA169210 and R01CA187040 to E.S.

REFERENCES

Adams H, Fritzsche FR, Dirnhofer S, Kristiansen G, Tzankov A. 2010. Class I histone deacetylases 1, 2 and 3 are highly expressed in classical Hodgkin's lymphoma. *Expert Opin Ther Targets* **14**: 577–584.

Aghdassi A, Sendler M, Guenther A, Mayerle J, Behn CO, Heidecke CD, Friess H, Buchler M, Evert M, Lerch MM, et al. 2012. Recruitment of histone deacetylases HDAC1 and HDAC2 by the transcriptional repressor ZEB1 downregulates E-cadherin expression in pancreatic cancer. *Gut* **61**: 439–448.

Ahn MY, Ahn JW, Kim HS, Lee J, Yoon JH. 2015. Apicidin inhibits cell growth by downregulating IGF-1R in salivary mucoepidermoid carcinoma cells. *Oncol Rep* **33**: 1899–1907.

Ahrens TD, Timme S, Hoeppner J, Ostendorp J, Hembach S, Follo M, Hopt UT, Werner M, Busch H, Boerries M, et al. 2015. Selective inhibition of esophageal cancer cells by combination of HDAC inhibitors and Azacytidine. *Epigenetics* **10**: 431–445.

Aldana-Masangkay GI, Rodriguez-Gonzalez A, Lin T, Ikeda AK, Hsieh YT, Kim YM, Lomenick B, Okemoto K, Landaw EM, Wang D, et al. 2011. Tubacin suppresses proliferation and induces apoptosis of acute lymphoblastic leukemia cells. *Leuk Lymphoma* **52:** 1544–1555.

Amengual JE, Johannet PM, Lombardo M, Zullo KM, Hoehn D, Bhagat G, Scotto L, Jirau-Serrano X, Radeski D, Heinen J, et al. 2015. Dual targeting of protein degradation pathways with the selective HDAC6 inhibitor, ACY-1215, and bortezomib is synergistic in lymphoma. *Clin Cancer Res* **21:** 4663–4675.

Bajbouj K, Mawrin C, Hartig R, Schulze-Luehrmann J, Wilisch-Neumann A, Roessner A, Schneider-Stock R. 2012. P53-dependent antiproliferative and pro-apoptotic effects of trichostatin A (TSA) in glioblastoma cells. *J Neurooncol* **107:** 503–516.

Balasubramanian S, Ramos J, Luo W, Sirisawad M, Verner E, Buggy JJ. 2008. A novel histone deacetylase 8 (HDAC8)-specific inhibitor PCI-34051 induces apoptosis in T-cell lymphomas. *Leukemia* **22:** 1026–1034.

Banerji U, van Doorn L, Papadatos-Pastos D, Kristeleit R, Debnam P, Tall M, Stewart A, Raynaud F, Garrett MD, Toal M, et al. 2012. A phase I pharmacokinetic and pharmacodynamic study of CHR-3996, an oral class I selective histone deacetylase inhibitor in refractory solid tumors. *Clin Cancer Res* **18:** 2687–2694.

Bangert A, Cristofanon S, Eckhardt I, Abhari BA, Kolodziej S, Hacker S, Vellanki SH, Lausen J, Debatin KM, Fulda S. 2012. Histone deacetylase inhibitors sensitize glioblastoma cells to TRAIL-induced apoptosis by c-myc-mediated downregulation of cFLIP. *Oncogene* **31:** 4677–4688.

Bantscheff M, Hopf C, Savitski MM, Dittmann A, Grandi P, Michon AM, Schlegl J, Abraham Y, Becher I, Bergamini G, et al. 2011. Chemoproteomics profiling of HDAC inhibitors reveals selective targeting of HDAC complexes. *Nat Biotechnol* **29:** 255–265.

Beagle BR, Nguyen DM, Mallya S, Tang SS, Lu M, Zeng Z, Konopleva M, Vo TT, Fruman DA. 2015. mTOR kinase inhibitors synergize with histone deacetylase inhibitors to kill B-cell acute lymphoblastic leukemia cells. *Oncotarget* **6:** 2088–2100.

Bergman JA, Woan K, Perez-Villarroel P, Villagra A, Sotomayor EM, Kozikowski AP. 2012. Selective histone deacetylase 6 inhibitors bearing substituted urea linkers inhibit melanoma cell growth. *J Med Chem* **55:** 9891–9899.

Bhaskara S, Knutson SK, Jiang G, Chandrasekharan MB, Wilson AJ, Zheng S, Yenamandra A, Locke K, Yuan JL, Bonine-Summers AR, et al. 2010. Hdac3 is essential for the maintenance of chromatin structure and genome stability. *Cancer Cell* **18:** 436–447.

Bilen MA, Fu S, Falchook GS, Ng CS, Wheler JJ, Abdelrahim M, Erguvan-Dogan B, Hong DS, Tsimberidou AM, Kurzrock R, et al. 2015. Phase I trial of valproic acid and lenalidomide in patients with advanced cancer. *Cancer Chemother Pharmacol* **75:** 869–874.

Bradbury CA, Khanim FL, Hayden R, Bunce CM, White DA, Drayson MT, Craddock C, Turner BM. 2005. Histone deacetylases in acute myeloid leukaemia show a distinctive pattern of expression that changes selectively in response to deacetylase inhibitors. *Leukemia* **19:** 1751–1759.

Bradner JE, West N, Grachan ML, Greenberg EF, Haggarty SJ, Warnow T, Mazitschek R. 2010. Chemical phylogenetics of histone deacetylases. *Nat Chem Biol* **6:** 238–243.

Brunetto AT, Ang JE, Lal R, Olmos D, Molife LR, Kristeleit R, Parker A, Casamayor I, Olaleye M, Mais A, et al. 2013. First-in-human, pharmacokinetic and pharmacodynamic phase I study of Resminostat, an oral histone deacetylase inhibitor, in patients with advanced solid tumors. *Clin Cancer Res* **19:** 5494–5504.

Butler KV, Kalin J, Brochier C, Vistoli G, Langley B, Kozikowski AP. 2010. Rational design and simple chemistry yield a superior, neuroprotective HDAC6 inhibitor, tubastatin A. *J Am Chem Soc* **132:** 10842–10846.

Buurman R, Gurlevik E, Schaffer V, Eilers M, Sandbothe M, Kreipe H, Wilkens L, Schlegelberger B, Kuhnel F, Skawran B. 2012. Histone deacetylases activate hepatocyte growth factor signaling by repressing microRNA-449 in hepatocellular carcinoma cells. *Gastroenterology* **143:** 811–820.

Byles V, Zhu L, Lovaas JD, Chmilewski LK, Wang J, Faller DV, Dai Y. 2012. SIRT1 induces EMT by cooperating with EMT transcription factors and enhances prostate cancer cell migration and metastasis. *Oncogene* **31:** 4619–4629.

Cai X, Zhai HX, Wang J, Forrester J, Qu H, Yin L, Lai CJ, Bao R, Qian C. 2010. Discovery of 7-(4-(3-ethynylphenylamino)-7-methoxyquinazolin-6-yloxy)-*N*-hydroxyheptanamide (CUDc-101) as a potent multi-acting HDAC, EGFR, and HER2 inhibitor for the treatment of cancer. *J Med Chem* **53:** 2000–2009.

Cao K, Wang G, Li W, Zhang L, Wang R, Huang Y, Du L, Jiang J, Wu C, He X, et al. 2015. Histone deacetylase inhibitors prevent activation-induced cell death and promote anti-tumor immunity. *Oncogene* **34:** 5960–5970.

Cha TL, Chuang MJ, Wu ST, Sun GH, Chang SY, Yu DS, Huang SM, Huan SK, Cheng TC, Chen TT, et al. 2009. Dual degradation of aurora A and B kinases by the histone deacetylase inhibitor LBH589 induces G_2-M arrest and apoptosis of renal cancer cells. *Clin Cancer Res* **15:** 840–850.

Chen IC, Chiang WF, Huang HH, Chen PF, Shen YY, Chiang HC. 2014. Role of SIRT1 in regulation of epithelial-to-mesenchymal transition in oral squamous cell carcinoma metastasis. *Mol Cancer* **13:** 254.

Chen S, Yin C, Lao T, Liang D, He D, Wang C, Sang N. 2015. AMPK-HDAC5 pathway facilitates nuclear accumulation of HIF-1α and functional activation of HIF-1 by deacetylating Hsp70 in the cytosol. *Cell Cycle* **14:** 2520–2536.

Choudhary C, Kumar C, Gnad F, Nielsen ML, Rehman M, Walther TC, Olsen JV, Mann M. 2009. Lysine acetylation targets protein complexes and co-regulates major cellular functions. *Science* **325:** 834–840.

Choy E, Flamand Y, Balasubramanian S, Butrynski JE, Harmon DC, George S, Cote GM, Wagner AJ, Morgan JA, Sirisawad M, et al. 2015. Phase 1 study of oral abexinostat, a histone deacetylase inhibitor, in combination with doxorubicin in patients with metastatic sarcoma. *Cancer* **121:** 1223–1230.

Christiansen AJ, West A, Banks KM, Haynes NM, Teng MW, Smyth MJ, Johnstone RW. 2011. Eradication of solid tumors using histone deacetylase inhibitors combined with immune-stimulating antibodies. *Proc Natl Acad Sci* **108:** 4141–4146.

Chun P. 2015. Histone deacetylase inhibitors in hematological malignancies and solid tumors. *Arch Pharm Res* **38**: 933–949.

Colarossi L, Memeo L, Colarossi C, Aiello E, Iuppa A, Espina V, Liotta L, Mueller C. 2014. Inhibition of histone deacetylase 4 increases cytotoxicity of docetaxel in gastric cancer cells. *Proteomics Clin Appl* **8**: 924–931.

Conrad E, Polonio-Vallon T, Meister M, Matt S, Bitomsky N, Herbel C, Liebl M, Greiner V, Kriznik B, Schumacher S, et al. 2015. HIPK2 restricts SIRT1 activity upon severe DNA damage by a phosphorylation-controlled mechanism. *Cell Death Differ* **23**: 110–122.

Conti C, Leo E, Eichler GS, Sordet O, Martin MM, Fan A, Aladjem MI, Pommier Y. 2010. Inhibition of histone deacetylase in cancer cells slows down replication forks, activates dormant origins, and induces DNA damage. *Cancer Res* **70**: 4470–4480.

de Cremoux P, Dalvai M, N'Doye O, Moutahir F, Rolland G, Chouchane-Mlik O, Assayag F, Lehmann-Che J, Kraus-Berthie L, Nicolas A, et al. 2015. HDAC inhibition does not induce estrogen receptor in human triple-negative breast cancer cell lines and patient-derived xenografts. *Breast Cancer Res Treat* **149**: 81–89.

Dong M, Ning ZQ, Xing PY, Xu JL, Cao HX, Dou GF, Meng ZY, Shi YK, Lu XP, Feng FY. 2012. Phase I study of chidamide (CS055/HBI-8000), a new histone deacetylase inhibitor, in patients with advanced solid tumors and lymphomas. *Cancer Chemother Pharmacol* **69**: 1413–1422.

Dong LH, Cheng S, Zheng Z, Wang L, Shen Y, Shen ZX, Chen SJ, Zhao WL. 2013. Histone deacetylase inhibitor potentiated the ability of MTOR inhibitor to induce autophagic cell death in Burkitt leukemia/lymphoma. *J Hematol Oncol* **6**: 53.

Dowling DP, Gantt SL, Gattis SG, Fierke CA, Christianson DW. 2008. Structural studies of human histone deacetylase 8 and its site-specific variants complexed with substrate and inhibitors. *Biochemistry* **47**: 13554–13563.

Duan W, Hou J, Chu X, Li X, Zhang J, Li J, Xu W, Zhang Y. 2015. Synthesis and biological evaluation of novel histone deacetylases inhibitors with nitric oxide releasing activity. *Bioorg Med Chem* **23**: 4481–4488.

Duong V, Bret C, Altucci L, Mai A, Duraffourd C, Loubersac J, Harmand PO, Bonnet S, Valente S, Maudelonde T, et al. 2008. Specific activity of class II histone deacetylases in human breast cancer cells. *Mol Cancer Res* **6**: 1908–1919.

Eades G, Yao Y, Yang M, Zhang Y, Chumsri S, Zhou Q. 2011. miR-200a regulates SIRT1 expression and epithelial to mesenchymal transition (EMT)-like transformation in mammary epithelial cells. *J Biol Chem* **286**: 25992–26002.

Ecker J, Oehme I, Mazitschek R, Korshunov A, Kool M, Hielscher T, Kiss J, Selt F, Konrad C, Lodrini M, et al. 2015. Targeting class I histone deacetylase 2 in MYC amplified group 3 medulloblastoma. *Acta Neuropathol Commun* **3**: 22.

Ellis L, Bots M, Lindemann RK, Bolden JE, Newbold A, Cluse LA, Scott CL, Strasser A, Atadja P, Lowe SW, et al. 2009. The histone deacetylase inhibitors LAQ824 and LBH589 do not require death receptor signaling or a functional apoptosome to mediate tumor cell death or therapeutic efficacy. *Blood* **114**: 380–393.

Falkenberg KJ, Johnstone RW. 2014. Histone deacetylases and their inhibitors in cancer, neurological diseases and immune disorders. *Nat Rev Drug Discov* **13**: 673–691.

Fan J, Lou B, Chen W, Zhang J, Lin S, Lv FF, Chen Y. 2014. Down-regulation of HDAC5 inhibits growth of human hepatocellular carcinoma by induction of apoptosis and cell cycle arrest. *Tumour Biol* **35**: 11523–11532.

Feng L, Pan M, Sun J, Lu H, Shen Q, Zhang S, Jiang T, Liu L, Jin W, Chen Y, et al. 2013. Histone deacetylase 3 inhibits expression of PUMA in gastric cancer cells. *J Mol Med (Berl)* **91**: 49–58.

Feng GW, Dong LD, Shang WJ, Pang XL, Li JF, Liu L, Wang Y. 2014. HDAC5 promotes cell proliferation in human hepatocellular carcinoma by up-regulating Six1 expression. *Eur Rev Med Pharmacol Sci* **18**: 811–816.

Fischer C, Leithner K, Wohlkoenig C, Quehenberger F, Bertsch A, Olschewski A, Olschewski H, Hrzenjak A. 2015. Panobinostat reduces hypoxia-induced cisplatin resistance of non-small cell lung carcinoma cells via HIF-1α destabilization. *Mol Cancer* **14**: 4.

Fiskus W, Sharma S, Shah B, Portier BP, Devaraj SG, Liu K, Iyer SP, Bearss D, Bhalla KN. 2014. Highly effective combination of LSD1 (KDM1A) antagonist and pan-histone deacetylase inhibitor against human AML cells. *Leukemia* **28**: 2155–2164.

Fleming CL, Ashton TD, Gaur V, McGee SL, Pfeffer FM. 2014. Improved synthesis and structural reassignment of MC1568: A class IIa selective HDAC inhibitor. *J Med Chem* **57**: 1132–1135.

Fraga MF, Ballestar E, Villar-Garea A, Boix-Chornet M, Espada J, Schotta G, Bonaldi T, Haydon C, Ropero S, Petrie K, et al. 2005. Loss of acetylation at Lys16 and trimethylation at Lys20 of histone H4 is a common hallmark of human cancer. *Nat Genet* **37**: 391–400.

Fritzsche FR, Weichert W, Roske A, Gekeler V, Beckers T, Stephan C, Jung K, Scholman K, Denkert C, Dietel M, et al. 2008. Class I histone deacetylases 1, 2 and 3 are highly expressed in renal cell cancer. *BMC Cancer* **8**: 381.

Fritsche P, Seidler B, Schuler S, Schnieke A, Gottlicher M, Schmid RM, Saur D, Schneider G. 2009. HDAC2 mediates therapeutic resistance of pancreatic cancer cells via the BH3-only protein NOXA. *Gut* **58**: 1399–1409.

Frumm SM, Fan ZP, Ross KN, Duvall JR, Gupta S, VerPlank L, Suh BC, Holson E, Wagner FF, Smith WB, et al. 2013. Selective HDAC1/HDAC2 inhibitors induce neuroblastoma differentiation. *Chem Biol* **20**: 713–725.

Frye R, Myers M, Axelrod KC, Ness EA, Piekarz RL, Bates SE, Booher S. 2012. Romidepsin: A new drug for the treatment of cutaneous T-cell lymphoma. *Clin J Oncol Nurs* **16**: 195–204.

Furumai R, Matsuyama A, Kobashi N, Lee KH, Nishiyama M, Nakajima H, Tanaka A, Komatsu Y, Nishino N, Yoshida M, et al. 2002. FK228 (depsipeptide) as a natural prodrug that inhibits class I histone deacetylases. *Cancer Res* **62**: 4916–4921.

Galli M, Salmoiraghi S, Golay J, Gozzini A, Crippa C, Pescosta N, Rambaldi A. 2010. A phase II multiple dose clinical trial of histone deacetylase inhibitor ITF2357 in patients with relapsed or progressive multiple myeloma. *Ann Hematol* **89**: 185–190.

Galloway TJ, Wirth LJ, Colevas AD, Gilbert J, Bauman JE, Saba NF, Raben D, Mehra R, Ma AW, Atoyan R, et al. 2015.

A Phase I Study of CUDC-101, a multitarget inhibitor of HDACs, EGFR, and HER2, in combination with chemoradiation in patients with head and neck squamous cell carcinoma. *Clin Cancer Res* **21:** 1566–1573.

Geng H, Harvey CT, Pittsenbarger J, Liu Q, Beer TM, Xue C, Qian DZ. 2011. HDAC4 protein regulates HIF1α protein lysine acetylation and cancer cell response to hypoxia. *J Biol Chem* **286:** 38095–38102.

Gorospe M, de Cabo R. 2008. AsSIRTing the DNA damage response. *Trends Cell Biol* **18:** 77–83.

Gruhn B, Naumann T, Gruner D, Walther M, Wittig S, Becker S, Beck JF, Sonnemann J. 2013. The expression of histone deacetylase 4 is associated with prednisone poor-response in childhood acute lymphoblastic leukemia. *Leuk Res* **37:** 1200–1207.

Gryder BE, Rood MK, Johnson KA, Patil V, Raftery ED, Yao LP, Rice M, Azizi B, Doyle DF, Oyelere AK. 2013. Histone deacetylase inhibitors equipped with estrogen receptor modulation activity. *J Med Chem* **56:** 5782–5796.

Guerrant W, Patil V, Canzoneri JC, Yao LP, Hood R, Oyelere AK. 2013. Dual-acting histone deacetylase-topoisomerase I inhibitors. *Bioorg Med Chem Lett* **23:** 3283–3287.

Guha M. 2015. HDAC inhibitors still need a home run, despite recent approval. *Nat Rev Drug Discov* **14:** 225–226.

Gupta M, Han JJ, Stenson M, Wellik L, Witzig TE. 2012. Regulation of STAT3 by histone deacetylase-3 in diffuse large B-cell lymphoma: Implications for therapy. *Leukemia* **26:** 1356–1364.

Guzman ML, Yang N, Sharma KK, Balys M, Corbett CA, Jordan CT, Becker MW, Steidl U, Abdel-Wahab O, Levine RL, et al. 2014. Selective activity of the histone deacetylase inhibitor AR-42 against leukemia stem cells: A novel potential strategy in acute myelogenous leukemia. *Mol Cancer Ther* **13:** 1979–1990.

Haggarty SJ, Koeller KM, Wong JC, Grozinger CM, Schreiber SL. 2003. Domain-selective small-molecule inhibitor of histone deacetylase 6 (HDAC6)-mediated tubulin deacetylation. *Proc Natl Acad Sci* **100:** 4389–4394.

Halsall JA, Turan N, Wiersma M, Turner BM. 2015. Cells adapt to the epigenomic disruption caused by histone deacetylase inhibitors through a coordinated, chromatin-mediated transcriptional response. *Epigenetics Chromatin* **8:** 29.

Hayashi A, Horiuchi A, Kikuchi N, Hayashi T, Fuseya C, Suzuki A, Konishi I, Shiozawa T. 2010. Type-specific roles of histone deacetylase (HDAC) overexpression in ovarian carcinoma: HDAC1 enhances cell proliferation and HDAC3 stimulates cell migration with downregulation of E-cadherin. *Int J Cancer* **127:** 1332–1346.

Hu E, Dul E, Sung CM, Chen Z, Kirkpatrick R, Zhang GF, Johanson K, Liu R, Lago A, Hofmann G, et al. 2003. Identification of novel isoform-selective inhibitors within class I histone deacetylases. *J Pharmacol Exp Ther* **307:** 720–728.

Huang WJ, Wang YC, Chao SW, Yang CY, Chen LC, Lin MH, Hou WC, Chen MY, Lee TL, Yang P, et al. 2012. Synthesis and biological evaluation of ortho-aryl *N*-hydroxycinnamides as potent histone deacetylase (HDAC) 8 isoform-selective inhibitors. *ChemMedChem* **7:** 1815–1824.

Huang R, Xu Y, Wan W, Shou X, Qian J, You Z, Liu B, Chang C, Zhou T, Lippincott-Schwartz J, et al. 2015a. Deacety-

lation of nuclear LC3 drives autophagy initiation under starvation. *Mol Cell* **57:** 456–466.

Huang Z, Peng S, Knoff J, Lee SY, Yang B, Wu TC, Hung CF. 2015b. Combination of proteasome and HDAC inhibitor enhances HPV16 E7-specific CD8$^+$ T cell immune response and antitumor effects in a preclinical cervical cancer model. *J Biomed Sci* **22:** 7.

Hug BA, Lazar MA. 2004. ETO interacting proteins. *Oncogene* **23:** 4270–4274.

Hui KF, Chiang AK. 2014. Combination of proteasome and class I HDAC inhibitors induces apoptosis of NPC cells through an HDAC6-independent ER stress-induced mechanism. *Int J Cancer* **135:** 2950–2961.

Iannitti T, Palmieri B. 2011. Clinical and experimental applications of sodium phenylbutyrate. *Drugs R D* **11:** 227–249.

Ishikawa S, Hayashi H, Kinoshita K, Abe M, Kuroki H, Tokunaga R, Tomiyasu S, Tanaka H, Sugita H, Arita T, et al. 2014. Statins inhibit tumor progression via an enhancer of *zeste* homolog 2-mediated epigenetic alteration in colorectal cancer. *Int J Cancer* **135:** 2528–2536.

Jazirehi AR, Kurdistani SK, Economou JS. 2014. Histone deacetylase inhibitor sensitizes apoptosis-resistant melanomas to cytotoxic human T lymphocytes through regulation of TRAIL/DR5 pathway. *J Immunol* **192:** 3981–3989.

Jin Z, Jiang W, Jiao F, Guo Z, Hu H, Wang L, Wang L. 2014. Decreased expression of histone deacetylase 10 predicts poor prognosis of gastric cancer patients. *Int J Clin Exp Pathol* **7:** 5872–5879.

Juengel E, Nowaz S, Makarevi J, Natsheh I, Werner I, Nelson K, Reiter M, Tsaur I, Mani J, Harder S, et al. 2014. HDAC-inhibition counteracts everolimus resistance in renal cell carcinoma in vitro by diminishing cdk2 and cyclin A. *Mol Cancer* **13:** 152.

Juergens RA, Wrangle J, Vendetti FP, Murphy SC, Zhao M, Coleman B, Sebree R, Rodgers K, Hooker CM, Franco N, et al. 2011. Combination epigenetic therapy has efficacy in patients with refractory advanced non-small cell lung cancer. *Cancer Discov* **1:** 598–607.

Jung KH, Noh JH, Kim JK, Eun JW, Bae HJ, Xie HJ, Chang YG, Kim MG, Park H, Lee JY, et al. 2012. HDAC2 overexpression confers oncogenic potential to human lung cancer cells by deregulating expression of apoptosis and cell cycle proteins. *J Cell Biochem* **113:** 2167–2177.

Kaidi A, Weinert BT, Choudhary C, Jackson SP. 2010. Human SIRT6 promotes DNA end resection through CtIP deacetylation. *Science* **329:** 1348–1353.

Kaliszczak M, Trousil S, Aberg O, Perumal M, Nguyen QD, Aboagye EO. 2013. A novel small molecule hydroxamate preferentially inhibits HDAC6 activity and tumour growth. *Br J Cancer* **108:** 342–350.

Kaluza D, Kroll J, Gesierich S, Yao TP, Boon RA, Hergenreider E, Tjwa M, Rossig L, Seto E, Augustin HG, et al. 2011. Class IIb HDAC6 regulates endothelial cell migration and angiogenesis by deacetylation of cortactin. *EMBO J* **30:** 4142–4156.

Kaluza D, Kroll J, Gesierich S, Manavski Y, Boeckel JN, Doebele C, Zelent A, Rossig L, Zeiher AM, Augustin HG, et al. 2013. Histone deacetylase 9 promotes angiogenesis by targeting the antiangiogenic microRNA-17–92 cluster in endothelial cells. *Arterioscler Thromb Vasc Biol* **33**: 533–543.

Kang Y, Nian H, Rajendran P, Kim E, Dashwood WM, Pinto JT, Boardman LA, Thibodeau SN, Limburg PJ, Lohr CV, et al. 2014. HDAC8 and STAT3 repress BMF gene activity in colon cancer cells. *Cell Death Dis* **5**: e1476.

Kao GD, McKenna WG, Guenther MG, Muschel RJ, Lazar MA, Yen TJ. 2003. Histone deacetylase 4 interacts with 53BP1 to mediate the DNA damage response. *J Cell Biol* **160**: 1017–1027.

Karthik S, Sankar R, Varunkumar K, Ravikumar V. 2014. Romidepsin induces cell cycle arrest, apoptosis, histone hyperacetylation and reduces matrix metalloproteinases 2 and 9 expression in bortezomib sensitized non-small cell lung cancer cells. *Biomed Pharmacother* **68**: 327–334.

Kato H, Tamamizu-Kato S, Shibasaki F. 2004. Histone deacetylase 7 associates with hypoxia-inducible factor 1α and increases transcriptional activity. *J Biol Chem* **279**: 41966–41974.

Khan N, Jeffers M, Kumar S, Hackett C, Boldog F, Khramtsov N, Qian X, Mills E, Berghs SC, Carey N, et al. 2008. Determination of the class and isoform selectivity of small-molecule histone deacetylase inhibitors. *Biochem J* **409**: 581–589.

Kikuchi S, Suzuki R, Ohguchi H, Yoshida Y, Lu D, Cottini F, Jakubikova J, Bianchi G, Harada T, Gorgun G, et al. 2015. Class IIa HDAC inhibition enhances ER stress-mediated cell death in multiple myeloma. *Leukemia* **29**: 1918–1927.

Kim JK, Noh JH, Eun JW, Jung KH, Bae HJ, Shen Q, Kim MG, Chang YG, Kim SJ, Park WS, et al. 2013. Targeted inactivation of HDAC2 restores p16INK4a activity and exerts antitumor effects on human gastric cancer. *Mol Cancer Res* **11**: 62–73.

Knipstein J, Gore L. 2011. Entinostat for treatment of solid tumors and hematologic malignancies. *Expert Opin Investig Drugs* **20**: 1455–1467.

Koeneke E, Witt O, Oehme I. 2015. HDAC family members intertwined in the regulation of autophagy: A druggable vulnerability in aggressive tumor entities. *Cells* **4**: 135–168.

Kong X, Lin Z, Liang D, Fath D, Sang N, Caro J. 2006. Histone deacetylase inhibitors induce VHL and ubiquitin-independent proteasomal degradation of hypoxia-inducible factor 1α. *Mol Cell Biol* **26**: 2019–2028.

Kong HS, Tian S, Kong Y, Du G, Zhang L, Jung M, Dritschilo A, Brown ML. 2012. Preclinical studies of YK-4-272, an inhibitor of class II histone deacetylases by disruption of nucleocytoplasmic shuttling. *Pharm Res* **29**: 3373–3383.

Koprinarova M, Botev P, Russev G. 2011. Histone deacetylase inhibitor sodium butyrate enhances cellular radiosensitivity by inhibiting both DNA nonhomologous end joining and homologous recombination. *DNA Repair (Amst)* **10**: 970–977.

Kotian S, Liyanarachchi S, Zelent A, Parvin JD. 2011. Histone deacetylases 9 and 10 are required for homologous recombination. *J Biol Chem* **286**: 7722–7726.

Kraker AJ, Mizzen CA, Hartl BG, Miin J, Allis CD, Merriman RL. 2003. Modulation of histone acetylation by [4-(acetylamino)-*N*-(2-amino-phenyl) benzamide] in HCT-8 colon carcinoma. *Mol Cancer Ther* **2**: 401–408.

Krennhrubec K, Marshall BL, Hedglin M, Verdin E, Ulrich SM. 2007. Design and evaluation of "Linkerless" hydroxamic acids as selective HDAC8 inhibitors. *Bioorg Med Chem Lett* **17**: 2874–2878.

Lachenmayer A, Toffanin S, Cabellos L, Alsinet C, Hoshida Y, Villanueva A, Minguez B, Tsai HW, Ward SC, Thung S, et al. 2012. Combination therapy for hepatocellular carcinoma: Additive preclinical efficacy of the HDAC inhibitor panobinostat with sorafenib. *J Hepatol* **56**: 1343–1350.

Lamblin M, Dabbas B, Spingarn R, Mendoza-Sanchez R, Wang TT, An BS, Huang DC, Kremer R, White JH, Gleason JL. 2010. Vitamin D receptor agonist/histone deacetylase inhibitor molecular hybrids. *Bioorg Med Chem* **18**: 4119–4137.

Lee H, Rezai-Zadeh N, Seto E. 2004. Negative regulation of histone deacetylase 8 activity by cyclic AMP-dependent protein kinase A. *Mol Cell Biol* **24**: 765–773.

Lee IH, Cao L, Mostoslavsky R, Lombard DB, Liu J, Bruns NE, Tsokos M, Alt FW, Finkel T. 2008. A role for the NAD-dependent deacetylase Sirt1 in the regulation of autophagy. *Proc Natl Acad Sci* **105**: 3374–3379.

Lee JY, Koga H, Kawaguchi Y, Tang W, Wong E, Gao YS, Pandey UB, Kaushik S, Tresse E, Lu J, et al. 2010a. HDAC6 controls autophagosome maturation essential for ubiquitin-selective quality-control autophagy. *EMBO J* **29**: 969–980.

Lee JY, Nagano Y, Taylor JP, Lim KL, Yao TP. 2010b. Disease-causing mutations in parkin impair mitochondrial ubiquitination, aggregation, and HDAC6-dependent mitophagy. *J Cell Biol* **189**: 671–679.

Lee JH, Mahendran A, Yao Y, Ngo L, Venta-Perez G, Choy ML, Kim N, Ham WS, Breslow R, Marks PA. 2013. Development of a histone deacetylase 6 inhibitor and its biological effects. *Proc Natl Acad Sci* **110**: 15704–15709.

Lee HY, Tsai AC, Chen MC, Shen PJ, Cheng YC, Kuo CC, Pan SL, Liu YM, Liu JF, Yeh TK, et al. 2014a. Azaindolylsulfonamides, with a more selective inhibitory effect on histone deacetylase 6 activity, exhibit antitumor activity in colorectal cancer HCT116 cells. *J Med Chem* **57**: 4009–4022.

Lee SH, Yoo C, Im S, Jung JH, Choi HJ, Yoo J. 2014b. Expression of histone deacetylases in diffuse large B-cell lymphoma and its clinical significance. *Int J Med Sci* **11**: 994–1000.

Li Z, Zhu WG. 2014. Targeting histone deacetylases for cancer therapy: from molecular mechanisms to clinical implications. *Int J Biol Sci* **10**: 757–770.

Li Y, Kao GD, Garcia BA, Shabanowitz J, Hunt DF, Qin J, Phelan C, Lazar MA. 2006. A novel histone deacetylase pathway regulates mitosis by modulating Aurora B kinase activity. *Genes Dev* **20**: 2566–2579.

Li D, Sun X, Zhang L, Yan B, Xie S, Liu R, Liu M, Zhou J. 2014. Histone deacetylase 6 and cytoplasmic linker protein 170 function together to regulate the motility of pancreatic cancer cells. *Protein Cell* **5**: 214–223.

Li S, Liu X, Chen X, Zhang L, Wang X. 2015a. Histone deacetylase 6 promotes growth of glioblastoma through inhibition of SMAD2 signaling. *Tumour Biol* **36**: 9661–9665.

Li Y, Peng L, Seto E. 2015b. HDAC10 regulates cell cycle G_2/M phase transition via a novel Let-7-HMGA2-Cyclin A2 pathway. *Mol Cell Biol* **35**: 3547–3565.

Liao W, Jordaan G, Srivastava MK, Dubinett S, Sharma S, Sharma S. 2013. Effect of epigenetic histone modifications on E-cadherin splicing and expression in lung cancer. *Am J Cancer Res* **3**: 374–389.

Liffers K, Kolbe K, Westphal M, Lamszus K, Schulte A. 2015. Histone deacetylase inhibitors resensitize EGFR/EGFRvIII-overexpressing, Erlotinib-resistant glioblastoma cells to tyrosine kinase inhibition. *Target Oncol* **11**: 29–40.

Lim JH, Lee YM, Chun YS, Chen J, Kim JE, Park JW. 2010. Sirtuin 1 modulates cellular responses to hypoxia by deacetylating hypoxia-inducible factor 1α. *Mol Cell* **38**: 864–878.

Lin YH, Yuan J, Pei H, Liu T, Ann DK, Lou Z. 2015. KAP1 Deacetylation by SIRT1 promotes non-homologous end-joining repair. *PLoS ONE* **10**: e0123935.

Liu KP, Zhou D, Ouyang DY, Xu LH, Wang Y, Wang LX, Pan H, He XH. 2013. LC3B-II deacetylation by histone deacetylase 6 is involved in serum-starvation-induced autophagic degradation. *Biochem Biophys Res Commun* **441**: 970–975.

Lobera M, Madauss KP, Pohlhaus DT, Wright QG, Trocha M, Schmidt DR, Baloglu E, Trump RP, Head MS, Hofmann GA, et al. 2013. Selective class IIa histone deacetylase inhibition via a nonchelating zinc-binding group. *Nat Chem Biol* **9**: 319–325.

Locatelli SL, Cleris L, Stirparo GG, Tartari S, Saba E, Pierdominici M, Malorni W, Carbone A, Anichini A, Carlo-Stella C. 2014. BIM upregulation and ROS-dependent necroptosis mediate the antitumor effects of the HDACi Givinostat and Sorafenib in Hodgkin lymphoma cell line xenografts. *Leukemia* **28**: 1861–1871.

Lopez G, Bill KL, Bid HK, Braggio D, Constantino D, Prudner B, Zewdu A, Batte K, Lev D, Pollock RE. 2015. HDAC8, A potential therapeutic target for the treatment of malignant peripheral nerve sheath tumors (MPNST). *PLoS ONE* **10**: e0133302.

Luna A, Aladjem MI, Kohn KW. 2013. SIRT1/PARP1 crosstalk: Connecting DNA damage and metabolism. *Genome Integr* **4**: 6.

Luo J, Nikolaev AY, Imai S, Chen D, Su F, Shiloh A, Guarente L, Gu W. 2001. Negative control of p53 by Sir2α promotes cell survival under stress. *Cell* **107**: 137–148.

Lv Z, Weng X, Du C, Zhang C, Xiao H, Cai X, Ye S, Cheng J, Ding C, Xie H, et al. 2015. Downregulation of HDAC6 promotes angiogenesis in hepatocellular carcinoma cells and predicts poor prognosis in liver transplantation patients. *Mol Carcinog* **55**: 1024–1033.

Mai A, Massa S, Pezzi R, Simeoni S, Rotili D, Nebbioso A, Scognamiglio A, Altucci L, Loidl P, Brosch G. 2005. Class II (IIa)-selective histone deacetylase inhibitors. 1: Synthesis and biological evaluation of novel (aryloxopropenyl)pyrrolyl hydroxyamides. *J Med Chem* **48**: 3344–3353.

Malvaez M, McQuown SC, Rogge GA, Astarabadi M, Jacques V, Carreiro S, Rusche JR, Wood MA. 2013. HDAC3-selective inhibitor enhances extinction of cocaine-seeking behavior in a persistent manner. *Proc Natl Acad Sci* **110**: 2647–2652.

Mann BS, Johnson JR, Cohen MH, Justice R, Pazdur R. 2007. FDA approval summary: Vorinostat for treatment of advanced primary cutaneous T-cell lymphoma. *Oncologist* **12**: 1247–1252.

Mao Z, Hine C, Tian X, Van Meter M, Au M, Vaidya A, Seluanov A, Gorbunova V. 2011. SIRT6 promotes DNA repair under stress by activating PARP1. *Science* **332**: 1443–1446.

Marek L, Hamacher A, Hansen FK, Kuna K, Gohlke H, Kassack MU, Kurz T. 2013. Histone deacetylase (HDAC) inhibitors with a novel connecting unit linker region reveal a selectivity profile for HDAC4 and HDAC5 with improved activity against chemoresistant cancer cells. *J Med Chem* **56**: 427–436.

Marquard L, Gjerdrum LM, Christensen IJ, Jensen PB, Sehested M, Ralfkiaer E. 2008. Prognostic significance of the therapeutic targets histone deacetylase 1, 2, 6 and acetylated histone H4 in cutaneous T-cell lymphoma. *Histopathology* **53**: 267–277.

Marquard L, Poulsen CB, Gjerdrum LM, de Nully Brown P, Christensen IJ, Jensen PB, Sehested M, Johansen P, Ralfkiaer E. 2009. Histone deacetylase 1, 2, 6 and acetylated histone H4 in B- and T-cell lymphomas. *Histopathology* **54**: 688–698.

Matsuba S, Niwa S, Muraki K, Kanatsuka S, Nakazono Y, Hatano N, Fujii M, Zhan P, Suzuki T, Ohya S. 2014. Downregulation of Ca^{2+}-activated Cl^- channel TMEM16A by the inhibition of histone deacetylase in TMEM16A-expressing cancer cells. *J Pharmacol Exp Ther* **351**: 510–518.

McDermott J, Jimeno A. 2014. Belinostat for the treatment of peripheral T-cell lymphomas. *Drugs Today (Barc)* **50**: 337–345.

McGraw AL. 2013. Romidepsin for the treatment of T-cell lymphomas. *Am J Health Syst Pharm* **70**: 1115–1122.

Meidhof S, Brabletz S, Lehmann W, Preca BT, Mock K, Ruh M, Schuler J, Berthold M, Weber A, Burk U, et al. 2015. ZEB1-associated drug resistance in cancer cells is reversed by the class I HDAC inhibitor mocetinostat. *EMBO Mol Med* **7**: 831–847.

Methot JL, Chakravarty PK, Chenard M, Close J, Cruz JC, Dahlberg WK, Fleming J, Hamblett CL, Hamill JE, Harrington P, et al. 2008. Exploration of the internal cavity of histone deacetylase (HDAC) with selective HDAC1/HDAC2 inhibitors (SHI-1:2). *Bioorg Med Chem Lett* **18**: 973–978.

Milde T, Oehme I, Korshunov A, Kopp-Schneider A, Remke M, Northcott P, Deubzer HE, Lodrini M, Taylor MD, von Deimling A, et al. 2010. HDAC5 and HDAC9 in medulloblastoma: Novel markers for risk stratification and role in tumor cell growth. *Clin Cancer Res* **16**: 3240–3252.

Miller KM, Tjeertes JV, Coates J, Legube G, Polo SE, Britton S, Jackson SP. 2010. Human HDAC1 and HDAC2 function in the DNA-damage response to promote DNA non-homologous end-joining. *Nat Struct Mol Biol* **17**: 1144–1151.

Min SK, Koh YH, Park Y, Kim HJ, Seo J, Park HR, Cho SJ, Kim IS. 2012. Expression of HAT1 and HDAC1, 2, 3 in diffuse large B-cell lymphomas, peripheral T-cell lymphomas, and NK/T-cell lymphomas. *Korean J Pathol* **46**: 142–150.

Minami J, Suzuki R, Mazitschek R, Gorgun G, Ghosh B, Cirstea D, Hu Y, Mimura N, Ohguchi H, Cottini F, et al. 2014. Histone deacetylase 3 as a novel therapeutic target in multiple myeloma. *Leukemia* **28**: 680–689.

Minamiya Y, Ono T, Saito H, Takahashi N, Ito M, Motoyama S, Ogawa J. 2010. Strong expression of HDAC3 correlates with a poor prognosis in patients with adenocarcinoma of the lung. *Tumour Biol* **31**: 533–539.

Minamiya Y, Ono T, Saito H, Takahashi N, Ito M, Mitsui M, Motoyama S, Ogawa J. 2011. Expression of histone deacetylase 1 correlates with a poor prognosis in patients with adenocarcinoma of the lung. *Lung Cancer* **74**: 300–304.

Mishima Y, Santo L, Eda H, Cirstea D, Nemani N, Yee AJ, O'Donnell E, Selig MK, Quayle SN, Arastu-Kapur S, et al. 2015. Ricolinostat (ACY-1215) induced inhibition of aggresome formation accelerates carfilzomib-induced multiple myeloma cell death. *Br J Haematol* **169**: 423–434.

Mithraprabhu S, Kalff A, Chow A, Khong T, Spencer A. 2014. Dysregulated class I histone deacetylases are indicators of poor prognosis in multiple myeloma. *Epigenetics* **9**: 1511–1520.

Moffat D, Patel S, Day F, Belfield A, Donald A, Rowlands M, Wibawa J, Brotherton D, Stimson L, Clark V, et al. 2010. Discovery of 2-(6-{[(6-fluoroquinolin-2-yl)methyl]amino}bicyclo[3.1.0]hex-3-yl)-*N*-hydroxypyrimidine-5-carboxamide (CHR-3996), a class I selective orally active histone deacetylase inhibitor. *J Med Chem* **53**: 8663–8678.

Moreno DA, Scrideli CA, Cortez MA, de Paula Queiroz R, Valera ET, da Silva Silveira V, Yunes JA, Brandalise SR, Tone LG. 2010. Differential expression of HDAC3, HDAC7 and HDAC9 is associated with prognosis and survival in childhood acute lymphoblastic leukaemia. *Br J Haematol* **150**: 665–673.

Moresi V, Carrer M, Grueter CE, Rifki OF, Shelton JM, Richardson JA, Bassel-Duby R, Olson EN. 2012. Histone deacetylases 1 and 2 regulate autophagy flux and skeletal muscle homeostasis in mice. *Proc Natl Acad Sci* **109**: 1649–1654.

Morschhauser F, Terriou L, Coiffier B, Bachy E, Varga A, Kloos I, Lelievre H, Sarry AL, Depil S, Ribrag V. 2015. Phase 1 study of the oral histone deacetylase inhibitor abexinostat in patients with Hodgkin lymphoma, non-Hodgkin lymphoma, or chronic lymphocytic leukaemia. *Invest New Drugs* **33**: 423–431.

Mostoslavsky R, Chua KF, Lombard DB, Pang WW, Fischer MR, Gellon L, Liu P, Mostoslavsky G, Franco S, Murphy MM, et al. 2006. Genomic instability and aging-like phenotype in the absence of mammalian SIRT6. *Cell* **124**: 315–329.

Muller BM, Jana L, Kasajima A, Lehmann A, Prinzler J, Budczies J, Winzer KJ, Dietel M, Weichert W, Denkert C. 2013. Differential expression of histone deacetylases HDAC1, 2 and 3 in human breast cancer—overexpression of HDAC2 and HDAC3 is associated with clinicopathological indicators of disease progression. *BMC Cancer* **13**: 215.

Munster PN, Thurn KT, Thomas S, Raha P, Lacevic M, Miller A, Melisko M, Ismail-Khan R, Rugo H, Moasser M, et al. 2011. A phase II study of the histone deacetylase inhibitor vorinostat combined with tamoxifen for the treat-

ment of patients with hormone therapy-resistant breast cancer. *Br J Cancer* **104**: 1828–1835.

Murakami T, Sato A, Chun NA, Hara M, Naito Y, Kobayashi Y, Kano Y, Ohtsuki M, Furukawa Y, Kobayashi E. 2008. Transcriptional modulation using HDACi depsipeptide promotes immune cell-mediated tumor destruction of murine B16 melanoma. *J Invest Dermatol* **128**: 1506–1516.

Nakashima H, Nguyen T, Goins WF, Chiocca EA. 2015. Interferon-stimulated gene 15 (ISG15) and ISG15-linked proteins can associate with members of the selective autophagic process, histone deacetylase 6 (HDAC6) and SQSTM1/p62. *J Biol Chem* **290**: 1485–1495.

Newbold A, Matthews GM, Bots M, Cluse LA, Clarke CJ, Banks KM, Cullinane C, Bolden JE, Christiansen AJ, Dickins RA, et al. 2013. Molecular and biologic analysis of histone deacetylase inhibitors with diverse specificities. *Mol Cancer Ther* **12**: 2709–2721.

Niegisch G, Knievel J, Koch A, Hader C, Fischer U, Albers P, Schulz WA. 2013. Changes in histone deacetylase (HDAC) expression patterns and activity of HDAC inhibitors in urothelial cancers. *Urol Oncol* **31**: 1770–1779.

Oehme I, Deubzer HE, Wegener D, Pickert D, Linke JP, Hero B, Kopp-Schneider A, Westermann F, Ulrich SM, von Deimling A, et al. 2009. Histone deacetylase 8 in neuroblastoma tumorigenesis. *Clin Cancer Res* **15**: 91–99.

Oehme I, Linke JP, Bock BC, Milde T, Lodrini M, Hartenstein B, Wiegand I, Eckert C, Roth W, Kool M, et al. 2013. Histone deacetylase 10 promotes autophagy-mediated cell survival. *Proc Natl Acad Sci* **110**: E2592–E2601.

Olsen CA, Montero A, Leman LJ, Ghadiri MR. 2012. Macrocyclic peptoid–peptide hybrids as inhibitors of class I histone deacetylases. *ACS Med Chem Lett* **3**: 749–753.

Osada H, Tatematsu Y, Saito H, Yatabe Y, Mitsudomi T, Takahashi T. 2004. Reduced expression of class II histone deacetylase genes is associated with poor prognosis in lung cancer patients. *Int J Cancer* **112**: 26–32.

Ou X, Lee MR, Huang X, Messina-Graham S, Broxmeyer HE. 2014. SIRT1 positively regulates autophagy and mitochondria function in embryonic stem cells under oxidative stress. *Stem Cells* **32**: 1183–1194.

Ouaissi M, Sielezneff I, Silvestre R, Sastre B, Bernard JP, Lafontaine JS, Payan MJ, Dahan L, Pirro N, Seitz JF, et al. 2008. High histone deacetylase 7 (HDAC7) expression is significantly associated with adenocarcinomas of the pancreas. *Ann Surg Oncol* **15**: 2318–2328.

Ouaissi M, Silvy F, Loncle C, Ferraz da Silva D, Martins Abreu C, Martinez E, Berthezene P, Cadra S, Le Treut YP, Hardwigsen J, et al. 2014. Further characterization of HDAC and SIRT gene expression patterns in pancreatic cancer and their relation to disease outcome. *PLoS ONE* **9**: e108520.

Palacios JA, Herranz D, De Bonis ML, Velasco S, Serrano M, Blasco MA. 2010. SIRT1 contributes to telomere maintenance and augments global homologous recombination. *J Cell Biol* **191**: 1299–1313.

Pandey UB, Nie Z, Batlevi Y, McCray BA, Ritson GP, Nedelsky NB, Schwartz SL, DiProspero NA, Knight MA, Schuldiner O, et al. 2007. HDAC6 rescues neurodegeneration and provides an essential link between autophagy and the UPS. *Nature* **447**: 859–863.

Papa L, Germain D. 2014. SirT3 regulates the mitochondrial unfolded protein response. *Mol Cell Biol* **34:** 699–710.

Parbin S, Kar S, Shilpi A, Sengupta D, Deb M, Rath SK, Patra SK. 2014. Histone deacetylases: A saga of perturbed acetylation homeostasis in cancer. *J Histochem Cytochem* **62:** 11–33.

Park J, Thomas S, Munster PN. 2015. Epigenetic modulation with histone deacetylase inhibitors in combination with immunotherapy. *Epigenomics* **7:** 641–652.

Pauer LR, Olivares J, Cunningham C, Williams A, Grove W, Kraker A, Olson S, Nemunaitis J. 2004. Phase I study of oral CI-994 in combination with carboplatin and paclitaxel in the treatment of patients with advanced solid tumors. *Cancer Invest* **22:** 886–896.

Peng L, Yuan Z, Li Y, Ling H, Izumi V, Fang B, Fukasawa K, Koomen J, Chen J, Seto E. 2015. Ubiquitinated sirtuin 1 (SIRT1) function is modulated during DNA damage-induced cell death and survival. *J Biol Chem* **290:** 8904–8912.

Poyet C, Jentsch B, Hermanns T, Schweckendiek D, Seifert HH, Schmidtpeter M, Sulser T, Moch H, Wild PJ, Kristiansen G. 2014. Expression of histone deacetylases 1, 2 and 3 in urothelial bladder cancer. *BMC Clin Pathol* **14:** 10.

Prebet T, Sun Z, Figueroa ME, Ketterling R, Melnick A, Greenberg PL, Herman J, Juckett M, Smith MR, Malick L, et al. 2014. Prolonged administration of azacitidine with or without entinostat for myelodysplastic syndrome and acute myeloid leukemia with myelodysplasia-related changes: results of the US Leukemia Intergroup trial E1905. *J Clin Oncol* **32:** 1242–1248.

Qian C, Lai CJ, Bao R, Wang DG, Wang J, Xu GX, Atoyan R, Qu H, Yin L, Samson M, et al. 2012. Cancer network disruption by a single molecule inhibitor targeting both histone deacetylase activity and phosphatidylinositol 3-kinase signaling. *Clin Cancer Res* **18:** 4104–4113.

Quint K, Agaimy A, Di Fazio P, Montalbano R, Steindorf C, Jung R, Hellerbrand C, Hartmann A, Sitter H, Neureiter D, et al. 2011. Clinical significance of histone deacetylases 1, 2, 3, and 7: HDAC2 is an independent predictor of survival in HCC. *Virchows Arch* **459:** 129–139.

Radhakrishnan R, Li Y, Xiang S, Yuan F, Yuan Z, Telles E, Fang J, Coppola D, Shibata D, Lane WS, et al. 2015. Histone deacetylase 10 regulates DNA mismatch repair and may Involve the deacetylation of MutS homolog 2. *J Biol Chem* **290:** 22795–22804.

Raha P, Thomas S, Thurn KT, Park J, Munster PN. 2015. Combined histone deacetylase inhibition and tamoxifen induces apoptosis in tamoxifen-resistant breast cancer models, by reversing Bcl-2 overexpression. *Breast Cancer Res* **17:** 26.

Ray A, Alalem M, Ray BK. 2013. Loss of epigenetic Kruppel-like factor 4 histone deacetylase (KLF-4-HDAC)-mediated transcriptional suppression is crucial in increasing vascular endothelial growth factor (VEGF) expression in breast cancer. *J Biol Chem* **288:** 27232–27242.

Reid T, Valone F, Lipera W, Irwin D, Paroly W, Natale R, Sreedharan S, Keer H, Lum B, Scappaticci F, et al. 2004. Phase II trial of the histone deacetylase inhibitor pivaloyloxymethyl butyrate (Pivanex, AN-9) in advanced non-small cell lung cancer. *Lung Cancer* **45:** 381–386.

Rettig I, Koeneke E, Trippel F, Mueller WC, Burhenne J, Kopp-Schneider A, Fabian J, Schober A, Fernekorn U, von Deimling A, et al. 2015. Selective inhibition of HDAC8 decreases neuroblastoma growth in vitro and in vivo and enhances retinoic acid-mediated differentiation. *Cell Death Dis* **6:** e1657.

Rhodes LV, Tate CR, Segar HC, Burks HE, Phamduy TB, Hoang V, Elliott S, Gilliam D, Pounder FN, Anbalagan M, et al. 2014. Suppression of triple-negative breast cancer metastasis by pan-DAC inhibitor panobinostat via inhibition of ZEB family of EMT master regulators. *Breast Cancer Res Treat* **145:** 593–604.

Richardson PG, Laubach JP, Lonial S, Moreau P, Yoon SS, Hungria VT, Dimopoulos MA, Beksac M, Alsina M, San-Miguel JF. 2015. Panobinostat: A novel pan-deacetylase inhibitor for the treatment of relapsed or relapsed and refractory multiple myeloma. *Expert Rev Anticancer Ther* **15:** 737–748.

Riley JS, Hutchinson R, McArt DG, Crawford N, Holohan C, Paul I, Van Schaeybroeck S, Salto-Tellez M, Johnston PG, Fennell DA, et al. 2013. Prognostic and therapeutic relevance of FLIP and procaspase-8 overexpression in non-small cell lung cancer. *Cell Death Dis* **4:** e951.

Ropero S, Fraga MF, Ballestar E, Hamelin R, Yamamoto H, Boix-Chornet M, Caballero R, Alaminos M, Setien F, Paz MF, et al. 2006. A truncating mutation of HDAC2 in human cancers confers resistance to histone deacetylase inhibition. *Nat Genet* **38:** 566–569.

Saijo K, Katoh T, Shimodaira H, Oda A, Takahashi O, Ishioka C. 2012. Romidepsin (FK228) and its analogs directly inhibit phosphatidylinositol 3-kinase activity and potently induce apoptosis as histone deacetylase/phosphatidylinositol 3-kinase dual inhibitors. *Cancer Sci* **103:** 1994–2001.

Saijo K, Imamura J, Narita K, Oda A, Shimodaira H, Katoh T, Ishioka C. 2015. Biochemical, biological and structural properties of romidepsin (FK228) and its analogs as novel HDAC/PI3K dual inhibitors. *Cancer Sci* **106:** 208–215.

Santo L, Hideshima T, Kung AL, Tseng JC, Tamang D, Yang M, Jarpe M, van Duzer JH, Mazitschek R, Ogier WC, et al. 2012. Preclinical activity, pharmacodynamic, and pharmacokinetic properties of a selective HDAC6 inhibitor, ACY-1215, in combination with bortezomib in multiple myeloma. *Blood* **119:** 2579–2589.

Santoro F, Botrugno OA, Dal Zuffo R, Pallavicini I, Matthews GM, Cluse L, Barozzi I, Senese S, Fornasari L, Moretti S, et al. 2013. A dual role for Hdac1: Oncosuppressor in tumorigenesis, oncogene in tumor maintenance. *Blood* **121:** 3459–3468.

Scholz C, Weinert BT, Wagner SA, Beli P, Miyake Y, Qi J, Jensen LJ, Streicher W, McCarthy AR, Westwood NJ, et al. 2015. Acetylation site specificities of lysine deacetylase inhibitors in human cells. *Nat Biotechnol* **33:** 415–423.

Schroeder FA, Lewis MC, Fass DM, Wagner FF, Zhang YL, Hennig KM, Gale J, Zhao WN, Reis S, Barker DD, et al. 2013. A selective HDAC 1/2 inhibitor modulates chromatin and gene expression in brain and alters mouse behavior in two mood-related tests. *PLoS ONE* **8:** e71323.

Schuler S, Fritsche P, Diersch S, Arlt A, Schmid RM, Saur D, Schneider G. 2010. HDAC2 attenuates TRAIL-induced apoptosis of pancreatic cancer cells. *Mol Cancer* **9:** 80.

Senese S, Zaragoza K, Minardi S, Muradore I, Ronzoni S, Passafaro A, Bernard L, Draetta GF, Alcalay M, Seiser C, et al. 2007. Role for histone deacetylase 1 in human tumor cell proliferation. *Mol Cell Biol* **27**: 4784–4795.

Seo HW, Kim EJ, Na H, Lee MO. 2009. Transcriptional activation of hypoxia-inducible factor-1α by HDAC4 and HDAC5 involves differential recruitment of p300 and FIH-1. *FEBS Lett* **583**: 55–60.

Seo J, Min SK, Park HR, Kim DH, Kwon MJ, Kim LS, Ju YS. 2014. Expression of histone deacetylases HDAC1, HDAC2, HDAC3, and HDAC6 in invasive ductal carcinomas of the breast. *J Breast Cancer* **17**: 323–331.

Seto E, Yoshida M. 2014. Erasers of histone acetylation: The histone deacetylase enzymes. *Cold Spring Harb Perspect Biol* **6**: a018713.

Shi W, Lawrence YR, Choy H, Werner-Wasik M, Andrews DW, Evans JJ, Judy KD, Farrell CJ, Moshel Y, Berger AC, et al. 2014. Vorinostat as a radiosensitizer for brain metastasis: A phase I clinical trial. *J Neurooncol* **118**: 313–319.

Shi Y, Dong M, Hong X, Zhang W, Feng J, Zhu J, Yu L, Ke X, Huang H, Shen Z, et al. 2015a. Results from a multicenter, open-label, pivotal phase II study of chidamide in relapsed or refractory peripheral T-cell lymphoma. *Ann Oncol* **26**: 1766–1771.

Shi Y, Huang J, Zhou J, Liu Y, Fu X, Li Y, Yin G, Wen J. 2015b. MicroRNA-204 inhibits proliferation, migration, invasion and epithelial-mesenchymal transition in osteosarcoma cells via targeting Sirtuin 1. *Oncol Rep* **34**: 399–406.

Shimizu T, LoRusso PM, Papadopoulos KP, Patnaik A, Beeram M, Smith LS, Rasco DW, Mays TA, Chambers G, Ma A, et al. 2015. Phase I first-in-human study of CUDC-101, a multitargeted inhibitor of HDACs, EGFR, and HER2 in patients with advanced solid tumors. *Clin Cancer Res* **20**: 5032–5040.

Shukla S, Khan S, Kumar S, Sinha S, Farhan M, Bora HK, Maurya R, Meeran SM. 2015. Cucurbitacin B alters the expression of tumor-related genes by epigenetic modifications in NSCLC and inhibits NNK-induced lung tumorigenesis. *Cancer Prev Res (Phila)* **8**: 552–562.

Simic P, Williams EO, Bell EL, Gong JJ, Bonkowski M, Guarente L. 2013. SIRT1 suppresses the epithelial-to-mesenchymal transition in cancer metastasis and organ fibrosis. *Cell Rep* **3**: 1175–1186.

Sodji QH, Kornacki JR, McDonald JF, Mrksich M, Oyelere AK. 2015. Design and structure activity relationship of tumor-homing histone deacetylase inhibitors conjugated to folic and pteroic acids. *Eur J Med Chem* **96**: 340–359.

Son CH, Keum JH, Yang K, Nam J, Kim MJ, Kim SH, Kang CD, Oh SO, Kim CD, Park YS, et al. 2014. Synergistic enhancement of NK cell-mediated cytotoxicity by combination of histone deacetylase inhibitor and ionizing radiation. *Radiat Oncol* **9**: 49.

Song C, Zhu S, Wu C, Kang J. 2013. Histone deacetylase (HDAC) 10 suppresses cervical cancer metastasis through inhibition of matrix metalloproteinase (MMP) 2 and 9 expression. *J Biol Chem* **288**: 28021–28033.

Sonnemann J, Marx C, Becker S, Wittig S, Palani CD, Kramer OH, Beck JF. 2014. p53–dependent and p53-independent anticancer effects of different histone deacetylase inhibitors. *Br J Cancer* **110**: 656–667.

Stark M, Hayward N. 2007. Genome-wide loss of heterozygosity and copy number analysis in melanoma using high-density single-nucleotide polymorphism arrays. *Cancer Res* **67**: 2632–2642.

Stronach EA, Alfraidi A, Rama N, Datler C, Studd JB, Agarwal R, Guney TG, Gourley C, Hennessy BT, Mills GB, et al. 2011. HDAC4-regulated STAT1 activation mediates platinum resistance in ovarian cancer. *Cancer Res* **71**: 4412–4422.

Stypula-Cyrus Y, Damania D, Kunte DP, Cruz MD, Subramanian H, Roy HK, Backman V. 2013. HDAC up-regulation in early colon field carcinogenesis is involved in cell tumorigenicity through regulation of chromatin structure. *PLoS ONE* **8**: e64600.

Sudo T, Mimori K, Nishida N, Kogo R, Iwaya T, Tanaka F, Shibata K, Fujita H, Shirouzu K, Mori M. 2011. Histone deacetylase 1 expression in gastric cancer. *Oncol Rep* **26**: 777–782.

Sun L, Kokura K, Izumi V, Koomen JM, Seto E, Chen J, Fang J. 2015. MPP8 and SIRT1 crosstalk in E-cadherin gene silencing and epithelial-mesenchymal transition. *EMBO Rep* **16**: 689–699.

Suzuki T, Ota Y, Ri M, Bando M, Gotoh A, Itoh Y, Tsumoto H, Tatum PR, Mizukami T, Nakagawa H, et al. 2012. Rapid discovery of highly potent and selective inhibitors of histone deacetylase 8 using click chemistry to generate candidate libraries. *J Med Chem* **55**: 9562–9575.

Suzuki T, Kasuya Y, Itoh Y, Ota Y, Zhan P, Asamitsu K, Nakagawa H, Okamoto T, Miyata N. 2013. Identification of highly selective and potent histone deacetylase 3 inhibitors using click chemistry-based combinatorial fragment assembly. *PLoS ONE* **8**: e68669.

Suzuki T, Muto N, Bando M, Itoh Y, Masaki A, Ri M, Ota Y, Nakagawa H, Iida S, Shirahige K, et al. 2014. Design, synthesis, and biological activity of NCC149 derivatives as histone deacetylase 8-selective inhibitors. *ChemMedChem* **9**: 657–664.

Takasaka N, Araya J, Hara H, Ito S, Kobayashi K, Kurita Y, Wakui H, Yoshii Y, Yumino Y, Fujii S, et al. 2014. Autophagy induction by SIRT6 through attenuation of insulin-like growth factor signaling is involved in the regulation of human bronchial epithelial cell senescence. *J Immunol* **192**: 958–968.

Tang C, Li C, Zhang S, Hu Z, Wu J, Dong C, Huang J, Zhou HB. 2015. Novel bioactive hybrid compound dual targeting estrogen receptor and histone deacetylase for the treatment of breast cancer. *J Med Chem* **58**: 4550–4572.

Taylor BS, DeCarolis PL, Angeles CV, Brenet F, Schultz N, Antonescu CR, Scandura JM, Sander C, Viale AJ, Socci ND, et al. 2011. Frequent alterations and epigenetic silencing of differentiation pathway genes in structurally rearranged liposarcomas. *Cancer Discov* **1**: 587–597.

Thurn KT, Thomas S, Raha P, Qureshi I, Munster PN. 2013. Histone deacetylase regulation of ATM-mediated DNA damage signaling. *Mol Cancer Ther* **12**: 2078–2087.

Toiber D, Erdel F, Bouazoune K, Silberman DM, Zhong L, Mulligan P, Sebastian C, Cosentino C, Martinez-Pastor B, Giacosa S, et al. 2013. SIRT6 recruits SNF2H to DNA break sites, preventing genomic instability through chromatin remodeling. *Mol Cell* **51**: 454–468.

Tomosugi M, Sowa Y, Yasuda S, Tanaka R, te Riele H, Ikawa H, Koyama M, Sakai T. 2012. Retinoblastoma gene-independent G$_1$ phase arrest by flavone, phosphatidylinositol

3-kinase inhibitor, and histone deacetylase inhibitor. *Cancer Sci* **103**: 2139–2143.

Tong ZT, Cai MY, Wang XG, Kong LL, Mai SJ, Liu YH, Zhang HB, Liao YJ, Zheng F, Zhu W, et al. 2012. EZH2 supports nasopharyngeal carcinoma cell aggressiveness by forming a co-repressor complex with HDAC1/HDAC2 and Snail to inhibit E-cadherin. *Oncogene* **31**: 583–594.

Tseng AH, Shieh SS, Wang DL. 2013. SIRT3 deacetylates FOXO3 to protect mitochondria against oxidative damage. *Free Radic Biol Med* **63**: 222–234.

Turtoi A, Mottet D, Matheus N, Dumont B, Peixoto P, Hennequiere V, Deroanne C, Colige A, De Pauw E, Bellahcene A, et al. 2012. The angiogenesis suppressor gene AKAP12 is under the epigenetic control of HDAC7 in endothelial cells. *Angiogenesis* **15**: 543–554.

Ungerstedt JS, Sowa Y, Xu WS, Shao Y, Dokmanovic M, Perez G, Ngo L, Holmgren A, Jiang X, Marks PA. 2005. Role of thioredoxin in the response of normal and transformed cells to histone deacetylase inhibitors. *Proc Natl Acad Sci* **102**: 673–678.

Urbich C, Rossig L, Kaluza D, Potente M, Boeckel JN, Knau A, Diehl F, Geng JG, Hofmann WK, Zeiher AM, et al. 2009. HDAC5 is a repressor of angiogenesis and determines the angiogenic gene expression pattern of endothelial cells. *Blood* **113**: 5669–5679.

Van Damme M, Crompot E, Meuleman N, Mineur P, Bron D, Lagneaux L, Stamatopoulos B. 2012. HDAC isoenzyme expression is deregulated in chronic lymphocytic leukemia B-cells and has a complex prognostic significance. *Epigenetics* **7**: 1403–1412.

Vasilatos SN, Katz TA, Oesterreich S, Wan Y, Davidson NE, Huang Y. 2013. Crosstalk between lysine-specific demethylase 1 (LSD1) and histone deacetylases mediates antineoplastic efficacy of HDAC inhibitors in human breast cancer cells. *Carcinogenesis* **34**: 1196–1207.

Vaziri H, Dessain SK, Ng Eaton E, Imai SI, Frye RA, Pandita TK, Guarente L, Weinberg RA. 2001. hSIR2(SIRT1) functions as an NAD-dependent p53 deacetylase. *Cell* **107**: 149–159.

Venugopal B, Baird R, Kristeleit RS, Plummer R, Cowan R, Stewart A, Fourneau N, Hellemans P, Elsayed Y, McClue S, et al. 2013. A phase I study of quisinostat (JNJ-26481585), an oral hydroxamate histone deacetylase inhibitor with evidence of target modulation and antitumor activity, in patients with advanced solid tumors. *Clin Cancer Res* **19**: 4262–4272.

Venza I, Visalli M, Oteri R, Cucinotta M, Teti D, Venza M. 2013. Class II-specific histone deacetylase inhibitors MC1568 and MC1575 suppress IL-8 expression in human melanoma cells. *Pigment Cell Melanoma Res* **26**: 193–204.

Vidal-Laliena M, Gallastegui E, Mateo F, Martinez-Balbas M, Pujol MJ, Bachs O. 2013. Histone deacetylase 3 regulates cyclin A stability. *J Biol Chem* **288**: 21096–21104.

von Burstin J, Eser S, Paul MC, Seidler B, Brandl M, Messer M, von Werder A, Schmidt A, Mages J, Pagel P, et al. 2009. E-cadherin regulates metastasis of pancreatic cancer in vivo and is suppressed by a SNAIL/HDAC1/HDAC2 repressor complex. *Gastroenterology* **137**: 361–371.

Wang RH, Sengupta K, Li C, Kim HS, Cao L, Xiao C, Kim S, Xu X, Zheng Y, Chilton B, et al. 2008. Impaired DNA damage response, genome instability, and tumorigenesis in SIRT1 mutant mice. *Cancer Cell* **14**: 312–323.

Wang JC, Kafeel MI, Avezbakiyev B, Chen C, Sun Y, Rathnasabapathy C, Kalavar M, He Z, Burton J, Lichter S. 2011. Histone deacetylase in chronic lymphocytic leukemia. *Oncology* **81**: 325–329.

Wang G, He J, Zhao J, Yun W, Xie C, Taub JW, Azmi A, Mohammad RM, Dong Y, Kong W, et al. 2012. Class I and class II histone deacetylases are potential therapeutic targets for treating pancreatic cancer. *PLoS ONE* **7**: e52095.

Wang RH, Lahusen TJ, Chen Q, Xu X, Jenkins LM, Leo E, Fu H, Aladjem M, Pommier Y, Appella E, et al. 2014. SIRT1 deacetylates TopBP1 and modulates intra-S-phase checkpoint and DNA replication origin firing. *Int J Biol Sci* **10**: 1193–1202.

Wang L, Chen G, Chen K, Ren Y, Li H, Jiang X, Jia L, Fu S, Li Y, Liu X, et al. 2015. Dual targeting of retinoid X receptor and histone deacetylase with DW22 as a novel antitumor approach. *Oncotarget* **6**: 9740–9755.

Webster BR, Scott I, Han K, Li JH, Lu Z, Stevens MV, Malide D, Chen Y, Samsel L, Connelly PS, et al. 2013. Restricted mitochondrial protein acetylation initiates mitochondrial autophagy. *J Cell Sci* **126**: 4843–4849.

Weichert W, Denkert C, Noske A, Darb-Esfahani S, Dietel M, Kalloger SE, Huntsman DG, Kobel M. 2008a. Expression of class I histone deacetylases indicates poor prognosis in endometrioid subtypes of ovarian and endometrial carcinomas. *Neoplasia* **10**: 1021–1027.

Weichert W, Roske A, Gekeler V, Beckers T, Ebert MP, Pross M, Dietel M, Denkert C, Rocken C. 2008b. Association of patterns of class I histone deacetylase expression with patient prognosis in gastric cancer: A retrospective analysis. *Lancet Oncol* **9**: 139–148.

Weichert W, Roske A, Gekeler V, Beckers T, Stephan C, Jung K, Fritzsche FR, Niesporek S, Denkert C, Dietel M, et al. 2008c. Histone deacetylases 1, 2 and 3 are highly expressed in prostate cancer and HDAC2 expression is associated with shorter PSA relapse time after radical prostatectomy. *Br J Cancer* **98**: 604–610.

Weichert W, Roske A, Niesporek S, Noske A, Buckendahl AC, Dietel M, Gekeler V, Boehm M, Beckers T, Denkert C. 2008d. Class I histone deacetylase expression has independent prognostic impact in human colorectal cancer: Specific role of class I histone deacetylases in vitro and in vivo. *Clin Cancer Res* **14**: 1669–1677.

Wells CE, Bhaskara S, Stengel KR, Zhao Y, Sirbu B, Chagot B, Cortez D, Khabele D, Chazin WJ, Cooper A, et al. 2013. Inhibition of histone deacetylase 3 causes replication stress in cutaneous T cell lymphoma. *PLoS ONE* **8**: e68915.

West AC, Johnstone RW. 2014. New and emerging HDAC inhibitors for cancer treatment. *J Clin Invest* **124**: 30–39.

West AC, Mattarollo SR, Shortt J, Cluse LA, Christiansen AJ, Smyth MJ, Johnstone RW. 2013. An intact immune system is required for the anticancer activities of histone deacetylase inhibitors. *Cancer Res* **73**: 7265–7276.

West AC, Smyth MJ, Johnstone RW. 2014. The anticancer effects of HDAC inhibitors require the immune system. *Oncoimmunology* **3**: e27414.

Wilmott JS, Colebatch AJ, Kakavand H, Shang P, Carlino MS, Thompson JF, Long GV, Scolyer RA, Hersey P. 2015.

Expression of the class 1 histone deacetylases HDAC8 and 3 are associated with improved survival of patients with metastatic melanoma. *Mod Pathol* **28:** 884–894.

Xiao W, Chen X, Liu X, Luo L, Ye S, Liu Y. 2014. Trichostatin A, a histone deacetylase inhibitor, suppresses proliferation and epithelial-mesenchymal transition in retinal pigment epithelium cells. *J Cell Mol Med* **18:** 646–655.

Xie HJ, Noh JH, Kim JK, Jung KH, Eun JW, Bae HJ, Kim MG, Chang YG, Lee JY, Park H, et al. 2012. HDAC1 inactivation induces mitotic defect and caspase-independent autophagic cell death in liver cancer. *PLoS ONE* **7:** e34265.

Xue K, Gu JJ, Zhang Q, Mavis C, Hernandez-Ilizaliturri FJ, Czuczman MS, Guo Y. 2015. Vorinostat, a histone deacetylase (HDAC) inhibitor, promotes cell cycle arrest and re-sensitizes rituximab- and chemo-resistant lymphoma cells to chemotherapy agents. *J Cancer Res Clin Oncol.*

Yamaguchi T, Cubizolles F, Zhang Y, Reichert N, Kohler H, Seiser C, Matthias P. 2010. Histone deacetylases 1 and 2 act in concert to promote the G_1-to-S progression. *Genes Dev* **24:** 455–469.

Yang FC, Tan BC, Chen WH, Lin YH, Huang JY, Chang HY, Sun HY, Hsu PH, Liou GG, Shen J, et al. 2013. Reversible acetylation regulates salt-inducible kinase (SIK2) and its function in autophagy. *J Biol Chem* **288:** 6227–6237.

Yoo YG, Kong G, Lee MO. 2006. Metastasis-associated protein 1 enhances stability of hypoxia-inducible factor-1α protein by recruiting histone deacetylase 1. *EMBO J* **25:** 1231–1241.

Younes A, Oki Y, Bociek RG, Kuruvilla J, Fanale M, Neelapu S, Copeland A, Buglio D, Galal A, Besterman J, et al. 2011. Mocetinostat for relapsed classical Hodgkin's lymphoma: An open-label, single-arm, phase 2 trial. *Lancet Oncol* **12:** 1222–1228.

Zhang J, Zhong Q. 2014. Histone deacetylase inhibitors and cell death. *Cell Mol Life Sci* **71:** 3885–3901.

Zhang Z, Yamashita H, Toyama T, Sugiura H, Omoto Y, Ando Y, Mita K, Hamaguchi M, Hayashi S, Iwase H. 2004. HDAC6 expression is correlated with better survival in breast cancer. *Clin Cancer Res* **10:** 6962–6968.

Zhang Z, Yamashita H, Toyama T, Sugiura H, Ando Y, Mita K, Hamaguchi M, Hara Y, Kobayashi S, Iwase H. 2005. Quantitation of HDAC1 mRNA expression in invasive carcinoma of the breast. *Breast Cancer Res Treat* **94:** 11–16.

Zhang Y, Adachi M, Kawamura R, Imai K. 2006. Bmf is a possible mediator in histone deacetylase inhibitors FK228 and CBHA-induced apoptosis. *Cell Death Differ* **13:** 129–140.

Zhang X, Su M, Chen Y, Li J, Lu W. 2013. The design and synthesis of a new class of RTK/HDAC dual-targeted inhibitors. *Molecules* **18:** 6491–6503.

Zhang M, Xiang S, Joo HY, Wang L, Williams KA, Liu W, Hu C, Tong D, Haakenson J, Wang C, et al. 2014. HDAC6 deacetylates and ubiquitinates MSH2 to maintain proper levels of MutSα. *Mol Cell* **55:** 31–46.

Zhang L, Zhang Y, Mehta A, Boufraqech M, Davis S, Wang J, Tian Z, Yu Z, Boxer MB, Kiefer JA, et al. 2015a. Dual inhibition of HDAC and EGFR signaling with CUDC-101 induces potent suppression of tumor growth and metastasis in anaplastic thyroid cancer. *Oncotarget* **6:** 9073–9085.

Zhang X, Kong Y, Zhang J, Su M, Zhou Y, Zang Y, Li J, Chen Y, Fang Y, Zhang X, et al. 2015b. Design, synthesis and biological evaluation of colchicine derivatives as novel tubulin and histone deacetylase dual inhibitors. *Eur J Med Chem* **95:** 127–135.

Zhao J, Lawless MW. 2015. Resminostat: Opening the door on epigenetic treatments for liver cancer. *Hepatology* **63:** 668–669.

Zhi X, Zhong Q. 2015. Autophagy in cancer. *F1000Prime Rep* **7:** 18.

Zhou Y, Tolstov Y, Arslan A, Roth W, Grullich C, Pahernik S, Hohenfellner M, Duensing S. 2014. Harnessing the p53–PUMA axis to overcome DNA damage resistance in renal cell carcinoma. *Neoplasia* **16:** 1028–1035.

Zhou ZW, Li XX, He ZX, Pan ST, Yang Y, Zhang X, Chow K, Yang T, Qiu JX, Zhou Q, et al. 2015. Induction of apoptosis and autophagy via sirtuin1- and PI3K/Akt/mTOR-mediated pathways by plumbagin in human prostate cancer cells. *Drug Des Devel Ther* **9:** 1511–1554.

Zibelman M, Wong YN, Devarajan K, Malizzia L, Corrigan A, Olszanski AJ, Denlinger CS, Roethke SK, Tetzlaff CH, Plimack ER. 2015. Phase I study of the mTOR inhibitor ridaforolimus and the HDAC inhibitor vorinostat in advanced renal cell carcinoma and other solid tumors. *Invest New Drugs* **33:** 1040–1047.

Zorzi AP, Bernstein M, Samson Y, Wall DA, Desai S, Nicksy D, Wainman N, Eisenhauer E, Baruchel S. 2013. A phase I study of histone deacetylase inhibitor, pracinostat (SB939), in pediatric patients with refractory solid tumors: IND203 a trial of the NCIC IND program/C17 pediatric phase I consortium. *Pediatr Blood Cancer* **60:** 1868–1874.

Zupkovitz G, Grausenburger R, Brunmeir R, Senese S, Tischler J, Jurkin J, Rembold M, Meunier D, Egger G, Lagger S, et al. 2010. The cyclin-dependent kinase inhibitor p21 is a crucial target for histone deacetylase 1 as a regulator of cellular proliferation. *Mol Cell Biol* **30:** 1171–1181.

Zwergel C, Valente S, Jacob C, Mai A. 2015. Emerging approaches for histone deacetylase inhibitor drug discovery. *Expert Opin Drug Discov* **10:** 599–613.

DNA Hypomethylating Drugs in Cancer Therapy

Takahiro Sato,[1] Jean-Pierre J. Issa,[1,2] and Patricia Kropf[2]

[1]Fels Institute for Cancer Research and Molecular Biology, Temple University School of Medicine, Philadelphia, Pennsylvania 19140

[2]Fox Chase Cancer Center, Temple Health, Philadelphia, Pennsylvania 19111

Correspondence: jpissa@temple.edu

Aberrant DNA methylation is a critically important modification in cancer cells, which, through promoter and enhancer DNA methylation changes, use this mechanism to activate oncogenes and silence of tumor-suppressor genes. Targeting DNA methylation in cancer using DNA hypomethylating drugs reprograms tumor cells to a more normal-like state by affecting multiple pathways, and also sensitizes these cells to chemotherapy and immunotherapy. The first generation hypomethylating drugs azacitidine and decitabine are routinely used for the treatment of myeloid leukemias and a next-generation drug (guadecitabine) is currently in clinical trials. This review will summarize preclinical and clinical data on DNA hypomethylating drugs as a cancer therapy.

DNA methylation occurs through covalent addition of a methyl group by DNA methyltransferases (DNMTs) to the C5 position of cytosine to form 5-methylcytosine (5mC). There are three members of DNMTs that have catalytic activity for DNA methylation— DNMT1, DNMT3A, and DNMT3B. DNMT1 is a maintenance methyltransferase that methylates preexisting hemimethylated DNA, whereas DNMT3A and 3B are de novo methyltransferases that establish methylation on unmethylated DNA (Baylin and Jones 2011). Other related members include DNMT3L, which lacks a catalytic domain and can modulate the activity of other DNMTs (Wienholz et al. 2010), and DNMT2, which has activity as a RNA methyltransferase (Goll et al. 2006). Demethylation of 5mC occurs through a reaction mediated by the three members of the ten-eleven trans location (TET) family, TET1, TET2, and TET3. TETs catalyze the conversion of 5mC to 5-hydroxymethylcytosine (5hmC), and 5hmC can be converted back to unmethylated cytosines either through active demethylation by the base excision repair pathway or passive demethylation by loss of 5hmC during cell division (Williams et al. 2012; Jin et al. 2015). DNA methylation can be recognized and bound by the methyl-binding proteins MBD1, MBD2, MBD4, and MeCP2. These MBDs form complexes with other epigenetic enzymes such as histone deacetylases (HDAC) and histone methyltransferases that catalyze the addition of histone modifications, which lead to compaction of the chromatin and silencing of gene expression (Parry and Clarke 2011).

The majority of DNA methylation occurs at cytosines that are followed by guanines (CpG sites), and there are certain regions in the genome where there is a high density of CpG sites that are termed CpG islands (CGIs). Fifty percent of gene promoters contain CGIs, and methylation of these CGIs has been associated with repression of gene expression (Taby and Issa 2010). In normal tissues, CpG sites throughout the genome are usually methylated, whereas promoter CGIs are usually unmethylated, with exceptions being the inactive X-chromosome, the silenced alleles of imprinted genes, and tissue-specific genes (Shen et al. 2007a; Smith and Meissner 2013).

CANCER AND DNA METHYLATION

Although DNA methylation in normal adult tissues is relatively stable, extensive DNA methylation changes occur in cancer that lead to global hypomethylation of the genome along with focal hypermethylation at promoter CGIs (Jones and Baylin 2007; Taby and Issa 2010; Jelinek et al. 2012). Five to ten percent of promoter CGIs become hypermethylated in most cancers, leading to silencing of many critical tumor-suppressor genes such as VHL, RB1, CDKN2A, MGMT, GATA4, and MLH1 (Taby and Issa 2010; Baylin and Jones 2011). In any given cancer, the number of genes showing promoter DNA methylation-associated gene silencing in cancer is substantially higher than those genes inactivated by genetic mutations, indicating that the majority of tumor-suppressor gene silencing in cancer is through epigenetic mechanisms (Plass et al. 2013). Global hypomethylation of the genome usually occurs in intergenic areas and does not have large effects on gene expression, but, in some cases, profound hypomethylation can lead to genomic instability through increased frequency of mutations, deletions, amplifications, inversions, and translocations (Chen et al. 1998).

One of the most well-known examples of the effect of DNA methylation on cancer occurs in patients categorized as having the CGI methylator phenotype (CIMP). CIMP was originally identified in colon cancer and indicates an abundance of cancer-specific hypermethylated promoter CGIs (Toyota et al. 1999). In colon cancer, CIMP is tightly correlated with BRAF mutations (Weisenberger et al. 2006; Shen et al. 2007b; Yagi et al. 2010; Hinoue et al. 2012; TCGA 2012), whereas in glioblastoma, CIMP tumors are associated with a high rate of IDH1 mutations and a better outcome (Noushmehr et al. 2010; Brennan et al. 2013). Further supporting the critical role of DNA methylation in cancer development, mutations in enzymes that regulate DNA methylation frequently occur in many hematological malignancies. DNMT3A has been found to be one of the most frequently mutated genes in hematological cancers (Yang et al. 2015). Loss-of-function mutations of TET2 or gain-of-function mutations of IDH1 or IDH2 frequently occur in hematological malignancies and are mutually exclusive (Abdel-Wahab et al. 2009; Delhommeau et al. 2009; Langemeijer et al. 2009; Gaidzik et al. 2012; Plass et al. 2013). IDH usually converts isocitrate to α-ketoglutarate, but mutant IDH will convert α-ketoglutarate to 2-hydroxyglutarate (2-HG) (Cohen et al. 2013), and 2-HG is an inhibitor of the TET proteins that mediates DNA demethylation (Xu et al. 2011). Patients with mutations in TET2 or IDH have specific hypermethylation signatures in both acute myelogenous leukemia (AML) (Figueroa et al. 2010; TCGA 2013; Yamazaki et al. 2015, 2016) and glioblastoma (Noushmehr et al. 2010; Turcan et al. 2012; Brennan et al. 2013).

DNA HYPOMETHYLATING DRUGS IN THE CLINIC

Given the reversibility of epigenetic modifications and the substantial DNA methylation changes that occur in cancer, it was hypothesized that DNA hypermethylation at promoter CGIs could be reversed to reexpress silenced genes and reprogram cancer cells to a more normal-like state.

This led to the pursuit of DNMT inhibitors for the treatment of cancer, and two DNMT inhibitors have had significant success in the clinic. They are the nucleoside analogs 5-aza-2′-deoxycytidine (decitabine) and 5-azacitidine (azacitidine) (Table 1). These inhibitors incorporate into newly synthesized DNA where

Table 1. Hypomethylating drugs

Chemical name	Generic name	Mechanism	Drug status
5-Azacytidine	Azacitidine	Cytosine analog	FDA approved for treatment of MDS
5-Aza-2′-deoxycytidine	Decitabine	Cytosine analog	FDA approved for treatment of MDS
SGI-110	Guadecitabine	Cytosine analog	Phase III clinical trial in AML
5-Fluro-2′-deoxycytidine	FdCyd	Cytosine analog	Phase II clinical trial in refractory solid tumors
Zebularine	–	Cytosine analog	Preclinical
CP-4200	–	Cytosine analog	Preclinical
RG108	–	Small molecule inhibitor	Preclinical
Nanaomycin A	–	Small molecule inhibitor	Preclinical

MDS, Myelodysplastic syndrome; AML, acute myelogenous leukemia.

they form a covalent bond with DNMTs, leading to the degradation of these DNMTs and hypomethylation of the genome through passive demethylation as the cells replicate and DNA methylation is not maintained (Issa and Kantarjian 2009). The main difference between these drugs is that decitabine incorporates into DNA, whereas azacitidine can incorporate into both DNA and RNA.

Decitabine and azacitidine were originally developed as cytotoxic anticancer agents in the 1960s, where they were used at high doses without clinical success (Taby and Issa 2010). However, further studies with these drugs led to the observation that they could cause differentiation of cells by inhibiting methylation of DNA (Jones and Taylor 1980). Importantly, it was discovered that hypomethylation after decitabine or azacitidine treatment did not occur at very high doses like the ones used previously in the clinic (Issa and Kantarjian 2009). These effects were attributed to the fact that high doses of decitabine and azacitidine would inhibit cell proliferation and DNA synthesis, and the incorporation of these drugs into DNA as well as their passive DNA demethylation effect were dependent on cell replication (Fig. 1) (Qin et al. 2007). Therefore, it was hypothesized that giving lower doses of the drug would lead to more effective DNA demethylation and improved clinical results (Issa and Kantarjian 2009). This hypothesis was supported by early phase clinical trials, which showed that repeated exposure with low doses of the inhibitors led to DNA demethylation and better responses than using the drugs at high doses (Wijermans et al. 2000; Issa et al. 2004).

These results eventually led to a phase III clinical trial of azacitidine in treating myelodysplastic syndrome (MDS) (Silverman et al. 2002). Patients that received azacitidine had an increased response rate (60% vs. 5%) and delayed progression to leukemia (21 mo vs. 13 mo) when compared with patients receiving supportive care (Silverman et al. 2002). Following these results, the FDA approved the use of azacitidine in 2004 for the treatment of patients with MDS (Taby and Issa 2010). A follow-up international phase III clinical trial confirmed the efficacy of azacitidine in treating MDS, as patients treated with azacitidine had higher median overall survival (24.5 mo) compared with patients receiving conventional care (15 mo) (Fenaux et al. 2009).

Meanwhile, a phase III clinical trial of decitabine in treating MDS resulted in higher response rate (17% vs. 0%) and longer median time to leukemia or death (12.1 mo vs. 7.8 mo) in patients treated with decitabine compared with best supportive care (Kantarjian et al. 2006). These results led to the 2006 FDA approval of decitabine in the treatment of patients with MDS (Taby and Issa 2010). Follow-up studies that optimized the dosing schedule showed a very high response rate (>70% objective response) of MDS patients treated with decitabine and increased survival compared with patients treated with chemotherapy (22 mo vs. 12 mo) (Kantarjian et al. 2007a,b). A more recent phase III trial performed in elderly patients with MDS that are ineligible for intensive chemotherapy showed that patients treated with decitabine had increased progression-free sur-

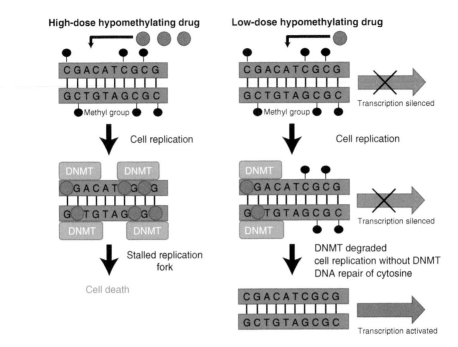

Figure 1. Comparison of the effects of high dose and low dose of hypomethylating drugs on cancer cells. When cells are treated with high doses of hypomethylating drugs, the drug gets incorporated in the DNA in a cell-replication-dependent manner, then binds and traps DNA methyltransferases (DNMTs). This causes the formation of bulky adducts, leading to a stalled replication fork and inhibition of DNA replication, which causes cell death. When cells are treated with low doses of hypomethylating drugs, the drugs still incorporate into DNA and bind the DNMTs, but the DNMTs end up being degraded. Without DNMTs to maintain DNA methylation, CpG sites lose their methylation after cell replication, and transcription of genes silenced by promoter methylation is restored.

vival (6.6 mo vs. 3 mo) and increased overall response rate (34% vs. 2%) when compared with patients receiving the best supportive care (Lübbert et al. 2011).

DNMT inhibitors have also shown effectiveness in treating patients with other leukemias. A phase II study in patients with chronic myelogenous leukemia (CML) showed complete response in 34% and partial response in 20% of patients treated with decitabine (Issa et al. 2005). Activity was also seen in AML, as a multicenter phase III trial performed on AML patients treated with decitabine showed an increase in overall survival (7.7 mo vs. 5.0 mo) and CR rate (17.8% vs. 7.8%) when compared with treatment choice (Kantarjian et al. 2012). These results led to the approval of decitabine in AML in Europe.

The effect of decitabine and azacitidine has a slower onset than traditional cytotoxic thera-pies, which usually require only one cycle of treatment to achieve a complete remission. For example, one study reported that although median time to first response to azacitidine was two cycles, continued treatment improved response in 48% of patients (Silverman et al. 2011). Another study reported that patients that show no evidence of response to decitabine after 3 mo can end up achieving complete remission with continued treatment (Oki et al. 2008). The delayed onset of response is consistent with the mechanism of action of decitabine and azacitidine, as the hypomethylation that occurs in response to inhibition of DNMTs is dependent on cell replication.

Treating solid tumors with DNMT inhibitors remains a challenge because solid cancers tend to have lower drug penetrance in larger tumors and often proliferate at slower rate than hematological cancers, which is problematic for DNMT inhibitors that are known to be S-phase-dependent and unstable in solution (Issa and Kantarjian 2009). However, there is evidence that DNMT inhibitors can be effective in inhibiting the growth of solid tumors in vitro (Bender et al. 1998; Qin et al. 2009; Tsai et al. 2012). Recent clinical trials of decitabine in solid tumors (Stewart et al. 2009) and azacitidine in combination with an HDAC inhibitor on non-small-cell lung cancer patients showed some objective responses (Juergens et al. 2011). There are ongoing clinical trials to determine whether using hypomethylating drugs as a monotherapy or in combination with other agents can be an effective therapy for solid tumors.

Collectively, these data indicate that hypomethylating drugs are an important therapy that acts through epigenetic mechanisms to target cancer instead of inducing general toxicity as seen with other chemotherapies. Although the mechanism of action of decitabine and azacitidine are proposed to be very similar, it is yet to be determined whether these two drugs will lead to different clinical outcomes because they have never been directly compared. There are some indications that the two drugs can have differing effects, as one study showed that some MDS patients resistant to azacitidine could be treated with decitabine and achieve objective responses (Borthakur et al. 2008). Regardless, both drugs remain important agents in the clinical treatment of MDS and leukemia, and quite possibly for solid tumors in the future.

RESISTANCE TO DNA HYPOMETHYLATING DRUGS

Despite the initial success of DNMT inhibitors in the treatment of MDS, lack of initial response to therapy (primary resistance), or acquired resistance after robust initial responses (secondary resistance) remains a major obstacle (Issa and Kantarjian 2009). Only ∼50% of MDS or AML patients treated with hypomethylating drugs achieve a clinical response (Treppendahl et al. 2014). Therefore, efforts have been made to find predictive biomarkers to DNMT inhibitors, but there has been limited correlation ob-

served with biomarkers such as DNA methylation before therapy or hypomethylation of tumor-suppressor genes after therapy (Oki et al. 2007; Fandy et al. 2009). Because all cancer patients have methylation defects, the key to correlating DNA methylation with response may be to find the right set of methylated gene promoters to use as biomarkers or to study methylated sequences outside the promoter region that may control gene expression, such as enhancers. Interestingly, a recent study found responders and nonresponders to decitabine could be distinguished in 40 chronic myelomonocytic leukemia (CMML) patients by using 167 differentially methylated regions that were primarily located in distal intergenic regions and enhancers (Meldi et al. 2015). It will be important to determine whether this observation can be replicated in a larger set of patients and in other leukemias. On the other hand, measuring hypomethylation of genes after therapy may not be the best approach to use as biomarkers because cells that are most sensitive to therapy and become hypomethylated may be the first to die off, leaving behind only the cells resistant to hypomethylation (Qin et al. 2007).

Mutations in proteins involved in regulation of DNA methylation have been tested as biomarkers for response to hypomethylating drugs, but the results have been mixed (Treppendahl et al. 2014). For example, two studies have found that MDS or AML patients with TET2 mutations were more likely to respond to hypomethylating drugs (Itzykson et al. 2011; Bejar et al. 2014), but two other studies have found limited correlation between TET2 mutations and response in MDS and CMML patients (Braun et al. 2011; Voso et al. 2011). Another recent study on patients with MDS and related disorders found a correlation between response to azacitidine or decitabine with TET2 or DNMT3A mutations (Traina et al. 2014). Because of to these conflicting reports, more studies are needed to establish whether TET2 or DNMT3A mutations can serve as a biomarker for hypomethylating drugs.

One cause of primary resistance to hypomethylating drugs was discovered when it was observed that cell lines that were resistant to decitabine in vitro had low deoxycytidine kinase (DCK), which activates decitabine through phosphorylation, and high cytosine deaminase (CDA), which inactivates decitabine through deamination (Qin et al. 2009). When resistance to decitabine was investigated in MDS patients, it was found that patients with primary resistance had a threefold higher CDA/DCK ratio (Qin et al. 2011). Another factor for the primary resistance may be the instability of DNMT inhibitors, because decitabine and azacitidine are rapidly cleared from the body and have a half-life less than 1 h (Derissen et al. 2013). In terms of secondary resistance, it was observed that a stable clone that initially responded to decitabine but gained secondary resistance had mutations in DCK (Qin et al. 2009). However, this mutation was not observed in any of the MDS patients that were treated with decitabine and developed secondary resistance (Qin et al. 2011).

EXPERIMENTAL STUDIES ON THE EFFECT OF DNA HYPOMETHYLATING DRUGS

There has been major progress made in experimental settings to better understand the effects of DNMT inhibitors. Studies using 27K or 450K methylation arrays (Tsai et al. 2012; Klco et al. 2013; Pandiyan et al. 2013), as well as with whole genome bisulfite sequencing (Lund et al. 2014) have shown that DNA demethylating agents cause global hypomethylation throughout the entire genome. However, these studies have also shown that only a small percentage of genes that undergo promoter hypomethylation actually become reactivated (Tsai et al. 2012; Klco et al. 2013; Pandiyan et al. 2013; Lund et al. 2014). This is most likely because other factors such as expression of the right transcription factor, changes in histone acetylation, changes in histone methylation, or chromatin remodeling are required on top of DNA demethylation for reactivation of gene expression, giving DNMT inhibitors a degree of specificity (Kondo et al. 2004; Paul et al. 2010; Si et al. 2010; Pandiyan et al. 2013). In fact, genes that gained chromatin accessibility after decitabine treatment were more likely to be genes that are expressed in normal tissue and down-regulated in cancer (Pandiyan et al. 2013).

In contrast to DNA methylation in the promoter, methylation in the gene body is usually associated with activation of gene expression (Maunakea et al. 2010; Jones 2012; Kulis et al. 2012; Varley et al. 2013). Because decitabine and azacitidine lead to global hypomethylation of the genome, these drugs also affect gene body methylation. This was shown by a recent study showing that decitabine can down-regulate expression of genes regulated by c-MYC as well as genes involved in metabolic processes through demethylation of gene bodies (Yang et al. 2014). This study indicated that, in addition to the therapeutic benefit gained from reactivation of silenced tumor-suppressor genes, hypomethylating drugs can also have a dual benefit by down-regulating oncogenes and metabolic genes (Yang et al. 2014).

Hypomethylating drugs can have very long-term effects on gene expression even after removal of the drug. The expression of a silenced, stably integrated GFP could still be detected 3 mo after initial decitabine treatment in a colon cancer model (Raynal et al. 2012). Similarly, the expression of a subset of alleles remained demethylated for more than 3 mo after treatment with decitabine in a breast cancer model (Kagey et al. 2010). These findings were also observed in a third study, in which it was found that low-dose DNMT inhibitors cause long-term gene-expression changes of hypermethylated genes involved in key antitumor pathways such as apoptosis, increased lineage commitment, and down-regulation of cell cycling (Tsai et al. 2012). Importantly, these long-term gene changes led to inhibition of cancer-initiating cells and tumor growth (Tsai et al. 2012).

In addition to the effect that hypomethylating drugs have on cancer cells, these drugs also affect the tumor microenvironment. For example, one study observed decreased angiogenesis after decitabine treatment because of inhibition of endothelial cell proliferation (Hellebrekers et al. 2006). Another study found that decitabine treatment increased expression of THBS1, which is an antiangiogenesis factor commonly methylated and silenced in various tumors

(Li et al. 1999). Consistent with these findings, combining azacitidine with lenalidomide, an FDA-approved therapy for MDS that has been shown to mediate some of its effects through affecting the tumor microenvironment, has shown promise in a phase II clinical trial on higher risk MDS patients (Sekeres et al. 2012).

IMMUNE RESPONSE MEDIATED BY DNA HYPOMETHYLATING DRUGS

Cancer cells avoid detection by the host immune system through altered expression of tumor-associated antigens and secretion of cytokines, which leads to deficient antigen presenting cells and cytolytic T cells (Heninger et al. 2015). DNA methylation is responsible for silencing of many immune-related genes, and DNMT inhibitors have been shown to induce expression of these genes in various experimental settings and clinical trial studies (Karpf et al. 1999; James et al. 2013; Wrangle et al. 2013; Odunsi et al. 2014; Heninger et al. 2015). For example, in a study looking at 63 cancer cell lines derived from breast, colorectal, and ovarian cancers that were treated with low-dose azacitidine for 3 days, it was found that genes commonly up-regulated by azacitidine were enriched for immunomodulatory genes and pathways such as cancer testis antigens (CTAs), interferon signaling, antigen presentation, inflammation, and cytokine signaling (Li et al. 2014). These immunomodulatory pathway genes were grouped together and termed the azacitidine immune gene set (AIMs). Patients in clinical trials that were treated with azacitidine and the HDAC inhibitor entinostat for 8 weeks had a higher expression of these AIM genes (Li et al. 2014).

Two recent studies have clarified the mechanism of the DNMT inhibitor-mediated immune response by showing that DNMT inhibitors could demethylate and induce endogenous retroviral sequences (ERVs), which led to activation of the cellular antiviral response (Chiappinelli et al. 2015; Roulois et al. 2015). One study looked at colorectal cancer cells treated with low-dose azacitidine, and found that there was a group of genes that were still highly expressed 42 days after drug withdrawal (Roulois et al. 2015). These genes were enriched in the interferon response and MDA5/MAVS RNA recognition pathway, and the up-regulation of these genes was mediated by hypomethylation of endogenous retroviral elements, which led to the expression of viral dsRNAs (Roulois et al. 2015). This viral recognition pathway was essential for the ability of azacitidine to inhibit cancer-initiating cells (Roulois et al. 2015). A second study studying ovarian cancer cells also found that the RNA-sensing pathways MAVS and TLR3 were up-regulated after azacitidine treatment, and this was a result of increased demethylation and expression of ERVs (Chiappinelli et al. 2015). Importantly, melanoma patients treated with immune checkpoint therapy with a high viral defense gene-expression signature had better clinical response, and mice pretreated with azacitidine had an amplified response to anti-CTLA4 immune checkpoint therapy, indicating that patients that lack the viral defense gene-expression signature may benefit from treatment with hypomethylating drugs before treatment with immunotherapy (Chiappinelli et al. 2015).

Given these findings relating cancer immune response to DNMT inhibitors, there is an ongoing phase II clinical trial to study the efficacy of the PD-1 inhibitor nivolumab on lung cancer patients pretreated with DNMT inhibitors (NCT01928576). These findings are also intriguing because biomarkers to predict response to DNMT inhibitors have been lacking as mentioned previously. Incorporating the expression of immune response genes affected by DNMT inhibitors with the expression of tumor-suppressor genes reactivated by DNMT inhibitors may give us a better ability to predict patients and cancers that will have the optimal response to hypomethylating drugs.

COMBINATION THERAPY WITH DNA HYPOMETHYLATING DRUGS

HDAC inhibitors have had success in the clinic for treating cutaneous T-cell lymphomas (Duvic et al. 2007; Whittaker et al. 2010), and combining DNMT inhibitors with HDAC inhibitors led to synergistic effects on reactivation of silenced tumor-suppressor genes (Cameron et al. 1999; Kalac et al. 2011). These observations led to the testing of this combination in numerous clinical trials. Because DNMT inhibitors are dependent on cell replication for its hypomethylating action and HDAC inhibitors lead to cell-cycle arrest, the effect of the combination is maximized when cells are treated with DNMT inhibitors first then with HDAC inhibitors, and clinical trials have mainly been tested using this dosing schedule. However, clinical trials have shown limited evidence of this combination being effective in improving patient survival (Garcia-Manero et al. 2006; Gore et al. 2006; Blum et al. 2007; Lin et al. 2009; Prebet et al. 2014; Issa et al. 2015a). For example, a recent phase II clinical trial in patients with MDS or AML showed that patients receiving only azacitidine had a median overall survival of 18 mo compared with 13 mo for the patients receiving the combination of azacitidine and entinostat (Prebet et al. 2014). Another phase II clinical trial in patients with MDS or AML showed that patients receiving only decitabine had a CR rate of 31% with overall response rate of 51%, compared with patients receiving decitabine along with valproic acid who had a CR rate of 37% and an overall response rate of 58%, neither of which were significantly improved over decitabine alone (Issa et al. 2015a). The median survival of patients receiving decitabine compared with decitabine plus valproic acid also did not have a significant improvement (11.2 mo vs. 11.9 mo). The lack of clinical success of combining these two agents may be because of the large, nonspecific gene-expression changes that occur when treating cells with HDAC inhibitors (Peart et al. 2005; LaBonte et al. 2009; Chueh et al. 2015).

Other agents that DNMT inhibitors have been combined with in clinical trials include lenalidomide, carboplatin, cisplatin, gemtuzumab ozogamicin, erythropoietin, filgrastim, romiplostim, bortezomib, arsenic trioxide, and sorafinib (Blum et al. 2012; Sekeres et al. 2012; Greenberg et al. 2013; Ravandi et al. 2013; Glasspool et al. 2014; Navada et al. 2014; Daver et al. 2015). These trials have produced mixed results about whether the combinations are beneficial and require further investigation, although the trials investigating the sensitization of cancers to platinum compounds by pretreatment with hypomethylating drugs have been particularly promising (Matei et al. 2012). The combination of decitabine with platinum compounds such as carboplatin is made even more intriguing by the fact that this combination showed increased epigenetic activity in reactivation of silenced tumor-suppressor genes such as MLH1 and PDLIM4 (Qin et al. 2015). This epigenetic synergy was found to be mediated through inhibition of HP1α expression by the platinum compounds, which led to reduced binding by MeCP2 and MBD2 (Qin et al. 2015). Similarly, combining arsenic trioxide or cardiac glycosides with hypomethylating drugs may have potential because a recent study showed that these drugs can also reactivate epigenetically silenced tumor-suppressor genes (Raynal et al. 2015). There are other epigenetic therapies in clinical development such as EZH2 inhibitors (McCabe et al. 2012; Knutson et al. 2013), LSD1 inhibitors (Mohammad et al. 2015), and BET inhibitors (Filippakopoulos et al. 2010), which have shown promising results in early clinical trials. It will be intriguing to explore whether these agents could have additive or synergistic effects in the clinic when combined with DNMT inhibitors because they may lead to synergistic gene reactivation as seen with the combination of hypomethylating drugs with HDAC inhibitors.

DEVELOPMENT OF NOVEL DNA HYPOMETHYLATING DRUGS

After the success of decitabine and azacitidine in the clinic, efforts were made to develop more effective hypomethylating drugs (Table 1). Zebularine is a cytidine analog similar to decitabine and azacitidine that incorporates into DNA and forms a reversible complex with DNMTs, effectively preventing methylation (Zhou et al. 2002; Champion et al. 2010). RG108 is a small molecular inhibitor discovered through virtual screening that binds to the catalytic site of DNMT1 and shows reactivation of silenced tumor-suppressor genes such as P16, SFRP1, and TIMP3 (Brueckner et al. 2005). CP-4200 is an elaidic acid derivative of azacitidine that allows entry into the cell independent of nucleoside transporters (Brueckner et al. 2010). Nanaomycin A is a small molecular inhibitor that has been shown to be selective for DNMT3B (Kuck et al. 2010). FdCyd is a nucleoside analog that has shown the ability to inhibit DNMTs and is undergoing early-stage clinical trials but requires simultaneous administration of tetahyrdrouridine to reduce enzymatic deamination (NCT00978250) (Newman et al. 2015). Further studies are needed to determine whether any of these inhibitors are more effective than azacitidine or decitabine in their ability to demethylate and reduce growth of tumors.

The most advanced of these novel DNA hypomethylating drugs has been SGI-110 (guadecitabine). Guadecitabine was designed to overcome one of the main limitations of decitabine and azacitidine, which is its instability in the human body because of degradation by CDA. This was accomplished by developing a compound with the structure of decitabine linked to a deoxyguanosine, which made it into a more stable compound resistant to this degradation (Griffiths et al. 2013). In experimental settings, guadecitabine had improved bioavailability and increased half-life while still maintaining the ability to act as a hypomethylating drug (Griffiths et al. 2013). In clinical trials, guadecitabine has shown promising results in a phase I clinical trial involving 93 patients with either AML or MDS (Issa et al. 2015b). Guadecitabine was well

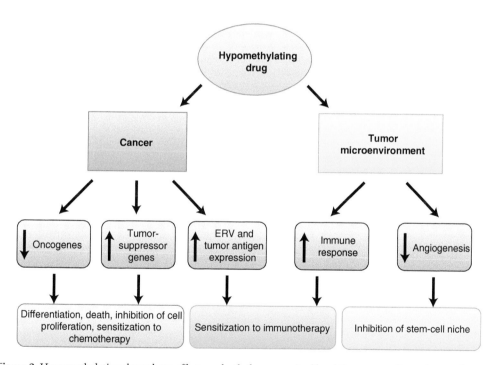

Figure 2. Hypomethylating drugs have effects on both the cancer itself and the surrounding microenvironment. In cancer cells, hypomethylating drugs can lead to reactivation of tumor-suppressor genes through demethylation of gene promoters, reactivation of oncogenes through demethylation of gene bodies, and reactivation of endogenous retroviral sequence (ERV) and tumor antigen expression through demethylation of these elements. These events lead to cell differentiation, death, inhibition of cell proliferation, sensitization to chemotherapy, and sensitization to immunotherapy. In the tumor microenvironment, hypomethylating drugs lead to increases in immune response and decreases in angiogenesis, which can subsequently lead to sensitization to immunotherapy and inhibition of the stem-cell niche.

tolerated and maximal demethylation occurred at 60 mg/m^2 with no additional demethylation at higher doses, similar to previous clinical data from decitabine and azacitidine. Six out of 19 MDS patients and six out of 74 AML patients had a clinical response to SGI-110. Importantly, there was a strong correlation between response and demethylation, indicating that guadecitabine is reducing tumor growth by acting as a hypomethylating drug (Issa et al. 2015b). Guadecitabine is now in phase II–III clinical trials in MDS and AML (NCT02096055 and NCT01261312).

CONCLUDING REMARKS

Modifications in DNA methylation is a critically important process in tumor initiation as well as progression. Targeting DNA methylation using DNMT inhibitors has changed the way we treat cancer, as it is an established therapy for MDS and AML. Hypomethylating drugs can be used to target cancer through effects on both the cancer itself and the surrounding tumor microenvironment (Fig. 2). Our increase in understanding of the mechanisms of the DNMT inhibitors as well as the degree and type of aberrant DNA methylation in different cancers should lead to even more clinical success in the future as we switch from using cytotoxic therapies to targeted therapies that can reverse the cancers back to a more normal-like state.

REFERENCES

Abdel-Wahab O, Mullally A, Hedvat C, Garcia-Manero G, Patel J, Wadleigh M, Malinge S, Yao J, Kilpivaara O, Bhat R, et al. 2009. Genetic characterization of TET1, TET2, and TET3 alterations in myeloid malignancies. *Blood* 114: 144–147.

Baylin SB, Jones PA. 2011. A decade of exploring the cancer epigenome—Biological and translational implications. *Nat Rev Cancer* 11: 726–734.

Bejar R, Lord A, Stevenson K, Bar-Natan M, Perez-Ladaga A, Zaneveld J, Wang H, Caughey B, Stojanov P, Getz G, et al. 2014. *TET2* mutations predict response to hypomethylating agents in myelodysplastic syndrome patients. *Blood* 124: 2705–2712.

Bender CM, Pao MM, Jones PA. 1998. Inhibition of DNA methylation by 5-aza-2′-deoxycytidine suppresses the growth of human tumor cell lines. *Cancer Res* 58: 95–101.

Blum W, Klisovic RB, Hackanson B, Liu Z, Liu S, Devine H, Vukosavljevic T, Huynh L, Lozanski G, Kefauver C, et al. 2007. Phase I study of decitabine alone or in combination with valproic acid in acute myeloid leukemia. *J Clin Oncol* 25: 3884–3891.

Blum W, Schwind S, Tarighat SS, Geyer S, Eisfeld AK, Whitman S, Walker A, Klisovic R, Byrd JC, Santhanam R, et al. 2012. Clinical and pharmacodynamic activity of bortezomib and decitabine in acute myeloid leukemia. *Blood* 119: 6025–6031.

Borthakur G, Ahdab SE, Ravandi F, Faderl S, Ferrajoli A, Newman B, Issa JP, Kantarjian H. 2008. Activity of decitabine in patients with myelodysplastic syndrome previously treated with azacitidine. *Leuk Lymphoma* 49: 690–695.

Braun T, Itzykson R, Renneville A, de Renzis B, Dreyfus F, Laribi K, Bouabdallah K, Vey N, Toma A, Recher C, et al. 2011. Molecular predictors of response to decitabine in advanced chronic myelomonocytic leukemia: A phase 2 trial. *Blood* 118: 3824–3831.

Brennan CW, Verhaak RG, McKenna A, Campos B, Noushmehr H, Salama SR, Zheng S, Chakravarty D, Sanborn JZ, Berman SH, et al. 2013. The somatic genomic landscape of glioblastoma. *Cell* 155: 462–477.

Brueckner B, Garcia Boy R, Siedlecki P, Musch T, Kliem HC, Zielenkiewicz P, Suhai S, Wiessler M, Lyko F. 2005. Epigenetic reactivation of tumor suppressor genes by a novel small-molecule inhibitor of human DNA methyltransferases. *Cancer Res* 65: 6305–6311.

Brueckner B, Rius M, Markelova MR, Fichtner I, Hals PA, Sandvold ML, Lyko F. 2010. Delivery of 5-azacytidine to human cancer cells by elaidic acid esterification increases therapeutic drug efficacy. *Mol Cancer Ther* 9: 1256–1264.

Cameron EE, Bachman KE, Myohanen S, Herman JG, Baylin SB. 1999. Synergy of demethylation and histone deacetylase inhibition in the re-expression of genes silenced in cancer. *Nat Genet* 21: 103–107.

Champion C, Guianvarc'h D, Senamaud-Beaufort C, Jurkowska RZ, Jeltsch A, Ponger L, Arimondo PB, Guieysse-Peugeot AL. 2010. Mechanistic insights on the inhibition of c5 DNA methyltransferases by zebularine. *PLoS ONE* 5: e12388.

Chen RZ, Pettersson U, Beard C, Jackson-Grusby L, Jaenisch R. 1998. DNA hypomethylation leads to elevated mutation rates. *Nature* 395: 89–93.

Chiappinelli KB, Strissel PL, Desrichard A, Li H, Henke C, Akman B, Hein A, Rote NS, Cope LM, Snyder A, et al. 2015. Inhibiting DNA methylation causes an interferon response in cancer via dsRNA including endogenous retroviruses. *Cell* 162: 974–986.

Chueh AC, Tse JW, Togel L, Mariadason JM. 2015. Mechanisms of histone deacetylase inhibitor-regulated gene expression in cancer cells. *Antioxid Redox Signal* 23: 66–84.

Cohen AL, Holmen SL, Colman H. 2013. IDH1 and IDH2 mutations in gliomas. *Curr Neurol Neurosci Rep* 13: 345.

Daver N, Kantarjian H, Ravandi F, Estey E, Wang X, Garcia-Manero G, Jabbour E, Konopleva M, O'Brien S, Verstovsek S, et al. 2015. A phase II study of decitabine and gemtuzumab ozogamicin in newly diagnosed and relapsed acute myeloid leukemia and high-risk myelodysplastic syndrome. *Leukemia* 30: 268–273.

Delhommeau F, Dupont S, Della Valle V, James C, Trannoy S, Masse A, Kosmider O, Le Couedic JP, Robert F, Alberdi A, et al. 2009. Mutation in *TET2* in myeloid cancers. *N Engl J Med* **360:** 2289–2301.

Derissen EJ, Beijnen JH, Schellens JH. 2013. Concise drug review: Azacitidine and decitabine. *Oncologist* **18:** 619–624.

Duvic M, Talpur R, Ni X, Zhang C, Hazarika P, Kelly C, Chiao JH, Reilly JF, Ricker JL, Richon VM, et al. 2007. Phase 2 trial of oral vorinostat (suberoylanilide hydroxamic acid, SAHA) for refractory cutaneous T-cell lymphoma (CTCL). *Blood* **109:** 31–39.

Fandy TE, Herman JG, Kerns P, Jiemjit A, Sugar EA, Choi SH, Yang AS, Aucott T, Dauses T, Odchimar-Reissig R, et al. 2009. Early epigenetic changes and DNA damage do not predict clinical response in an overlapping schedule of 5-azacytidine and entinostat in patients with myeloid malignancies. *Blood* **114:** 2764–2773.

Fenaux P, Mufti GJ, Hellstrom-Lindberg E, Santini V, Finelli C, Giagounidis A, Schoch R, Gattermann N, Sanz G, List A, et al. 2009. Efficacy of azacitidine compared with that of conventional care regimens in the treatment of higher-risk myelodysplastic syndromes: A randomised, open-label, phase III study. *Lancet Oncol* **10:** 223–232.

Figueroa ME, Abdel-Wahab O, Lu C, Ward PS, Patel J, Shih A, Li Y, Bhagwat N, Vasanthakumar A, Fernandez HF, et al. 2010. Leukemic IDH1 and IDH2 mutations result in a hypermethylation phenotype, disrupt TET2 function, and impair hematopoietic differentiation. *Cancer Cell* **18:** 553–567.

Filippakopoulos P, Qi J, Picaud S, Shen Y, Smith WB, Fedorov O, Morse EM, Keates T, Hickman TT, Felletar I, et al. 2010. Selective inhibition of BET bromodomains. *Nature* **468:** 1067–1073.

Gaidzik VI, Paschka P, Spath D, Habdank M, Kohne CH, Germing U, von Lilienfeld-Toal M, Held G, Horst HA, Haase D, et al. 2012. *TET2* mutations in acute myeloid leukemia (AML): Results from a comprehensive genetic and clinical analysis of the AML study group. *J Clin Oncol* **30:** 1350–1357.

Garcia-Manero G, Kantarjian HM, Sanchez-Gonzalez B, Yang H, Rosner G, Verstovsek S, Rytting M, Wierda WG, Ravandi F, Koller C, et al. 2006. Phase 1/2 study of the combination of 5-aza-2′-deoxycytidine with valproic acid in patients with leukemia. *Blood* **108:** 3271–3279.

Glasspool RM, Brown R, Gore ME, Rustin GJ, McNeish IA, Wilson RH, Pledge S, Paul J, Mackean M, Hall GD, et al. 2014. A randomised, phase II trial of the DNA-hypomethylating agent 5-aza-2′-deoxycytidine (decitabine) in combination with carboplatin vs carboplatin alone in patients with recurrent, partially platinum-sensitive ovarian cancer. *Br J Cancer* **110:** 1923–1929.

Goll MG, Kirpekar F, Maggert KA, Yoder JA, Hsieh CL, Zhang X, Golic KG, Jacobsen SE, Bestor TH. 2006. Methylation of tRNAAsp by the DNA methyltransferase homolog Dnmt2. *Science* **311:** 395–398.

Gore SD, Baylin S, Sugar E, Carraway H, Miller CB, Carducci M, Grever M, Galm O, Dauses T, Karp JE, et al. 2006. Combined DNA methyltransferase and histone deacetylase inhibition in the treatment of myeloid neoplasms. *Cancer Res* **66:** 6361–6369.

Greenberg PL, Garcia-Manero G, Moore M, Damon L, Roboz G, Hu K, Yang AS, Franklin J. 2013. A randomized controlled trial of romiplostim in patients with low- or intermediate-risk myelodysplastic syndrome receiving decitabine. *Leuk Lymphoma* **54:** 321–328.

Griffiths EA, Choy G, Redkar S, Taverna P, Azab M, Karpf AR. 2013. SGI-110: DNA methyltransferase inhibitor oncolytic. *Drugs Future* **38:** 535–543.

Hellebrekers DM, Jair KW, Vire E, Eguchi S, Hoebers NT, Fraga MF, Esteller M, Fuks F, Baylin SB, van Engeland M, et al. 2006. Angiostatic activity of DNA methyltransferase inhibitors. *Mol Cancer Ther* **5:** 467–475.

Heninger E, Krueger TE, Lang JM. 2015. Augmenting antitumor immune responses with epigenetic modifying agents. *Front Immunol* **6:** 29.

Hinoue T, Weisenberger DJ, Lange CP, Shen H, Byun HM, Van Den Berg D, Malik S, Pan F, Noushmehr H, van Dijk CM, et al. 2012. Genome-scale analysis of aberrant DNA methylation in colorectal cancer. *Genome Res* **22:** 271–282.

Issa JP, Kantarjian HM. 2009. Targeting DNA methylation. *Clin Cancer Res* **15:** 3938–3946.

Issa JP, Garcia-Manero G, Giles FJ, Mannari R, Thomas D, Faderl S, Bayar E, Lyons J, Rosenfeld CS, Cortes J, et al. 2004. Phase 1 study of low-dose prolonged exposure schedules of the hypomethylating agent 5-aza-2′-deoxycytidine (decitabine) in hematopoietic malignancies. *Blood* **103:** 1635–1640.

Issa JP, Gharibyan V, Cortes J, Jelinek J, Morris G, Verstovsek S, Talpaz M, Garcia-Manero G, Kantarjian HM. 2005. Phase II study of low-dose decitabine in patients with chronic myelogenous leukemia resistant to imatinib mesylate. *J Clin Oncol* **23:** 3948–3956.

Issa JP, Garcia-Manero G, Huang X, Cortes J, Ravandi F, Jabbour E, Borthakur G, Brandt M, Pierce S, Kantarjian HM. 2015a. Results of phase 2 randomized study of low-dose decitabine with or without valproic acid in patients with myelodysplastic syndrome and acute myelogenous leukemia. *Cancer* **121:** 556–561.

Issa JP, Roboz G, Rizzieri D, Jabbour E, Stock W, O'Connell C, Yee K, Tibes R, Griffiths EA, Walsh K, et al. 2015b. Safety and tolerability of guadecitabine (SGI-110) in patients with myelodysplastic syndrome and acute myeloid leukaemia: A multicentre, randomised, dose-escalation phase 1 study. *Lancet Oncol* **16:** 1099–1110.

Itzykson R, Kosmider O, Cluzeau T, Mansat-De Mas V, Dreyfus F, Beyne-Rauzy O, Quesnel B, Vey N, Gelsi-Boyer V, Raynaud S, et al. 2011. Impact of *TET2* mutations on response rate to azacitidine in myelodysplastic syndromes and low blast count acute myeloid leukemias. *Leukemia* **25:** 1147–1152.

James SR, Cedeno CD, Sharma A, Zhang W, Mohler JL, Odunsi K, Wilson EM, Karpf AR. 2013. DNA methylation and nucleosome occupancy regulate the cancer germline antigen gene *MAGEA11*. *Epigenetics* **8:** 849–863.

Jelinek J, Liang S, Lu Y, He R, Ramagli LS, Shpall EJ, Estecio MR, Issa JP. 2012. Conserved DNA methylation patterns in healthy blood cells and extensive changes in leukemia measured by a new quantitative technique. *Epigenetics* **7:** 1368–1378.

Jin C, Qin T, Barton MC, Jelinek J, Issa JJ. 2015. Minimal role of base excision repair in TET-induced global DNA demethylation in HEK293T cells. *Epigenetics* 10: 1006–1013.

Jones PA. 2012. Functions of DNA methylation: Islands, start sites, gene bodies and beyond. *Nat Rev Genet* 13: 484–492.

Jones PA, Baylin SB. 2007. The epigenomics of cancer. *Cell* 128: 683–692.

Jones PA, Taylor SM. 1980. Cellular differentiation, cytidine analogs and DNA methylation. *Cell* 20: 85–93.

Juergens RA, Wrangle J, Vendetti FP, Murphy SC, Zhao M, Coleman B, Sebree R, Rodgers K, Hooker CM, Franco N, et al. 2011. Combination epigenetic therapy has efficacy in patients with refractory advanced non-small cell lung cancer. *Cancer Discov* 1: 598–607.

Kagey JD, Kapoor-Vazirani P, McCabe MT, Powell DR, Vertino PM. 2010. Long-term stability of demethylation after transient exposure to 5-aza-2′-deoxycytidine correlates with sustained RNA polymerase II occupancy. *Mol Cancer Res* 8: 1048–1059.

Kalac M, Scotto L, Marchi E, Amengual J, Seshan VE, Bhagat G, Ulahannan N, Leshchenko VV, Temkin AM, Parekh S, et al. 2011. HDAC inhibitors and decitabine are highly synergistic and associated with unique gene-expression and epigenetic profiles in models of DLBCL. *Blood* 118: 5506–5516.

Kantarjian H, Issa JP, Rosenfeld CS, Bennett JM, Albitar M, DiPersio J, Klimek V, Slack J, de Castro C, Ravandi F, et al. 2006. Decitabine improves patient outcomes in myelodysplastic syndromes: Results of a phase III randomized study. *Cancer* 106: 1794–1803.

Kantarjian H, Oki Y, Garcia-Manero G, Huang X, O'Brien S, Cortes J, Faderl S, Bueso-Ramos C, Ravandi F, Estrov Z, et al. 2007a. Results of a randomized study of 3 schedules of low-dose decitabine in higher-risk myelodysplastic syndrome and chronic myelomonocytic leukemia. *Blood* 109: 52–57.

Kantarjian HM, O'Brien S, Huang X, Garcia-Manero G, Ravandi F, Cortes J, Shan J, Davisson J, Bueso-Ramos CE, Issa JP. 2007b. Survival advantage with decitabine versus intensive chemotherapy in patients with higher risk myelodysplastic syndrome: Comparison with historical experience. *Cancer* 109: 1133–1137.

Kantarjian HM, Thomas XG, Dmoszynska A, Wierzbowska A, Mazur G, Mayer J, Gau JP, Chou WC, Buckstein R, Cermak J, et al. 2012. Multicenter, randomized, open-label, phase III trial of decitabine versus patient choice, with physician advice, of either supportive care or low-dose cytarabine for the treatment of older patients with newly diagnosed acute myeloid leukemia. *J Clin Oncol* 30: 2670–2677.

Karpf AR, Peterson PW, Rawlins JT, Dalley BK, Yang Q, Albertsen H, Jones DA. 1999. Inhibition of DNA methyltransferase stimulates the expression of signal transducer and activator of transcription 1, 2, and 3 genes in colon tumor cells. *Proc Natl Acad Sci* 96: 14007–14012.

Klco JM, Spencer DH, Lamprecht TL, Sarkaria SM, Wylie T, Magrini V, Hundal J, Walker J, Varghese N, Erdmann-Gilmore P, et al. 2013. Genomic impact of transient low-dose decitabine treatment on primary AML cells. *Blood* 121: 1633–1643.

Knutson SK, Warholic NM, Wigle TJ, Klaus CR, Allain CJ, Raimondi A, Porter Scott M, Chesworth R, Moyer MP, Copeland RA, et al. 2013. Durable tumor regression in genetically altered malignant rhabdoid tumors by inhibition of methyltransferase EZH2. *Proc Natl Acad Sci* 110: 7922–7927.

Kondo Y, Shen L, Yan PS, Huang TH, Issa JP. 2004. Chromatin immunoprecipitation microarrays for identification of genes silenced by histone H3 lysine 9 methylation. *Proc Natl Acad Sci* 101: 7398–7403.

Kuck D, Caulfield T, Lyko F, Medina-Franco JL. 2010. Nanaomycin A selectively inhibits DNMT3B and reactivates silenced tumor suppressor genes in human cancer cells. *Mol Cancer Ther* 9: 3015–3023.

Kulis M, Heath S, Bibikova M, Queiros AC, Navarro A, Clot G, Martinez-Trillos A, Castellano G, Brun-Heath I, Pinyol M, et al. 2012. Epigenomic analysis detects widespread gene-body DNA hypomethylation in chronic lymphocytic leukemia. *Nat Genet* 44: 1236–1242.

LaBonte MJ, Wilson PM, Fazzone W, Groshen S, Lenz HJ, Ladner RD. 2009. DNA microarray profiling of genes differentially regulated by the histone deacetylase inhibitors vorinostat and LBH589 in colon cancer cell lines. *BMC Med Genomics* 2: 67.

Langemeijer SM, Kuiper RP, Berends M, Knops R, Aslanyan MG, Massop M, Stevens-Linders E, van Hoogen P, van Kessel AG, Raymakers RA, et al. 2009. Acquired mutations in *TET2* are common in myelodysplastic syndromes. *Nat Genet* 41: 838–842.

Li Q, Ahuja N, Burger PC, Issa JP. 1999. Methylation and silencing of the Thrombospondin-1 promoter in human cancer. *Oncogene* 18: 3284–3289.

Li H, Chiappinelli KB, Guzzetta AA, Easwaran H, Yen RW, Vatapalli R, Topper MJ, Luo J, Connolly RM, Azad NS, et al. 2014. Immune regulation by low doses of the DNA methyltransferase inhibitor 5-azacitidine in common human epithelial cancers. *Oncotarget* 5: 587–598.

Lin J, Gilbert J, Rudek MA, Zwiebel JA, Gore S, Jiemjit A, Zhao M, Baker SD, Ambinder RF, Herman JG, et al. 2009. A phase I dose-finding study of 5-azacytidine in combination with sodium phenylbutyrate in patients with refractory solid tumors. *Clin Cancer Res* 15: 6241–6249.

Lübbert M, Suciu S, Baila L, Rüter BH, Platzbecker U, Giagounidis A, Selleslag D, Labar B, Germing U, Salih HR, et al. 2011. Low-dose decitabine versus best supportive care in elderly patients with intermediate- or high-risk myelodysplasticcontrast syndrome (MDS) ineligible for intensive chemotherapy: Final results of the randomized phase III study of the European Organisation for Research and Treatment of Cancer Leukemia Group and the German MDS Study Group. *J Clin Oncol* 29: 1987–1996.

Lund K, Cole JJ, VanderKraats ND, McBryan T, Pchelintsev NA, Clark W, Copland M, Edwards JR, Adams PD. 2014. DNMT inhibitors reverse a specific signature of aberrant promoter DNA methylation and associated gene silencing in AML. *Genome Biol* 15: 406.

Matei D, Fang F, Shen C, Schilder J, Arnold A, Zeng Y, Berry WA, Huang T, Nephew KP. 2012. Epigenetic resensitization to platinum in ovarian cancer. *Cancer Res* 72: 2197–2205.

Maunakea AK, Nagarajan RP, Bilenky M, Ballinger TJ, D'Souza C, Fouse SD, Johnson BE, Hong C, Nielsen C, Zhao Y, et al. 2010. Conserved role of intragenic DNA methylation in regulating alternative promoters. *Nature* **466:** 253–257.

McCabe MT, Ott HM, Ganji G, Korenchuk S, Thompson C, Van Aller GS, Liu Y, Graves AP, Della Pietra A III, Diaz E, et al. 2012. EZH2 inhibition as a therapeutic strategy for lymphoma with EZH2-activating mutations. *Nature* **492:** 108–112.

Meldi K, Qin T, Buchi F, Droin N, Sotzen J, Micol JB, Selimoglu-Buet D, Masala E, Allione B, Gioia D, et al. 2015. Specific molecular signatures predict decitabine response in chronic myelomonocytic leukemia. *J Clin Invest* **125:** 1857–1872.

Mohammad HP, Smitheman KN, Kamat CD, Soong D, Federowicz KE, Van Aller GS, Schneck JL, Carson JD, Liu Y, Butticello M, et al. 2015. A DNA hypomethylation signature predicts antitumor activity of LSD1 inhibitors in SCLC. *Cancer Cell* **28:** 57–69.

Navada SC, Steinmann J, Lubbert M, Silverman LR. 2014. Clinical development of demethylating agents in hematology. *J Clin Invest* **124:** 40–46.

Newman EM, Morgan RJ, Kummar S, Beumer JH, Blanchard MS, Ruel C, El-Khoueiry AB, Carroll MI, Hou JM, Li C, et al. 2015. A phase I, pharmacokinetic, and pharmacodynamic evaluation of the DNA methyltransferase inhibitor 5-fluoro-2′-deoxycytidine, administered with tetrahydrouridine. *Cancer Chemother Pharmacol* **75:** 537–546.

Noushmehr H, Weisenberger DJ, Diefes K, Phillips HS, Pujara K, Berman BP, Pan F, Pelloski CE, Sulman EP, Bhat KP, et al. 2010. Identification of a CpG island methylator phenotype that defines a distinct subgroup of glioma. *Cancer Cell* **17:** 510–522.

Odunsi K, Matsuzaki J, James SR, Mhawech-Fauceglia P, Tsuji T, Miller A, Zhang W, Akers SN, Griffiths EA, Miliotto A, et al. 2014. Epigenetic potentiation of NY-ESO-1 vaccine therapy in human ovarian cancer. *Cancer Immunol Res* **2:** 37–49.

Oki Y, Aoki E, Issa JP. 2007. Decitabine—Bedside to bench. *Crit Rev Oncol Hematol* **61:** 140–152.

Oki Y, Jelinek J, Shen L, Kantarjian HM, Issa JP. 2008. Induction of hypomethylation and molecular response after decitabine therapy in patients with chronic myelomonocytic leukemia. *Blood* **111:** 2382–2384.

Pandiyan K, You JS, Yang X, Dai C, Zhou XJ, Baylin SB, Jones PA, Liang G. 2013. Functional DNA demethylation is accompanied by chromatin accessibility. *Nucleic Acids Res* **41:** 3973–3985.

Parry L, Clarke AR. 2011. The roles of the methyl-CpG binding proteins in cancer. *Genes Cancer* **2:** 618–630.

Paul TA, Bies J, Small D, Wolff L. 2010. Signatures of polycomb repression and reduced H3K4 trimethylation are associated with p15INK4b DNA methylation in AML. *Blood* **115:** 3098–3108.

Peart MJ, Smyth GK, van Laar RK, Bowtell DD, Richon VM, Marks PA, Holloway AJ, Johnstone RW. 2005. Identification and functional significance of genes regulated by structurally different histone deacetylase inhibitors. *Proc Natl Acad Sci* **102:** 3697–3702.

Plass C, Pfister SM, Lindroth AM, Bogatyrova O, Claus R, Lichter P. 2013. Mutations in regulators of the epigenome and their connections to global chromatin patterns in cancer. *Nat Rev Genet* **14:** 765–780.

Prebet T, Sun Z, Figueroa ME, Ketterling R, Melnick A, Greenberg PL, Herman J, Juckett M, Smith MR, Malick L, et al. 2014. Prolonged administration of azacitidine with or without entinostat for myelodysplastic syndrome and acute myeloid leukemia with myelodysplasia-related changes: Results of the US Leukemia Intergroup trial E1905. *J Clin Oncol* **32:** 1242–1248.

Qin T, Youssef EM, Jelinek J, Chen R, Yang AS, Garcia-Manero G, Issa JP. 2007. Effect of cytarabine and decitabine in combination in human leukemic cell lines. *Clin Cancer Res* **13:** 4225–4232.

Qin T, Jelinek J, Si J, Shu J, Issa JP. 2009. Mechanisms of resistance to 5-aza-2′-deoxycytidine in human cancer cell lines. *Blood* **113:** 659–667.

Qin T, Castoro R, El Ahdab S, Jelinek J, Wang X, Si J, Shu J, He R, Zhang N, Chung W, et al. 2011. Mechanisms of resistance to decitabine in the myelodysplastic syndrome. *PLoS ONE* **6:** e23372.

Qin T, Si J, Raynal NJ, Wang X, Gharibyan V, Ahmed S, Hu X, Jin C, Lu Y, Shu J, et al. 2015. Epigenetic synergy between decitabine and platinum derivatives. *Clin Epigenet* **7:** 97.

Ravandi F, Alattar ML, Grunwald MR, Rudek MA, Rajkhowa T, Richie MA, Pierce S, Daver N, Garcia-Manero G, Faderl S, et al. 2013. Phase 2 study of azacytidine plus sorafenib in patients with acute myeloid leukemia and FLT-3 internal tandem duplication mutation. *Blood* **121:** 4655–4662.

Raynal NJ, Si J, Taby RF, Gharibyan V, Ahmed S, Jelinek J, Estecio MR, Issa JP. 2012. DNA methylation does not stably lock gene expression but instead serves as a molecular mark for gene silencing memory. *Cancer Res* **72:** 1170–1181.

Raynal NJ, Lee JT, Wang Y, Beaudry A, Madireddi P, Garriga J, Malouf GG, Dumont S, Dettman EJ, Gharibyan V, et al. 2015. Targeting calcium signaling induces epigenetic reactivation of tumor suppressor genes in cancer. *Cancer Res* **76:** 1494–1505.

Roulois D, Loo Yau H, Singhania R, Wang Y, Danesh A, Shen SY, Han H, Liang G, Jones PA, Pugh TJ, et al. 2015. DNA-demethylating agents target colorectal cancer cells by inducing viral mimicry by endogenous transcripts. *Cell* **162:** 961–973.

Sekeres MA, Tiu RV, Komrokji R, Lancet J, Advani AS, Afable M, Englehaupt R, Juersivich J, Cuthbertson D, Paleveda J, et al. 2012. Phase 2 study of the lenalidomide and azacitidine combination in patients with higher-risk myelodysplastic syndromes. *Blood* **120:** 4945–4951.

Shen L, Kondo Y, Guo Y, Zhang J, Zhang L, Ahmed S, Shu J, Chen X, Waterland RA, Issa JP. 2007a. Genome-wide profiling of DNA methylation reveals a class of normally methylated CpG island promoters. *PLoS Genet* **3:** 2023–2036.

Shen L, Toyota M, Kondo Y, Lin E, Zhang L, Guo Y, Hernandez NS, Chen X, Ahmed S, Konishi K, et al. 2007b. Integrated genetic and epigenetic analysis identifies three different subclasses of colon cancer. *Proc Natl Acad Sci* **104:** 18654–18659.

Si J, Boumber YA, Shu J, Qin T, Ahmed S, He R, Jelinek J, Issa JP. 2010. Chromatin remodeling is required for gene reactivation after decitabine-mediated DNA hypomethylation. *Cancer Res* **70:** 6968–6977.

Silverman LR, Demakos EP, Peterson BL, Kornblith AB, Holland JC, Odchimar-Reissig R, Stone RM, Nelson D, Powell BL, DeCastro CM, et al. 2002. Randomized controlled trial of azacitidine in patients with the myelodysplastic syndrome: A study of the cancer and leukemia group B. *J Clin Oncol* **20:** 2429–2440.

Silverman LR, Fenaux P, Mufti GJ, Santini V, Hellstrom-Lindberg E, Gattermann N, Sanz G, List AF, Gore SD, Seymour JF. 2011. Continued azacitidine therapy beyond time of first response improves quality of response in patients with higher-risk myelodysplastic syndromes. *Cancer* **117:** 2697–2702.

Smith ZD, Meissner A. 2013. DNA methylation: Roles in mammalian development. *Nat Rev Genet* **14:** 204–220.

Stewart DJ, Issa JP, Kurzrock R, Nunez MI, Jelinek J, Hong D, Oki Y, Guo Z, Gupta S, Wistuba II. 2009. Decitabine effect on tumor global DNA methylation and other parameters in a phase I trial in refractory solid tumors and lymphomas. *Clin Cancer Res* **15:** 3881–3888.

Taby R, Issa JP. 2010. Cancer epigenetics. *CA Cancer J Clin* **60:** 376–392.

TCGA. 2012. Comprehensive molecular characterization of human colon and rectal cancer. *Nature* **487:** 330–337.

TCGA. 2013. Genomic and epigenomic landscapes of adult de novo acute myeloid leukemia. *N Engl J Med* **368:** 2059–2074.

Toyota M, Ahuja N, Ohe-Toyota M, Herman JG, Baylin SB, Issa JP. 1999. CpG island methylator phenotype in colorectal cancer. *Proc Natl Acad Sci* **96:** 8681–8686.

Traina F, Visconte V, Elson P, Tabarroki A, Jankowska AM, Hasrouni E, Sugimoto Y, Szpurka H, Makishima H, O'Keefe CL, et al. 2014. Impact of molecular mutations on treatment response to DNMT inhibitors in myelodysplasia and related neoplasms. *Leukemia* **28:** 78–87.

Treppendahl MB, Kristensen LS, Gronbaek K. 2014. Predicting response to epigenetic therapy. *J Clin Invest* **124:** 47–55.

Tsai HC, Li H, Van Neste L, Cai Y, Robert C, Rassool FV, Shin JJ, Harbom KM, Beaty R, Pappou E, et al. 2012. Transient low doses of DNA-demethylating agents exert durable antitumor effects on hematological and epithelial tumor cells. *Cancer Cell* **21:** 430–446.

Turcan S, Rohle D, Goenka A, Walsh LA, Fang F, Yilmaz E, Campos C, Fabius AW, Lu C, Ward PS, et al. 2012. IDH1 mutation is sufficient to establish the glioma hypermethylator phenotype. *Nature* **483:** 479–483.

Varley KE, Gertz J, Bowling KM, Parker SL, Reddy TE, Pauli-Behn F, Cross MK, Williams BA, Stamatoyannopoulos JA, Crawford GE, et al. 2013. Dynamic DNA methylation across diverse human cell lines and tissues. *Genome Res* **23:** 555–567.

Voso MT, Fabiani E, Piciocchi A, Matteucci C, Brandimarte L, Finelli C, Pogliani E, Angelucci E, Fioritoni G, Musto P, et al. 2011. Role of BCL2L10 methylation and TET2 mutations in higher risk myelodysplastic syndromes treated with 5-azacytidine. *Leukemia* **25:** 1910–1913.

Weisenberger DJ, Siegmund KD, Campan M, Young J, Long TI, Faasse MA, Kang GH, Widschwendter M, Weener D, Buchanan D, et al. 2006. CpG island methylator phenotype underlies sporadic microsatellite instability and is tightly associated with *BRAF* mutation in colorectal cancer. *Nat Genet* **38:** 787–793.

Whittaker SJ, Demierre MF, Kim EJ, Rook AH, Lerner A, Duvic M, Scarisbrick J, Reddy S, Robak T, Becker JC, et al. 2010. Final results from a multicenter, international, pivotal study of romidepsin in refractory cutaneous T-cell lymphoma. *J Clin Oncol* **28:** 4485–4491.

Wienholz BL, Kareta MS, Moarefi AH, Gordon CA, Ginno PA, Chedin F. 2010. DNMT3L modulates significant and distinct flanking sequence preference for DNA methylation by DNMT3A and DNMT3B in vivo. *PLoS Genet* **6:** e1001106.

Wijermans P, Lubbert M, Verhoef G, Bosly A, Ravoet C, Andre M, Ferrant A. 2000. Low-dose 5-aza-2'-deoxycytidine, a DNA hypomethylating agent, for the treatment of high-risk myelodysplastic syndrome: A multicenter phase II study in elderly patients. *J Clin Oncol* **18:** 956–962.

Williams K, Christensen J, Helin K. 2012. DNA methylation: TET proteins-guardians of CpG islands? *EMBO Rep* **13:** 28–35.

Wrangle J, Wang W, Koch A, Easwaran H, Mohammad HP, Vendetti F, Vancriekinge W, Demeyer T, Du Z, Parsana P, et al. 2013. Alterations of immune response of non-small cell lung cancer with azacytidine. *Oncotarget* **4:** 2067–2079.

Xu W, Yang H, Liu Y, Yang Y, Wang P, Kim SH, Ito S, Yang C, Xiao MT, Liu LX, et al. 2011. Oncometabolite 2-hydroxyglutarate is a competitive inhibitor of α-ketoglutarate-dependent dioxygenases. *Cancer Cell* **19:** 17–30.

Yagi K, Akagi K, Hayashi H, Nagae G, Tsuji S, Isagawa T, Midorikawa Y, Nishimura Y, Sakamoto H, Seto Y, et al. 2010. Three DNA methylation epigenotypes in human colorectal cancer. *Clin Cancer Res* **16:** 21–33.

Yamazaki J, Jelinek J, Lu Y, Cesaroni M, Madzo J, Neumann F, He R, Taby R, Vasanthakumar A, Macrae T, et al. 2015. *TET2* mutations affect non-CpG island DNA methylation at enhancers and transcription factor binding sites in chronic myelomonocytic leukemia. *Cancer Res* **75:** 2833–2843.

Yamazaki J, Taby R, Jelinek J, Raynal NJ, Cesaroni M, Pierce SA, Kornblau SM, Bueso-Ramos CE, Ravandi F, Kantarjian HM, et al. 2016. Hypomethylation of TET2 target genes identifies a curable subset of acute myeloid leukemia. *J Natl Cancer Inst* **108**.

Yang X, Han H, De Carvalho DD, Lay FD, Jones PA, Liang G. 2014. Gene body methylation can alter gene expression and is a therapeutic target in cancer. *Cancer Cell* **26:** 577–590.

Yang L, Rau R, Goodell MA. 2015. DNMT3A in haematological malignancies. *Nat Rev Cancer* **15:** 152–165.

Zhou L, Cheng X, Connolly BA, Dickman MJ, Hurd PJ, Hornby DP. 2002. Zebularine: A novel DNA methylation inhibitor that forms a covalent complex with DNA methyltransferases. *J Mol Biol* **321:** 591–599.

Targeting Cancer Cells with BET Bromodomain Inhibitors

Yali Xu and Christopher R. Vakoc

Cold Spring Harbor Laboratory, Cold Spring Harbor, New York 11724

Correspondence: vakoc@cshl.edu

Cancer cells are often hypersensitive to the targeting of transcriptional regulators, which may reflect the deregulated gene expression programs that underlie malignant transformation. One of the most prominent transcriptional vulnerabilities in human cancer to emerge in recent years is the bromodomain and extraterminal (BET) family of proteins, which are coactivators that link acetylated transcription factors and histones to the activation of RNA polymerase II. Despite unclear mechanisms underlying the gene specificity of BET protein function, small molecules targeting these regulators preferentially suppress the transcription of cancer-promoting genes. As a consequence, BET inhibitors elicit anticancer activity in numerous malignant contexts at doses that can be tolerated by normal tissues, a finding supported by animal studies and by phase I clinical trials in human cancer patients. In this review, we will discuss the remarkable, and often perplexing, therapeutic effects of BET bromodomain inhibition in cancer.

In eukaryotic cells, sequence-specific DNA-binding transcription factors (TFs) activate their target genes by recruiting multisubunit coactivator complexes, which use diverse biochemical mechanisms to activate RNA polymerase II (Pol II). One important class of coactivators possess lysine acetyltransferase (KAT) activity, which transfers acetyl groups from acetyl-coenzyme A to the epsilon amino group of lysine residues of various substrate proteins. Many KAT enzymes (e.g., p300/CBP) have permissive substrate specificity and will acetylate unstructured, lysine-rich peptides found on TFs, histones, and various other components of the transcription apparatus (Dancy and Cole 2015). The pervasiveness of TF–KAT interactions in transcriptional regulation leads to a global partitioning of eukaryotic genomes into hyperacetylated and hypoacetylated domains, which strongly correlate with active and inactive *cis*-regulatory regulatory elements, respectively (Wang et al. 2008).

One mechanism by which acetylation influences transcription is by neutralizing the positive charge of lysine side chains to disrupt electrostatic interactions (e.g., between histones and DNA), which can lead to chromatin decompaction (Roth et al. 2001). In an alternative mechanism, lysine side-chain acetylation of many transcriptional regulators will create docking sites for proteins possessing acetyllysine binding/reader domains. In this setting, acetyllysine serves a vital function in the assembly of the transcriptional apparatus at enhancer and promoter elements. The most well established acetyllysine reader is the bromodomain,

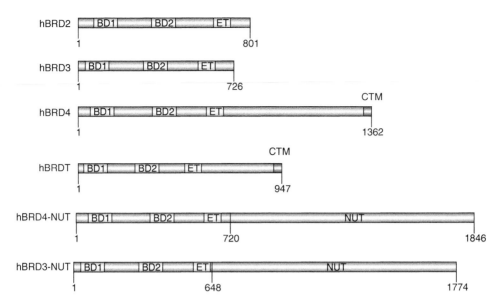

Figure 1. Domain structure of the bromodomain and extraterminal (BET) protein family and BET–NUT fusion proteins. Each BET protein contains two bromodomains (BD1 and BD2) and an extraterminal (ET) domain. BRD4 and BRDT have an additional carboxy-terminal motif (CTM). BRD3/BRD4–NUT fusion proteins found in NUT midline carcinoma (NMC) patients fuse the amino terminus of BRD4 (or BRD3) with almost the entire NUT protein.

which is present on 46 different proteins encoded in the human genome (Dhalluin et al. 1999; Filippakopoulos et al. 2012). A bromodomain is composed of a left-handed bundle of four α helices, with interhelical loops forming a hydrophobic binding pocket that engages in acetyllysine recognition (Dhalluin et al. 1999). Studies spanning nearly two decades have implicated bromodomain-containing coactivator proteins as integral components of TF-mediated gene regulation by linking lysine acetylation to downstream effects on chromatin structure and transcription (Sanchez and Zhou 2009). Moreover, the functional diversity and emerging "drugability" of bromodomain modules with small molecules has motivated a widespread interest in this class of proteins as therapeutic targets (Filippakopoulos and Knapp 2014).

BET PROTEIN FAMILY OF TRANSCRIPTIONAL COACTIVATORS

The mammalian BET (bromodomain and extraterminal domain-containing) protein family consists of four members, including the ubiq-

uitously expressed BRD2, BRD3, BRD4, and the germ-cell-specific BRDT (Fig. 1) (Wu and Chiang 2007; Shi and Vakoc 2014). All four BET proteins have two conserved bromodomains that preferentially bind to multiacetylated peptides (Fig. 1) (Dey et al. 2003; Moriniere et al. 2009; Gamsjaeger et al. 2011; Filippakopoulos et al. 2012). The preferred ligand of the first bromodomain (BD1) is $K^{ac}XXK^{ac}$, with the intervening X amino acids having small side chains (e.g., glycine or alanine), whereas the second bromodomain (BD2) is more permissive to binding multiacetylated peptides in diverse sequence contexts (Dey et al. 2003; Moriniere et al. 2009; Gamsjaeger et al. 2011; Filippakopoulos et al. 2012). The acetylated tails of core histones H3 and H4 and acetylated regions of TFs are the most well-validated binding partners of BET bromodomains, which are generated as consequence of TF-mediated KAT recruitment (Dey et al. 2003; Lamonica et al. 2011; Shi et al. 2014; Roe et al. 2015). ChIP-seq studies in several cell types have shown that BRD4 localizes preferentially to the nucleosome-free site occupied by TFs at enhancers and promoters, which is consistent with acetylated TFs being important recruiters of BET proteins (Roe et al. 2015; Stonestrom et al. 2015). It is likely that a multitude of acetylated peptides contribute to BET protein recruitment to a particular DNA element, with different acetylated peptides being relevant at different *cis* elements. However, it is challenging to pinpoint the complete repertoire of acetylated peptides responsible for recruiting BET proteins to chromatin.

In addition to two bromodomains, all four BET proteins possess a conserved extraterminal (ET) domain that performs an effector role in transcriptional activation and in chromatin remodeling. The ET domain interacts with several different cofactors, including the demethylase protein JMJD6, the methyltransferase/adaptor protein NSD3, and the chromatin remodeling ATPases CHD4 and BRG1 (Rahman et al. 2011; Shen et al. 2015). The ET domain also interacts with the virally encoded proteins, such as murine leukemia virus (MLV) integrase and latency-associated nuclear antigen (LANA) peptide of Kaposi's sarcoma-associated herpesvirus (KSHV) (Hellert et al. 2013; Crowe et al. 2016). All of these interactions are mediated by a hydrophobic groove on the ET domain that recognizes a consensus motif of alternating lysine and hydrophobic residues found on several of the above-mentioned cofactors (e.g., an IKLKI motif on NSD3 and a LKIKL motif on CHD4) (Hellert et al. 2013; Shen et al. 2015; Crowe et al. 2016; Zhang et al. 2016). The available evidence indicates that BRD4 relies on a unique subset of these ET-interacting proteins for transcriptional activation in particular cell types. In HEK293T cells, BRD4 uses JMJD6 as its ET-domain effector, which will demethylate histones and noncoding RNA to promote enhancer-mediated gene activation (Liu et al. 2013). In acute myeloid leukemia (AML) cells, the ET domain of BRD4 activates transcription by interacting with NSD3, which functions as a scaffold to recruit the chromatin-remodeling enzyme CHD8 (Shen et al. 2015). This apparent context-specificity of ET domain function is not well understood at present.

BRD4 and BRDT possess a unique carboxy-terminal motif (CTM), which binds to the serine/threonine kinase P-TEFb as an additional mechanism of gene activation (Bisgrove et al. 2007; Krueger et al. 2010). P-TEFb is a heterodimer of the kinase Cdk9 and a K, T1, or T2-type cyclin, which together can phosphorylate the serine 2 position of the Pol II carboxy-terminal domain (CTD), as well as serine and threonine residues on the pausing factors DSIF and NELF (Krueger et al. 2010; Jonkers and Lis 2015). Thus, BRD4-mediated P-TEFb recruitment will drive a variety of local phosphorylation events to bypass the paused state of Pol II and promote transcription elongation. At the biochemical level, BRD4 uses multiple mechanisms to regulate P-TEFb activity. Using purified proteins, the interaction with BRD4 is sufficient to stimulate P-TEFb kinase activity (Itzen et al. 2014). In cells, this BRD4 interaction is competitive with the interaction of P-TEFb with HEXIM1/7SK RNA, which are inhibitors of its kinase activity (Jang et al. 2005; Yang et al. 2005). BRD4 also contributes to the localization of P-TEFb to hyperacetylated enhancers and promoters across the genome, thus guiding P-TEFb to it relevant substrates near TF-bound sites (Jang et al. 2005; Yang et al. 2005). Another key player in the functional linkage between BRD4 and P-TEFb is the Mediator complex, which is a 30 subunit coactivator complex that physically associates with BRD4 and with P-TEFb (Jiang et al. 1998; Jang et al. 2005; Donner et al. 2010; Allen and Taatjes 2015). Although the precise interaction surface between BRD4 and Mediator has yet to be mapped, the MED23 subunit has been implicated in this interaction (Wang et al. 2013). BRD4 and Mediator stabilize each other's occupancy at specific sites across the genome, and these two machineries cooperate in recruiting P-TEFb (Jang et al. 2005; Donner et al. 2010; Bhagwat et al. 2016). It should be noted that the CTM region is found on BRD4 (and BRDT), but not on BRD2 and BRD3, which may explain why BRD4 performs a broader nonredundant role in transcriptional activation than the other BET proteins. For example, genetic inactivation of BRD4 will lead to slow growth phenotypes in essentially all mammalian cell lines, whereas BRD2 and BRD3 lead to only subtle phenotypes when targeted (Vakoc CR, unpubl.). Only a few cellular contexts have been identified in which redundancy exists among the BET proteins, as was recently shown for BRD2 and BRD3 in hematopoietic cells (Stonestrom et al. 2015).

Recent studies have found that BRD4 protein possesses intrinsic kinase and KAT activity in in vitro assays (Devaiah et al. 2012, 2016). Similar to P-TEFb, BRD4 can directly phosphorylate the CTD of Pol II at the serine 2 position and can acetylate multiple residues on histone H3 and H4, including H3K112 found on the globular region of the nucleosome (Devaiah et al. 2012, 2016). Earlier work had also identified an intrinsic kinase activity in purified BRD2 and in FSH, the *Drosophila melanogaster* ortholog of BRD4 (Denis and Green 1996; Chang et al. 2007). The presence of these activities in biochemical assays is difficult to reconcile with the lack of an obvious kinase or KAT domain in the BRD4 polypeptide, and hence these activities should be considered provisional at present and await further validation to confirm their importance in vivo.

SMALL-MOLECULE INHIBITORS OF BET BROMODOMAINS

The above description of BET proteins suggests a general role of these regulators in transcriptional control, particularly because acetylated TFs and histones are found at all active promoters and enhancers in the genomes. It is only recently that the attention of the field has turned toward identifying biological processes that are disproportionately BET protein-dependent. This avenue of research was invigorated by two studies published in 2010 describing the first selective small-molecule inhibitors of BET bromodomains (Filippakopoulos et al. 2010; Nicodeme et al. 2010). The potency, specificity, and in vivo activity of these molecules in modulating BET proteins has allowed numerous studies in a myriad of animal models of disease. This work has exposed a remarkable gene specificity of transcriptional effects of BET inhibition that underpins a broad interest in BET proteins as therapeutic targets.

The first class of BET bromodomain inhibitors, which are a series of thienotriazolodiazepines, were originally filed as patents by the Mitsubishi Tanabe Pharmaceutical Corporation (Adachi et al. 2006; Miyoshi et al. 2010, 2013). The compounds belong to the diazepine family and are analogs of benzodiazepine, which has been used extensively in the clinic as psychoactive drug (Smith et al. 2014). Thereafter, the Bradner laboratory and researchers at GlaxoSmithKline independently published the highly potent and selective BET bromodomain inhibitors JQ1 (a thienotriazolodiazepine) and iBET (a benzodiazepine), respectively (Filippakopoulos et al. 2010; Nicodeme et al. 2010). Of note, both JQ1 and iBET are pan-BET bromodomain inhibitors, which do not discriminate between the two bromodomains within the same BET protein, nor among the four BET family members (Filippakopoulos et al. 2010; Nicodeme et al. 2010). Because the tandem bromodomains within the same BET protein have distinct functions and binding affinities toward acetylated peptides, the lack of specificity of JQ1 and iBET limits the potential use of these chemical probes to study the roles of individual BET proteins. Nonetheless, the paninhibitory activity of these compounds may contribute to their high potency in modulating biological processes in vivo. Because BRD4 tends to be the dominant transcriptional regulator within the BET protein family in somatic cell types, most studies have linked the transcriptional effects of BET inhibitors to BRD4 inhibition, with BRD2 and BRD3 contributing to a lesser degree.

Chemists have continued to optimize these compounds with the aim to improve the selectivity among the BET protein family and enhance drug potency and in vivo pharmacodynamics, in an effort to make these compounds suitable for clinical investigation. This includes multiple pan inhibitors, some of which have entered clinical trials (see below), BD1 selective inhibitors, such as MS-436, Olinone, and BI-2536, as well as the BD2 selective inhibitors RVX-208 and RVX-297 (Steegmaier et al. 2007; Park et al. 2013; Picaud et al. 2013; Zhang et al. 2013; McLure et al. 2014; Kharenko et al. 2016). As expected, inhibition of individual BET bromodomains will lead to different transcriptional and phenotypic outcomes. For example, the BD1-specific inhibitor Olineone induces differentiation of mouse primary oligodendrocytes, whereas pan BET bromodomain inhibitors have the opposite effect (Gacias et al. 2014). The BD2-selective inhibitor RVX-208 was originally identified in a cell-based chemical screen to enhance the production of apoA-1, and only recently was discovered to target BET proteins after completing phase III clinical trials for treatment of atherosclerosis (Bailey et al. 2010; Nicholls et al. 2011, 2012; Picaud et al. 2013; McLure et al. 2014). Interestingly, RVX-208 causes a milder effect on the transcriptome of cells when compared with inhibitors that bind to both BET bromodomains (Picaud et al. 2013).

All of the compounds described above bind to BET bromodomains in a competitive manner with acetyllysine to displace BET-containing protein complexes from chromatin. A more recent innovation in BET inhibitor design has been to conjugate JQ1 with chemical moieties that promote recruitment of E3 ubiquitin ligases, which leads to polyubiquitylation of BET proteins and proteasome-dependent degradation, a strategy known as proteolysis targeting chimera (PROTAC) (Lu et al. 2015; Winter et al. 2015; Zengerle et al. 2015). This new generation of inhibitors (known as dBET1, ARV-825, or MZ1) leads to more potent suppression of BET proteins in cells, and may provide an additional strategy for therapeutic targeting.

Recent studies have shown that many clinical-stage kinase inhibitors, including the CDK inhibitor dinaciclib, the JAK2 inhibitor TG101209, and the PLK1 inhibitor BI-2536, potently inhibit BET bromodomains as an unintended off-target effect (Martin et al. 2013; Ciceri et al. 2014; Dittmann et al. 2014; Ember et al. 2014). These findings raise the possibility that the off-target effect on BET proteins might contribute to the therapeutic effect of these kinase inhibitors. Moreover, these findings reveal an opportunity for the rational design of drugs that simultaneously target specific kinases and BET proteins to augment anticancer activity or prevent drug resistance (Ciceri et al. 2014).

TARGETING THE BRD4–NUT FUSION ONCOPROTEIN IN NUT MIDLINE CARCINOMA

The first malignant context in which BET proteins were proposed as therapeutic targets is in a rare cancer called NUT midline carcinoma (NMC), which is an aggressive subtype of squamous cell carcinoma with a median survival of only 6.7 months (French et al. 2003; French 2010). Most cases of NMC possess a chromosomal translocation that generates a fusion of the amino terminus of BRD4 (or less commonly BRD3 or NSD3) to the carboxyl terminus of NUT, which is normally only expressed in testes (French et al. 2007, 2014; French 2010). The resulting BRD4–NUT oncoprotein retains the two bromodomains and the ET domain fused to a region of NUT that binds to p300, a protein with KAT activity (French et al. 2003, 2007; Reynoird et al. 2010; Wang and You 2015). It is interesting to note that BRD4–NUT (and presumably BRD3–NUT) requires an interaction with NSD3 via its ET domain for its oncogenic function, whereas the NSD3–NUT fusion requires its BRD4-binding motif for its oncogenic function (French et al. 2014). This highlights a remarkable convergence of molecular functions among BRD3–, BRD4–, and NSD3–NUT fusion proteins, and implies that each of these proteins function through similar multisubunit complexes to regulate transcription.

The tumorigenic properties of BRD4–NUT stems from the coupling of its bromodomains with the p300 binding site on NUT, which leads to a positive feedback loop that generates hyperacetylation-driven nuclear foci (Fig. 2) (French et al. 2007, 2014; Yan et al. 2011; Grayson et al. 2014; Alekseyenko et al. 2015; Wang and You 2015). At a genomic level, this positive feedback loop of acetylation and bromodomain-mediated binding generates large contiguous "megadomains" of active chromatin, which are enriched for BRD4–NUT, p300, and histone hyperacetylation, which span >1 Mb of the genome (Alekseyenko et al. 2015). The BRD4–NUT-induced active chromatin domain seems to propagate unfettered until it

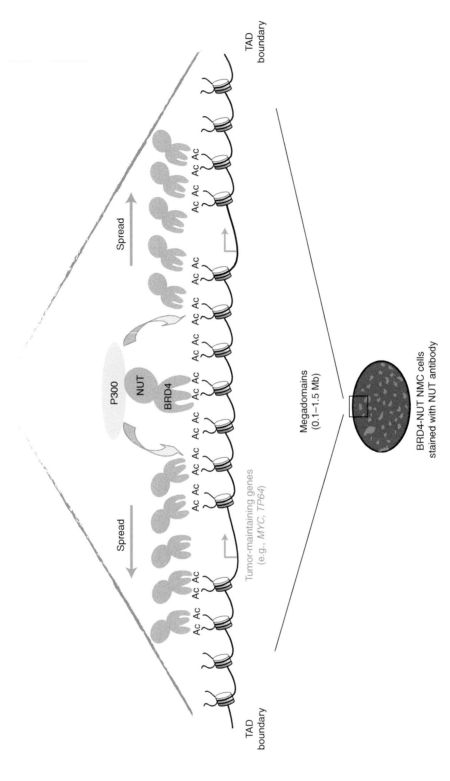

Figure 2. Oncogenic mechanism of the BRD4–NUT fusion protein. In NUT midline carcinoma (NMC) cells, BRD4–NUT forms 80 to 100 large nuclear foci, due to the aggregation of BRD4–NUT on the chromatin. On a molecular level, BRD4–NUT occupies a stretch of highly acetylated chromatin, spanning 0.1–1.5 Mb, termed "mega-domains." The NUT portion of BRD4–NUT recruits p300, which is a histone acetyltransferases that drives further increases in local histone acetylation, thus forming a feed forward loop. The size and spreading of megadomains tends to be restricted to specific topologically associated domains (TADs). Megadomains can occur over genes that are important for the survival of NMC cells, like *MYC* and *TP63*.

encounters the edge of a toplogical domain defined by CTCF/cohesion (Alekseyenko et al. 2015). Active chromatin domains of this size are not observed in normal mammalian chromatin, highlighting a unique chromatin-based mechanism of transformation used by BRD4−NUT. Despite the enormity of these megadomains, only a specific program of genes becomes aberrantly expressed by BRD4−NUT function, which includes *MYC* and the epithelial fate determinant *TP63* (Alekseyenko et al. 2015).

Because the bromodomains of BRD4−NUT are essential to its oncogenic function, NMC provides a clear rationale for evaluating the therapeutic activity of BET bromodomain inhibitors. Exposing cultured NMC cell lines to JQ1 leads to a rapid eviction of BRD4−NUT from chromatin, followed by a rapid suppression of its direct target genes, such as *MYC*, and the induction of terminal squamous cell differentiation (Filippakopoulos et al. 2010; Yan et al. 2011; Grayson et al. 2014; Alekseyenko et al. 2015). Moreover, JQ1 treatment of mice bearing subcutaneous NMC patient-derived xenograft leads to a pronounced inhibition of tumor growth in vivo, with minimal toxicity to normal tissues (Filippakopoulos et al. 2010). This remarkable study provided the first evidence for BET inhibition having a therapeutic index in treating cancer, and provided rationale for subsequent clinical studies of BET inhibition in NMC patients (see below).

A WIDESPREAD SENSITIVITY OF CANCER CELLS TO BET BROMODOMAIN INHIBITION

Given the rarity of NMC, a key question arises as to whether malignancies that lack *BRD4* rearrangements would also be sensitive to BET inhibitors. Because the original description of NMC sensitivity to BET inhibition, numerous studies have shown that BRD4 is a non-oncogene dependency in several forms of cancer. Two studies published in 2011 implicated BRD4 as a vulnerability in the MLL-rearranged subtype of AML (Dawson et al. 2011; Zuber et al. 2011). One used shRNA screening to reveal that AML cells were hypersensitive to genetic knockdown of BRD4, whereas the other study used a proteomic approach to link BRD4 with MLL-fusion cofactors (Dawson et al. 2011; Zuber et al. 2011). These studies, as well as others using multiple myeloma and lymphoma models, showed that BET inhibitors show therapeutic effects in diverse genetic contexts of hematological malignancy, with effects comparable to observations in NMC models (Dawson et al. 2011; Delmore et al. 2011; Mertz et al. 2011; Zuber et al. 2011). Although BET inhibitors show broad efficacy in the blood malignancies, the precise pattern of gene expression changes incurred by BET inhibitor treatment is remarkably heterogeneous among different cancer cell lines. In addition, some cancer cell types will terminally differentiate in response to JQ1/iBET exposure, whereas others undergo apoptosis (Dawson et al. 2011; Delmore et al. 2011; Mertz et al. 2011; Zuber et al. 2011). Although there is generally a lack of consistent gene-expression alterations following BET inhibitor treatment in these malignancies, the well-established oncogenes *MYC*, *BCL2*, and *CDK6* are often suppressed by these drugs, whereas housekeeping genes tend to be unaffected (Shi and Vakoc 2014).

Over the past 5 years, the efficacy of BET inhibitors has been shown in numerous preclinical solid tumor models, including tumors of the prostate, breast, colon, intestine, pancreas, liver, and brain (Sahai et al. 2016). Large-scale profiling studies in human cancer cell lines have suggested that a specific subset of these different tumors harbor exceptional sensitivity to BET inhibitors (Rathert et al. 2015). However, it has been challenging to identify biomarkers that predict hypersensitivity to BET inhibition that might guide patient enrollment into clinical trials. Nonetheless, we can now appreciate that sensitivity to BET inhibition is pervasive across different malignancies.

Although the broad anticancer activity of BET inhibitors is remarkable, it should be emphasized that normal cell types are also affected in unique ways by these agents. This leads to pleiotropic phenotypes in the normal tissues of mice, and presumably in humans. Some of these effects may present therapeutic opportunities in areas outside of oncology. For example, iBET will attenuate cytokine production in innate immune cells, which can allow iBET-treated mice to survive septic shock (Nicodeme et al. 2010). BET inhibitors will also suppress the pathological remodeling of cardiomyoctyes in response to pressure overload (Anand et al. 2013). It has also been proposed that BET inhibitors might be used as a male contraceptive, owing to the reversible impairment in spermatogenesis consequent to BRDT inhibition in the testes (Matzuk et al. 2012; Berkovits and Wolgemuth 2013). BET inhibition will reactivate latent HIV infection, which might prove useful to eliminate viral reservoirs in infected patients (Banerjee et al. 2012; Zhang et al. 2012; Zhu et al. 2012; Li et al. 2013). However, it remains to be determined whether these indications for BET inhibition will be translated into human clinical investigation.

There are also more concerning effects of BET inhibition, such as an impairment in memory formation and an autism-like syndrome in the central nervous system and a worsening of viral/bacterial infections consequent to immunosuppression (Marazzi et al. 2012; Korb et al. 2015; Sullivan et al. 2015). Transgenic *Brd4* shRNA mice, in which BRD4 levels are reduced conditionally in adult tissues, show stem cell depletion in the small intestine and hyperplasia of epidermal tissues (Bolden et al. 2014). The latter phenotype may reflect the emerging role for BRD4 in tumor suppression, and hence BET inhibition might be expected to worsen certain malignancies (Alsarraj and Hunter 2012; Fernandez et al. 2014; Tasdemir et al. 2016). Notably, the side effects of BRD4 knockdown in vivo are known to be reversible after restoring BET protein function (Matzuk et al. 2012; Bolden et al. 2014; Nagarajan et al. 2014).

PHASE I CLINICAL STUDIES OF BET INHIBITORS IN HUMAN CANCER PATIENTS

The efficacy of BET inhibitors in preclinical cancer models provided the rationale for a multitude of ongoing human clinical trials, which includes patients with hematological malignancies, BRD4–NUTexpressing NMC, and various solid tumors. A summary of these ongoing trials can be found in Table 1. Although we cannot determine at the present time the ultimate impact BET inhibitors will have in oncology, there a few key observations than have been made from the initial administration of these agents to human patients thus far. Importantly, the toxicities in humans have been determined for three clinical BET inhibitors (OTX015, TEN-010, and CPI-0610) (Abramson et al. 2015; Shapiro et al. 2015; Amorim et al. 2016; Berthon et al. 2016; Stathis et al. 2016). In one set of phase 1 trials in hematological cancers, OTX015 was administered orally once or twice a day for 21-day cycles. At the higher doses (120–160 mg/d), OTX015 resulted in a substantial, yet reversible thrombocytopenia (drop in platelet counts), severe gastrointestinal events, and fatigue (Amorim et al. 2016; Berthon et al. 2016). The recommended dose for leukemia and lymphoma patients was identified in this study as 80 mg/d given in repetitive cycles of 14 days on followed by 7 days off (Amorim et al. 2016; Berthon et al. 2016). At different doses in these trials, evidence was reported of disease reduction in five out of 37 acute leukemia patients and five out of 17 diffuse large B cell lymphoma patients, whereas no responses were observed in any of the 12 multiple myeloma patients treated (Amorim et al. 2016; Berthon et al. 2016). Consistent with observations in mouse models, OTX015 was found to cause the terminal differentiation of myeloid leukemia cells, as indicated by an increase in peripheral neutrophil counts during treatment (Dawson et al. 2011; Zuber et al. 2011; Berthon et al. 2016). Unfortunately, a specific genetic mutation in leukemia patients has yet to be identified that correlates with responses to OTX015. This highlights the formidable challenge of identifying a predictive biomarker to guide patient enrollment in future studies. In NMC patients, rapid responses to OTX015 have also been identified, in association with tumor regression (Shapiro et al. 2015; Stathis et al. 2016). Although these findings are encouraging and will motivate further phase II evaluation, many of the patients that initially responded later relapsed several months after initiating

Table 1. Clinical trials of BET bromdomain inhibitors

Compound	Sponsor	NCT identifier	Conditions	Clinical phase
ABBV-075	AbbVie	NCT02391480	Advanced cancer; breast cancer; non-small-cell lung cancer (NSCLC); acute myeloid leukemia (AML); multiple myeloma	Phase I (recruiting)
BAY 1238097	Bayer	NCT02369029	Neoplasms	Phase I (terminated)
BI 894999	Boehringer Ingelheim	NCT02516553	Neoplasms	Phase I (recruiting)
BMS-986158	Bristol-Myers Squibb	NCT02419417	Multiple indications cancer	Phase I/IIa (recruiting)
		NCT01949883	Lymphoma	Phase I (recruiting)
CPI-0610	Constellation Pharmaceuticals	NCT02157636	Multiple myeloma	Phase I (recruiting)
		NCT02158858	Leukemia, myelocytic, acute; myelodysplastic syndrome (MDS); meylodusplastic/ myeloproliferative neoplasm, unclassifiable; meylofibrosis	Phase I (recruiting)
FT-1101	Forma Therapeutics	NCT02543879	AML; acute myelogenous leukemia; myelodysplastic syndrome	Phase I (recruiting)
GS-5829	Gilead Sciences	NCT02607228	Metastatic castration-resistant prostate cancer (CRPC) (as a single agent or in combination with enzalutamide)	Phase I (recruiting)
GSK2820151	GlaxoSmithKline	NCT02630251	Cancer	Phase I (not yet open for recruiting)
GSK525762/I-BET762	GlaxoSmithKline	NCT01587703	Carcinoma, midline	Phase I (recruiting)
		NCT01943851	Cancer	Phase I (recruiting)
INCB054329	Incyte Corporation	NCT02431260	Advanced cancer	Phase I/II (recruiting)
		NCT02698189	AML including AML de novo and AML secondary to MDSs; diffuse large B-cell lymphoma (DLBCL)	Phase I (recruiting)
MK-8628	Merck Sharp & Dohme Corp.	NCT02698176	NUT-midline carcinoma (NMC); triple-negative breast cancer (TNBC); NSCLC; CRPC	Phase I (recruiting)
N-methyl-2-pyrrolidone	Peter MacCallum Cancer Centre, Australia	NCT02468687	Multiple myeloma	Phase I (recruiting)
		NCT01713582	AML; DLBCL; acute lymphoblastic leukemia; multiple myeloma	Phase I (active, not recruiting)
OTX015/MK-8628	OncoEthix GmbH/Merck	NCT02259114	NMC; TNBC; NSCLC with rearranged ALK gene/fusion protein or KRAS mutation; CRPC; pancreatic ductal adenocarcinoma	Phase I (active, not recruiting)

Continued

Table 1. *Continued*

Compound	Sponsor	NCT identifier	Conditions	Clinical phase
		NCT02296476	Glioblastoma multiforme	Phase I (terminated)
RVX-208/ RVX 000222	Resverlogix Corp.	NCT02586155	Diabetes mellitus, type 2; coronary artery disease; cardiovascular diseases	Phase III (recruiting)
TEN 010	Tensha Therapeutics/ Roche	NCT01987362	Solid tumors; advanced solid tumors	Phase I (recruiting)
		NCT02308761	MDSs; AML	Phase I (recruiting)
ZEN003694	Zenith Epigenetics	NCT02711956	Metastatic CRPC (in combination with enzalutamide)	Phase I (not yet open for recruiting)
		NCT02705469	Metastatic CRPC	Phase I (recruiting)

See clinicaltrials.gov.

treatment (Stathis et al. 2016). This indicates the importance of studying mechanisms of resistance and the potential of combining BET inhibitors with other agents to provide more durable responses. Taken together, these initial studies have generated sufficient enthusiasm within the pharmaceutical industry to justify a continuing of phase II clinical investigation.

WHY ARE CANCER GENES HYPERSENSITIVE TO BET INHIBITION?

When considering the basic molecular function of BRD4 described above, it is difficult to understand why chemical inhibition of BET proteins would lead to preferential impairment to cancer cells versus nontransformed cell types. Transcriptome-level studies have revealed that BET inhibitors suppress hundreds of genes in each cell type (Anand et al. 2013; Chapuy et al. 2013; Lovén et al. 2013; Asangani et al. 2014). The identity of BET-dependent genes varies dramatically from cell type to cell type, which poses a major challenge in proposing a unified mechanism to explain the anticancer effects of BET inhibitors. At present, our understanding of these effects is limited to correlative observations that these compounds will preferentially suppress expression of cancer-promoting genes versus that of housekeeping genes (Chapuy et al.

2013; Lovén et al. 2013; Nagarajan et al. 2014; Roe et al. 2015; Bhagwat et al. 2016; Henssen et al. 2016). An alternative summation of the available evidence is that JQ1 preferentially suppresses "highly regulated" genes, that is, genes that are dynamically expressed in response to exogenous stimuli or those genes that are expressed in a lineage-specific manner, and hence are influenced by numerous *trans-* and *cis-*acting regulators. Indeed, many growth/cancer-promoting genes (e.g., *MYC* and *BCL2*) fall into this broad category (Lovén et al. 2013; Asangani et al. 2014; Roe et al. 2015; Shu et al. 2016). Such a model would also explain why JQ1 suppresses cytokine genes in immune cells and the immediate-early genes in cardiomyocytes and neurons (Nicodeme et al. 2010; Anand et al. 2013; Brown et al. 2014; Korb et al. 2015; Toniolo et al. 2015). It is important to note that the effects of JQ1 on transcription have been shown to be reversible, that is, withdrawing JQ1 leads to a rapid restoration of the preexisting transcription level (Mertz et al. 2011). This is likely to account for why normal tissues are able to recover in the setting of 14-day-on, 7-day-off BET inhibition treatment-cycles in humans (Amorim et al. 2016; Berthon et al. 2016). Cancer cells are perhaps less able to recover following BET inhibitor treatment, owing to their "addiction" to oncogenes like MYC

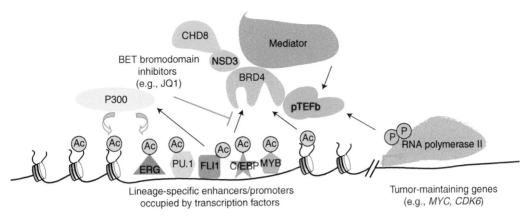

Figure 3. The BRD4 pathway in acute myeloid leukemia (AML) cells. In AML cells, BRD4 is recruited to lineage-specific enhancers and promoters by acetylated histones and transcription factors (TFs), which are acetylated by p300. BRD4 will then recruit several proteins to regions through direct physical interaction, including NSD3/CHD8, the Mediator complex, and p-TEFb, which promotes transcriptional activation. BET bromodomain inhibitors will release BRD4 from chromatin along with its cofactors to suppress transcription. Because many BRD4-occupied enhancers are located at distal upstream or downstream sites, several of these activities are occurring at a distance from the target gene promoter.

(Arvanitis and Felsher 2006). Nevertheless, the specific pattern of genes that are suppressed by BET inhibition in each cell type is clearly of central importance to the therapeutic efficacy of these agents. Below, we describe studies that have defined molecular mechanisms that underlie the gene-specific transcriptional effects of BET bromodomain inhibition.

One explanation for the context-specific effects of BET inhibition is that each cell type expresses a different complement of acetylated TFs that bind to BET bromodomains. In AML cells, for example, it has been found that BRD4 binding across the genome is highly correlated with a set of hematopoietic lineage TFs (ERG, FLI1, PU.1, MYB, C/EBPα, and C/EBPβ), which can each physically associate BRD4 in either a bromodomain-dependent or independent manner (Roe et al. 2015). In JQ1-treated leukemia cells, the downstream target genes of this set of TFs are rapidly suppressed, whereas TF occupancy on DNA remains unaffected (Fig. 3) (Roe et al. 2015). Moreover, ectopic expression of these hematopoietic TFs in fibroblasts can recapitulate the JQ1 transcriptional response seen in leukemia cells, indicated that these TFs are sufficient to specify the effects of BET inhibition (Roe et al. 2015). An expanding

body of literature is continuing to link the anticancer effects of BET inhibition to the functional suppression key TFs—the androgen receptor in prostate cancer (Asangani et al. 2014), the estrogen receptor and TWIST in breast cancer (Nagarajan et al. 2014; Shi et al. 2014), NF-κB in lymphoma and lung cancers (Asangani et al. 2014; Nagarajan et al. 2014; Zou et al. 2014; Gao et al. 2015). The broad suppression of TF function in normal cell types is also likely to be related to the on-target toxicities of BET inhibitors seen in human patients. As an example, the demonstrated interaction between the diacetylated TF GATA-1 and BET proteins provides a potential explanation for the thrombocytopenia observed in BET inhibitor-treated patients (Gamsjaeger et al. 2011; Lamonica et al. 2011; Stonestrom et al. 2015).

It has been observed that many of the genes that are sensitive to BET inhibition have an unusually large repertoire of enhancer elements in their vicinity, which might be interpreted as another indicator of genes that are "highly regulated." Indeed, genes with a large number of enhancers tend to encode lineage-specific and growth-regulatory factors (Hnisz et al. 2013; Whyte et al. 2013). Such enhancers have been given numerous labels (superenhancers, locus control regions, stretch enhancers) and often exist in clusters and harbor high levels of BRD4 enrichment (Li et al. 2002; Chapuy et al. 2013; Lovén et al. 2013; Parker et al. 2013). Using genomic approaches applied to several different cell types, it has been observed that genes with "superenhancers" nearby tend to be more suppressed by JQ1 than randomly chosen expressed genes (Chapuy et al. 2013; Lovén et al. 2013; Peeters et al. 2015). More recent evidence indicates that only a minority of superenhancers are in fact targeted by JQ1, as indicated by measurements of Mediator eviction following BET inhibitor exposure (Bhagwat et al. 2016).

From these studies, it is clear that specific cis-elements in the genome are more suppressed by BET inhibition than others, and this contributes to the biased effects of JQ1 on certain genes. The mechanistic basis underlying these heterogeneous effects is still not understood. One possibility is that enhancer-binding proteins like BRD4 have variable on–off rates at each cis-regulatory element, and that perhaps BET inhibitors will preferentially evict BRD4-containing proteins complexes from chromatin at sites that are more dynamic. Taken together, the context-specific consequences of BET inhibition can be attributed, at least in part, to the specific complement of TF-bound cis-elements (enhancers and promoters) that are suppressed by these small molecules.

MECHANISMS OF RESISTANCE TO BET INHIBITION

All cancer monotherapies are limited by the emergence of drug-resistant cell populations, with the ongoing clinical trials indicating that BET inhibitors are not an exception. Hence, an important area of ongoing investigation has been to define mechanisms of resistance to BET bromodomain inhibition. Several recent studies have shown that acquired resistance to BET inhibition is associated with nongenetic mechanisms, in association with a global alteration of gene expression that compensates for the effects of BET inhibition (Tang et al. 2014; Fong et al. 2015; Rathert et al. 2015; Shu et al. 2016). In MLL-fusion AML, this compensatory

change in gene expression is linked to activation of the WNT signaling pathway, which can restore MYC expression despite chemical blockade of BRD4 (Fong et al. 2015; Rathert et al. 2015). In breast cancer cells, JQ1 resistance is associated with an elevated level of BRD4 phosphorylation, which in turn will bind more tightly to the Mediator complex to achieve bromodomain-independent recruitment to chromatin (Shu et al. 2016). Across a panel of heterogeneous cell lines, it has been found that the sensitivity to BET inhibitors can be correlated with the preexisting expression level of genes that encode apoptosis regulators (Conery et al. 2016). For example, leukemia cell lines with high BCL2 expression and low expression of BCL2L1 (also called BCL-xl) or BAD are generally correlated with higher sensitivity to BET inhibition (Conery et al. 2016). Consistent with this observation, acquired resistance to BET inhibition can be linked to an increase in the expression of BCL2L1 by gaining super enhancers upstream of the BCL2L1 gene (Shu et al. 2016). What is notable across these studies is that resistance to BET inhibition is not associated with BRD2/BRD3/BRD4 mutations, but instead is associated with selection for a rare (presumably preexisting) cell population harboring a pattern of gene expression that bypasses BET inhibition (Fong et al. 2015; Rathert et al. 2015; Conery et al. 2016).

A clear objective for future clinical investigation is to identify drugs that synergize with BET inhibitors in causing anticancer effects, but without having a synergistic increase in toxicity. One promising area of drug combinations is to use BET inhibition as a means to eliminate drug resistance to other targeted agents. In T-cell leukemia driven by activating NOTCH1 mutations, drug-tolerant cells are able to survive NOTCH pathway inhibition (Knoechel et al. 2014). These resistant T-ALL cells are more sensitive to BET bromodomain inhibition than the parental population, thus providing a rationale to combine NOTCH and BET-targeting agents in this disease (Knoechel et al. 2014). A similar scenario is found in breast cancer driven by PI3 kinase mutations, in which sensitivity to PI3K inhibitors is attenuated by feedback pathways that bypass PI3K through the activation of tyrosine kinase pathways (Stratikopoulos et al. 2015). Remarkably, these bypass pathways can be suppressed transcriptionally via BET bromodomain inhibition. Hence, combinations of BET and PI3K inhibitors are a promising therapeutic approach in breast cancer. Resistance to estrogen receptor modulation, Sonic hedgehog, and androgen receptor blockade can also be overcome by BET bromodomain inhibition, thus providing numerous opportunities to explore drug combinations in the clinic (Asangani et al. 2014; Nagarajan et al. 2014; Tang et al. 2014; Shu et al. 2016).

CONCLUDING REMARKS

There is a long history of treating cancer patients with agents that disrupt fundamental cellular processes (e.g., antimetabolites and DNA alkylating agents), which can cause more severe cell death responses in cancer cells than in normal tissues. A similar description could be applied to BET bromodomain inhibitors, which target a set of transcriptional coactivators to disrupt an important hub for numerous TF pathways. As we have outlined in this review, there is clearly specificity in the transcriptional effects of BET inhibition, but only to a degree. Hence, these agents can cause detrimental effects to cancer cells in association with tolerable pleiotropic biological effects in normal tissues. The basic and preclinical research performed in this field has provided a roadmap for the implementation of BET inhibition in the clinic. In our view, the success of BET inhibition in clinical studies will rest squarely on our ability to find predictive biomarkers of therapeutic response and the most effective drug combinations for achieving durable disease remissions.

ACKNOWLEDGMENTS

C.R.V. is supported by the Leukemia and Lymphoma Society, the Burroughs-Wellcome Fund, the Pershing Square Sohn Cancer Research Alliance, the Starr Cancer Consortium, and the National Institutes of Health/National Cancer Institute (NIH/NCI) Grant RO1 CA174793.

REFERENCES

Abramson JS, Blum KA, Flinn IW, Gutierrez M, Goy A, Maris M, Cooper M, Meara M, Borger D, Mertz J, et al. 2015. BET inhibitor CPI-0610 is well tolerated and induces responses in diffuse large B-cell lymphoma and follicular lymphoma: Preliminary analysis of an ongoing phase 1 study. *Blood* **126:** 1491.

Adachi K, Hikawa H, Hamada M, Endoh JI, Ishibuchi S, Fujie N, Tanaka M, Sugahara K, Oshita K, Murata M. 2006. Thienotriazolodiazepine compound and a medicinal use thereof. WPO Patent No. WO2006129623A1.

Alekseyenko AA, Walsh EM, Wang X, Grayson AR, Hsi PT, Kharchenko PV, Kuroda MI, French CA. 2015. The oncogenic BRD4–NUT chromatin regulator drives aberrant transcription within large topological domains. *Genes Dev* **29:** 1507–1523.

Allen BL, Taatjes DJ. 2015. The mediator complex: A central integrator of transcription. *Nat Rev Mol Cell Biol* **16:** 155–166.

Alsarraj J, Hunter KW. 2012. Bromodomain-containing protein 4: A dynamic regulator of breast cancer metastasis through modulation of the extracellular matrix. *Int J Breast Cancer* **2012:** 670632.

Amorim S, Stathis A, Gleeson M, Iyengar S, Magarotto V, Leleu X, Morschhauser F, Karlin L, Broussais F, Rezai K, et al. 2016. Bromodomain inhibitor OTX015 in patients with lymphoma or multiple myeloma: A dose-escalation, open-label, pharmacokinetic, phase 1 study. *Lancet Haematol* **3:** e196–e204.

Anand P, Brown JD, Lin CY, Qi J, Zhang R, Artero PC, Alaiti MA, Bullard J, Alazem K, Margulies KB, et al. 2013. BET bromodomains mediate transcriptional pause release in heart failure. *Cell* **154:** 569–582.

Arvanitis C, Felsher DW. 2006. Conditional transgenic models define how MYC initiates and maintains tumorigenesis. *Semin Cancer Biol* **16:** 313–317.

Asangani IA, Dommeti VL, Wang X, Malik R, Cieslik M, Yang R, Escara-Wilke J, Wilder-Romans K, Dhanireddy S, Engelke C, et al. 2014. Therapeutic targeting of BET bromodomain proteins in castration-resistant prostate cancer. *Nature* **510:** 278–282.

Bailey D, Jahagirdar R, Gordon A, Hafiane A, Campbell S, Chatur S, Wagner GS, Hansen HC, Chiacchia FS, Johansson J, et al. 2010. RVX-208: A small molecule that increases apolipoprotein A-I and high-density lipoprotein cholesterol in vitro and in vivo. *J Am Coll Cardiol* **55:** 2580–2589.

Banerjee C, Archin N, Michaels D, Belkina AC, Denis GV, Bradner J, Sebastiani P, Margolis DM, Montano M. 2012. BET bromodomain inhibition as a novel strategy for reactivation of HIV-1. *J Leukoc Biol* **92:** 1147–1154.

Berkovits BD, Wolgemuth DJ. 2013. The role of the double bromodomain-containing BET genes during mammalian spermatogenesis. *Curr Top Dev Biol* **102:** 293–326.

Berthon C, Raffoux E, Thomas X, Vey N, Gomez-Roca C, Yee K, Taussig DC, Rezai K, Roumier C, Herait P, et al. 2016. Bromodomain inhibitor OTX015 in patients with acute leukaemia: A dose-escalation, phase 1 study. *Lancet Haematol* **3:** e186–e195.

Bhagwat AS, Roe JS, Mok BA, Hohmann AF, Shi J, Vakoc CR. 2016. BET bromodomain inhibition releases the Mediator complex from select *cis*-regulatory elements. *Cell Rep* **15**: 519–530.

Bisgrove DA, Mahmoudi T, Henklein P, Verdin E. 2007. Conserved P-TEFb-interacting domain of BRD4 inhibits HIV transcription. *Proc Natl Acad Sci* **104**: 13690–13695.

Bolden JE, Tasdemir N, Dow Lukas E, van Es Johan H, Wilkinson John E, Zhao Z, Clevers H, Lowe Scott W. 2014. Inducible in vivo silencing of Brd4 identifies potential toxicities of sustained BET protein inhibition. *Cell Rep* **8**: 1919–1929.

Brown JD, Lin CY, Duan Q, Griffin G, Federation A, Paranal RM, Bair S, Newton G, Lichtman A, Kung A, et al. 2014. NF-κB directs dynamic super enhancer formation in inflammation and atherogenesis. *Mol Cell* **56**: 219–231.

Chang YL, King B, Lin SC, Kennison JA, Huang DH. 2007. A double-bromodomain protein, FSH-S, activates the homeotic gene ultrabithorax through a critical promoter-proximal region. *Mol Cell Biol* **27**: 5486–5498.

Chapuy B, McKeown Michael R, Lin Charles Y, Monti S, Roemer Margaretha GM, Qi J, Rahl Peter B, Sun Heather H, Yeda Kelly T, Doench John G, et al. 2013. Discovery and characterization of super-enhancer-associated dependencies in diffuse large B cell lymphoma. *Cancer Cell* **24**: 777–790.

Ciceri P, Müller S, O'Mahony A, Fedorov O, Filippakopoulos P, Hunt JP, Lasater EA, Pallares G, Picaud S, Wells C, et al. 2014. Dual kinase-bromodomain inhibitors for rationally designed polypharmacology. *Nat Chem Biol* **10**: 305–312.

Conery AR, Centore RC, Spillane KL, Follmer NE, Bommi-Reddy A, Hatton C, Bryant BM, Greninger P, Amzallag A, Benes CH, et al. 2016. Preclinical anticancer efficacy of BET bromodomain inhibitors is determined by the apoptotic response. *Cancer Res* **76**: 1313–1319.

Crowe BL, Larue RC, Yuan C, Hess S, Kvaratskhelia M, Foster MP. 2016. Structure of the Brd4 ET domain bound to a C-terminal motif from γ-retroviral integrases reveals a conserved mechanism of interaction. *Proc Natl Acad Sci* **113**: 2086–2091.

Dancy BM, Cole PA. 2015. Protein lysine acetylation by p300/CBP. *Chem Rev* **115**: 2419–2452.

Dawson MA, Prinjha RK, Dittman A, Giotopoulos G, Bantscheff M, Chan W-I, Robson SC, Chung Cw, Hopf C, Savitski MM, et al. 2011. Inhibition of BET recruitment to chromatin as an effective treatment for MLL-fusion leukaemia. *Nature* **478**: 529–533.

Delmore JE, Issa GC, Lemieux ME, Rahl PB, Shi J, Jacobs HM, Kastritis E, Gilpatrick T, Paranal RM, Qi J, et al. 2011. BET bromodomain inhibition as a therapeutic strategy to target c-Myc. *Cell* **146**: 904–917.

Denis GV, Green MR. 1996. A novel, mitogen-activated nuclear kinase is related to a *Drosophila* developmental regulator. *Genes Dev* **10**: 261–271.

Devaiah BN, Lewis BA, Cherman N, Hewitt MC, Albrecht BK, Robey PG, Ozato K, Sims RJ, Singer DS. 2012. BRD4 is an atypical kinase that phosphorylates serine2 of the RNA polymerase II carboxy-terminal domain. *Proc Natl Acad Sci* **109**: 6927–6932.

Devaiah BN, Case-Borden C, Gegonne A, Hsu CH, Chen Q, Meerzaman D, Dey A, Ozato K, Singer DS. 2016. BRD4 is a histone acetyltransferase that evicts nucleosomes from chromatin. *Nat Struct Mol Biol* **23**: 540–548.

Dey A, Chitsaz F, Abbasi A, Misteli T, Ozato K. 2003. The double bromodomain protein Brd4 binds to acetylated chromatin during interphase and mitosis. *Proc Natl Acad Sci* **100**: 8758–8763.

Dhalluin C, Carlson JE, Zeng L, He C, Aggarwal AK, Zhou MM, Zhou MM. 1999. Structure and ligand of a histone acetyltransferase bromodomain. *Nature* **399**: 491–496.

Dittmann A, Werner T, Chung CW, Savitski MM, Fälth Savitski M, Grandi P, Hopf C, Lindon M, Neubauer G, Prinjha RK, et al. 2014. The commonly used PI3-kinase probe LY294002 is an inhibitor of BET bromodomains. *ACS Chem Biol* **9**: 495–502.

Donner AJ, Ebmeier CC, Taatjes DJ, Espinosa JM. 2010. CDK8 is a positive regulator of transcriptional elongation within the serum response network. *Nat Struct Mol Biol* **17**: 194–201.

Ember SWJ, Zhu JY, Olesen SH, Martin MP, Becker A, Berndt N, Georg GI, Schönbrunn E. 2014. Acetyl-lysine binding site of bromodomain-containing protein 4 (BRD4) interacts with diverse kinase inhibitors. *ACS Chem Biol* **9**: 1160–1171.

Fernandez P, Scaffidi P, Markert E, Lee JH, Rane S, Misteli T. 2014. Transformation resistance in a premature aging disorder identifies a tumor-protective function of BRD4. *Cell Rep* **9**: 248–260.

Filippakopoulos P, Knapp S. 2014. Targeting bromodomains: Epigenetic readers of lysine acetylation. *Nat Rev Drug Discov* **13**: 337–356.

Filippakopoulos P, Qi J, Picaud S, Shen Y, Smith WB, Fedorov O, Morse EM, Keates T, Hickman TT, Felletar I, et al. 2010. Selective inhibition of BET bromodomains. *Nature* **468**: 1067–1073.

Filippakopoulos P, Picaud S, Mangos M, Keates T, Lambert JP, Barsyte-Lovejoy D, Felletar I, Volkmer R, Müller S, Pawson T, et al. 2012. Histone recognition and large-scale structural analysis of the human bromodomain family. *Cell* **149**: 214–231.

Fong CY, Gilan O, Lam EYN, Rubin AF, Ftouni S, Tyler D, Stanley K, Sinha D, Yeh P, Morison J, et al. 2015. BET inhibitor resistance emerges from leukaemia stem cells. *Nature* **525**: 538–542.

French CA. 2010. NUT midline carcinoma. *Cancer Genet Cytogenet* **203**: 16–20.

French CA, Miyoshi I, Kubonishi I, Grier HE, Perez-Atayde AR, Fletcher JA. 2003. BRD4–NUT fusion oncogene: A novel mechanism in aggressive carcinoma. *Cancer Res* **63**: 304.

French CA, Ramirez CL, Kolmakova J, Hickman TT, Cameron MJ, Thyne ME, Kutok JL, Toretsky JA, Tadavarthy AK, Kees UR, et al. 2007. BRD–NUT oncoproteins: A family of closely related nuclear proteins that block epithelial differentiation and maintain the growth of carcinoma cells. *Oncogene* **27**: 2237–2242.

French CA, Rahman S, Walsh EM, Kühnle S, Grayson AR, Lemieux ME, Grunfeld N, Rubin BP, Antonescu CR, Zhang S, et al. 2014. NSD3–NUT fusion oncoprotein in NUT midline carcinoma: Implications for a novel oncogenic mechanism. *Cancer Discov* **4**: 928–941.

Gacias M, Gerona-Navarro G, Plotnikov AN, Zhang G, Zeng L, Kaur J, Moy G, Rusinova E, Rodriguez Y, Matikainen B, et al. 2014. Selective chemical modulation of gene tran-

Cite this article as *Cold Spring Harb Perspect Med* doi: 10.1101/cshperspect.a026674

scription favors oligodendrocyte lineage progression. *Chem Biol* **21:** 841–854.

Gamsjaeger R, Webb SR, Lamonica JM, Billin A, Blobel GA, Mackay JP. 2011. Structural basis and specificity of acetylated transcription factor GATA1 recognition by BET family bromodomain protein Brd3. *Mol Cell Biol* **31:** 2632–2640.

Gao F, Yang Y, Wang Z, Gao X, Zheng B. 2015. BRAD4 plays a critical role in germinal center response by regulating Bcl-6 and NF-κB activation. *Cell Immunol* **294:** 1–8.

Grayson AR, Walsh EM, Cameron MJ, Godec J, Ashworth T, Ambrose JM, Aserlind AB, Wang H, Evan GI, Kluk MJ, et al. 2014. MYC, a downstream target of BRD–NUT, is necessary and sufficient for the blockade of differentiation in NUT midline carcinoma. *Oncogene* **33:** 1736–1742.

Hellert J, Weidner-Glunde M, Krausze J, Richter U, Adler H, Fedorov R, Pietrek M, Rückert J, Ritter C, Schulz TF, et al. 2013. A structural basis for BRD2/4-mediated host chromatin interaction and oligomer assembly of Kaposi sarcoma-associated herpesvirus and murine γ-herpesvirus LANA proteins. *PLoS Pathog* **9:** e1003640.

Henssen A, Althoff K, Odersky A, Beckers A, Koche R, Speleman F, Schäfers S, Bell E, Nortmeyer M, Westermann F, et al. 2016. Targeting MYCN-driven transcription by BET-bromodomain inhibition. *Clin Cancer Res* **22:** 2470–2481.

Hnisz D, Abraham BJ, Lee TI, Lau A, Saint-André V, Sigova AA, Hoke H, Young RA. 2013. Super-enhancers in the control of cell identity and disease. *Cell* **155:** 934–947.

Itzen F, Greifenberg AK, Bösken CA, Geyer M. 2014. Brd4 activates P-TEFb for RNA polymerase II CTD phosphorylation. *Nucleic Acids Res* **42:** 7577–7590.

Jang MK, Mochizuki K, Zhou M, Jeong HS, Brady JN, Ozato K. 2005. The bromodomain protein Brd4 is a positive regulatory component of P-TEFb and stimulates RNA polymerase II-dependent transcription. *Mol Cell* **19:** 523–534.

Jiang YW, Veschambre P, Erdjument-Bromage H, Tempst P, Conaway JW, Conaway RC, Kornberg RD. 1998. Mammalian mediator of transcriptional regulation and its possible role as an end-point of signal transduction pathways. *Proc Natl Acad Sci* **95:** 8538–8543.

Jonkers I, Lis JT. 2015. Getting up to speed with transcription elongation by RNA polymerase II. *Nat Rev Mol Cell Biol* **16:** 167–177.

Kharenko OA, Gesner EM, Patel RG, Norek K, White A, Fontano E, Suto RK, Young PR, McLure KG, Hansen HC. 2016. RVX-297—A novel BD2 selective inhibitor of BET bromodomains. *Biochem Biophys Res Commun* **477:** 62–67.

Knoechel B, Roderick JE, Williamson KE, Zhu J, Lohr JG, Cotton MJ, Gillespie SM, Fernandez D, Ku M, Wang H, et al. 2014. An epigenetic mechanism of resistance to targeted therapy in T-cell acute lymphoblastic leukemia. *Nat Genet* **46:** 364–370.

Korb E, Herre M, Zucker-Scharff I, Darnell RB, Allis CD. 2015. BET protein Brd4 activates transcription in neurons and BET inhibitor Jq1 blocks memory in mice. *Nat Neurosci* **18:** 1464–1473.

Krueger BJ, Varzavand K, Cooper JJ, Price DH. 2010. The mechanism of release of P-TEFb and HEXIM1 from the 7SK snRNP by viral and cellular activators includes conformational change in 7SK. *PLoS ONE* **5:** e12335.

Lamonica JM, Deng W, Kadauke S, Campbell AE, Gamsjaeger R, Wang H, Cheng Y, Billin AN, Hardison RC, Mackay JP, et al. 2011. Bromodomain protein Brd3 associates with acetylated GATA1 to promote its chromatin occupancy at erythroid target genes. *Proc Natl Acad Sci* **108:** E159–E168.

Li Q, Peterson KR, Fang X, Stamatoyannopoulos G. 2002. Locus control regions. *Blood* **100:** 3077–3086.

Li Z, Guo J, Wu Y, Zhou Q. 2013. The BET bromodomain inhibitor JQ1 activates HIV latency through antagonizing Brd4 inhibition of Tat-transactivation. *Nucleic Acids Res* **41:** 277–287.

Liu W, Ma Q, Wong K, Li W, Ohgi K, Zhang J, Aggarwal A K, Rosenfeld M G. 2013. Brd4 and JMJD6-associated anti-pause enhancers in regulation of transcriptional pause release. *Cell* **155:** 1581–1595.

Lovén J, Hoke HA, Lin CY, Lau A, Orlando DA, Vakoc CR, Bradner JE, Lee TI, Young RA. 2013. Selective inhibition of tumor oncogenes by disruption of super-enhancers. *Cell* **153:** 320–334.

Lu J, Qian Y, Altieri M, Dong H, Wang J, Raina K, Hines J, Winkler J D, Crew Andrew P, Coleman K, et al. 2015. Hijacking the E3 ubiquitin ligase Cereblon to efficiently target BRD4. *Chem Biol* **22:** 755–763.

Marazzi I, Ho JSY, Kim J, Manicassamy B, Dewell S, Albrecht RA, Seibert CW, Schaefer U, Jeffrey KL, Prinjha RK, et al. 2012. Suppression of the antiviral response by an influenza histone mimic. *Nature* **483:** 428–433.

Martin MP, Olesen SH, Georg GI, Schönbrunn E. 2013. Cyclin-dependent kinase inhibitor dinaciclib interacts with the acetyl-lysine recognition site of bromodomains. *ACS Chem Biol* **8:** 2360–2365.

Matzuk MM, McKeown MR, Filippakopoulos P, Li Q, Ma L, Agno JE, Lemieux ME, Picaud S, Yu RN, Qi J, et al. 2012. Small-molecule inhibition of BRDT for male contraception. *Cell* **150:** 673–684.

McLure KG, Gesner EM, Tsujikawa L, Kharenko OA, Attwell S, Campeau E, Wasiak S, Stein A, White A, Fontano E, et al. 2014. RVX-208, an inducer of ApoA-I in humans, is a BET bromodomain antagonist. *PLoS ONE* **8:** e83190.

Mertz JA, Conery AR, Bryant BM, Sandy P, Balasubramanian S, Mele DA, Bergeron L, Sims RJ. 2011. Targeting MYC dependence in cancer by inhibiting BET bromodomains. *Proc Natl Acad Sci* **108:** 16669–16674.

Miyoshi S, Ooike S, Iwata K, Hikawa H, Sugahara K. 2010. Antitumor agent. U.S. Patent 20100286127 A1.

Miyoshi S, Ooike S, Iwata K, Hikawa H, Sugahara K. 2013. Antitumor agent. U.S. Patent 8476260 B2.

Moriniere J, Rousseaux S, Steuerwald U, Soler-Lopez M, Curtet S, Vitte AL, Govin J, Gaucher J, Sadoul K, Hart DJ, et al. 2009. Cooperative binding of two acetylation marks on a histone tail by a single bromodomain. *Nature* **461:** 664–668.

Nagarajan S, Hossan T, Alawi M, Najafova Z, Indenbirken D, Bedi U, Taipaleenmäki H, Ben-Batalla I, Scheller M, Loges S, et al. 2014. Bromodomain protein BRD4 is required for estrogen receptor-dependent enhancer activation and gene transcription. *Cell Rep* **8:** 460–469.

Nicholls SJ, Gordon A, Johansson J, Wolski K, Ballantyne CM, Kastelein JJP, Taylor A, Borgman M, Nissen SE. 2011. Efficacy and safety of a novel oral inducer of apolipoprotein A-I synthesis in statin-treated patients with stable coronary artery disease: A randomized controlled trial. *J Am College Cardiol* **57**: 1111–1119.

Nicholls SJ, Gordon A, Johannson J, Ballantyne CM, Barter PJ, Brewer HB, Kastelein JJP, Wong NC, Borgman MRN, Nissen SE. 2012. ApoA-I induction as a potential cardio-protective strategy: Rationale for the SUSTAIN and ASSURE studies. *Cardiovasc Drugs Ther* **26**: 181–187.

Nicodeme E, Jeffrey KL, Schaefer U, Beinke S, Dewell S, Chung Cw, Chandwani R, Marazzi I, Wilson P, Coste H, et al. 2010. Suppression of inflammation by a synthetic histone mimic. *Nature* **468**: 1119–1123.

Park SR, Speranza G, Piekarz R, Wright JJ, Kinders RJ, Wang L, Pfister T, Trepel JB, Lee MJ, Alarcon S, et al. 2013. A multi-histology trial of fostamatinib in patients with advanced colorectal, non-small cell lung, head and neck, thyroid, and renal cell carcinomas, and pheochromocytomas. *Cancer Chemother Pharmacol* **71**: 981.

Parker SCJ, Stitzel ML, Taylor DL, Orozco JM, Erdos MR, Akiyama JA, van Bueren KL, Chines PS, Narisu N, Program NCS, et al. 2013. Chromatin stretch enhancer states drive cell-specific gene regulation and harbor human disease risk variants. *Proc Natl Acad Sci* **110**: 17921–19726.

Peeters JGC, Vervoort SJ, Tan Sander C, Mijnheer G, de Roock S, Vastert SJ, Nieuwenhuis EES, van Wijk F, Prakken BJ, Creyghton MP, et al. 2015. Inhibition of super-enhancer activity in autoinflammatory site-derived T cells reduces disease-associated gene expression. *Cell Rep* **12**: 1986–1996.

Picaud S, Wells C, Felletar I, Brotherton D, Martin S, Savitsky P, Diez-Dacal B, Philpott M, Bountra C, Lingard H, et al. 2013. RVX-208, an inhibitor of BET transcriptional regulators with selectivity for the second bromodomain. *Proc Natl Acad Sci* **110**: 19754–19759.

Rahman S, Sowa ME, Ottinger M, Smith JA, Shi Y, Harper JW, Howley PM. 2011. The Brd4 extraterminal domain confers transcription activation independent of pTEFb by recruiting multiple proteins, including NSD3. *Mol Cell Biol* **31**: 2641–2652.

Rathert P, Roth M, Neumann T, Muerdter F, Roe JS, Muhar M, Deswal S, Cerny-Reiterer S, Peter B, Jude J, et al. 2015. Transcriptional plasticity promotes primary and acquired resistance to BET inhibition. *Nature* **525**: 543–547.

Reynoird N, Schwartz BE, Delvecchio M, Sadoul K, Meyers D, Mukherjee C, Caron C, Kimura H, Rousseaux S, Cole PA, et al. 2010. Oncogenesis by sequestration of CBP/p300 in transcriptionally inactive hyperacetylated chromatin domains. *EMBO J* **29**: 2943–2952.

Roe JS, Mercan F, Rivera K, Pappin DJ, Vakoc CR. 2015. BET bromodomain inhibition suppresses the function of hematopoietic transcription factors in acute myeloid leukemia. *Mol Cell* **58**: 1028–1039.

Roth SY, Denu JM, Allis CD. 2001. Histone acetyltransferases. *Annu Rev Biochem* **70**: 81–120.

Sahai V, Redig AJ, Collier KA, Eckerdt FD, Munshi HG. 2016. Targeting bet bromodomain proteins in solid tumors. *Oncotarget* doi: 10.18632/oncotarget.9804.

Sanchez R, Zhou MM. 2009. The role of human bromodomains in chromatin biology and gene transcription. *Curr Opin Drug Discov Dev* **12**: 659–665.

Shapiro GI, Dowlati A, LoRusso PM, Eder JP, Anderson A, Do KT, Kagey MH, Sirard C, Bradner JE, Landau SB. 2015. Clinically efficacy of the BET bromodomain inhibitor TEN-010 in an open-label substudy with patients with documented NUT-midline carcinoma (NMC). *AACR Annual Meeting, Abstract A49.* Philadelphia, PA, April 18–22.

Shen C, Ipsaro JJ, Shi J, Milazzo JP, Wang E, Roe JS, Suzuki Y, Pappin DJ, Joshua-Tor L, Vakoc CR. 2015. NSD3-Short is an adaptor protein that couples BRD4 to the CHD8 chromatin remodeler. *Mol Cell* **60**: 847–859.

Shi J, Vakoc CR. 2014. The mechanisms behind the therapeutic activity of BET bromodomain inhibition. *Mol Cell* **54**: 728–736.

Shi J, Wang Y, Zeng L, Wu Y, Deng J, Zhang Q, Lin Y, Li J, Kang T, Tao M, et al. 2014. Disrupting the interaction of BRD4 with di-acetylated twist suppresses tumorigenesis in basal-like breast cancer. *Cancer Cell* **25**: 210–225.

Shu S, Lin CY, He HH, Witwicki RM, Tabassum DP, Roberts JM, Janiszewska M, Jin Huh S, Liang Y, Ryan J, et al. 2016. Response and resistance to BET bromodomain inhibitors in triple-negative breast cancer. *Nature* **529**: 413–417.

Smith SG, Sanchez R, Zhou MM. 2014. Privileged diazepine compounds and their emergence as bromodomain inhibitors. *Chem Biol* **21**: 573–583.

Stathis A, Zucca E, Bekradda M, Gomez-Roca C, Delord JP, de La Motte Rouge T, Uro-Coste E, de Braud F, Pelosi G, French CA. 2016. Clinical response of carcinomas harboring the BRD4–NUT oncoprotein to the targeted bromodomain inhibitor OTX015/MK-8628. *Cancer Discov* **6**: 492–500.

Steegmaier M, Hoffmann M, Baum A, Lénárt P, Petronczki M, Krššák M, Gürtler U, Garin-Chesa P, Lieb S, Quant J, et al. 2007. BI 2536, a potent and selective inhibitor of polo-like kinase 1, inhibits tumor growth in vivo. *Curr Biol* **17**: 316–322.

Stonestrom AJ, Hsu SC, Jahn KS, Huang P, Keller CA, Giardine BM, Kadauke S, Campbell AE, Evans P, Hardison RC, et al. 2015. Functions of BET proteins in erythroid gene expression. *Blood* **125**: 2825–2834.

Stratikopoulos E E, Dendy M, Szabolcs M, Khaykin Alan J, Lefebvre C, Zhou MM, Parsons R. 2015. Kinase and BET inhibitors together clamp inhibition of PI3K signaling and overcome resistance to therapy. *Cancer Cell* **27**: 837–851.

Sullivan JM, Badimon A, Schaefer U, Ayata P, Gray J, Chung CW, von Schimmelmann M, Zhang F, Garton N, Smithers N, et al. 2015. Autism-like syndrome is induced by pharmacological suppression of BET proteins in young mice. *J Exp Med* **212**: 1771–1781.

Tang Y, Gholamin S, Schubert S, Willardson MI, Lee A, Bandopadhayay P, Bergthold G, Masoud S, Nguyen B, Vue N, et al. 2014. Epigenetic targeting of Hedgehog pathway transcriptional output through BET bromodomain inhibition. *Nat Med* **20**: 732–740.

Tasdemir N, Banito A, Roe JS, Alonso-Curbelo D, Camiolo M, Tschaharganeh DF, Huang C-H, Aksoy O, Bolden JE, Chen CC, et al. 2016. BRD4 connects enhancer remodeling to senescence immune surveillance. *Cancer Discov* **6:** 612–629.

Toniolo PA, Liu S, Yeh JE, Moraes-Vieira PM, Walker SR, Vafaizadeh V, Barbuto JAM, Frank DA. 2015. Inhibiting STAT5 by the BET bromodomain inhibitor JQ1 disrupts human dendritic cell maturation. *J Immunol* **194:** 3180–3190.

Wang R, You J. 2015. Mechanistic analysis of the role of bromodomain-containing protein 4 (BRD4) in BRD4–NUT oncoprotein-induced transcriptional activation. *J Biol Chem* **290:** 2744–2758.

Wang Z, Zang C, Rosenfeld JA, Schones DE, Barski A, Cuddapah S, Cui K, Roh TY, Peng W, Zhang MQ, et al. 2008. Combinatorial patterns of histone acetylations and methylations in the human genome. *Nat Genet* **40:** 897–903.

Wang W, Yao X, Huang Y, Hu X, Liu R, Hou D, Chen R, Wang G. 2013. Mediator MED23 regulates basal transcription in vivo via an interaction with P-TEFb. *Transcription* **4:** 39–51.

Whyte WA, Orlando DA, Hnisz D, Abraham BJ, Lin CY, Kagey MH, Rahl PB, Lee TI, Young RA. 2013. Master transcription factors and Mediator establish superenhancers at key cell identity genes. *Cell* **153:** 307–319.

Winter GE, Buckley DL, Paulk J, Roberts JM, Souza A, Dhe-Paganon S, Bradner JE. 2015. Phthalimide conjugation as a strategy for in vivo target protein degradation. *Science* **348:** 1376–1381.

Wu SY, Chiang CM. 2007. The double bromodomain-containing chromatin adaptor Brd4 and transcriptional regulation. *J Biol Chem* **282:** 13141–13145.

Yan J, Diaz J, Jiao J, Wang R, You J. 2011. Perturbation of BRD4 protein function by BRD4–NUT protein abrogates cellular differentiation in NUT midline carcinoma. *J Biol Chem* **286:** 27663–27675.

Yang Z, Yik JHN, Chen R, He N, Jang MK, Ozato K, Zhou Q. 2005. Recruitment of P-TEFb for stimulation of transcriptional elongation by the bromodomain protein Brd4. *Mol Cell* **19:** 535–545.

Zengerle M, Chan KH, Ciulli A. 2015. Selective small molecule induced degradation of the BET bromodomain protein BRD4. *ACS Chem Biol* **10:** 1770–1777.

Zhang G, Liu R, Zhong Y, Plotnikov AN, Zhang W, Zeng L, Rusinova E, Gerona-Nevarro G, Moshkina N, Joshua J, et al. 2012. Down-regulation of NF-κB transcriptional activity in HIV-associated kidney disease by BRD4 inhibition. *J Biol Chem* **287:** 28840–28851.

Zhang G, Plotnikov AN, Rusinova E, Shen T, Morohashi K, Joshua J, Zeng L, Mujtaba S, Ohlmeyer M, Zhou MM. 2013. Structure-guided design of potent diazobenzene inhibitors for the BET bromodomains. *J Med Chem* **56:** 9251–9264.

Zhang Q, Zeng L, Shen C, Ju Y, Konuma T, Zhao C, Vakoc Christopher R, Zhou MM. 2016. Structural mechanism of transcriptional regulator NSD3 recognition by the ET domain of BRD4. *Structure* **24:** 1201–1208.

Zhu J, Gaiha GD, Sinu JP, Pertel T, Chin CR, Gao G, Qu H, Walker BD, Elledge SJ, Brass AL. 2012. Reactivation of latent HIV-1 by inhibition of BRD4. *Cell Rep* **2:** 807–816.

Zou Z, Huang B, Wu X, Zhang H, Qi J, Bradner J, Nair S, Chen LF. 2014. Brd4 maintains constitutively active NF-κB in cancer cells by binding to acetylated RelA. *Oncogene* **33:** 2395–2404.

Zuber J, Shi J, Wang E, Rappaport AR, Herrmann H, Sison EA, Magoon D, Qi J, Blatt K, Wunderlich M, et al. 2011. RNAi screen identifies Brd4 as a therapeutic target in acute myeloid leukaemia. *Nature* **478:** 524–528.

Histone Lysine Demethylase Inhibitors

Ashwini Jambhekar,[1] Jamie N. Anastas,[1,2] and Yang Shi[1,2]

[1]Division of Newborn Medicine and Epigenetics Program, Department of Medicine, Boston Children's Hospital, Boston, Massachusetts 02115

[2]Department of Cell Biology, Harvard Medical School, Boston, Massachusetts 02115

Correspondence: yang.shi@childrens.harvard.edu

The dynamic regulation of covalent modifications to histones is essential for maintaining genomic integrity and cell identity and is often compromised in cancer. Aberrant expression of histone lysine demethylases has been documented in many types of blood and solid tumors, and thus demethylases represent promising therapeutic targets. Recent advances in high-throughput chemical screening, structure-based drug design, and structure–activity relationship studies have improved both the specificity and the in vivo efficacy of demethylase inhibitors. This review will briefly outline the connection between demethylases and cancer and will provide a comprehensive overview of the structure, specificity, and utility of currently available demethylase inhibitors. To date, a select group of demethylase inhibitors is being evaluated in clinical trials, and additional compounds may soon follow from the bench to the bedside.

THE DYNAMIC NATURE OF HISTONE METHYLATION

Chromatin, which is mainly composed of DNA and histone proteins, is the template on which many important nuclear processes take place, including transcription and DNA replication. Chromatin consists of repeating units of nucleosomes, comprising 147 base pairs of DNA wrapped around an octamer of histones (typically two each of histones H2A, H2B, H3, and H4) (Luger et al. 1997), which are then further compacted into higher order structures. The histones themselves, particularly H3 and H4, are subject to extensive chemical modifications such as phosphorylation, ubiquitination, acetylation, and methylation (Jenuwein and Al-lis 2001), which have profound effects on gene expression. Consequently, the mechanisms that regulate these modifications are relevant to many areas of biology. The effects of histone methylation, which occurs primarily on arginines and lysines, depend on the site of modification, the extent of methylation, as well as on additional modifications on the same or neighboring histones (Kouzarides 2007). Patterns of nucleosome methylation affect gene expression, replication, the maintenance of genome stability, and other DNA metabolic processes; thus, the mechanisms that regulate histone methylation are relevant to both normal development and diseases like cancer.

As methylation marks are quite stable, they were initially considered to be irreversible. Early models for reversal of histone methylation invoked clipping of modified histone tails or replacement of entire histones, although both failed to explain the rapid changes in histone modifications observed in vivo (Bannister et al. 2002). However, an early study measured formaldehyde production as an indication for possible histone demethylase activity and found potential activities primarily in the kidney (Paik and Kim 1973). However, it was unclear whether formaldehyde production was the direct action of a demethylase, and no evidence was provided for the resulting demethylated histones, or for the molecular nature/mechanism of the demethylase enzyme. The first irrefutable evidence that methylation could be dynamically

regulated came in 2004 with the discovery of the lysine-specific demethylase LSD1 (also known as KDM1A) (Shi et al. 2004). Similar to monoamine oxidases (MAOs), LSD1 uses FAD as a cofactor to oxidize the methyl group and hydrolyze it to formaldehyde (Fig. 1A). This mechanism precludes the use of trimethylated lysine as a substrate, which does not contain a free electron pair required for the first step of the reaction. Accordingly, LSD1 demethylates H3K4me1/2, but not H3K4me3, or other methylated lysines in H3 such as H3K20me2 (Shi et al. 2004). In prostate cancer cells, LSD1 also demethylates H3K9me1/2 when complexed to the androgen receptor (Metzger et al. 2005), and other LSD1 variants have shown different

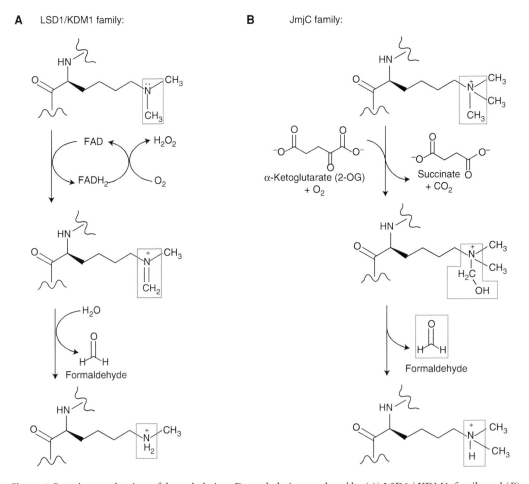

Figure 1. Reaction mechanism of demethylation. Demethylation catalyzed by (*A*) LSD1/ KDM1-family and (*B*) JmjC-family enzymes.

Cite this article as *Cold Spring Harb Perspect Med* doi: 10.1101/cshperspect.a026484

substrate specificities (Laurent et al. 2015; Wang et al. 2015a). Later, multiple groups discovered additional histone demethylases with various substrate requirements (both in terms of lysine residues as well as extent of methylation), revealing the dynamic nature of multiple types of histone methylations. With the exception of LSD2, a close homolog of LSD1 (Karytinos et al. 2009; Fang et al. 2010), the other demethylases fall into the Jumonji C (JmjC) class, which uses Fe(II) and 2-oxoglutarate (2-OG, or α-ketoglutarate) as cofactors to hydroxylate the methyl groups via a free-radical mechanism (Fig. 1B), which is then released as formaldehyde (Tsukada et al. 2006). Importantly, this reaction mechanism allows the reversal of trimethylations (Cloos et al. 2006; Klose et al. 2006; Whetstine et al. 2006), which LSD1 and LSD2 cannot catalyze. The discoveries of these enzymes highlight the specific and dynamic regulation of methylation at various histone lysine residues.

ABERRANT HISTONE REGULATION IN CANCER

Misregulation of lysine demethylases is frequently observed in cancer, and the diverse natures of the regulatory defects indicate that cellular homeostasis requires a precise balance of histone methylation and demethylation. Mutations or translocations of genes encoding demethylases are relatively infrequent, but variations in the expression levels of demethylases are more common (Chen et al. 2014). Because the combination of histone modifications with other regulatory processes influences the overall biological outcomes, the misregulation of demethylases in tumors can result in different consequences depending on the tissue of origin, the presence of other mutations, and the activities of other gene expression networks. For example, LSD1 is down-regulated in some breast cancers (Wang et al. 2009), but up-regulated in many other solid tumors and leukemias (Hayami et al. 2011; Schildhaus et al. 2011). A single type of cancer can show misregulation of different demethylases, such as up-regulation of KDM5C (Wang et al. 2015c), KDM5B (Yamane et al. 2007), KDM4A, or KDM4B (Berry et al.

2012) in breast cancers. Although co-occurring mutations in chromatin regulators have been documented in bladder cancer (Gui et al. 2011), the extent to which such events occur has not been comprehensively analyzed, leaving open the possibility that defects in multiple chromatin regulators may be operative in some cancers. Although current cancer genome sequencing efforts have not found frequent mutations and translocations of histone demethylases in cancers, there are a few notable examples. KDM6A (UTX), an H3K27me2/3 demethylase (Agger et al. 2007; Lan et al. 2007), is often truncated or mutated in a wide range of cancers, including cancers of the breast, pancreas, lung, and kidney and in leukemias (van Haaften et al. 2009; Dalgliesh et al. 2010). Additionally, a fusion of the H3K4me2/3 demethylase KDM5A (JARID1A) to the nuclear pore component Nup98 was described in one acute myeloid leukemia (AML) case (van Zutven et al. 2006), and this fusion caused leukemias when expressed in mice (Gough et al. 2014). Thus, changes in expression levels, mutations, or translocations of histone demethylase genes can all contribute to carcinogenesis.

Although the precise effects of the misregulation of histone demethylases have not been determined for many cancers, the carcinogenic potential of these events appears to rest on direct or indirect regulation of tumor suppressors or oncogenes. Increased LSD1 expression, induced by expressing a genetic MLL-AF9 fusion in murine leukemia model, promoted transcription of MYC and the associated "core module" of embryonic stem cell (ESC) genes (Harris et al. 2012), leading to excessive proliferation and a loss of normal differentiation. In AML, LSD1 represses differentiation markers such as E-cadherin (Murray-Stewart et al. 2014). Consistently, chemical or genetic inhibition of LSD1 results in differentiation of leukemia cells in vitro (Schenk et al. 2012), and LSD1 has been suggested to regulate cancer stem cells (Harris et al. 2012). Loss of LSD1 expression in some breast cancers up-regulates transforming growth factor β (TGF-β), promoting cellular invasion (Wang et al. 2009), whereas overexpression of KDM5B (PLU-1) represses genes that promote differentiation and maintenance of genome integrity, such as CAV-1 and BRCA-1 (Yamane et al. 2007). In prostate cancer, KDM4B overexpression promotes cell proliferation by targeting cell cycle regulators such as PLK and Aurora kinase A (Duan et al. 2015). KDM5A (JARID1A) promotes the maintenance of a drug-resistant state in a subset of cancer cells, although the essential KDM5A targets responsible for this phenotype have yet to be identified (Sharma et al. 2010). These studies highlight the context-dependent effects of demethylases in cancer, and suggest that re-establishing the balance of methylation levels could be an effective therapeutic strategy.

Given the therapeutic potential of targeting histone methylation in cancer and other diseases, many research groups are pursuing the development of demethylase inhibitors. Because the catalytic amine oxidase domain of LSD1 is homologous to that of MAOs, the first drugs developed against LSD1 were based on MAO inhibitors (MAOIs), which suffered from a lack of specificity for LSD1 (Lee et al. 2006). Drug development for the JmjC class of demethylases has focused on optimizing mimics of the 2-OG cofactor (Chen et al. 2007) or developing inhibitors that interfere with metal binding (e.g., Sekirnik et al. 2009). Additionally, focused screens of natural compounds (Willmann et al. 2012; Wu et al. 2012) or other classes of molecules such as polyphenols (Abdulla et al. 2013) have yielded effective histone demethylase inhibitors. Over the years, a plethora of compounds targeting histone lysine demethylases has been developed; we will focus on inhibitors reported in the academic literature to have shown activity in cellular or animal model studies of cancer, as these compounds show the most potential for development into therapeutics.

LSD1/KDM1A

LSD1 is a flavin-dependent monoamine oxidase homolog that demethylates H3K4me2 in vitro and in vivo (Shi et al. 2004), and can demethylate H3K9me2 in vivo when associated with nuclear hormone receptor complexes (Metzger et al. 2005; Garcia-Bassets et al. 2007; Perillo et al. 2008). LSD1-dependent demethylation is conducted by the catalytic amine oxidase domain, in conjunction with a SWIRM domain that mediates LSD1 stability (Chen et al. 2006; Stavropoulos et al. 2006) and histone recognition (Stavropoulos et al. 2006), and a Tower domain that interacts with the transcriptional repressor CoREST (Chen et al. 2006), which is important for LSD1 to access nucleosomal substrates (Lee et al. 2005; Shi et al. 2005). LSD1 can both activate and repress transcription (Wang et al. 2007), depending on the surrounding chromatin landscape (Perillo et al. 2008), its association with cofactors (Garcia-Bassets et al. 2007), and whether it targets H3K4me2 (Martin and Zhang 2005) or H3K9me2 (Barski et al. 2007). The specificity of LSD1 for H3K4me2 over H3K9me2 is in part determined by the LSD1 splice isoform being expressed (Laurent et al. 2015; Wada et al. 2015; Wang et al. 2015a). We found that the inclusion of an eight-amino-acid exon switched its specificity from H3K4me2 to H3K9me2 (Laurent et al. 2015), and others have reported substrate preferences of H3K9me2 (Wada et al. 2015) or H4K20me1/2 for different splice variants of LSD1 (Wang et al. 2015a). LSD1 is typically overexpressed in cancers (Hayami et al. 2011; Schildhaus et al. 2011), and even though the critical targets that promote carcinogenesis vary between cancer types, LSD1 inhibitors have been actively pursued. LSD1 inhibitors fall into four major classes: irreversible derivatives of the monoamine oxidase inhibitors (Fig. 2), reversible polyamine or peptide inhibitors, rationally designed fusions of active molecules, and novel compounds not known to inhibit MAOIs (Fig. 3). These classes are discussed in detail below (Table 1; Figs. 2 and 3).

IRREVERSIBLE LSD1 INHIBITORS

The earliest inhibitors tested against LSD1 were irreversible MAOIs, selected based on the homology of the catalytic domain of LSD1 with that of the monoamine oxidases (MAOs, of which there are two, MAO-A and MAO-B). Pargyline (Fig. 2), which forms a covalent adduct with FAD (Oreland et al. 1973), was effective in cell culture (Metzger et al. 2005; Lv et al. 2012) and tumor xenografts models (Cortez et al.

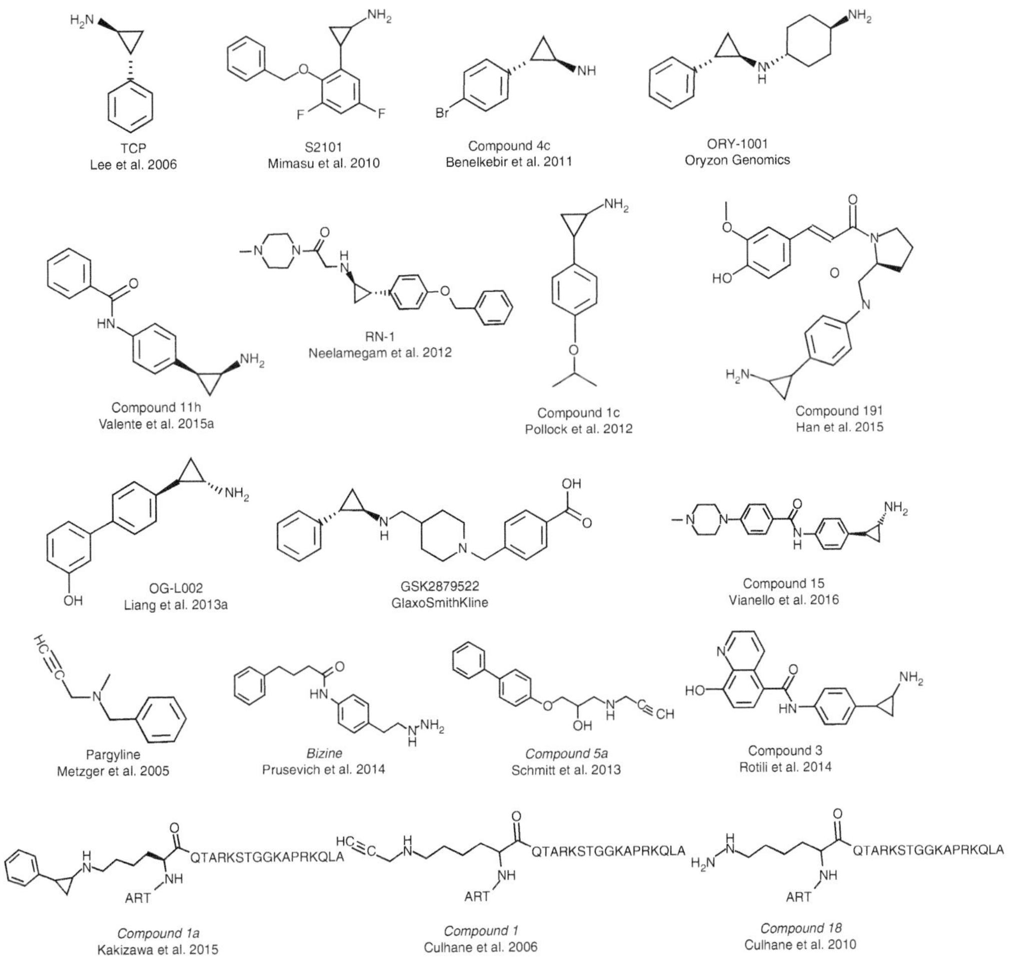

Figure 2. Chemical structures of irreversible LSD1 inhibitors, and references describing their development or use. Compounds listed in italics have not shown cytotoxic activity.

2012; Sareddy et al. 2013; Wang et al. 2015b) on its own, or when used in combination therapies (Rose et al. 2008; Huang et al. 2012; Vasilatos et al. 2013). However, pargyline showed low potency for inhibiting LSD1 in vitro (Schmidt and McCafferty 2007) and in vivo (Kauffman et al. 2011), suggesting that certain transcriptional or chromatin regulatory events might be sensitive to incomplete inhibition of LSD1, or the cellular effects of pargyline may arise from off-target interactions. In contrast, bizine (Fig. 2), a derivative of the irreversible MAOI phenelzine, promisingly showed specificity for LSD1 over MAO-A and MAO-B, but has not been extensively tested in vivo (Prusevich et al. 2014). A third irrevers-

ible inhibitor, tranylcypromine (TCP) (Fig. 2), long used for treating psychiatric diseases (Agin 1960), also efficiently inhibited LSD1 (Lee et al. 2006) by forming a covalent adduct with the FAD cofactor (Schmidt and McCafferty 2007; Yang et al. 2007). Despite a lack of specificity for LSD1 over other MAOs, TCP inhibited growth in several cell-culture-based models of cancer (Ding et al. 2013; Ferrari-Amorotti et al. 2013), mouse xenograft models of breast cancer and oral squamous cell carcinoma (Ferrari-Amorotti et al. 2014; Wang et al. 2016), and a genetic model of MLL-AF9-driven driven acute myeloid leukemia (AML) (Harris et al. 2012). Additionally, TCP has been effective in

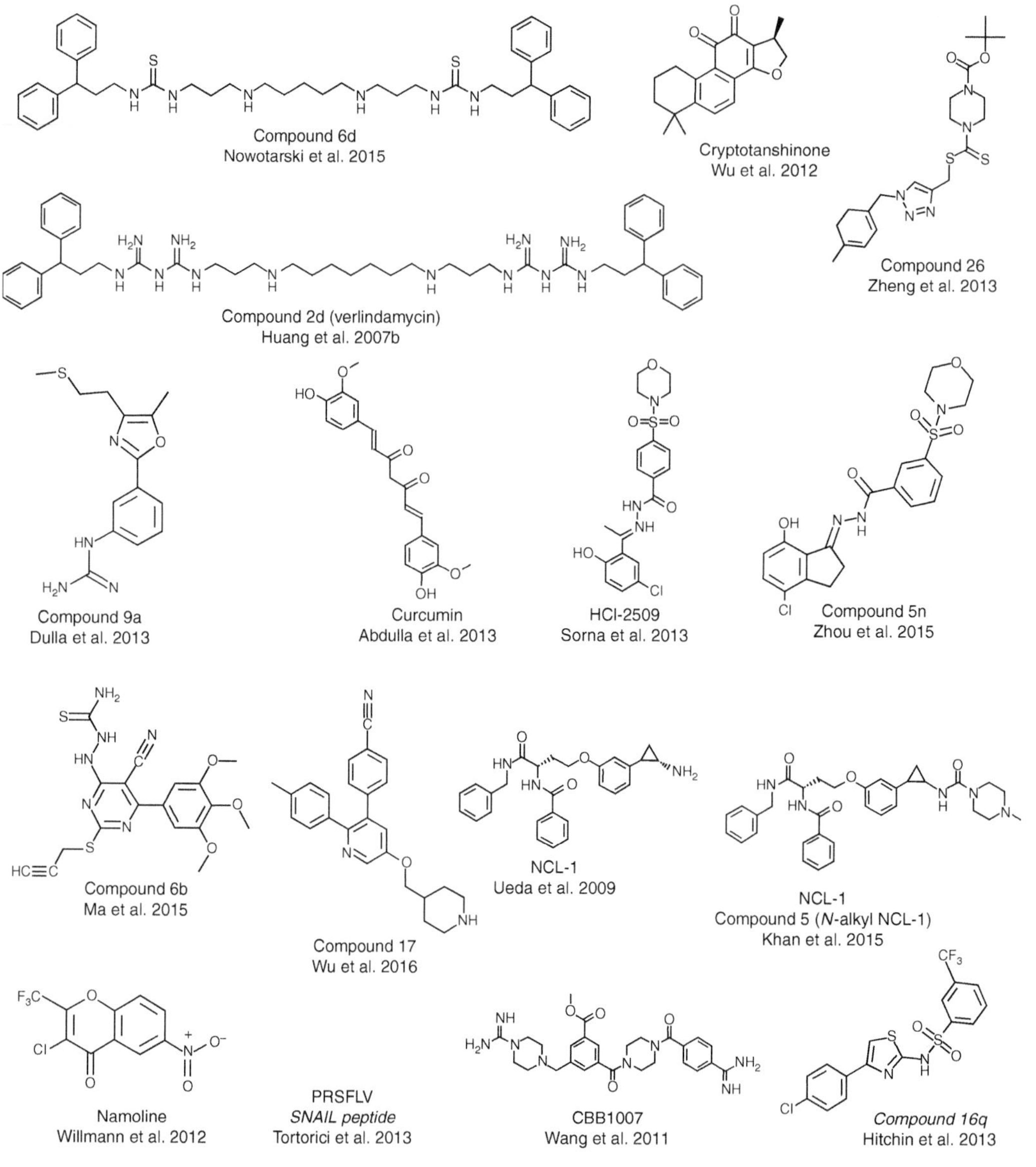

Figure 3. Chemical structures of reversible LSD1 inhibitors, and references describing their development or use. Letters in large font denote single-letter amino acid codes. Compounds listed in italics have not shown cytotoxic activity.

combination therapy approaches (Singh et al. 2011; Schenk et al. 2012), notably promoting AML differentiation in combination with all-*trans* retinoic acid (ATRA) (Schenk et al. 2012). Two clinical trials are underway to evaluate this combination therapy in AML and myelodysplastic syndrome (MDS). The success of combination therapy approaches indicates that

this strategy may be a viable alternative to the development of highly specific compounds, as it can allow each drug to be used at a lower dose, thus minimizing off-target effects.

Despite the successes of TCP, multiple groups have attempted to increase its specificity for LSD1, resulting in compounds with small (e.g., compounds 4c [Benelkebir et al. 2011),

Table 1. LSD1 inhibitors displaying cytotoxic effects

Name	In vitro			Mouse tumor studies	Additional references
	Where first described	Inhibits	Does not inhibit		
TCP[*]	Lee et al. 2006	LSD1, MAO-A, MAO-B		Leukemia and oral squamous cell carcinoma xenografts	Harris et al. 2012; Ferrari-Amorotti et al. 2014; Wang et al. 2016
ORY-1001[*]	Oryzon Genomics	LSD1			
GSK2879553[*]	GlaxoSmithKline	LSD1			
Bizine	Prusevich et al. 2014	LSD1			
Compound 1	Culhane et al. 2006	LSD1			
Compound 1a	Kakizawa et al. 2015	LSD1			
Compound 1c	Pollock et al. 2012	LSD1			
Compound 3	Rotili et al. 2014	LSD1, KDM2, KDM3, KDM4C, KDM4E, KDM5, KDM6, PHD2	MAO-A, MAO-B, FIH		
Compound 4c	Benelkebir et al. 2011	LSD1			
Compound 5a	Schmitt et al. 2013	LSD1, MAO-A, MAO-B			
Compound 11h	Valente et al. 2015a	LSD1, MAO-A	MAO-B		Binda et al. 2010
Compound 15	Vianello et al. 2016	LSD1	LSD2, MAO-A, MAO-B	Genetic APL mouse model	Binda et al. 2010
Compound 18	Culhane et al. 2010	LSD1			
Compound 19l	Han et al. 2015	LSD1	MAO-A, MAO-B		
OG-L002	Liang et al. 2013a	LSD1	MAO-A, MAO-B		
Pargyline	Metzger et al. 2005	LSD1, MAO-A, MAO-B		Breast cancer and glioma xenograft	Huang et al. 2012; Sareddy et al. 2013
RN-1	Neelamegam et al. 2012	LSD1, MAO-A (weakly)	MAO-B		Konovalov and Garcia-Bassets 2013
S2101	Mimasu et al. 2010	LSD1	MAO-A, MAO-B		Konovalov and Garcia-Bassets 2013; Suva et al. 2014
CBB1007	Wang et al. 2011	LSD1			
Compound 2d (verlindamycin)	Huang et al. 2007b	LSD1		AML and colon cancer xenograft	Huang et al. 2009; Schenk et al. 2012
Compound 5	Khan et al. 2015	LSD1			

Continued

Table 1. *Continued*

Name	In vitro			Mouse tumor studies	Additional references
	Where first described	Inhibits	Does not inhibit		
Compound 5n (HCI-2509 derivative)	Zhou et al. 2015	LSD1			
Compound 6b	Ma et al. 2015	LSD1		Gastric cancer xenograft	
Compound 6d	Nowotarski et al. 2015	LSD1	MAO-A, MAO-B		
Compound 9a	Dulla et al. 2013	LSD1			
Compound 16q	Hitchin et al. 2013	LSD1	MAO-A		
Compound 17	Wu et al. 2016	LSD1, MAO-B	MAO-A		
Compound 26	Zheng et al. 2013	LSD1	MAO-A, MAO-B		
Cryptotanshinone	Wu et al. 2012	LSD1			
Curcumin	Abdulla et al. 2013	LSD1			
HCI-2509	Sorna et al. 2013	LSD1	MAO-A, MAO-B	Ewing's sarcoma xenograft	Fiskus et al. 2014; Sankar et al. 2014; Theisen et al. 2014; Zhou et al. 2015
Namoline	Willmann et al. 2012	LSD1	MAO-A, MAO-B	Prostate cancer xenograft	
NCL-1	Ogasawara et al. 2011	LSD1	MAO-A, MAO-B	Prostate cancer xenograft	Ueda et al. 2009; Cortez et al. 2012; Sareddy et al. 2013; Etani et al. 2015; Hoshino et al. 2016
SNAIL peptide	Tortorici et al. 2013	LSD1			

Note: Inhibitors marked with an asterisk are in clinical trials, and those in bold are irreversible. Italics indicate that the compounds have not been tested in vivo. In cases in which a single report described multiple, related compounds, only the most potent and/or selective one is listed. Specificity is reported based on the original authors' interpretation of their data; additional information can be found in the cited references.

APL, acute promyelocytic leukemia; AML, acute myeloid leukemia.

1c (Pollock et al. 2012), and OG-L002 (Liang et al. 2013a]) or large (e.g., compounds 11h [Binda et al. 2010; Valente et al. 2015a,b], 15 [Binda et al. 2010; Vianello et al. 2016], and 17 [Wu et al. 2016]) substitutions at the *para* position, and/or at the primary amine (e.g., RN-1 [Neelamegam et al. 2012] and compound 17 [Wu et al. 2016]) (Figs. 2 and 3). These compounds effectively inhibited cancer cell growth in culture (Binda et al. 2010; Benelkebir et al. 2011; Pollock et al. 2012; Konovalov and Garcia-Bassets 2013; Valente et al. 2015a,b; Wu et al. 2016). Notably, compound 17 (Fig. 3) (Wu et al. 2016) functioned as a competitor of the methylated H3 peptide, having lost its ability to form a covalent bond with FAD during the derivatization process (Wu et al. 2016). Two compounds (13b and 14e) developed as specific

LSD1 inhibitors (Binda et al. 2010) were further derivatized and optimized with respect to stereochemistry, yielding compounds 11h (Valente et al. 2015a) and 15 (Fig. 2) (Vianello et al. 2016). The latter compound increased survival in an acute promyelocytic leukemia mouse model when delivered orally as a single agent (Vianello et al. 2016). Guided by predictions of the docking of TCP derivatives into the active site of LSD1, Mimasu et al. (2010) capitalized on the slightly larger catalytic cleft of LSD1 to develop an analog (S2101) (Fig. 2) that inhibited LSD1 at least 50-fold more effectively than it did either MAO, and reduced the viability of glioma (Suva et al. 2014) and ovarian cancer cells (Konovalov and Garcia-Bassets 2013) in culture. Although some of these studies systematically compared the activity of TCP or its derivatives against LSD1 and MAOs in vitro (Binda et al. 2010; Mimasu et al. 2010; Neelamegam et al. 2012; Valente et al. 2015a,b; Wu et al. 2016), it is not clear whether any in vivo effects arise from a combination of LSD1 inactivation and inhibition of other enzymes. In fact, RN-1 (Fig. 2) affects memory in mice through mechanisms that are not fully understood. Depletion of a neuronal LSD1 isoform (Wang et al. 2015a) was reported to impair learning and memory, and inhibition of other MAOs is also expected to inhibit memory storage (Neelamegam et al. 2012) by affecting the expression of glucocorticoid receptors (Heydendael and Jacobson 2009), which play key roles in memory consolidation (Quirarte et al. 1997). Other indirect effects that could contribute to the memory phenotypes caused by RN-1 include impaired production of insulin on MAO inhibition (Feldman and Chapman 1975). Thus, the effects of RN-1 could arise from LSD1 inhibition, or off-target effects on other enzymes, including the two other MAOs (Shih et al. 2011). Whether any of the TCP derivatives also reacts with any other FAD-utilizing enzymes—of which there are more than 75, participating in processes such as sugar catabolism, electron transport, and lipid metabolism (Lienhart et al. 2013)—has not been thoroughly tested, and would be important to determine before application in the clinic. Systematic compari-

sons of the effects of TCP analog treatment, with or without additional LSD1 inhibition, would yield insight into the extent to which off-target effects contribute to in vivo phenotypes. In general, bulky substituents that capitalize on the larger binding pocket of LSD1 compared with the other MAOs seem to confer some degree of specificity. Notably, the structure-guided designs of Mimasu et al. (2010) achieved specificity for LSD1 over the MAOs in vitro, suggesting that further development of specific TCP derivatives could be achieved by rational design. Given the variety of TCP derivatives reported in the literature, a systematic comparison of their chemical features may yield insight into the types and locations of substitutions that confer specificity over each MAO, aiding further rational design attempts.

REVERSIBLE INHIBITORS

Reversible poly- and monoamine oxidase inhibitor scaffolds—such as polyamines (Bianchi et al. 2006) and pyrimidines (George et al. 1971)—have also been used to generate LSD1 inhibitors. Derivatization of these scaffolds (Huang et al. 2007b; Ma et al. 2015; Nowotarski et al. 2015) yielded molecules (2d, 6b, and 6d, respectively) (Fig. 3) that displayed improved specificity for LSD1 in vitro and inhibited the growth of several types of cancer cells in culture. A pyrimidine thiourea-containing compound (6b) potently inhibited LSD1, and repressed the growth of gastric cancer cell lines in culture and in mouse xenografts with no overt side effects (Ma et al. 2015) (Fig. 3). To improve specificity, peptide-based molecules have also been used, which take advantage of the specific binding of LSD1 to its substrates or other interaction partners. A 6-mer peptide derived from SNAIL (Fig. 3), a competitive inhibitor of LSD1 in vivo (Baron et al. 2011), showed efficacy in vitro (Tortorici et al. 2013). In a second study, CBB1003 and its methyl ester CBB1007 (Fig. 3), both small molecules designed to mimic the H3 substrate, reduced the proliferation of multiple cancers derived from pluripotent cells (e.g., embryonic carcinomas) in culture (Wang et al. 2011). Few studies have compared the effects of reversible and irreversible inhibitors in parallel on cell or tumor growth. In one case, polyamine compound 2d (verlindamycin) (Fig. 3) (Huang et al. 2009) proved less potent than TCP at inducing AML differentiation in combination with ATRA; but on its own, 2d was surprisingly more effective (Schenk et al. 2012). These results suggest that polyamines and TCP may affect different aspects of LSD1 function in addition to its catalytic activity (e.g., its assembly into complexes or its localization to H3K4me2 sites). In fact, both TCP and pargyline were shown to reduce LSD1 protein levels, in addition to inactivating the enzyme (Wang et al. 2016). A comparative assessment of the effects of reversible and irreversible LSD1 inhibitors on its expression levels, localization, and association with binding partners could reveal additional mechanisms of LSD1 inhibition, and novel strategies for developing specific inhibitors. Just as unique applications for irreversible and reversible cyclooxygenases inhibitors have been established (for suppressing clotting [Schror 1997] and inflammation [Simon and Mills 1980], respectively), further studies may reveal similar nonredundant uses for irreversible and reversible LSD1 inhibitors.

BIFUNCTIONAL LSD1 INHIBITORS

Several groups have designed inhibitors by fusing two or more features of known LSD1 inhibitors into a single molecule, as multifunctional drugs can improve the efficacy and/or reduce off-target effects of individual drugs (Mai et al. 2008). Histone H3 peptides (compounds 1 [Culhane et al. 2006], 18 [Culhane et al. 2010], and 1a [Kakizawa et al. 2015]) (Fig. 2), or peptide mimics, fused at the ε-amine of K4 to cyclic groups (mimicking the cyclopropane group of TCP) (Ueda et al. 2009) or propargylamines (mimicking pargyline) (e.g., compound 5a [Schmitt et al. 2013]), inhibited LSD1 in vitro. The peptide mimic NCL-1 (Fig. 3) (Ueda et al. 2009; Ogasawara et al. 2011) inhibited the growth of multiple types of cancer cells in culture (Cortez et al. 2012; Sareddy et al. 2013; Hoshino et al. 2016), as well as prostate cancer xenografts in mice with minimal toxicity (Etani

et al. 2015). Surprisingly, an *N*-alkylated derivative (compound 5), fusing features of ORY-1001 to NCL-1, displayed improved potency only in vitro, but not in vivo (Fig. 3) (Khan et al. 2015). A separate study combined elements of TCP, polyamine analog inhibitors, and peptide inhibitors to generate compound 9a, which suppressed breast cancer cell proliferation in culture (Fig. 3) (Dulla et al. 2013).

Two groups generated bifunctional inhibitors by combining MAOIs with compounds displaying anti-proliferative activities—either hydroxylcinnamic acid (compound 19l [Han et al. 2015]) (Fig. 2) or carbamates (compound 26 [Zheng et al. 2013]) (Fig. 3)—and both sets suppressed proliferation in a variety of cancer lines in culture. Based on the finding that LSD1 and the JmjC demethylase KDM4 colocalize at androgen-dependent promoters (Wissmann et al. 2007), Rotili and colleagues fused TCP to JmjC inhibitors bipyridine (compound 2) or hydroxyquinoline (compound 3). Both fusions effectively inhibited LSD1 and all tested JmjC demethylases with minimal effects on MAO-B (compound 2) or both MAO-A and MAO-B (compound 3), and preferentially suppressed the growth of prostate cancer lines over a noncancerous cell line in culture (Rotili et al. 2014). Intriguingly, the fusion compounds induced apoptosis more effectively than did a combination of the two inhibitors alone at the same concentration (Rotili et al. 2014) (Fig. 2). Like combination therapies, these strategies aimed to maximize potency and specificity combinatorially, although the ORY-1001-NCL-1 fusion (Khan et al. 2015) shows that this approach is not foolproof.

Although these bifunctional compounds have shown remarkable success, their mechanisms of action remain unresolved. The compounds reported by Rotili et al. (2014) raise the question of how the demethylase targets are arranged spatially in the cell, and whether their geometries (in addition to their catalytic activities) are important for their function. For example, do the fusion inhibitors trap the demethylases in a configuration that is incompatible with any catalytic-independent function(s) (e.g., recruitment of additional regulators), in a way that the individual inhibitors cannot accomplish? Biochemical and structural studies of the demethylase complexes with each inhibitor would be particularly informative in understanding the functions of the inhibitors and their demethylase targets in cells. The fusion compounds with antiproliferatives raise additional questions, such as the identities of the molecules targeted by these compounds. The targets of the antiproliferatives remain unknown, and it is likely that each compound has multiple targets, given that a fusion of hydroxylcinnamic acid to a Ras inhibitor increased cytotoxicity (Ling et al. 2014), despite the fact that Ras and LSD1 localize to different cellular compartments. These results suggest that hydroxylcinnamic acid could have both plasma membrane-localized and nuclear targets, or it may simply increase the permeability of its fusion partners into cells. Investigations into these areas will be important for understanding the mechanisms of action of these molecules, and will provide insight into the interaction of target molecules with each other, which could be used to inform further development of potent and specific inhibitors.

OTHER INHIBITORS

Several other groups have developed inhibitors specific to LSD1 by identifying compounds not already known to be MAOIs. Screening-based approaches identified aminothiazoles (compound 16q [Hitchin et al. 2013]) (Fig. 3), the natural compounds resveratrol and curcumin (Fig. 3) (Abdulla et al. 2013) (the latter also reported as a KDM4 inhibitor [Kim et al. 2014]); cryptotanshinone (Fig. 3) (Wu et al. 2012), and Namoline (Fig. 3) (Willmann et al. 2012) (a γ-pyrone that the same group had recently identified as an MAOI [Wetzel et al. 2010]), all of which (with the exception of aminothiazoles [Hitchin et al. 2013]) inhibited the growth of cancer cells in culture. Namoline showed some toxicity when tested in a prostate cancer xenograft (Willmann et al. 2012), highlighting a potential disadvantage of developing new inhibitors not previously used therapeutically. A virtual screen of compounds that can

dock against LSD1 led to the rational design of a phenylethylidene-benzohydride (HCI-2509, aka SP2509) (Fig. 3), which suppressed the growth of cancer cells in culture (Sorna et al. 2013), or in xenograft models of Ewing's sarcoma (Sankar et al. 2014) or endometrial cancer (Theisen et al. 2014) on its own, or AML xenografts in combination with the histone deacetylase (HDAC) inhibitor panobinostat (Fiskus et al. 2014). A sterically constrained derivative of HCI-2509, compound 5n, inhibited the growth of multiple cancer cell lines in culture more potently than HCI-2509 did (Fig. 3) (Zhou et al. 2015). Collectively, these studies suggest that compounds not already known to be MAOIs could have LSD1-specific inhibitory properties, and could be optimized for clinical use.

Despite the numerous classes of LSD1 inhibitors developed, few have been tested systematically for toxicity or efficacy in mouse models, and so far only three—TCP, ORY-1001, and GSK 2879552 (Fig. 2) (Maes et al. 2015)—are in clinical trials. Achieving specific inhibition of LSD1 over the other MAOs, and potentially other FAD-utilizing enzymes (in the case of irreversible inhibitors), remains a significant challenge, but the recent advances suggest the potential for further optimization. The in vivo and clinical applications of many of the LSD1 inhibitors described in the literature remain to be fulfilled, and could yield effective targeted therapeutics in the future.

JmjC FAMILY

The JmjC family of histone demethylases uses 2-OG and Fe(II) as cofactors to demethylate mono-, di-, and tri-methylated lysines. The substrate specificities of the enzymes vary, with some accepting multiple methylated lysines (e.g., KDM4A-C, which act on both H3K9 and H3K36 methylations [Woon et al. 2012; Labbe et al. 2013]), whereas others recognize only a single substrate (e.g., KDM2, which specifically demethylates H3K36me1-2 [Tsukada et al. 2006]). KDM4A, although it recognizes both H3K9 and H3K36 (Couture et al. 2007), shows a preference for trimethylated lysines (Klose et al. 2006; Whetstine et al. 2006), displaying more specificity for the methylation state than for the surrounding protein sequence. Like LSD1, JmjC enzymes display complex and context-dependent effects on gene expression. Overexpression of these enzymes in cancers is a common theme, whereas loss of function occurs less frequently (Hojfeldt et al. 2013) (with the exception of KDM6A, which is frequently mutated in cancers, as described above). In fact, KDM4C (GASC-1), had been initially found amplified in esophageal cancers (Yang et al. 2000), and was later implicated in medulloblastoma (Ehrbrecht et al. 2006) and breast cancer (Liu et al. 2009). As observed for LSD1, the KDM4 family exerts its effects in breast and prostate cancers by associating with nuclear hormone receptors (Yamane et al. 2007; Shi et al. 2011; Berry et al. 2012), leading to either overexpression of pro-proliferation genes or repression of tumor suppressors. Despite the similarities between enzymes of the same family, their roles in carcinogenesis appear distinct, as only some family members are found to be misregulated in each cancer type (Berry et al. 2012). Although the mechanisms underlying demethylase-promoted oncogenesis vary, demethylases represent good targets for drug therapies because of their widespread up-regulation across cancers, and in some cases showed roles in cancer stem cells (e.g., Nakamura et al. 2013). Many JmjC inhibitors have been identified, but as was the case for LSD1, achieving specificity has been challenging because the catalytic sites of these enzymes are not only homologous to each other, but also to other 2-OG dependent oxidases, such as PHD1 and FIH (Elkins et al. 2003). The major classes of inhibitors include cofactor mimics, substrate mimics, as well as compounds whose mechanism of action is poorly defined (Figs. 4 and 5; Table 2).

INHIBITORS OF 2-OG-DEPENDENT ENZYMES

As for LSD1, the first inhibitors tested for JmjC demethylase inhibition were those that had been developed for other related enzymes that use similar catalytic mechanisms. A number of researchers have exploited the 2-OG dependence of JmjC family demethylases to design inhibitors that interfere with 2-OG function. Because of their structural similarity to N-methyl and RNA demethylases, and to nucleic acid oxygenases, JmjC family demethylases have been targeted by inhibitors of 2-OG oxygenases, such as hydroxamate derivatives (Hamada et al. 2009; Itoh et al. 2015), N-oxalyl amino acid derivatives (Rose et al. 2008, 2010; Hamada et al. 2009), pyridine dicarboxylates (Rose et al. 2008), and agents such as disulfiram (Sekirnik et al. 2009) (a drug used to treat alcoholism [Ellis and Dronsfield 2013]) that interfere with metal binding.

Two primary scaffolds for 2-OG-dependent enzyme inhibitors are N-oxalylglycine (NOG) (Fig. 4), a 2-OG cofactor mimic (Baader et al. 1994) that binds Fe(II) but is resistant to superoxide attack (Elkins et al. 2003), and $para$ 2,4 dicarboxylic acid (2,4-PDCA) (Fig. 4), another 2-OG mimic that occupies the 2-OG binding site but cannot complete catalysis (Tschank et al. 1987). NOG was first shown to inhibit KDM4C (Cloos et al. 2006), but showed little specificity for demethylases over other 2-OG oxygenases such as PHD2 and FIH (Hopkinson et al. 2013). Both rational (Rose et al. 2010) and screen-based (Mannironi et al. 2014) approaches identified NOG derivatives, of which the latter (compound 3195) (Fig. 5) weakly inhibited HeLa cell proliferation in culture (Mannironi et al. 2014) (the other, an N-oxalyl-D-tyrosine derivative, compound 7f [Fig. 4], was not tested in vivo [Rose et al. 2010]). 2,4-PDCA (Fig. 4) also proved to be an effective inhibitor of JmjC demethylase activity in vitro (Rose et al. 2008; Mackeen et al. 2010; Hopkinson et al. 2013) and in vivo (Mackeen et al. 2010; Kristensen et al. 2012; Hopkinson et al. 2013), showing a preference for KDM4C over KDM6A (Kristensen et al. 2012), but generally inhibiting other 2-OG-dependent enzymes with equal potency (Hopkinson et al. 2013). An effective 2,4-PDCA derivative more specific for KDM4E than for PHD2 was generated by including a fluorophenyl substituted amine at the 3 position, compound 47 (Fig. 4) (Thalhammer et al. 2011), but this com-

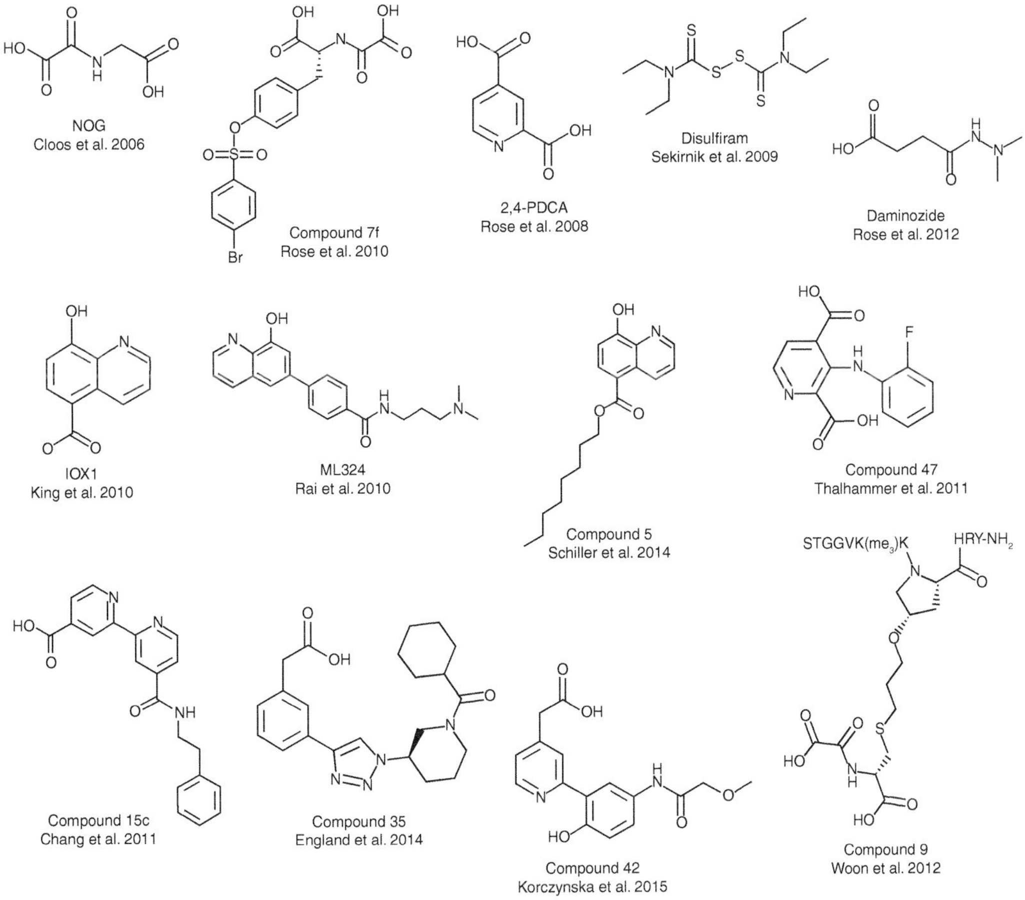

Figure 4. Chemical structures of select JmjC-family inhibitors that are effective in vitro, but have not shown JmjC-mediated cytotoxic activity. References describing their development or use are provided. Letters in large font denote single-letter amino acid codes.

pound has not been tested in vivo. Further development of the 2,4-PDCA scaffold could generate potent and selective inhibitors of histone demethylases, although a further challenge will be improving its cell permeability, as even the ester analogs require high concentrations in culture (Mackeen et al. 2010; Hopkinson et al. 2013) and have not been shown to inhibit cell proliferation. Systematic structural studies of the various JmjC members (as well as other 2-OG-dependent enzymes) bound to 2-OG mimics, in the presence and absence of 2-OG, could reveal any differences in binding modes that can be exploited to develop inhibitors specific to individual family members.

BIFUNCTIONAL JmjC INHIBITORS

Other groups took the strategy of improving specificity of JmjC family inhibitors by generating bifunctional compounds, as was done for LSD1. Because the in vivo specificity of JmjC enzymes for their substrates stems from the sequence of the peptide surrounding the substrate lysine, bifunctional molecules relying on peptides (or their mimics) fused to 2-OG mimics have been synthesized (NCDM-32B [Hamada et al. 2009, 2010], methylstat [Luo et al. 2011], and compound 9 [Woon et al. 2012]) (Figs. 4 and 5). Surprisingly, this strategy led to unintended effects; an early attempt yielded an

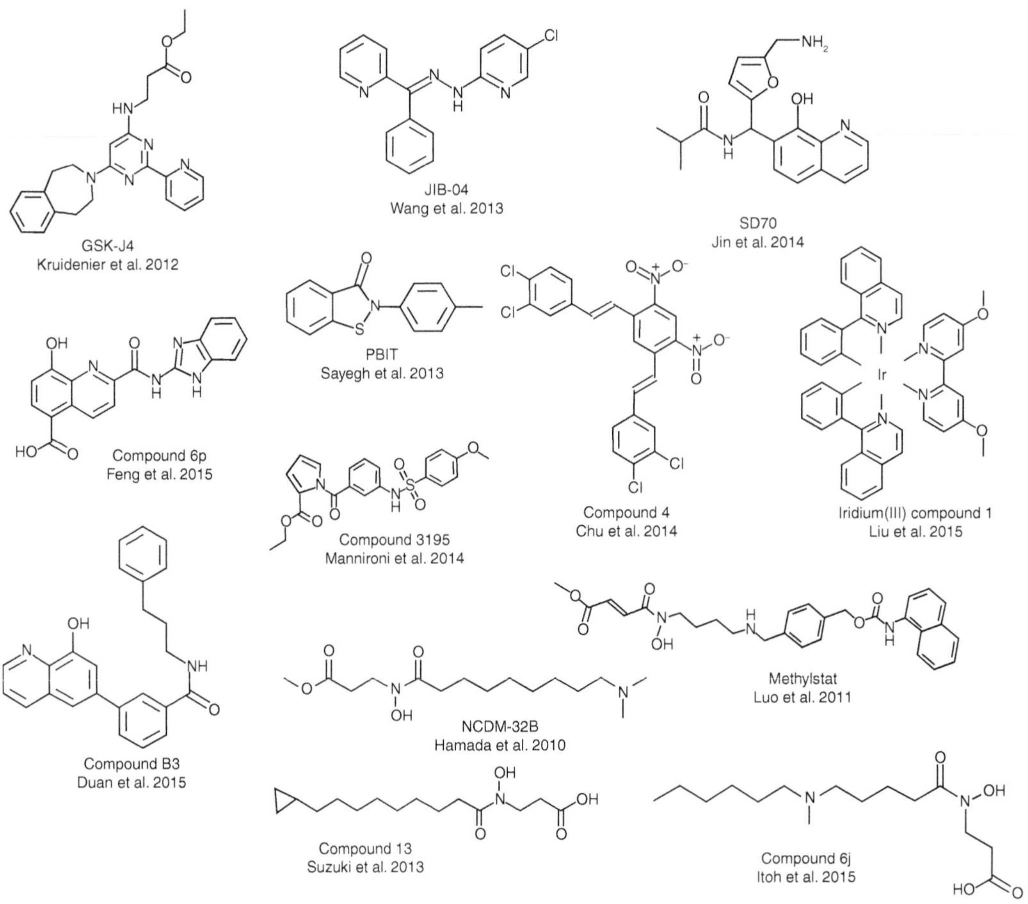

Figure 5. Chemical structures of JmjC-family inhibitors that have been shown to display cytotoxic activities. References describing their development or use are provided.

NOG derivative that inhibited enzyme(s) other than the one it was designed against (Hamada et al. 2009). Substitution of a hydroxamate at the amide linkage of NOG (which improved metal chelation, and was later used to develop the KDM5A inhibitor compound 6j [Itoh et al. 2015]), along with simplification of the peptide mimic, generated NCDM-32B (Fig. 5), which restored specificity for KDM4 in vitro (Hamada et al. 2010), and suppressed growth of breast (Ye et al. 2015) and prostate cancer cells (Hamada et al. 2010) (the latter in combination with an LSD1 inhibitor). Yet another compound, methylstat (Fig. 5), featuring a similar hydroxamate group and a bulky substrate mimic, was specific for KDM4 and KDM6 demeth-

ylases over PHF8 and FIH and inhibited KDM4C-mediated myogenesis in culture (Luo et al. 2011). These results reveal the power of inhibiting metal binding by chelation with hydroxamate, and more generally show that whereas peptide scaffolds can be exploited for specificity, the outcomes are difficult to predict a priori.

Although the bifunctional LSD1 and JmjC inhibitors share some similarities, it is noteworthy that the two classes have used conceptually different approaches. The JmjC inhibitors have aimed to displace both the substrate and the cofactor (either 2-OG and/or Fe(II)), whereas the LSD1 inhibitors have typically displaced only the substrate while irreversibly reacting

Table 2. JmjC family inhibitors

Name	Where first described	In vitro		Mouse tumor studies	Additional references
		Inhibits	Does not inhibit		
GSK-J4[*]	Kruidenier et al. 2012	KDM5B, KDM5C, KDM6A, KDM6B	KDM4A, KDM4C, KDM4D, KDM4E	Ovarian cancer xenograft	Heinemann et al. 2014; Ntziachristos et al. 2014; Sakaki et al. 2015; Horton et al. 2016
JIB-04[*]	Wang et al. 2013	KDM4A, KDM4B, KDM4C, KDM4E, KDM6B		Non-small-cell lung cancer xenograft	Horton et al. 2016
SD70[*]	Jin et al. 2014	KDM4C		Prostate cancer xenograft	
Compound 4	Chu et al. 2014	KDM4A, KDM4B			
Compound 6p	Feng et al. 2015	KDM4A	PHD2		
Compound 6j	Itoh et al. 2015	KDM5A	KDM4C, KDM3A		
Compound 13	Suzuki et al. 2013	KDM2A, KDM7A, KDM7B	KDM4A, KDM4C, KDM5A, KDM6A		
Compound 3195	Mannironi et al. 2014	KDM5A			
Compound B3	Duan et al. 2015	KDM4B, KDM4D	KDM4A, KDM4C		
Iridium(III) compound 1	Liu et al. 2015	KDM4D	KDM5A, KDM6B, HDAC		
Methylstat	Luo et al. 2011	KDM4A, KDM4C, KDM4E, KDM6	PHF8, FIH, LSD1, HDAC		
NCDM-32B	Hamada et al. 2010	KDM2A, KDM2C	PHD1, PHD2		
PBIT	Sayegh et al. 2013	KDM5A, KDM5B, KDM5C	KDM6A, KDM6B		
2,4 PDCA	Rose et al. 2008	KDM2A, KDM3A, KDM4A, KDM4C, KDM4D, KDM4E, PHD2, FIH	KDM6A, KDM6B (weak inhibition)		Hopkinson et al. 2013
Compound 5	Schiller et al. 2014	KDM4C, KDM4E	KDM2A, KDM3A, KDM5C, KDM6B, PHD2		
Compound 7f	Rose et al. 2010	KDM4A, KDM4E, FIH	PHD2		
Compound 9	Woon et al. 2012	KDM4A	KDM2A, KDM4E, PHF8		

Continued

Table 2. *Continued*

Name	Where first described	In vitro		Mouse tumor studies	Additional references
		Inhibits	Does not inhibit		
Compound 15c	Chang et al. 2011	KDM4E			
Compound 35	England et al. 2014	KDM2A	KDM3A, KDM4A, KDM4C, KDM4E, KDM5C, KDM6B		
Compound 42	Korczynska et al. 2015	KDM3A, KDM4C, KDM4D, KDM5B	KDM2A, KDM6B, FIH		
Compound 47	Thalhammer et al. 2011	KDM4E	PHD2		
Daminozide	Rose et al. 2012	KDM2A, KDM7A, PHF8	KDM3A, KDM4E, KDM5C, KDM6B, FIH, PHD2, BBOX1		
Disulfiram	Sekirnik et al. 2009	KDM4A, aldehyde dehydrogenase, HIV nucleocapsid p7			
IOX1	King et al. 2010	KDM3A, KDM4A, KDM4C, KDM4D, KDM4E, KDM6A, KDM6B	KDM2A, KDM5C, PHF8, FIH, PHD2		
ML324	Rai et al. 2010	KDM4E			
NOG	Cloos et al. 2006	KDM2A, KDM4A, KDM4C, KDM4D, KDM5C, KDM6A, KDM6B, PHD2	KDM4E, PHF8, FIH		Hopkinson et al. 2013

Inhibitors in bold have shown cytotoxic effects, and those marked with an asterisk have shown efficacy in animal models. In cases in which a single report described multiple, related compounds, only the most potent and/or selective one is listed. Specificity is reported based on the original authors' interpretation of their data; additional information can be found in the cited references.

with the cofactor. Second, with the exception of the compounds described by Rotili et al. (2014), none of the JmjC bifunctional inhibitors was designed to target two independent molecules simultaneously. Targeting other molecules in association with JmjC enzymes could be an effective approach to improving potency and specificity, and could open up novel strategies of inhibition, such as imposing steric constraints rather than simply inhibiting catalytic activity. The success of bifunctional inhibitors for both LSD1 and JmjC demethylases suggests that this strategy might be widely used to achieve specific inhibition, thus allowing use of compounds previously developed for targeting other enzymes.

OTHER JmjC INHIBITORS

To identify novel scaffolds for JmjC demethylase inhibitors, several targeted and large-scale screens, including virtual screens, have been conducted, yielding molecules such as IOX1 (King et al. 2010) and compounds 35 (England et al. 2014) and 42 (Korczynska et al. 2015). In many screens, 8-hydroxyquinoline derivatives emerged as lead compounds. 5-carboxy-8-hydroxyquinoline (later named IOX1) (Fig. 4) was initially identified in a large screen as a potent and somewhat specific inhibitor of the KDM4 family (King et al. 2010; Hopkinson et al. 2013), and was later derivatized to improve cell permeability (compound 5 [Schiller et al. 2014]) (Fig. 4) and specificity (Feng et al. 2015), with a 2-1*H*-benzo[*d*]imidazole substituted compound (6p) (Fig. 5) inhibiting the proliferation of multiple types of cancer cells in culture (Feng et al. 2015). Other modifications of the parent 8HQ have improved both its potency in vivo and selectivity (e.g., ML324 [Rai et al. 2010; Liang et al. 2013b] and compound B3 [Duan et al. 2015]). Additional compounds identified by screening that inhibited JmjC enzymes in vitro and/or cancer cell growth in culture included a novel heterocycle, PBIT (Fig. 5) (Sayegh et al. 2013); catechols (Sakurai et al. 2010; Nielsen et al. 2012); bipyridyl compounds with carboxylates such as 2,4-PDCA (Rose et al. 2008) and compound 15c (Fig. 4) (Chang et al. 2011); the plant growth retardant daminozide, which chelates metal cofactors (Fig. 4) (Rose et al. 2012); and a dinitrobenzene derivative, compound 4 (Fig. 5) (Chu et al. 2014). Because daminozide was known to be toxic (Fan and Jackson 1989), a derivative (compound 13) containing a hydroxamate but lacking the potentially toxic 1,1-dimethylhydrazine structure was developed and shown to retain daminozide's preference for KDM2 and KDM7, and was effective in culture (Fig. 5) (Suzuki et al. 2013). Although the vast majority of inhibitors reported to date have been organic molecules, recently an iridium(III)-bipyridine complex (Fig. 5) was shown to inhibit KDM4 enzymes preferentially over KDM5A, KDM6A, or HDACs, and to inhibit lung cancer cell proliferation in culture (Liu et al. 2015). Together, these studies highlight the diversity of compounds capable of inhibiting JmjC family demethylases, and suggest that further screening may yield additional inhibitors. As is the case for the novel LSD1 inhibitors, many questions remain regarding the mechanisms of action of the JmjC inhibitors. For many inhibitors, their in vivo specificities, genome-wide effects on the distribution of methylated histones, and the expression and localizations of target and nontarget JmjC enzymes remain poorly defined. Furthermore, the efficacies of these inhibitors in animal models of cancer have not yet been reported; this information will be crucial for the development of JmjC inhibitors as therapeutics.

Three molecules identified by screening approaches have shown in vivo efficacy in mouse tumor models. The first, GSK-J4 (Fig. 5), was derived from a weak hit generated by screening the GlaxoSmithKline collection of two million compounds (Kruidenier et al. 2012). Guided by the crystal structure of KDM6A (JMJD3), Kruidenier et al. (2012) optimized interactions of the lead compound with the KDM6A active site to generate GSK-J1, which was reported to be specific for the KDM6 family of demethylases (Kruidenier et al. 2012). The cell-permeable analog, GSK-J4, suppressed KDM6A-mediated proinflammatory responses in macrophages, inhibited growth of breast cancer cells (Horton et al. 2016) in culture, and reduced tumor volume of ovarian cancer cells in mouse xenografts (Sakaki et al. 2015). GSK-J4 was also effective against T-cell acute lymphocytic leukemia (T-ALL) (Ntziachristos et al. 2014; Benyoucef et al. 2016). One study described an oncogenic role for KDM6B, and inhibition of leukemic cell growth was achieved in culture by treatment with GSK-J4 (Ntziachristos et al. 2014). Although this study also reported a tumor suppressive role for KDM6A (Ntziachristos et al. 2014), a subsequent study found that KDM6A functioned as an oncogene in T-ALLs driven by Tal1 (Benyoucef et al. 2016), a transcription factor commonly associated with this disease (Chen et al. 1990). Treatment of these cells with GSK-J4 inhibited proliferation and induced transcriptional programs similar to those observed on KDM6A knockdown, suggesting that KDM6A was the primary target of GSK-J4, and also inhibited T-ALL xenograft growth in a mouse model (Benyoucef et al. 2016). GSK-J1/4s reported specificity for the KDM6 family generated enthusiasm (Harrison 2012), but soon Heinemann et al. (2014) showed that the molecules inhibited other JmjC family demethylases not initially tested. These results reveal the importance of testing potential specific inhibitors against a broad panel of 2-OG-dependent enzymes, as some inhibitors may react with a more distantly related family member rather than a close one. Given that KDM6A cooperates with other histone methylation regulators such as LSD1 (in the case of Tal1-driven transcription [Hu et al. 2009]) and the H3K4 methyltransferse MLL2 (Cho et al. 2007; Issaeva et al. 2007), it is possible that some of the effects of GSK-J1/4 on T-ALL could arise from inhibition of other demethylases operating in the same pathways. Nevertheless, the effectiveness of GSKJ1/4 in animal tumor models suggests potential therapeutic uses.

Two other screening approaches have yielded additional inhibitors of JmjC-family enzymes that are active. One screening approach, based on the expression of a GFP reporter, yielded JIB-04 (Fig. 5), which was identified as a demethylase inhibitor based on the similarity of the expression profiles of JIB-04-treated cells with those of cells treated with inhibitors of 2-OG-dependent enzymes (Wang et al. 2013). JIB-04 prolonged the survival of mice bearing orthotopic mammary tumors (Horton et al. 2016). A more targeted screen of molecules regulating epigenetic or nuclear factors yielded SD70 (Fig. 5), an 8-HQ derivative that localized to enhancers and inhibited KDM4C in vitro, and decreased the tumor volume of prostate cancer xenografts in mice (Jin et al. 2014). Importantly, "Chem-Seq" pull-down experiments (Anders et al. 2014)—which used a biotin tag on SD70 to isolate the molecule from cells, followed by high-throughput sequencing to identify the genomic regions to which it localized (via binding to its target proteins or DNA)—revealed that the genomic localization sites of SD70 overlapped with androgen receptor (AR)

enhancer sites, and that SD70 inhibited expression of the AR target genes. These results provide valuable information on the molecular effects of the inhibitor, suggesting that the drug might interfere with the catalytic activity of KDM4C, but have minimal effects on its targeting to enhancers. Chem-Seq could be a powerful approach for investigating the molecular mechanisms of other inhibitors, provided that the linker and biotin tag are tolerated by the target enzymes.

CONCLUDING REMARKS

The success of histone demethylase inhibitors in decreasing the growth of many types of cancer cells in culture points to the potential for further developing these molecules for clinical use. A major obstacle toward this goal is achieving specific inhibition of key demethylases, as most inhibitors thus far act on related enzymes (other demethylases as well as enzymes using the same catalytic strategy). Although specific inhibitors are less likely to induce toxicity or side effects, pan- (or multi-) demethylase inhibitors could also be useful (as observed for HDAC inhibitors), as several demethylases show functional redundancies. A second challenge lies in predicting the transcriptional and cellular outcomes of demethylase inhibition. As discussed earlier, demethylase activity can affect transcriptional outputs in different ways, depending on cell type, and the key demethylase target genes that contribute to unrestrained cell proliferation vary between cancers. Therefore, predicting which demethylases to inhibit, and to what extent, may be challenging, as achieving a fine balance of histone modifications genome-wide is critical for normal cell function. Furthermore, recent work has revealed nonhistone substrates for histone demethylases, including regulation of p53 (Huang et al. 2007a) and E2F1 (Kontaki and Talianidis 2010) by LSD1-mediated demethylation, and a cytoplasmic role for KDM4A in promoting translation (Van Rechem et al. 2015). In accord with their homology with 2-OG oxygenases, some JmjC enzymes have also been reported to hydroxylate proteins, such as the splicing factor U2AF65 (Webby et al.

2009). Thus, it is possible that some of the cellular and in vivo effects of demethylase inhibitors arise from altered methylation of nonhistone substrates, instead of (or in addition to) effects on histone methylation. Given the pleiotropic roles of demethylases, it is remarkable that studies targeting overexpressed demethylases in cancer have met with success, and these studies hold promise for successful application of this strategy. Finally, demethylase inhibitors are not immune to the usual challenges of drug design such as achieving stability and cell permeability; in fact, some strong inhibitors such as NOG derivatives (Hamada et al. 2009) have already shown poor efficacy in vivo despite their inhibitory activity in vitro. Nevertheless, epigenetic modifiers represent unique drug targets, as their inhibition can be effective at blocking cell proliferation even if other mutations underlie the cancer phenotype. Therefore, further development of the demethylase inhibitors holds the promise of generating effective therapies for a wide range of cancers, either singly or in combination with other chromatin-targeting agents (e.g., Huang et al. 2009; Singh et al. 2011; Huang et al. 2012; Vasilatos et al. 2013; Fiskus et al. 2014), differentiation therapies (Schenk et al. 2012), or immunotherapeutics (Maio 2015).

ACKNOWLEDGMENTS

This work is supported by grants from the National Institutes of Health (CA118487 and MH096066) and an Ellison Foundation Senior Scholar Award to Y.S., and grants from the National Cancer Institute (5F32CA189741-02), Rally Foundation for Pediatric Cancer Research, and the Vs. Cancer Foundation to J.N.A. Y.S. is an American Cancer Society Research Professor. Y.S. is a cofounder of Constellation Pharmaceuticals and a member of its scientific advisory board and is a consultant for Active Motif.

REFERENCES

Abdulla A, Zhao X, Yang F. 2013. Natural polyphenols inhibit lysine-specific demethylase-1. *J Biochem Pharmacol Res* **1**: 56–63.

Agger K, Cloos PA, Christensen J, Pasini D, Rose S, Rappsilber J, Issaeva I, Canaani E, Salcini AE, Helin K. 2007. UTX and JMJD3 are histone H3K27 demethylases involved in HOX gene regulation and development. *Nature* **449**: 731–734.

Agin HV. 1960. Tranylcypromine in depression: A clinical report. *Am J Psychiatry* **117**: 150–151.

Anders L, Guenther MG, Qi J, Fan ZP, Marineau JJ, Rahl PB, Loven J, Sigova AA, Smith WB, Lee TI, et al. 2014. Genome-wide localization of small molecules. *Nat Biotechnol* **32**: 92–96.

Baader E, Tschank G, Baringhaus KH, Burghard H, Gunzler V. 1994. Inhibition of prolyl 4-hydroxylase by oxalyl amino acid derivatives in vitro, in isolated microsomes and in embryonic chicken tissues. *Biochem J* **300**: 525–530.

Bannister AJ, Schneider R, Kouzarides T. 2002. Histone methylation: Dynamic or static? *Cell* **109**: 801–806.

Baron R, Binda C, Tortorici M, McCammon JA, Mattevi A. 2011. Molecular mimicry and ligand recognition in binding and catalysis by the histone demethylase LSD1-CoREST complex. *Structure* **19**: 212–220.

Barski A, Cuddapah S, Cui K, Roh TY, Schones DE, Wang Z, Wei G, Chepelev I, Zhao K. 2007. High-resolution profiling of histone methylations in the human genome. *Cell* **129**: 823–837.

Benelkebir H, Hodgkinson C, Duriez PJ, Hayden AL, Bulleid RA, Crabb SJ, Packham G, Ganesan A. 2011. Enantioselective synthesis of tranylcypromine analogues as lysine demethylase (LSD1) inhibitors. *Bioorg Med Chem* **19**: 3709–3716.

Benyoucef A, Palii CG, Wang C, Porter CJ, Chu A, Dai F, Tremblay V, Rakopoulos P, Singh K, Huang S, et al. 2016. UTX inhibition as selective epigenetic therapy against TAL1-driven T-cell acute lymphoblastic leukemia. *Genes Dev* **30**: 508–521.

Berry WL, Shin S, Lightfoot SA, Janknecht R. 2012. Oncogenic features of the JMJD2A histone demethylase in breast cancer. *Int J Oncol* **41**: 1701–1706.

Bianchi M, Polticelli F, Ascenzi P, Botta M, Federico R, Mariottini P, Cona A. 2006. Inhibition of polyamine and spermine oxidases by polyamine analogues. *FEBS J* **273**: 1115–1123.

Binda C, Valente S, Romanenghi M, Pilotto S, Cirilli R, Karytinos A, Ciossani G, Botrugno OA, Forneris F, Tardugno M, et al. 2010. Biochemical, structural, and biological evaluation of tranylcypromine derivatives as inhibitors of histone demethylases LSD1 and LSD2. *J Am Chem Soc* **132**: 6827–6833.

Chang KH, King ON, Tumber A, Woon EC, Heightman TD, McDonough MA, Schofield CJ, Rose NR. 2011. Inhibition of histone demethylases by 4-carboxy-2,2′-bipyridyl compounds. *ChemMedChem* **6**: 759–764.

Chen Q, Cheng JT, Tasi LH, Schneider N, Buchanan G, Carroll A, Crist W, Ozanne B, Siciliano MJ, Baer R. 1990. The *tal* gene undergoes chromosome translocation in T cell leukemia and potentially encodes a helix–loop–helix protein. *EMBO J* **9**: 415–424.

Chen Y, Yang Y, Wang F, Wan K, Yamane K, Zhang Y, Lei M. 2006. Crystal structure of human histone lysine-specific demethylase 1 (LSD1). *Proc Natl Acad Sci* **103**: 13956–13961.

Chen Z, Zang J, Kappler J, Hong X, Crawford F, Wang Q, Lan F, Jiang C, Whetstine J, Dai S, et al. 2007. Structural basis of the recognition of a methylated histone tail by JMJD2A. *Proc Natl Acad Sci* **104:** 10818–10823.

Chen QW, Zhu XY, Li YY, Meng ZQ. 2014. Epigenetic regulation and cancer (review). *Oncol Rep* **31:** 523–532.

Cho YW, Hong T, Hong S, Guo H, Yu H, Kim D, Guszczynski T, Dressler GR, Copeland TD, Kalkum M, et al. 2007. PTIP associates with MLL3- and MLL4-containing histone H3 lysine 4 methyltransferase complex. *J Biol Chem* **282:** 20395–20406.

Chu CH, Wang LY, Hsu KC, Chen CC, Cheng HH, Wang SM, Wu CM, Chen TJ, Li LT, Liu R, et al. 2014. KDM4B as a target for prostate cancer: Structural analysis and selective inhibition by a novel inhibitor. *J Med Chem* **57:** 5975–5985.

Cloos PA, Christensen J, Agger K, Maiolica A, Rappsilber J, Antal T, Hansen KH, Helin K. 2006. The putative oncogene GASC1 demethylates tri- and dimethylated lysine 9 on histone H3. *Nature* **442:** 307–311.

Cortez V, Mann M, Tekmal S, Suzuki T, Miyata N, Rodriguez-Aguayo C, Lopez-Berestein G, Sood AK, Vadlamudi RK. 2012. Targeting the PELP1-KDM1 axis as a potential therapeutic strategy for breast cancer. *Breast Cancer Res* **14:** R108.

Couture JF, Collazo E, Ortiz-Tello PA, Brunzelle JS, Trievel RC. 2007. Specificity and mechanism of JMJD2A, a trimethyllysine-specific histone demethylase. *Nat Struct Mol Biol* **14:** 689–695.

Culhane JC, Szewczuk LM, Liu X, Da G, Marmorstein R, Cole PA. 2006. A mechanism-based inactivator for histone demethylase LSD1. *J Am Chem Soc* **128:** 4536–4537.

Culhane JC, Wang D, Yen PM, Cole PA. 2010. Comparative analysis of small molecules and histone substrate analogues as LSD1 lysine demethylase inhibitors. *J Am Chem Soc* **132:** 3164–3176.

Dalgliesh GL, Furge K, Greenman C, Chen L, Bignell G, Butler A, Davies H, Edkins S, Hardy C, Latimer C, et al. 2010. Systematic sequencing of renal carcinoma reveals inactivation of histone modifying genes. *Nature* **463:** 360–363.

Ding J, Zhang ZM, Xia Y, Liao GQ, Pan Y, Liu S, Zhang Y, Yan ZS. 2013. LSD1-mediated epigenetic modification contributes to proliferation and metastasis of colon cancer. *Br J Cancer* **109:** 994–1003.

Duan L, Rai G, Roggero C, Zhang QJ, Wei Q, Ma SH, Zhou Y, Santoyo J, Martinez ED, Xiao G, et al. 2015. KDM4/JMJD2 histone demethylase inhibitors block prostate tumor growth by suppressing the expression of AR and BMYB-regulated genes. *Chem Biol* **22:** 1185–1196.

Dulla B, Kirla KT, Rathore V, Deora GS, Kavela S, Maddika S, Chatti K, Reiser O, Iqbal J, Pal M. 2013. Synthesis and evaluation of 3-amino/guanidine substituted phenyl oxazoles as a novel class of LSD1 inhibitors with anti-proliferative properties. *Org Biomol Chem* **11:** 3103–3107.

Ehrbrecht A, Muller U, Wolter M, Hoischen A, Koch A, Radlwimmer B, Actor B, Mincheva A, Pietsch T, Lichter P, et al. 2006. Comprehensive genomic analysis of desmoplastic medulloblastomas: Identification of novel amplified genes and separate evaluation of the different histological components. *J Pathol* **208:** 554–563.

Elkins JM, Hewitson KS, McNeill LA, Seibel JF, Schlemminger I, Pugh CW, Ratcliffe PJ, Schofield CJ. 2003. Structure of factor-inhibiting hypoxia-inducible factor (HIF) reveals mechanism of oxidative modification of HIF-1α. *J Biol Chem* **278:** 1802–1806.

Ellis PM, Dronsfield AT. 2013. Antabuse's diamond anniversary: Still sparkling on? *Drug Alcohol Rev* **32:** 342–344.

England KS, Tumber A, Krojer T, Scozzafava G, Ng SS, Daniel M, Szykowska A, Che K, von Delft F, Burgess-Brown NA, et al. 2014. Optimisation of a triazolopyridine based histone demethylase inhibitor yields a potent and selective KDM2A (FBXL11) inhibitor. *MedChemComm* **5:** 1879–1886.

Etani T, Suzuki T, Naiki T, Naiki-Ito A, Ando R, Iida K, Kawai N, Tozawa K, Miyata N, Kohri K, et al. 2015. NCL1, a highly selective lysine-specific demethylase 1 inhibitor, suppresses prostate cancer without adverse effect. *Oncotarget* **6:** 2865–2878.

Fan AM, Jackson RJ. 1989. Pesticides and food safety. *Regul Toxicol Pharmacol* **9:** 158–174.

Fang R, Barbera AJ, Xu Y, Rutenberg M, Leonor T, Bi Q, Lan F, Mei P, Yuan GC, Lian C, et al. 2010. Human LSD2/KDM1b/AOF1 regulates gene transcription by modulating intragenic H3K4me2 methylation. *Mol Cell* **39:** 222–233.

Feldman JM, Chapman B. 1975. Monoamine oxidase inhibitors: Nature of their interaction with rabbit pancreatic islets to alter insulin secretion. *Diabetologia* **11:** 487–494.

Feng T, Li D, Wang H, Zhuang J, Liu F, Bao Q, Lei Y, Chen W, Zhang X, Xu X, et al. 2015. Novel 5-carboxy-8-HQ based histone demethylase JMJD2A inhibitors: Introduction of an additional carboxyl group at the C-2 position of quinoline. *Eur J Med Chem* **105:** 145–155.

Ferrari-Amorotti G, Fragliasso V, Esteki R, Prudente Z, Soliera AR, Cattelani S, Manzotti G, Grisendi G, Dominici M, Pieraccioli M, et al. 2013. Inhibiting interactions of lysine demethylase LSD1 with snail/slug blocks cancer cell invasion. *Cancer Res* **73:** 235–245.

Ferrari-Amorotti G, Chiodoni C, Shen F, Cattelani S, Soliera AR, Manzotti G, Grisendi G, Dominici M, Rivasi F, Colombo MP, et al. 2014. Suppression of invasion and metastasis of triple-negative breast cancer lines by pharmacological or genetic inhibition of slug activity. *Neoplasia* **16:** 1047–1058.

Fiskus W, Sharma S, Shah B, Portier BP, Devaraj SG, Liu K, Iyer SP, Bearss D, Bhalla KN. 2014. Highly effective combination of LSD1 (KDM1A) antagonist and pan-histone deacetylase inhibitor against human AML cells. *Leukemia* **28:** 2155–2164.

Garcia-Bassets I, Kwon YS, Telese F, Prefontaine GG, Hutt KR, Cheng CS, Ju BG, Ohgi KA, Wang J, Escoubet-Lozach L, et al. 2007. Histone methylation-dependent mechanisms impose ligand dependency for gene activation by nuclear receptors. *Cell* **128:** 505–518.

George T, Kaul CL, Grewal RS, Tahilramani R. 1971. Antihypertensive and monoamine oxidase inhibitory activity of some derivatives of 3-formyl-4-oxo-4H-pyrido (1,2-a) pyrimidine. *J Med Chem* **14:** 913–915.

Gough SM, Lee F, Yang F, Walker RL, Zhu YJ, Pineda M, Onozawa M, Chung YJ, Bilke S, Wagner EK, et al. 2014. NUP98-PHF23 is a chromatin-modifying oncoprotein that causes a wide array of leukemias sensitive to inhibition of PHD histone reader function. *Cancer Discov* **4:** 564–577.

Gui Y, Guo G, Huang Y, Hu X, Tang A, Gao S, Wu R, Chen C, Li X, Zhou L, et al. 2011. Frequent mutations of chromatin remodeling genes in transitional cell carcinoma of the bladder. *Nat Genet* **43:** 875–878.

Hamada S, Kim TD, Suzuki T, Itoh Y, Tsumoto H, Nakagawa H, Janknecht R, Miyata N. 2009. Synthesis and activity of *N*-oxalylglycine and its derivatives as Jumonji C-domain-containing histone lysine demethylase inhibitors. *Bioorg Med Chem Lett* **19:** 2852–2855.

Hamada S, Suzuki T, Mino K, Koseki K, Oehme F, Flamme I, Ozasa H, Itoh Y, Ogasawara D, Komaarashi H, et al. 2010. Design, synthesis, enzyme-inhibitory activity, and effect on human cancer cells of a novel series of Jumonji domain-containing protein 2 histone demethylase inhibitors. *J Med Chem* **53:** 5629–5638.

Han Y, Wu C, Lv H, Liu N, Deng H. 2015. Novel tranylcypromine/hydroxylcinnamic acid hybrids as lysine-specific demethylase 1 inhibitors with potent antitumor activity. *Chem Pharm Bull (Tokyo)* **63:** 882–889.

Harris WJ, Huang X, Lynch JT, Spencer GJ, Hitchin JR, Li Y, Ciceri F, Blaser JG, Greystoke BF, Jordan AM, et al. 2012. The histone demethylase KDM1A sustains the oncogenic potential of MLL-AF9 leukemia stem cells. *Cancer Cell* **21:** 473–487.

Harrison C. 2012. Structure-based drug design: Opening the door to an epigenetic target. *Nat Rev Drug Discov* **11:** 672.

Hayami S, Kelly JD, Cho HS, Yoshimatsu M, Unoki M, Tsunoda T, Field HI, Neal DE, Yamaue H, Ponder BA, et al. 2011. Overexpression of LSD1 contributes to human carcinogenesis through chromatin regulation in various cancers. *Int J Cancer* **128:** 574–586.

Heinemann B, Nielsen JM, Hudlebusch HR, Lees MJ, Larsen DV, Boesen T, Labelle M, Gerlach LO, Birk P, Helin K. 2014. Inhibition of demethylases by GSK-J1/J4. *Nature* **514:** E1–E2.

Heydendael W, Jacobson L. 2009. Glucocorticoid status affects antidepressant regulation of locus coeruleus tyrosine hydroxylase and dorsal raphe tryptophan hydroxylase gene expression. *Brain Res* **1288:** 69–78.

Hitchin JR, Blagg J, Burke R, Burns S, Cockerill MJ, Fairweather EE, Hutton C, Jordan AM, McAndrew C, Mirza A, et al. 2013. Development and evaluation of selective, reversible LSD1 inhibitors derived from fragments. *Medchemcomm* **4:** 1513–1522.

Hojfeldt JW, Agger K, Helin K. 2013. Histone lysine demethylases as targets for anticancer therapy. *Nat Rev Drug Discov* **12:** 917–930.

Hopkinson RJ, Tumber A, Yapp C, Chowdhury R, Aik W, Che KH, Li XS, Kristensen JB, King ON, Chan MC, et al. 2013. 5-Carboxy-8-hydroxyquinoline is a broad spectrum 2-oxoglutarate oxygenase inhibitor which causes iron translocation. *Chem Sci* **4:** 3110–3117.

Horton JR, Engstrom A, Zoeller EL, Liu X, Shanks JR, Zhang X, Johns MA, Vertino PM, Fu H, Cheng X. 2016. Characterization of a linked Jumonji domain of the KDM5/JARID1 family of histone H3 lysine 4 demethylases. *J Biol Chem* **291:** 2631–2646.

Hoshino I, Akutsu Y, Murakami K, Akanuma N, Isozaki Y, Maruyama T, Toyozumi T, Matsumoto Y, Suito H, Taka-

hashi M, et al. 2016. Histone demethylase LSD1 inhibitors prevent cell growth by regulating gene expression in esophageal squamous cell carcinoma cells. *Ann Surg Oncol* **23:** 312–320.

Hu X, Li X, Valverde K, Fu X, Noguchi C, Qiu Y, Huang S. 2009. LSD1-mediated epigenetic modification is required for TAL1 function and hematopoiesis. *Proc Natl Acad Sci* **106:** 10141–10146.

Huang J, Sengupta R, Espejo AB, Lee MG, Dorsey JA, Richter M, Opravil S, Shiekhattar R, Bedford MT, Jenuwein T, et al. 2007a. p53 is regulated by the lysine demethylase LSD1. *Nature* **449:** 105–108.

Huang Y, Greene E, Murray Stewart T, Goodwin AC, Baylin SB, Woster PM, Casero RA Jr. 2007b. Inhibition of lysine-specific demethylase 1 by polyamine analogues results in reexpression of aberrantly silenced genes. *Proc Natl Acad Sci* **104:** 8023–8028.

Huang Y, Stewart TM, Wu Y, Baylin SB, Marton LJ, Perkins B, Jones RJ, Woster PM, Casero RA Jr. 2009. Novel oligoamine analogues inhibit lysine-specific demethylase 1 and induce reexpression of epigenetically silenced genes. *Clin Cancer Res* **15:** 7217–7228.

Huang Y, Vasilatos SN, Boric L, Shaw PG, Davidson NE. 2012. Inhibitors of histone demethylation and histone deacetylation cooperate in regulating gene expression and inhibiting growth in human breast cancer cells. *Breast Cancer Res Treat* **131:** 777–789.

Issaeva I, Zonis Y, Rozovskaia T, Orlovsky K, Croce CM, Nakamura T, Mazo A, Eisenbach L, Canaani E. 2007. Knockdown of ALR (MLL2) reveals ALR target genes and leads to alterations in cell adhesion and growth. *Mol Cell Biol* **27:** 1889–1903.

Itoh Y, Sawada H, Suzuki M, Tojo T, Sasaki R, Hasegawa M, Mizukami T, Suzuki T. 2015. Identification of Jumonji AT-rich interactive domain 1A inhibitors and their effect on cancer cells. *ACS Med Chem Lett* **6:** 665–670.

Jenuwein T, Allis CD. 2001. Translating the histone code. *Science* **293:** 1074–1080.

Jin C, Yang L, Xie M, Lin C, Merkurjev D, Yang JC, Tanasa B, Oh S, Zhang J, Ohgi KA, et al. 2014. Chem-seq permits identification of genomic targets of drugs against androgen receptor regulation selected by functional phenotypic screens. *Proc Natl Acad Sci* **111:** 9235–9240.

Kakizawa T, Ota Y, Itoh Y, Tsumoto H, Suzuki T. 2015. Histone H3 peptide based LSD1-selective inhibitors. *Bioorg Med Chem Lett* **25:** 1925–1928.

Karytinos A, Forneris F, Profumo A, Ciossani G, Battaglioli E, Binda C, Mattevi A. 2009. A novel mammalian flavin-dependent histone demethylase. *J Biol Chem* **284:** 17775–17782.

Kauffman EC, Robinson BD, Downes MJ, Powell LG, Lee MM, Scherr DS, Gudas LJ, Mongan NP. 2011. Role of androgen receptor and associated lysine-demethylase coregulators, LSD1 and JMJD2A, in localized and advanced human bladder cancer. *Mol Carcinog* **50:** 931–944.

Khan MNA, Tsumoto H, Itoh Y, Ota Y, Suzuki M, Ogasawara D, Nakagawa H, Mizukami T, Miyata N, Suzuki T. 2015. Design, synthesis, and biological activity of *N*-alkylated analogue of NCL1, a selective inhibitor of lysine-specific demethylase 1. *Med Chem Commun* **6:** 407–412.

Kim TD, Fuchs JR, Schwartz E, Abdelhamid D, Etter J, Berry WL, Li C, Ihnat MA, Li PK, Janknecht R. 2014. Progrowth role of the JMJD2C histone demethylase in HCT-116 colon cancer cells and identification of curcuminoids as JMJD2 inhibitors. *Am J Transl Res* **6**: 236–247.

King ON, Li XS, Sakurai M, Kawamura A, Rose NR, Ng SS, Quinn AM, Rai G, Mott BT, Beswick P, et al. 2010. Quantitative high-throughput screening identifies 8-hydroxyquinolines as cell-active histone demethylase inhibitors. *PLoS ONE* **5**: e15535.

Klose RJ, Yamane K, Bae Y, Zhang D, Erdjument-Bromage H, Tempst P, Wong J, Zhang Y. 2006. The transcriptional repressor JHDM3A demethylates trimethyl histone H3 lysine 9 and lysine 36. *Nature* **442**: 312–316.

Konovalov S, Garcia-Bassets I. 2013. Analysis of the levels of lysine-specific demethylase 1 (LSD1) mRNA in human ovarian tumors and the effects of chemical LSD1 inhibitors in ovarian cancer cell lines. *J Ovarian Res* **6**: 75.

Kontaki H, Talianidis I. 2010. Lysine methylation regulates E2F1-induced cell death. *Mol Cell* **39**: 152–160.

Korczynska M, Le DD, Younger N, Gregori-Puigjane E, Tumber A, Krojer T, Velupillai S, Gileadi C, Nowak RP, Iwasa E, et al. 2015. Docking and linking of fragments to discover Jumonji histone demethylase inhibitors. *J Med Chem* **59**: 1580–1598.

Kouzarides T. 2007. Chromatin modifications and their function. *Cell* **128**: 693–705.

Kristensen LH, Nielsen AL, Helgstrand C, Lees M, Cloos P, Kastrup JS, Helin K, Olsen L, Gajhede M. 2012. Studies of H3K4me3 demethylation by KDM5B/Jarid1B/PLU1 reveals strong substrate recognition in vitro and identifies 2,4-pyridine-dicarboxylic acid as an in vitro and in cell inhibitor. *FEBS J* **279**: 1905–1914.

Kruidenier L, Chung CW, Cheng Z, Liddle J, Che K, Joberty G, Bantscheff M, Bountra C, Bridges A, Diallo H, et al. 2012. A selective Jumonji H3K27 demethylase inhibitor modulates the proinflammatory macrophage response. *Nature* **488**: 404–408.

Labbe RM, Holowatyj A, Yang ZQ. 2013. Histone lysine demethylase (KDM) subfamily 4: Structures, functions and therapeutic potential. *Am J Transl Res* **6**: 1–15.

Lan F, Bayliss PE, Rinn JL, Whetstine JR, Wang JK, Chen S, Iwase S, Alpatov R, Issaeva I, Canaani E, et al. 2007. A histone H3 lysine 27 demethylase regulates animal posterior development. *Nature* **449**: 689–694.

Laurent B, Ruitu L, Murn J, Hempel K, Ferrao R, Xiang Y, Liu S, Garcia BA, Wu H, Wu F, et al. 2015. A specific LSD1/KDM1A isoform regulates neuronal differentiation through H3K9 demethylation. *Mol Cell* **57**: 957–970.

Lee MG, Wynder C, Cooch N, Shiekhattar R. 2005. An essential role for CoREST in nucleosomal histone 3 lysine 4 demethylation. *Nature* **437**: 432–435.

Lee MG, Wynder C, Schmidt DM, McCafferty DG, Shiekhattar R. 2006. Histone H3 lysine 4 demethylation is a target of nonselective antidepressive medications. *Chem Biol* **13**: 563–567.

Liang Y, Quenelle D, Vogel JL, Mascaro C, Ortega A, Kristie TM. 2013a. A novel selective LSD1/KDM1A inhibitor epigenetically blocks herpes simplex virus lytic replica-

tion and reactivation from latency. *MBio* **4**: e00558–e00512.

Liang Y, Vogel JL, Arbuckle JH, Rai G, Jadhav A, Simeonov A, Maloney DJ, Kristie TM. 2013b. Targeting the JMJD2 histone demethylases to epigenetically control herpesvirus infection and reactivation from latency. *Sci Transl Med* **5**: 167ra165.

Lienhart WD, Gudipati V, Macheroux P. 2013. The human flavoproteome. *Arch Biochem Biophys* **535**: 150–162.

Ling Y, Wang Z, Wang X, Zhao Y, Zhang W, Wang X, Chen L, Huang Z, Zhang Y. 2014. Synthesis and biological evaluation of hybrids from farnesylthiosalicylic acid and hydroxylcinnamic acid with dual inhibitory activities of Ras-related signaling and phosphorylated NF-κB. *Org Biomol Chem* **12**: 4517–4530.

Liu G, Bollig-Fischer A, Kreike B, van de Vijver MJ, Abrams J, Ethier SP, Yang ZQ. 2009. Genomic amplification and oncogenic properties of the GASC1 histone demethylase gene in breast cancer. *Oncogene* **28**: 4491–4500.

Liu LJ, Lu L, Zhong HJ, He B, Kwong DW, Ma DL, Leung CH. 2015. An iridium(III) complex inhibits JMJD2 activities and acts as a potential epigenetic modulator. *J Med Chem* **58**: 6697–6703.

Luger K, Mader AW, Richmond RK, Sargent DF, Richmond TJ. 1997. Crystal structure of the nucleosome core particle at 2.8 A resolution. *Nature* **389**: 251–260.

Luo X, Liu Y, Kubicek S, Myllyharju J, Tumber A, Ng S, Che KH, Podoll J, Heightman TD, Oppermann U, et al. 2011. A selective inhibitor and probe of the cellular functions of Jumonji C domain-containing histone demethylases. *J Am Chem Soc* **133**: 9451–9456.

Lv T, Yuan D, Miao X, Lv Y, Zhan P, Shen X, Song Y. 2012. Over-expression of LSD1 promotes proliferation, migration and invasion in non-small cell lung cancer. *PLoS ONE* **7**: e35065.

Ma LY, Zheng YC, Wang SQ, Wang B, Wang ZR, Pang LP, Zhang M, Wang JW, Ding L, Li J, et al. 2015. Design, synthesis, and structure–activity relationship of novel LSD1 inhibitors based on pyrimidine-thiourea hybrids as potent, orally active antitumor agents. *J Med Chem* **58**: 1705–1716.

Mackeen MM, Kramer HB, Chang KH, Coleman ML, Hopkinson RJ, Schofield CJ, Kessler BM. 2010. Small-molecule-based inhibition of histone demethylation in cells assessed by quantitative mass spectrometry. *J Proteome Res* **9**: 4082–4092.

Maes T, Carceller E, Salas J, Ortega A, Buesa C. 2015. Advances in the development of histone lysine demethylase inhibitors. *Curr Opin Pharmacol* **23**: 52–60.

Mai A, Cheng D, Bedford MT, Valente S, Nebbioso A, Perrone A, Brosch G, Sbardella G, De Bellis F, Miceli M, et al. 2008. Epigenetic multiple ligands: Mixed histone/protein methyltransferase, acetyltransferase, and class III deacetylase (sirtuin) inhibitors. *J Med Chem* **51**: 2279–2290.

Maio M, Covre A, Fratta E, Di Giacomo AM, Taverna P, Natali PG, Coral S, Sigalotti L. 2015. Molecular pathways: At the crossroads of cancer epigenetics and immunotherapy. *Clin Cancer Res* **21**: 4040–4047.

Mannironi C, Proietto M, Bufalieri F, Cundari E, Alagia A, Danovska S, Rinaldi T, Famiglini V, Coluccia A, La Regina G, et al. 2014. An high-throughput in vivo screening system to select H3K4-specific histone demethylase inhibitors. *PLoS ONE* **9**: e86002.

Martin C, Zhang Y. 2005. The diverse functions of histone lysine methylation. *Nat Rev Mol Cell Biol* **6**: 838–849.

Metzger E, Wissmann M, Yin N, Muller JM, Schneider R, Peters AH, Gunther T, Buettner R, Schule R. 2005. LSD1 demethylates repressive histone marks to promote androgen-receptor-dependent transcription. *Nature* **437**: 436–439.

Mimasu S, Umezawa N, Sato S, Higuchi T, Umehara T, Yokoyama S. 2010. Structurally designed trans-2-phenylcyclopropylamine derivatives potently inhibit histone demethylase LSD1/KDM1. *Biochemistry* **49**: 6494–6503.

Murray-Stewart T, Woster PM, Casero RA Jr. 2014. The reexpression of the epigenetically silenced e-cadherin gene by a polyamine analogue lysine-specific demethylase-1 (LSD1) inhibitor in human acute myeloid leukemia cell lines. *Amino Acids* **46**: 585–594.

Nakamura S, Tan L, Nagata Y, Takemura T, Asahina A, Yokota D, Yagyu T, Shibata K, Fujisawa S, Ohnishi K. 2013. JmjC-domain containing histone demethylase 1B-mediated p15(Ink4b) suppression promotes the proliferation of leukemic progenitor cells through modulation of cell cycle progression in acute myeloid leukemia. *Mol Carcinog* **52**: 57–69.

Neelamegam R, Ricq EL, Malvaez M, Patnaik D, Norton S, Carlin SM, Hill IT, Wood MA, Haggarty SJ, Hooker JM. 2012. Brain-penetrant LSD1 inhibitors can block memory consolidation. *ACS Chem Neurosci* **3**: 120–128.

Nielsen AL, Kristensen LH, Stephansen KB, Kristensen JB, Helgstrand C, Lees M, Cloos P, Helin K, Gajhede M, Olsen L. 2012. Identification of catechols as histone-lysine demethylase inhibitors. *FEBS Lett* **586**: 1190–1194.

Nowotarski SL, Pachaiyappan B, Holshouser SL, Kutz CJ, Li Y, Huang Y, Sharma SK, Casero RA Jr, Woster PM. 2015. Structure–activity study for (bis)ureidopropyl- and (bis)thioureidopropyldiamine LSD1 inhibitors with 3-5-3 and 3-6-3 carbon backbone architectures. *Bioorg Med Chem* **23**: 1601–1612.

Ntziachristos P, Tsirigos A, Welstead GG, Trimarchi T, Bakogianni S, Xu L, Loizou E, Holmfeldt L, Strikoudis A, King B, et al. 2014. Contrasting roles of histone 3 lysine 27 demethylases in acute lymphoblastic leukaemia. *Nature* **514**: 513–517.

Ogasawara D, Suzuki T, Mino K, Ueda R, Khan MN, Matsubara T, Koseki K, Hasegawa M, Sasaki R, Nakagawa H, et al. 2011. Synthesis and biological activity of optically active NCL-1, a lysine-specific demethylase 1 selective inhibitor. *Bioorg Med Chem* **19**: 3702–3708.

Oreland L, Kinemuchi H, Yoo BY. 1973. The mechanism of action of the monoamine oxidase inhibitor pargyline. *Life Sci* **13**: 1533–1541.

Paik WK, Kim S. 1973. Enzymatic demethylation of calf thymus histones. *Biochem Biophys Res Commun* **51**: 781–788.

Perillo B, Ombra MN, Bertoni A, Cuozzo C, Sacchetti S, Sasso A, Chiariotti L, Malorni A, Abbondanza C, Avvedimento EV. 2008. DNA oxidation as triggered by H3K9me2 demethylation drives estrogen-induced gene expression. *Science* **319**: 202–206.

Pollock JA, Larrea MD, Jasper JS, McDonnell DP, McCafferty DG. 2012. Lysine-specific histone demethylase 1

inhibitors control breast cancer proliferation in ERα-dependent and -independent manners. *ACS Chem Biol* **7**: 1221–1231.

Prusevich P, Kalin JH, Ming SA, Basso M, Givens J, Li X, Hu J, Taylor MS, Cieniewicz AM, Hsiao PY, et al. 2014. A selective phenelzine analogue inhibitor of histone demethylase LSD1. *ACS Chem Biol* **9**: 1284–1293.

Quirarte GL, Roozendaal B, McGaugh JL. 1997. Glucocorticoid enhancement of memory storage involves noradrenergic activation in the basolateral amygdala. *Proc Natl Acad Sci* **94**: 14048–14053.

Rai G, Kawamura A, Tumber A, Liang Y, Vogel JL, Arbuckle JH, Rose NR, Dexheimer TS, Foley TL, King ON, et al. 2010. Discovery of ML324, a JMJD2 demethylase inhibitor with demonstrated antiviral activity. *Probe Reports from the NIH Molecular Libraries Program*, Bethesda, MD.

Rose NR, Ng SS, Mecinovic J, Lienard BM, Bello SH, Sun Z, McDonough MA, Oppermann U, Schofield CJ. 2008. Inhibitor scaffolds for 2-oxoglutarate-dependent histone lysine demethylases. *J Med Chem* **51**: 7053–7056.

Rose NR, Woon EC, Kingham GL, King ON, Mecinovic J, Clifton IJ, Ng SS, Talib-Hardy J, Oppermann U, McDonough MA, et al. 2010. Selective inhibitors of the JMJD2 histone demethylases: Combined nondenaturing mass spectrometric screening and crystallographic approaches. *J Med Chem* **53**: 1810–1818.

Rose NR, Woon EC, Tumber A, Walport LJ, Chowdhury R, Li XS, King ON, Lejeune C, Ng SS, Krojer T, et al. 2012. Plant growth regulator daminozide is a selective inhibitor of human KDM2/7 histone demethylases. *J Med Chem* **55**: 6639–6643.

Rotili D, Tomassi S, Conte M, Benedetti R, Tortorici M, Ciossani G, Valente S, Marrocco B, Labella D, Novellino E, et al. 2014. Pan-histone demethylase inhibitors simultaneously targeting Jumonji C and lysine-specific demethylases display high anticancer activities. *J Med Chem* **57**: 42–55.

Sakaki H, Okada M, Kuramoto K, Takeda H, Watarai H, Suzuki S, Seino S, Seino M, Ohta T, Nagase S, et al. 2015. GSKJ4, a selective Jumonji H3K27 demethylase inhibitor, effectively targets ovarian cancer stem cells. *Anticancer Res* **35**: 6607–6614.

Sakurai M, Rose NR, Schultz L, Quinn AM, Jadhav A, Ng SS, Oppermann U, Schofield CJ, Simeonov A. 2010. A miniaturized screen for inhibitors of Jumonji histone demethylases. *Mol Biosyst* **6**: 357–364.

Sankar S, Theisen ER, Bearss J, Mulvihill T, Hoffman LM, Sorna V, Beckerle MC, Sharma S, Lessnick SL. 2014. Reversible LSD1 inhibition interferes with global EWS/ETS transcriptional activity and impedes Ewing sarcoma tumor growth. *Clin Cancer Res* **20**: 4584–4597.

Sareddy GR, Nair BC, Krishnan SK, Gonugunta VK, Zhang QG, Suzuki T, Miyata N, Brenner AJ, Brann DW, Vadlamudi RK. 2013. KDM1 is a novel therapeutic target for the treatment of gliomas. *Oncotarget* **4**: 18–28.

Sayegh J, Cao J, Zou MR, Morales A, Blair LP, Norcia M, Hoyer D, Tackett AJ, Merkel JS, Yan Q. 2013. Identification of small molecule inhibitors of Jumonji AT-rich interactive domain 1B (JARID1B) histone demethylase by a sensitive high throughput screen. *J Biol Chem* **288**: 9408–9417.

Schenk T, Chen WC, Gollner S, Howell L, Jin L, Hebestreit K, Klein HU, Popescu AC, Burnett A, Mills K, et al. 2012. Inhibition of the LSD1 (KDM1A) demethylase reactivates the all-trans-retinoic acid differentiation pathway in acute myeloid leukemia. *Nat Med* **18:** 605–611.

Schildhaus HU, Riegel R, Hartmann W, Steiner S, Wardelmann E, Merkelbach-Bruse S, Tanaka S, Sonobe H, Schule R, Buettner R, et al. 2011. Lysine-specific demethylase 1 is highly expressed in solitary fibrous tumors, synovial sarcomas, rhabdomyosarcomas, desmoplastic small round cell tumors, and malignant peripheral nerve sheath tumors. *Hum Pathol* **42:** 1667–1675.

Schiller R, Scozzafava G, Tumber A, Wickens JR, Bush JT, Rai G, Lejeune C, Choi H, Yeh TL, Chan MC, et al. 2014. A cell-permeable ester derivative of the JmjC histone demethylase inhibitor IOX1. *ChemMedChem* **9:** 566–571.

Schmidt DM, McCafferty DG. 2007. *trans*-2-Phenylcyclopropylamine is a mechanism-based inactivator of the histone demethylase LSD1. *Biochemistry* **46:** 4408–4416.

Schmitt ML, Hauser AT, Carlino L, Pippel M, Schulz-Fincke J, Metzger E, Willmann D, Yiu T, Barton M, Schule R, et al. 2013. Nonpeptidic propargylamines as inhibitors of lysine specific demethylase 1 (LSD1) with cellular activity. *J Med Chem* **56:** 7334–7342.

Schror K. 1997. Aspirin and platelets: The antiplatelet action of aspirin and its role in thrombosis treatment and prophylaxis. *Semin Thromb Hemost* **23:** 349–356.

Sekirnik R, Rose NR, Thalhammer A, Seden PT, Mecinovic J, Schofield CJ. 2009. Inhibition of the histone lysine demethylase JMJD2A by ejection of structural Zn(II). *Chem Commun (Camb)* **42:** 6376–6378.

Sharma SV, Lee DY, Li B, Quinlan MP, Takahashi F, Maheswaran S, McDermott U, Azizian N, Zou L, Fischbach MA, et al. 2010. A chromatin-mediated reversible drug-tolerant state in cancer cell subpopulations. *Cell* **141:** 69–80.

Shi Y, Lan F, Matson C, Mulligan P, Whetstine JR, Cole PA, Casero RA, Shi Y. 2004. Histone demethylation mediated by the nuclear amine oxidase homolog LSD1. *Cell* **119:** 941–953.

Shi YJ, Matson C, Lan F, Iwase S, Baba T, Shi Y. 2005. Regulation of LSD1 histone demethylase activity by its associated factors. *Mol Cell* **19:** 857–864.

Shi L, Sun L, Li Q, Liang J, Yu W, Yi X, Yang X, Li Y, Han X, Zhang Y, et al. 2011. Histone demethylase JMJD2B coordinates H3K4/H3K9 methylation and promotes hormonally responsive breast carcinogenesis. *Proc Natl Acad Sci* **108:** 7541–7546.

Shih JC, Wu JB, Chen K. 2011. Transcriptional regulation and multiple functions of MAO genes. *J Neur Trans* **118:** 979–986.

Simon LS, Mills JA. 1980. Nonsteroidal antiinflammatory drugs (second of two parts). *N Engl J Med* **302:** 1237–1243.

Singh MM, Manton CA, Bhat KP, Tsai WW, Aldape K, Barton MC, Chandra J. 2011. Inhibition of LSD1 sensitizes glioblastoma cells to histone deacetylase inhibitors. *Neuro Oncol* **13:** 894–903.

Sorna V, Theisen ER, Stephens B, Warner SL, Bearss DJ, Vankayalapati H, Sharma S. 2013. High-throughput virtual screening identifies novel N'-(1-phenylethylidene)-benzohydrazides as potent, specific, and reversible LSD1 inhibitors. *J Med Chem* **56:** 9496–9508.

Stavropoulos P, Blobel G, Hoelz A. 2006. Crystal structure and mechanism of human lysine-specific demethylase-1. *Nat Struct Mol Biol* **13:** 626–632.

Suva ML, Rheinbay E, Gillespie SM, Patel AP, Wakimoto H, Rabkin SD, Riggi N, Chi AS, Cahill DP, Nahed BV, et al. 2014. Reconstructing and reprogramming the tumor-propagating potential of glioblastoma stem-like cells. *Cell* **157:** 580–594.

Suzuki T, Ozasa H, Itoh Y, Zhan P, Sawada H, Mino K, Walport L, Ohkubo R, Kawamura A, Yonezawa M, et al. 2013. Identification of the KDM2/7 histone lysine demethylase subfamily inhibitor and its antiproliferative activity. *J Med Chem* **56:** 7222–7231.

Thalhammer A, Mecinovic J, Loenarz C, Tumber A, Rose NR, Heightman TD, Schofield CJ. 2011. Inhibition of the histone demethylase JMJD2E by 3-substituted pyridine 2,4-dicarboxylates. *Org Biomol Chem* **9:** 127–135.

Theisen ER, Gajiwala S, Bearss J, Sorna V, Sharma S, Janat-Amsbury M. 2014. Reversible inhibition of lysine specific demethylase 1 is a novel anti-tumor strategy for poorly differentiated endometrial carcinoma. *BMC Cancer* **14:** 752.

Tortorici M, Borrello MT, Tardugno M, Chiarelli LR, Pilotto S, Ciossani G, Vellore NA, Bailey SG, Cowan J, O'Connell M, et al. 2013. Protein recognition by short peptide reversible inhibitors of the chromatin-modifying LSD1/CoREST lysine demethylase. *ACS Chem Biol* **8:** 1677–1682.

Tschank G, Raghunath M, Gunzler V, Hanauske-Abel HM. 1987. Pyridinedicarboxylates, the first mechanism-derived inhibitors for prolyl 4-hydroxylase, selectively suppress cellular hydroxyprolyl biosynthesis. Decrease in interstitial collagen and Clq secretion in cell culture. *Biochem J* **248:** 625–633.

Tsukada Y, Fang J, Erdjument-Bromage H, Warren ME, Borchers CH, Tempst P, Zhang Y. 2006. Histone demethylation by a family of JmjC domain-containing proteins. *Nature* **439:** 811–816.

Ueda R, Suzuki T, Mino K, Tsumoto H, Nakagawa H, Hasegawa M, Sasaki R, Mizukami T, Miyata N. 2009. Identification of cell-active lysine specific demethylase 1-selective inhibitors. *J Am Chem Soc* **131:** 17536–17537.

Valente S, Rodriguez V, Mercurio C, Vianello P, Saponara B, Cirilli R, Ciossani G, Labella D, Marrocco B, Monaldi D, et al. 2015a. Pure enantiomers of benzoylamino-tranylcypromine: LSD1 inhibition, gene modulation in human leukemia cells and effects on clonogenic potential of murine promyelocytic blasts. *Eur J Med Chem* **94:** 163–174.

Valente S, Rodriguez V, Mercurio C, Vianello P, Saponara B, Cirilli R, Ciossani G, Labella D, Marrocco B, Ruoppolo G, et al. 2015b. Pure diastereomers of a tranylcypromine-based LSD1 inhibitor: Enzyme selectivity and in-cell studies. *ACS Med Chem Lett* **6:** 173–177.

van Haaften G, Dalgliesh GL, Davies H, Chen L, Bignell G, Greenman C, Edkins S, Hardy C, O'Meara S, Teague J, et al. 2009. Somatic mutations of the histone H3K27 demethylase gene UTX in human cancer. *Nat Genet* **41:** 521–523.

Van Rechem C, Black JC, Boukhali M, Aryee MJ, Graslund S, Haas W, Benes CH, Whetstine JR. 2015. Lysine demethylase KDM4A associates with translation machinery and regulates protein synthesis. *Cancer Discov* **5:** 255–263.

van Zutven LJ, Onen E, Velthuizen SC, van Drunen E, von Bergh AR, van den Heuvel-Eibrink MM, Veronese A, Mecucci C, Negrini M, de Greef GE, et al. 2006. Identification of NUP98 abnormalities in acute leukemia: JARID1A (12p13) as a new partner gene. *Genes Chromosomes Cancer* **45**: 437–446.

Vasilatos SN, Katz TA, Oesterreich S, Wan Y, Davidson NE, Huang Y. 2013. Crosstalk between lysine-specific demethylase 1 (LSD1) and histone deacetylases mediates antineoplastic efficacy of HDAC inhibitors in human breast cancer cells. *Carcinogenesis* **34**: 1196–1207.

Vianello P, Botrugno OA, Cappa A, Dal Zuffo R, Dessanti P, Mai A, Marrocco B, Mattevi A, Meroni G, Minucci S, et al. 2016. Discovery of a novel inhibitor of histone lysine-specific demethylase 1A (KDM1A/LSD1) as orally active antitumor agent. *J Med Chem* **59**: 1501–1517.

Wada T, Koyama D, Kikuchi J, Honda H, Furukawa Y. 2015. Overexpression of the shortest isoform of histone demethylase LSD1 primes hematopoietic stem cells for malignant transformation. *Blood* **125**: 3731–3746.

Wang J, Scully K, Zhu X, Cai L, Zhang J, Prefontaine GG, Krones A, Ohgi KA, Zhu P, Garcia-Bassets I, et al. 2007. Opposing LSD1 complexes function in developmental gene activation and repression programmes. *Nature* **446**: 882–887.

Wang Y, Zhang H, Chen Y, Sun Y, Yang F, Yu W, Liang J, Sun L, Yang X, Shi L, et al. 2009. LSD1 is a subunit of the NuRD complex and targets the metastasis programs in breast cancer. *Cell* **138**: 660–672.

Wang L, Chang J, Varghese D, Dellinger M, Kumar S, Best AM, Ruiz J, Bruick R, Pena-Llopis S, Xu J, et al. 2013. A small molecule modulates Jumonji histone demethylase activity and selectively inhibits cancer growth. *Nat Commun* **4**: 2035.

Wang J, Telese F, Tan Y, Li W, Jin C, He X, Basnet H, Ma Q, Merkurjev D, Zhu X, et al. 2015a. LSD1n is an H4K20 demethylase regulating memory formation via transcriptional elongation control. *Nat Neurosci* **18**: 1256–1264.

Wang M, Liu X, Guo J, Weng X, Jiang G, Wang Z, He L. 2015b. Inhibition of LSD1 by Pargyline inhibited process of EMT and delayed progression of prostate cancer in vivo. *Biochem Biophys Res Commun* **467**: 310–315.

Wang Q, Wei J, Su P, Gao P. 2015c. Histone demethylase JARID1C promotes breast cancer metastasis cells via down regulating BRMS1 expression. *Biochem Biophys Res Commun* **464**: 659–666.

Wang Y, Zhu Y, Wang Q, Hu H, Li Z, Wang D, Zhang W, Qi B, Ye J, Wu H, et al. 2016. The histone demethylase LSD1 is a novel oncogene and therapeutic target in oral cancer. *Cancer Lett.* **374**: 12–21.

Webby CJ, Wolf A, Gromak N, Dreger M, Kramer H, Kessler B, Nielsen ML, Schmitz C, Butler DS, Yates JR 3rd, et al. 2009. Jmjd6 catalyses lysyl-hydroxylation of U2AF65, a protein associated with RNA splicing. *Science* **325**: 90–93.

Wetzel S, Wilk W, Chammaa S, Sperl B, Roth AG, Yektaoglu A, Renner S, Berg T, Arenz C, Giannis A, et al. 2010. A scaffold-tree-merging strategy for prospective bioactivity annotation of γ-pyrones. *Angew Chem Int Ed Engl* **49**: 3666–3670.

Whetstine JR, Nottke A, Lan F, Huarte M, Smolikov S, Chen Z, Spooner E, Li E, Zhang G, Colaiacovo M, et al. 2006. Reversal of histone lysine trimethylation by the JMJD2 family of histone demethylases. *Cell* **125**: 467–481.

Willmann D, Lim S, Wetzel S, Metzger E, Jandausch A, Wilk W, Jung M, Forne I, Imhof A, Janzer A, et al. 2012. Impairment of prostate cancer cell growth by a selective and reversible lysine-specific demethylase 1 inhibitor. *Int J Cancer* **131**: 2704–2709.

Wissmann M, Yin N, Muller JM, Greschik H, Fodor BD, Jenuwein T, Vogler C, Schneider R, Gunther T, Buettner R, et al. 2007. Cooperative demethylation by JMJD2C and LSD1 promotes androgen receptor-dependent gene expression. *Nat Cell Biol* **9**: 347–353.

Woon EC, Tumber A, Kawamura A, Hillringhaus L, Ge W, Rose NR, Ma JH, Chan MC, Walport LJ, Che KH, et al. 2012. Linking of 2-oxoglutarate and substrate binding sites enables potent and highly selective inhibition of JmjC histone demethylases. *Angew Chem Int Ed Engl* **51**: 1631–1634.

Wu CY, Hsieh CY, Huang KE, Chang C, Kang HY. 2012. Cryptotanshinone down-regulates androgen receptor signaling by modulating lysine-specific demethylase 1 function. *Int J Cancer* **131**: 1423–1434.

Wu F, Zhou C, Yao Y, Wei L, Feng Z, Deng L, Song Y. 2016. 3-(Piperidin-4-ylmethoxy)pyridine containing compounds are potent inhibitors of lysine specific demethylase 1. *J Med Chem* **59**: 253–263.

Yamane K, Tateishi K, Klose RJ, Fang J, Fabrizio LA, Erdjument-Bromage H, Taylor-Papadimitriou J, Tempst P, Zhang Y. 2007. PLU-1 is an H3K4 demethylase involved in transcriptional repression and breast cancer cell proliferation. *Mol Cell* **25**: 801–812.

Yang ZQ, Imoto I, Fukuda Y, Pimkhaokham A, Shimada Y, Imamura M, Sugano S, Nakamura Y, Inazawa J. 2000. Identification of a novel gene, GASC1, within an amplicon at 9p23-24 frequently detected in esophageal cancer cell lines. *Cancer Res* **60**: 4735–4739.

Yang M, Culhane JC, Szewczuk LM, Jalili P, Ball HL, Machius M, Cole PA, Yu H. 2007. Structural basis for the inhibition of the LSD1 histone demethylase by the antidepressant trans-2-phenylcyclopropylamine. *Biochemistry* **46**: 8058–8065.

Ye Q, Holowatyj A, Wu J, Liu H, Zhang L, Suzuki T, Yang ZQ. 2015. Genetic alterations of KDM4 subfamily and therapeutic effect of novel demethylase inhibitor in breast cancer. *Am J Cancer Res* **5**: 1519–1530.

Zheng YC, Duan YC, Ma JL, Xu RM, Zi X, Lv WL, Wang MM, Ye XW, Zhu S, Mobley D, et al. 2013. Triazole-dithiocarbamate based selective lysine specific demethylase 1 (LSD1) inactivators inhibit gastric cancer cell growth, invasion, and migration. *J Med Chem* **56**: 8543–8560.

Zhou Y, Li Y, Wang WJ, Xiang P, Luo XM, Yang L, Yang SY, Zhao YL. 2015. Synthesis and biological evaluation of novel (E)-N′-(2, 3-dihydro-1H-inden-1-ylidene) benzohydrazides as potent LSD1 inhibitors. *Bioorg Med Chem Lett* doi: 10.1016/j.bmcl2015.06.054.

Mechanisms of Nucleosome Dynamics In Vivo

Steven Henikoff

Howard Hughes Medical Institute, Fred Hutchinson Cancer Research Center, Seattle, Washington 98109

Correspondence: steveh@fredhutch.org

Nucleosomes function to tightly package DNA into chromosomes, but the nucleosomal landscape becomes disrupted during active processes such as replication, transcription, and repair. The realization that many proteins responsible for chromatin regulation are frequently mutated in cancer has drawn attention to chromatin dynamics; however, the basic mechanisms whereby nucleosomes are disrupted and reassembled is incompletely understood. Here, I present an overview of chromatin dynamics as has been elucidated in model organisms, in which our understanding is most advanced. A basic understanding of chromatin dynamics during normal developmental processes can provide the context for understanding how this machinery can go awry during oncogenesis.

Sequencing of tumor DNA has uncovered mutations and rearrangements of well-known tumor suppressor genes and oncogenes in a wide variety of human cancers, confirming much of what had been learned from decades of cancer research (Vogelstein et al. 2013). Such genetic insights verify the long-held assumption that cancer is not a single disease, but rather many diseases. Nevertheless, tumor DNA sequencing has also revealed a surprising number of likely driver mutations in a variety of shared chromatin components, sometimes seen in cancers that otherwise have little else in common (Pon and Marra 2015). From this epigenetic perspective, cancer might be viewed as a complex syndrome in which normal mechanisms that maintain chromatin homeostasis become disrupted in such a way that they may be subject to selection for uncontrolled proliferation. On the one hand, finding a chromatin basis for so many cancers has led to the hope for therapeutic

intervention to reverse the cancer phenotype, and other reviews discuss progress on these fronts. On the other hand, the complexity of the chromatin landscape makes it difficult to explain a wide variety of observations, some of which almost seem to lack a rational basis. For example, a histone modification that is present in all eukaryotic life, H3K79 methylation, can be essentially eliminated in the hematopoietic system in which it efficiently kills leukemia cells but has little effect on normal stem/progenitor cells (Daigle et al. 2011). Another example is a lysine-to-methionine mutation in the histone H3 tail that promotes a highly aggressive tumor when it occurs in a specific site in a child's brain (Wu et al. 2012), but a nearby mutation in the same histone tail results in a benign tumor when it occurs in a specific bone-forming cell of an adult (Behjati et al. 2013). To make sense of these and other issues raised by the discovery of mutations in chromatin regulators, we need to consider the molecular and developmental context in which the various components of the chromatin machinery normally function.

Here, I will survey chromatin dynamics during normal processes, focusing on those proteins and complexes that are most frequently mutated in cancer (Fig. 1). Most of our understanding of these processes comes from studies in model organisms, in which powerful genetic, biochemical, and genomic tools have long been applied. With the advent of new technologies, such as genome editing (Laufer and Singh 2015) and live super-resolution microscopy (Liu et al. 2015), the impact of traditional genetic studies of model organisms on understanding cancer genetics and epigenetics is likely to continue.

DISRUPTING AND REMODELING NUCLEOSOMES

The tight wrapping of DNA around the octameric core of nucleosomes requires their mobilization or eviction to make the DNA accessible for replication, transcription, and repair to occur. Nucleosomes are completely disrupted every cell cycle before the DNA duplex passes through the replicative helicase, which separates the Watson and Crick strands for templated DNA synthesis, followed by reassembly on leading and lagging strands (Ramachandran and Henikoff 2015). Nucleosomes are also disrupted during passage of RNA polymerases, although the high density of nucleosomes over all but the most actively transcribed genes (Weintraub and Groudine 1976) implies that nucleosome disruption and reassembly during transcription must be very efficient. Considering that a nucleosome is an impenetrable barrier to the large RNA polymerase II (RNAPII) holoenzyme complex in vitro, how it can transcribe through a nucleosome in vivo remains incompletely understood (Teves et al. 2014). Another dynamic process that disrupts nucleosomes is remodeling by a class of DNA translocases related to the yeast switch-2/sucrose nonfermenting-2 (SWI2/SNF2) ATPase (Clapier and Cairns 2009).

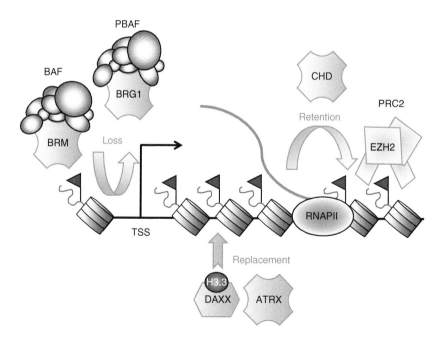

Figure 1. Regulators of chromatin dynamics implicated in cancer. (P)BAF complexes catalyze nucleosome sliding and/or eviction, ATRX and DAXX promote replacement with H3.3 nucleosomes, CHD ATPases facilitate transcriptional elongation, and the PRC2 complex methylates the H3 amino-terminal tail at lysine-27 to stabilize nucleosomes.

Cite this article as *Cold Spring Harb Perspect Med* doi: 10.1101/cshperspect.a026666

Phylogenetic analysis of human and yeast members of the SWI/SNF superfamily reveals 11 distinct subfamilies, all with at least one human and one yeast member (Fig. 2). Four subfamilies, RAD54, RAD26, RAD16, and FUN30, are DNA translocases that function in DNA repair and/or recombination (Chen et al. 2012; Costelloe et al. 2012; Hinz and Czaja 2015; Li 2015; Waters et al. 2015), and MOT1 is dedicated to regulate TATA-binding protein by removing it from high-affinity sites (Wollmann et al. 2011; Zentner and Henikoff 2013). The other six subfamilies use nucleosomes as substrates for remodeling, and except for IRC5, which is as-yet uncharacterized, all have

been shown to perform distinct nucleosome mobilization reactions. CHD (chromo-helicase-DNA-binding), ISWI (imitation switch), and SNF2 subfamily translocases slide histone cores along the DNA duplex. Biochemical characterization of SNF2 and ISWI remodelers in yeast and *Drosophila* led to the realization that these different phylogenetic subgroups have rather different actions on nucleosomes: SNF2-class remodelers disrupt nucleosomes to facilitate activation, ISWI-class remodelers reposition and regularly space nucleosomes, and CHD-class remodelers help RNAPII transcribe through a nucleosome. These differences led to the realization that nucleosome dynam-

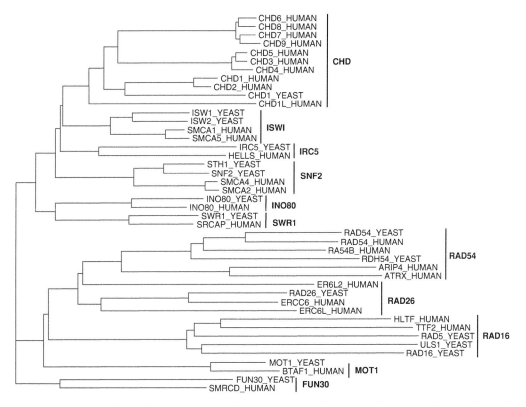

Figure 2. Phylogeny of SWI/SNF ATPase superfamily members in humans and yeast. Dendrogram shows that each subfamily is represented in both species. For example, the SNF2 subfamily includes the ATPase subunit for yeast RSC (Sth1) and SWI/SNF (Snf2) and for PBAF (SMARCA4=BRG1) and BAF (SMARCA2=BRM). For consistency, the SWISS-PROT name for each protein is used in the phylogeny. To obtain human and yeast members of the superfamily, ATRX_HUMAN was used as query versus human- and yeast-annotated SWISS-PROT amino acid sequences. Multiple alignment was performed using MAFFT with defaults, pruning with the MaxAlin option to maximize alignment quality, and a neighbor-joining tree was constructed using all 364 gap-free residues with the JTT amino acid substitution matrix. Ten distinct subfamilies were identified, each with at least one human and one yeast protein.

ics can have a profound effect on gene expression.

Biochemical characterization of nucleosome remodelers has also distinguished them based on differences in subunit composition. Whereas the yeast Chd1 remodeler is a single subunit enzyme, Swr1 and Ino80 ATPases are each part of 14-subunit complexes. The SWR1 complex is dedicated to replacing canonical histone H2A with the H2A.Z histone variant, and the INO80 complex is capable of performing the reverse reaction (Watanabe and Peterson 2010). It is thought that these opposing reactions constitute a futile cycle in which H2A/H2B dimers, which flank the central $(H3/H4)_2$ tetramer in the nucleosome, are dynamically exchanged during transcription. Dynamic exchange of histone dimers is promoted by acetylation of histone H3K56 (Watanabe et al. 2013), an example of how chromatin remodeling, histone variants, and histone modifications can work together to dynamically maintain the chromatin landscape.

In the budding yeast, there are two SNF2-class remodeler complexes, RSC and SWI/SNF (Clapier and Cairns 2009). RSC is abundant, broadly distributed, and essential, whereas SWI/SNF is much less abundant than RSC and is nonessential. RSC is similar to the Polybromo (PBAF) complex, in having multiple subunits with bromodomains, a class of modules that bind to acetylated histone tails. In fact, eight of the 15 bromodomains encoded in the yeast genome are present in the RSC complex. SNF2 is similar to the BRM subunit of the BAF complex, which in mammals shares most of its subunits with PBAF. Yeast RSC and SWI/SNF have different catalytic subunits (Sth1 for RSC and Snf2 for SWI/SNF), and mammalian PBAF and BAF also have different catalytic subunits (BRG1 for PBAF and BRM for BAF), which suggest that the two complexes play different chromatin remodeling roles. However, *Drosophila* has only a single SNF2-class catalytic subunit, BRM, which suggests that these complexes are redundant at the basic biochemical level.

Most (P)BAF cancer mutations are found in subunits shared between the two complexes, which suggests that some common feature of SWI/SNF-class remodelers is responsible for chromatin alterations characteristic of the oncogenic state (Kadoch et al. 2013). Insofar as >20% of all human cancers are found to harbor loss-of-function mutations in P-BAF or BAF complex subunits, elucidating the molecular mechanism of SNF2-class remodeling is crucial for understanding oncogenesis in these cancers.

The yeast RSC complex has been intensively studied both biochemically and structurally. RSC is several-fold larger than the nucleosome, and RSC is a powerful DNA translocase that hydrolyzes 3 ATPs per base pair with a step size of ~1 base pair (Eastlund et al. 2013). A model for RSC action is that it engulfs and unwraps the nucleosome up to the dyad axis by electrostatic attraction of DNA to its inner surface (Chaban et al. 2008; Lorch et al. 2010). Then Sth1 hydrolyzes ATP in a power stroke that releases the remaining histone–DNA contacts to pump through DNA, resulting in directional sliding of the histone octamer along DNA. This model is supported by the in vivo observation that RSC-bound nucleosomes are unwrapped up to the dyad axis, a putative RSC/nucleosome remodeling intermediate (Ramachandran et al. 2015). In some circumstances, RSC can evict a nucleosome core from the DNA entirely (Lorch et al. 2006), and the evidence that RSC facilitates the loss and replacement (turnover) of nucleosomes in vivo (Dion et al. 2007; Hartley and Madhani 2009) is consistent with eviction being an extreme manifestation of sliding in the context of a chromatin fiber that is densely packed with nucleosomes. These sliding and eviction capabilities of RSC can account for the well-established role of RSC in the generation of nucleosome-depleted regions genome-wide in vivo (Ganguli et al. 2014), which is a prerequisite for formation of a preinitiation complex that recruits RNAPII. We might view the SNF2-class remodelers as machines that clear nucleosomes from promoters so that transcription can initiate.

It is probable that *Drosophila* Brahma, the only SNF2-class remodeler in the fly genome (Tamkun et al. 1992), and mammalian PBAF and BAF complexes (Wilson and Roberts 2011) are likewise responsible for clearing nucleosomes from promoters. Moreover, the evidence that enhancers are also sites of enhancer RNA (eRNA) transcriptional initiation (Lai and Shiekhattar 2014) suggests that PBAF and BAF may serve a similar function at *cis*-regulatory sites in general. Thus, the central role of SNF2-class remodelers in transcription initiation throughout the genome may have made cells especially vulnerable to transcriptional misregulation when their function is altered by loss of a regulatory subunit (Kim and Roberts 2014). The fact that some individual subunits have been implicated in cell-type-specific functions of PBAF/BAF complexes (Hargreaves and Crabtree 2011) fits with the notion that cancer loss-of-function mutations in various subunits result in cell-type-specific misregulation of this powerful nucleosome clearing apparatus.

Tumor DNA sequencing has also led to the identification of driver mutations in ATRX (α-thalassaemia/mental retardation X-linked) (Picketts et al. 1996), which belongs to the RAD54 branch of the SWI/SNF superfamily of DNA translocases (Fig. 2). Germline homozygous loss-of-function ATRX mutations cause ATRX syndrome, in which an α-thalassaemia is associated with improper chromatin packaging of the α-globin CpG island promoter (Law et al. 2010). Little was understood about the mechanism of action of ATRX syndrome until tumor sequencing revealed that mutations in ATRX and in the associated DAXX histone chaperone are mutually exclusive driver mutations in ∼4% of human cancers (Heaphy et al. 2011a; Jiao et al. 2011). Indeed, the study of these cancer mutations led to elucidation of the ATRX/DAXX nucleosome assembly pathway, a reversal of the usual paradigm in which basic science advances help to inform the mechanism of cancer drivers.

The Rad54 DNA translocase does not remodel nucleosomes, but studies of the Rad54-related ATRX ortholog in *Drosophila*, XNP, have suggested that ATRX translocation prepares the DNA substrate for assembly of histone H3.3 nucleosomes (Schneiderman et al. 2009, 2012). DAXX is a histone chaperone that is dedicated to the assembly of nucleosomes containing the

histone H3.3 variant, by depositing successive dimers of H3.3/H4 to form the central tetramer of octameric $(H2A/H2B/H3.3/H4)_2$ nucleosomes (Ray-Gallet et al. 2002; Drane et al. 2010; Lewis et al. 2010; Elsasser et al. 2012). The normal function of the ATRX-DAXX-H3.3 nucleosome assembly pathway is thought to be filling of gaps in the nucleosome landscape at sites where periodicities and base-compositional biases disfavor tight wrapping of nucleosomes (Fig. 2) (Schneiderman et al. 2009).

A unique feature of ATRX and DAXX mutations is that they are hallmarks of the alternative lengthening of telomeres (ALT) phenotype (Lovejoy et al. 2012). ALT is recognized as a less frequent alternative to telomerase activation that cancer cells can adopt to escape senescence, which likely accounts for the occurrence of ALT in a wide variety of cancers. ALT is readily detected using fluorescence in situ hybridization (FISH) with telomere DNA probes, seen as sporadic cells with extraordinarily long telomeres (Heaphy et al. 2011b). The precise mechanism whereby mutations in the H3.3 assembly pathway lead to the ALT phenotype has not been fully elucidated. However, a recent finding that CHK1-mediated phosphorylation of H3.3 serine-31, one of only four H3.3 residues not found in replication-coupled (RC) variants H3.1 and H3.2, is essential for the full ALT phenotype (Chang et al. 2015) raises the possibility that CHK1 kinase inhibitors might be used to specifically kill cancer cells driven by ATRX and DAXX mutations. ALT also sensitizes cells to ATR (ataxia telangiectasia and Rad3-related) protein kinase inhibitors (Flynn et al. 2015), presumably by reducing phosphorylation of histone H2A.X as part of the DNA damage response pathway. Thus, histone variant phosphorylation promises to be a more general regulatory paradigm that might be exploited in treating ALT-associated cancers.

MODIFYING NUCLEOSOMES

The surface of the nucleosome core and its histone tails are rich in basic residues that neutralize the strong negative charges of the DNA phosphate backbone. Whereas core surface arginine side chains sit in the DNA minor groove to tightly wrap the duplex around the particle, lysines on the tails can undergo dynamic cycles of acetylation and deacetylation that neutralize and recharge the tails to facilitate active processes such as transcription (Wolffe 1992; Waterborg 2002). Charge neutralization is thus a universal function of acetylation. In addition, the acetyltransferases and deacetylases that mediate this process are often mutated or aberrantly expressed in cancer, and histone deacetylase (HDAC) inhibitors have been widely used in cancer chemotherapy (Ma et al. 2016). But whether histones themselves are the therapeutic targets of HDAC inhibitors is unclear, in that many nonhistone proteins, including metabolic enzymes, are regulated by acetylation (Downey and Baetz 2015). Furthermore, it has been proposed that a bulk function of histone tail acetylation is pH homeostasis, in which release of acetate and export from the cell reduces intracellular pH (McBrian et al. 2013). These observations complicate interpretation of HDAC inhibition studies.

Histone lysines also undergo cycles of methylation and demethylation, although on much slower time scales from cycles of acetylation and deacetylation. Whereas the half-life of an acetyl on a histone might be measured in seconds or minutes, the average methyl has a half-life that is nearly the same as the histone (Waterborg 1993, 2002). From a kinetic perspective, the stability of histone methylation makes it a strong candidate for a modification that perpetuates chromatin memory, one that can be selectively removed to allow for gene expression to occur. For example, ~70% of H3K27 lysines in the *Drosophila* genome are dimethylated by Polycomb repressive complex 2 (PRC2), which serves to prevent global unscheduled transcription (Lee et al. 2015), but only H3K27me3 is associated with regulated developmental silencing (Schwartz and Pirrotta 2013). To selectively derepress a gene without removing the histone, the UTX H3K27-specific demethylase may be targeted to demethylate H3K27 residues down to di- and monomethylation (Agger et al. 2007).

It is often asserted that histone methylation is "epigenetic" based on the analogy to DNA

methylation (Ptashne 2007). DNA methyls are inherited by the action of the Dnmt1 DNA methyltransferase, which faithfully methylates the cytosine of a CG dinucleotide on the newly replicated strand at the replication fork, but only when the parental cytosine on the opposite strand is methylated (Bestor 1996). However, unlike modifications of DNA, all DNA-binding proteins are removed in advance of DNA polymerization by the action of the replicative helicase. Therefore, all histones, whether old or new, must be deposited de novo on leading and lagging strands behind the replication fork. This means that inheritance of a modification cannot be assumed, and without direct evidence, it remains formally possible that all histone modifications are reestablished de novo as are the nucleosomes that they modify. Direct evidence for inheritance of a histone modification has been obtained in the case of the *Caenorhabditis elegans* germline, in which loss of the PRC2 complex that methylates H3K27 nevertheless allows for retention of H3K27me3 (Gaydos et al. 2014). Just how this feat of legerdemain at the replication fork is accomplished remains the subject of speculation (Ramachandran and Henikoff 2015).

EZH2 (enhancer-of-zeste homolog 2), the catalytic component of the mammalian PRC2 complex, is up-regulated in many cancers, and effective and safe inhibition of EZH2 is a major goal of pharmacological research (Koppens and van Lohuizen 2015). Other histone modifications, including acetylation and H3K4, H3K36, and H3K79 methylation, are associated with the transcriptionally active state, and some of the enzymes responsible for these modifications have been implicated in driving tumorigenesis (Bernt et al. 2011; Colon-Bolea and Crespo 2014; Riedel et al. 2015). From the perspective of cancer therapeutics, these activation-associated modifications represent attractive targets, insofar as inhibition of the enzymes that are involved in the gene activation process might be expected to down-regulate oncogenes that are aberrantly expressed in cancer. However, from a mechanistic perspective, possible roles of these "activating" modifications in mediating gene expression is more poorly understood than in the case of "silencing" modifications. Specifically, the high-affinity binding of histone binder ("reader") proteins, including Polycomb to H3K27me3, and heterochromatin-associated protein 1 (HP-1) to H3K9me2 and H3K9me3, helps to immobilize nucleosomes (Canzio et al. 2011; Schwartz and Pirrotta 2013). Reducing nucleosome dynamics likely impedes transcriptional activation, which requires accessible DNA for transcription factors to bind. In contrast, histone-modifying enzymes that are enriched at active genes most likely act during the dynamic process of transcription, being associated with the carboxy-terminal domain of the large subunit of RNA polymerase II (Henikoff and Shilatifard 2011). Thus, these enzymes act on nucleosomes as they are being disrupted and reassembled by the machinery that moves a denaturation bubble forward as it adds RNA bases onto the growing RNA chain. As is the case for disruption of nucleosomes by replication fork passage, our understanding of nucleosome dynamics during transcription is far from complete.

An additional uncertainty in understanding the role of histone modifications in cancer is whether the modification of histones, as opposed to nonhistone proteins, is relevant to the cancer phenotype (Carlson and Gozani 2016). Moreover, cancer cell lines in which overexpression of the NSD2 SET domain-containing histone methyltransferase causes increases in H3K36 methylation also causes decreases in H3K27 methylation (Popovic et al. 2014). This anticorrelation could result from interference along the H3 tail, but might also be an indirect effect. As lysine methylation is a common regulatory modification of many proteins, including transcription factors (Levy et al. 2011; Carlson et al. 2015), it is also possible that the therapeutic target of NSD2 is instead a nonhistone protein, in which case these changes that occur on the lysine tail represent collateral damage with no physiological consequences. It should be kept in mind that calling these enzymes "histone" methyltransferases, demethylases, etc., does not necessarily reflect their in vivo activities, but rather the historical fact that the lysine richness of histones made them

convenient substrates for use in biochemical purification. To avoid this misunderstanding, the accepted nomenclature for what was previously referred to as histone methyltransferases (HMTs) has been changed to "lysine" methyltransferases (KMTs) and so on for other chromatin-modifying enzymes (Allis et al. 2007).

A similar uncertainty as to physiological function applies to histone reader proteins, which bind their modified substrates sometimes with nanomolar affinities. In the case of Polycomb and HP-1, there is both genetic and biochemical evidence that their action includes preferential binding of methylated H3K27 and H3K9 lysines, respectively, including similar phenotypes for the orthologous modifying enzymes and reader proteins in *Drosophila* (Schotta et al. 2002; Schwartz and Pirrotta 2013). However, in other cases, the situation is ambiguous, for example, whether the various bromodomains found to preferentially bind acetylated histone tails actually bind these tails in vivo, and if they do, whether or not the binding is simply a consequence of the extraordinarily high concentration of acetylated histone lysines in the nucleus (Shi and Vakoc 2014). The development of reader protein inhibitors (Filippakopoulos et al. 2010) and ascertainment of their safety in the clinic will likely benefit from a better understanding of their physiologically relevant targets.

REPLACING CANONICAL HISTONES WITH VARIANTS

Most histones are rapidly synthesized from multicopy genes during S phase by specialized mRNA processing machineries and are deposited into nucleosomes immediately behind the replication fork (Marzluff et al. 2002). Other histones are synthesized throughout the cell cycle from ordinary genes and are incorporated into nucleosomes by dedicated histone chaperones (Henikoff and Smith 2015). These replication-independent (RI) histones include the aforementioned H3.3 histone variant, which is the exclusive substrate for RI assembly on chromosome arms, as are H3.1 and H3.2 the exclusive substrates for RC nucleosome assembly

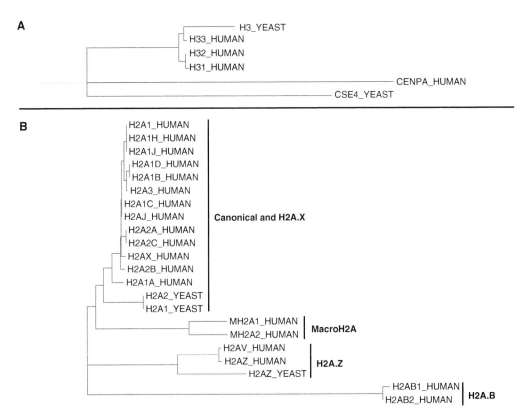

Figure 3. Phylogenies of histone variants in humans and yeast. All H3 (*A*), and H2A (*B*) variants annotated in SWISS-PROT are included (see the legend to Fig. 1 for details).

(Fig. 3A). Another notable histone variant is cenH3 (called CENP-A in mammals), which is the defining component of nucleosomes that form the chromatin foundation for the centromere. Most other RI histones are variants of H2A, including H2A.X, which is rapidly phosphorylated on a carboxy-terminal serine when DNA breaks in its vicinity, H2A.Z, which acts to weaken the nucleosome barrier to transcription, and macroH2A, H2A.B, and H2A.L, which are mammalian histone variants of uncertain function (Fig. 3B). The substitution of a histone variant for a canonical histone represents a profound change in nucleosome properties that can potentially affect nucleosome dynamics. For example, H2A.Z has a very different "docking domain" from H2A, which contacts H3/H4 in the nucleosome core and can affect its stability (Suto et al. 2000). Also, H2A.B forms a "short wrapper" nucleosome core that

makes fewer contacts with DNA, potentially reducing stability (Bao et al. 2004). macroH2A, which is enriched in silent chromatin, is unique among core histones in having a globular domain of uncertain function that protrudes from the canonical core (Costanzi and Pehrson 1998).

Until recently, histone variants have received much less attention in the cancer and chromatin field than histone modifications, which are catalyzed by many different modifying and demodifying enzymes and bound by reader proteins that provide attractive targets for small molecule drug development (Copeland et al. 2010; Filippakopoulos et al. 2010). However, tumor DNA sequencing studies have revealed unexpected roles for histone variants in promoting tumorigenesis. Specific driver mutations have been discovered in H3.3 genes that are characteristic of specific cancer types to an extraordinary degree: H3.3K27M mutations are found in most diffuse midline gliomas (DIPGs) and many pediatric glioblastomas (Wu et al. 2012), and H3.3K36M mutations are driver mutations in nearly all chondroblastomas and H3.3G34W/L mutations in nearly all giant cell tumors of bone (Behjati et al. 2013). Whereas DIPGs are highly aggressive pediatric cancers, the bone tumors are less invasive and sometimes benign, and are also found in adults. All are likely gain-of-function mutations in that these tumors also express normal H3.3 copies. For example, in mammalian cells, expression of the H3.3K27M protein contributes to neoplastic transformation (Funato et al. 2014), and a similar neoplastic effect is seen for H3K36M/I mutations found in chondroblastomas (Lu et al. 2016). A gain-of-function interpretation of these phenotypes is also supported by studies in *Drosophila*, whereby introduction of the K27M mutation in an H3.3 transgene results in dominant Polycomb phenotypes, consistent with methionine-27 titrating the PRC2 complex (Lewis et al. 2013; Herz et al. 2014). Indeed, elucidation of the structure of a PRC2 complex homolog binding a histone H3K27M amino-terminal tail peptide shows that the mutation results in occlusion of the enzymatic active site by the neighboring H3R26 arginine side chain (Jiao and Liu 2015). Another candidate for mediating at least some of these cancers is the ZMYND11 zinc-finger tumor suppressor protein that specifically binds to both lysine-36 and (H3.3-specific) serine-31 (Wen et al. 2014).

Tumor sequencing and gene expression studies have also drawn attention to the possibility that RI nucleosome assembly pathway components might be potential drug targets, for example, the inhibition of the CHK1 and ATR kinases, which preferentially phosphorylate H3.3 and H2A.X in cancer cells (Chang et al. 2015). In the case of H2A.Z, overexpression is associated with poor prognosis in ERα breast cancer (Hua et al. 2008), and hyperacetylation of H2A.Z is a feature of deregulated promoters in prostate cancer (Valdes-Mora et al. 2012). Changes in macroH2A levels are also common in cancer. For example, the degree of malignancy in melanoma is inversely corre-lated with macroH2A levels, and knockdown of macroH2A increases malignancy (Kapoor et al. 2010). As macroH2A nucleosomes are more compacted, it is attractive to consider that its presence in a nucleosome reduces dynamics (Chakravarthy et al. 2005), and so its loss can promote gene misexpression. However, just what role that global changes in macroH2A levels might have in tumor progression is complicated by the evidence that different splicing isoforms produced by two different macroH2A genes are associated with very different outcomes (Sporn and Jung 2012). It is possible that these complications arise from the differential abundances of these isoforms in different cancer cell types, rather than any difference in the action of the isoforms in compacting chromatin. Interestingly, macroH2A is negatively regulated by ATRX (Ratnakumar et al. 2012), and this raises the possibility that misregulation of macroH2A contributes to the ALT phenotype.

Whereas most RI histone variants are of interest because of their potential involvement in transcriptional regulation, cenH3 nucleosomes form the foundation of the centromere in the large majority of eukaryotes, and so are crucial for genome stability (Quenet and Dalal 2012). As cenH3 mislocalization can lead to the appearance of ectopic "neocentromeres," and bridge–breakage–fusion cycles can result from a second centromere on a chromosome, overproduction of CENP-A in some cancers makes it a potential causative factor in aneuploidy (Lacoste et al. 2014; Athwal et al. 2015). Reducing CENP-A levels in cancer, for example, by targeting expression of its gene or inhibiting its incorporation into chromatin, may be a potential therapeutic strategy that would have little if any impact on nondividing cells.

CONCLUDING REMARKS

This survey of chromatin components involved in nucleosome dynamics of potential relevance to cancer underscores the intricate interrelationships between the major components, including DNA translocases, histone modifications and their reader proteins, and histone variants and their chaperones. For example, the

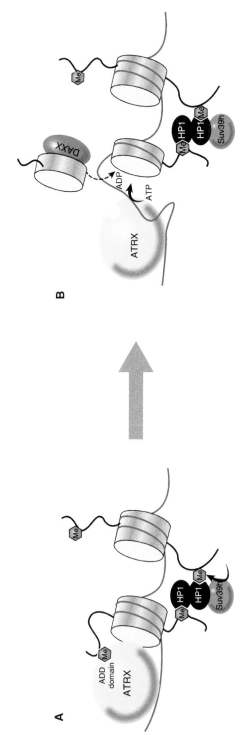

Figure 4. Model for the concerted action of multiple chromatin regulators. (A) The Suv39h H3K9 methyltransferase is recruited by HP-1 protein, which binds preferentially to methylated H3K9. To perpetuate this mark when the nucleosome turns over, the ATRX ATPase is concentrated at the site via its ATRX-DNMT3-DNMT3L (ADD) "reader" domain, which binds with high specificity to methylated H3K9 on tails that entirely lack H3K4 methylation (because there are no H3K4 methyltransferases in this region of the genome). (B) ATRX provides the energy of ATP and works together with the H3.3-specific DAXX histone chaperone complex to incorporate the new nucleosome. The high local concentration of Suv39h results in a new nucleosome with the same H3K9 methylation as the nucleosome that was lost. (From Henikoff and Smith 2015; reprinted, with permission, from the authors.)

ATRX translocase, which is associated with the H3.3-specific DAXX chaperone has a dual-reader module that tightly binds di- or trimethylated lysine-9 and unmethylated lysine-4 on the histone 3 amino-terminal tail (Fig. 4) (Eustermann et al. 2011). Teasing out these interdependencies to understand how ATRX and DAXX mutations can drive ALT or how different H3.3 mutations can drive such different tumors as aggressive DIPGs and benign chondroblastomas requires a better understanding of the dynamic processes that these components normally carry out. Increases in nucleosome dynamics can also result in genome instability, as loss of nucleosomes exposes DNA and can result in double-strand breaks (Yang et al. 2015).

In this regard, studies of model organisms have the potential of providing insight into how disruption of normal developmental processes can result in cancer. For example, an important simplification of what otherwise seems to be a byzantine network of chromatin regulators involved in tumor progression originated with fly genetics. According to the central paradigm for *Drosophila* development, genes are heritably maintained in the silent state in *trans* by the product of the Polycomb locus (Lewis 1978). This paradigm was further elaborated with the discovery of additional loci in the Polycomb group (PcGs) that maintain developmental silencing (Schwartz and Pirrotta 2013), and in the trithorax group (trxGs) that maintain gene activity (Steffen and Ringrose 2014). Although originally described for maintaining silencing and expression of "homoeotic" master regulators of the anterior–posterior morphological axis, the paradigm applies generally to maintaining developmental memory in complex animals. Biochemical studies of PcGs showed them to consist of subunits of PRC1 complexes, which act to compact chromatin (Grau et al. 2011), and of the PRC2 complex, which methylates H3K27 to heritably maintain silencing (Schwartz and Pirrotta 2013). Biochemical studies of trxGs showed them to be chromatin remodelers, histone-modifying enzymes and other chromatin regulators that are associated with gene activity (Kingston and Tamkun 2006). Upsetting the balance between trxGs

and PcGs that normally regulate developmental processes in *Drosophila* uncovers homoeotic phenotypes. Likewise, mutations in and silencing or misexpression of mammalian orthologs of trxGs and PcGs upsets the balance between them that normally maintains developmental homeostasis. A fuller understanding of the mechanisms whereby trxGs and PcGs affect nucleosome dynamics is needed to address the question of how mutation and misregulation of these proteins can drive tumorigenesis.

ACKNOWLEDGMENTS

I thank Paul Talbert for comments on the manuscript. Work in my laboratory is supported by the Howard Hughes Medical Institute and a grant from the National Institutes of Health (R01 ES020116).

REFERENCES

*Reference is also in this collection.

Agger K, Cloos PA, Christensen J, Pasini D, Rose S, Rappsilber J, Issaeva I, Canaani E, Salcini AE, Helin K. 2007. UTX and JMJD3 are histone H3K27 demethylases involved in HOX gene regulation and development. *Nature* **449:** 731–734.

Allis CD, Berger SL, Cote J, Dent S, Jenuwien T, Kouzarides T, Pillus L, Reinberg D, Shi Y, Shiekhattar R, et al. 2007. New nomenclature for chromatin-modifying enzymes. *Cell* **131:** 633–636.

Athwal RK, Walkiewicz MP, Baek S, Fu S, Bui M, Camps J, Ried T, Sung MH, Dalal Y. 2015. CENP-A nucleosomes localize to transcription factor hotspots and subtelomeric sites in human cancer cells. *Epigenetics Chromatin* **8:** 2.

Bao Y, Konesky K, Park YJ, Rosu S, Dyer PN, Rangasamy D, Tremethick DJ, Laybourn PJ, Luger K. 2004. Nucleosomes containing the histone variant H2A.Bbd organize only 118 base pairs of DNA. *EMBO J* **23:** 3314–3324.

Behjati S, Tarpey PS, Presneau N, Scheipl S, Pillay N, Van Loo P, Wedge DC, Cooke SL, Gundem G, Davies H, et al. 2013. Distinct H3F3A and H3F3B driver mutations define chondroblastoma and giant cell tumor of bone. *Nat Genet* **45:** 1479–1482.

Bernt KM, Zhu N, Sinha AU, Vempati S, Faber J, Krivtsov AV, Feng Z, Punt N, Daigle A, Bullinger L, et al. 2011. MLL-rearranged leukemia is dependent on aberrant H3K79 methylation by DOT1L. *Cancer Cell* **20:** 66–78.

Bestor TH. 1996. DNA methyltransferases in mammalian development and genome defense. In *Epigenetic mechanisms of gene regulation* (ed. Russo VEA, Martienssen RA, Riggs AD), pp. 61–76. Cold Spring Harbor Laboratory Press, Cold Spring Harbor, NY.

Canzio D, Chang EY, Shankar S, Kuchenbecker KM, Simon MD, Madhani HD, Narlikar GJ, Al-Sady B. 2011. Chromodomain-mediated oligomerization of HP1 suggests a nucleosome-bridging mechanism for heterochromatin assembly. *Mol Cell* **41:** 67–81.

* Carlson SM, Gozani O. 2016. Non-histone lysine methylation in the regulation of cancer pathways. *Cold Spring Harb Perpect Med* doi: 10.1101/cshperspect.a026435.

Carlson SM, Moore KE, Sankaran SM, Reynoird N, Elias JE, Gozani O. 2015. A proteomic strategy identifies lysine methylation of splicing factor snRNP70 by the SETMAR enzyme. *J Biol Chem* **290:** 12040–12047.

Chaban Y, Ezeokonkwo C, Chung WH, Zhang F, Kornberg RD, Maier-Davis B, Lorch Y, Asturias FJ. 2008. Structure of a RSC-nucleosome complex and insights into chromatin remodeling. *Nat Struct Mol Biol* **15:** 1272–1277.

Chakravarthy S, Gundimella SK, Caron C, Perche PY, Pehrson JR, Khochbin S, Luger K. 2005. Structural characterization of the histone variant macroH2A. *Mol Cell Biol* **25:** 7616–7624.

Chang FT, Chan FL, JD RM, Udugama M, Mayne L, Collas P, Mann JR, Wong LH. 2015. CHK1-driven histone H3.3 serine 31 phosphorylation is important for chromatin maintenance and cell survival in human ALT cancer cells. *Nucleic Acids Res* **43:** 2603–2614.

Chen X, Cui D, Papusha A, Zhang X, Chu CD, Tang J, Chen K, Pan X, Ira G. 2012. The Fun30 nucleosome remodeller promotes resection of DNA double-strand break ends. *Nature* **489:** 576–580.

Clapier CR, Cairns BR. 2009. The biology of chromatin remodeling complexes. *Annu Rev Biochem* **78:** 273–304.

Colon-Bolea P, Crespo P. 2014. Lysine methylation in cancer: SMYD3-MAP3K2 teaches us new lessons in the Ras-ERK pathway. *Bioessays* **36:** 1162–1169.

Copeland RA, Olhava EJ, Scott MP. 2010. Targeting epigenetic enzymes for drug discovery. *Curr Opin Chem Biol* **14:** 505–510.

Costanzi C, Pehrson JR. 1998. Histone macroH2A1 is concentrated in the inactive X chromosome of female mammals. *Nature* **393:** 599–601.

Costelloe T, Louge R, Tomimatsu N, Mukherjee B, Martini E, Khadaroo B, Dubois K, Wiegant WW, Thierry A, Burma S, et al. 2012. The yeast Fun30 and human SMARCAD1 chromatin remodellers promote DNA end resection. *Nature* **489:** 581–584.

Daigle SR, Olhava EJ, Therkelsen CA, Majer CR, Sneeringer CJ, Song J, Johnston LD, Scott MP, Smith JJ, Xiao Y, et al. 2011. Selective killing of mixed lineage leukemia cells by a potent small-molecule DOT1L inhibitor. *Cancer Cell* **20:** 53–65.

Dion M, Kaplan T, Friedman N, Rando OJ. 2007. Dynamics of replication-independent histone turnover in budding yeast. *Science* **315:** 1405–1408.

Downey M, Baetz K. 2015. Building a KATalogue of acetyllysine targeting and function. *Brief Funct Genomics* **15:** 109–118.

Drane P, Ouararhni K, Depaux A, Shuaib M, Hamiche A. 2010. The death-associated protein DAXX is a novel histone chaperone involved in the replication-independent deposition of H3.3. *Genes Dev* **24:** 1253–1265.

Eastlund A, Malik SS, Fischer CJ. 2013. Kinetic mechanism of DNA translocation by the RSC molecular motor. *Arch Biochem Biophys* **532:** 73–83.

Elsasser SJ, Huang H, Lewis PW, Chin JW, Allis CD, Patel DJ. 2012. DAXX envelops a histone H3.3-H4 dimer for H3.3-specific recognition. *Nature* **491:** 560–565.

Eustermann S, Yang JC, Law MJ, Amos R, Chapman LM, Jelinska C, Garrick D, Clynes D, Gibbons RJ, Rhodes D, et al. 2011. Combinatorial readout of histone H3 modifications specifies localization of ATRX to heterochromatin. *Nat Struct Mol Biol* **18:** 777–782.

Filippakopoulos P, Qi J, Picaud S, Shen Y, Smith WB, Fedorov O, Morse EM, Keates T, Hickman TT, Felletar I, et al. 2010. Selective inhibition of BET bromodomains. *Nature* **468:** 1067–1073.

Flynn RL, Cox KE, Jeitany M, Wakimoto H, Bryll AR, Ganem NJ, Bersani F, Pineda JR, Suva ML, Benes CH, et al. 2015. Alternative lengthening of telomeres renders cancer cells hypersensitive to ATR inhibitors. *Science* **347:** 273–277.

Funato K, Major T, Lewis PW, Allis CD, Tabar V. 2014. Use of human embryonic stem cells to model pediatric gliomas with H3.3K27M histone mutation. *Science* **346:** 1529–1533.

Ganguli D, Chereji RV, Iben JR, Cole HA, Clark DJ. 2014. RSC-dependent constructive and destructive interference between opposing arrays of phased nucleosomes in yeast. *Genome Res* **24:** 1637–1649.

Gaydos LJ, Wang W, Strome S. 2014. Gene repression. H3K27me and PRC2 transmit a memory of repression across generations and during development. *Science* **345:** 1515–1518.

Grau DJ, Chapman BA, Garlick JD, Borowsky M, Francis NJ, Kingston RE. 2011. Compaction of chromatin by diverse Polycomb group proteins requires localized regions of high charge. *Genes Dev* **25:** 2210–2221.

Hargreaves DC, Crabtree GR. 2011. ATP-dependent chromatin remodeling: Genetics, genomics and mechanisms. *Cell Res* **21:** 396–420.

Hartley PD, Madhani HD. 2009. Mechanisms that specify promoter nucleosome location and identity. *Cell* **137:** 445–458.

Heaphy CM, de Wilde RF, Jiao Y, Klein AP, Edil BH, Shi C, Bettegowda C, Rodriguez FJ, Eberhart CG, Hebbar S, et al. 2011a. Altered telomeres in tumors with ATRX and DAXX mutations. *Science* **333:** 425.

Heaphy CM, Subhawong AP, Hong SM, Goggins MG, Montgomery EA, Gabrielson E, Netto GJ, Epstein JI, Lotan TL, Westra WH, et al. 2011b. Prevalence of the alternative lengthening of telomeres telomere maintenance mechanism in human cancer subtypes. *Am J Pathol* **179:** 1608–1615.

Henikoff S, Shilatifard A. 2011. Histone modification: Cause or cog? *Trends Genet* **27:** 389–396.

Henikoff S, Smith MM. 2015. Histone variants and epigenetics. *Cold Spring Harb Perspect Biol* **7:** a019364.

Herz HM, Morgan M, Gao X, Jackson J, Rickels R, Swanson SK, Florens L, Washburn MP, Eissenberg JC, Shilatifard A. 2014. Histone H3 lysine-to-methionine mutants as a paradigm to study chromatin signaling. *Science* **345:** 1065–1070.

Hinz JM, Czaja W. 2015. Facilitation of base excision repair by chromatin remodeling. *DNA Repair (Amst)* **36**: 91–97.

Hua S, Kallen CB, Dhar R, Baquero MT, Mason CE, Russell BA, Shah PK, Liu J, Khramtsov A, Tretiakova MS, et al. 2008. Genomic analysis of estrogen cascade reveals histone variant H2A.Z associated with breast cancer progression. *Mol Syst Biol* **4**: 188.

Jiao L, Liu X. 2015. Structural basis of histone H3K27 trimethylation by an active Polycomb repressive complex 2. *Science* **350**: aac4383.

Jiao Y, Shi C, Edil BH, de Wilde RF, Klimstra DS, Maitra A, Schulick RD, Tang LH, Wolfgang CL, Choti MA, et al. 2011. DAXX/ATRX, MEN1, and mTOR pathway genes are frequently altered in pancreatic neuroendocrine tumors. *Science* **331**: 1199–1203.

Kadoch C, Hargreaves DC, Hodges C, Elias L, Ho L, Ranish J, Crabtree GR. 2013. Proteomic and bioinformatic analysis of mammalian SWI/SNF complexes identifies extensive roles in human malignancy. *Nat Genet* **45**: 592–601.

Kapoor A, Goldberg MS, Cumberland LK, Ratnakumar K, Segura MF, Emanuel PO, Menendez S, Vardabasso C, Leroy G, Vidal CI, et al. 2010. The histone variant macroH2A suppresses melanoma progression through regulation of CDK8. *Nature* **468**: 1105–1109.

Kim KH, Roberts CW. 2014. Mechanisms by which SMARCB1 loss drives rhabdoid tumor growth. *Cancer Genet* **207**: 365–372.

Kingston RE, Tamkun J. 2014. Transcriptional regulation by trithorax-group proteins. *Cold Spring Harb Perspect Biol* **6**: 019349.

Koppens M, van Lohuizen M. 2015. Context-dependent actions of Polycomb repressors in cancer. *Oncogene* **17**: 1341–1352.

Lacoste N, Woolfe A, Tachiwana H, Garea AV, Barth T, Cantaloube S, Kurumizaka H, Imhof A, Almouzni G. 2014. Mislocalization of the centromeric histone variant CenH3/CENP-A in human cells depends on the chaperone DAXX. *Mol Cell* **53**: 631–644.

Lai F, Shiekhattar R. 2014. Enhancer RNAs: The new molecules of transcription. *Curr Opin Genet Dev* **25**: 38–42.

Laufer BI, Singh SM. 2015. Strategies for precision modulation of gene expression by epigenome editing: An overview. *Epigenetics Chromatin* **8**: 34.

Law MJ, Lower KM, Voon HP, Hughes JR, Garrick D, Viprakasit V, Mitson M, De Gobbi M, Marra M, Morris A, et al. 2010. ATR-X syndrome protein targets tandem repeats and influences allele-specific expression in a size-dependent manner. *Cell* **143**: 367–378.

Lee HG, Kahn TG, Simcox A, Schwartz YB, Pirrotta V. 2015. Genome-wide activities of polycomb complexes control pervasive transcription. *Genome Res* **25**: 1170–1181.

Levy D, Liu CL, Yang Z, Newman AM, Alizadeh AA, Utz PJ, Gozani O. 2011. A proteomic approach for the identification of novel lysine methyltransferase substrates. *Epigenetics Chromatin* **4**: 19.

Lewis EB. 1978. A gene complex controlling segmentation in *Drosophila*. *Nature* **276**: 565–570.

Lewis PW, Elsaesser SJ, Noh KM, Stadler SC, Allis CD. 2010. Daxx is an H3.3-specific histone chaperone and cooperates with ATRX in replication-independent chromatin assembly at telomeres. *Proc Natl Acad Sci* **107**: 14075–14080.

Lewis PW, Muller MM, Koletsky MS, Cordero F, Lin S, Banaszynski LA, Garcia BA, Muir TW, Becher OJ, Allis CD. 2013. Inhibition of PRC2 activity by a gain-of-function H3 mutation found in pediatric glioblastoma. *Science* **340**: 857–861.

Li S. 2015. Transcription coupled nucleotide excision repair in the yeast Saccharomyces cerevisiae: The ambiguous role of Rad26. *DNA Repair (Amst)* **36**: 43–48.

Liu Z, Lavis LD, Betzig E. 2015. Imaging live-cell dynamics and structure at the single-molecule level. *Mol Cell* **58**: 644–659.

Lorch Y, Maier-Davis B, Kornberg RD. 2006. Chromatin remodeling by nucleosome disassembly in vitro. *Proc Natl Acad Sci* **103**: 3090–3093.

Lorch Y, Maier-Davis B, Kornberg RD. 2010. Mechanism of chromatin remodeling. *Proc Natl Acad Sci* **107**: 3458–3462.

Lovejoy CA, Li W, Reisenweber S, Thongthip S, Bruno J, de Lange T, De S, Petrini JH, Sung PA, Jasin M, et al. 2012. Loss of ATRX, genome instability, and an altered DNA damage response are hallmarks of the alternative lengthening of telomeres pathway. *PLoS Genet* **8**: e1002772.

Lu C, Jain SU, Hoelper D, Bechet D, Molden RC, Ran L, Murphy D, Venneti S, Hameed M, Pawel BR, et al. 2016. Histone H3K36 mutations promote sarcomagenesis through altered histone methylation landscape. *Science* **352**: 844–849.

Ma N, Luo Y, Wang Y, Liao C, Ye WC, Jiang S. 2016. Selective histone deacetylase inhibitors with anticancer activity. *Curr Top Med Chem* **16**: 415–426.

Marzluff WF, Gongidi P, Woods KR, Jin J, Maltais LJ. 2002. The human and mouse replication-dependent histone genes. *Genomics* **80**: 487–498.

McBrian MA, Behbahan IS, Ferrari R, Su T, Huang TW, Li K, Hong CS, Christofk HR, Vogelauer M, Seligson DB, et al. 2013. Histone acetylation regulates intracellular pH. *Mol Cell* **49**: 310–321.

Morishita M, Mevius D, di Luccio E. 2014. In vitro histone lysine methylation by NSD1, NSD2/MMSET/WHSC1 and NSD3/WHSC1L. *BMC Struct Biol* **14**: 25.

Picketts DJ, Higgs DR, Bachoo S, Blake DJ, Quarrell OW, Gibbons RJ. 1996. ATRX encodes a novel member of the SNF2 family of proteins: Mutations point to a common mechanism underlying the ATR-X syndrome. *Hum Mol Genet* **5**: 1899–1907.

Pon JR, Marra MA. 2015. Driver and passenger mutations in cancer. *Annu Rev Pathol* **10**: 25–50.

Popovic R, Martinez-Garcia E, Giannopoulou EG, Zhang Q, Zhang Q, Ezponda T, Shah MY, Zheng Y, Will CM, Small EC, et al. 2014. Histone methyltransferase MMSET/NSD2 alters EZH2 binding and reprograms the myeloma epigenome through global and focal changes in H3K36 and H3K27 methylation. *PLoS Genet* **10**: e1004566.

Ptashne M. 2007. On the use of the word "epigenetic." *Curr Biol* **17**: R233–R236.

Quenet D, Dalal Y. 2012. The CENP-A nucleosome: A dynamic structure and role at the centromere. *Chromosome Res* **20**: 465–479.

Ramachandran S, Henikoff S. 2015. Replicating nucleo-somes. *Sci Adv* **1:** e1500587.

Ramachandran S, Zentner GE, Henikoff S. 2015. Asymmetric nucleosomes flank promoters in the budding yeast genome. *Genome Res* **25:** 381–390.

Ratnakumar K, Duarte LF, LeRoy G, Hasson D, Smeets D, Vardabasso C, Bonisch C, Zeng T, Xiang B, Zhang DY, et al. 2012. ATRX-mediated chromatin association of histone variant macroH2A1 regulates α-globin expression. *Genes Dev* **26:** 433–438.

Ray-Gallet D, Quivy J-P, Scamps C, Martini EM, Lipinski M, Almouzni G. 2002. HIRA is critical for a nucleosome assembly pathway independent of DNA synthesis. *Mol Cell* **9:** 1091–1100.

Riedel SS, Neff T, Bernt KM. 2015. Histone profiles in cancer. *Pharmacol Ther* **154:** 87–109.

Schneiderman JI, Sakai A, Goldstein S, Ahmad K. 2009. The XNP remodeler targets dynamic chromatin in *Drosophila. Proc Natl Acad Sci* **106:** 14472–14477.

Schneiderman JI, Orsi GA, Hughes KT, Loppin B, Ahmad K. 2012. Nucleosome-depleted chromatin gaps recruit assembly factors for the H3.3 histone variant. *Proc Natl Acad Sci* **109:** 19721–19726.

Schotta G, Ebert A, Krauss V, Fischer A, Hoffmann J, Rea S, Jenuwein T, Dorn R, Reuter G. 2002. Central role of *Drosophila* SU(VAR)3-9 in histone H3-K9 methylation and heterochromatic gene silencing. *EMBO J* **21:** 1121–1131.

Schwartz YB, Pirrotta V. 2013. A new world of polycombs: Unexpected partnerships and emerging functions. *Nat Rev Genet* **14:** 853–864.

Shi J, Vakoc CR. 2014. The mechanisms behind the therapeutic activity of BET bromodomain inhibition. *Mol Cell* **54:** 728–736.

Sporn JC, Jung B. 2012. Differential regulation and predictive potential of MacroH2A1 isoforms in colon cancer. *Am J Pathol* **180:** 2516–2526.

Steffen PA, Ringrose L. 2014. What are memories made of? How polycomb and trithorax proteins mediate epigenetic memory. *Nat Rev Mol Cell Biol* **15:** 340–356.

Suto RK, Clarkson MJ, Tremethick DJ, Luger K. 2000. Crystal structure of a nucleosome core particle containing the variant histone H2A.Z. *Nat Struct Biol* **7:** 1121–1124.

Tamkun JW, Deuring R, Scott MP, Kissinger M, Pattatucci AM, Kaufman TC, Kennison JA. 1992. brahma: A regulator of *Drosophila* homeotic genes structurally related to the yeast transcriptional activator SNF2/SWI2. *Cell* **68:** 561–572.

Teves SS, Weber CM, Henikoff S. 2014. Transcribing through the nucleosome. *Trends Biochem Sci* **39:** 577–586.

Valdes-Mora F, Song JZ, Statham AL, Strbenac D, Robinson MD, Nair SS, Patterson KI, Tremethick DJ, Stirzaker C,

Clark SJ. 2012. Acetylation of H2A.Z is a key epigenetic modification associated with gene deregulation and epigenetic remodeling in cancer. *Genome Res* **22:** 307–321.

Vogelstein B, Papadopoulos N, Velculescu VE, Zhou S, Diaz LA Jr, Kinzler KW. 2013. Cancer genome landscapes. *Science* **339:** 1546–1558.

Watanabe S, Peterson CL. 2010. The INO80 family of chromatin-remodeling enzymes: Regulators of histone variant dynamics. *Cold Spring Harb Symp Quant Biol* **75:** 35–42.

Watanabe S, Radman-Livaja M, Rando OJ, Peterson CL. 2013. A histone acetylation switch regulates H2A.Z deposition by the SWR-C remodeling enzyme. *Science* **340:** 195–199.

Waterborg JH. 1993. Histone synthesis and turnover in alfalfa. Fast loss of highly acetylated replacement histone variant H3.2. *J Biol Chem* **268:** 4912–4917.

Waterborg JH. 2002. Dynamics of histone acetylation in vivo. A function for acetylation turnover? *Biochem Cell Biol* **80:** 363–378.

Waters R, van Eijk P, Reed S. 2015. Histone modification and chromatin remodeling during NER. *DNA Repair (Amst)* **36:** 105–113.

Weintraub H, Groudine M. 1976. Chromosomal subunits in active genes have an altered conformation. *Science* **193:** 848–856.

Wen H, Li Y, Xi Y, Jiang S, Stratton S, Peng D, Tanaka K, Ren Y, Xia Z, Wu J, et al. 2014. ZMYND11 links histone H3.3K36me3 to transcription elongation and tumour suppression. *Nature* **508:** 263–268.

Wilson BG, Roberts CW. 2011. SWI/SNF nucleosome remodellers and cancer. *Nat Rev Cancer* **11:** 481–492.

Wolffe AP. 1992. *Chromatin: Structure and function.* Academic, San Diego.

Wollmann P, Cui S, Viswanathan R, Berninghausen O, Wells MN, Moldt M, Witte G, Butryn A, Wendler P, Beckmann R, et al. 2011. Structure and mechanism of the Swi2/Snf2 remodeller Mot1 in complex with its substrate TBP. *Nature* **475:** 403–407.

Wu G, Broniscer A, McEachron TA, Lu C, Paugh BS, Becksfort J, Qu C, Ding L, Huether R, Parker M, et al. 2012. Somatic histone H3 alterations in pediatric diffuse intrinsic pontine gliomas and non-brainstem glioblastomas. *Nat Genet* **44:** 251–253.

Yang F, Kemp CJ, Henikoff S. 2015. Anthracyclines induce double-strand DNA breaks at active gene promoters. *Mutat Res* **773:** 9–15.

Zentner GE, Henikoff S. 2013. Mot1 redistributes TBP from TATA-containing to TATA-less promoters. *Mol Cell Biol* **33:** 4996–5004.

Nonhistone Lysine Methylation in the Regulation of Cancer Pathways

Scott M. Carlson and Or Gozani

Department of Biology, Stanford University, Stanford, California 94305

Correspondence: scottmc@stanford.edu; ogozani@stanford.edu

Proteins are regulated by an incredible array of posttranslational modifications (PTMs). Methylation of lysine residues on histone proteins is a PTM with well-established roles in regulating chromatin and epigenetic processes. The recent discovery that hundreds and likely thousands of nonhistone proteins are also methylated at lysine has opened a tremendous new area of research. Major cellular pathways involved in cancer, such as growth signaling and the DNA damage response, are regulated by lysine methylation. Although the field has developed quickly in recent years many fundamental questions remain to be addressed. We review the history and molecular functions of lysine methylation. We then discuss the enzymes that catalyze methylation of lysine residues, the enzymes that remove lysine methylation, and the cancer pathways known to be regulated by lysine methylation. The rest of the article focuses on two open questions that we suggest as a roadmap for future research. First is understanding the large number of candidate methyltransferase and demethylation enzymes whose enzymatic activity is not yet defined and which are potentially associated with cancer through genetic studies. Second is investigating the biological processes and cancer mechanisms potentially regulated by the multitude of lysine methylation sites that have been recently discovered.

Lysine methylation is the addition of one, two, or three methyl groups to the ε-nitrogen of a lysine sidechain (Fig. 1). Methylation is generated by lysine methyltransferase enzymes (KMTs) and removed by lysine demethylases (KDMs). Lysine methylation was first described in 1959 by R.P Ambler and M.W. Rees on a *Salmonella typhimurium* flagellar protein (Ambler and Rees 1959). Lysine methylation in mammals was first reported 5 years later when Kenneth Murray (1964) found methylated lysine on bovine histone proteins. Over the next few years lysine methylation was found on a variety of other proteins (Paik et al. 2007). The biological purpose of lysine methylation remained enigmatic for several decades and it was only in the first decade of this century that an understanding began to emerge of how lysine methylation contributes to regulation of processes, including epigenetics, chromatin function, and cellular signaling.

Most research on lysine methylation has focused on histone methylation because of its early discovery and clear importance in chromatin biology and gene regulation. Thomas Jenuwein and colleagues found that the human enzyme

Cite this article as *Cold Spring Harb Perspect Med* doi: 10.1101/cshperspect.a026435

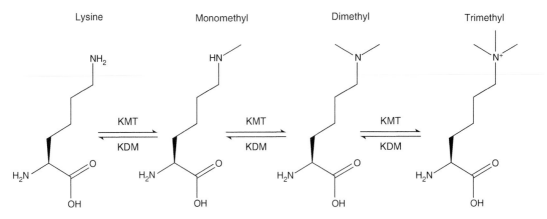

Figure 1. Lysine residues can be modified by the addition of up to three methyl groups at the ε-nitrogen. Lysine methylation is catalyzed by lysine methyltransferase enzymes (KMTs) and removed by lysine demethylase enzymes (KDMs). Methylation states of lysine are recognized by protein "reader domains" that bind to specific methyl or nonmethyl states of their target proteins.

SUV39H1 is able to methylate histone H3 at lysine 9 (H3K9) (Rea et al. 2000). The next year the groups of Thomas Jenuwein and Tony Kouzarides independently discovered that the chromodomain of HP1 is a methyllysine "reader domain" able to suppress transcription by selectively binding to methylated H3K9 (Bannister et al. 2001; Lachner et al. 2001). The first lysine demethylase enzyme was discovered in 2004 by the group of Yang Shi (Shi et al. 2004). The combination of enzymes to "write" and "erase" methylation with proteins that can "read" methylation of their interacting partners established lysine methylation as a dynamic signaling process similar in principle to phosphorylation.

Many KMTs, KDMs, and reader domains have now been identified that act on or recognize histone proteins (Black et al. 2012). A smaller number of enzymes have been identified with the ability to methylate lysine on nonhistone proteins, although lysine methylation has been established as an important player in cellular signaling processes (see Biggar and Li 2015 for a detailed history). Similar to its role on histones, the primary function of nonhistone lysine methylation is to regulate protein–protein interactions. Such interactions control a wide range of downstream processes such as protein stability, subcellular localization, and DNA binding (Hamamoto et al. 2015). It is not yet

known whether the same protein domain families that recognize histone lysine methylation are also responsible for reading nonhistone lysine methylation events, or whether there are yet to be discovered families that recognize nonhistone lysine methylation.

Until recently, research into nonhistone lysine methylation was limited because there were no strategies to identify lysine methylation across the entire proteome. Starting in 2013, we and several other groups developed techniques to identify methylated proteins across the entire proteome. We recently reviewed these techniques (Carlson and Gozani 2014) and they will be discussed in more detail later in the text. These proteomic studies have identified many hundreds of new methylated proteins and methylated lysine residues, but there is little known about the functional relevance of these modifications, the enzymes responsible for generating these methylation events, and the potential roles for nonhistone lysine methylation in cancer or other diseases. However, a striking number of methylated proteins have been discovered in pathways related to oncogenesis and cancer progression. We will discuss one example of nonhistone lysine methylation regulating growth pathways central to cancer and examine some of the other cancer pathways potentially regulated by lysine methylation.

Herein we will review the nonhistone KMT and KDM enzymes associated with cancer, and we will discuss what is currently known about the molecular mechanisms underlying these associations. In the rest of the article, we will focus on areas of ongoing and future research. In particular, we will highlight cancer-associated KMTs and KDMs that lack known enzymatic functions and lysine methylation in cancer pathways where the possibility of a regulatory function has not yet been investigated.

SET DOMAIN ENZYMES IN CANCER

There are two proteins families known to methylate lysine: SET domain family and the seven β-strand enzymes (7β-strand enzymes). Most research has focused on the SET domain family, which includes about 55 human proteins (named for three *Drosophila* proteins originally recognized to contain the domain: Su(var)3-9, enhancer of zeste, trithorax) (Petrossian and Clarke 2011). The human genome encodes approximately 125 members of the 7β-strand family (Petrossian and Clarke 2011). Different members of the 7β-strand family are able to methylate a wide range of substrates, including lysine, arginine, other amino acid side chains,

DNA, RNA, metabolites, and arsenic (Schubert et al. 2003). The 7β-strand family is discussed later in the chapter.

Most research into SET domain proteins has focused on their role in methylating histones. This seems to be the primary activity for about a third of the family. A smaller number target both histones and nonhistone proteins, and a few characterized enzymes are likely to act primarily on nonhistone substrates (Fig. 2). Close to half of the SET domain proteins have no known substrates and it is not clear whether they have enzymatic activity at all.

Experimental evidence for the activity of many SET domain proteins is incomplete, making it difficult to sort them by activity on histones or nonhistone substrates. One challenge comes from nonspecific activity when these enzymes are expressed as isolated proteins and probed for activity on candidate substrates. Many enzymes will methylate recombinant substrates in vitro, which are not substrates in a natural context. This is especially true of histone proteins because their tail regions contain many lysine residues that are excellent substrates for some of the commonly characterized KMTs. Thus, some SET domain proteins will methylate free histones but have little or no activity on

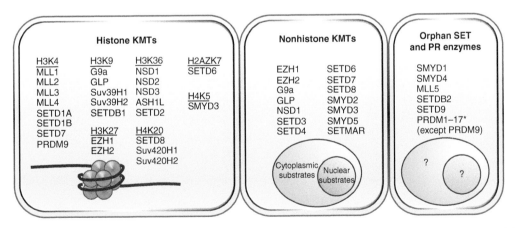

Figure 2. SET domain proteins categorized by their established substrate specificity for (1) histone proteins, (2) nonhistone proteins, or (3) orphan enzymes with no well-established enzymatic activity. Note that there are enzymes that are primarily histone lysine methyltransferase enzymes (KMTs) but also have reported nonhistone substrates. In addition, there are proteins that are primarily nonhistone KMTs, with reports of histone methylation activity. We note that several PRDMs (e.g., 1, 2, 3, and 16) have been reported to have activity on H3K9 and more work is needed to understand the catalytic activities of these enyzmes.

histones packaged into nucleosomes that contain a full histone octomer wrapped with DNA (the physiologic context in which histones commonly exist) (Levy et al. 2011; Carlson et al. 2015). Other enzymes methylate many sites on free histones but become specific for a single residue when presented with nucleosome substrates (our unpublished observations). We anticipate that an in vitro activity seen only on free histones but not nucleosomes needs to be carefully interpreted and rigorously evaluated in cell culture to conclude that it is physiologic. Similarly, many SET domain enzymes have non-specific activity on nonhistone substrates in vitro so candidate nonhistone substrates need to be supported by extensive characterization in cell culture.

A second challenge is overinterpretation of data from Western blotting with methyl-specific antibodies. These antibodies are often incompletely characterized or else bind nonspecifically to other methylated residues, methylation on other proteins, other methylation states (e.g., trimethyl instead of dimethyl), or other post-translational modifications of lysine (Fuchs et al. 2011). This is especially true when candidate KMT substrates contain multiple methylated residues, or when the methylation state being probed is much less abundant than other methyl marks, a frequent issue when analyzing histone modifications.

To frame the discussion of nonhistone KMTs in cancer, we have consolidated evidence from literature reports as well as unpublished experiments in our own laboratory to sort the SET domain enzymes into enzymes that are well established to act (1) on histones, (2) on nonhistone substrates, or else (3) are "orphans," which have no activity that has been widely observed—we note that there are some enzymes that are present in both the first and second groups (Fig. 2). With regard to the orphan SET domains, they may not be enzymatic, may only be active in the presence of other unidentified factors, or may target substrates that have not been identified. In some cases there are individual reports of enzymatic activity that await independent corroboration and for the sake of caution we have included them as orphans.

These and other orphan KMTs are discussed later in the text. The literature is too extensive on the topic to cite it thoroughly here; in lieu we refer to excellent reviews of SET domain activities (Dillon et al. 2005; Herz et al. 2013).

MOLECULAR MECHANISMS OF SET DOMAIN NONHISTONE METHYLATION IN CANCER

It is beyond the scope of this article to discuss all the nonhistone KMTs that touch on pathways involved in cancer. Comprehensive reviews of nonhistone lysine methylation have recently been provided by Nakamura and colleagues (Hamamoto et al. 2015) and Biggar and Li (2015). Here we will focus on establishing a framework to understand KMTs and their role in cancer. With this in mind, we will discuss two examples of nonhistone KMTs with clear roles in cancer. First is SMYD3, a primarily cytoplasmic enzyme that cooperates with active oncogenic K-Ras to regulate growth signaling in lung and pancreatic carcinomas and likely other types of cancer. Second is the pair of enzymes G9a and GLP, closely related KMTs that are strongly associated with cancer progression and metastasis. These enzymes methylate both histone and nonhistone substrates, which raises important questions about how to discern the biological and pathologic role of these activities. Other nonhistone KMTs that affect cancer pathways will be discussed briefly and we will finish the section by reviewing the orphan SET domain proteins that are strongly associated with cancer.

SMYD3: A Cytoplasmic KMT that Promotes Ras-Driven Cancer Pathways

SMYD3 was first identified based on an expressed sequence tag that was overexpressed in colorectal and hepatocellular carcinomas (Hamamoto et al. 2004). It was grouped into the SMYD family based on its combination of its split SET and zf-MYND domains. Its cancer relevance was quickly established based on its ability to enhance growth of NIH3T3 fibroblasts. Since the initial report, overexpression of SMYD3 has been found in many other types of cancer, especially those carrying mutations in K-Ras that lead to constitutively active growth signaling (Hamamoto et al. 2004; Gaedcke et al. 2010; Watanabe et al. 2011). The link to K-Ras is clinically important because mutations in the K-Ras gene occur in more than half of all human cancers (Forrester et al. 1987; Prior et al. 2012). K-Ras itself has been the target of extensive efforts to develop targeted inhibitors but the protein itself remains difficult to target directly with pharmacological treatments (McCormick 2015).

SMYD3 was initially described as an enzyme that generates trimethylation of histone H3 at lysine 4 (H3K4me3), a hallmark of active transcription (Hamamoto et al. 2004). However, subsequent work has shown that SMYD3 has no detectable activity on H3K4 (Van Aller et al. 2012; Mazur et al. 2014). SMYD3 does methylate histones, specifically histone H4 at lysine 5 (H4K5) in vitro, and SMYD3 knockdown reduced H4K5 methylation in cell culture, although H4K5me is a very lowly abundant species of histone modification (Van Aller et al. 2012). Indeed, SMYD3 largely resides in the cytoplasm and the first evidence that SMYD3 acts on nonhistone proteins was reported by Furukawa and colleagues in 2007 (Kunizaki et al. 2007). They found that SMYD3 methylates vascular endothelial growth factor receptor 1 (VEGFR1) both in vitro and when the proteins are overexpressed in cell culture. VEGFR1 methylation increased its kinase activity, although the molecular mechanism for this is not known. In addition, whether this methylation event is linked to cell growth or cancer progression has yet to be determined.

SMYD3 was recently shown to trimethylate the kinase MAP3K2 (Mazur et al. 2014). MAP3K2 is one of several kinases within the K-Ras-regulated mitogen-activated protein kinase-signaling cascade (MAPK pathway) that regulates cell growth. MAPK pathways are composed of a three-kinase signaling cascade, where each kinase in the pathway becomes activated, then phosphorylates and activates the next kinase down. The pathway is deactivated when these kinases are dephosphorylated by a phosphatase. One mechanism by which oncogenic mutations in K-Ras and other *Ras* genes promote cancer is through activation of a MAPK pathway leading to activation of the kinases ERK1 and ERK2. Activated ERK1 and ERK2 phosphorylate a wide range of effector proteins, and especially transcription factors, that contribute to proliferation, cell survival, and metastasis (Plotnikov et al. 2011).

The methylation of MAP3K2 by SMYD3 was shown to promote activation of ERK1/2 in mouse models of Ras-driven pancreatic and lung adenocarcinoma. The methylation event does not intrinsically affect the kinase activity of MAP3K2 and rather prevents the protein phosphatase 2A complex (PP2A) from binding to MAP3K2. PP2A regulates the MAPK cascade by dephosphorylating and inactivating kinases in the cascade. The interaction with PP2A explains how SMYD3 cooperates with mutant K-Ras to promote cancer. In this model SMYD3 relieves repression by PP2A and increases mutant K-Ras signaling to drive inappropriate growth signals through the MAPK cascade. Chemical inhibition of PP2A phenocopies overexpression of SMYD3 in lung and pancreatic cancer cell and xenograft models, suggesting that this is important mechanism by which SMYD3 promotes these cancers.

MAP3K2 is not highly expressed in all types of cancers, all cancers that carry K-Ras mutations, or even all cancers that overexpress SMYD3. It is, therefore, likely that SMYD3 acts through different mechanisms depending on the cellular context and it will be important to elucidate the relevant substrates in other cancers regulated by SMYD3. Establishing how SMYD3 contributes to other types of cancer is an important area for future study.

G9a and GLP: Coassociated Mixed Histone/Nonhistone KMTs that Promote Cancer

G9a (also called EHMT2) was discovered in 2001 as a "hyperactive" methyltransferase enzyme able to mono- and dimethylate lysines 9 and 27 of histone H3 (Tachibana et al. 2001). G9a and the closely related enzyme G9a-like protein (GLP) are responsible for the bulk of

H3K9 dimethylation (H3K9me2) in human cells (Tachibana et al. 2002, 2005). These enzymes are crucial for early development and high G9a expression is associated with aggressive cancer, metastasis, and poor prognosis in cancers of the breast, lung, head and neck, brain, and ovaries among others (Casciello et al. 2015).

The potential importance of G9a in so many types of cancer led in part to development of the inhibitor BIX-01294 (Kubicek et al. 2007). Treatment with this inhibitor blocked epithelial-to-mesenchymal transition in cell culture and in in vivo models of breast cancer (Dong et al. 2012), and reduced proliferation rates of some leukemia cell lines (Savickiene et al. 2014). BIX-01294 causes significant cellular toxicity separate from its function inhibiting G9a and GLP, limiting its usefulness in preclinical research. Improved inhibitors UNC0638, BRD4770, UNC0642, and A-366 have now achieved cellular toxicity, target specificity, and pharmacokinetics that are suitable for animal studies (Vedadi et al. 2011; Yuan et al. 2012; Liu et al. 2013a; Sweis et al. 2014).

In cells G9a and GLP are responsible for a large fraction of all the H3K9me1 and H3K9me2. H3K9me2 can be bound by a number of reader domain-containing proteins that are linked to transcriptional repression, such as the HP1 proteins, G9a and GLP themselves, and others (Yun et al. 2011). Transcriptional repression through H3K9me2 is certainly an important function of G9a and GLP but these enzymes are also able to methylate a variety of nonhistone proteins.

The first example of G9a acting as a nonhistone KMT was shown by the group of Alexander Tarakhovsky in 2007 (Sampath et al. 2007). They reported an automethylation site on G9a, which, like H3K9me2, is bound by HP1 and potentially contributes to silencing chromatin. In 2008 Jeltsch and colleagues used peptide arrays to identify additional candidate nonhistone substrates for G9a, and verified that several of these can be methylated by G9a in cells (CDYL1, WIZ, ACINUS, and confirmed G9a itself) (Rathert et al. 2008). Since then, many additional nonhistone substrate for G9a have

been reported, suggesting that it has important functions modifying both histone and nonhistone proteins, although the biological role of nonhistone substrates is not well understood (Shinkai and Tachibana 2011). These enzymes are extremely promiscuous in vitro and overexpressed enzymes may not be properly targeted in cell culture, making it difficult to determine which substrates contribute to their physiological effects. Further research using G9a inhibitors, genetic perturbations, or cell-based chemical biology will be important to identify or validate physiological targets for these and other enzymes (we recently reviewed these approaches in detail (Carlson and Gozani 2014)).

One nonhistone G9a substrate clearly linked to cancer is the tumor suppressor p53. G9a and GLP methylate p53 at lysine 373 (Huang et al. 2010). The function of this methylation site is not known but methylation appears to be reduced during repair of DNA damage, suggesting that it may limit p53 activity. Indeed, the first example of a histone KMT with dual activity was shown by Reinberg and colleagues, who showed that SETD7 monomethylates p53 at K372 to positively regulate p53 protein stability (Chuikov et al. 2004). The methyltransferase enzymes SMYD2 and SETD8 also target p53 at other lysine residues (Huang et al. 2006; Shi et al. 2007). With many KMTs targeting p53, it is clear that lysine methylation has an important role in coordinating the tightly regulated response of p53 to genotoxic stress. Our understanding of methylation events on p53 has been reviewed in detail elsewhere (Huang and Berger 2008; West and Gozani 2011).

G9a and GLP have another feature that shows the role of reader domains in methylation signal transduction. Both proteins include an ankyrin repeat domain that binds specifically to mono- and dimethylated lysine (Collins et al. 2008). A major target for this activity is histone H3K9me2. There are several proposed functions for G9a/GLP binding to their own enzymatic product. It may serve to protect the modification from conversion to trimethylation or from demethylation (Collins and Cheng 2010). Alternatively, recent evidence indicates that H3K9 methylation promotes similar methylation at adjacent nucleosomes (Liu et al. 2015). This would establish positive feedback allowing H3K9me2 to "spread" along a chromatin until it encounters opposing regulatory processes that establish a boundary between regulatory regions. Finally, this binding event can help maintain chromatin regions containing H3K9me2 on DNA replication (Margueron and Reinberg 2010).

In addition to recognizing histones, the ankryin repeats are able to bind to methylated nonhistone proteins. Specifically, the NF-κB transcription factor RelA is methylated by SETD6, which creates a docking site recognized by the ankryin repeats of GLP (Levy et al. 2011). This brings GLP to NF-κB target genes where it represses gene activity by methylating H3K9. Understanding the roles of histone and nonhistone binding for these and other methyllysine reader domains is a crucial and exciting area for ongoing research.

Other SET Domain Enzymes Affecting Cancer Pathways

There are several enzymes with intriguing links to cancer or cancer cell phenotypes in which specific enzyme/substrate relationships in this disease have yet to be elucidated. For example, the enzyme SMYD2 is overexpressed in several types of cancer and associated with poor clinical prognoses (Komatsu et al. 2009; Sakamoto et al. 2014). SMYD2 is a promiscuous enzyme in vitro and it is often referred to as a histone KMT based on its robust in vitro activity on recombinant histone substrates. The enzyme has many reported nonhistone substrates that have been validated in cell culture and could be relevant in cancer, including the tumor suppressors p53, Rb and PTEN, protein-folding chaperone HSP90, and estrogen receptor α (Huang et al. 2006; Abu-Farha et al. 2011; Cho et al. 2012; Zhang et al. 2013; Nakakido et al. 2015).

SETD6, SETD7, SETD8, and SETMAR are all SET domain enzymes that do not appear to act directly as oncogenes or tumor suppressors but which methylate proteins involved in cancer pathways. SETD6 targets the RelA subunit of the transcription factor NF-κB to regulate inflam-

mation (Levy et al. 2011); SETD7 has many cancer-associated substrates, including p53, the DNA methyltransferase DNMT1, the phosphatase PPP1R12A, which regulates cell cycle through Rb, and many others (Pradhan et al. 2009); SETD8 methylates histone H4 as well as p53 and proliferating cell nuclear antigen (PCNA) (Nishioka et al. 2002; Shi et al. 2007; Takawa et al. 2012); SETMAR is involved in repairing DNA damage and methylates the splicing factor snRNP70 (Fnu et al. 2011; Carlson et al. 2015).

Orphan SET Domain Proteins Associated with Cancer

Many other SET domain proteins do not have known molecular catalytic activity but are still associated with cancer through genome or exome sequencing, experiments in cell culture, or experiments in animal models. Table 1 lists these proteins and selected major associations with cancer. It may be that these proteins are pseudoenzymes and instead work through nonenzymatic mechanisms, such as binding DNA or by protein interactions. There are clear examples of KMT proteins with nonenzymatic functions, including KMTs expressed as isoforms lacking their catalytic domain or else truncated by chromosomal rearrangement. For example, acute myeloid leukemia is frequently driven by translations involving the *MLL* gene truncated so that it lacks the catalytic SET domain (McCabe et al. 1992; Meyer et al. 2013). Similarly, an isoform of NSD3 that lacks the SET domain has recently been shown to act by bridging a protein interaction between histones and other chromatin-associated proteins (Shen et al. 2015). With all of the hundreds of lysine methylation sites recently discovered across the proteome, we expect that many of these enzymes have still unknown activity as nonhistone KMTs and we simply have not developed the right tools and methods to characterize many of these orphan enzymes.

The PRDM subfamily deserves to be highlighted as a major area for future research. These 17 human proteins contain an amino-terminal SET domain and a variable number of carboxy-

Table 1. Examples of selected "orphan" SET domain proteins associated with cancer

Enzyme	Selected cancer associations	Enzyme	Selected cancer associations
SMYD4	Deleted/silenced in medulloblastoma (Northcott et al. 2009)	PRDM5	Silenced in gastric cancer (Watanabe et al. 2007)
PRDM1	Deleted in diffuse large B-cell lymphoma (Pasqualucci et al. 2006)	PRDM11	Deleted or silenced in diffuse large B-cell lymphoma (Fog et al. 2015)
PRDM2 (RIZ1)	Silenced in diverse solid tumors (Du et al. 2001)	PRDM14	Overexpressed in breast cancer (Hu et al. 2005; Nishikawa et al. 2007)
PRDM3 (MECOM, MDS1/EVI1)	Chromosomal rearrangement in myelodysplastic syndrome and myeloid leukemia (Morishita et al. 1992)		

terminal zinc fingers (Fog et al. 2012). The SET domain is often called a PR or PR/SET domain because it diverges significantly from the other SET domain proteins. So far, only PRDM9 is clearly shown to have enzymatic activity, having the greatest activity on H3K4 and lower activity on H3K9 and H3K36 in vitro (Hayashi et al. 2005). The H3K4me3 activity is thought to mark meiotic hotspots and be critical for speciation (Baudat et al. 2010; Myers et al. 2010; Parvanov et al. 2010); whether H3K9 and H3K36 methylation also play a role in these meiosis-specific functions remains to be determined.

Many of the PRDM family proteins have important functions in cellular differentiation and development, and many are frequently deleted, mutated, or overexpressed in cancer (Fog et al. 2012). For example, PRDM1 is a master regulator of terminal B-cell differentiation, and loss of PRDM1 expression has been found in a majority of activated B-cell-like diffuse large B-cell lymphomas (Pasqualucci et al. 2006). Loss of PRDM1 promotes reactivation of the cell cycle in terminally differentiated B cells. Similarly, overexpression of a specific splicing isoform of PRDM2 called RIZ1 acts as a tumor suppressor in cancers of the breast, liver, and colon (Du et al. 2001).

With such central roles in development and disease, identifying methylation substrates or other molecular functions for the PRDM family is a crucial area for future research. There have been reports of histone methylation activity by several PRDM proteins in addition to PRDM9

(Derunes et al. 2005; Eom et al. 2009; Pinheiro et al. 2012); however, structural studies have suggested that these PR/SET domains lack residues necessary for binding to S-adenosyl methionine (Wu et al. 2013), the cofactor that donates methyl groups for KMT activity. Extensive research by several groups, including our own, has failed to establish enzymatic activity for these enzymes. Several groups have found that PRDM proteins are involved in recruitment of other chromatin-modifying factors, suggesting that their function may be independent of any enzymatic activity (Hohenauer and Moore 2012). Except for PRDM9 it is not yet clear whether enzymatic activities attributed to the PRDM family are intrinsic to proteins themselves or if they belong to copurifying enzymes. Another possibility is that PRDM proteins are responsible for methylation of nonhistone substrates that have yet to be identified. Given their involvement in cancer and other diseases it will be important to determine if some or all of the PRDM proteins have intrinsic enzymatic activity, or whether noncatalyic biological functions exist for these PR/SET domains.

7β-STRAND ENZYMES: A LARGE AND ENIGMATIC FAMILY ASSOCIATED WITH CANCER

Members of the 7β-strand methyltransferase family are the second group of enzymes able to methylate lysine. The best studied of these enzymes, which act on lysine in humans, is DOT1L, which methylates histone H3 at lysine 79 strictly in a nucleosome context (Ng et al. 2002). Although DOT1L is the only 7β-strand enzyme known to methylate a lysine residue on histones, other members of this family are linked to lysine methylation of nonhistone proteins, arginine residues on histone and nonhistone proteins, and nucleic acids (Schubert et al. 2003).

In 2010 Robert Houtz and colleagues discovered that human calmodulin is methylated by a 7β-strand enzyme now named calmodulin-lysine *N*-methyltransferase (CAMKMT) (Magnani et al. 2010). Soon after, the group of Pål Falnes found that the 7β-strand enzyme METTL21D (renamed VCPKMT) methylates VCP, an ATPase enzyme (Kernstock et al. 2012). Several groups have since identified additional 7β-strand enzymes that methylate lysine residues of nonhistone substrates. Although research into 7β-strand KMTs is progressing quickly, there is still very little about their role in biology or human disease (Cloutier et al. 2013, 2014; Jakobsson et al. 2013; Davydova et al. 2014; Małecki et al. 2015). Dozens of human 7β-strand enzymes are still entirely uncharacterized and we expect that more of them will turn out to be KMTs.

To understand the potential scope of KMT enzymes in cancer, we reviewed data from The Cancer Genome Atlas (TCGA) to find candidate KMTs that are associated with cancer (Cerami et al. 2012; Gao et al. 2013). We included only enzymes that methylate lysine on nonhistone proteins or else have no known function. Table 2 lists the candidate 7β-strand KMTs that are most strongly associated with cancer in the TCGA dataset or in studies indexed by the TCGA. This list is subjective and only meant to serve as a guide for future work. Hypothesis-driven research will be needed to determine if any of these enzymes contribute to cancer or present any therapeutic opportunity.

THE LSD1 LYSINE DEMETHYLASE IN CANCER

The first lysine demethylase (KDM), lysine-specific demethylase 1A (LSD1) was discovered 2004 by Yang Shi and colleagues (Shi et al. 2004). It belongs to the family of FAD-dependent monoamine oxidases. LSD1 removes mono- and dimethylation from histone H3K4, and under specific circumstances from H3K9 (Metzger et al. 2005; Laurent et al. 2015). In 2007, the group of Shelley Berger found that LSD1 regulates p53 by removing dimethylation from lysine 370, providing the first evidence of reversible lysine methylation on human nonhistone proteins (Huang et al. 2007). Other nonhistone substrates of LSD1 have since been discovered, including E2F1, DNMT1, and MYPT1 (Zheng et al. 2015). Many of these targets belong to pathways that contribute to oncogenesis or tumor suppression, although their individual roles in cancer progression have not been established. LSD2 is another member of the FAD-dependent monoamine oxidases, which can demethylate histones, but it has not been reported to remove methylation from nonhistone proteins (Black et al. 2012).

LSD1 is frequently overexpressed in cancer, especially of the prostate, bladder, lung, and colon (Shi 2007; Hayami et al. 2011). High LSD1 expression is generally associated with invasive disease and a poor prognosis, and knockdown or inhibition of LSD1 reduces invasive phenotypes in several cell culture systems (Wang et al. 2011). These roles in cancer make LSD1 a promising target for new therapeutics and there has been extensive work on developing small molecule inhibitors of LSD1 (Zheng et al. 2015). In particular, LSD1 inhibition has been shown to sensitize acute myeloid leukemia to treatment with all-*trans* retinoic acid (Schenk et al. 2012), and phase I clinical trials were initiated in 2014 to exploit this effects (clinicaltrials.gov, identifiers NCT02273102 and NCT02261779). GlaxoSmithKline has also begun phase I clinical trials to test the effect of the LSD1 inhibitor GSK2879552 in small-cell lung cancer and acute myeloid leukemia (clinicaltrials.gov, identifiers NCT02177812 and NCT02034123).

The second family of KDM is the Jumonji C domain (JmjC)-containing enzymes. The first member of this family was JHDM1, discovered by the group of Yi Zhang in 2006 (Tsukada et al. 2006). Several of these enzymes have well-characterized activity on histones both in vitro and in cells (Black et al. 2012). Demethylation of

Table 2. Selected uncharacterized 7β-strand methyltransferase proteins frequently deleted or amplified cancer in datasets from The Cancer Genome Atlas (TCGA)

Enzyme	Selected cancer associations	Enzyme	Selected cancer associations
FAM86B1	Deletion: Uterine carcinosarcoma Prostate adenocarcinoma Bladder urothelial carcinoma	METTL11B	Amplification: Bladder urothelial carcinoma Invasive breast carcinoma Liver hepatocellular carcinoma
FAM86B2	Deletion: Uterine carcinosarcoma Prostate adenocarcinoma Bladder urothelial carcinoma	METTL13	Amplification: Bladder urothelial carcinoma Liver hepatocellular carcinoma Pancreatic cancer Lung adenocarcinoma
FAM173B	Amplification: Lung squamous cell carcinoma Lung adenocarcinoma Bladder urothelial carcinoma	METTL18	Amplification: Bladder urothelial carcinoma Liver hepatocellular carcinoma Breast invasive carcinoma Lung adenocarcinoma
FTSJ2	Amplification: Pancreatic cancer Stomach adenocarcinoma Lung adenocarcinoma	METTL21B	Amplification: Sarcoma Glioblastoma
FTSJ3	Amplification: Breast invasive carcinoma	METTL24	Deletion: Diffuse large B-cell lymphoma Prostate adenocarcinoma
KIAA1456	Deletion: Uterine sarcoma Prostate adenocarcinoma Bladder urothelial carcinoma	TGS1	Amplification: Metastatic prostate adenocarcinoma Uveal melanoma Breast invasive carcinoma
METTL7A	Deletion: Adenoid cystic carcinoma	THUMPD3	Amplification: Bladder urothelial carcinoma
METTL7B	Deletion: Adenoid cystic carcinoma		

nonhistone proteins by the JmjC-family of enzymes has not been well characterized. With hundreds of methylated proteins discovered in recent years, we believe it likely that more nonhistone KDMs will soon be discovered.

CONCLUDING REMARKS: UNDERSTANDING LYSINE METHYLATION ACROSS CANCER PATHWAYS

Proteomic techniques allow many posttranslational modifications to be routinely identified and measured across hundreds or thousands of proteins. In these approaches, modified proteins or peptides are captured using chemical affinity or immunoprecipitation. Enriched proteins or peptides are then identified using mass spectrometry. Early attempts to analyze the posttranslational "lysine methylome" were limited by the antibodies available to immunoprecipitate methylated lysine (Ong et al. 2004; Ong and Mann 2006).

In 2013, the authors and the group of Shawn Li independently used methyllysine reader domains to enrich and identify methylated proteins (Liu et al. 2013b; Moore et al. 2013). Soon after the group of Ben Garcia and the company Cell Signaling Technologies, each used antimethyl antibodies with broad sequence specificity to perform proteomic studies that identified lysine methylation at hundreds of new sites (Cao et al. 2013; Guo et al. 2014). There are currently 1324 proteins with lysine methylation appearing the PhosphoSitePlus database

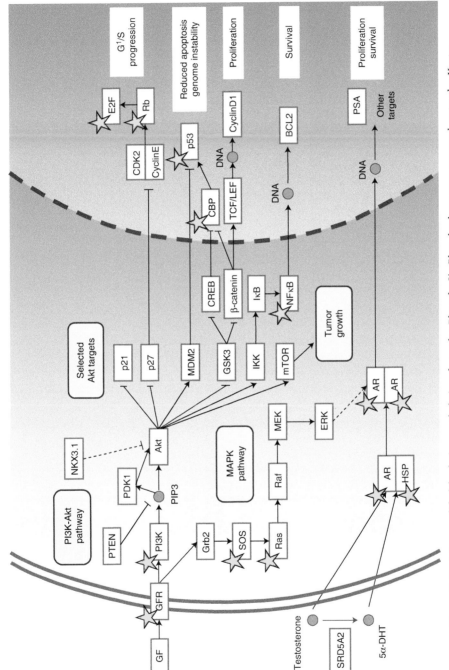

Figure 3. Proteins modified by lysine methylation from the PhosphoSitePlus database were mapped onto the Kyoto Encyclopedia of Genes and Genomes (KEGG) prostate cancer pathway. The figure shows a simplified pathway diagram; methylated proteins are indicated by a star.

(www.phosphosite.org) (Hornbeck et al. 2012). It is important to be aware that of factors that cause high-throughput identification of methylated residues to be subject to a higher rate of false positives than other posttranslational modifications (Hart-Smith et al. 2015). This is largely because of the fact that several pairs of amino acid substitutions introduce the same 14 Da mass shift as methylation (for example, glycine to alanine). It is also common for samples treated with methanol to be chemically methylated at acidic residues. These potential artifacts can only be eliminated by careful evaluation of peptide fragmentation spectra. We recommend that researchers investigating a methylation site identified by high-throughput proteomics purify the protein from their particular biological system. LC-MS/MS can then be used to confirm that the modification exists and to measure the stoichiometry of mono-, di- and trimethylated forms.

To visualize cancer pathways that may be regulated by lysine methylation we mapped the list of methylated proteins onto cancer pathways annotated in the Kyoto Encyclopedia of Genes and Genomes (KEGG) (Kanehisa et al. 2015). Many core pathways are shared among these cancer pathways so we have visualized only the "prostate cancer" pathway (Fig. 3). Earlier we discussed how the MAPK pathway is regulated by methylation of MAP3K2. Methylation also occurs on core members of the pathway including growth factor receptors, SOS and Ras proteins, as well as in the androgen receptor pathway, the NF-κB pathway, on proteins that control cell cycle, and on p53.

With over 1000 methylated lysine residues reported so far, there is clearly a tremendous opportunity for future research. Techniques to identify these sites have existed for less than 3 years so many more methylated residues likely remain to be identified. Many of the nonhistone methylation sites studied in detail so far have revealed important new regulatory processes, and especially processes affecting cancer-related pathways. It will be of great scientific and clinical importance to understand the roles of nonhistone lysine methylation in normal physiology, to discover which sites and whether they act

in cancer, and to map the pathways that they regulate.

ACKNOWLEDGMENTS

O.G. is supported by a grant from the National Institutes of Health (NIH) (RO1 CA172560). S.M.C. is supported by an NIH K99/R00 Pathway to Independence Award (K99CA190803). O.G. is a co-founder of EpiCypher.

REFERENCES

Abu-Farha M, Lanouette S, Elisma F, Tremblay V, Butson J, Figeys D, Couture JF. 2011. Proteomic analyses of the SMYD family interactomes identify HSP90 as a novel target for SMYD2. *J Mol Cell Biol* **3:** 301–308.

Ambler RP, Rees MW. 1959. ε-*N*-Methyl-lysine in bacterial flagellar protein. *Nature* **184:** 56–57.

Bannister AJ, Zegerman P, Partridge JF, Miska EA, Thomas JO, Allshire RC, Kouzarides T. 2001. Selective recognition of methylated lysine 9 on histone H3 by the HP1 chromo domain. *Nature* **410:** 120–124.

Baudat F, Buard J, Grey C, Fledel-Alon A, Ober C, Przeworski M, Coop G, de Massy B. 2010. PRDM9 is a major determinant of meiotic recombination hotspots in humans and mice. *Science* **327:** 836–840.

Biggar KK, Li SS. 2015. Non-histone protein methylation as a regulator of cellular signalling and function. *Nat Rev Mol Cell Biol* **16:** 5–17.

Black JC, Van Rechem C, Whetstine JR. 2012. Histone lysine methylation dynamics: Establishment, regulation, and biological impact. *Mol Cell* **48:** 491–507.

Cao XJ, Arnaudo AM, Garcia BA. 2013. Large-scale global identification of protein lysine methylation in vivo. *Epigenetics* **8:** 477-485.

Carlson SM, Gozani O. 2014. Emerging technologies to map the protein methylome. *J Mol Biol* **426:** 3350–3362.

Carlson SM, Moore KE, Sankaran SM, Reynoird N, Elias JE, Gozani O. 2015. A proteomic strategy identifies lysine methylation of splicing factor snRNP70 by the SETMAR enzyme. *J Biol Chem* **290:** 12040–12047.

Casciello F, Windloch K, Gannon F, Lee JS. 2015. Functional role of G9a histone methyltransferase in cancer. *Front Immunol* **6:** 487.

Cerami E, Gao J, Dogrusoz U, Gross BE, Sumer SO, Aksoy BA, Jacobsen A, Byrne CJ, Heuer ML, Larsson E, et al. 2012. The cBio cancer genomics portal: An open platform for exploring multidimensional cancer genomics data. *Cancer Discov* **2:** 401–404.

Cho HS, Hayami S, Toyokawa G, Maejima K, Yamane Y, Suzuki T, Dohmae N, Kogure M, Kang D, Neal DE, et al. 2012. RB1 methylation by SMYD2 enhances cell cycle progression through an increase of RB1 phosphorylation. *Neoplasia* **14:** 476–486.

Chuikov S, Kurash JK, Wilson JR, Xiao B, Justin N, Ivanov GS, McKinney K, Tempst P, Prives C, Gamblin SJ, et al.

2004. Regulation of p53 activity through lysine methylation. *Nature* **432**: 353–360.

Cloutier P, Lavallée-Adam M, Faubert D, Blanchette M, Coulombe B. 2013. A newly uncovered group of distantly related lysine methyltransferases preferentially interact with molecular chaperones to regulate their activity. *PLoS Genet* **9**: e1003210.

Cloutier P, Lavallée-Adam M, Faubert D, Blanchette M, Coulombe B. 2014. Methylation of the DNA/RNA-binding protein Kin17 by METTL22 affects its association with chromatin. *J Proteomics* **100**: 115–124.

Collins R, Cheng X. 2010. A case study in cross-talk: The histone lysine methyltransferases G9a and GLP. *Nucleic Acids Res* **38**: 3503–3511.

Collins RE, Northrop JP, Horton JR, Lee DY, Zhang X, Stallcup MR, Cheng X. 2008. The ankyrin repeats of G9a and GLP histone methyltransferases are mono- and dimethyllysine binding modules. *Nat Struct Mol Biol* **15**: 245–250.

Davydova E, Ho AY, Malecki J, Moen A, Enserink JM, Jakobsson ME, Loenarz C, Falnes P. 2014. Identification and characterization of a novel evolutionarily conserved lysine-specific methyltransferase targeting eukaryotic translation elongation factor 2 (eEF2). *J Biol Chem* **289**: 30499–30510.

Derunes C, Briknarová K, Geng L, Li S, Gessner CR, Hewitt K, Wu S, Huang S, Woods VI, Ely KR. 2005. Characterization of the PR domain of RIZ1 histone methyltransferase. *Biochem Biophys Res Commun* **333**: 925–934.

Dillon SC, Zhang X, Trievel RC, Cheng X. 2005. The SET-domain protein superfamily: Protein lysine methyltransferases. *Genome Biol* **6**: 227.

Dong C, Wu Y, Yao J, Wang Y, Yu Y, Rychahou PG, Evers BM, Zhou BP. 2012. G9a interacts with Snail and is critical for Snail-mediated E-cadherin repression in human breast cancer. *J Clin Invest* **122**: 1469–1486.

Du Y, Carling T, Fang W, Piao Z, Sheu JC, Huang S. 2001. Hypermethylation in human cancers of the RIZ1 tumor suppressor gene, a member of a histone/protein methyltransferase superfamily. *Cancer Res* **61**: 8094–8099.

Eom GH, Kim K, Kim SM, Kee HJ, Kim JY, Jin HM, Kim JR, Kim JH, Choe N, Kim KB, et al. 2009. Histone methyltransferase PRDM8 regulates mouse testis steroidogenesis. *Biochem Biophys Res Commun* **388**: 131–136.

Fnu S, Williamson EA, De Haro LP, Brenneman M, Wray J, Shaheen M, Radhakrishnan K, Lee SH, Nickoloff JA, Hromas R. 2011. Methylation of histone H3 lysine 36 enhances DNA repair by nonhomologous end-joining. *Proc Natl Acad Sci* **108**: 540–545.

Fog CK, Galli GG, Lund AH. 2012. PRDM proteins: Important players in differentiation and disease. *Bioessays* **34**: 50–60.

Fog CK, Asmar F, Côme C, Jensen KT, Johansen JV, Kheir TB, Jacobsen L, Friis C, Louw A, Rosgaard L, et al. 2015. Loss of PRDM11 promotes MYC-driven lymphomagenesis. *Blood* **125**: 1272–1281.

Forrester K, Almoguera C, Han K, Grizzle WE, Perucho M. 1987. Detection of high incidence of K-ras oncogenes during human colon tumorigenesis. *Nature* **327**: 298–303.

Fuchs SM, Krajewski K, Baker RW, Miller VL, Strahl BD. 2011. Influence of combinatorial histone modifications

on antibody and effector protein recognition. *Curr Biol* **21**: 53–58.

Gaedcke J, Grade M, Jung K, Camps J, Jo P, Emons G, Gehoff A, Sax U, Schirmer M, Becker H, et al. 2010. Mutated KRAS results in overexpression of DUSP4, a MAP-kinase phosphatase, and SMYD3, a histone methyltransferase, in rectal carcinomas. *Genes Chromosomes Cancer* **49**: 1024–1034.

Gao J, Aksoy BA, Dogrusoz U, Dresdner G, Gross B, Sumer SO, Sun Y, Jacobsen A, Sinha R, Larsson E, et al. 2013. Integrative analysis of complex cancer genomics and clinical profiles using the cBioPortal. *Sci Signal* **6**: pl1.

Guo A, Gu H, Zhou J, Mulhern D, Wang Y, Lee KA, Yang V, Aguiar M, Kornhauser J, Jia X, et al. 2014. Immunoaffinity enrichment and mass spectrometry analysis of protein methylation. *Mol Cell Proteomics* **13**: 372–387.

Hamamoto R, Furukawa Y, Morita M, Iimura Y, Silva FP, Li M, Yagyu R, Nakamura Y. 2004. SMYD3 encodes a histone methyltransferase involved in the proliferation of cancer cells. *Nat Cell Biol* **6**: 731–740.

Hamamoto R, Saloura V, Nakamura Y. 2015. Critical roles of non-histone protein lysine methylation in human tumorigenesis. *Nat Rev Cancer* **15**: 110–124.

Hart-Smith G, Yagoub D, Tay AP, Pickford R, Wilkins MR. 2015. Large-scale mass spectrometry-based identifications of enzyme-mediated protein methylation are subject to high false discovery rates. *Mol Cell Proteomics* **15**: 989–1006.

Hayami S, Kelly JD, Cho HS, Yoshimatsu M, Unoki M, Tsunoda T, Field HI, Neal DE, Yamaue H, Ponder BA, et al. 2011. Overexpression of LSD1 contributes to human carcinogenesis through chromatin regulation in various cancers. *Int J Cancer* **128**: 574–586.

Hayashi K, Yoshida K, Matsui Y. 2005. A histone H3 methyltransferase controls epigenetic events required for meiotic prophase. *Nature* **438**: 374–378.

Herz HM, Garruss A, Shilatifard A. 2013. SET for life: Biochemical activities and biological functions of SET domain-containing proteins. *Trends Biochem Sci* **38**: 621–639.

Hohenauer T, Moore AW. 2012. The Prdm family: Expanding roles in stem cells and development. *Development* **139**: 2267–2282.

Hornbeck PV, Kornhauser JM, Tkachev S, Zhang B, Skrzypek E, Murray B, Latham V, Sullivan M. 2012. PhosphoSitePlus: A comprehensive resource for investigating the structure and function of experimentally determined post-translational modifications in man and mouse. *Nucleic Acids Res* **40**: D261–D270.

Hu M, Yao J, Cai L, Bachman KE, van den Brûle F, Velculescu V, Polyak K. 2005. Distinct epigenetic changes in the stromal cells of breast cancers. *Nat Genet* **37**: 899–905.

Huang J, Berger SL. 2008. The emerging field of dynamic lysine methylation of non-histone proteins. *Curr Opin Genet Dev* **18**: 152–158.

Huang J, Perez-Burgos L, Placek BJ, Sengupta R, Richter M, Dorsey JA, Kubicek S, Opravil S, Jenuwein T, Berger SL. 2006. Repression of p53 activity by Smyd2-mediated methylation. *Nature* **444**: 629–632.

Huang J, Sengupta R, Espejo AB, Lee MG, Dorsey JA, Richter M, Opravil S, Shiekhattar R, Bedford MT, Jenuwein T, et al. 2007. p53 is regulated by the lysine demethylase LSD1. *Nature* **449:** 105–108.

Huang J, Dorsey J, Chuikov S, Pérez-Burgos L, Zhang X, Jenuwein T, Reinberg D, Berger SL. 2010. G9a and Glp methylate lysine 373 in the tumor suppressor p53. *J Biol Chem* **285:** 9636–9641.

Jakobsson ME, Moen A, Bousset L, Egge-Jacobsen W, Kernstock S, Melki R, Falnes P. 2013. Identification and characterization of a novel human methyltransferase modulating Hsp70 protein function through lysine methylation. *J Biol Chem* **288:** 27752–27763.

Kanehisa M, Sato Y, Kawashima M, Furumichi M, Tanabe M. 2015. KEGG as a reference resource for gene and protein annotation. *Nucleic Acids Res* **44:** D457–D462.

Kernstock S, Davydova E, Jakobsson M, Moen A, Pettersen S, Mælandsmo GM, Egge-Jacobsen W, Falnes P. 2012. Lysine methylation of VCP by a member of a novel human protein methyltransferase family. *Nat Commun* **3:** 1038.

Komatsu S, Imoto I, Tsuda H, Kozaki KI, Muramatsu T, Shimada Y, Aiko S, Yoshizumi Y, Ichikawa D, Otsuji E, et al. 2009. Overexpression of SMYD2 relates to tumor cell proliferation and malignant outcome of esophageal squamous cell carcinoma. *Carcinogenesis* **30:** 1139–1146.

Kubicek S, O'Sullivan RJ, August EM, Hickey ER, Zhang Q, Teodoro ML, Rea S, Mechtler K, Kowalski JA, Homon CA, et al. 2007. Reversal of H3K9me2 by a small-molecule inhibitor for the G9a histone methyltransferase. *Mol Cell* **25:** 473–481.

Kunizaki M, Hamamoto R, Silva FP, Yamaguchi K, Nagayasu T, Shibuya M, Nakamura Y, Furukawa Y. 2007. The lysine 831 of vascular endothelial growth factor receptor 1 is a novel target of methylation by SMYD3. *Cancer Res* **67:** 10759–10765.

Lachner M, O'Carroll D, Rea S, Mechtler K, Jenuwein T. 2001. Methylation of histone H3 lysine 9 creates a binding site for HP1 proteins. *Nature* **410:** 116–120.

Laurent B, Ruitu L, Murn J, Hempel K, Ferrao R, Xiang Y, Liu S, Garcia BA, Wu H, Wu F, et al. 2015. A specific LSD1/KDM1A isoform regulates neuronal differentiation through H3K9 demethylation. *Mol Cell* **57:** 957–970.

Levy D, Kuo AJ, Chang Y, Schaefer U, Kitson C, Cheung P, Espejo A, Zee BM, Liu CL, Tangsombatvisit S, et al. 2011. Lysine methylation of the NF-κB subunit RelA by SETD6 couples activity of the histone methyltransferase GLP at chromatin to tonic repression of NF-κB signaling. *Nat Immunol* **12:** 29–36.

Liu F, Barsyte-Lovejoy D, Li F, Xiong Y, Korboukh V, Huang XP, Allali-Hassani A, Janzen WP, Roth BL, Frye SV, et al. 2013a. Discovery of an in vivo chemical probe of the lysine methyltransferases G9a and GLP. *J Med Chem* **56:** 8931–8942.

Liu H, Galka M, Mori E, Liu X, Lin YF, Wei R, Pittock P, Voss C, Dhami G, Li X, et al. 2013b. A method for systematic mapping of protein lysine methylation identifies functions for HP1β in DNA damage response. *Mol Cell* **50:** 723–735.

Liu N, Zhang Z, Wu H, Jiang Y, Meng L, Xiong J, Zhao Z, Zhou X, Li J, Li H, et al. 2015. Recognition of H3K9 methylation by GLP is required for efficient establishment of H3K9 methylation, rapid target gene repression, and mouse viability. *Genes Dev* **29:** 379–393.

Magnani R, Dirk LM, Trievel RC, Houtz RL. 2010. Calmodulin methyltransferase is an evolutionarily conserved enzyme that trimethylates Lys-115 in calmodulin. *Nat Commun* **1:** 43.

Margueron R, Reinberg D. 2010. Chromatin structure and the inheritance of epigenetic information. *Nat Rev Genet* **11:** 285–296.

Mazur PK, Reynoird N, Khatri P, Jansen PW, Wilkinson AW, Liu S, Barbash O, Van Aller GS, Huddleston M, Dhanak D, et al. 2014. SMYD3 links lysine methylation of MAP3K2 to Ras-driven cancer. *Nature* **510:** 283–287.

Małecki J, Ho AY, Moen A, Dahl HA, Falnes P. 2015. Human METTL20 is a mitochondrial lysine methyltransferase that targets the β subunit of electron transfer flavoprotein (ETFβ) and modulates its activity. *J Biol Chem* **290:** 423–434.

McCabe NR, Burnett RC, Gill HJ, Thirman MJ, Mbangkollo D, Kipiniak M, van Melle E, Ziemin-van der Poel S, Rowley JD, Diaz MO. 1992. Cloning of cDNAs of the *MLL* gene that detect DNA rearrangements and altered RNA transcripts in human leukemic cells with 11q23 translocations. *Proc Natl Acad Sci* **89:** 11794–11798.

McCormick F. 2015. KRAS as a therapeutic target. *Clin Cancer Res* **21:** 1797–1801.

Metzger E, Wissmann M, Yin N, Müller JM, Schneider R, Peters AH, Günther T, Buettner R, Schüle R. 2005. LSD1 demethylates repressive histone marks to promote androgen-receptor-dependent transcription. *Nature* **437:** 436–439.

Meyer C, Hofmann J, Burmeister T, Gröger D, Park TS, Emerenciano M, Pombo de Oliveira M, Renneville A, Villarese P, Macintyre E, et al. 2013. The *MLL* recombinome of acute leukemias in 2013. *Leukemia* **27:** 2165–2176.

Moore KE, Carlson SM, Camp ND, Cheung P, James RG, Chua KF, Wolf-Yadlin A, Gozani O. 2013. A general molecular affinity strategy for global detection and proteomic analysis of lysine methylation. *Mol Cell* **50:** 444–456.

Morishita K, Parganas E, William CL, Whittaker MH, Drabkin H, Oval J, Taetle R, Valentine MB, Ihle JN. 1992. Activation of *EVI1* gene expression in human acute myelogenous leukemias by translocations spanning 300–400 kilobases on chromosome band 3q26. *Proc Natl Acad Sci* **89:** 3937–3941.

Murray K. 1964. The occurrence of iε-*N*-Methyl lysine in histones. *Biochemistry* **3:** 10–15.

Myers S, Bowden R, Tumian A, Bontrop RE, Freeman C, MacFie TS, McVean G, Donnelly P. 2010. Drive against hotspot motifs in primates implicates the *PRDM9* gene in meiotic recombination. *Science* **327:** 876–879.

Nakakido M, Deng Z, Suzuki T, Dohmae N, Nakamura Y, Hamamoto R. 2015. Dysregulation of AKT pathway by SMYD2-mediated lysine methylation on PTEN. *Neoplasia* **17:** 367–373.

Ng HH, Feng Q, Wang H, Erdjument-Bromage H, Tempst P, Zhang Y, Struhl K. 2002. Lysine methylation within the globular domain of histone H3 by Dot1 is important for telomeric silencing and Sir protein association. *Genes Dev* **16:** 1518–1527.

Nishikawa N, Toyota M, Suzuki H, Honma T, Fujikane T, Ohmura T, Nishidate T, Ohe-Toyota M, Maruyama R, Sonoda T, et al. 2007. Gene amplification and overexpression of PRDM14 in breast cancers. *Cancer Res* **67**: 9649–9657.

Nishioka K, Rice JC, Sarma K, Erdjument-Bromage H, Werner J, Wang Y, Chuikov S, Valenzuela P, Tempst P, Steward R, et al. 2002. PR-Set7 is a nucleosome-specific methyltransferase that modifies lysine 20 of histone H4 and is associated with silent chromatin. *Mol Cell* **9**: 1201–1213.

Northcott PA, Nakahara Y, Wu X, Feuk L, Ellison DW, Croul S, Mack S, Kongkham PN, Peacock J, Dubuc A, et al. 2009. Multiple recurrent genetic events converge on control of histone lysine methylation in medulloblastoma. *Nat Genet* **41**: 465–472.

Ong S, Mann M. 2006. Identifying and quantifying sites of protein methylation by heavy methyl SILAC. *Curr Protoc Protein Sci* doi: 10.1002/0471140864.ps1409s46.

Ong SE, Mittler G, Mann M. 2004. Identifying and quantifying in vivo methylation sites by heavy methyl SILAC. *Nat Methods* **1**: 119–126.

Paik WK, Paik DC, Kim S. 2007. Historical review: The field of protein methylation. *Trends Biochem Sci* **32**: 146–152.

Parvanov ED, Petkov PM, Paigen K. 2010. Prdm9 controls activation of mammalian recombination hotspots. *Science* **327**: 835.

Pasqualucci L, Compagno M, Houldsworth J, Monti S, Grunn A, Nandula SV, Aster JC, Murty VV, Shipp MA, Dalla-Favera R. 2006. Inactivation of the PRDM1/BLIMP1 gene in diffuse large B cell lymphoma. *J Exp Med* **203**: 311–317.

Petrossian TC, Clarke SG. 2011. Uncovering the human methyltransferasome. *Mol Cell Proteomics* **10**: M110. 000976.

Pinheiro I, Margueron R, Shukeir N, Eisold M, Fritzsch C, Richter FM, Mittler G, Genoud C, Goyama S, Kurokawa M, et al. 2012. Prdm3 and Prdm16 are H3K9me1 methyltransferases required for mammalian heterochromatin integrity. *Cell* **150**: 948–960.

Plotnikov A, Zehorai E, Procaccia S, Seger R. 2011. The MAPK cascades: Signaling components, nuclear roles and mechanisms of nuclear translocation. *Biochim Biophys Acta* **1813**: 1619–1633.

Pradhan S, Chin HG, Estève PO, Jacobsen SE. 2009. SET7/9 mediated methylation of non-histone proteins in mammalian cells. *Epigenetics* **4**: 383–387.

Prior IA, Lewis PD, Mattos C. 2012. A comprehensive survey of Ras mutations in cancer. *Cancer Res* **72**: 2457–2467.

Rathert P, Dhayalan A, Murakami M, Zhang X, Tamas R, Jurkowska R, Komatsu Y, Shinkai Y, Cheng X, Jeltsch A. 2008. Protein lysine methyltransferase G9a acts on nonhistone targets. *Nat Chem Biol* **4**: 344–346.

Rea S, Eisenhaber F, O'Carroll D, Strahl BD, Sun ZW, Schmid M, Opravil S, Mechtler K, Ponting CP, Allis CD, et al. 2000. Regulation of chromatin structure by site-specific histone H3 methyltransferases. *Nature* **406**: 593–599.

Sakamoto LH, Andrade RV, Felipe MS, Motoyama AB, Pittella Silva F. 2014. SMYD2 is highly expressed in pediatric acute lymphoblastic leukemia and constitutes a bad prognostic factor. *Leuk Res* **38**: 496–502.

Sampath SC, Marazzi I, Yap KL, Krutchinsky AN, Mecklenbräuker I, Viale A, Rudensky E, Zhou MM, Chait BT, Tarakhovsky A. 2007. Methylation of a histone mimic within the histone methyltransferase G9a regulates protein complex assembly. *Mol Cell* **27**: 596–608.

Savickiene J, Treigyte G, Stirblyte I, Valiuliene G, Navakauskiene R. 2014. Euchromatic histone methyltransferase 2 inhibitor, BIX-01294, sensitizes human promyelocytic leukemia HL-60 and NB4 cells to growth inhibition and differentiation. *Leuk Res* **38**: 822–829.

Schenk T, Chen WC, Göllner S, Howell L, Jin L, Hebestreit K, Klein HU, Popescu AC, Burnett A, Mills K, et al. 2012. Inhibition of the LSD1 (KDM1A) demethylase reactivates the all-*trans*-retinoic acid differentiation pathway in acute myeloid leukemia. *Nat Med* **18**: 605–611.

Schubert HL, Blumenthal RM, Cheng X. 2003. Many paths to methyltransfer: A chronicle of convergence. *Trends Biochem Sci* **28**: 329–335.

Shen C, Ipsaro JJ, Shi J, Milazzo JP, Wang E, Roe JS, Suzuki Y, Pappin DJ, Joshua-Tor L, Vakoc CR. 2015. NSD3-short is an adaptor protein that couples BRD4 to the CHD8 chromatin remodeler. *Mol Cell* **60**: 847–859.

Shi Y. 2007. Histone lysine demethylases: Emerging roles in development, physiology and disease. *Nat Rev Genet* **8**: 829–833.

Shi Y, Lan F, Matson C, Mulligan P, Whetstine JR, Cole PA, Casero RA. 2004. Histone demethylation mediated by the nuclear amine oxidase homolog LSD1. *Cell* **119**: 941–953.

Shi X, Kachirskaia I, Yamaguchi H, West LE, Wen H, Wang EW, Dutta S, Appella E, Gozani O. 2007. Modulation of p53 function by SET8-mediated methylation at lysine 382. *Mol Cell* **27**: 636–646.

Shinkai Y, Tachibana M. 2011. H3K9 methyltransferase G9a and the related molecule GLP. *Genes Dev* **25**: 781–788.

Sweis RF, Pliushchev M, Brown PJ, Guo J, Li F, Maag D, Petros AM, Soni NB, Tse C, Vedadi M, et al. 2014. Discovery and development of potent and selective inhibitors of histone methyltransferase g9a. *ACS Med Chem Lett* **5**: 205–209.

Tachibana M, Sugimoto K, Fukushima T, Shinkai Y. 2001. Set domain-containing protein, G9a, is a novel lysine-preferring mammalian histone methyltransferase with hyperactivity and specific selectivity to lysines 9 and 27 of histone H3. *J Biol Chem* **276**: 25309–25317.

Tachibana M, Sugimoto K, Nozaki M, Ueda J, Ohta T, Ohki M, Fukuda M, Takeda N, Niida H, Kato H, et al. 2002. G9a histone methyltransferase plays a dominant role in euchromatic histone H3 lysine 9 methylation and is essential for early embryogenesis. *Genes Dev* **16**: 1779–1791.

Tachibana M, Ueda J, Fukuda M, Takeda N, Ohta T, Iwanari H, Sakihama T, Kodama T, Hamakubo T, Shinkai Y. 2005. Histone methyltransferases G9a and GLP form heteromeric complexes and are both crucial for methylation of euchromatin at H3-K9. *Genes Dev* **19**: 815–826.

Takawa M, Cho HS, Hayami S, Toyokawa G, Kogure M, Yamane Y, Iwai Y, Maejima K, Ueda K, Masuda A, et al. 2012. Histone lysine methyltransferase SETD8 promotes carcinogenesis by deregulating PCNA expression. *Cancer Res* **72**: 3217–3227.

Tsukada Y, Fang J, Erdjument-Bromage H, Warren ME, Borchers CH, Tempst P, Zhang Y. 2006. Histone demethylation by a family of JmjC domain-containing proteins. *Nature* **439**: 811–816.

Van Aller GS, Reynoird N, Barbash O, Huddleston M, Liu S, Zmoos AF, McDevitt P, Sinnamon R, Le B, Mas G, et al. 2012. Smyd3 regulates cancer cell phenotypes and catalyzes histone H4 lysine 5 methylation. *Epigenetics* **7**: 340–343.

Vedadi M, Barsyte-Lovejoy D, Liu F, Rival-Gervier S, Allali-Hassani A, Labrie V, Wigle TJ, Dimaggio PA, Wasney GA, Siarheyeva A, et al. 2011. A chemical probe selectively inhibits G9a and GLP methyltransferase activity in cells. *Nat Chem Biol* **7**: 566–574.

Wang J, Lu F, Ren Q, Sun H, Xu Z, Lan R, Liu Y, Ward D, Quan J, Ye T, et al. 2011. Novel histone demethylase LSD1 inhibitors selectively target cancer cells with pluripotent stem cell properties. *Cancer Res* **71**: 7238–7249.

Watanabe Y, Toyota M, Kondo Y, Suzuki H, Imai T, Ohe-Toyota M, Maruyama R, Nojima M, Sasaki Y, Sekido Y, et al. 2007. PRDM5 identified as a target of epigenetic silencing in colorectal and gastric cancer. *Clin Cancer Res* **13**: 4786–4794.

Watanabe T, Kobunai T, Yamamoto Y, Matsuda K, Ishihara S, Nozawa K, Iinuma H, Ikeuchi H, Eshima K. 2011. Differential gene expression signatures between colorectal cancers with and without KRAS mutations: Crosstalk between the KRAS pathway and other signalling pathways. *Eur J Cancer* **47**: 1946–1954.

West LE, Gozani O. 2011. Regulation of p53 function by lysine methylation. *Epigenomics* **3**: 361–369.

Wu H, Mathioudakis N, Diagouraga B, Dong A, Dombrovski L, Baudat F, Cusack S, de Massy B, Kadlec J. 2013. Molecular basis for the regulation of the H3K4 methyltransferase activity of PRDM9. *Cell Rep* **5**: 13–20.

Yuan Y, Wang Q, Paulk J, Kubicek S, Kemp MM, Adams DJ, Shamji AF, Wagner BK, Schreiber SL. 2012. A small-molecule probe of the histone methyltransferase G9a induces cellular senescence in pancreatic adenocarcinoma. *ACS Chem Biol* **7**: 1152–1157.

Yun M, Wu J, Workman JL, Li B. 2011. Readers of histone modifications. *Cell Res* **21**: 564–578.

Zhang X, Tanaka K, Yan J, Li J, Peng D, Jiang Y, Yang Z, Barton MC, Wen H, Shi X. 2013. Regulation of estrogen receptor α by histone methyltransferase SMYD2-mediated protein methylation. *Proc Natl Acad Sci* **110**: 17284–17289.

Zheng YC, Ma J, Wang Z, Li J, Jiang B, Zhou W, Shi X, Wang X, Zhao W, Liu HM. 2015. A systematic review of histone lysine-specific demethylase 1 and its inhibitors. *Med Res Rev* **35**: 1032–1071.

Cite this article as *Cold Spring Harb Perspect Med* doi: 10.1101/cshperspect.a026435

Long Noncoding RNAs: At the Intersection of Cancer and Chromatin Biology

Adam M. Schmitt[1] and Howard Y. Chang[2]

[1]Department of Radiation Oncology, Memorial Sloan Kettering Cancer Center, New York, New York 10065

[2]Center for Personal Dynamic Regulomes, Stanford University School of Medicine, Stanford, California 94305

Correspondence: howchang@stanford.edu

Although only 2% of the genome encodes protein, RNA is transcribed from the majority of the genetic sequence, suggesting a massive degree of cellular functionality is programmed in the noncoding genome. The mammalian genome contains tens of thousands of long noncoding RNAs (lncRNAs), many of which occur at disease-associated loci or are specifically expressed in cancer. Although the vast majority of lncRNAs have no known function, recurring molecular mechanisms for lncRNAs are now being observed in chromatin regulation and cancer pathways and emerging technologies are now providing tools to interrogate lncRNA molecular interactions and determine function of these abundant cellular macromolecules.

It has become clear only recently that alterations within the noncoding genome are major contributors to cancer and other human diseases (Maurano et al. 2012). Indeed recurrent somatic noncoding mutations (Melton et al. 2015), epigenetic alterations (Roadmap Epigenomics et al. 2015), or somatic copy number alterations (Beroukhim et al. 2010) are implicated in multiple cancer types. One of the most unexpected findings of the genomics era of biology is that the noncoding genome is the source of extensive transcription of noncoding RNA (Rinn and Chang 2012; Morris and Mattick 2014). Remarkably, several cancer-associated loci contain ultraconserved noncoding RNA sequences suggesting that these transcripts play important roles in cancer pathogenesis (Calin et al. 2007).

The catalog of noncoding genes has grown tremendously over the past few years in large part because of the identification of extensive long noncoding RNA (lncRNA) transcription (Djebali et al. 2012; Harrow et al. 2012; Iyer et al. 2015), arising from active enhancers (Kim et al. 2010), promoters (Seila et al. 2008), and intergenic sequences. Long noncoding RNAs (lncRNAs) are functionally defined as transcripts >200 nucleotides in length with no protein coding potential. These transcripts are biochemically similar to messenger RNA (mRNA) and are typically transcribed by RNA polymerase II, polyadenylated, and are frequently spliced, although they often contain fewer exons than coding genes. It is now recognized that lncRNAs are exquisitely regulated and are restricted to specific cell types to a greater degree than mRNA (Cabili et al. 2011). The catalog of identified lncRNAs now includes tens of thousands of genes and many are uniquely expressed in differentiated tissues or specific cancer types (Iyer et al. 2015). Indeed, the number of lncRNA genes outnumbers protein-coding genes (Derrien et al. 2012) with more than 90% having no appreciable peptide products (Banfai et al. 2012; Guttman et al. 2013).

Supporting an important functional role for lncRNAs, some lncRNA genes are evolutionarily conserved, although may display limited sequence similarity. Syntenic genomic sites frequently encode evolutionarily conserved, functional lncRNAs (Chodroff et al. 2010; Ulitsky et al. 2011; Hezroni et al. 2015). In a recent report, identification of regions of conserved microhomology within lncRNAs arising from syntenic sites was used to identify a set of 47 orthologs of the *Drosophila* lncRNA roX in diverse Drosophilid species (Quinn et al. 2016). Despite drastic sequence divergence, roX orthologs maintain regions of conserved secondary structure, retain X-chromosome interactions, and can rescue male lethality in roX-null *Drosophila melanogaster*.

Despite the rapid pace of discovery in the field, the vast majority of annotated lncRNAs have yet to be functionally characterized. However, several recurring lncRNA molecular mechanisms have been observed, particularly in chromatin biology and gene regulation. The emergence of several technologies over the last few years have expanded investigators abilities to functionally annotate cancer-associated lncRNAs and resulted in an exponential increase in our understanding of these enigmatic molecules (Chu et al. 2015). In this article, we provide an overview of the current state of lncRNA function in cellular molecular mechanisms and review the known roles for lncRNAs in cancer pathophysiology. Portions of this review are abbreviated from recent articles by the investigators (Chu et al. 2015; Schmitt and Chang 2016), and readers are referred to these articles for more in depth treatment of lncRNA as cancer biomarkers or lncRNA-centric technologies, respectively.

LONG NONCODING RNA: EMERGENCE OF A DISTINCT CLASS OF REGULATORY RNA

In this section we highlight several of the recurring molecular roles of lncRNAs in cellular processes (Fig. 1). Functionality in part begins with lncRNA cellular localization: nuclear lncRNAs are enriched for functionality involving nuclear structure, chromatin interactions, transcriptional regulation, and RNA processing, whereas cytoplasmic lncRNAs can modulate mRNA stability or translation and influence cellular signaling cascades (Batista and Chang 2013). LncRNA transcriptional regulation at the level of chromatin is a widely observed mechanism that can involve activities affecting neighboring intrachromosomal genes in *cis* or targeting of gene in *trans* on different chromosomes (Huarte et al. 2010; Huarte 2015; Sahu et al. 2015). LncRNAs are widely known to regulate genes in *cis* as enhancer-associated RNAs (Orom et al. 2010), through transcription activation or silencing (Huarte et al. 2010; Zhu et al. 2013; Dimitrova et al. 2014; Trimarchi et al. 2014), transcription factor trapping (Sigova et al. 2015), chromatin looping (Wang et al. 2011), and gene methylation (Schmitz et al. 2010) to name just a few mechanisms. LncRNAs also regulate distant genes through modulation of transcription factor recruitment (Hung et al. 2011; Yang et al. 2013b), chromatin modification (Wang and Chang 2011), and serving as a scaffold for assembly of multiple regulatory molecules at single locus (Tsai et al. 2010).

Nuclear Organization

Deranged nuclear architecture has been a defining feature of cancer cells for >100 years since Beale described abnormal nuclear structure in pharyngeal carcinoma cells in the mid-19th century. Nuclear atypia remains an important cytopathologic feature in diagnosing and grading of malignant cells such as in the case of Pap test for cervical neoplasia and in breast carcinoma. Nuclear organization relies heavily on noncoding nuclear RNA (Quinodoz and Guttman 2014). RNase treatment and transcription inhibitors substantially disrupt nuclear organiza-

 Cite this article as *Cold Spring Harb Perspect Med* doi: 10.1101/cshperspect.a026492

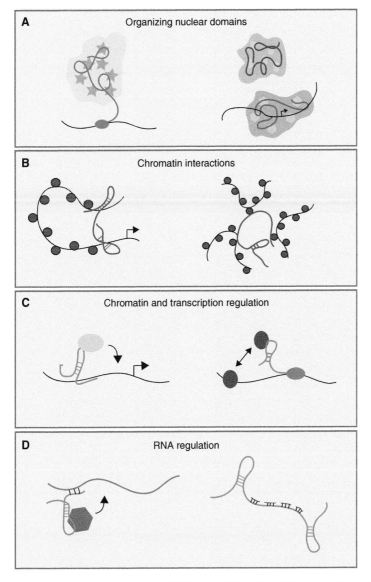

Figure 1. Molecular mechanisms of long noncoding RNAs (lncRNAs). LncRNAs interactions with DNA, RNA, and protein mediate a wide array of cellular functions including (*A*) nuclear organization, (*B*) chromatin interactions, (*C*) chromatin and transcriptional regulation, and (*D*) RNA regulation.

tion, whereas translation inhibitors did not (Nickerson et al. 1989), pointing to important structural roles for nuclear RNA. It is now clear that nuclear lncRNAs interact with protein, chromatin, and RNA to organize nuclear domains and maintain structure.

Formation of the nuclear paraspeckle, a poorly understood nuclear domain, is dependent on the ncRNA Nuclear enriched abundant transcript 1 (NEAT1). NEAT1 serves as a nidus for formation of the paraspeckle where, on NEAT1 transcription, soluble protein components of the paraspeckle localize to regions of NEAT1 accumulation thus forming the nuclear domain (Clemson et al. 2009; Mao et al. 2011; Shevtsov and Dundr 2011). This domain has been implicated in nuclear retention of A-to-I hyperedited transcripts (Chen and Carmichael

2009) and the sequestration of a transcription repressor from its chromatin targets (Imamura et al. 2014). Overexpression of NEAT1 in breast cancer specimens is associated with poor patient survival in which a hypoxic tumor microenvironment induces NEAT1 transcription, increasing paraspeckle formation, enhancing cancer cell survival and stimulating proliferation (Choudhry et al. 2015).

LncRNAs associated with functional nuclear domains can provide anchorage points for the localization of genes undergoing dynamic regulation. The serum response, the activation of which in breast cancer is associated with poor survival (Chang et al. 2004), regulates the localization of dynamically regulated genes between transcriptionally inactive and active nuclear domains. Serum stimulation recruits the demethylase KDM4C to select promoter regions that are regulated by E2F where it demethylates human Polycomb 2 homolog (Pc2), also known as Chromobox 4 (Yang et al. 2011). Methylated Pc2 interacts with the lncRNA TUG1 in Polycomb bodies, whereas demethylated Pc2 binds MALAT1/NEAT2 localized in interchromatin granules, essentially creating a posttranslation switch for Pc2 interaction with domain defining lncRNAs.

Remarkably, mouse genetic loss of function models of lncRNAs Neat1 and Malat1 have revealed no clear phenotypes in normal animal development (Nakagawa et al. 2011, 2012; Eissmann et al. 2012; Zhang et al. 2012). The contrast of these results with observed in vitro cellular phenotypes suggests that redundancy or compensatory mechanisms may be at work. Yet, Malat1 knockout or depletion by antisense oligonucleotides in a mouse MMTV-PyMT breast cancer model slows tumor growth and inhibits metastasis. Malat1 loss of function promotes cystic differentiation of breast cancer cells and enhances adhesion and inhibits migration (Arun et al. 2016).

Chromatin Interactions

The three-dimensional organization of chromatin facilitates contacts between distantly spaced intra- or interchromosomal sites allowing for transmittal of regulatory information between distant loci. Chromatin looping facilitates the interaction of the *cis* interaction of the regulatory lncRNA HOTTIP, which is transcribed from the margin of the HOXA locus, bringing HOTTIP into proximity with other HOXA locus genes and supporting transcription at the locus. HOTTIP, transcribed from the 5′ end of the locus, coordinates transcription along the multigene locus by interacting with WDR5 and recruiting the WDR5/MLL complex to target genes within the locus, increasing H3K4 trimethylation and gene transcription (Wang et al. 2011). Similarly, a class of lncRNAs referred to as noncoding RNA-activating recruit Mediator to the site of lncRNA transcription and thereby promote gene transcription of neighboring genes in close proximity because of chromatin looping (Lai et al. 2013).

Distant intra- and interchromosomal chromatin interactions are also mediated by lncRNAs. Enhancer RNAs are a subset of lncRNA transcribed from active enhancer sites. In response to estrogen exposure, enhancer RNAs transcribed from estrogen receptor α (ERα) bound enhancers facilitate the enhancer-promoter looping induced by estrogen signaling (Li et al. 2013). Similarly, the prostate cancer-specific lncRNA PCGEM1 enables chromatin looping between androgen receptor occupied enhancer regions and the promoters of target genes to facilitate their regulation. PCGEM1 bridges the enhancer and promoter regions by binding separately to the androgen receptor (AR) at enhancer sites and pygopus 2, a H3K4 trimethyl "reader" protein bound to the promoters of AR-regulated genes (Yang et al. 2013b). The X-linked lncRNA Firre interacts with hnRNPU to organize interactions between five chromosomes (Hacisuleyman et al. 2014).

Regulation of Chromatin

One particularly well-characterized mechanism by which lncRNAs regulate gene expression both in *cis* and in *trans* involves interaction with chromatin to facilitate histone modification (Khalil et al. 2009). X-inactive specific transcript (Xist), one of the first functionally annotated lncRNAs, regulates dosage compensation in female mammals by localizing to the X chromosome and recruiting multiple factors directly and indirectly, including the polycomb repressive complex 2 (PRC2), to execute X chromosome inactivation (XCI) (Gendrel and Heard 2011; Lee and Bartolomei 2013). That human malignancies frequently have X aneuploidy suggested an important role for dosage compensation of X-linked genes in preventing malignant transformation (Pageau et al. 2007). Confirming this, female mice with Xist deletion in the hematopoietic compartment develop an aggressive myeloproliferative disorder with full penetrance (Yildirim et al. 2013).

Although Xist coating of the inactive chromosome is required for X silencing, annotation of Xist interaction domains is a necessary prerequisite for clarifying the role of Xist. To define the genomic interactions of lncRNAs, Chu and colleagues developed chromatin isolation by RNA purification (ChIRP), in which short complementary biotinylated oligonucleotides hybridize to target RNAs and copurify chromatin cross-linked to the target RNA (Chu et al. 2011). Similar methods including capture hybridization analysis of RNA targets (CHART) and RNA antisense purification (RAP) showed that Xist initially binds gene-rich islands on the X chromosome that are in close three-dimensional proximity, then spreads to gene-poor regions during de novo X chromosome inactivation (Simon et al. 2011, 2013; Engreitz et al. 2014).

Transcriptional Regulation

Cancer transcriptional programs are also modulated by lncRNA recruitment to distant promoters and enhancers. ChIRP revealed that Paupar, a central nervous system (CNS) restricted lncRNA located adjacent to Pax6, interacts with numerous promoters to regulate the cell cycle and maintain the dedifferentiated state of neuroblastoma (Vance et al. 2014). NEAT1, a component of nuclear paraspeckles (Clemson et al. 2009), is associated with an increased risk of biochemical progression and metastasis of prostate cancer and is a downstream transcriptional target of ERα (Chakravarty et al. 2014).

ChIRP revealed that NEAT1 localizes to promoters of genes involved in prostate cancer growth and increased chromatin marks of active transcription at these sites, driving androgen-independent prostate cancer growth. T helper 17 cells are proinflammatory T cells, and the lncRNA RMRP was shown by ChIRP to coassociate with the key Th17 transcription factor ROR-γt for inflammatory effector function (Huang et al. 2015). Nascent RNA transcripts in the vicinity of enhancers and promoters can also bind the transcription factor YY1, promoting local accumulation of YY1 in the vicinity of the regulatory sites, supporting the engagement of YY1 with these elements, and driving a positive feedback loop that maintains active transcription (Sigova et al. 2015).

Regulation of RNA Processes

LncRNA modulation of RNA metabolism is an emerging theme with described roles in the control of mRNA stability, splicing, and translation. Staufen-1 (STAU-1) mediates mRNA decay through interaction with a double-stranded RNA in the 3′ UTR of select mRNAs. However, a subset of STAU-1 regulated mRNAs can only form the requisite double-stranded RNA stem necessary for STAU-1 binding by duplexing with an Alu element of a cytoplasmic lncRNA (Gong and Maquat 2011), thus a lncRNA can confer specificity on the STAU-1-mediated mRNA decay. STAU-1–mRNA interactions are also modulated by terminal differentiation-induced noncoding RNA (TINCR), which drives epidermal differentiation and is down-regulated in poorly differentiated cutaneous squamous-cell carcinomas (Kretz et al. 2013). TINCR recruits STAU-1 to mRNA bearing the 25 nucleotide TINCR Box, somewhat unexpectedly, stabilizing these messages and facilitating the translation of several genes involved in keratinocyte differentiation. DNA damage-induced lncRNA-p21 has been shown to inhibit translation of certain mRNAs such as JUNB and CTNNB1 through a direct RNA–RNA interaction mechanism (Yoon et al. 2012).

Multiple lncRNAs have also been implicated in regulating posttranscriptional mRNA processing. The prostate cancer-specific lncRNA PCA3 (Bussemakers et al. 1999), which is the first lncRNA assayed in an FDA-approved clinical test for prostate cancer, is an intronic lncRNA transcribed antisense to the protein coding gene *PRUNE2*. Expression of PCA3 in prostate cancer cells leads to formation of a double-stranded RNA heteroduplex with PRUNE2 that is a template for ADAR to perform adenosine-to-ionosine RNA editing and down-regulation of this tumor suppressor (Salameh et al. 2015). In line with prior observations that MALAT1 can regulate alternative splicing through interactions with serine/arginine splicing factors and pre-mRNA, Malat1 binds to nascent strands of pre-mRNA, localizing Malat1 to localize at the proximal chromatin region of transcriptionally active genes (Tripathi et al. 2010; Engreitz et al. 2014).

LncRNAs and pseudogene RNA have also been postulated to act as competing endogenous RNA (ceRNA) or "RNA sponges," interacting with microRNAs in a manner that can sequester these molecules and reduce their regulatory effect on target mRNA (Tay et al. 2014). Transcription of the *PTEN* pseudogene, *PTENP1* and other transcripts, although unable to serve as a template for PTEN translation, nonetheless can increase PTEN protein levels through an RNA-dependent mechanism that involves binding and sequestering PTEN targeting miRNA (Poliseno et al. 2010; Karreth et al. 2011; Tay et al. 2011). Indeed, lncRNAs serving as RNA sponges appear to regulate a diverse array of prostate cancer driver genes (Du et al. 2016). Similar ceRNA mechanisms have been reported for lncRNAs H19 and HULC (Wang et al. 2010; Keniry et al. 2012; Kallen et al. 2013). However, quantitative analyses of miRNA and target mRNA abundance suggest that the ceRNA mechanism could only operate given appropriate stoichiometry between ceRNA, mRNA, and miRNAs. Denzler and colleagues provided compelling quantitative evidence that very high ceRNA copy number thresholds exist for derepression of mRNAs targeted by a miRNA. Their data supports a model in which the number of endogenous miRNA target sites exceeds the number of miRNA copies per cell, resulting in

a high fraction of miRNAs already bound to target sites. This model implies that derepression of mRNAs by a ceRNA mechanism requires the number of decoy target sites encoded in a ceRNA to equal or exceed the total number of all miRNA target sites in the cell, which in the case of miR-122 is $\sim 10^5$ (Denzler et al. 2014). Thus, quantitative experimental analyses and knockout studies are needed beyond the presence of matching miRNA seed sequences to validate ceRNA mechanisms.

LncRNA DRIVERS OF CANCER PHENOTYPES

Cancer cells modulate intracellular signaling networks to proliferate, overcome cytostatic and tumor-suppressor pathways, enhance viability, and promote invasion and metastasis (Hanahan and Weinberg 2000; Hanahan and Weinberg 2011). LncRNAs are key regulators of pathways involving each of the hallmarks of cancer, as described by Hanahan and Weinberg (2000) (Fig. 2). In this section, we review the myriad of cellular activities of lncRNAs in driving classic cancer phenotypes.

Cell Growth and Tumor Suppression

Normal tissue homeostasis requires a careful balance of the imperative of cellular proliferation and repopulation on the one hand with the need to limit the expansion of deranged cell populations on the other. Multiple lncRNAs are downstream targets of chemokine and hormonal pathways (Xing et al. 2014). In T-cell acute lymphoblastic leukemia, the Notch1 oncogene drives growth in part by inducing the lncRNA LUNAR1 to up-regulate insulin like growth factor 1 receptor expression and signaling (Trimarchi et al. 2014). Estrogen responsive genes depend on enhancer lncRNAs to facilitate local chromatin interactions that promote transcriptional regulation by ERα (Li et al. 2013). Furthermore, androgen signaling in prostate cancer also relies on a number of lncRNAs that are implicated in prostate cancer proliferation that act through direct interactions with the androgen receptor (Yang et al. 2013b; Zhang

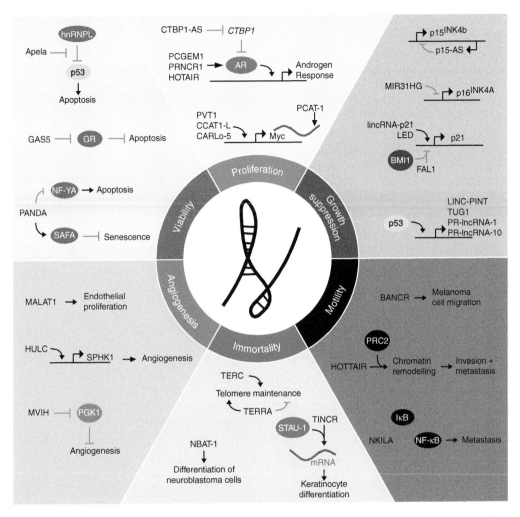

Figure 2. Long noncoding RNAs (lncRNAs) in cancer phenotypes. LncRNAs contribute to each of the six hallmarks of cancer defined by Hanahan and Weinberg. Selected examples of lncRNAs and their molecular partners or genomic targets are shown for proliferation, growth suppression, motility, immortality, angiogenesis, and viability cancer phenotypes. (From Schmitt and Chang 2016; reprinted, with permission, from the authors.)

et al. 2015) or by inhibiting repressors of the androgen receptor (Takayama et al. 2013).

The myc proto-oncogene is the subject of extensive regulation by lncRNAs. Myc transcription is activated in *cis* by the colon cancer–associated lncRNA CCAT-1, also known as CARLo-5, which promotes long-range interactions between myc and an enhancer element (Kim et al. 2014; Xiang et al. 2014). The prostate specific lncRNA PCGEM1 binds to myc enhancing myc's transcriptional activation of sev-

eral genes that regulate metabolic processes in prostate cancer cells (Hung et al. 2014). Myc also targets numerous lncRNAs for transcriptional regulation (Zheng et al. 2014), which can, in turn, regulate cell-cycle progression (Kim et al. 2015).

Amplification of the 8q24 locus is a well-characterized oncogenic event in human malignancies that amplifies *MYC* copy number. *PVT1* is a lncRNA gene that is at the breakpoint of the t(2:8) translocation in Burkitt lymphoma that brings the human immunoglobulin enhancer to *PVT1-MYC* locus. In a mouse model of Myc oncogensis, single-copy amplification of *Myc* alone was insufficient to enhance tumor formation, whereas amplification of a multigene segment encompassing *Myc* and the lncRNA *Pvt1* promoted efficient tumor development (Tseng et al. 2014). Coamplification of *Pvt1* and *Myc* increased myc protein levels, whereas depletion of PVT1 in myc-driven human colon cancer cells impaired proliferation.

Recently several lncRNAs have been found to play an extensive role in modulating tumor-suppressor and growth arrest pathways. The complement of lncRNA transcription is dynamically regulated under differing cell-cycling conditions (Hung et al. 2011) and during senescence (Abdelmohsen et al. 2013). Several tumor-suppressor genes are regulated in *cis* by their antisense transcripts. For example, the expression of the tumor suppressor TCF21 is activated by its antisense RNA TARID, which recruits the GADD45a to the *TCF21* promoter to facilitate demethylation (Arab et al. 2014).

Several lncRNAs regulate the expression of key tumor suppressors from the *CDKN2a/CDKN2b* locus, which encodes $p15^{INK4b}$, $p16^{INK4a}$, and $p14^{ARF}$. The antisense noncoding transcript p15AS induces silencing of $p15^{INK4b}$ through heterochromatin formation (Yu et al. 2008) and elevated p15AS expression is associated with low $p15^{INK4b}$ expression in leukemic cells. The lncRNA MIR31HG recruits Polycomb group proteins to the *INK4A* locus to repress its transcription during normal growth but is sequestered away from the *INK4A* locus during senescence (Montes et al. 2015).

LncRNAs in the p53 Pathway

Regulation of the p53 pathway by lncRNAs has been a topic of especially intense interest. The maternally imprinted RNA MEG3 binds to p53 and activates p53-dependent transcription of a subset of p53-regulated genes (Zhou et al. 2007). Furthermore, the known complement of p53-regulated lncRNAs is growing rapidly pointing to widespread involvement of lncRNAs downstream of p53 activation (Huarte et al. 2010;

Hung et al. 2011; Marin-Bejar et al. 2013; Sanchez et al. 2014; Zhang et al. 2014; Younger et al. 2015). Distant p53 bound enhancer regions generate enhancer RNAs that are required for p53's regulation of multiple genes from a single enhancer site (Melo et al. 2013). Genome wide profiling of p53-regulated enhancer RNAs identified the p53-induced lncRNA LED, which interacts with and activates strong enhancers including a *CKDN1A* enhancer to support cell-cycle arrest by p53. LED is silenced in a subset of p53-wildtype leukemia cells indicating a possible tumor-suppressor role for this lncRNA (Leveille et al. 2015).

LncRNA-p21, which is induced by DNA damage in a p53-dependent manner, interacts with hnRNA-K to regulate *CDKN1A* in *cis* and arrest the cell cycle in a p21-dependent manner (Huarte et al. 2010; Dimitrova et al. 2014). Apela RNA positively regulates DNA damage-induced apoptosis in mouse embryonic stem cells by binding to hnRNPL, inhibiting its ability to sequester p53 away from the mitochondria (Li et al. 2015). The lncRNA FAL-1, located within a region of chromosome 1 with frequent amplification in cancer, recruits the chromatin repressor protein BMI-1 to multiple genes including *CDKN1A* to promote tumor cell proliferation (Hu et al. 2014a). Additional p53 pathway activities are also regulated by the p53 and DNA damage inducible lncRNA PANDA, which inhibits DNA damage induced apoptosis by binding to the transcription factor NF-YA and blocking its recruitment to proapoptotic genes (Hung et al. 2011). PANDA has also been shown to regulate senescence through and interaction with SAFA and PRC1 (Puvvula et al. 2014).

Enhancing Cellular Survival

The selective advantage of tumor cells is driven in some cancers by preservation of an undifferentiated tumor cell population. LncRNAs are extensively involved in differentiation circuits and modulation of this functional class of lncRNAs may contribute to the maintenance of cancer stem-cell populations (Flynn and Chang 2014). LncTCF7 recruits the SWI/SNF complex to the TCF7 promoter to activate Wnt signaling to maintain liver cancer stem-cell self-renewal (Wang et al. 2015). Cutaneous squamous-cell carcinoma cells repress the lncRNA TINCR, which is required for keratinocyte differentiation by stabilizing differentiation-associated mRNAs through recruitment of STAU-1 (Kretz et al. 2013). The lncRNA NBAT-1 has also been observed to promote neuronal differentiation in neuroblastoma cells through regulation of the neuron specific transcription factor NRSF/REST, and repression of this lncRNA is associated with high-risk neuroblastoma (Pandey et al. 2014).

The replicative potential of differentiated somatic mammalian cells is typically limited by progressive telomere shortening following replication. However, tumor cells adapt mechanisms to maintain telomeres and effectively obtain immortality. The telomere RNA component (TERC) is a critical subunit of the ribonucleoprotein telomerase complex and encodes the template for the hexanucleotide repeats that compose the telomere sequence (Feng et al. 1995). Single-nucleotide polymorphisms at the TERC locus are associated with telomere lengthening and an increased risk of developing high-grade glioma (Walsh et al. 2014). Meanwhile, Marek's disease virus efficiently induces T-cell lymphomas in chickens by expressing a viral telomerase RNA to promote telomere lengthening (Trapp et al. 2006).

The telomeric repeat containing RNA (TERRA) transcribed from subtelomeric and telomeric DNA sequences exerts both telomerase-dependent and telomerase-independent effects on telomere maintenance (Rippe and Luke 2015). One role for TERRA involves its dynamic regulation during the cell cycle, which promotes the exchange of single-strand DNA-binding protein RPA by POT-1 and thus telomere capping (Flynn et al. 2011). Cancer cells lacking the SWI/SNF component ATRX maintain persistent TERRA loci at telomeres as cells transition from S phase to G_2. This results in persistent RPA occupancy on the single-stranded telomeric DNA preventing telomerase-dependent telomere lengthening. These cells instead rely on the recombination-dependent alternative pathway of telomere lengthening,

which requires ATR, rendering ATRX-deficient cancer cells highly sensitive to ATR inhibitors (Flynn et al. 2015).

Cancer cells that overcome intrinsic checks on growth are yet constantly subjected to conditions of nutrient scarcity and hypoxia and must develop compensatory mechanisms to survive under conditions of stress. Growth arrest specific 5 (Gas5) is induced in cells arrested by nutrient deprivation or withdrawal of growth factors. Gas5 blocks glucocorticoid responsive gene expression by binding to the glucocorticoid receptor's DNA-binding domain and acting as a decoy (Kino et al. 2010; Hudson et al. 2014). This blockade of glucocorticoid receptor decreases expression of the cellular inhibitor of apoptosis 2 (Kino et al. 2010), thereby enhancing apoptosis under stressed conditions in normal cells. Suppression of Gas5 in human breast cancer cells relative to adjacent normal breast tissue has been observed and may support the enhanced viability of breast cancer cells in the nutrient-poor tumor microenvironment (Mourtada-Maarabouni et al. 2009).

Tumor Invasion and Metastasis

Cancer metastasis was one of the first cancer phenotypes associated with dysregulated lncRNAs. Overexpression of MALAT1 in early stage nonsmall cell lung cancer patients was found to predict a high risk of metastatic progression (Ji et al. 2003). Although MALAT1 loss of function in mouse revealed that it is a nonessential gene in development or for adult normal tissue homeostasis (Nakagawa et al. 2012; Zhang et al. 2012), depletion of MALAT1 in lung carcinoma cells impairs cellular motility in vitro and metastasis in mice (Gutschner et al. 2013), suggesting that MALAT1 overexpression in cancer may drive gain-of-function phenotypes not observed during normal tissue development or homeostasis.

Multiple cancer-associated lncRNAs have been implicated in regulating cancer invasion and metastases (Flockhart et al. 2012; Hu et al. 2014b; Huarte 2015). Overexpression of the HOX-associated lncRNA HOX transcript antisense RNA (HOTAIR) in breast cancer reprograms the chromatin landscape genome-wide via recruitment of PRC2, enforcing a mesenchymal cellular phenotype which promotes breast cancer metastasis (Gupta et al. 2010), and is associated with poor prognosis in other malignancies as well (Kogo et al. 2011). The prostate cancer lncRNA SChLAP1, associated with poor prognosis and metastatic progression (Prensner et al. 2014), promotes prostate cancer invasion and metastasis by disrupting the metastasis suppressing activity of the SWI/SNF complex (Prensner et al. 2013). Transforming growth factor β (TGF-β) was found to induce the expression of lncRNA-ATB in hepatocellular carcinoma (HCC) cells, which facilitated the epithelial to mesenchymal transition (EMT), cellular invasion, and organ colonization by HCC cells by two distinct RNA–RNA interactions (Yuan et al. 2014). LncRNA-ATB competitively binds miR-200 to activate the expression of ZEB1 and ZEB2 during EMT, although interactions with IL-11 mRNA enhances Stat3 signaling to promote metastasis. The breast cancer–associated lncRNA BCAR4 connects extracellular CCL22 to noncanonical Gli2 signaling by binding the transcription factors SNIP1 and PNUTS, stimulating cell migration and metastasis (Xing et al. 2014).

Recent identification of metastasis suppressing lncRNAs has opened a new perspective into a link between the tumor microenvironment regulation of lncRNAs that promote or impair metastasis. The lncRNA LET connects the hypoxia response to the metastasis phenotype. Hypoxia-induced histone deacetylase 3 suppresses the LET promoter, impairing its expression and facilitating nuclear factor 90 accumulation and hypoxia-induced cellular invasion (Yang et al. 2013a). On the other hand, NKILA is a metastasis suppressor lncRNA that is induced by NF-κB in response to inflammatory signaling. NKILA mediates a negative feedback loop that suppresses NF-κB signaling by binding the cytoplasmic NF-κB/IκB complex and preventing IκB phosphorylation, NF-κB release, and nuclear localization (Liu et al. 2015). Downregulation of NKILA expression in human breast cancer is associated to metastatic dissemination of disease and poor prognosis.

FUTURE DIRECTIONS

LncRNAs occupy a unique niche at the interface of chromatin biology and the regulation of cancer phenotypes. That thousands of these transcripts are differentially expressed in cancer suggests that lncRNAs may play a far greater role in cancer pathways that currently appreciated. However, additional layers of lncRNA regulation are clearly at work that may increase the diversity of lncRNA function far beyond what is currently appreciated. The extensive posttranscriptional modifications of RNA, such as N^6-methyladenosine, are dynamically regulated in mammalian cells and influence cell fate decisions (Batista et al. 2014; Zhou et al. 2015). Furthermore, RNA structural approaches to evaluating lncRNA functions have revealed that even single-nucleotide polymorphisms can alter local RNA structure at functionally relevant sites involved in miRNA or protein binding (Wan et al. 2014). Application of whole transcriptome RNA structural analyses in cancer may reveal structural consequences of SNPs or somatic mutations within cancer-associated lncRNAs that may alter lncRNA function (Chu et al. 2015; Spitale et al. 2015). Furthermore, the emerging evidence that some putative lncRNAs may encode short, translated open reading frames (Ingolia et al. 2014; Anderson et al. 2015), and that coding RNAs can exert translation-independent functional roles (Li et al. 2015), suggests that distinction between mRNA and lncRNA may be less absolute than once thought.

Cataloging the physiologic role of cancer-associated lncRNAs will require an integrated approaches utilizing molecular and cellular characterization as well as animal genetic models (Sauvageau et al. 2013), as recently suggested (Li and Chang 2014). Animal models will be especially critical for elucidating the emerging physiologic roles for noncoding RNA in intercellular signaling, inflammation, angiogenesis, and immune modulation (Bernard et al. 2012; Yuan et al. 2012; Michalik et al. 2014; Lu et al. 2015; Satpathy and Chang 2015) and for identification of therapeutic targets in cancer. Multiple RNA-directed technologies, including antisense oligonucleotides, now provide an opportunity to rationally design targeted drugs to modulate lncRNAs involved in cancer pathophysiology for therapeutic gain (Gaudet et al. 2014; Buller et al. 2015; Meng et al. 2015).

ACKNOWLEDGMENTS

We apologize to colleagues whose work was not included in this review because of space limitations. This work is supported by the National Institutes of Health (NIH) (R35-CA209919 and R01-ES023168 to H.Y.C.). H.Y.C. is a founder of Epinomics and was a member of the scientific advisory board of RaNa Therapeutics.

REFERENCES

Abdelmohsen K, Panda A, Kang MJ, Xu J, Selimyan R, Yoon JH, Martindale JL, De S, Wood WH III, Becker KG, et al. 2013. Senescence-associated lncRNAs: senescence-associated long noncoding RNAs. *Aging Cell* **12:** 890–900.

Anderson DM, Anderson KM, Chang CL, Makarewich CA, Nelson BR, McAnally JR, Kasaragod P, Shelton JM, Liou J, Bassel-Duby R, et al. 2015. A micropeptide encoded by a putative long noncoding RNA regulates muscle performance. *Cell* **160:** 595–606.

Arab K, Park YJ, Lindroth AM, Schafer A, Oakes C, Weichenhan D, Lukanova A, Lundin E, Risch A, Meister M, et al. 2014. Long noncoding RNA TARID directs demethylation and activation of the tumor suppressor TCF21 via GADD45A. *Mol Cell* **55:** 604–614.

Arun G, Diermeier S, Akerman M, Chang KC, Wilkinson JE, Hearn S, Kim Y, MacLeod AR, Krainer AR, Norton L, et al. 2016. Differentiation of mammary tumors and reduction in metastasis upon Malat1 lncRNA loss. *Genes Dev* **30:** 34–51.

Banfai B, Jia H, Khatun J, Wood E, Risk B, Gundling WE Jr, Kundaje A, Gunawardena HP, Yu Y, Xie L, et al. 2012. Long noncoding RNAs are rarely translated in two human cell lines. *Genome Res* **22:** 1646–1657.

Batista PJ, Chang HY. 2013. Long noncoding RNAs: Cellular address codes in development and disease. *Cell* **152:** 1298–1307.

Batista PJ, Molinie B, Wang J, Qu K, Zhang J, Li L, Bouley DM, Lujan E, Haddad B, Daneshvar K, et al. 2014. m⁶A RNA modification controls cell fate transition in mammalian embryonic stem cells. *Cell Stem Cell* **15:** 707–719.

Bernard JJ, Cowing-Zitron C, Nakatsuji T, Muehleisen B, Muto J, Borkowski AW, Martinez L, Greidinger EL, Yu BD, Gallo RL. 2012. Ultraviolet radiation damages self noncoding RNA and is detected by TLR3. *Nat Med* **18:** 1286–1290.

Beroukhim R, Mermel CH, Porter D, Wei G, Raychaudhuri S, Donovan J, Barretina J, Boehm JS, Dobson J, Urashima M, et al. 2010. The landscape of somatic copy-number alteration across human cancers. *Nature* **463:** 899–905.

Buller HR, Bethune C, Bhanot S, Gailani D, Monia BP, Raskob GE, Segers A, Verhamme P, Weitz JI, Investigators F-AT. 2015. Factor XI antisense oligonucleotide for prevention of venous thrombosis. *N Engl J Med* **372:** 232–240.

Bussemakers MJ, van Bokhoven A, Verhaegh GW, Smit FP, Karthaus HF, Schalken JA, Debruyne FM, Ru N, Isaacs WB. 1999. DD3: A new prostate-specific gene, highly overexpressed in prostate cancer. *Cancer Res* **59:** 5975–5979.

Cabili MN, Trapnell C, Goff L, Koziol M, Tazon-Vega B, Regev A, Rinn JL. 2011. Integrative annotation of human large intergenic noncoding RNAs reveals global properties and specific subclasses. *Genes Dev* **25:** 1915–1927.

Calin GA, Liu CG, Ferracin M, Hyslop T, Spizzo R, Sevignani C, Fabbri M, Cimmino A, Lee EJ, Wojcik SE, et al. 2007. Ultraconserved regions encoding ncRNAs are altered in human leukemias and carcinomas. *Cancer Cell* **12:** 215–229.

Chakravarty D, Sboner A, Nair SS, Giannopoulou E, Li R, Hennig S, Mosquera JM, Pauwels J, Park K, Kossai M, et al. 2014. The oestrogen receptor α-regulated lncRNA NEAT1 is a critical modulator of prostate cancer. *Nat Commun* **5:** 5383.

Chang HY, Sneddon JB, Alizadeh AA, Sood R, West RB, Montgomery K, Chi JT, van de Rijn M, Botstein D, Brown PO. 2004. Gene expression signature of fibroblast serum response predicts human cancer progression: Similarities between tumors and wounds. *PLoS Biol* **2:** E7.

Chen LL, Carmichael GG. 2009. Altered nuclear retention of mRNAs containing inverted repeats in human embryonic stem cells: Functional role of a nuclear noncoding RNA. *Mol Cell* **35:** 467–478.

Chodroff RA, Goodstadt L, Sirey TM, Oliver PL, Davies KE, Green ED, Molnar Z, Ponting CP. 2010. Long noncoding RNA genes: Conservation of sequence and brain expression among diverse amniotes. *Genome Biol* **11:** R72.

Choudhry H, Albukhari A, Morotti M, Haider S, Moralli D, Smythies J, Schodel J, Green CM, Camps C, Buffa F, et al. 2015. Tumor hypoxia induces nuclear paraspeckle formation through HIF-2α dependent transcriptional activation of NEAT1 leading to cancer cell survival. *Oncogene* **34:** 4482–4490.

Chu C, Qu K, Zhong FL, Artandi SE, Chang HY. 2011. Genomic maps of long noncoding RNA occupancy reveal principles of RNA-chromatin interactions. *Mol Cell* **44:** 667–678.

Chu C, Spitale RC, Chang HY. 2015. Technologies to probe functions and mechanisms of long noncoding RNAs. *Nat Struct Mol Biol* **22:** 29–35.

Clemson CM, Hutchinson JN, Sara SA, Ensminger AW, Fox AH, Chess A, Lawrence JB. 2009. An architectural role for a nuclear noncoding RNA: NEAT1 RNA is essential for the structure of paraspeckles. *Mol Cell* **33:** 717–726.

Denzler R, Agarwal V, Stefano J, Bartel DP, Stoffel M. 2014. Assessing the ceRNA hypothesis with quantitative measurements of miRNA and target abundance. *Mol Cell* **54:** 766–776.

Derrien T, Johnson R, Bussotti G, Tanzer A, Djebali S, Tilgner H, Guernec G, Martin D, Merkel A, Knowles DG, et al. 2012. The GENCODE v7 catalog of human long noncoding RNAs: Analysis of their gene structure, evolution, and expression. *Genome Res* **22:** 1775–1789.

Dimitrova N, Zamudio JR, Jong RM, Soukup D, Resnick R, Sarma K, Ward AJ, Raj A, Lee JT, Sharp PA, et al. 2014. *LincRNA-p21* activates *p21* in *cis* to promote Polycomb target gene expression and to enforce the G_1/S checkpoint. *Mol Cell* **54:** 777–790.

Djebali S, Davis CA, Merkel A, Dobin A, Lassmann T, Mortazavi A, Tanzer A, Lagarde J, Lin W, Schlesinger F, et al. 2012. Landscape of transcription in human cells. *Nature* **489:** 101–108.

Du Z, Sun T, Hacisuleyman E, Fei T, Wang X, Brown M, Rinn JL, Lee MG, Chen Y, Kantoff PW, et al. 2016. Integrative analyses reveal a long noncoding RNA-mediated sponge regulatory network in prostate cancer. *Nat Commun* **7:** 10982.

Eissmann M, Gutschner T, Hammerle M, Gunther S, Caudron-Herger M, Gross M, Schirmacher P, Rippe K, Braun T, Zornig M, et al. 2012. Loss of the abundant nuclear noncoding RNA MALAT1 is compatible with life and development. *RNA Biol* **9:** 1076–1087.

Engreitz JM, Sirokman K, McDonel P, Shishkin AA, Surka C, Russell P, Grossman SR, Chow AY, Guttman M, Lander ES. 2014. RNA–RNA interactions enable specific targeting of noncoding RNAs to nascent Pre-mRNAs and chromatin sites. *Cell* **159:** 188–199.

Feng J, Funk WD, Wang SS, Weinrich SL, Avilion AA, Chiu CP, Adams RR, Chang E, Allsopp RC, Yu J, et al. 1995. The RNA component of human telomerase. *Science* **269:** 1236–1241.

Flockhart RJ, Webster DE, Qu K, Mascarenhas N, Kovalski J, Kretz M, Khavari PA. 2012. BRAFV600E remodels the melanocyte transcriptome and induces BANCR to regulate melanoma cell migration. *Genome Res* **22:** 1006–1014.

Flynn RA, Chang HY. 2014. Long noncoding RNAs in cell-fate programming and reprogramming. *Cell Stem Cell* **14:** 752–761.

Flynn RL, Centore RC, O'Sullivan RJ, Rai R, Tse A, Songyang Z, Chang S, Karlseder J, Zou L. 2011. TERRA and hnRNPA1 orchestrate an RPA-to-POT1 switch on telomeric single-stranded DNA. *Nature* **471:** 532–536.

Flynn RL, Cox KE, Jeitany M, Wakimoto H, Bryll AR, Ganem NJ, Bersani F, Pineda JR, Suva ML, Benes CH, et al. 2015. Alternative lengthening of telomeres renders cancer cells hypersensitive to ATR inhibitors. *Science* **347:** 273–277.

Gaudet D, Brisson D, Tremblay K, Alexander VJ, Singleton W, Hughes SG, Geary RS, Baker BF, Graham MJ, Crooke RM, et al. 2014. Targeting APOC3 in the familial chylomicronemia syndrome. *N Engl J Med* **371:** 2200–2206.

Gendrel AV, Heard E. 2011. Fifty years of X-inactivation research. *Development* **138:** 5049–5055.

Gong C, Maquat LE. 2011. lncRNAs transactivate STAU1-mediated mRNA decay by duplexing with 3′ UTRs via Alu elements. *Nature* **470:** 284–288.

Gupta RA, Shah N, Wang KC, Kim J, Horlings HM, Wong DJ, Tsai MC, Hung T, Argani P, Rinn JL, et al. 2010. Long noncoding RNA HOTAIR reprograms chromatin state to promote cancer metastasis. *Nature* **464:** 1071–1076.

Gutschner T, Hammerle M, Eissmann M, Hsu J, Kim Y, Hung G, Revenko A, Arun G, Stentrup M, Gross M, et al. 2013. The noncoding RNA MALAT1 is a critical regulator of the metastasis phenotype of lung cancer cells. *Cancer Res* **73:** 1180–1189.

Guttman M, Russell P, Ingolia NT, Weissman JS, Lander ES. 2013. Ribosome profiling provides evidence that large noncoding RNAs do not encode proteins. *Cell* **154:** 240–251.

Hacisuleyman E, Goff LA, Trapnell C, Williams A, Henao-Mejia J, Sun L, McClanahan P, Hendrickson DG, Sauvageau M, Kelley DR, et al. 2014. Topological organization of multichromosomal regions by the long intergenic noncoding RNA Firre. *Nat Struct Mol Biol* **21:** 198–206.

Hanahan D, Weinberg RA. 2000. The hallmarks of cancer. *Cell* **100:** 57–70.

Hanahan D, Weinberg RA. 2011. Hallmarks of cancer: The next generation. *Cell* **144:** 646–674.

Harrow J, Frankish A, Gonzalez JM, Tapanari E, Diekhans M, Kokocinski F, Aken BL, Barrell D, Zadissa A, Searle S, et al. 2012. GENCODE: The reference human genome annotation for The ENCODE Project. *Genome Res* **22:** 1760–1774.

Hezroni H, Koppstein D, Schwartz MG, Avrutin A, Bartel DP, Ulitsky I. 2015. Principles of long noncoding RNA evolution derived from direct comparison of transcriptomes in 17 species. *Cell Rep* **11:** 1110–1122.

Hu X, Feng Y, Zhang D, Zhao SD, Hu Z, Greshock J, Zhang Y, Yang L, Zhong X, Wang LP, et al. 2014a. A functional genomic approach identifies FAL1 as an oncogenic long noncoding RNA that associates with BMI1 and represses p21 expression in cancer. *Cancer Cell* **26:** 344–357.

Hu Y, Wang J, Qian J, Kong X, Tang J, Wang Y, Chen H, Hong J, Zou W, Chen Y, et al. 2014b. Long noncoding RNA GAPLINC regulates CD44-dependent cell invasiveness and associates with poor prognosis of gastric cancer. *Cancer Res* **74:** 6890–6902.

Huang W, Thomas B, Flynn RA, Gavzy SJ, Wu L, Kim SV, Hall JA, Miraldi ER, Ng CP, Rigo FW, et al. 2015. DDX5 and its associated lncRNA Rmrp modulate TH17 cell effector functions. *Nature* **528:** 517–522.

Huarte M. 2015. The emerging role of lncRNAs in cancer. *Nat Med* **21:** 1253–1261.

Huarte M, Guttman M, Feldser D, Garber M, Koziol MJ, Kenzelmann-Broz D, Khalil AM, Zuk O, Amit I, Rabani M, et al. 2010. A large intergenic noncoding RNA induced by p53 mediates global gene repression in the p53 response. *Cell* **142:** 409–419.

Hudson WH, Pickard MR, de Vera IM, Kuiper EG, Mourtada-Maarabouni M, Conn GL, Kojetin DJ, Williams GT, Ortlund EA. 2014. Conserved sequence-specific lincRNA-steroid receptor interactions drive transcriptional repression and direct cell fate. *Nat Commun* **5:** 5395.

Hung T, Wang Y, Lin MF, Koegel AK, Kotake Y, Grant GD, Horlings HM, Shah N, Umbricht C, Wang P, et al. 2011. Extensive and coordinated transcription of noncoding RNAs within cell-cycle promoters. *Nat Genet* **43:** 621–629.

Hung CL, Wang LY, Yu YL, Chen HW, Srivastava S, Petrovics G, Kung HJ. 2014. A long noncoding RNA connects c-Myc to tumor metabolism. *Proc Natl Acad Sci* **111:** 18697–18702.

Imamura K, Imamachi N, Akizuki G, Kumakura M, Kawaguchi A, Nagata K, Kato A, Kawaguchi Y, Sato H, Yoneda M, et al. 2014. Long noncoding RNA NEAT1-dependent SFPQ relocation from promoter region to paraspeckle mediates IL8 expression upon immune stimuli. *Mol Cell* **53:** 393–406.

Ingolia NT, Brar GA, Stern-Ginossar N, Harris MS, Talhouarne GJ, Jackson SE, Wills MR, Weissman JS. 2014. Ribosome profiling reveals pervasive translation outside of annotated protein-coding genes. *Cell Rep* **8:** 1365–1379.

Iyer MK, Niknafs YS, Malik R, Singhal U, Sahu A, Hosono Y, Barrette TR, Prensner JR, Evans JR, Zhao S, et al. 2015. The landscape of long noncoding RNAs in the human transcriptome. *Nat Genet* **47:** 199–208.

Ji P, Diederichs S, Wang W, Boing S, Metzger R, Schneider PM, Tidow N, Brandt B, Buerger H, Bulk E, et al. 2003. MALAT-1, a novel noncoding RNA, and thymosin β4 predict metastasis and survival in early-stage nonsmall cell lung cancer. *Oncogene* **22:** 8031–8041.

Kallen AN, Zhou XB, Xu J, Qiao C, Ma J, Yan L, Lu L, Liu C, Yi JS, Zhang H, et al. 2013. The imprinted H19 lncRNA antagonizes let-7 microRNAs. *Mol Cell* **52:** 101–112.

Karreth FA, Tay Y, Perna D, Ala U, Tan SM, Rust AG, DeNicola G, Webster KA, Weiss D, Perez-Mancera PA, et al. 2011. In vivo identification of tumor-suppressive PTEN ceRNAs in an oncogenic BRAF-induced mouse model of melanoma. *Cell* **147:** 382–395.

Keniry A, Oxley D, Monnier P, Kyba M, Dandolo L, Smits G, Reik W. 2012. The H19 lincRNA is a developmental reservoir of miR-675 that suppresses growth and Igf1r. *Nat Cell Biol* **14:** 659–665.

Khalil AM, Guttman M, Huarte M, Garber M, Raj A, Rivea Morales D, Thomas K, Presser A, Bernstein BE, van Oudenaarden A, et al. 2009. Many human large intergenic noncoding RNAs associate with chromatin-modifying complexes and affect gene expression. *Proc Natl Acad Sci* **106:** 11667–11672.

Kim TK, Hemberg M, Gray JM, Costa AM, Bear DM, Wu J, Harmin DA, Laptewicz M, Barbara-Haley K, Kuersten S, et al. 2010. Widespread transcription at neuronal activity-regulated enhancers. *Nature* **465:** 182–187.

Kim T, Cui R, Jeon YJ, Lee JH, Lee JH, Sim H, Park JK, Fadda P, Tili E, Nakanishi H, et al. 2014. Long-range interaction and correlation between MYC enhancer and oncogenic long noncoding RNA CARLo-5. *Proc Natl Acad Sci* **111:** 4173–4178.

Kim T, Cui R, Jeon YJ, Fadda P, Alder H, Croce CM. 2015. MYC-repressed long noncoding RNAs antagonize MYC-induced cell proliferation and cell cycle progression. *Oncotarget* **6:** 18780–18789.

Kino T, Hurt DE, Ichijo T, Nader N, Chrousos GP. 2010. Noncoding RNA gas5 is a growth arrest- and starvation-associated repressor of the glucocorticoid receptor. *Sci Signal* **3:** ra8.

Kogo R, Shimamura T, Mimori K, Kawahara K, Imoto S, Sudo T, Tanaka F, Shibata K, Suzuki A, Komune S, et al. 2011. Long noncoding RNA HOTAIR regulates polycomb-dependent chromatin modification and is associated with poor prognosis in colorectal cancers. *Cancer Res* **71:** 6320–6326.

Kretz M, Siprashvili Z, Chu C, Webster DE, Zehnder A, Qu K, Lee CS, Flockhart RJ, Groff AF, Chow J, et al. 2013. Control of somatic tissue differentiation by the long noncoding RNA TINCR. *Nature* **493:** 231–235.

Lai F, Orom UA, Cesaroni M, Beringer M, Taatjes DJ, Blobel GA, Shiekhattar R. 2013. Activating RNAs associate with Mediator to enhance chromatin architecture and transcription. *Nature* **494:** 497–501.

Lee JT, Bartolomei MS. 2013. X-inactivation, imprinting, and long noncoding RNAs in health and disease. *Cell* **152:** 1308–1323.

Leveille N, Melo CA, Rooijers K, Diaz-Lagares A, Melo SA, Korkmaz G, Lopes R, Akbari Moqadam F, Maia AR, Wijchers PJ, et al. 2015. Genome-wide profiling of p53-regulated enhancer RNAs uncovers a subset of enhancers controlled by a lncRNA. *Nat Commun* **6:** 6520.

Li L, Chang HY. 2014. Physiological roles of long noncoding RNAs: Insight from knockout mice. *Trends Cell Biol* **24:** 594–602.

Li W, Notani D, Ma Q, Tanasa B, Nunez E, Chen AY, Merkurjev D, Zhang J, Ohgi K, Song X, et al. 2013. Functional roles of enhancer RNAs for oestrogen-dependent transcriptional activation. *Nature* **498:** 516–520.

Li M, Gou H, Tripathi BK, Huang J, Jiang S, Dubois W, Waybright T, Lei M, Shi J, Zhou M, et al. 2015. An Apela RNA-containing negative feedback loop regulates p53-mediated apoptosis in embryonic stem cells. *Cell Stem Cell* **16:** 669–683.

Liu B, Sun L, Liu Q, Gong C, Yao Y, Lv X, Lin L, Yao H, Su F, Li D, et al. 2015. A cytoplasmic NF-κB interacting long noncoding RNA blocks IκB phosphorylation and suppresses breast cancer metastasis. *Cancer Cell* **27:** 370–381.

Lu Z, Xiao Z, Liu F, Cui M, Li W, Yang Z, Li J, Ye L, Zhang X. 2015. Long noncoding RNA HULC promotes tumor angiogenesis in liver cancer by up-regulating sphingosine kinase 1 (SPHK1). *Oncotarget* **7:** 241–254.

Mao YS, Sunwoo H, Zhang B, Spector DL. 2011. Direct visualization of the cotranscriptional assembly of a nuclear body by noncoding RNAs. *Nat Cell Biol* **13:** 95–101.

Marin-Bejar O, Marchese FP, Athie A, Sanchez Y, Gonzalez J, Segura V, Huang L, Moreno I, Navarro A, Monzo M, et al. 2013. Pint lincRNA connects the p53 pathway with epigenetic silencing by the Polycomb repressive complex 2. *Genome Biol* **14:** R104.

Maurano MT, Humbert R, Rynes E, Thurman RE, Haugen E, Wang H, Reynolds AP, Sandstrom R, Qu H, Brody J, et al. 2012. Systematic localization of common disease-associated variation in regulatory DNA. *Science* **337:** 1190–1195.

Melo CA, Drost J, Wijchers PJ, van de Werken H, de Wit E, Oude Vrielink JA, Elkon R, Melo SA, Leveille N, Kalluri R, et al. 2013. eRNAs are required for p53-dependent enhancer activity and gene transcription. *Mol Cell* **49:** 524–535.

Melton C, Reuter JA, Spacek DV, Snyder M. 2015. Recurrent somatic mutations in regulatory regions of human cancer genomes. *Nat Genet* **47:** 710–716.

Meng L, Ward AJ, Chun S, Bennett CF, Beaudet AL, Rigo F. 2015. Towards a therapy for Angelman syndrome by targeting a long noncoding RNA. *Nature* **518:** 409–412.

Michalik KM, You X, Manavski Y, Doddaballapur A, Zornig M, Braun T, John D, Ponomareva Y, Chen W, Uchida S, et al. 2014. Long noncoding RNA MALAT1 regulates endothelial cell function and vessel growth. *Circ Res* **114:** 1389–1397.

Montes M, Nielsen MM, Maglieri G, Jacobsen A, Hojfeldt J, Agrawal-Singh S, Hansen K, Helin K, van de Werken HJ, Pedersen JS, et al. 2015. The lncRNA MIR31HG regulates p16(INK4A) expression to modulate senescence. *Nat Commun* **6:** 6967.

Morris KV, Mattick JS. 2014. The rise of regulatory RNA. *Nat Rev Genet* **15:** 423–437.

Mourtada-Maarabouni M, Pickard MR, Hedge VL, Farzaneh F, Williams GT. 2009. GAS5, a nonprotein-coding RNA, controls apoptosis and is downregulated in breast cancer. *Oncogene* **28:** 195–208.

Nakagawa S, Naganuma T, Shioi G, Hirose T. 2011. Paraspeckles are subpopulation-specific nuclear bodies that are not essential in mice. *J Cell Biol* **193:** 31–39.

Nakagawa S, Ip JY, Shioi G, Tripathi V, Zong X, Hirose T, Prasanth KV. 2012. Malat1 is not an essential component of nuclear speckles in mice. *RNA* **18:** 1487–1499.

Nickerson JA, Krochmalnic G, Wan KM, Penman S. 1989. Chromatin architecture and nuclear RNA. *Proc Natl Acad Sci* **86:** 177–181.

Orom UA, Derrien T, Beringer M, Gumireddy K, Gardini A, Bussotti G, Lai F, Zytnicki M, Notredame C, Huang Q, et al. 2010. Long noncoding RNAs with enhancer-like function in human cells. *Cell* **143:** 46–58.

Pageau GJ, Hall LL, Ganesan S, Livingston DM, Lawrence JB. 2007. The disappearing Barr body in breast and ovarian cancers. *Nat Rev Cancer* **7:** 628–633.

Pandey GK, Mitra S, Subhash S, Hertwig F, Kanduri M, Mishra K, Fransson S, Ganeshram A, Mondal T, Bandaru S, et al. 2014. The risk-associated long noncoding RNA NBAT-1 controls neuroblastoma progression by regulating cell proliferation and neuronal differentiation. *Cancer Cell* **26:** 722–737.

Poliseno L, Salmena L, Zhang J, Carver B, Haveman WJ, Pandolfi PP. 2010. A coding-independent function of gene and pseudogene mRNAs regulates tumour biology. *Nature* **465:** 1033–1038.

Prensner JR, Iyer MK, Sahu A, Asangani IA, Cao Q, Patel L, Vergara IA, Davicioni E, Erho N, Ghadessi M, et al. 2013. The long noncoding RNA SChLAP1 promotes aggressive prostate cancer and antagonizes the SWI/SNF complex. *Nat Genet* **45:** 1392–1398.

Prensner JR, Zhao S, Erho N, Schipper M, Iyer MK, Dhanasekaran SM, Magi-Galluzzi C, Mehra R, Sahu A, Siddiqui J, et al. 2014. RNA biomarkers associated with metastatic progression in prostate cancer: A multi-institutional high-throughput analysis of SChLAP1. *Lancet Oncol* **15:** 1469–1480.

Puvvula PK, Desetty RD, Pineau P, Marchio A, Moon A, Dejean A, Bischof O. 2014. Long noncoding RNA PANDA and scaffold-attachment-factor SAFA control senescence entry and exit. *Nat Commun* **5:** 5323.

Quinn JJ, Zhang QC, Georgiev P, Ilik IA, Akhtar A, Chang HY. 2016. Rapid evolutionary turnover underlies conserved lncRNA–genome interactions. *Genes Dev* **30:** 191–207.

Quinodoz S, Guttman M. 2014. Long noncoding RNAs: An emerging link between gene regulation and nuclear organization. *Trends Cell Biol* **24:** 651–663.

Rinn JL, Chang HY. 2012. Genome regulation by long noncoding RNAs. *Annu Rev Biochem* **81:** 145–166.

Rippe K, Luke B. 2015. TERRA and the state of the telomere. *Nat Struct Mol Biol* **22:** 853–858.

Roadmap Epigenomics C, Kundaje A, Meuleman W, Ernst J, Bilenky M, Yen A, Heravi-Moussavi A, Kheradpour P, Zhang Z, Wang J, et al. 2015. Integrative analysis of 111 reference human epigenomes. *Nature* **518:** 317–330.

Sahu A, Singhal U, Chinnaiyan AM. 2015. Long noncoding RNAs in cancer: From function to translation. *Trends Cancer* **1:** 93–109.

Salameh A, Lee AK, Cardo-Vila M, Nunes DN, Efstathiou E, Staquicini FI, Dobroff AS, Marchio S, Navone NM, Hosoya H, et al. 2015. PRUNE2 is a human prostate cancer suppressor regulated by the intronic long noncoding RNA PCA3. *Proc Natl Acad Sci* **112:** 8403–8408.

Sanchez Y, Segura V, Marin-Bejar O, Athie A, Marchese FP, Gonzalez J, Bujanda L, Guo S, Matheu A, Huarte M. 2014. Genome-wide analysis of the human p53 transcriptional network unveils an lncRNA tumour suppressor signature. *Nat Commun* **5:** 5812.

Satpathy AT, Chang HY. 2015. Long noncoding RNA in hematopoiesis and immunity. *Immunity* **42:** 792–804.

Sauvageau M, Goff LA, Lodato S, Bonev B, Groff AF, Gerhardinger C, Sanchez-Gomez DB, Hacisuleyman E, Li E, Spence M, et al. 2013. Multiple knockout mouse models reveal lincRNAs are required for life and brain development. *eLife* **2:** e01749.

Schmitt AM, Chang HY. 2016. Long noncoding RNAs in cancer pathways. *Cancer Cell* **29:** 452–463.

Schmitz KM, Mayer C, Postepska A, Grummt I. 2010. Interaction of noncoding RNA with the rDNA promoter mediates recruitment of DNMT3b and silencing of rRNA genes. *Genes Dev* **24:** 2264–2269.

Seila AC, Calabrese JM, Levine SS, Yeo GW, Rahl PB, Flynn RA, Young RA, Sharp PA. 2008. Divergent transcription from active promoters. *Science* **322:** 1849–1851.

Shevtsov SP, Dundr M. 2011. Nucleation of nuclear bodies by RNA. *Nat Cell Biol* **13:** 167–173.

Sigova AA, Abraham BJ, Ji X, Molinie B, Hannett NM, Guo YE, Jangi M, Giallourakis CC, Sharp PA, Young RA. 2015. Transcription factor trapping by RNA in gene regulatory elements. *Science* **350:** 978–981.

Simon MD, Wang CI, Kharchenko PV, West JA, Chapman BA, Alekseyenko AA, Borowsky ML, Kuroda MI, Kingston RE. 2011. The genomic binding sites of a noncoding RNA. *Proc Natl Acad Sci* **108:** 20497–20502.

Simon MD, Pinter SF, Fang R, Sarma K, Rutenberg-Schoenberg M, Bowman SK, Kesner BA, Maier VK, Kingston RE, Lee JT. 2013. High-resolution Xist binding maps reveal two-step spreading during X-chromosome inactivation. *Nature* **504:** 465–469.

Spitale RC, Flynn RA, Zhang QC, Crisalli P, Lee B, Jung JW, Kuchelmeister HY, Batista PJ, Torre EA, Kool ET, et al.

2015. Structural imprints in vivo decode RNA regulatory mechanisms. *Nature* 519: 486–490.

Takayama K, Horie-Inoue K, Katayama S, Suzuki T, Tsutsumi S, Ikeda K, Urano T, Fujimura T, Takagi K, Takahashi S, et al. 2013. Androgen-responsive long noncoding RNA CTBP1-AS promotes prostate cancer. *EMBO J* 32: 1665–1680.

Tay Y, Kats L, Salmena L, Weiss D, Tan SM, Ala U, Karreth F, Poliseno L, Provero P, Di Cunto F, et al. 2011. Coding-independent regulation of the tumor suppressor PTEN by competing endogenous mRNAs. *Cell* 147: 344–357.

Tay Y, Rinn J, Pandolfi PP. 2014. The multilayered complexity of ceRNA crosstalk and competition. *Nature* 505: 344–352.

Trapp S, Parcells MS, Kamil JP, Schumacher D, Tischer BK, Kumar PM, Nair VK, Osterrieder N. 2006. A virus-encoded telomerase RNA promotes malignant T cell lymphomagenesis. *J Exp Med* 203: 1307–1317.

Trimarchi T, Bilal E, Ntziachristos P, Fabbri G, Dalla-Favera R, Tsirigos A, Aifantis I. 2014. Genome-wide mapping and characterization of Notch-regulated long noncoding RNAs in acute leukemia. *Cell* 158: 593–606.

Tripathi V, Ellis JD, Shen Z, Song DY, Pan Q, Watt AT, Freier SM, Bennett CF, Sharma A, Bubulya PA, et al. 2010. The nuclear-retained noncoding RNA MALAT1 regulates alternative splicing by modulating SR splicing factor phosphorylation. *Mol Cell* 39: 925–938.

Tsai MC, Manor O, Wan Y, Mosammaparast N, Wang JK, Lan F, Shi Y, Segal E, Chang HY. 2010. Long noncoding RNA as modular scaffold of histone modification complexes. *Science* 329: 689–693.

Tseng YY, Moriarity BS, Gong W, Akiyama R, Tiwari A, Kawakami H, Ronning P, Reuland B, Guenther K, Beadnell TC, et al. 2014. PVT1 dependence in cancer with MYC copy-number increase. *Nature* 512: 82–86.

Ulitsky I, Shkumatava A, Jan CH, Sive H, Bartel DP. 2011. Conserved function of lincRNAs in vertebrate embryonic development despite rapid sequence evolution. *Cell* 147: 1537–1550.

Vance KW, Sansom SN, Lee S, Chalei V, Kong L, Cooper SE, Oliver PL, Ponting CP. 2014. The long noncoding RNA Paupar regulates the expression of both local and distal genes. *EMBO J* 33: 296–311.

Walsh KM, Codd V, Smirnov IV, Rice T, Decker PA, Hansen HM, Kollmeyer T, Kosel ML, Molinaro AM, McCoy LS, et al. 2014. Variants near TERT and TERC influencing telomere length are associated with high-grade glioma risk. *Nat Genet* 46: 731–735.

Wan Y, Qu K, Zhang QC, Flynn RA, Manor O, Ouyang Z, Zhang J, Spitale RC, Snyder MP, Segal E, et al. 2014. Landscape and variation of RNA secondary structure across the human transcriptome. *Nature* 505: 706–709.

Wang KC, Chang HY. 2011. Molecular mechanisms of long noncoding RNAs. *Mol Cell* 43: 904–914.

Wang J, Liu X, Wu H, Ni P, Gu Z, Qiao Y, Chen N, Sun F, Fan Q. 2010. CREB up-regulates long noncoding RNA, HULC expression through interaction with microRNA-372 in liver cancer. *Nucleic Acids Res* 38: 5366–5383.

Wang KC, Yang YW, Liu B, Sanyal A, Corces-Zimmerman R, Chen Y, Lajoie BR, Protacio A, Flynn RA, Gupta RA, et al. 2011. A long noncoding RNA maintains active chromatin to coordinate homeotic gene expression. *Nature* 472: 120–124.

Wang Y, He L, Du Y, Zhu P, Huang G, Luo J, Yan X, Ye B, Li C, Xia P, et al. 2015. The long noncoding RNA lncTCF7 promotes self-renewal of human liver cancer stem cells through activation of Wnt signaling. *Cell Stem Cell* 16: 413–425.

Xiang JF, Yin QF, Chen T, Zhang Y, Zhang XO, Wu Z, Zhang S, Wang HB, Ge J, Lu X, et al. 2014. Human colorectal cancer-specific CCAT1-L lncRNA regulates long-range chromatin interactions at the MYC locus. *Cell Res* 24: 513–531.

Xing Z, Lin A, Li C, Liang K, Wang S, Liu Y, Park PK, Qin L, Wei Y, Hawke DH, et al. 2014. lncRNA directs cooperative epigenetic regulation downstream of chemokine signals. *Cell* 159: 1110–1125.

Yang L, Lin C, Liu W, Zhang J, Ohgi KA, Grinstein JD, Dorrestein PC, Rosenfeld MG. 2011. ncRNA- and Pc2 methylation-dependent gene relocation between nuclear structures mediates gene activation programs. *Cell* 147: 773–788.

Yang F, Huo XS, Yuan SX, Zhang L, Zhou WP, Wang F, Sun SH. 2013a. Repression of the long noncoding RNA-LET by histone deacetylase 3 contributes to hypoxia-mediated metastasis. *Mol Cell* 49: 1083–1096.

Yang L, Lin C, Jin C, Yang JC, Tanasa B, Li W, Merkurjev D, Ohgi KA, Meng D, Zhang J, et al. 2013b. lncRNA-dependent mechanisms of androgen-receptor-regulated gene activation programs. *Nature* 500: 598–602.

Yildirim E, Kirby JE, Brown DE, Mercier FE, Sadreyev RI, Scadden DT, Lee JT. 2013. Xist RNA is a potent suppressor of hematologic cancer in mice. *Cell* 152: 727–742.

Yoon JH, Abdelmohsen K, Srikantan S, Yang X, Martindale JL, De S, Huarte M, Zhan M, Becker KG, Gorospe M. 2012. LincRNA-p21 suppresses target mRNA translation. *Mol Cell* 47: 648–655.

Younger ST, Kenzelmann-Broz D, Jung H, Attardi LD, Rinn JL. 2015. Integrative genomic analysis reveals widespread enhancer regulation by p53 in response to DNA damage. *Nucleic Acids Res* 43: 4447–4462.

Yu W, Gius D, Onyango P, Muldoon-Jacobs K, Karp J, Feinberg AP, Cui H. 2008. Epigenetic silencing of tumour suppressor gene p15 by its antisense RNA. *Nature* 451: 202–206.

Yuan SX, Yang F, Yang Y, Tao QF, Zhang J, Huang G, Yang Y, Wang RY, Yang S, Huo XS, et al. 2012. Long noncoding RNA associated with microvascular invasion in hepatocellular carcinoma promotes angiogenesis and serves as a predictor for hepatocellular carcinoma patients' poor recurrence-free survival after hepatectomy. *Hepatology* 56: 2231–2241.

Yuan JH, Yang F, Wang F, Ma JZ, Guo YJ, Tao QF, Liu F, Pan W, Wang TT, Zhou CC, et al. 2014. A long noncoding RNA activated by TGF-β promotes the invasion-metastasis cascade in hepatocellular carcinoma. *Cancer Cell* 25: 666–681.

Zhang B, Arun G, Mao YS, Lazar Z, Hung G, Bhattacharjee G, Xiao X, Booth CJ, Wu J, Zhang C, et al. 2012. The lncRNA Malat1 is dispensable for mouse development but its transcription plays a *cis*-regulatory role in the adult. *Cell Rep* 2: 111–123.

Zhang EB, Yin DD, Sun M, Kong R, Liu XH, You LH, Han L, Xia R, Wang KM, Yang JS, et al. 2014. P53-regulated long noncoding RNA TUG1 affects cell proliferation in human nonsmall cell lung cancer, partly through epigenetically regulating HOXB7 expression. *Cell Death Dis* **5**: e1243.

Zhang A, Zhao JC, Kim J, Fong KW, Yang YA, Chakravarti D, Mo YY, Yu J. 2015. LncRNA HOTAIR enhances the androgen-receptor-mediated transcriptional program and drives castration-resistant prostate cancer. *Cell Rep* **13**: 209–221.

Zheng GX, Do BT, Webster DE, Khavari PA, Chang HY. 2014. Dicer-microRNA-Myc circuit promotes transcrip-

tion of hundreds of long noncoding RNAs. *Nat Struct Mol Biol* **21**: 585–590.

Zhou Y, Zhong Y, Wang Y, Zhang X, Batista DL, Gejman R, Ansell PJ, Zhao J, Weng C, Klibanski A. 2007. Activation of p53 by MEG3 noncoding RNA. *J Biol Chem* **282**: 24731–24742.

Zhou J, Wan J, Gao X, Zhang X, Jaffrey SR, Qian SB. 2015. Dynamic m^{6}A mRNA methylation directs translational control of heat shock response. *Nature* **526**: 591–594.

Zhu Y, Rowley MJ, Bohmdorfer G, Wierzbicki AT. 2013. A SWI/SNF chromatin-remodeling complex acts in noncoding RNA-mediated transcriptional silencing. *Mol Cell* **49**: 298–309.

Index